2015 IEEE 33rd VLSI Test Symposium

(VTS 2015)

Napa, California, USA
27-29 April 2015

IEEE Catalog Number:	CFP15029-POD
ISBN:	978-1-4799-7598-3

Copyright © 2015 by the Institute of Electrical and Electronic Engineers, Inc
All Rights Reserved

Copyright and Reprint Permissions: Abstracting is permitted with credit to the source. Libraries are permitted to photocopy beyond the limit of U.S. copyright law for private use of patrons those articles in this volume that carry a code at the bottom of the first page, provided the per-copy fee indicated in the code is paid through Copyright Clearance Center, 222 Rosewood Drive, Danvers, MA 01923.

For other copying, reprint or republication permission, write to IEEE Copyrights Manager, IEEE Service Center, 445 Hoes Lane, Piscataway, NJ 08854. All rights reserved.

***This publication is a representation of what appears in the IEEE Digital Libraries. Some format issues inherent in the e-media version may also appear in this print version.**

IEEE Catalog Number: CFP15029-POD
ISBN 13:   978-1-4799-7598-3

Additional Copies of This Publication Are Available From:

Curran Associates, Inc
57 Morehouse Lane
Red Hook, NY 12571 USA
Phone: (845) 758-0400
Fax: (845) 758-2633
E-mail: curran@proceedings.com
Web: www.proceedings.com

TABLE OF CONTENTS

FAULT DIAGNOSIS FOR FLOW-BASED MICROFLUIDIC BIOCHIPS 1
Hu, Kai ; Bhattacharya, Bhargab B. ; Chakrabarty, Krishnendu

RAPID ONLINE FAULT RECOVERY FOR CYBER-PHYSICAL DIGITAL MICROFLUIDIC
BIOCHIPS 7
Jaress, Christopher ; Brisk, Philip ; Grissom, Daniel

FAULT MODELING AND TESTING OF 1T1R MEMRISTOR MEMORIES 13
Chen, Yong-Xiao ; Li, Jin-Fu

A LOW COST JITTER SEPARATION AND CHARACTERIZATION METHOD 19
Xu, Li ; Duan, Yan ; Chen, Degang

ULTRAFAST STIMULUS ERROR REMOVAL ALGORITHM FOR ADC LINEARITY TEST 24
Chen, Tao ; Chen, Degang

DISTURBANCE-FREE BIST FOR LOOP CHARACTERIZATION OF DC-DC BUCK
CONVERTERS 29
Beohar, Navankur ; Bakliwal, Priyanka ; Roy, Sidhanto ; Mandal, Debashis ; Adell, Philippe ; Vermeire, Bert ;
Bakkaloglu, Bertan ; Ozev, Sule

INNOVATIVE PRACTICES SESSION 1C: NEW TECHNOLOGIES, NEW CHALLENGES - 1 [3
PRESENTATIONS] 35
Tracey, Paul

A MULTI-LAYERED METHODOLOGY FOR DEFECT-TOLERANCE OF DATAPATH
MODULES IN PROCESSORS 36
Hsiung, Hsunwei ; Gupta, Sandeep K.

PPB: PARTIALLY-WORKING PROCESSORS BINNING FOR MAXIMIZING WAFER
UTILIZATION 42
Cheng, Da ; Gupta, Sandeep K.

IN-DEPTH SOFT ERROR VULNERABILITY ANALYSIS USING SYNTHETIC BENCHMARKS 48
Mirkhani, Shahrzad ; Samynathan, Balavinayagam ; Abraham, Jacob A.

TMO: A NEW CLASS OF ATTACK ON CIPHER MISUSING TEST INFRASTRUCTURE 54
Ali, Sk Subidh ; Sinanoglu, Ozgur

A CALL TO ACTION: SECURING IEEE 1687 AND THE NEED FOR AN IEEE TEST SECURITY
STANDARD 58
Dworak, Jennifer ; Crouch, Al

ENABLING UNAUTHORIZED RF TRANSMISSION BELOW NOISE FLOOR WITH NO
DETECTABLE IMPACT ON PRIMARY COMMUNICATION PERFORMANCE 62
Chang, Doohwang ; Bakkaloglu, Bertan ; Ozev, Sule

INNOVATIVE PRACTICES SESSION 2C: NEW TECHNOLOGIES, NEW CHALLENGES - 2 66
Sindia, Suraj

EXTRACTING EFFECTIVE FUNCTIONAL TESTS FROM COMMERCIAL PROGRAMS 67
Kodakara, Sreekumar Vadakke ; Sagar, Mehul V. ; Yuen, Joel

STATISTICAL TECHNIQUES FOR PREDICTING SYSTEM-LEVEL FAILURE USING STRESS-
TEST DATA 73
Chen, Harry H. ; Kuo, Shih-Hua ; Tung, Jonathan ; Chao, Mango C.-T.

YIELD PROGNOSIS FOR FAB-TO-FAB PRODUCT MIGRATION 79
Ahmadi, Ali ; Huang, Ke ; Nahar, Amit ; Orr, Bob ; Pas, Michael ; Carulli, John M. ; Makris, Yiorgos

3D MICROELECTRONIC WITH BEOL COMPATIBLE DEVICES 85
Drouin, D ; A-Bounouar, M ; Droulers, G ; Labalette, M ; Pioro-Ladriere, M ; Souifi, A ; Ecoffey, S

INNOVATIVE PRACTICES SESSION 3C: ADVANCES IN SILICON DEBUG & DIAGNOSIS 86
Ricchetti, Mike

PANEL: WHEN WILL THE COST OF DEPENDABILITY END INNOVATION IN COMPUTER
DESIGN? 87
Bertacco, Valeria

SPECIAL SESSION: HOT TOPICS: STATISTICAL TEST METHODS 88
Barragan, Manuel J. ; Leger, Gildas ; Azais, Florence ; Blanton, R.D. ; Singh, Adit D. ; Sunter, Stephen

EXTEST SCHEDULING FOR 2.5D SYSTEM-ON-CHIP INTEGRATED CIRCUITS 90
Wang, Ran ; Guoliang Li ; Rui Li ; Jun Qian ; Chakrabarty, Krishnendu

PULSE SHRINKAGE BASED PRE-BOND THROUGH SILICON VIAS TEST IN 3D IC 96
Chang Hao ; Liang Huaguo

TESTING OF 3D-STACKED ICS WITH HARD- AND SOFT-DIES - A PARTICLE SWARM OPTIMIZATION BASED APPROACH 102
Karmakar, Rajit ; Agarwal, Aditya ; Chattopadhyay, Santanu

IMPROVING DIAGNOSIS RESOLUTION OF A FAULT DETECTION TEST SET 108
Riefert, Andreas ; Sauer, Matthias ; Reddy, Sudhakar ; Becker, Bernd

IMPROVING THE ACCURACY OF DEFECT DIAGNOSIS BY CONSIDERING REDUCED DIAGNOSTIC INFORMATION 114
Pomeranz, Irith

SIGNATURE ORIENTED MODEL PRUNING TO FACILITATE MULTI-THREADED PROCESSORS DEBUGGING 120
Refan, Fatemeh ; Alizadeh, Bijan ; Navabi, Zainalabedin

INNOVATIVE PRACTICES SESSION 5C: ADVANCEMENTS IN TEST -KEEPING MOORE MOVING! 126
Amyeen, Enamul

AT-PRODUCT-TEST DEDICATED ADAPTIVE SUPPLY-RESONANCE SUPPRESSION 127
Taniguchi, Kohki ; Miura, Noriyuki ; Hayashi, Taisuke ; Nagata, Makoto

LOW COST HIGH FREQUENCY SIGNAL SYNTHESIS: APPLICATION TO RF CHANNEL INTERFERENCE TESTING 133
Wang, Xian ; Banerjee, Debashis ; Chatterjee, Abhijit

AUTOMATED TESTING OF MIXED-SIGNAL INTEGRATED CIRCUITS BY TOPOLOGY MODIFICATION 139
Coyette, Anthony ; Esen, Baris ; Vanhooren, Ronny ; Dobbelaere, Wim ; Gielen, Georges

IMPACT OF PARAMETER VARIATIONS ON FINFET FAULTS 145
Harutyunyan, G. ; Tshagharyan, G. ; Zorian, Y.

MEMORY REPAIR FOR HIGH DEFECT DENSITIES 149
Nicolaidis, Michael ; Papavramidou, Panagiota

HORIZONTAL-FPN FAULT COVERAGE IMPROVEMENT IN PRODUCTION TEST OF CMOS IMAGERS 153
Fei, R. ; Moreau, J. ; Mir, S. ; Marcellin, A. ; Mandier, C. ; Huss, E. ; Palmigiani, G. ; Vitrou, P. ; Droniou, T.

CAPACITIVE COUPLING MITIGATION FOR TSV-BASED 3D ICS 159
Eghbal, Ashkan ; Yaghini, Pooria M. ; Bagherzadeh, Nader

IMPROVING ACCURACY OF ON-CHIP DIAGNOSIS VIA INCREMENTAL LEARNING 165
Ren, Xuanle ; Martin, Mitchell ; Blanton, R.D.

RESILIENCY CHALLENGES IN SUB-10NM TECHNOLOGIES 171
Aitken, Rob ; Cannon, Ethan H. ; Pant, Mondira ; Tahoori, Mehdi B.

INNOVATIVE PRACTICES SESSION 7C: MIXED SIGNAL TEST AND DEBUG 175
Natarajan, Suriya

PANEL: ANALOG/RF BIST: ARE WE THERE YET? 176
Ozev, Sule ; Milor, Linda

NO FAULT FOUND: THE ROOT CAUSE 177
Larsson, Erik ; Eklow, Bill ; Davidsson, Scott ; Aitken, Rob ; Jutman, Artur ; Lotz, Christophe

SPECIAL SESSION 8C: E.J. MCCLUSKEY DOCTORAL THESIS AWARD SEMI-FINAL 178
Portolan, M. ; Huang, K.

ABSTRACTION-BASED RELATION MINING FOR FUNCTIONAL TEST GENERATION 180
Gent, Kelson ; Hsiao, Michael S.

RANDOM PATTERN GENERATION FOR POST-SILICON VALIDATION OF DDR3 SDRAM 186
Yang, Hao-Yu ; Kuo, Shih-Hua ; Huang, Tzu-Hsuan ; Chen, Chi-Hung ; Lin, Chris ; Chao, Mango C.-T.

UPF-BASED FORMAL VERIFICATION OF LOW POWER TECHNIQUES IN MODERN PROCESSORS 192
Sharafinejad, Reza ; Alizadeh, Bijan ; Fujita, Masahiro

MBIST AND STATISTICAL HYPOTHESIS TEST FOR TIME DEPENDENT DIELECTRIC BREAKDOWNS DUE TO GOBD VS. BTDDB IN AN SRAM ARRAY 198
Kim, Woongrae ; Chen, Chang-Chih ; Cha, Soonyoung ; Milor, Linda

AN EARLY PREDICTION METHODOLOGY FOR AGING SENSOR INSERTION TO ASSURE SAFE CIRCUIT OPERATION DUE TO NBTI AGING 204
Gomez, Andres ; Poehls, Leticia ; Vargas, Fabian ; Champac, Victor

INTEGRAL IMPACT OF BTI AND VOLTAGE TEMPERATURE VARIATION ON SRAM SENSE AMPLIFIER 210
Agbo, Innocent ; Taouil, Mottaqiallah ; Hamdioui, Said ; Kukner, Halil ; Weckx, Pieter ; Raghavan, Praveen ; Catthoor, Francky

A ROBUST DIGITAL SENSOR IP AND SENSOR INSERTION FLOW FOR IN-SITU PATH TIMING SLACK MONITORING IN SOCS 216
Sadi, M. ; Winemberg, L. ; Tehranipoor, M.

SCALABILITY STUDY OF PSANDE: POWER SUPPLY ANALYSIS FOR NOISE AND DELAY ESTIMATION 222
Rao, Sushmita Kadiyala ; Shivashankar, Bharath ; Robucci, Ryan ; Banerjee, Nilanjan ; Patel, Chintan

ROBUST COUNTERFEIT PCB DETECTION EXPLOITING INTRINSIC TRACE IMPEDANCE VARIATIONS 228
Zhang, Fengchao ; Hennessy, Andrew ; Bhunia, Swarup

FIELD, EXPERIMENTAL, AND ANALYTICAL DATA ON LARGE-SCALE HPC SYSTEMS AND EVALUATION OF THE IMPLICATIONS FOR EXASCALE SYSTEM DESIGN 234
DeBardeleben, Nathan ; Blanchard, Sean ; Kaeli, David ; Rech, Paolo

MULTI-CYCLE CIRCUIT PARAMETER INDEPENDENT ATPG FOR INTERCONNECT OPEN DEFECTS 236
Erb, Dominik ; Scheibler, Karsten ; Sauer, Matthias ; Reddy, Sudhakar M. ; Becker, Bernd

TEST VECTOR OMISSION WITH MINIMAL SETS OF SIMULATED FAULTS 242
Pomeranz, Irith

TEST COMPACTION BY TEST CUBE MERGING FOR FOUR-WAY BRIDGING FAULTS 248
Pomeranz, Irith

PANEL: IS DESIGN-FOR-SECURITY THE NEW DFT? 254
Aitken, Rob

INNOVATIVE PRACTICES SESSION 11C: ADVANCED SCAN METHODOLOGIES [3 PRESENTATIONS] 255
Rajski, Janusz ; Mukherjee, Nilanjan

TESTING CROSS WIRE OPENS WITHIN COMPLEX GATES 256
Han, Chao ; Singh, Adit D.

A DEFINITION OF THE NUMBER OF DETECTIONS FOR FAULTS WITH SINGLE TESTS IN A COMPACT SCAN-BASED TEST SET 262
Pomeranz, Irith

EFFICIENT BUILT-IN SELF TEST OF REGULAR LOGIC CHARACTERIZATION VEHICLES 268
Niewenhuis, Ben ; Blanton, R.D.

SPECIAL SESSION 12B: PANEL: IOT - RELIABLE? SECURE? OR DEATH BY A BILLION CUTS? 274
Kodakara, Sreekumar V. ; Natarajan, Suriya

Author Index

PROCEEDINGS

2015 IEEE 33rd
VLSI TEST SYMPOSIUM
(VTS)

PROCEEDINGS

2015 IEEE 33rd VLSI Test Symposium (VTS)

—— VTS 2015 ——

April 27th – 29th 2015

Napa, California (USA)

2015 IEEE 33rd VLSI Test Symposium (VTS)

Foreword

Welcome to VTS 2015, the thirty third in a series of annual symposia that focus on innovation in the field of testing of integrated circuits and systems.

The core of VTS 2015, the three day technical program, responds to the many trends and challenges in the semiconductor design and manufacturing industries, with papers covering a diverse and seminal set of topics including: Analog, Mixed-Signal & RF Test; ATPG & Test Compression; Delay Testing; Dependability and Reliability; Design for Testability (DFT); Design Verification & Validation; Diagnosis & Debug; Emerging Technologies Test and New Directions in Testing; Hardware Security; Power Issues in Test; Processor, System Testing and Yield; 3D IC Test.

In addition to the three-day technical program, VTS 2015 features several special sessions including five panels, one new topic session, one student activity session, three hot topic sessions and two embedded tutorials. VTS 2015 continues the tradition of featuring the Innovative Practices track. The sessions that make up this track highlight cutting-edge challenges faced by test practitioners, and innovative solutions employed to address them. A special Keynote talk speculating on the future of the test is given by a prominent speaker, Kenneth Wagner, VP of Engineering at PMC Sierra.

The social program at VTS provides an opportunity for informal technical discussions among participants. Napa area provides a highly attractive backdrop for all VTS 2015 activities.

VTS is the result of the work of many dedicated volunteers: the reviewers, the best paper award judges, the Program Committee, the Organizing Committee, and the Steering Committee. We wholeheartedly thank them all. We also wish to thank all the authors who submitted their works to VTS 2015, and the program participants for their contribution at the Symposium. We thank the IEEE Computer Society and the IEEE Computer Society Test Technology Technical Council for the continued sponsorship and support. Finally, we thank the Corporate Supporters of VTS 2015.

We hope that you will find VTS 2015 enlightening, thought-provoking, rewarding, and enjoyable. We wish you all a fun-filled and productive week in the Napa Valley area and hope that you will keep making VTS a success by actively participating in it, assisting in its organization, and letting us always know when we can do something better. Thank you all for coming.

General Chair
Claude Thibeault

Program Chair
Lorena Anghel

VTS 2015 ORGANIZING COMMITTEE

General Chair

Claude THIBEAULT
E. Tech. Sup. Montreal, CA

Program Chair

Lorena ANGHEL
Grenoble Alpes University, TIMA Laboratory, FR

Vice General Chair

Chen-Huan CHIANG
Alcatel-Lucent, USA

Vice General Chair

Yiorgos MAKRIS
University of Texas at Dallas, USA

Vice Program Chair

Rohit KAPUR
Synopsys, USA

Vice Program Chair

Srivaths RAVI
Texas Instruments, USA

New Topics

Bernard COURTOIS
CMP, FR

New Topics

Bozena KAMINSKA
Simon Fraser U., USA

Special Sessions

Sule OZEV
Arizona State University, USA

Innovative Practices Track

Amit MAJUMDAR
XILINX, US

Publication

Giorgio DI NATALE
LIRMM/CNRS, FR

Publicity

Zainalabedin NAVABI
Worcester Polytechnic Finance, USA

Registration

Jennifer DWORAK
SMU, USA

Web

Alessandro SAVINO
Politecnico di Torino, IT

Special Sessions

Matteo SONZA REORDA
Politecnico di Torino, IT

Innovative Practices Track

Rafic MAKKI
GLOBALFOUNDRIES, USA

Publication

Elena Ioana VATAJELU
Politecnico di Torino, IT

Publicity

Kazumi HATAYAMA
NAIST, JP

Web

Stefano DI CARLO
Politecnico di Torino, IT

Finances

Peilin SONG
IBM, USA

Local Arrangements

William EKLOW
Cisco, USA

Audio/Visual

Chintan PATEL
*University of Maryland,
Baltimore County (UMBC)*

Ex-Officio

Yervant ZORIAN
Synopsys, USA

VTS 2015 STEERING AND PROGRAM COMMITTEES

Steering Committee

M. Abadir - Freescale
J. Figueras - U Pol Catalunya
A. Ivanov - UBC
M. Nicolaidis - Grenoble Alpes University, TIMA Laboratory
P. Prinetto - Polit di Torino
A. Singh - Auburn U
P. Varma - Apache Design
Y. Zorian - Synopsys

Program Committee

J. Abraham – Univ. of Texas at Austin
V. Agrawal – Auburn Univ.
B. Becker – Univ. of Freiburg
S. Bernard – LIRMM
R. Blanton – CMU
A. Chatterjee – Georgia Tech
V. Champac – INAOE
C.J. Clark – Intellitech
H. Chen – Mediatek
B. Cory – Nvidia
K. Chung – Samsung
C. Dixit – Avagotech
R. Galivanche – Intel
D. Gizopoulos – Univ. of Athens
S. Gupta – Univ. of Southern California
S. Hamidioui – TU Delft
I. Hartanto – Xilinx
S. Hellebrand – Univ. of Paderborn
R. Karri – NYU-Poly
E. Larsson – Lund University
X. Li – Chinese Academy of Science

P. Maxwell – ON Semiconductor
S. Mitra – Stanford Univ.
S. Makar
S. Natarajan – Intel
A. Orailoglu – UC San Diego
J. Rajski – Mentor Graphics
S. Reddy – Univ. of Iowa
M. Renovell – LIRMM
M. Richetti – Synopsys
S. Shaikh – Broadcom
S. Shoukourian – Synopsys
O. Sinanoglu – NYU Abu Dhabi
H. Stratigopoulos – Grenoble Alpes U., TIMA Lab.
S. Sunter – Mentor Graphics
M. Tehranipoor – Univ. of Connecticut
J. Tyszer – Poznan Univ.
L. Wang – UC Santa Barbara
C. W. Wu – NTHU
H. J. Wunderlich – Univ. of Stuttgart
O. Yaglioglu – FormFactor

Acknowledgments

VTS, like any complex organization, is the result of the efforts of a large number of volunteers, who selflessly have volunteered their time and energy, with their only reward being the satisfaction of seeing a job well done, and the consciousness to have contributed to the dissemination of scientific knowledge through the continued success of a forum dedicated to the exchange of advances in both research and practice in VLSI Test. No words would compare to the magnitude of the efforts displayed by these volunteers. However, we would nonetheless like to register herein our personal note of thanks to the whole body of volunteers, who made it possible the organization of VTS 2015.

A special thanks goes to each member of the Organizing Committee, who excellently played a leading role in each aspect of VTS 2015 organization, with an enormous expenditure of energy and time: without their contribution VTS 2015 could have never taken place. Among all VTS 2015 volunteers, we would like to thank all members of the VTS 2015 Technical Committee and Steering Committee.

Also, we would like to thank all of you, the VTS 2015 participants, the paper submitters, authors and speakers, reviewers, moderators, and IP and special session participants for making the VLSI Test Symposium a continued success and establishing it as the preeminent forum for the exchange of innovative ideas in all aspects of VLSI Test.

General Chair
Claude Thibeault

Program Chair
Lorena Anghel

Each year, VTS proudly presents the Best Paper Award to the author(s) of the most outstanding paper from those presented at the previous year's symposium. The candidates for this honor are initially selected based solely on the numerical ratings of the reviewers and symposium attendees, as recorded on the review forms and the session rating cards. The Best Paper Award Judges then carefully review the candidate papers as published in the proceedings. The judges provide numerical scores and comments for each candidate paper. The scores and comments are compiled to select the best paper.

The paper selected by VTS 2014 Best Paper Award Judges for the **Best Paper Award** is:

3.A.1: A Built-In Self-Test Technique for Load Inductance and Lossless Current Sensing of DC-DC Converters

Tao Liu, Chao Fu, Sule Ozev, Bertan Bakkaloglu *(Arizona State University)*

The VTS 2014 Best Paper Award selection committee is listed below. VTS extends special thanks to these individuals for reviewing the papers and offering invaluable comments.

Jacob Abraham, *University of Texas at Austin*
Amitava Majumdar, *Xilinx*
Salvador Mir, *Grenoble Alpes University, TIMA Laboratory*
Janusz Rajski, *Mentor Graphics*
Adit Singh, *Auburn University*

Each year, VTS recognizes the organizers and presenters of the Best Innovative Practices Session at the previous year's symposium. The selection is based entirely on audience feedback, as recorded on the attendee feedback forms.

For VTS 2014, the Best Innovative Practices Session Award goes to:

IP Session 7.C: Reduced Pin-Count Testing: How Low Can We Go?

Organizer & Moderator:

Stephen SUNTER *(Mentor Graphics)*

Presenters:

Steve COMEN *(Texas Instruments)*

Paul BERNDT *(Cypress Semiconductor)*

Ram RAJAMANI *(Intel)*

Each year, VTS recognizes the organizers and presenters of the Best Special Session at the previous year's symposium. The selection is based entirely on audience feedback, as recorded on the attendee feedback forms.

For VTS 2014, the Best Special Session Award goes to:

**Panel Session 8.B: In-Field Testing of SoC Devices:
Which Solutions by Which Players?**

Organizers:

Matteo SONZA-REORDA (*Politecnico di Torino*)

Dimitris GIZOPOULOS (*University of Athens*)

Presenters:

Jacob ABRAHAM (*University of Texas at Austin*)

Xinli GU (*Huawei*)

Teresa MCLAURIN (*ARM*)

Janusz RAJSKI (*Mentor Graphics*)

Paul RYAN (*Intel*)

IEEE Computer Society

TTTC: Test Technology Technical Council

TTTC IN GENERAL

PURPOSE: The Test Technology Technical Council is a volunteer professional organization sponsored by the IEEE Computer Society. The goals of TTTC are to contribute to members' professional development and advancement and to help them solve engineering problems in electronic test, and help advance the state-of-the art. In particular, TTTC aims at facilitating the knowledge flow in an integrated manner, to ensure overall quality in terms of technical excellence, fairness, openness, and equal opportunities.

MEMBERSHIP: Membership is open to all individuals interested in test engineering at a professional level.

DUES: There are NO dues for TTTC membership and no parent-organization membership requirements.

BENEFITS: The TTTC members benefit from personal association with other test professionals. They may have the opportunity to be involved on a wide range of committees. They receive appropriate and updated information and announcements. There are substantial reductions in fees for TTTC-sponsored meetings and tutorials for members of IEEE and/or IEEE Computer Society.

TTTC ACTIVITIES

TECHNICAL MEETINGS: To spread technical knowledge and advance the state-of-the art, TTTC sponsors many well-known conferences and symposia and holds numerous regional and topical workshops worldwide.

STANDARDS: TTTC initiates, nurtures and encourages new test standards. TTTC-initiated Working Groups have produced numerous IEEE standards, including the 1149 series used throughout the industry.

TECHNICAL ACTIVITIES: TTTC sponsors a number of Technical Activity Committees (TACs) that address emerging test technology topics and guide a wide range of activities.

TUTORIALS and EDUCATION: TTTC sponsors a comprehensive *Test Technology Educational Program (TTEP)*. This program provides opportunities for design and test professionals to update and expand their knowledge base in test technology, and to earn official accreditation from IEEE TTTC, upon the completion of four full day tutorials proposed by TTEP.

TTTC CONTACT

TTTC On-Line: The TTTC Web Site at http://tab.computer.org/tttc offers samples of the TTTC Newsletter, information about technical activities, conferences, workshops and standards, and links to the Web pages of a number of TTTC-sponsored technical meetings.

Becoming a MEMBER: Becoming a TTTC member is extremely simple. You may either contact by phone or e-mail the TTTC office, or fill out and submit a TTTC application form, or visit the membership section of the TTTC web site.

TTTC OFFICE: 1 Marsh Elder Lane, Savannah, GA 31411, USA
　　　　　　　Phone: +1-540-937-5066 Fax: +1-540-937-7848 E-mail:tttc@computer.org

TTTC Officers for 2015

Chair	**Michael NICOLAIDIS** TIMA Laboratory - France	michael.nicolaidis@imag.fr
1st Vice Chair	**Chen-Huan CHIANG** Alcatel-Lucent - USA	chen-huan.chiang@alcatel-lucent.com
2nd Vice Chair	**Rohit KAPUR** Synopsys, Inc. - USA	rkapur@synopsys.com
President of Board	**Yervant ZORIAN** Synopsys Inc. - USA	yervant.zorian@synopsis.com
Past Chair	**Adit D. SINGH** Auburn Univ. - USA	adsingh@eng.auburn.edu
Senior Past Chair	**André IVANOV** U. of British Columbia - Canada	ivanov@ece.ubc.ca
IEEE Design & Test EIC	**André IVANOV** U. of British Columbia - Canada	ivanov@ece.ubc.ca
ITC General Chair	**Mike Purtell** - Intersil	m.purtell@ieee.org
Test Week Coordinator	**Yervant ZORIAN** Synopsys Inc. - USA	Yervant.Zorian@synopsys.com
Secretary	**Joan Figueras** UPC Barcelona Tech - Spain	figueras@eel.upc.edu
Vice Secretary	**Adam OSSEIRAN** Edith Cowan U. – Australia	a.osseiran@ecu.edu.au
Finance Chair	**Chen-Huan CHIANG** Alcatel-Lucent - USA	chen-huan.chiang@alcatel-lucent.com
Finance Vice-Chair	**Bill EKLOW** Cisco Systems, Inc. - USA	beklow@cisco.com

Group Chairs

Technical Meetings	**Chen-Huan CHIANG** Alcatel-Lucent- USA	chen-huan.chiang@alcatel-lucent.com
Technical Activities	**Patrick Girard** LIRMM – France	patrick.girard@lirmm.fr
Tutorials & Education	**Dimitris GIZOPOULOS** University of Piraeus - Greece	dgizop@unipi.gr
Standards	**Rohit KAPUR** Synopsys, Inc. - USA	rkapur@synopsys.com
Communications	**Cecilia METRA** U. of Bologna - Italy	cmetra@deis.unibo.it
Standing Committees	**André IVANOV** U. of British Columbia - Canada	ivanov@ece.ubc.ca
Industry Advisory Board	**Yervant ZORIAN** Synopsys Inc. - USA	Yervant.Zorian@synopsys.com
Electronic Media	**Alfredo BENSO** Politecnico di Torino - Italy	alfredo.benso@polito.it
Asia & Pacific	**Kazumi HATAYAMA** STARC - Japan	hatayama.kazumi@starc.or.jp
Europe	**Matteo SONZA REORDA** Politecnico di Torino – Italy	matteo.sonzareorda@polito.it
Latin America	**Victor Hugo CHAMPAC** Inst. Natl. de Astrofisica - Mexico	champac@inaoep.mx
North America	**André IVANOV** U. of British Columbia - Canada	ivanov@ece.ubc.ca
Middle East & Africa	**Ibrahim HAJJ** American U. of Beirut - Lebanon	ihajj@aub.edu.lb

Technical Activity Committees

Board Testing	**Bill Bill EKLOW** Cisco Systems, Inc. - USA	beklow@cisco.com
Defect Tolerance	**Vincenzo PIURI** Politecnico di Milano - Italy	piuri@elet.polimi.it
Economics of Test	**Magdy S. ABADIR** Freescale, Inc. - USA	m.abadir@freescale.com
Embedded Core Test	**Yervant ZORIAN** Synopsys Inc. - USA	Yervant.Zorian@synopsys.com
FPGA Testing	**Michel RENOVELL** LIRMM - France	renovell@lirmm.fr
Infrastructure IP	**Yervant ZORIAN** Synopsys Inc. - USA	Yervant.Zorian@synopsys.com
Memory Testing	**Yervant ZORIAN** Synopsys Inc. - USA	Yervant.Zorian@synopsys.com
MEMs Testing	**Ronald D. BLANTON** Carnegie-Mellon U. - USA	blanton@ece.cmu.edu
	Bernard COURTOIS TIMA - France	bernard.courtois@imag.fr
Mixed-Signal Testing	**Bozena KAMINSKA** IMS Pultronics, Inc. - USA	bozena@pultronics.com
Nanometer Testing	**Jaume SEGURA** U. of the Balearic Islands - Spain	dfsjsf4@clust.uib.es
Nanotechnology Test	**Fabrizio LOMBARDI** Northeastern U. - USA	lombardi@ece.neu.edu
Network-On-Chip Test	**Erik Jan MARINISSEN** IMEC - Belgium	erik.jan.marinissen@imec.be
On-Line Testing	**Michael NICOLAIDIS** iRoC Technologies - France	michael.nicolaidis@iroctech.com
RF Testing	**Iboun Taimiya SYLLA** Texas Instruments - USA	isylla@ti.com
Silicon Debug and Diagnosis	**Michael RICHETTI** ATI Research, Inc. - USA	mike_ricchetti@ieee.org
System Test	**Ian HARRIS** UC Irvine - USA	harris@ics.uci.edu
3D chips & SiP Testing	**Yervant ZORIAN** Virage Logic Corp. - USA	zorian@viragelogic.com
Test Compression	**Rohit KAPUR** Synopsys, Inc. - USA	rkapur@synopsys.com
Test & Verification	**Magdy S. ABADIR** Freescale, Inc. - USA	m.abadir@freescale.com
Test Education	**Sule OZEV** Duke U. - USA	sule@ee.duke.edu
Test Synthesis	**Scott DAVIDSON** Sun Microsystems - USA	scott.davidson@eng.sun.com
Thermal Testing	**Bernard COURTOIS** TIMA - France	bernard.courtois@imag.fr

Standards Working Groups

IEEE 1149.1	**Christopher J. CLARK** Intellitech Corporation - USA	cclark@intellitech.com
IEEE 1149.4	**Bambang SUPARJO** Mentor Graphics - USA	bambang_suparjo@mentor.com
IEEE 1149.6	**Bill EKLOW** Cisco Systems, Inc. - USA	beklow@cisco.com
IEEE P1149.7	**Robert OSHANA** Texas Instruments – USA	roshana@ti.com
IEEE 1450-1999	**Gregory MASTON** Synopsys, Inc. - USA	gmaston@synopsys.com
IEEE 1450.1	**Tony TAYLOR**	t.taylor@ieee.org
IEEE 1450.2-2002	**Gregg WILDER** Texas Instruments - USA	gwilder@ti.com
IEEE P1450.3	**Tony TAYLOR**	t.taylor@ieee.org
IEEE P1450.4	**Doug SPRAGUE** IBM - USA	dsprague@us.ibm.com
	Jim O'REILLY Analog Devices - USA	jim_oreilly@ieee.org
IEEE P1450.6-1	**Bruce CORY** NVIDIA – USA	bcory@nvidia.com
IEEE P1450.6-2	**Saman ADHAM** LogicVision, Inc. - Canada	saman@logicvision.com

IEEE 1450.6-2005	**Rohit KAPUR** Synopsys, Inc. - USA	rkapur@synopsys.com
IEEE P1450.7	**Jean-Louis CARBONERO** STMicroelectronics - France	jean-louis.carbonero@st.com
IEEE 1500	**Yervant ZORIAN** Virage Logic Corp. - USA	zorian@viragelogic.com
IEEE 1532	**Neil JACOBSON** Xilinx Corp. - USA	neil.jacobson@xilinx.com
IEEE P1581	**Heiko EHRENBERG** GOEPEL Electronics - USA	h.ehrenberg@goepel.com
IEEE P1687	**Kenneth POSSE** AMD - USA	kepos@comcast.net
	Alfred CROUCH Asset InterTech - USA	al.crouch@asset-intertech.com
IEEE P1838	**Erik Jan MARINISSEN** IMEC - Belgium	erik.jan.marinissen@imec.be

TTTC-Sponsored Technical Meetings in 2015

*For the most current information, please visit the TTTC website (http://tab.computer.org/tttc)
or TTTC Events website (http://www.tttc-events.org)*

3/9-3/13	Design, Automation and Test in Europe (DATE), Grenoble, France	W. Nebel
3/25-3/27	Latin American Test Symposium (LATS), Puerto Vallarta, Mexico	V. Champac, Y. Zorian
3/31-4/1	Workshop on Silicon Errors in Logic - System Effects (SELSE), Austin, TX, USA	S. Michalak, H. Naemi
4/22-4/24	Design & Diagnosis of Electronic Circuits & Systems Symposium (DDECS), Belgrade, Serbia	Z. Stamenkovic
4/27-4/29	VLSI Test Symposium (VTS), Napa Valley, CA, USA	C. Thibeault
5/5-5/7	Int'l Symposium on Hardware-Oriented Security and Trust (HOST), McLean, VA, USA	W.H. Robinson
5/22-5/29	European Test Symposium (ETS), Cluj-Napoca, Romania	L. Miclea
7/6-7/8	International On-Line Testing Symposium (IOLTS), Halkidiki, Greece	M. Nicolaidis, A. Paschalis
7/15-7/16	ATE: Vision 2020, San Francisco, CA, USA	S. Tilden
9/TBA	Board Test Workshop (BTW), Austin, TX, USA	W. Eklow
9/26-9/29	East-West Design and Test Symposium (EWDTS), Batumi, Georgia	V. Hahanov, Y. Zorian
10/6-10/8	International Test Conference (ITC), Anaheim, CA, USA	D. Young
10/8-10/9	Int'l Defect and Adaptive Testing Workshop (DATA), Anaheim, CA, USA	A. Sinha
10/8-10/9	Wkshop. on Testing Three-Dimensional Stacked Integrated Circuits (3D-Test), Anaheim, CA, USA	Y. Zorian
10/12-10/14	Int'l Symp. on Defect & Fault Tolerance in VLSI and Nanotechnolohy Systems (DFT), Amherst, MA, USA	S. Kundu, S. Pontarelli
11/22-11/25	Asian Test Symposium (ATS), Bombay, India	M. Inoue, V. Singh
11/25-11/26	Workshop on RTL and High Level Testing (WRTLT), TBA	
12/TBA	International Workshop on Microprocessor Test and Verification (MTV), Austin, TX, USA	M. Abadir
12/TBA	International Design & Test Symposium (IDT), Jordan	Y. Zorian, K. Mhaidat

TTTC Office

1 Marsh Elder Lane
Savannah, GA 31411
USA

Phone: +1-540-937-5066
Fax: +1-540-937-7848
E-mail: tttc@computer.org

http://tab.computer.org/tttc

Keynote Address

New Opportunities in the Internet of Things
Yankin Tanuhan (Synopsys)

Dr. Yankin Tanurhan is Vice President of Engineering for DesignWare Processor Cores, IP Subsystems and Non-Volatile Memory at Synopsys. He leads the low power and high performance ARC embedded Processor developments targeted from Mobile, IoT, Digital Home, Automotive/Industrial to Storage markets, ASIP tool development with products like Processor Designer and IP Designer, IP Subsystems products like Sensor IP Subsystems and Audio Subsystems and CMOS based Non Volatile IP development. Before joining Synopsys, Dr. Tanurhan was Vice President and General Manager of Virage Logic's Processors, SoC Infrastructure and NVM Solutions business units. Virage Logic was acquired by Synopsys in September 2010. Prior to this, Dr. Tanurhan served as Vice President of Actel's Advanced Applications and System Solutions, where he lead Actel's new architecture design, IP and MPU business units, system and hardware tools and product validation departments. He was also responsible for leading Actel's embedded FPGA, embedded processor and DSP activities.

Previously in his research career he served as the director of the department of electronic systems and microsystems of FZI (Forschungszentrum Informatik) a German contract research institute attached to the University of Karlsruhe. Dr. Tanurhan has authored more than 100 papers in refereed publications. He holds a B.S. and M.S. in Electrical and Computer Engineering from Rheinisch Westfaellische Technische Hochschule (RWTH) in Aachen, Germany and a Dr. Ing. degree summa cum laude in Electrical Engineering from the University of Karlsruhe (TH) in Karlsruhe, Germany.

Fault Diagnosis for Flow-Based Microfluidic Biochips

Kai Hu, Bhargab B. Bhattacharya and Krishnendu Chakrabarty

Abstract—Advances in flow-based microfluidics allow biochemistry-on-a-chip for DNA sequencing, drug discovery, and point-of-care disease diagnosis. However, the adoption of flow-based biochips is hampered by defects that frequently occur in chips fabricated using soft lithography techniques. Fault diagnosis methods are now needed to improve fabrication processes and facilitate the (partial) use of chips that have defects. We present the first approach for the automated diagnosis of flow-based microfluidic biochips. The proposed method facilitates the identification of defects through syndrome analysis and a hitting-set problem formulation. The proposed technique is evaluated using three fabricated biochips, and exact defect localization and identification of the defect type is achieved in all cases.

I. INTRODUCTION

Flow-based microfluidic biochips constitute an emerging technology for the automation of biochemical procedures [1]. These devices allow the manipulation of small volumes of liquids, typically at the nanoliter scale, through microchannels and thousands of integrated microvalves [2]. To increase throughput and provide more functionality, the typical size of an integrated microvalve has been reduced ninefold over the past 10 years [3]. Such a tremendous increase in the integration level and valve density has resulted in high defect rates for these devices [5], [6]. Although the chip itself is inexpensive, an incorrect experimental outcome caused by a defective biochip could lead to erroneous clinical diagnosis, which is unacceptable and must be prevented.

Recent advances in the testing of flow-based microfluidic biochips include a behavioral abstraction of physical defects in microchannels and microvalves as stuck at and bridging faults [5], [6]. The flow paths and flow control are modeled as a logic circuit composed of Boolean gates, which enables test generation using typical automatic test-pattern generation (ATPG) tools.

Despite these promising developments in pass/fail testing [5], [6], the diagnosis problem for flow-based microfluidic biochips has yet to be addressed. To date, no automated solution has been proposed and visual inspection under microscopes is the only method available [9]. This diagnosis method is impractical because it is too labor-intensive to facilitate timely process improvements or design refinements. In addition, defects are small enough in size that even skilled observers cannot scan the entire chip at high resolution to identify all defects. As in the semiconductor industry, a manufacturing test procedure should not only detect defects, but also identify physical defects in order to improve yield, reliability, and design practices. Moreover, the discarding of defective biochips without any attempt to identify defects can lead to significant yield loss because a defective chip may be partially functional, and thereby usable for some applications. For example, a cell culture chip contains dozens of identical culture chambers connected in parallel [10]. Such a chip can be used (with lower throughout or parallelism) even if some of the chambers are defective.

In this paper, we present the first approach for defect diagnosis in flow-based microfluidic biochips. The proposed method targets the identification of defect types and defect locations. It can significantly reduce the set of likely locations of defects, and in many cases, exactly identify the defects through syndrome analysis and hitting-set analysis. Three fabricated biochips are used to evaluate the proposed diagnosis technique; for these chips, the proposed approach is able to exactly locate and identify all defects.

The remainder of this paper is organized as follows. Section II provides an overview of a flow-based microfluidic biochip. Related prior work is described in Section III. Section IV describes the diagnosis problem and the proposed solution. The details of the diagnosis algorithm are presented in Section V. Three fabricated flow-based devices are used to demonstrate the proposed approach in Section VI. Conclusions are drawn in Section VII.

II. FLOW-BASED MICROFLUIDIC BIOCHIPS

A flow-based microfluidic biochip utilizes thousands of on-chip microvalves to manipulate pressure-driven flows in a complex network of etched microchannels [2]. A basic microfluidic device is composed of two elastomer layers, each with its own channel network; see Fig. 1(a). Microchannels in the flow layer, called flow channels, are connected to fluid reservoirs through pumps that generates fluid flow. Microchannels in the control layer, called control channels, are connected to an external air pressure source through control pins, which are holes punched on the chip to connect the pressure source to on-chip control channels and microvalves. A flexible membrane, working as a microvalve, is formed at the overlapping area between a flow channel and a control channel. When the pressure source is activated, high pressure in the control channels deflects the membrane of the valve to restrict the fluidic flow in the underlying flow channel. When the pressure source is not active, the fluid is permitted to pass freely. A simple layout of a flow-based biochip is shown in Fig. 1(b), which contains eleven valves and four control pins (Components $n1$, $O1$, $O2$ and $O3$).

Kai Hu and Krishnendu Chakrabarty are with the Department of Electrical & Computer Engineering, Duke University, Durham, NC, 27708, USA (e-mail: {kh131, krish}@duke.edu).

Bhargab B. Bhattacharya is with the Advanced Computing & Microelectronics Unit, Indian Statistical Institute, Kolkata-700108, India (email: bhargab@isical.ac.in).

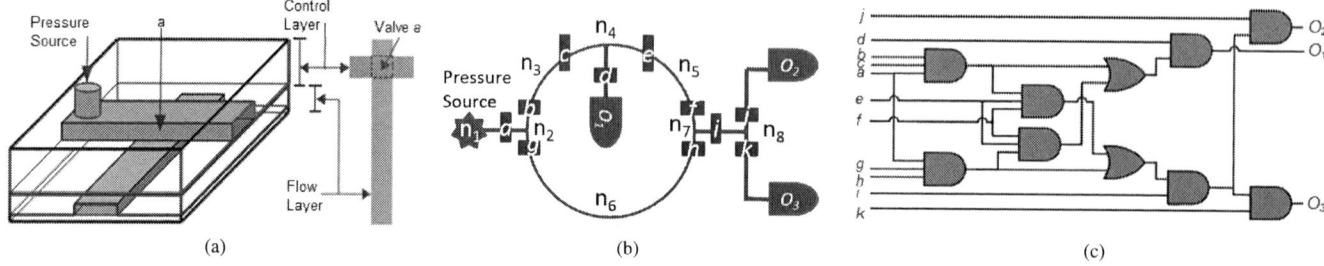

Fig. 1: (a) A schematic of a two-layer flow-based biochip; (b) Layout of a simple microfluidic biochip with a mixer (the circle) and a branch. The junctions of flow channels are labeled as n1– n6. Rectangles indicate the positions of valves, which are connected to pumps via control channels (not shown in the figure); (c) Logic circuit model of the biochip shown in (b).

Table I: Behavioral-level fault model for flow-based biochips.

	Flow Layer	Control Layer
Block	stuck-at-0	stuck-at-1
Leak	OR bridge (1-dominant)	AND bridge (0-dominant)

III. FAULT MODELING AND TESTING

Prior work has shown that, despite the complexity of flow-based microfluidic biochips, the consequences of defects in these devices can be simply described as either a blockage or a leakage [5], [6]. Four types of faults have been defined to represent different defect types in the different layers:

1) Blockage in flow channels: A blockage in the flow channel between Valve g and Valve h in Fig. 1(b) leads to the behavior that Valve g cannot be deactivated. We use a "0" to indicate that the valve is activated, i.e., high pressure in the corresponding control channel. Hence, the faulty behavior of blockage in flow channels is similar to that of a stuck-at-0 fault in integrated circuits.

2) Blockage in control channels: A blockage in the control channel of Valve g in Fig. 1(b) leads to the behavior that Valve g cannot be activated. A "1" at a valve indicates that the valve is deactivated, i.e., there is low pressure in the corresponding control channel. Hence, the faulty behavior due to blockage in control channels is similar to that of a stuck-at-1 fault in integrated circuits.

3) Leakage in the flow channel: Similar to an OR bridging fault in integrated circuits, if a leak occurs between flow channels $g - h$ and $b - c$ in Fig. 1(b), the liquid in channel $g - h$ infiltrates channel $b - c$.

4) Leakage in the control channel: Similar to an AND bridging fault in integrated circuits, if a leak occurs between control channels of g and b in Fig. 1(b), the two shorted valves effectively form one valve. When either valve is activated, both valves become activated.

Table I summarizes behavioral-level fault models for different defect types in the two layers. For example, the chip layout in Fig. 1(a) can be modeled as a logic circuit composed of AND and OR Boolean gates only, as shown in Fig. 1(c). Based on Table I and the logic-circuit model, each physical defect in the chip layout can be mapped to a logic-level fault. Hence, test generation can be realized with a traditional ATPG method [5], [6], [7]. Test patterns are mapped to pressure actuation at input parts and test responses correspond to pressure readings at output ports.

IV. FAULT DIAGNOSIS: PROBLEM DESCRIPTION

During diagnosis, a list of possible defects is examined until the root cause is identified. In this section, we first demonstrate the validity of the single-defect assumption for flow-based microfluidic biochips.

A. Single-defect type assumption

We first assume that only one type of defect may exist in a chip. This assumption can be justified as follows:

1). Each root cause of a defect typically produces the same type of defect. For example, a particle pollutant is likely to block channels, whereas a fiber pollutant is likely to cause leakage between channels. Because the feature size of microfluidic fabrication (~5 micron) is large compared to that of semiconductor chips, it is rare that a defect blocks a channel and also connects adjacent channels. Even if such a situation (i.e., simultaneous blockage and leakage due to a single defect) occurred, the defect is likely to be so large that it can be easily located by visual inspection.

2). There are no inter-layer defects because the control and flow layers of these devices are fabricated separately. As a result, even though different fault models represent the defects in the different layers, it is still acceptable to assume that a faulty chip contains only one type of defect in either the control layer or the flow layer.

Although the single-defect assumption simplifies biochip fault diagnosis, the same type of defect may occur multiple times in a layer. For example, the faulty chip may contain multiple blockage defects that resulted from a scratch on the silicon mold for the control layer.

For the purpose of diagnosis, we must not only directly compare the responses of test patterns but also analyze the erroneous responses in order to identify the locations and types of defects. Note that the logic-circuit models for flow-based biochips are composed only of AND gates and OR gates; there is no inverter in the model. This characteristic of the logic-circuit model causes errors due to the 1/0 (a fault-free value of 1 and a faulty value of 0) and 0/1 (a fault-free value of 0 and a faulty value of 1) to be propagated from the defect sites to the primary outputs without inversion. Hence, we can make a preliminary determination of the defect type by simply observing the type of erroneous responses. For example, if a fault 1/0 is observed at the Outlet O_1 in Fig. 1(a), we can exclude the possibility that the chip contains blockage in the

978-1-4799-7598-3/15 $31.00 © 2015 IEEE

#	Pathways	#	Pathways
1	a, b, c, d, O_1	4	a, g, h, i, j, O_2
2	a, g, h, f, e, d, O_1	5	a, b, c, e, f, i, k, O_3
3	a, b, c, e, f, i, j, O_2	6	a, g, h, i, k, O_3

Table II: Pathway dictionary corresponding to Fig. 1(b).

control channels or leakage in the flow channels under the single-defect assumption. Similarly, if a fault 0/1 is observed, it is impossible that the chip contains blockage in the flow channels or leakage in the control channels.

B. Diagnosis Method

In this subsection, we propose an efficient approach to reduce the number of defect candidates through (1) syndrome analysis, and (2) hitting-set formulation.

Definition 1. *Syndrome(A): Syndrome(A) is the set of erroneous observations at outlet ports for a given pattern if fault A occurs in the logic-circuit model.*

For example, in Fig. 1(c), Syndrome(stuck-at-0 fault at Input i) = {1/0 at O_2, 1/0 at O_3} if all valves are open.

Theorem 1. *Assuming a single defect in the biochip, $Syndrome(A) \cup Syndrome(B) = Syndrome(A, B)$ in the logic-circuit model, where $Syndrome(A, B)$ represents the erroneous observations if both Fault A and Fault B occur.*

The proof of the theorem can be found in [4]. Based on Theorem 1, we can reduce the number of possible defects for a chip by analyzing the syndrome for each single defect. The possibility that Fault A is one of the root causes of a defective chip can be eliminated unless Syndrome(A) is a subset of the observed syndrome. For example, in Fig. 1(b), if all of the valves are open and only the pressure sensor at O_1 reports an error, i.e., observed_syndrome= $\{O_1\}$, we can conclude that there is no blockage defect at Channel a because Syndrome(blockage at Channel a) = $\{O_1, O_2, O_3\}$ is not a subset of the observed_syndrome.

Once we have determined that a defect exists and determined the possible defect types through preliminary response comparison and syndrome analysis, we then proceed to identify the actual defect type and location. A *pathway* is a path that connects a pressure source to a pressure sensor, and a pathway is considered to be *active* for a certain test pattern if all of the microvalves along the pathway are open for the pattern of interest. We use a graph $\mathcal{G} = (V, E)$ to model the continuous fluid-flow topology for a given chip layout to facilitate the search for active paths. The vertices denoted by the set V in the graph represent channel junctions, while the edges denoted by the set E represent not only valves but also their downstream flow channels. Fig. 2 illustrates the graph model for the design shown in Fig. 1(b). A pathway p can be represented by a set of edges (valves/channels) along the path. Hence, $p \subseteq E$. Table II shows the pathway dictionary corresponding to Fig. 1(b).

We next discuss how to locate blockage defects. If there is an active pathway between pressure sources and pressure meters, the pressure meters will detect a high pressure injected

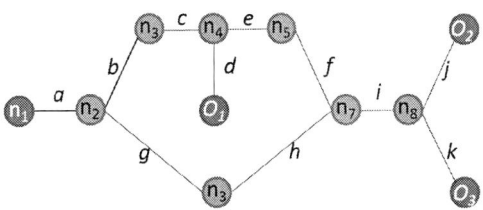

Fig. 2: Graph model corresponding to Fig. 1(b). An edge in the graph represents a valve, while a vertex represents a flow channel.

by the pumps. Conversely, if all active pathways are blocked, the pressure meters cannot sense high pressure. Hence, if a chip suffers from a blockage defect and the faulty observation is 1/0, we can conclude the following: (1) the blockage defects are in the flow layer, and (2) all of the active pathways are blocked, that is, there is at least one blockage defect in each active pathway, blocking the pressure propagation from the pressure source to the pressure sensor.

We next show that the problem of locating blockage defects in the flow layer can be modeled as a hitting-set problem. The proof of Theorem 2 can be found in [4].

Definition 2. A hitting set H of a collection $\mathbb{R}$ is a set having a non-empty intersection with each subset $S \in \mathbb{R}$; that is, $S \cap H \neq \emptyset, \forall S \in \mathbb{R}$ [13].

Theorem 2. *The set of blockage defects in the flow channel must be a hitting set of the active pathways between the pressure source and the pressure sensors reporting errors.*

Next, we describe how we can establish the active pathway collection $\mathbb{R}$. Before the start of the diagnosis procedure, we establish a path dictionary Ω by backtracking and then indexing pathways for a given chip. An active pathway collection $\mathbb{R}$ can therefore be obtained by removing the inactive pathways in the pathway dictionary Ω. Note that Theorem 2 holds even if the collection of active pathways is incomplete. In [11], [7] two path-search algorithms were presented for the generation of path dictionary; details are omitted here due to space limits.

Let T denote the set of test patterns for the biochip. For a test pattern $t \in T$ and an Outlet o, we let $\mathbb{R}_{t,o} \subseteq \Omega$ represent a collection that is composed of all the active pathways between the source and Outlet o for test pattern t. Active pathways for a test pattern can be identified by determining whether the pathway $p \in \Omega$ contains a valve $j \in E$ that should be closed. For a given test pattern, only pathways with all valves open are considered as active pathways. Hence, by screening (and thereby eliminating) paths in the path dictionary Ω, we can obtain a sub-collection $\mathbb{R}$ that includes the active pathways for each faulty observation for each test pattern, i.e., $\mathbb{R} = \cup_{t \in T, o \in p_t} \mathbb{R}_{t,o} = \cup S$, where p_t is the set of outlets where the pressure sensors report errors.

The number of possible blockage defect locations can be further reduced by determining the hitting set of Collection $\mathbb{R}$. If the chip has multiple defects, the hitting set contains multiple elements. A collection can have multiple hitting sets as long as the condition described in Definition 1 is satisfied; each hitting set of $\mathbb{R}$ can block all active pathways and produce

978-1-4799-7598-3/15 $31.00 © 2015 IEEE

the observed erroneous responses of pressure sensors.

Let us revisit Fig. 1(b). All valves are open except Valve i. An error 1/0 is observed at O_1. Hence, $\mathbb{R}$ consists of Pathway 1 and Pathway 2 in Table II. Because an error 1/0 cannot be observed unless all of the pathways in $\mathbb{R}$ are blocked, the defects must be located within the hitting set of $\mathbb{R}$, namely $\{a\}$, $\{d\}$, $\{b, h\}$, $\{c, e\}$, etc.

We can formulate the diagnosis problem for blockage in the control layer as a hitting-set problem as well. If there is a blockage in the control channel for a valve, the corresponding valve can no longer be closed and an error 0/1 (a fault-free state 0 and faulty state 1) may be observed.

Definition 3. A Cut c of Outlet o is a partitioning of the set of edges (valves) in the graph model that separates the pressure source and the Outlet o into two disjoint subsets, thereby cutting off all of the active pathways.

Hence, a '0' output at Outlet o implies that there is a Cut c of Outlet o that disconnects all of the active pathways in the graph model. If an error 0/1 is observed at Outlet o, we can conclude that one or more valves on the cuts of this outlet that are expected to be closed in the fault-free case are actually open, which results in an active pathway between the pressure source and the pressure sensor.

We next show that the problem of locating blockage defects in the control layer can be modeled as a hitting-set problem of cuts. Again, the proof of the theorem can be found in [4].

Theorem 3. *The locations of blockage defects in the control layer must be a hitting set of cuts of the outlets reporting errors.*

Next, as in the diagnosis procedure for blockage defects in the flow layer, once the cuts have been found for each erroneous response and test pattern, the number of possible blockage defects in the control layer can be reduced by determining the hitting set $\mathbb{R}$ of the cuts. If the chip has multiple defects, the hitting set should contain multiple elements. A collection $\mathbb{R}$ can have multiple hitting sets as long as Definition 2 is satisfied; each hitting set of $\mathbb{R}$ can produce the observed erroneous response. Hence, more effort is needed to further identify the defects. For example, in Fig. 3, a test pattern t with all valves open except Valve c and Valve h, is applied to the chip of Fig. 1(b). For this example, Edge c and Edge h should be removed in the graph shown in Fig. 2. The pressure in the flow channels cannot be sensed at O_1 because the removal of Edge c and Edge h forms a cut that prevents the high pressure injected at Inlet n_1 from reaching the pressure meter at Outlet O_1 (Fig. 3). If there is a blockage defect in the control channels of either Valve c or Valve h, the corresponding valve is prevented from closing, and an error 0/1 is observed at Outlet O_1. Hence, $\mathbb{R} = \{c, h\}$ and these defects must be included in the hitting set of $\mathbb{R}$, i.e., either Valve c or Valve h or both.

C. Fault simulation for defect-location identification

The method described above is applicable only for blockage defects. If the erroneous responses are caused by a leakage defect, it is necessary to enhance the method to differentiate

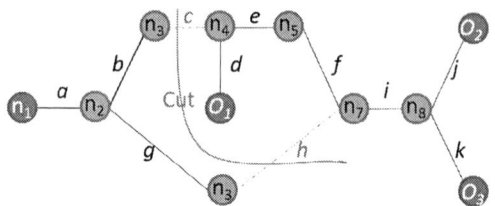

Fig. 3: Graph model for pattern t, where all valves are open except Valve c and Valve h. A cut is generated because Valve c and Valve h are closed.

leakage from blockage. Recall that leakage is caused by defective areas of the channel wall. Therefore, it is reasonable to exclude the possibility of leakage defects if all the candidate defective channels/valves in the hitting set are not adjacent to each other in the physical layout. It was shown in [5], [6] that leakage defects can be mapped to bridging faults, whose faulty behavior can be modeled as conditional stuck-at faults. Consequently, we can identify defect types by simulating fault behaviors and checking whether the necessary conditions for the corresponding bridging faults are satisfied for all test patterns. If the necessary conditions are not satisfied but an erroneous response is observed, we can conclude that the chip suffers from blockage defects rather than leakage defects.

On the other hand, Theorem 2 and Theorem 3 show that the hitting sets only provide candidates for the actual root cause; an element of a hitting set cannot be regarded as a root cause unless the fault-simulation results are identical to the observed responses at each pressure sensor for each test pattern. This validation must therefore be carried out for each hitting set of the active pathways/cuts.

V. Algorithm Design

The proposed diagnosis algorithm has two main parts: data preparation and diagnosis. The data-preparation stage (Step 1-3) is completed before the start of experiment, without any knowledge of the particular characteristics of a chip. This off-line data-preparation step needs to be performed only once but it significantly accelerates the diagnosis process. In the diagnosis stage (Step 4-9), a specific biochip is tested and the responses are reported by pressure meters. Syndrome analysis and the hitting-set problem formulation are used to reduce the number of defect candidates until the possible root causes are determined. We next explain the algorithm step-by-step.

Step 1: A graph model is generated to model the continuous fluid-flow topology for a given chip layout.

Step 2: We simulate and store the syndrome of each fault-candidate for each test pattern *a-priori*. As stated in Theorem 1, the likelihood of Fault A causing a defect can be eliminated unless Syndrome(A) is a subset of the observed syndrome.

Step 3: The path/cut dictionary is established before test application. For example, Table II is the path dictionary of the biochip shown in Fig. 1(c).

Step 4: If Fault 1/0 is observed, the defect type is either blockage in the flow channel (stuck-at-0) or leakage in the control channel (AND bridging). Otherwise, the defect type is either blockage in the control channel (stuck-at-1) or leakage in the flow channel (OR bridging).

978-1-4799-7598-3/15 $31.00 © 2015 IEEE

Step 5: See Subsection IV-B (Syndrome Analysis).

Step 6: Defect candidates are identified by determining the hitting set H of the active pathways and that of the cuts. The problem of finding a hitting set for a given collection is NP-complete [13], so we use an efficient heuristic solution. To avoid a local minimum, the hitting set is searched using simulating annealing (SA) [14], and the number of iterations in the cooling scheduling of SA i_1 is set to 500 for one run. Note that the first solution provided by the SA algorithm is one possible hitting set, and it may not necessary represent the actual defect set. Hence, we re-run the SA algorithm ten times to provide ten hitting sets, in one pass. We process these hitting sets by simulating their responses.

Step 8: Because a leakage is caused by defective areas of the channel wall, the possibility of leakage defects can be excluded if candidate defective channels/valves in the hitting set are not adjacent to each other in the physical layout.

Step 9: In this step, we validate the solutions of the hitting set using fault simulation, in ascending order of their sizes. If the simulated responses of blockage and leakage defects are identical to the observed responses at each pressure sensor for all applied test patterns, the corresponding hitting set (i.e., the fault-list) is said to pass the validation, and is announced as the source of defects. Otherwise, we continue to do further simulation with the remaining solutions. If all of them fail, we rerun the SA algorithm for one more pass (Step 6) to generate ten new solutions, and repeat the simulation steps.

Note that the diagnosis of the defect type is not necessary because, as long as the defect sites are located correctly, the defect types can be easily rapidly determined by visual inspection with a microscope. An analysis of computational complexity of the diagnosis can be found in [4]. For a design with v valves, p outlets, n test patterns, and w flow channels, the computational complexity is $O(v + w)pn$. Hence, the runtime of the proposed diagnosis algorithm grows slowly with an increase in chip complexity.

VI. RESULTS: APPLICATION TO FABRICATED BIOCHIPS

This section evaluates the effectiveness of the proposed diagnosis method using three fabricated biochips. The specification of each chip is listed in Table III. As ours is the first approach to fault diagnosis in flow-based microfluidic biochips, we compare it with a baseline solution based on exhaustive search. The two methods are described below.

- Baseline method: All possible combinations of candidate defects are obtained through exhaustive search. The symptoms for all of the combinations are simulated until the simulated result matches the observed ones. Hence, if the chip-under-test has n defects, the worst-case computational complexity is $O(2^n)$.
- Proposed method: The number of possible defect combinations is reduced by syndrome analysis and the search for hitting sets. Hence, significantly fewer fault simulations are needed in order to find the root cause.

A. Results for ChIP chip

The flow-based biochip used in the first example is designed for chromatin immunoprecipitation (ChIP), an assay used to

Chip Name	# Valves	# Pc	# Pf	# P	# AND	# OR
ChIP	28	28	15	30	28	5
WGA	235	23	19	12	235	49
Cell Culture	720	48	24	92	720	94

Table III: The specification of fabricated chips used as examples. "# Pc" denotes the number of ports in the control layer, "# Pf" denotes the number of ports in the flow layer, "# AND" and "# OR" denotes the number of AND gates and OR gates in the logic circuit model.

	#D	Blockage			Leakage			CPU Time	
		#E	#C	i_2	#E	#C	i_2	T_p	T_B
ChIP	2	6	10.8	1	5	14.5	1	3.38	7.9
	3	9	10.5	1	8	16.5	1	5.32	5.4e2
	4	8	11.6	1	6	18	1	6.44	1.7e4
	5	9	13.8	1	8	15	1	7.41	4.7e5
	6	10	13.1	1	9	19	1	8.38	1.4e7
WGA	2	4	11.8	1	3	13.8	2	61	8.0
	3	6	20.4	1	4	31.6	3	85	2.5e2
	4	6	21.2	1	5	38.4	3	148	1.9e4
	5	9	29.8	1	8	25.9	4	193	1.2e6
	6	11	36.6	1	10	30.6	3	215	5.2e7
Cell Culture	1	6	23.4	1	7	25.9	1	1151	1.2e5
	2	25	11.5	1	19	14.9	3	165	1.5e7
	3	13	49.2	1	8	112.3	5	1183	-
	4	18	65.2	1	15	98.8	4	1240	-
	5	48	55.4	1	15	144.6	7	1360	-

Table IV: Additional information about experimental results. #D: the number of defects in the chip, #E: the number of erroneous observations reported by pressure sensors, #C: denotes the number of fault candidates after syndrome analysis, i_2: the average number of passes for SA needed to find the root cause, T_p and T_b: the CPU time (Unit: ms) of the proposed method and the baseline method, respectively.

analyze DNA-protein interactions [15]. A total of 34 paths are found and then stored in the path dictionary Ω. For each defect, the corresponding syndrome is simulated and stored in the syndrome dictionary. A total of 30 test patterns reported in [5] are needed to achieve 100% fault coverage. Two tests are performed to evaluate the proposed method. In each test, we evaluate the proposed method using 50 independent simulations, where each simulation has n random defects and n ranges from 2 to 6. Because of the single-fault-type assumption, all defects in each simulation are the same type.

The proposed method successfully finds all blockage defects. Although the defects are correctly located in all cases, 7% of the leakage defects are misdiagnosed as blockage defects. This is not a concern since in practice, once the automatic diagnosis has been completed, candidate defect sites must be visually inspected. Details of experimental results including the number of erroneous observations, the number of fault-candidates, are shown in Table IV.

The average CPU time for the proposed method and the baseline method are shown in Table IV. The proposed method requires much less CPU time than the baseline method, especially when the chip under test has multiple defects. In the 6-defect case, the CPU time increases to 3.5 hours for the baseline method. If we the extrapolate results, a chip with 7 defects will require a diagnosis time of 12 days, which is obviously unacceptable.

B. Results for WGA chip

We use experimental data reported in [6]. The biochip-under-test is used for whole genome amplification (WGA).

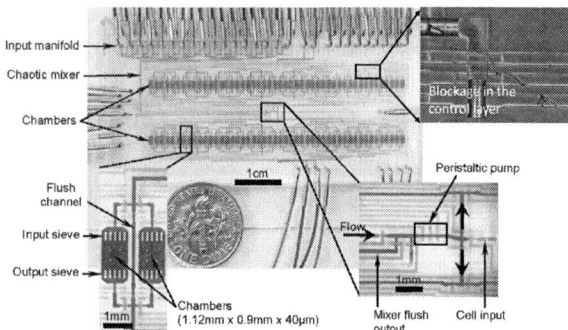

Fig. 4: Image of the faulty cell culture chip. The chip suffers from block defects in the control channels [10], [6].

The chip layout is shown in [4]. Control channels are shown in red. Port "Pressure" is connected to a pressure source.

The experimental test results of a defective chip with block defects were reported in [6]. During the testing of the defective chip, the pressure sensors report errors for two test patterns due to blockage defects [6]. The proposed diagnosis method further analyzes the experimental data for the two test patterns. A total of 144 active pathways are found for these patterns, and then SA is used to search for hitting sets. Finally, the proposed method correctly identifies the defect in 85 ms, while the baseline method takes 249 ms.

Next, we evaluate the proposed methods for more general cases involving a larger number of defects. Two tests are performed to evaluate the proposed method. In each test, we evaluate the proposed method using 20 independent simulations. In each simulation, $n\%$ of the valves or flow channels are randomly selected as defect sites and n ranges from 1 to 5. Once again, the proposed method correctly locates the defects in all cases, though 18% of the leakage defects are misdiagnosed as blockage defects. More information about the proposed method is available in Table IV.

The average CPU time for the proposed method and baseline method are shown in Table IV. In the case that a chip has 5 defects, the CPU time of the baseline method has increased to 20 hours. According to extrapolated results, the diagnosis of a chip with 6 defects will take a year of CPU time.

C. Results for cell culture chip

The last example presented to evaluate the proposed diagnosis method is a cell culture chip shown in Fig. 4 [10]. The chip has a channel width of 0.13 mm, hence a 10x magnification is necessary for inspection. However, a 10x magnification corresponds to a 1 mm by 1 mm field of view. Because the chip size is 32.6 mm by 66.6 mm, this implies at least 2200 observations to inspect the entire chip. Thus, visual inspection is too labor-intensive to be practical.

For this chip, a total of 92 test patterns are generated and 305 pathways are found. The experimental test results of a defective chip with block defects were reported in [6]. Magnified images of the defects are shown in Fig. 4. The proposed method correctly identifies the defects in 2.77 s.

Next, five tests are performed to evaluate the proposed method. In each test, n defects are randomly chosen, where

the number n ranges from 1 to 5. Each test consists of 50 independent simulations. All defects in each simulation are the same type. The proposed method successfully locates the defects in all cases, though 25% of the leakage defects are misdiagnosed as blockage defects. Additional information is available in Table IV.

The CPU time of the proposed method and the baseline method (exhaustive search) are shown in Table IV. The CPU time of the baseline method increases to 4 hours for two defects. Because the CPU time of the baseline method increases exponentially, the diagnosis processes expected to take more than 15 days in the case of three defects.

VII. CONCLUSION

We have presented the first approach for defect diagnosis in flow-based microfluidic biochips. Given a set of erroneous responses reported by pressure meters for different test patterns (valve conditions), syndrome analysis and hitting-set theory are used to reduce the number of suspicious defects, possibly to the point of exact identification of defect locations. In this way, visual inspection is facilitated through the pinpointing of the possible locations of defects. Three fabricated biochips have been used to evaluate the proposed diagnosis method. We have shown that, compared to a baseline exhaustive-search method, the proposed approach can exactly locate the defects in much less time, and the benefits are especially obvious for larger chips and a larger number of defects.

REFERENCES

[1] D. Mark et al., "Microfluidic lab-on-a-chip platforms: requirements, characteristics and applications.," *Chemical Society Reviews*, vol. 39, Mar. 2010.

[2] T. Squires and S. Quake, "Microfluidics: Fluid physics at the nanoliter scale," *Reviews of Modern Physics*, vol. 77, pp. 977–1026, Oct. 2005.

[3] I. E. Araci and S. R. Quake, "Microfluidic very large scale integration (mVLSI) with integrated micromechanical valves." *Lab on a Chip*, vol.12, pp. 2803–6, Aug. 2012.

[4] https://drive.google.com/file/d/0B8kBOaUd8Fr0RFRjeGhzX0wxUVk/view?usp=sharing

[5] K. Hu, T.-Y. Ho, and K. Chakrabarty, "Testing of flow-based microfluidic biochips", *Proc. IEEE VLSI Test Symposium*, pp. 124-129, 2013.

[6] K. Hu, F. Yu, T.-Y. Ho and K. Chakrabarty, "Testing of flow-based microfluidic biochips: Fault modeling, test generation, and experimental demonstration", *IEEE Trans. CAD*, vol. 33, pp. 1463-1475, Oct. 2014.

[7] K. Hu, T.-Y. Ho and K. Chakrabarty, "Test generation and design-for-testability for flow-based mVLSI microfluidic biochips", *Proc. IEEE VLSI Test Symposium*, pp. 97-102, 2014.

[8] http://www.youtube.com/watch?v=xWdRczefirs

[9] H. Hassanin, A. Mohammadkhani, and K. Jiang, "Fabrication of Hybrid Nanostructured Arrays Using a PDMS/PDMS Replication Process.,"*Lab on a Chip*, vol. 12, no. 20, pp. 4160–7, Sep. 2012.

[10] R. Gomez-Sjoberg, et al., "Versatile, fully automated, microfluidic cell culture system", *Anal. Chem.*, vol. 79, pp. 8557–8563, 2007.

[11] K. Hu, T.-Y. Ho and K. Chakrabarty, "Wash optimization for cross-contamination removal in flow-based microfluidic biochips", *ASP-DAC*, pp. 244-249, 2014.

[12] B. Milic and M. Malek, "Adaptation of the breadth first search algorithm for cut-edge detection in wireless multihop networks," *Proc. of. ACM MSWiM*, pp. 377-386, 2007.

[13] R. Niedermeier, et al., "An Efficient Fixed Parameter Algorithm for 3-Hitting Set," *Journal of Discrete Algorithms*, 1: 89-102, 2003.

[14] S. Kirkpatrick , C. Gellat and M. Vecchi "Optimization by simulated annealing", *Science*, vol. 220, no. 4598, pp. 671-680, 1983

[15] A. Wu et al., "Automated Microfluidic Chromatin Immunoprecipitation from 2,000 Cells," *Lab on a Chip*, vol. 9, pp. 1365–1730, 2009.

[16] P. Blainey, S. Quake, "Digital MDA for enumeration of total nucleic acid contamination", *Nucleic Acids Research Advance Access*, Nov. 2010.

Rapid Online Fault Recovery for Cyber-physical Digital Microfluidic Biochips

Christopher Jaress, Philip Brisk
Department of Computer Science and Engineering
University of California, Riverside
Riverside, CA, USA
chrisjaress@gmail.com, philip@cs.ucr.edu

Daniel Grissom
Department of Engineering and Computer Science
Azusa Pacific University
Azusa, CA, USA
dgrissom@apu.edu

Abstract—**Microfluidic technologies offer benefits to the biological sciences by miniaturizing and automating chemical reactions. Software-controlled laboratories-on-a-chip (LoCs) execute biological protocols (assays) specified using high-level languages. Integrated sensors and video monitoring provide a closed feedback loop between the LoC and its control software, which provide timely information about the progress of an ongoing assay and the overall health of the LoC. This paper introduces a cyber-physical control algorithm that rectifies hard and soft faults that are detected dynamically while executing an assay on a digital microfluidic biochip (DMFB), one specific LoC technology. The approach is scalable (i.e., there is no fixed limit on the number of faults that may occur), and runs efficiently in practice, thereby limiting the performance overhead incurred when a hard or soft fault occurs during assay execution.**

Keywords—Digital Microfluidic Biochip, Error Recovery

I. INTRODUCTION

A digital microfluidic biochip (DMFB) [20] is a device that manipulates discrete droplets of liquid via electrostatic actuation atop a 2-dimensional grid of electrodes. Compared to existing laboratory-on-a-chip (LoC) technologies that manipulate continuous flows of fluid, DMFBs offer three key advantages: (1) the ability to manipulate fluids individually; (2) the ability to immerse solids within liquids without the risk of clogging one or more microchannels; and (3) compatibility with a wide variety of fluid volumes. DMFB applications include DNA sequencing, immunoassays, point-of-care diagnostics, and many others [11].

Recent DMFBs integrate devices such as heaters [13], photo-detectors [14, 29], impedance sensors [22], or magnetic separators [6], which provide feedback to a PC that controls the execution of an assay (biochemical reaction) running on the DMFB, forming a feedback-control loop as shown in Fig. 1. Such a *cyber-physical* DMFB can execute assays that incorporate sensory feedback and real-time decision-making into their specification. Historically, assays were specified as directed acyclic graphs (DAGs) without decision-making or control flow. A cyber-physical DMFB can now execute assays specified as control flow graphs (CFGs), as shown in Fig. 2, which include conditions and loops whose behavior is driven by sensor feedback. Each CFG node contains a DAG, and the last operation in each DAG is either a branch or the CFG exit point, signifying that the assay has terminated.

Each control flow operation signifies a reconfiguration point, as it is not possible to predict control behavior at compile-time, and the precise configuration of droplets at the start point of each DAG is not guaranteed to be the same each time that a CFG node is invoked for execution. Consequently, it is necessary to re-compile each DAG on-the-fly as the CFG executes, i.e., the system employs a just-in-time (JIT) compiler. Each call to the compiler must schedule, place, and route the assay in real-time, i.e., the assay pauses while the compiler solves these interdependent NP-complete problems. In this context, a premium must be placed on the runtime of the JIT compiler, as opposed to solution quality.

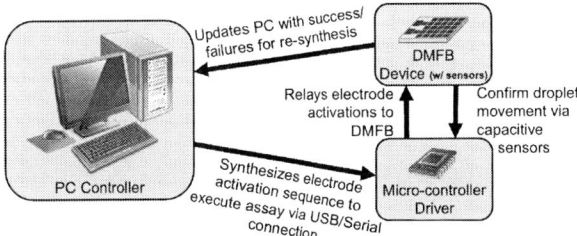

Fig. 1. A feedback-control loop for a cyber-physical DMFB with integrated capacitive-touch sensors.

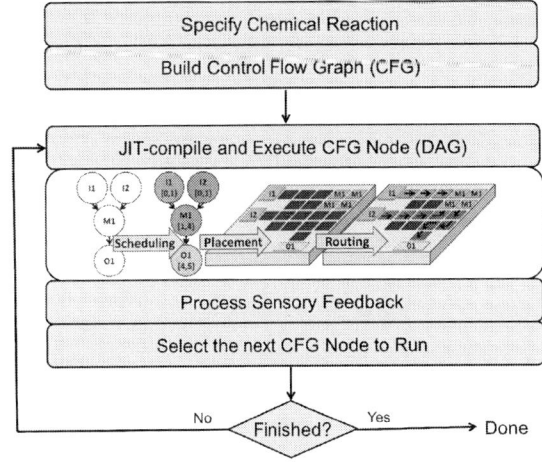

Fig. 2. Software architecture of a system that executes assays specified as control flow graphs (CFGs) in real-time.

This work was supported in part by NSF Grant CNS-1035603. Any opinions, findings, and conclusions or recommendations expressed in this material are those of the authors and do not necessarily reflect those of the NSF.

978-1-4799-7598-3/15 $31.00 © 2015 IEEE

Within the context of an online JIT compiler, this paper introduces a cyber-physical control model for DMFBs that enables recovery from faults that occur during assay execution. A hard fault refers to a device-level failure that renders a portion of the DMFB unusable. DMFBs offer abundant spatial parallelism, so a typical hard fault manifests itself as a loss of some parallelism. In the worst case, a hard fault could block the area that interfaces with an external device, such as a heater, or could block access to an I/O reservoir on the perimeter of the chip. Most hard faults are not catastrophic.

A cyber-physical DMFB can detect a hard fault in real-time by comparing the behavior of a droplet with its expected action based on a control signal sent to the device. For example, if an electrode adjacent to a droplet is activated, the expected action is droplet motion; if the droplet does not move in response, then we can assume that the droplet is stuck and the region of the chip surrounding the droplet is no longer usable. The remainder of the assay must be recompiled to avoid the faulty area. Fast re-compilation methods are needed to achieve high throughput in the presence of faults and to avoid spoilage of samples and reagents.

Soft faults, in contrast, represent erroneous assay operations that do not indicate device failure. For example, one droplet may be split into two droplets of significantly unequal volume [2, 3], or the concentration of a droplet may not be within the calibrated range of the sensor [29]. If so, the erroneous droplet must be discarded, and the part of the assay that produced the droplet is re-executed. Likewise, this entails (re-)compiling a portion of the assay to introduce new operations whose necessity could not be predicted statically.

Contribution: This paper contributes a compiler and runtime monitoring system for cyber-physical DMFBs to enable fast dynamic fault recovery. Compared to prior work, our system offers the following advantages: (1) This is the first control mechanism for cyber-physical DMFBs that handles hard and soft faults in a unified fashion; (2) the algorithm is scalable, i.e., there is no hard upper bound on the number of faults that can be tolerated; (3) the algorithm is faster than all prior scalable fault recovery algorithms that have been published to date; and (4) the general approach is intuitive and easy to implement, which favors rapid software development and a lower likelihood of errors and bug fixes later on.

II. BACKGROUND AND APPROACH

Checkpoints are automatically inserted into an assay to test for observable errors, [29]. Each checkpoint routes a droplet to a sensor/detector for assessment; if the assessment fails, an error recovery subgraph (a static program slice [27] containing all operations that may affect the droplet at the checkpoint) is inserted into the assay (Fig. 3), and the schedule, placement, and routing plan are updated. Checkpoint and error recovery subgraph insertion can be done manually or by a compiler.

The assay is initially specified as a DAG. Checkpoints and error recovery subgraphs are inserted into the assay, converting into an executable CFG, as shown in Fig. 2. Each checkpoint ends with a condition based on sensory feedback: if an error occurs, control transfers to the error recovery subgraph; if there is no error, control transfers to the next operation.

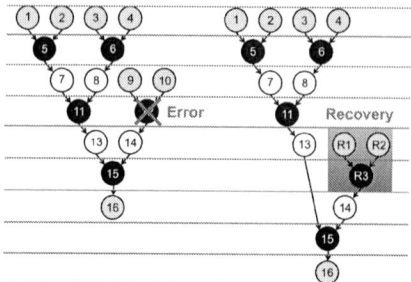

Fig. 3. A soft faults occurs at runtime in a scheduled DAG. An error recovery subgraph is introduced and the updated DAG is rescheduled.

III. RELATED WORK

Cyber-physical integration enables real-time detection and fault recovery; several algorithms have been introduced to reschedule assay operations and reconfigure the DMFB to recover from faults. Table 1 lists the algorithms and their relevant properties in comparison to algorithms introduced in this paper. With the exception of Luo et al. [14], all of these techniques focus explicitly on hard or soft faults, but not both.

Zhao et al. [29] pioneered real-time soft fault detection and recovery for DMFBs. Their approach had two limitations: (1) all operations stop during recovery, including those that do not depend on droplets involved in the fault; and (2) operations within the error recovery subgraph must be fault-free. Subsequent work has addressed these limitations [14, 15].

Maftei et al. [16] and Alistar et al. [2] detect hard faults offline; their compiler avoids the use of faulty DMFB regions; they do not detect or recover faults that occur online.

Alistar et al. [1] and Luo et al. [15] enumerate all combinations of soft faults that might occur during assay execution, and generate all of the error recovery subgraphs that could reconfigure the system at runtime. These approaches reduce recompilation times, but can only tolerate a small number of faults due to exponentially large storage costs. They are non-scalable and cannot tolerate hard faults.

Many assays produce intermediate droplets that are not used. Early DMFB compilers dispose of all unneeded droplets, as there was no motivation to store them. Hseih et al. [10] and Luo et al. [14] optimistically store some of these intermediate droplets, as they can reduce the overhead of the fault recovery process. Our approach can support droplet re-use if desired.

Table I shows that: (1) prior work has used dynamic recompilation to recover from soft, but not hard, faults; (2) list scheduling is preferred, presumably due to its efficiency; (3) only one paper has used a polynomial-time placement heuristic, and its runtime is quadratic [14]; and (4) prior work has not considered droplet routing on recompilation.

The contribution of our work is an online recompilation technique for hard and soft faults that is scalabile, achieves a linear time complexity for placement, and accounts for droplet routing. The time router's complexity is O(MN) [21], as it uses Soukup's routing algorithm internally [24]. The average case performance of Soukup's router is less than its worst-case time complexity, and our online alternative to placement guarantees routability and helps the router converge quickly.

TABLE I. COMPARISON BETWEEN THE FAULT RECOVERY TECHNIQUES INTRODUCED IN THIS PAPER AND PRIOR WORK; THE LIMITATIONS AND WEAKNESSES OF PRIOR WORK COMPARED TO OURS ARE HIGHLIGHTED.

Reference	Overview of Fault Detection and Recovery				During Recovery		Online Recompilation Algorithms		
	Fault Type	Fault Detection	Scalable	Droplet Re-use	Assay Pauses	Tolerance to further faults	Scheduling	Placement	Routing
Zhao et al. [28]	**Soft**	Online	Yes	Yes	**Yes**	**No**	Computed Offline		
Alistar et al. [1]	**Soft**	Online	**No**	**No**	No	Yes	Enumerated Offline		**No**
Alistar et al. [3]	**Soft**	Online	Yes	**No**	No	Yes	O(nlogn)*	**No**	**No**
Maftei et al. [16]	**Hard**	**Offline**	Yes	**No**	N/A	N/A	N/A		
Luo et al. [15]	**Soft**	Online	**No**	**No**	No	Yes	Enumerated Offline		
Hsieh et al. [10]	**Soft**	Online	Yes	Yes	No	Yes	Ref. [14] or [27]		
Alistar et al. [2]	**Hard**	**Offline**	**No**	N/A	N/A	N/A	N/A		
Luo et al. [14]	Both	Online	Yes	Yes	No	Yes	O(nlogn)*	**O(MN)****	**No**
	Both	Online	Yes	Yes	No	Yes	**Iterative Improvement**		**No**
Our Work	Both	Online	Yes	Yes	No	Yes	O(nlogn)*	O(P)*** [12]	O(MN)**

* n is the number of assay operations (vertices in the DAG); list scheduling [7, 24] has an O(nlogn) time complexity.
** M and N are the DMFB length and width. O(MN) is the time complexity of Soukup's algorithm [24], used for path planning during routing [21].
*** P is the number of modules placed on the chip during reconfiguration; the O(P) time complexity is reported in ref. [12].

IV. ONLINE FAULT RECOVERY

A. Virtual Topology

The key to enable fast and efficient JIT compilation is the notion of a virtual topology [7], as shown in Fig. 4. A virtual topology segregates specific regions of the chip (work modules) to perform assay operations (mixing, dilution, storage, etc.), while leaving space (streets) between modules for droplet transport. External devices (heaters, detectors, etc.) may enhance the functionality of a work module, but do not affect droplets transported on a street.

Virtual topologies eliminate certain mistakes that arise from the interdependence between schedulers, placers, and routers. In Fig. 5(a), the schedule dictates that seven concurrent modules execute: in principle, there is enough free space on the chip to perform all operations, however, a 4x6 contiguous region cannot be found for module M7; this is an established problem called fragmentation, which occurs in dynamic placement for runtime reconfigurable FPGAs [5, 12]. In Fig. 5(b), a legal placement has been found, but the placed modules abut one another, blocking the path that droplet D would like to take to reach the detector on the other side of the chip. Lastly, Fig. 5(c) illustrates the complex and chaotic nature of the routing in the presence of many droplets [26].

Since the JIT compiler places a premium on runtime, the time spent to detect and correct the problems shown in Fig. 5(a) and (b) is unacceptable. The virtual topology eliminates these problems completely: the number of on-chip resources is clearly articulated to the scheduler. For example, in Fig. 4, there are four work modules that can perform mixing, splitting, and storage; one can perform heating, and another can perform detection. The scheduler has exact knowledge of what resources are available for different assay operations. As all operations occur in work modules, placement becomes a conceptually simpler binding problem [7]. Routing path blockages cannot occur, since the virtual topology ensures that all droplet routing paths (input port-to-module; module-to-module; module-to-output port) are blockage-free. Lastly, the virtual topology eliminates the chaos depicted in Fig. 5(c) due to the orderly layout of streets, and prior work [7] has demonstrated provably deadlock-free routing algorithms.

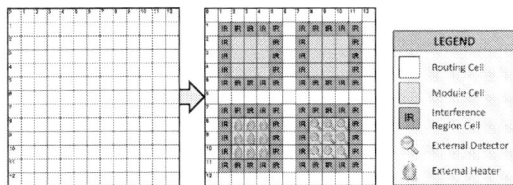

Fig. 4. Depiction of a virtual topology [7].

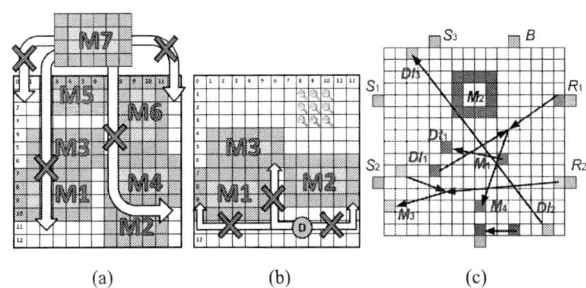

Fig. 5. Online DMFB placement is challenging because of: fragmentation (a); legal placements may have adverse affects on routing (b); and routing a large number of droplets at once can be chaotic (c) [26, Fig. 8(a)].

In a virtual topology, each work module has dedicated virtual I/O ports, which restrict the locations where droplets may enter or exit (Fig. 6). If a work module can store k droplets, then it requires k virtual input (and output) ports, along with an interference-free path from each virtual input port to its corresponding virtual output port within the module; this enables droplets to enter and leave independently without interfering with one another. The virtual I/O ports play an important role in ensuring provably deadlock-free routing at the point where droplets enter and exit work modules [7]. Droplets are allowed to wait in I/O cells as long as necessary, and spacing between them ensures that arriving droplets do not interfere with departing droplets during routing.

B. Fault Recovery Model

We assume that the assay is specified as a control flow graph with checkpoints and error recovery subgraphs inserted a-priori. We review our system's soft fault handing capabilities [8] and introduce techniques to handle hard faults.

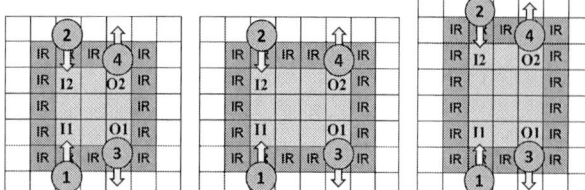

Fig. 6. Each work module in a virtual topology has dedicated virtual I/O ports where droplets can enter and leave without interfering with one another.

Soft Fault Recovery: Any erroneous droplets that are no longer usable are transported to a waste reservoir. Control flow transfers to the error recovery subgraph. The JIT compiler schedules, places, and routes the error recovery subgraph on the virtual topology using fast and efficient algorithms, e.g., list scheduling [7, 25], a binding algorithm (in lieu of placement) that selects a work module for each scheduled operation [7]; and a fast routing algorithm [7].

In contrast, Luo et al. [14] recompute the placement at each time step of the updated schedule. If the target chip is area-constrained, this approach may fail. Since Luo et al. do not include routing results, it is not possible to determine if the placements obtained by their tool are routable or not.

Hard Fault Recovery: Hard faults that occur on the DMFB surface can ruin a virtual topology. A fault in a module cell renders it unusable, while a relatively small number of faults that occur in routing cells could potentially block every pair of paths between two modules in the topology. To fix the situation, we reconfigure the virtual topology when a fault occurs to use smaller modules, as shown in Fig. 8.

The algorithm to repair the virtual topology in response to a hard fault is derived from an algorithm for online FPGA placement called Keep All Maximal Empty Rectangles (KAMER) [5]. KAMER represents the free space on the DMFB [12] using a set of overlapping maximal empty rectangles (MERs). An empty rectangle is maximal if no other rectangle encloses it. KAMER treats each work module (including its surrounding interference region [26]) as an operation; the MERs are shown in Fig. 7(a).

Assume that a hard fault affects one cell in the DMFB, as shown in Fig. 7(b). The 3x3 faulty region (FR) surrounding the cell must be avoided for the remaining lifetime of the chip. We treat the FR as a non-reconfigurable operation that persists through all future placements. Any modules that intersect the FR must be reconfigured; they are removed from the list of active modules and KAMER updates its set of MERs, as shown in Fig. 7(b).

The next step is to introduce smaller work modules with limited functionality (slower mix times [19] and reduced storage capacity), as shown in Fig. 7(c). The preferred strategy is to introduce the largest module that can fit into the reallocated space, to minimize the resulting increase in mixing time [19]. We then query the MER data structure to return the largest rectangle representing free space on the chip. If the MER is large enough to accommodate a new work module, then we add it to the chip. This repeats until no MER can accommodate any more work modules.

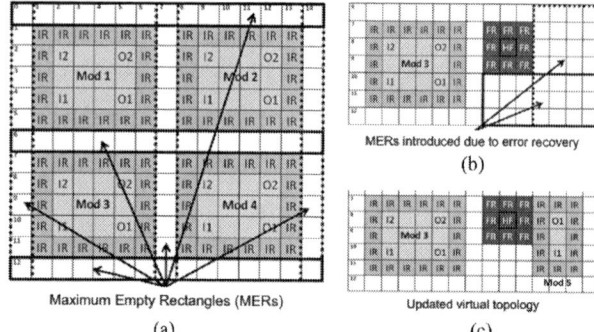

Fig. 7. On a 15x13 DMFB with 4x3 modules, (a) the MERs initially consist of the three horizontal and three vertical streets; (b) a hard fault (HF) and its surrounding faulty region (FR) makes Mod 4 unusable, resulting in two new MERs; (c) a smaller 1x3 module (Mod 5) with well-defined I/O ports is introduced and placed within the larger MER.

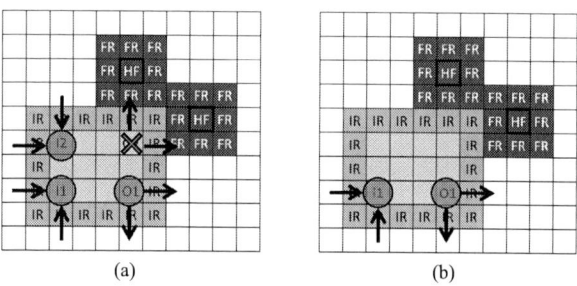

Fig. 8. Two hard faults (HFs) and their faulty regions (FRs) abut against a module. Although they will not interfere with an operation inside the module, (a) output cell O2 is unusable because there is no unobstructed path out of the module, thus (b) only one set of I/Os can be used in this module.

The last step is to determine the number of droplets that each new work module can accommodate, and select the location of the virtual I/O ports which restrict the locations where droplets may enter or exit the module (see Fig. 6).

There must be a path from an adjacent street to each virtual I/O port; otherwise, the port becomes inaccessible. In Fig. 8, a 3x4 module can store up to two droplets with two pairs of virtual I/O ports; however, two hard faults block access to virtual output port O2. As a result, the work module can only store one droplet. In principle, we could reconfigure virtual input port (I2) to be bi-directional; however, doing so would break the deadlock-free droplet routing property [7]. With less storage capacity, reducing the module size is feasible, but doing so would increase the latency of mixing operations [19].

Fig. 9 presents pseudocode for the fault recovery process.

V. SIMULATION RESULTS

We implemented cyber-physical error detection and recovery in a publicly available open source compiler and simulator for digital microfluidics [8]. The experiments compare the performance and effectiveness of error detection and recovery using the virtual topology to an approach similar to the work by Luo et al. [14], which we call *free placement (FP)*. Both approaches employ list scheduling [7, 25] and a greedy droplet routing algorithm [19]. FP uses KAMER for placement [5, 12], but not for dynamic fault recovery.

978-1-4799-7598-3/15 $31.00 © 2015 IEEE

```
ReconfigureFault( 3x3 Fault Region F )
//      Data structures
1.      MER data structure (includes pre-existing 3x3 fault regions): MER
2.      List of modules placed on the DMFB: L

//      Remove modules affected by the fault from the virtual topology
3.      If F occurs in a work module M or within M's interference region
            L.remove(M)
4.      ElseIf the center of F occurs in a street and F intersects the interference
        region of at least one work module
5.          Select a work module M whose interference region intersects F
6.          L.remove(M)
7.      EndIf

//      Update the MER data structure
8.      MER.insert(F)
9.      For each module M∈L
10.         MER.insert(M)
11.     EndFor

//      Introduce new, smaller work modules in the vicinity of the
//      fault, and reconstruct the virtual topology
12.     Do
13.         Boolean stop ← True
14.         For each max. empty rectangle R∈MER
15.             If R.area() >= Minimum module area
16.                 M ← CreateNewWorkModule(R)
17.                 MER.insert(M)
18.                 stop ← False
19.             EndIf
20.         EndFor
21.     While(stop = False)
22.     L←RebuildVirtualTopology(MER)

//      Add virtual inputs and output ports of each work module
//      introduced as part of the reconfiguration process.
23.     For each newly inserted module M
24.         Select the maximum number of droplets that M can store
25.         Add virtual input and output ports for each droplet
26.     EndFor

//      Remove virtual input and output ports (if necessary) from work
//      modules that abut the faulty region F.
27.     For each module M abutting F
28.         For each virtual I/O port P blocked by F
29.             Remove P and its partner port from M
30.             Reduce the max. number of droplets that M can store
31.         EndFor
32.     EndFor
```

Fig. 9. Pseudocode describing virtual topology reconfiguration in response to a hard fault.

The virtual topology employs a restricted variation of KAMER to reconfigure itself when hard faults are detected; however, this is done once per fault discovery, and should not be confused with the usage of KAMER as a free placer that reconfigures the placement when each operation starts/stops.

We consider an exponential protein dilution assay with 5 levels (Protein-Split 5 [7]). We target a 15x19 DMFB with a 2x2 virtual topology where modules store up to two droplets. We converted the assay to a CFG [8] by inserting checkpoints and error recovery subgraphs [29]. We assume that stuck droplet faults can be detected instantaneously [4, 18]. For each experiment, we compile the assay using the virtual topology (VT) and free placement (FP) approaches. The simulator steps through the protocol at 100 Hz (10ms per cycle).

Our experiments measure the recovery time in response to hard and soft faults, and whether or not hard fault recovery is successful. All experiments were performed on a desktop PC running an Intel i7 processor clocked at 3.4 GHz with 10GB of DDR3 DRAM running Windows 8.1.

First, the assay is compiled using VT and FP; we report the initial compilation time. After building the CFG, we randomly selected 5 operations and simulated 5 soft faults; Fig. 10(a) reports the recompilation time. We then randomly selected 5 operations and simulated 5 hard faults; Fig. 10(b) reports the recompilation time. Last, we randomly generated 102 fault scenarios and recompiled the assay using both approaches; in two cases, FP failed; Fig. 10(c) reports the average recovery times for the first two faults for VT and FP for the 100 cases where both approaches were successful.

Fig. 10(a) shows three trends: (1) VT is marginally faster than FP; (2) droplet routing, not scheduling or placement, dominates recompilation time; and (3) the recovery time is to less for faults that occur later in the schedule (since more of the assay has executed, the DAG to be recompiled becomes smaller as the simulation progresses toward completion). The difference in runtime shown in Fig. 10(a) is mostly due to placement, which shows that VT's binding approach [7] is more efficient than invocation of the KAMER placer.

In Fig. 10(b), FP fails to successfully recompile the assay after the 3rd hard fault, while VT successfully recovers after all 5 faults; for the initial compilation step and dynamic recovery from the first two hard faults, the results are similar to Fig. 10(a) for soft faults. Fig. 10(c) reports similar results as well. Altogether, VT is more efficient than FP in terms of spatial resource management as hard faults are introduced into the chip. FP suffers from fragmentation, as the number of hard faults increases, while VT does not.

These results clearly indicate that dynamic recompilation could benefit from faster droplet routing algorithms whose runtimes are comparable to the scheduler and placer (VT binding and FP's invocation of KAMER); since prior work has established that droplet routing does not significantly affect total assay execution time [7, 25, 26], there is reason to believe that significant benefits could be accrued by sacrificing routing solution quality to reduce runtime.

VI. CONCLUSION

Online error recovery for DMFBs necessitates fast and efficient algorithms. Existing approaches do not effectively deal with the interdependence between scheduling, placement, and routing. The approach to error recovery outlined in this paper sidesteps these issues by leveraging a virtual topology: placement is converted to a binding problem, and fast, provably deadlock-free routing algorithms can be used to quickly converge. This paper has shown how to reconfigure a virtual topology in response to hard faults, thus providing graceful degradation as the chip ages. Prior work has established the viability of virtual topologies for efficient detection and recovery from soft faults.

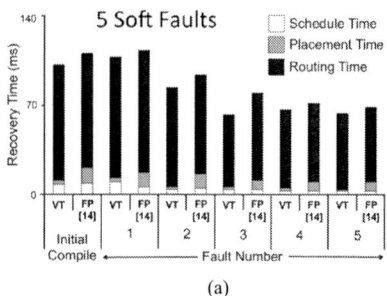 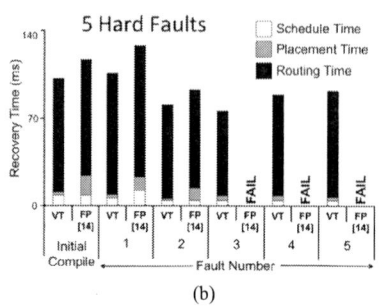 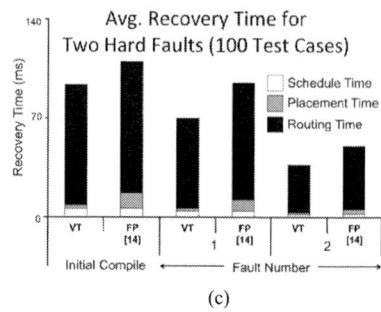

(a)   (b)   (c)

Fig. 10. Recovery time and success rate using the virtual topology (VT) and free placement (FP) [14] on the Protein-Split 5 assay running on a 15x19 DMFB with (a) five soft faults; (b) five hard faults; and (c) average recovery time for 100 simulated executions of VT and FP after compiling and re-compiling the Protein-Split 5 assay with two hard faults.

REFERENCES

[1] Alistar, M., Maftei, E., Pop, P., and Madsen, J. 2010. Synthesis of biochemical applications on digital microfluidics with operation variability. In *Proceedings of the IEEE Symposium on Design Test Integration and Packaging of MEMS/MOEMS* (Seville, Spain, May 05-07, 2010) DTIP '10, 350-357.

[2] Alistar, M., Pop, P., and Madsen, J. 2013. Application-specific fault tolerant architecture synthesis for digital microfluidic biochips. In *Proceedings of the Asia and South Pacific Design Automation Conference* (Yokohama, Japan, Jan. 22-25, 2013), ASPDAC '13, 794-800. DOI= http://dx.doi.org/10.1109/ASPDAC.2013.6509697

[3] Alistar, M., Pop, P., and Madsen, J. 2012. Online synthesis for error recovery in digital microfluidic biochips with operation variability. In *Proceedings of the IEEE Symposium on Design Test Integration and Packaging of MEMS/MOEMS* (Cannes, France, April 25-27, 2012) DTIP '12, 53-58.

[4] Basu, A. S. 2013. Droplet morphometry and velocimetry (DMV): a video processing software for time-resolved label-free tracking of droplet parameters, *Lab-on-a-Chip* 13, 10 (Mar. 2013), 1892-1901. DOI=http://dx.doi.org/10.1039/C3LC50074H

[5] Bazargan, K., Kastner, R., and Sarrafzadeh, M. 2000. Fast template placement for reconfigurable computing systems. *IEEE Design & Test of Computers*, 17, 1 (Jan.-Mar. 2000) 68-83. DOI= http://dx.doi.org/10.1109/54.825678

[6] Choi, K., et al. 2013. Automated digital microfluidic platform for magnetic-particle-based immunoassays with optimization by design of experiments. *Analytical Chemistry* 85, 20 (Aug. 2013) 9638-9646. DOI= http://dx.doi.org/10.1021/ac401847x

[7] Grissom, D., and Brisk, P. Fast online synthesis of generally programmable digital microfluidic biochips. *IEEE Trans CAD* 33, 3 (Mar. 2014), 356-369. DOI= http://dx.doi.org/10.1109/TCAD.2013.2290582

[8] Grissom, D., Curtis, C., and Brisk, P. 2014. Interpreting assays with control flow on digital microfluidic biochips. *ACM Journal on Emerging Technologies in Computing Systems* 10, 3 (Apr. 2014), article #24. DOI= http://dx.doi.org/10.1145/2567669

[9] Grissom, D., O'Neal, K., Preciado, B., Patel, H., Doherty, R., Liao, N., and Brisk, P. 2012. A digital microfluidic biochip synthesis framework. In *Proceedings of the IEEE/IFIP International Conference on VLSI and System-on-a-Chip* (Santa Cruz, CA, USA, October 07 - 10, 2012). VLSI-SOC '12, 177-182, DOI= http://dx.doi.org/10.1109/VLSI-SoC.2012.6379026

[10] Hsieh, Y-L., Ho, T-Y., and Chakrabarty, K. 2012. Design methodology for sample preparation on digital microfluidic biochips. In *Proceedings of the International Conference on Computer Design* (Montreal, Canada, Sep. 30 - Oct. 3, 2012) ICCD'12, 189-194. DOI= http://dx.doi.org/10.1109/ICCD.2012.6378639

[11] Jebrail, M. J., Bartsch, M. S., and Patel, K. D. 2012. Digital microfluidics: a versatile tool for applications in chemistry, biology, and medicine. *Lab-on-a-Chip* 12, 14 (Jul. 2012), 5452-2463. DOI= http://dx.doi.org/10.1039/C2LC40318H

[12] Lu, Y. Marconi, T., Gaydadjiev, G., and Bertels, K. 2008. An efficient algorithm for free resources management on the FPGA. In *Proceedings of Design Automation and Test in Europe* (Munich, Germany, March 10-14, 2008) DATE '08, 1095-1098. DOI= http://dx.doi.org/10.1109/DATE.2008.4484923

[13] Luo, Y., Bhattacharya, B. B., Ho, T-Y., and Chakrabarty, K. Optimization of polymerase chain reaction on a cyberphysical digital microfluidic biochip. In *Proceedings of the International Conference on Computer-Aided Design* (San Jose, CA, Nov. 18-21, 2013) ICCAD'13, 622-629. DOI= http://dx.doi.org/10.1109/ICCAD.2013.6691181

[14] Luo, Y., Chakrabarty, K., and Ho, T-Y. 2013. Error recovery in cyberphysical digital microfluidic biochips. *IEEE Trans CAD* 32, 1 (Jan. 2013), 59-72. DOI= http://dx.doi.org/10.1109/TCAD.2012.2211104

[15] Luo, Y. Chakrabarty, K., and Ho, T-Y. 2012. Dictionary-based error recovery in cyberphysical digital-microfluidic biochips. In *Proceedings of the International Conference on Computer-Aided Design* (San Jose, CA, USA, November 05-08, 2012) ICCAD '12, 369-376. DOI= http://dx.doi.org/10.1145/2429384.2429463

[16] Maftei, M., Pop, P., and Madsen, J. 2013. Droplet-aware module-based synthesis for fault-tolerance digital microfluidic biochips. In *Proceedings of the IEEE Symposium on Design Test Integration and Packaging of MEMS/MOEMS* (Cannes, France, April 25-27, 2012) DTIP '12, 47-52.

[17] Mitra, D., et al. 2012. Automated path planning for washing in digital microfluidic biochips. In *Proceedings of the International Conference on Automation Science and Engineering* (Seoul, Korea, Aug. 20-24, 2012) CASE'12, 115-120. DOI= http://dx.doi.org/10.1109/CoASE.2012.6386419

[18] Murran, M. A., and Najjaran, H. 2012. Capacitance-based droplet position estimator for digital microfluidic devices. *Lab-on-a-Chip* 12, 11 (Mar. 2012) 2053-2059. DOI= http://dx.doi.org/10.1039/c2lc21241b

[19] Paik, P., Pamula, V. K., and Fair, R. B. 2003. Rapid droplet mixers for digital microfluidic systems. *Lab-on-a-Chip* 3, 4 (Nov. 2003), 253-259.

[20] Pollack, M. G., Shenderov, A. D., and Fair, R. B. 2002. Electrowetting-based actuation of droplets for integrated microfluidics. *Lab-on-a-Chip* 2, 2 (Mar. 2002), 96-101. DOI=http://dx.doi.org/10.1039/b110474h

[21] Roy, P., Rahaman, H., and Dasgupta, P. 2010. A novel droplet routing algorithm for digital microfluidic biochips. In *Proceedings of the 20th Great Lakes Symposium on VLSI* (Providence, RI, USA, May 16 - 18, 2010) GLSVLSI '10, 441-446. DOI= http://dx.doi.org/10.1145/1785481.1785583

[22] Shih, S. C. C., et al., Digital microfluidics with impedance sensing for integrated cell culture and analysis. *Biosensors and Bioelectronics* 42, 4 (Apr. 2013) 314-320. DOI= http://dx.doi.org/10.1016/j.bios.2012.10.035

[23] Shih, S. C. C., Fobel, R., Kumar, P., and Wheeler, A. R. 2011. A feedback control system for high-fidelity digital microfluidics. *Lab-on-a-Chip* 11, 3 (Feb. 2011) 535-540. DOI= http://dx.doi.org/10.1039/C0LC00223B

[24] Soukup, J. Fast maze router. In *Proceedings of the Design Automation Conference* (Las Vegas, NV, USA, June 19-21, 1978) DAC '78, 100-102, DOI= http://dx.doi.org/10.1109/DAC.1978.1585154

[25] Su, F., and Chakrabarty, K. 2008. High-level synthesis of digital microfluidic biochips. *ACM Journal on Emerging Technologies in Computing Systems* 3, 4 (Jan. 2008), article #16. DOI= http://dx.doi.org/10.1145/1324177.1324178

[26] Su, F., Hwang, W., and Chakrabarty, K. 2006. Droplet routing in the synthesis of digital microfluidic biochips. In *Proceedings of Design Automation and Test in Europe* (Munich, Germany, March 06-10, 2006) DATE '06, 1-6. DOI= http://dx.doi.org/10.1109/DATE.2006.244177

[27] Weiser, M. 1984. Program slicing. *IEEE Trans. Software Engineering* 10, 4 (July 1984) 352-357. DOI= http://dx.doi.org/10.1109/TSE.1984.5010248

[28] Xu, T., and Chakrabarty, K. 2007. Functional testing of digital microfluidic biochips In *Proceedings of the IEEE International Test Conference* (Santa Clara, CA, USA, Oct. 21-26, 2007) ITC'07. DOI= http://dx.doi.org/10.1109/TEST.2007.4437614

[29] Zhao, Y., Xu, T., and Chakrabarty, K. 2010. Integrated control-path design and error recovery in the synthesis of digital microfluidic biochips. *ACM Journal on Emerging Technologies in Computing Systems* DOI= http://dx.doi.org/10.1145/1777401.1777404

Fault Modeling and Testing of 1T1R Memristor Memories

Yong-Xiao Chen and Jin-Fu Li
Advanced Reliable Systems (ARES) Lab.
Department of Electrical Engineering
National Central University
Taoyuan, Taiwan 320

Abstract—Memristor memory has attracted more attentions to act as one of future non-volatile memories. One access transistor and one memristor (1T1R) cell structure can be used to eliminate the issue of sneak path current of memristor memories with crossbar structure. In this paper, we propose several fault models for 1T1R memristor memories based on electrical defects, such as resistive bridge between two nodes, transistor stuck-on and stuck-open faults. In comparison with existing faults, two new faults, write disturbance fault (WDF) and dynamic write disturbance fault ($dWDF$), are found. In addition, a March test is proposed to cover the defined faults. The March test requires $(1+2a+2b)N$ write operations and $5N$ read operations for an N-bit memristor memory, where a and b are the number of consecutive Write-1 and Write-0 operations for activating a $dWDF$.

Index Terms—Non-volatile memory, memristor, resistive memory, test.

I. INTRODUCTION

Memristor is a non-linear device which has the characteristics of memory and resistor. It can be used in many applications, such as digital circuits, analog circuits, neuromorphic systems, and so on [1]. One important application of memristor is the memristor memory. In a memristor memory, the resistance of a memristor is used to represent logic 0 or logic 1 [2]–[4]. Thus, the state of memristor memory can be retained after the power is turned off. Therefore, the memristor memory is one possible future non-volatile memory [5].

The simplest method for designing a memristor memory array is the crossbar structure where only a memristor (1R) connects to the bitline and the wordline. The 1R memristor memory array can achieve the highest density. However, it has the problem of sneak path. The sneak path current may result in an unsuccessful write operation on the selected cell, an undesirable write operation on an unselected cell, and a read operation failure. Several biasing schemes were proposed to alleviate the sneak path impact [3], [6]. However, these techniques cannot eliminate the sneak path current, and the current incurs intolerable power consumption if the array size is large [7]. Sneak paths can be avoided by using selector device in connection with the memristor at each node. One possible selector device is diode [8], i.e., each node of the memristor memory array consists of a diode and a memristor (1D1R). Since the diode is a unipolar device, only unipolar memristor can be used. Another possible selector device is the

transistor, i.e., each node consists of an access transistor and a memristor (1T1R). A memristor memory with 1T1R cells can eliminate the sneak-path current and bipolar memristor can be used. However, it is challenging to provide high current levels with both polarities [9].

Several works have addressed the defect analysis and testing of memristor memories [10]–[13]. In [10], a framework of defect oriented testing in 1R memristor memory based on electrical simulation was proposed. In [11], fault modeling for 1R 3D resistive RAMs by injecting open defects was investigated. In addition, two design-for-testability techniques by using short write time and low write voltage were proposed to detect undefined state faults. In [12], the authors examined the fault models and proposed an efficient testing technique for the 1R memristor memories. Fault models incurred by parametric variation of a memristor are also considered. In [13], fault models of 1D1R memristor memories were investigated by considering bridge defects between two nodes.

In this paper, a comprehensive electrical fault modeling for 1T1R memristor memories by injecting transistor stuck-on, stuck-open, open, and bridge defects is investigated. Furthermore, the bridge defects between neighboring cells are considered. In addition, a test algorithm is proposed to cover the defined faults. The remainder of this paper is organized as follows. Section II briefly reviews the 1T1R memristor memory. Section III presents electrical defects and fault models. Section IV introduces the proposed march test for the defined faults. Finally, Section V concludes this paper.

II. 1T1R MEMRISTOR MEMORY

Fig. 1(a) shows a 1T1R memristor cell. The cell has three terminals connected to the word line (WL), source line (SL), and bit line (BL) of the memory array. The memristor cell operations depend on the duration of access time and the value of supply voltage applied on WL and SL. Fig. 1(b) shows the bias conditions when a write operation is executed. If a Write-0 operation is executed, V_{w0} is applied to BL and SL is connected to ground which changes the resistance of memristor to low resistance state (LRS). If a Write-1 operation is executed, V_{w1} is applied to SL and BL is connected to ground which changes the resistance of memristor to high resistance state (HRS). To read the data from a selected cell, a

small read voltage V_{read} is applied to BL and SL is connected to ground as shown in Fig. 1(c) such that the read current is smaller than the threshold current for changing the state of the memristor.

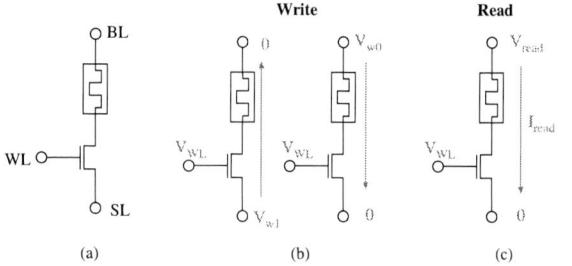

Fig. 1. (a) 1T1R cell. (b) Bias conditions of write operations. (c) Bias conditions of read operation.

Fig. 2 shows the block diagram of a $m \times n$ 1T1R memristor memory. In the memory cell array, the cells in the same column share a SL and a BL, and the cells in the same row share a WL. If a write or a read operation is performed, the write and read circuit generate the corresponding bias of SL and BL for the selected cell. Other peripheral circuits in 1T1R memristor memory are similar to that of in conventional memory.

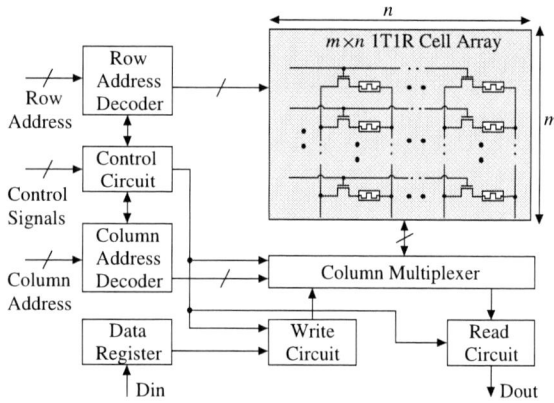

Fig. 2. Organization of a $m \times n$-bit memristor memory.

III. ELECTRICAL DEFECTS AND FAULTS

A. Electrical Defects

We analyze possible fault behaviors of a memristor memory by injecting the following electrical defects. 1) Transistor stuck-on—a transistor is always turned on regardless of its gate voltage. 2) Transistor stuck-open—a transistor is always turned off regardless of its gate voltage. 3) Open—represents unwanted resistance within a connection. 4) Bridge—represents unwanted resistance between two signal lines.

Fig. 3(a) shows possible defects and defect locations in a 1T1R cell, where AT_{on} and AT_{open} denote the transistor stuck-on and stuck-open defects at the access transistor. Also, R_{o1}, R_{o2}, and R_{o3} denote possible open defects within the

1T1R cell. Since the density of memristor memory is high, we consider the bridge defects within a 1T1R cell and between neighboring 1T1R cells. Therefore, a 2×2 1T1R memristor memory array as shown in Fig. 3(b) is considered for defect analysis. Possible bridge defects for the 2×2 1T1R memristor memory are listed in Table I.

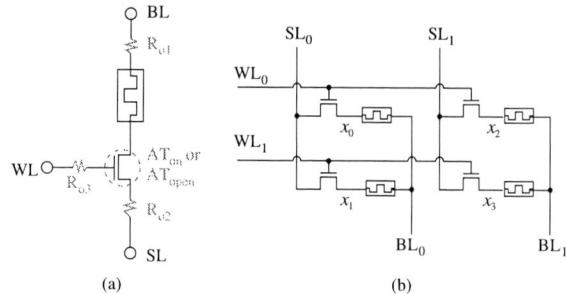

Fig. 3. (a) Possible open, transistor stuck-on, transistor stuck-open defects in a 1T1R cell. (b) A 2×2 1T1R cell array.

TABLE I
POSSIBLE BRIDGE DEFECTS IN THE 2×2 1T1R MEMRISTOR MEMORY
SHOWN IN FIG. 3(B)

Bridge	Location	Bridge	Location
R_{B1}	WL_0-SL_0	R_{B9}	BL_0-SL_1
R_{B2}	WL_0-x_0	R_{B10}	BL_0-BL_1
R_{B3}	WL_0-BL_0	R_{B11}	x_0-x_1
R_{B4}	SL_0-x_0	R_{B12}	x_0-WL_1
R_{B5}	BL_0-x_0	R_{B13}	x_0-SL_1
R_{B6}	SL_0-BL_0	R_{B14}	x_0-BL_1
R_{B7}	WL_0-WL_1	R_{B15}	x_0-x_2
R_{B8}	SL_0-SL_1	R_{B16}	x_0-x_3

B. Simulation Model

Fig. 4 shows the simulation model of a 2×2 1T1R memristor memory array for Hspice simulation. Here TSMC 180nm CMOS technology is used for the access transistor of 1T1R cell and the peripheral circuitry. The memristor Verilog-A model reported in [14] is used for the simulation of memristor. We design the peripheral circuitry by referring to the resistive random access memory design reported in [15]. The specification of the 2×2 memristor memory is summarized in Table II.

TABLE II
SPECIFICATION OF THE 2×2 1T1R MEMRISTOR MEMORY

Item	Parameters	Description
R_{on}	10KΩ	Value of high resistance state
R_{off}	1000KΩ	Value of low resistance state
T_{write}	5 ns	Write time
V_{w0}/V_{w1}	1.6/1.8V	Voltages of Write-0 and Write-1 operations
T_{read}	8.5 ns	Read time
$V_{th,M}$	0.5V	Threshold voltage of memristor
V_{OH}	1.6V	Minimum Logic 1 output voltage
V_{OL}	0.2V	Maximum Logic 0 output voltage

Before the simulation of the memristor with defect injection, the correctness of the 2×2 memristor memory has to be verified. Fig. 5 shows a sample defect-free simulation waveform

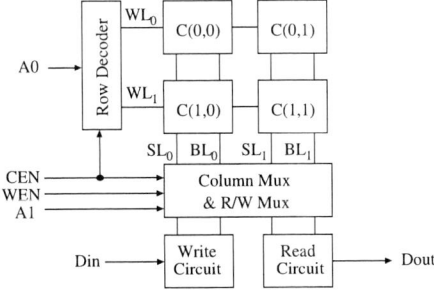

Fig. 4. Simulation model of a 2×2 1T1R memristor memory for Hspice simulation.

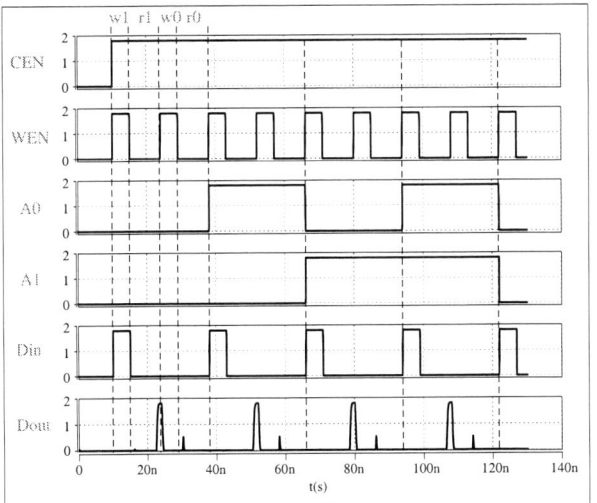

Fig. 5. Sample of defect-free simulation waveform of the 2×2 memristor memory.

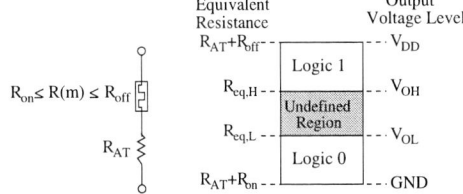

Fig. 6. Relationship between the equivalent resistance and the output voltage level.

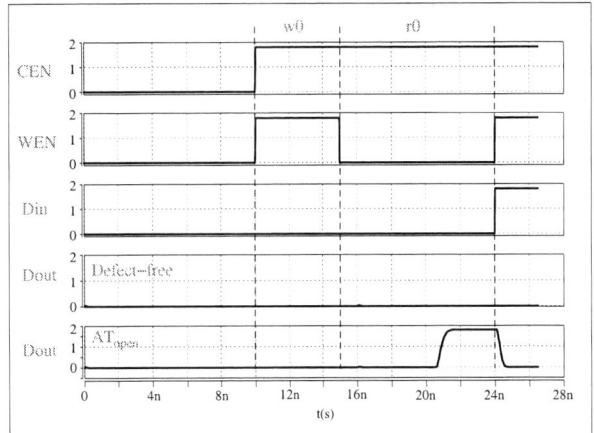

Fig. 7. Waveform sample if the transistor has a stuck-open defect under the $w0r0$ operation.

when the memristor memory executes the operation sequence $S=w1, r1, w0, r0$ for each cell, where wy and ry denote a write operation with data y and a read operation with expected data y, respectively. The write operation is applied after the selected cell is initialized to 0 or 1. As Fig. 4 shows, the CEN signal is used to enable the memristor memory. If $CEN = 1$, then the read/write operation can be executed. If $WEN = 1$, a write operation is executed and the data of D_{in} is written into the addressed cell. If $WEN = 0$, a read operation is executed and the data stored in the addressed cell is read out to the D_{out}.

In the sequel, analysis of different defects is executed by simulating the memristor memory with defect injection. We assume that the resistance of open defect and bridge defect are in the interval $[0, \infty)$. Also, only a single defect is injected each time, and then the circuit simulation is performed.

C. Faults Caused by Transistor Stuck-Open or Open Defects

Let the equivalent resistance of the access transistor and the memristor in a 1T1R cell be R_{AT} and $R(m)$ as the left figure of Fig. 6 shows. The value of $R(m)$ ranges from R_{on} to R_{off}, which is affected by the value and direction of current

flowing through the memristor. Assume that the relationship between the equivalent resistance of the 1T1R cell and the output voltage is as shown in right figure of Fig. 6. Here V_{OH} and V_{OL} denote the minimal voltage value of logic 1 and the maximal voltage value of logic 0, respectively. Also, $R_{eq,H}$ and $R_{eq,L}$ denote the minimal resistance value of logic 1 and the maximal resistance value of logic 0. The stuck-on and open defects only result in single cell faults. Therefore, the sensitizing sequences $\{0w1, 1w0, 0r0, 1r1\}$ are used to simulate the fault behaviors caused by those defects, where $0w1$ ($1w0$) denotes a Write-1 (Write-0) to selected cell initialized to 0 (1) and $0r0$ ($1r1$) denotes a Read-0 (Read-1) from selected cell initialized to 0 (1).

If the access transistor of a 1T1R cell has a stuck-open defect, the $R_{AT} = \infty$. Thus, the equivalent resistance will be ∞ and the read data will be logic 1 regardless of the written data. Therefore, the transistor stuck-open defect results in a stuck-at-1 fault. Fig. 7 shows a sample of waveform if the 1T1R cell has a transistor stuck-open defect under the $w0r0$ operations and the initial state of memristor is logic 1. As the figure shows, the read out data of the $r0$ operation.

As Fig. 3(a) shows, if an open defect is located between the SL and BL, e.g, R_{o1} or R_{o2}, then the equivalent resistance of the 1T1R cell is increased, which results in the voltage across the memristor is reduced. If the voltage across the memristor is smaller than the threshold voltage $V_{th,M}$, the cell cannot be programmed. Therefore, the resistance value of the memristor

978-1-4799-7598-3/15 $31.00 © 2015 IEEE

is dependent on the initial state of the memristor. When the resistance (R_o) of a open defect R_{o1} or R_{o2} in a cell is large, clearly, the open defect results in a stuck-at-1 fault ($SA1F$) regardless of the initial state of the cell. Thus, if the R_o is large enough, $(R_{AT} + R(m) + R_o) > R_{eq,H}$ regardless the value of $R(m)$. On the contrary, if the resistance of a open defect R_{o1} or R_{o2} in a cell is not large, the open defect may result in a $SA1F$, stuck-at-0 fault ($SA0F$), an up transition fault (TF_u), or a unknown read fault (URF) [12]. The reason is that if the R_o is small, the sum of equivalent resistance $R_{AT} + R(m) + R_o$ may larger than $R_{eq,H}$, within the range between $R_{eq,H}$ and $R_{eq,L}$, or smaller than $R_{eq,L}$ depending on the value of $R(m)$ and R_o. If $R_{eq,H} \geq R_{AT} + R(m) + R_o \geq R_{eq,L}$, this results in an undefined output, which may be interpreted by the sense amplifier as either a logic 0 or logic 1. Thus, the read data may be logic 0 or logic 1 randomly. This results in an URF [12].

For the open defect R_{o1}, the defect affects the equivalent resistance of cell indirectly. It results in additional delay of in the gate terminal such that the turn-on period of the access transistor is shortened within the write or read cycle period. Therefore, the the current through the memristor is not long enough. Thus, the value of $R(m)$ is related to the turn-on period of the access transistor which is affected by the defect size of R_{o3}. Similar to R_{o1} and R_{o2}, the R_o3 also may result in $SA0F$, $SA1F$, TF_u, and URF.

Table III summarizes possible fault caused by transistor stuck-open and open defects with respect to different defect sizes. We can see that the defined faults for 1T1R memristor memories caused by transistor stuck-open and open defects also appear in 1R memristor memories [10], [12].

TABLE III
POSSIBLE FAULTS WITH RESPECT TO DIFFERENT SIZES OF DEFECTS OF
TRANSISTOR STUCK-OPEN AND OPEN

Defect Location	Defect Size	Fault Models
AT_{open}	-	$SA1F$
R_{o1}	$R_o > 395K\Omega$	$SA1F$
	$64K\Omega < R_o \leq 395K\Omega$	$SA0F, SA1F, URF$
	$8K\Omega < R_o \leq 64K\Omega$	TF_u, URF
R_{o2}	$R_o > 360K\Omega$	$SA1F$
	$57K\Omega < R_o \leq 360K\Omega$	$SA0F, SA1F, URF$
	$12K\Omega < R_o \leq 57K\Omega$	TF_u, URF
R_{o3}	$R_o > 55.3M\Omega$	$SA1F$
	$7.02M\Omega < R_o \leq 55.3M\Omega$	$SA0F, SA1F, URF$
	$6.2M\Omega < R_o \leq 7.02M\Omega$	TF_u, URF

D. Faults Caused by Transistor Stuck-on or Bridge Defects

Transistor stuck-on and bridge defects may result in single-cell faults or two-cell faults. Similar to coupling faults [16], for a two-cell fault, one cell is the aggressor cell and the other cell is the victim cell. The state of the aggressor cell affects the state of the victim cell. To analyze the defects results in two-cell faults, the sensitizing sequences for the aggressor cell and the victim cell are listed as follows: $S_a = \{0, 1, 0w1, 1w0, 0r0, 1r1\}$; $S_v = \{0, 1\}$, where S_a and S_v denote the state of aggressor and victim, respectively. Here the Hspice simulation is executed on a 2×2 1T1R memristor

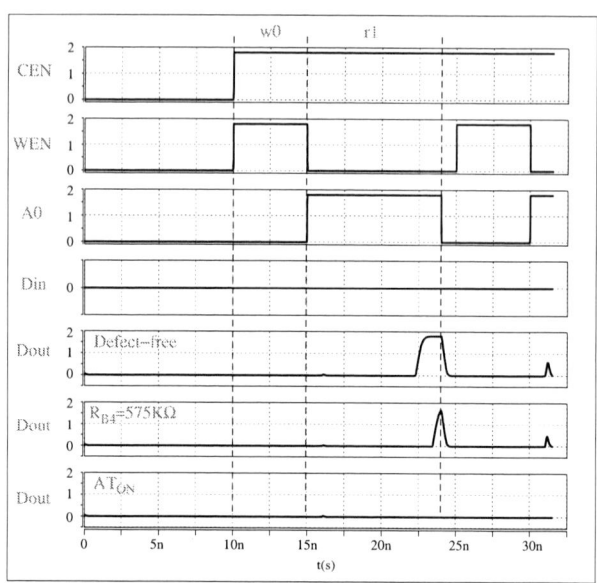

Fig. 8. Sample of waveform of a WDF when the aggressor executes a 1w0 operation and the initial state of the victim is 1.

memory as shown in Fig. 3(b) by injecting possible bridge defects listed in Table I and transistor stuck-on defects. Consider that the AT_{on} defect or bridge defect R_{B4} is injected in the cell $C(0, 0)$. Fig. 8 shows that the simulation results for those two defects if the aggressor cell $C(1, 0)$ executes $1w0$ sensitizing sequence and the victim cell $C(0, 0)$ has an initial value 1. We see that the read data of $C(0, 0)$ is logic 0 if the AT_{on} defect is injected. The reason is that when the $C(1, 0)$ executes a $w0$ operation, the $C(0, 0)$ is written as well due to the AT_{on} defect. Thus, the state of victim cell is disturbed due to a write operation executing in the aggressor cell. We call the two cells have a WDF. However, the state of victim cell is not logic 0 if the $R_{B4} = 575K\Omega$ is injected. The reason is that the resistance of R_{B4} is larger than the turn-on resistance the transistor such that the disturbance of victim is smaller. That is, the resistance value of bridge defect has impact on the influence of disturbance.

Since the resistance of memristor is affected by the duration and magnitude of the current through the memristor, the defect size and the number of write operations on the aggressor also has the impact on the influence of disturbance. Again, the defects AT_{on} and R_{B4} are illustrated to explain this in more details. As Fig. 9(a) shows, the cell $C(0, 0)$ has an AT_{on} defect. The turn-on resistance of the transistor is much smaller than $575K\Omega$. Assume that the initial state of the aggressor and victim is logic 1. If a defect size is small, then only one Write-0 operation in the aggressor $C(1, 0)$ will change the $R(m)$ of the $C(0, 0)$ from R_{HRS} to R_{LRS} such that the state of the victim is flip to logic 0. However, if the $R_{B4} = 575K\Omega$ defect is considered as shown in Fig. 9(b), one Write-0 operation in the aggressor cannot change the $R(m)$ of $C(0, 0)$ to R_{LRS} as shown in the right figure of Fig. 9(b). But, if two consecutive Write-0 operations are executed in the aggressor $C(1, 0)$, the

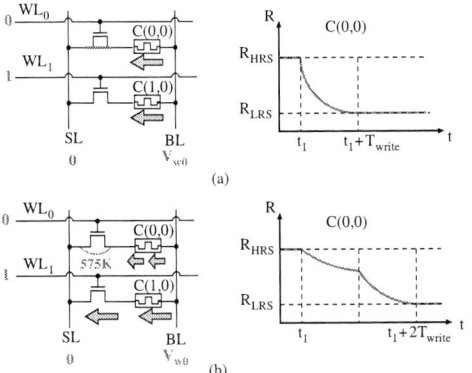

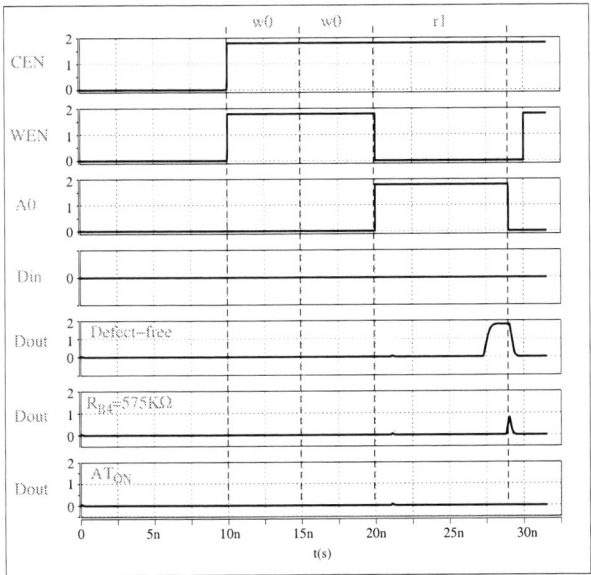

Fig. 9. (a) An example of WDF caused by AT_{on}. (b) An example of dWDF caused by R_{B4}.

Fig. 10. Sample of waveform of a $dWDF$ when the aggressor executes 1w0w0 operations and the initial state of the victim is 1.

$R(m)$ can be reduced to R_{LRS} and the state of victim is flipped. Fig. 10 shows sample of waveform when the aggressor cell executes $1w0w0$ sequence and the initial state of victim cell is 1 for the defects AT_{on} and R_{B4}. In comparison with the waveform shown in Fig. 8, we see that the data of the victim is also flipped for the defect $R_{B4}=575K\Omega$. Consequently, if a WDF is activated by larger than two consecutive write operations in the aggressor, we call the fault as a dynamic write disturbance fault ($dWDF$).

According to the discussion above, we see that the number of write operation in the aggressor affects what fault type is observed in the victim. We use $< S_a; S_v/F >$ to represent WDF and $dWDF$, where S_a, S_v, and F denote the sensitizing sequences of the aggressor, fault-free state of the victim, and faulty state of the victim, respectively. Assume that the initial states of aggressor and victim are x and z. WDF and $dWDF$ can be represented by $< x(wy)^k; z/\bar{z} >$, where k

denotes the number of required write operations sensitizing the fault. If $k = 1$, then the fault is WDF. If $k \geq 2$, then the fault is $dWDF$.

In the same way, we can analyze all the other possible bridge defects in a 2×2 memristor memory. Table IV summarizes possible faults caused by corresponding transistor stuck-on (AT_{on}) and bridge defects, where x and y in $WDF(x, y)$, $dWDF(x, y)$, and $CF_{st}(x, y)$ denote the state of aggressor and victim. Some of faults are also defined in existing works [10]–[13]. However, we found two new faults, write disturbance fault (WDF) and dynamic write disturbance fault ($dWDF$), existing in 1T1R memristor memories.

IV. PROPOSED MARCH TEST FOR 1T1R MEMRISTOR MEMORIES

A March test is proposed to cover the defined faults in 1T1R memristor memories. The March test, March-1T1R, is as below

$$\left\{ \begin{array}{c} \Updownarrow (w0); \Uparrow (r0, w1, r1, (w1)^{a-1}); \\ \Uparrow (r1, (w0)^b); \Downarrow (r0, (w1)^a); \Downarrow (r1, (w0)^b); \end{array} \right\}$$

where a and b denote the number of required consecutive Write-1 and Write-0 operations for detecting $dWDF$, respectively; $\Uparrow$, $\Downarrow$, or $\Updownarrow$ denotes the addressing sequence is ascending, descending, or either. The March-1T1R consists of five march elements, $M1 = \Updownarrow (w0)$, $M2 = \Uparrow (r0, w1, r1, (w1)^{a-1})$, ..., and $M5 = \Downarrow (r1, (w0)^b)$.

Subsequently, we explain how the March-1T1R can cover the defined faults as below.

1) $SA1F$: it can be sensitized by $M1$ and detected by the first read operation of $M2$.
2) $SA0F$: it can be sensitized and detected by $M2$.
3) TF_u: it can be sensitized by $\{M1, M2\}$ and detected by the second read operation of $M2$.
4) $CF_{st}(0, 0)$: the faults $CF_{st}(0, 0)$ whose aggressor position is higher than the victim position are sensitized and detected by $\{M2, M3\}$; the faults $CF_{st}(0, 0)$ whose aggressor position is lower than the victim position are sensitized and detected by $\{M4, M5\}$.
5) WDF and $dWDF$: the faults $WDF(1, 1)$ and $dWDF(1, 1)$ whose aggressor position is lower (higher) than the victim are sensitized and detected by $\{M1, M2\}$ ($\{M3, M4\}$); the faults $WDF(0, 0)$ and $dWDF(0, 0)$ whose aggressor position is lower (higher) than the victim are sensitized and detected by $\{M2, M3\}$ ($\{M4, M5\}$); the faults $WDF(1, 0)$ and $dWDF(1, 0)$ whose aggressor position is lower (higher) than the victim are sensitized and detected by $\{M2, M3\}$ ($\{M4, M5\}$).

Table V summarizes the comparison results of March-MOM [12] and the proposed March-1T1R. The March-MOM is used for the 1R memristor memory and the March-1T1R is used for 1T1R memristor. We see that the proposed March-1T1R can completely cover the faults in 1T1R memristor memory. However, March-MOM can provide full coverage on the SAF and TF_u. It only can cover partial CF_{st}, WDF, and $dWDF$,

TABLE IV
FAULT TYPES CAUSED BY TRANSISTOR STUCK-ON AND BRIDGE DEFECTS

Defect Location	Aggressor	Victim	Fault Models
AT_{on}	-	C(1,0)	$SA0F$
	C(1,0)	C(0,0)	$WDF(0,0), WDF(1,1)$
$R_{B5}, R_{B6}, R_{B8}, R_{B9}, R_{B10}\ R_{B15}$	-	C(0,0)	$SA0F$
R_{B1}, R_{B2}, R_{B3}	-	C(0,0)	$SA1F$
	-	C(1,0)	$SA0F$
R_{B2}	C(1,0)	C(0,0)	$SA1F, WDF(0,0), dWDF(0,0)$
	C(0,0)	C(1,0)	$SA0F, CF_{st}(0,0)$
R_{B4}	C(1,0)	C(0,0)	$WDF(0,0), WDF(1,1), dWDF(0,0), dWDF(1,1)$
	C(0,0)	C(1,0)	$CF_{st}(0,0)$
R_{B12}	C(1,0)	C(0,0)	$SA0F, WDF(1,1), dWDF(1,1)$
	C(0,0)	C(1,0)	$SA0F, CF_{st}(0,0)$
R_{B7}, R_{B11}	C(1,0)	C(0,0)	$SA0F, CF_{st}(0,0)$
R_{B13}	C(1,1)	C(0,0)	$SA1F, WDF(1,0), dWDF(1,0)$
R_{B14}, R_{B16}	C(1,1)	C(0,0)	$SA1F, WDF(0,0), dWDF(0,0)$

since those faults occur in 1T1R memristor memory. Please note that March-MOM can cover many faults that occur in 1R memristor which are not listed in the table since we focus on the 1T1R memristor memory in this paper. For an N-bit memristor memory, March-MOM needs $5N$ write operations and $4N$ read operations. March-1T1R needs $(1 + 2a + 2b)N$ write operations and $5N$ read operations, where we assume that the number of required consecutive Write-1 and Write-0 operations for $dWDF$ is a and b, respectively.

TABLE V
COMPARISON RESULTS OF DIFFERENT MARCH TESTS

Test Algorithm		March-MOM [12]	March-1T1R
Memristor Memory Type		1R Memristor	1T1R Memristor
Covered Faults	SAF	Full	Full
	TF_u	Full	Full
	CF_{st}	Partial	Full
	WDF	Partial	Full
	$dWDF$	Partial	Full
Test Complexity	Write	$5N$	$(1 + 2a + 2b)N$
	Read	$4N$	$5N$

V. CONCLUSIONS

In this paper, a comprehensive defect analysis for a 2×2 1T1R memristor memory has been executed. Electrical defects of transistor stuck-on, stuck-open, open, and bridge are simulated by Hspice simulation. Two new functional faults are found, which are write disturbance fault and dynamic write disturbance fault. In addition, a March test has been proposed to cover the defined faults of 1T1R memristor memories. The proposed March test requires $(1 + 2a + 2b)N$ write operations and $5N$ read operations for an N-bit memristor memory, where a and b are the number of consecutive Write-1 and Write-0 operations for activating a $dWDF$.

ACKNOWLEDGMENT

This work was supported in part by the National Science Council, R.O.C., under Contract NSC 101-2221-E-008-130-MY3 and NSC 102-2221-E-008-108-MY3.

REFERENCES

[1] P. Mazumder, S. Kang, and R. Waser, "Memristors: Devices, models and applications," *Proc. of the IEEE*, vol. 100, no. 6, pp. 1911–1919, Jun. 2012.

[2] M. D. Ventra, Y. V. Pershin, and L. O. Chua, "Circuit elements with memory: Memristors, memcapacitors, and meminductors," *Proc. of the IEEE*, vol. 97, no. 10, pp. 1717–1724, Oct. 2009.

[3] H. Manem, J. Rajendran, and G. S. Rose, "Design considerations for multilevel CMOS/nano memristive memory," *ACM Jour. on Emerging Tech. in Computing Systems*, vol. 8, no. 1, pp. 6:1–6:22, Feb. 2012.

[4] D. B. Strukov, G. S. Snider, D. R. Stewart, and R. S. Williams, "The missing memristor found," *Nature Letters*, vol. 453, pp. 80–83, May 2008.

[5] Semiconductor Industry Association, "International technology roadmap for semiconductor (ITRS)," http://www.itrs.net, 2013.

[6] C. Xu, X. Dong, N. P. Jouppi, and Y. Xie, "Design implications of memristor-based RRAM cross-point structures," in *Proc. Conf. Design, Automation, and Test in Europe (DATE)*, Mar. 2011, pp. 1–6.

[7] M. A. Zidan, H. A. Fahmy, M. M. Hussain, and K. N. Salama, "Memristor-based memory: The sneak paths problem and solutions," *Microelectronics Journal*, vol. 44, no. 2, pp. 176–183, Feb. 2013.

[8] M.-J. Lee and et al., "2-stack 1D-1R cross-point structure with oxide diodes as switch elements for high density resistance RAM applications," in *Proc. IEEE Int'l Electron Devices Meeting*, 2007, pp. 771–774.

[9] S. Hamdioui, H. Aziza, and G. C. Sirakoulis, "Memristor based memories: technology, design and test," in *IEEE Int'l Conf. on Design and Technology of Integrated Systems in Nanoscale Era (DTIS)*, Santorini, May 2014, pp. 1–7.

[10] N. Z. Haron and S. Hamdioui, "On defect-oriented testing for hybrid CMOS/memristor memory," in *IEEE Asian Test Symp. (ATS)*, Nov. 2011, pp. 353–358.

[11] ——, "DfT schemes for resistive open defects in RRAMs," in *Proc. Conf. Design, Automation, and Test in Europe (DATE)*, Mar. 2012, pp. 799–804.

[12] S. Kannan, J. Rajendran, R. Karri, and O. Sinaoglu, "Sneak-path testing of crossbar-based nonvolatile random access memories," *IEEE Trans. on Nanotech.*, vol. 12, no. 3, pp. 413–426, May 2013.

[13] O. Ginez, J. Portal, and C. Muller, "Design and test challenges in resistive switching RAM (ReRAM): An electrical model for defect injections," in *IEEE European Test Symp. (ETS)*, May 2009, pp. 61–66.

[14] S. Kvatinsky, E. Friedman, A. Kolodny, and U. Weiser, "TEAM: Threshold adaptive memristor model," *IEEE Trans. on Circuits and Systems I: Fundamental Theory and Applications*, vol. 60, no. 1, pp. 211–221, Jan. 2013.

[15] S.-S. Sheu, P.-C. Chiang, W.-P. Lin, H.-Y. Lee, P.-S. Chen, Y.-S. Chen, T.-Y. Wu, F. Chen, K.-L. Su, M.-J. Kuo, K.-H. Cheng, and M.-J. Tsai, "A 5ns fast write multi-level non-volatile 1 k bits rram memory with advance write scheme," in *IEEE Symp. on VLSI Circuits*, Jun. 2009, pp. 82–83.

[16] A. J. van de Goor and Z. Al-Ars, "Functional memory faults: a formal notation and a taxonomy," in *Proc. IEEE VLSI Test Symp. (VTS)*, 2000, pp. 281–289.

978-1-4799-7598-3/15 $31.00 © 2015 IEEE

A Low Cost Jitter Separation and Characterization Method

Li Xu, Yan Duan, Degang Chen

Department of Electrical and Computer Engineering
Iowa State University, Ames, IA, USA

Abstract- Clock jitter is a crucial factor in high speed and high performance application. Traditional jitter measurement method relies on precise and expensive instrumentations. This paper proposes a low cost jitter measurement and separation method. Instead of using traditional time internal analysis equipment, a simple Analog-to-Digital Converter (ADC) is used as the jitter measurement device. The clock under test is applied as the sampling clock of an ADC while the ADC is sampling a full scale sine wave. The ADC output contains the information of the clock jitter. The algorithm will separately detect the effects of Periodic Jitter, Dual-Dirac Jitter and Random Jitter, and accurately compute the rms value of each jitter component. This method offers great potential for wide use in low cost applications and especially in on-chip or on-board jitter measurement applications. Simulation results demonstrate the functionality, accuracy and robustness of the proposed low-cost jitter measurement method.

Keywords—Random jitter, Dual-Dirac jitter, jitter separation, jitter characterization, low-cost jitter testing.

I. INTRODUCTION

Jitter, defined as the variation in the sampling instant, is an important specification in high speed analog-mixed signal devices. Deterministic Jitter (DJ) and Random Jitter(RJ) are two important categories of clock jitter[1]. As the increasing of frequency and data rate, jitter can be the ultimate limit of the performance in some application. Clock jitter measurement is usually a difficult task. It usually relies on expensive Automatic Test Equipment (ATE) and can easily consume long test time[2]. Furthermore, the situation is getting worse as the System-on-Chip (SoC) is becoming a trend, because access to deeply embedded signals is not always possible.

The conventional jitter estimation methods always use expensive instrumentations to measure the clock instant and then analyze it to obtain the jitter information[1, 3]. One of widely used in jitter decomposition methods is based on Time Interval Error (TIE) [4], which is implemented as software in measurement instrument such as oscilloscope. However, the TIE measurement needs time reference (clock reference) which needs an extra clock recovery method (usually is PLL circuit). It is very expensive and complex for the instrument as the data rate increases. Many jitter separation and characterization methods have been proposed. An FFT-based jitter separation method was proposed to identify four kinds of DJ from collected time intervals[1]. A method for modeling and quantifying bounded Gaussian jitter(BGJ), as well as bounded Gaussian noise (BGN) was proposed in [5]. It can be used for jitter and noise estimation and testing. This paper proposes a low cost jitter measurement and separation method. Instead of using traditional time internal analysis equipment, a simple ADC is used as the jitter measurement device. The clock under test is applied as the sampling clock of an ADC while the ADC is sampling a full scale sine wave. The ADC output contains the information of the clock jitter.

In recent years, as the development of high performance Analog and Mixed Signal (AMS) application, the requirement on sampling clock is increasing. ADC is an important category of AMS products. Clock jitter plays a crucial role in ADC testing as the Signal-to-Noise Ratio (SNR) decreases if there is jitter in the sampling clock. Many researchers have proposed methodologies to measure clock jitter in ADC application. Conventional methods apply two input with sufficient separate frequencies with to the ADC under test to calculate the jitter information [6-8]. Dual frequencies method increases the test cost as the requirement of signal generator and synthesizers are high for ATE test. And for SoC test, the two frequency method increases the test cost because for low frequency on chip test needs large die area for capacitors. A fast and accurate jitter and noise measurement method with one frequency test signal was proposed in [9]. By setting certain number of harmonics of the ADC output to be zero in frequency domain, the residue of ADC output were separated to be two sets with different jitter power. The RMS of jitter was obtained by processing the two set of data. This method is accurate and efficient in jitter estimation. However, it requires knowing the number of harmonics before setting them to be zero, and non-harmonic spurs affect the testing result. An efficient method was proposed to accurately measure jitter and noise power based on one frequency measurement and it does not need to have prior knowledge of harmonics[10].Although methods mentioned above can separate random jitter from ADC intrinsic noise, they cannot handle it when there is DJ, which are common in AMS circuits.

The proposed algorithm will accurately remove the effects due to ADC harmonic distortion, additive noise and so on. The algorithm will then separately detect the effects of Periodic Jitter (PJ), Dual-Dirac Jitter (DDJ) and Random Jitter, and accurately compute the rms value of each jitter component. And this algorithm can also separate jitter and ADC additive noise, making it possible to calculate ADC specifications with poor quality clock. This method offers great potential for wide use in low cost applications and especially in on-chip or on-board jitter measurement applications.

This paper is organized as following: In section II, jitter and its effect on ADC output is introduced and the Expectation Maximization (EM) algorithm used for jitter separation is reviewed. Section III proposes the new method to separate PJ, DDJ, RJ and ADC intrinsic noise. Section IV shows the simulation results and accuracy of the method. The last section concludes the result.

II. REVIEW OF JITTER AND EM ALGORITHM

A. Basic concept of jitter

In this paper, jitter is discussed as the difference between the real clock edge and ideal clock edge. As shown in Fig.1, the ideal clock period is Ts and the sampling instant is at time 0, Ts, 2Ts and so on. However, there are deviations from the ideal sampling edge and the real ones. The deviations tt_1, tt_2 and tt_3 shown in Fig.1 are the jitter of the clock. This jitter is the total jitter at each sampling instant, and it can be the summation of several kinds of clock jitter. DJ and RJ are two important categories of clock jitter [11]. RJ is usually generated by thermal noise and it follows Gaussian distribution, as shown in Fig.2 (a). Periodic jitter and Dual-Dirac Jitter are two kinds of DJ existing in AMS. Sinusoidal Jitter (SJ) is one of the most common PJ, and Fig.2 (c) shows an example of its distribution. Fig.2(b) is the distribution of DDJ when the rms is 20ps. When there is both Random Jitter and Dual Dirac distributed jitter, the distribution of the total jitter is like shown in Fig.2(d).

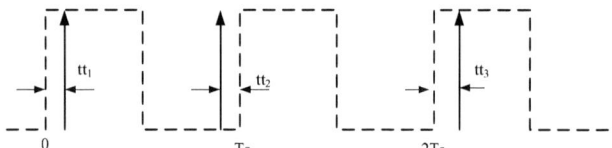

Figure 1. Clock jitter

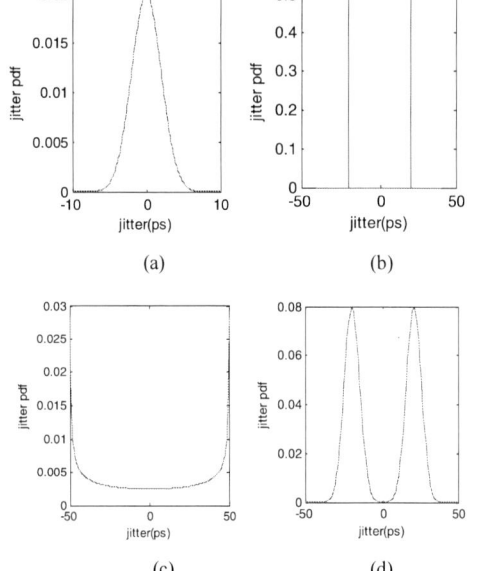

(a) (b)

(c) (d)

Figure 2.(a) Gaussian distribution (b) Dual Dirac distribution (d) Sinusoidal distribution (d) double peak Gaussian distribution

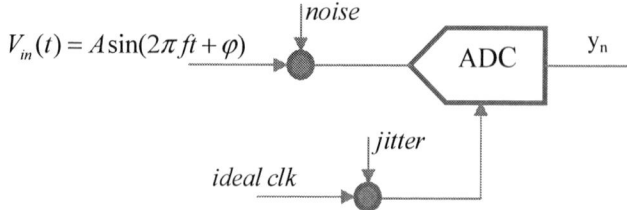

Figure 3. ADC test configuration

In the proposed method, an ADC device is used to test the jitter information. A sine wave is applied to the ADC as the input, and the sampling clock is the clock with jitter under test. The ADC output contains the jitter information. The model of jitter estimation using ADC is shown in Fig.3. A pure input sine wave is applied to the ADC, where the additive noise of ADC is modeled as random voltage added to the input signal. Jitter is modeled as random clock variation to the ideal clock instant. There are 3 components of the clock jitter: PJ, DDJ and RJ. The distribution of RJ and DDJ is shown as Fig.1 (a) and (b). The PJ can be sinusoidal jitter and square wave jitter in time domain, as long as it is periodic. A data sequence of the ADC output y_n is collected to analyze the jitter information.

Fig.4 shows spectra of an ADC with and without jitter effect. The blue spectrum is the one of the ADC output which is sampled by ideal clock jitter. The red spectrum when the ADC is sampled with clock jitter. In frequency domain, both the Random Jitter and Dual Dirac distributed jitter increase the noise floor as well as the additive noise in the ADC output spectrum, and the PJ generates non-harmonic spurs as shown in the red spectrum in Fig.4. Except for jitter information, the other conditions of the two spectra are the same. It can be seen that the noise floor of the spectrum with jitter is higher than that without jitter. If one calculates SNR and ENOB from the red spectrum directly, these two parameters will be less than the true ones (calculated from the blue spectrum).

From the spectra shown in Fig.4, it can be seen that the PJ is easy to be identified as it generates spurs in the spectrum. However, the RJ and DDJ increase the noise floor of the spectrum as well as the ADC intrinsic noise and quantization noise. It is difficult to separate RJ, DDJ from the ADC noise using existing methods. This paper uses an EM algorithm to calculate the RMS of RJ, DDJ and ADC noise. The EM algorithm is reviewed in the following.

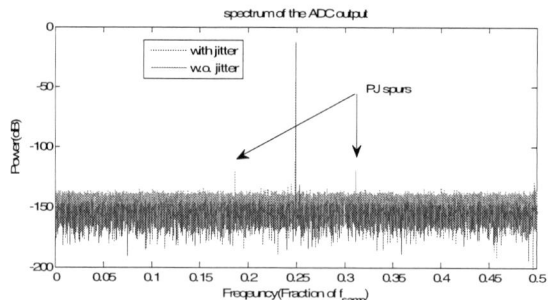

Figure 4. Spectra of ADC output with and without jitter effect

978-1-4799-7598-3/15 $31.00 © 2015 IEEE   20

B. Review of EM algorithm

The EM algorithm provides a systematic approach to find maximum likelihood estimates in cases where the model can be formulated in terms of "observed" and "unobserved" (missing) data. Here, "missing data" refers to quantities that, if we could measure them, would allow us to estimate easily the parameters of interest.

The following is the description of parameters used in the EM algorithm:

- Observations, measurements $x = [x_1, x_2, ... x_N]^T$;
- Parameter vector of interest θ;
- u, unobserved (missing) data;
- y = (u, x), complete data;

We seek the ML estimate

$$\hat{\theta} = \arg\max_{\theta} f_{(X|\Theta)}\left(x \mid \theta\right) \qquad (1)$$

The EM iteration that aims at computing $\hat{\theta}$ alternates between the following two steps:

Expectation (E) step:

$$Q\left(\theta \mid \theta^{(t)}\right) = E_{U|\Theta,X}[\ln f_{U,X|\Theta}\left(u,x \mid \theta\right) \mid \theta^{(t)}, x]$$
$$= \int \ln f_{U,X|\Theta}\left(u,x \mid \theta\right) f_{U|\Theta,X}\left(u \mid \theta^{(t)}, x\right) du \qquad (2)$$

Maximization (M) step:

$$\theta^{(t+1)} = \arg\max_{\theta} Q\left(\theta \mid \theta^{(t)}\right) \qquad (3)$$

After certain EM iterations, estimation of θ converges and the estimation results can be obtained.

III. THE PROPOSED METHOD

This section proposes a new method to separate PJ, DDJ and RJ using an ADC as measurement device. A pure sine wave is applied as the input of the ADC. The sampling clock is not ideal; it can contain PJ, DDJ and RJ. A set of ADC output is collected with the data length that is usually used for ADC spectral testing. First, the PJ spurs are identified from the ADC output spectrum and the rms of PJ is calculated. Then the fundamental and harmonics are set to be zero, the residue only contains jitter and noise information. EM algorithm is used to identify the rms or variance of DDJ, RJ and ADC intrinsic noise. The ADC intrinsic noise can also be used to calculate the ADC characteristics.

Assume the input signal of the ADC under test is a pure sine wave, which can be written as:

$$V_{in}(t) = A \sin(2\pi f t + \varphi) \qquad (4)$$

where A, f and φ are amplitude, frequency and phase of the input signal respectively. One set of ADC output with M sampling points are collected, the analog representation of the output is:

$$y_n = V_{in}(nT_s + tt_n) + h.d + w_n \qquad n = 0,1....M-1 \qquad (5)$$

where h.d is high order distortions, w_n is the summation of additive and quantization noise, tt_n is the total clock jitter at sampling instant nT_s and T_s is the sampling period. The total clock jitter consists of three components: PJ (p_n), DDJ (d_n) and RJ (δ_n).

A. Estimation of PJ

As tt_n is usually small compared to nT_s, we apply Taylor expansion to equation (5), the expression of y_n can be written as (6):

$$y_n \approx A \sin(2\pi f n T_s + \varphi) + 2\pi f A \cos(2\pi f n T_s + \varphi) p_n$$
$$+ 2\pi f A \cos(2\pi f n T_s + \varphi)(d_n + \delta_n) + h.d + w_n \qquad (6)$$

As p_n is a periodic term, it generates spurs in the spectrum if FFT is applied to y_n, as shown in Fig.4. For example, if p_n is sinusoidal jitter, it can be expressed as:

$$p_n = p_{\sin} \sin(2\pi f_{\sin} n T_s + \varphi_{\sin}) \qquad (7)$$

In equation (7), $p_{\sin}$, $f_{\sin}$ and $\varphi_{\sin}$ are the amplitude, frequency and initial phase of the sinusoidal jitter. As we want to identify the rms of the PJ, we just need to estimate $p_{\sin}$, the amplitude of the sinusoidal jitter.

First, the fundamental component of y_n is estimated, and the estimated values of $\hat{A}$, $\hat{f}$ and $\hat{\varphi}$ are obtained. Then, in the spectrum of the ADC output, as the red one shown in Fig.4, the height of the PJ spur and the height of the fundamental tone are estimated: h_1dB and h_2dB respectively. Then amplitude of the sinusoidal jitter is calculated as:

$$\widehat{p_{\sin}} = \frac{\sqrt{10^{\left(\frac{h_1 - h_2}{10}\right)}}}{\pi f} \qquad (8)$$

Then the rms of SJ can be obtained as $\widehat{p_{\sin}} / \sqrt{2}$.

B. Initial estimation of DDJ and RJ

Once the rms of PJ is calculated, the PJ spurs and harmonics are set to be 0 in the frequency domain. Inverse Fast Fourier Transform (IFFT) is applied to the new frequency domain data, and a new sequence y_{n1} is obtained where the PJ and high order harmonics are removed from y_n.

$$y_{n1} \approx A \sin(2\pi f \left(n T_s + tt_{1n}\right) + \varphi) + w_n \qquad (9)$$

In equation (9), tt_{1n} is the summation of DDJ and RJ at n^{th} sampling instant. If we ignore the noise effect, tt_{1n} can be roughly calculated as (10).

$$tt_{1n} \approx \frac{\sin^{-1}\left(\frac{y_{n1}}{\hat{A}}\right) - \hat{\varphi}}{2\pi \hat{f}} - nT_s \qquad (10)$$

The distribution of tt_{1n} is like the one shown in Fig.2 (d). One can estimate the approximate rms value of DDJ and RJ from tt_{1n}. Assume the rms of DDJ is d and the RJ follows a Gaussian distribution with 0 mean and variance of σ_1^2: $\delta_n \sim N(0, \sigma_1^2)$. And the noise w_n is modeled as a variable following Gaussian distribution: $w_n \sim N(0, \sigma_2^2)$. The initial value of d can be calculated as the mean value of the right bell shape and the initial value of σ_1^2 can be calculated as the variance of one bell shape of the tt_{1n} distribution. These are just the initial values and further steps are applied to calculate the accurate ones. And the initial value of σ_2^2 can be set as the same as that of σ_1^2.

C. Accurate estimation of DDJ and RJ

If we remove the fundamental component from (9), the residue will only contains jitter and noise information:

$$y_{n2} \approx 2\pi f A \cos(2\pi f n T_s + \varphi)(d_n + \delta_n) + w_n \quad (11)$$

For simplicity, we use a_n to represent the term $2\pi f A cos(2\pi f n T_s + \varphi)$ in equation (11), then a variable X_n is defined with M observations x_n as described in (11). Suppose $X_n = a_n Z_n S_n + w_n$, where Z_n follows a Gaussian distribution with a mean of d and a variance of σ_1^2: $Z_n \sim N(d, \sigma_1^2)$, and S_n follows Bernoulli distribution, the probability of S_n equals to be 1 or -1 is both 0.5. The parameters we want to solve in this case is $\theta = \langle d, \sigma_1^2, \sigma_2^2 \rangle$. For given S_n, d, σ_1^2 and σ_2^2, the distribution of x_n can be described as:

$$x_n \mid S_n, d, \sigma_1^2, \sigma_2^2 \sim N\left(a_n d S_n, a_n^2 \sigma_1^2 + \sigma_2^2\right) \quad (12)$$

Then the pdf of x_n can be given as:

$$P(x_n \mid S_n, d, \sigma_1^2, \sigma_2^2) = \frac{1}{\sqrt{2\pi\left(a_n^2\sigma_1^2 + \sigma_2^2\right)}} e^{-\frac{1}{2}\frac{1}{a_n^2\sigma_1^2+\sigma_2^2}(x_n - a_n d S_n)^2} \quad (13)$$

The joint pdf of all M x_n for given $S_{1:M}$ and θ is :

$$f\left(x_{1:N} \mid S_{1:N}, \theta\right) = \prod_{n=1}^{M} f\left(x_n \mid S_n, \theta\right) \quad (14)$$

And equation (14) can be rewritten as (15):

$$f\left(x_{1:M} \mid S_{1:M}, \theta\right)$$
$$= \prod_{n=1}^{M} \frac{1}{\sqrt{2\pi}} \frac{1}{\sqrt{a_n^2\sigma_1^2 + \sigma_2^2}} \exp\left[-\frac{1}{2}\frac{1}{a_n^2\sigma_1^2+\sigma_2^2}(x_n - a_n d S_n)^2\right] \quad (15)$$

As the joint pdf of S_n if $f(S_{1:M}|\theta) = \frac{1}{2^M}$, we can get the expectations of the likelihood function as shown in equation (16):

$$E\left(\ln f\left(x_{1:M}, S_{1:M} \mid \theta\right)\right) = -M \ln 2\sqrt{2\pi} - \frac{1}{2}\sum_{n=1}^{M} \ln\left(a_n^2\sigma_1^2 + \sigma_2^2\right)$$
$$-\frac{1}{2}\sum_{n=1}^{M}\frac{1}{a_n^2\sigma_1^2+\sigma_2^2}\left(x_n^2 - 2a_n d x_n ES_n + a_n^2 d^2\right) \quad (16)$$

From EM algorithm, the next step is to find parameters $\theta = \langle d, \sigma_1^2, \sigma_2^2 \rangle$ that can maximize the expression shown in (16). The parameters are updated after several iterations of maximization as shown in equation (17), (18) and (19), where k is the index of iteration. Usually after several

iterations the value of parameters will converge to a certain level that it can be stopped.

$$d^{(k+1)} = \frac{\sum \dfrac{a_i x_i \left(ES_i\right)^{(k+1)}}{a_i^2\sigma_1^{2(k)} + \sigma_2^{2(k)}}}{\sum \dfrac{a_i^2}{a_i^2\sigma_1^{2(k)} + \sigma_2^{2(k)}}} \quad (17)$$

$$\sigma_1^{2(k+1)} = \frac{\sum \dfrac{a_i^2}{\left(a_i^2\sigma_1^{2(k)} + \sigma_2^{2(k)}\right)^2}\left(x_i^2 - 2a_i d^{(k+1)} x_i \left(ES_i\right)^{(k+1)} + a_i^2\left(d^{(k+1)}\right)^2 - \sigma_2^{2(k)}\right)}{\sum \dfrac{a_i^4}{\left(a_i^2\sigma_1^{2(k)} + \sigma_2^{2(k)}\right)^2}} \quad (18)$$

$$\sigma_2^{2(k+1)} = \frac{\sum \dfrac{1}{\left(a_i^2\sigma_1^{2(k+1)} + \sigma_2^{2(k)}\right)^2}\left(x_i^2 - 2a_i d^{(k+1)} x_i \left(ES_i\right)^{(k+1)} + a_i^2\left(d^{(k+1)}\right)^2 - a_i^2\sigma_1^{2(k+1)}\right)}{\sum \dfrac{1}{\left(a_i^2\sigma_1^{2(k+1)} + \sigma_2^{2(k)}\right)^2}} \quad (19)$$

IV. SIMULATION RESULT

In this section, the proposed jitter measurement method is validated by simulation data generated in MATLAB. 14-bit, 16-bit and 18-bit ADCs are modeled respectively in MATLAB as a set of transition levels. The nonlinearity error is modeled as a set of Gaussian distributed variable with zero mean and a certain standard deviation for different resolution ADCs. The amplitude of the input sine wave is selected to be full scale of the ADC under test, and the input frequency is a little bit smaller than a half of sampling frequency. The additive noise is introduced with the input. It is Gaussian distributed random noise with zero mean and a certain standard deviation. The jitter is modeled as a random error added to the ideal sampling instant, which contains SJ, DDJ and RJ. Here we set the rms of SJ and DDJ at tens of ps level and RJ at several ps according to data from [1].

Table I. Testing result of ADCs with different resolution

	14 bit	16bit	18bit
Sampling frequency	5MHz	2MHz	2MHz
M	2^14	2^16	2^17
THD	-93.98dB	-102.2dB	-113.32dB
RMS of Dual Dirac Jitter Added	50ps	40ps	30ps
RMS of Dual Dirac Jitter Estimated	49.98ps	40.01ps	29.99ps
RMS of Random Jitter Added	4.99ps	2.993ps	1.996ps
RMS of Random Jitter Estimated	5.01ps	2.998ps	1.999ps
RMS of SJ Added	35.36ps	28.28ps	21.21ps
RMS of SJ Estimated	35.38ps	28.29ps	21.21ps

978-1-4799-7598-3/15 $31.00 © 2015 IEEE

A. Functionality Test

Table I shows the setup of the ADC under test. From the estimation results it can be seen that the jitter and noise estimation is accurate.

B. Robustness test

200 runs with different ADCs are applied to test the robustness of the proposed method. Table II shows the testing results of SJ, DDJ and RJ. It can be seen from the table that the testing result is accurate and the standard deviation of all three terms are very small.

Table II. Robustness testing results of SJ,DDJ and RJ (unit: ps)

Rms of Jitter added		Mean	Standard Deviation
RJ	2	2.01	0.006
	3	3.03	0.015
	5	5.01	0.009
DDJ	20	20	0.005
	30	30	0.01
	50	50	0.006
SJ	14.14	14.14	0.05
	21.21	21.21	0.05
	35.35	35.34	0.04

V. CONCLUSION

A simple, accurate and low-cost method that simultaneously extracts the Periodic Jitter, Dual Dirac Jitter and Random Jitter of the clock was presented. This method offers great potential for wide use in low cost applications and especially in on-chip or on-board jitter measurement applications. Simulation results demonstrated the functionality, accuracy and robustness of the proposed low-cost jitter measurement method. As the limitation of the ADC speed, the sampling clock used in this jitter characterization method is not high speed. If one needs to test high speed clock, one can use tens of the clock under test as one sampling clock to drive the ADC in order to increase the test

capability. In future work, the method needs to be justified by measurement results.

VI. ACKNOWLEDGEMENT

Materials presented in this paper are based upon the work supported in part by the National Science Foundation, the Semiconductor Research Corporation and Texas Instruments Inc. Any opinions, findings and conclusions or recommendations expressed in this paper are those of the authors and do not necessarily reflect the views of the sponsors.

REFERENCE

[1] T. J. Yamaguchi and H. X. Hou, "An FFT-based jitter separation method for high-frequency jitter testing with a 10x reduction in test time," in *IEEE International Test Conference*, 2007, pp. 1-8.

[2] H. Jui-Jer and H. Jiun-Lang, "A low-cost jitter measurement technique for BIST applications," in *Test Symposium, 2003. ATS 2003. 12th Asian*, 2003, pp. 336-339.

[3] T. J. Yamaguchi, M. Soma, M. Ishida, T. Watanabe, and T. Ohmi, "Extraction of instantaneous and RMS sinusoidal jitter using an analytic signal method," *Ieee Transactions on Circuits and Systems Ii-Analog and Digital Signal Processing*, vol. 50, pp. 288-298, Jun 2003.

[4] "Analyzing Jitter Using Agilent EZJIT Plus Software," *Application Note 1563.*

[5] M. Shimanouchi, M. P. Li, and D. Chow, "New modeling methods for bounded Gaussian jitter (BGJ)/noise (BGN) and their applications in jitter/noise estimation/testing," in *Test Conference, 2009. ITC 2009. International*, 2009, pp. 1-8.

[6] "IEEE Standard for Terminology and Test Methods for Analog-to-Digital Converters," *IEEE Std.1241*, 2010.

[7] S. Shariat-Panahi, F. A. C. Alegria, A. Manuel, and A. M. D. Serra, "IEEE 1057 Jitter Test of Waveform Recorders," *Ieee Transactions on Instrumentation and Measurement*, vol. 58, pp. 2234-2244, Jul 2009.

[8] W. Kester, "The Data Conversion Handbook," *Analog Device, Inc*, pp. 5.73-5.75, 2004.

[9] W. Minshun, C. Degang, and D. Jingbo, "Fast & accurate algorithm for jitter test with a single frequency test signal," in *2011 IEEE International Conference on Electro/Information Technology (EIT)*, 2011, pp. 1-5.

[10] L. Xu and D. Chen, "Accurate and efficient method of jitter and noise separation and its application to ADC testing," in *VLSI Test Symposium (VTS), 2014 IEEE 32nd*, 2014, pp. 1-5.

[11] T. J. Yamaguchi and K. Ichiyama, "A robust method for identifying a deterministic jitter model in a total jitter distribution," in *International Test Conference*, 2009, pp. 1-10.

Ultrafast Stimulus Error Removal Algorithm for ADC Linearity Test

Tao Chen, Degang Chen

Department of Electrical and Computer Engineering
Iowa State University, Ames, IA, USA
taoc@iastate.edu djchen@iastate.edu

Abstract— **Linearity test of an analog-to-digital converter (ADC) can be very challenging because it requires a signal generator substantially more linear than the ADC under test. For high performance ADCs, the overall manufacturing cost could be dominated by the long test time and the high-precision test instruments. This paper introduces the ultrafast stimulus error removal and segmented model identification of linearity errors (USER-SMILE) method for high resolution ADC linearity test, allowing the stimulus signal's linearity requirement to be significantly relaxed and the test time to be reduced by orders of magnitude compared to the state-of-art histogram method. The USER-SMILE algorithm uses two nonlinear but functionally related input signals as ADC excitations and uses a stimulus error removal technique to recover test accuracy. The USER-SMILE algorithm also uses the ultrafast segmented model identification of linearity errors (uSMILE) approach to dramatically reduce test time while achieving test accuracy and coverage superior to the histogram method. The USER-SMILE algorithm is validated by extensive simulation with different types of ADCs, different resolution levels, and different types of input signals including nonlinear ramps, nonlinear sine waves and even random input signals. Statistical simulation results show that for a 16-bit SAR ADC, with two 1 hit/code nonlinear ramp signals, the INL test error is within +/- 0.4LSB.**

Keywords—Analog-to-digital converter; integral nonlinearity; histogram; ultrafast stimulus error removal and segmented model identification of linearity errors (USER-SMILE); built-in-self-test

I. INTRODUCTION

The analog-to-digital converter (ADC) is one of the most important analog and mixed signal (AMS) products [1]. Accurate linearity test of ADC can be very challenging, especially for high resolution ADCs [2]. As the manufacturing cost goes down, the test cost becomes more and more dominant in the overall cost. The ADC test cost is mainly due to the test equipment cost and the test time. To test the ADC nonlinearity, the state-of-art histogram method uses a highly linear signal generated from the high-precision automated test equipment (ATE) [3-6]. The signal source is required to be substantially more linear than the device under test (DUT). It becomes more and more difficult to generate linear source as the ADC resolution goes high. Furthermore, the histogram method requires much more samples than the number of transitions in the ADC. As the industry standard, the histogram test usually uses tens or even hundreds hits per code to accurately test the

ADC nonlinearity, which results in a very long data acquisition time. For high resolution ADC (higher than 16-bit), it is usually not practical to fully test the ADC linearity in production test due to the extremely long test time.

The stringent requirement on the input signal linearity and the extremely long time test become the challenges in the AMS test. Significant works have been done to overcome these challenges. Recently, researchers have developed different ways to address the stringent linearity requirement of the input signal. In [7], the author employed the delta-sigma modulation technique to generate the highly linear input signal. However, it is not easy to design such a signal generator as the ADC's resolution or speed goes high. The design complexity often increases the cost. In the contrast, some researchers have put efforts on algorithms to relax the stimulus linearity requirement. In [8-10], stimulus error identification and removal (SEIR) algorithm is proposed to test precision ADC using nonlinear stimulus. It has been proved that 7-bit linear ramp signal can be used to test high resolution ADC and achieve more than 16 bits accuracy. A constant offset is required to identify the nonlinear components in the signal source. As the requirement on the input signal linearity is relaxed, built-in-self-test for ADC full code INL/DNL test becomes practical. However, SEIR is based on the histogram method, which means the data acquisition time is still very long.

Other than relaxing the input signal requirement, lots of efforts have been made to reduce the test time. In [11, 12], a method was proposed to use fast Fourier transform (FFT) test to estimate the ADC's INL. In [13], the INL can be estimated with a combined spectral and histogram method. A system identification approach is proposed in [14] to evaluate the nonlinearity of a pipeline ADC. In [15], the author uses the polynomial fitting method with low resolution input signal to test the ADC nonlinearity. In [16], Goyal, et al introduced a selective code measurement method to reduce the test time of SAR ADCs. However, all above methods or similar ones reduce the test time by sacrificing other test aspects, so that they cannot achieve similar coverage or test accuracy than the histogram method. Therefore, the application is very limited. An ultrafast segmented model identification of linearity errors (uSMILE) [17] algorithm was proposed recently to take a system identification approach to capture both linear and nonlinear errors in the ADC. With the concept of the segmented non-parametric model, the algorithm can reduce the

---

The work is supported in part by Freescale Semiconductor and Semiconductor Research Corporation.

test data by a factor of over 100 and achieve a test accuracy superior to the histogram method. However, it still requires highly linear input signal source.

In summary, the existing solutions have at least one of the following issues: long test time, highly linear stimulus, low accuracy or coverage. Currently there is no valid solution to resolve all these issues at the same time. Test time is an important factor in AMS test and accurate input signal cannot be easily implemented on chip. There is a strong need to test the ADC using an easy-to-implement signal generator with much less test time. In this paper, a new algorithm combing the concept of SEIR and uSMILE is proposed for accurate linearity test with dramatically reduced test time and also relaxes the requirement on source linearity. Two nonlinear input signals with constant offset between them are applied to the ADC. Two sets of ADC output codes will be generated. Segmented non-parametric model is used to represent the final INL. Rather than directly finding the INL of the ADC, the INL is indirectly evaluated from the difference of the segmented INL by subtracting the two sets of the output codes. The test accuracy and coverage is superior to the state-of-art histogram method.

The following of this paper is organized as follows. Section II reviews two fundamental algorithms: the SEIR and the uSMILE algorithm. Section III presents the proposed algorithm. Section IV shows the simulation results. And section V concludes the paper.

II. BACKGROUND

The SEIR algorithm relaxes the linearity requirement on the input signal by injecting a constant offset in the input signal. The uSMILE algorithm significantly reduces the test time with the segmented non-parametric model. These two algorithms are the basis of the USER-SMILE algorithm and will be reviewed below.

A. SEIR

Define the nonlinear ramp signal to be x(t) and normalize the time so that $t_0 = 0$ $t_{N-2} = 1$ and T_k is the transition level for output from code k-1 to code k. Then it can be expressed as

$$x(t) = T_0 + (T_{N-2} - T_0)t + F(t) \qquad (1)$$

where F(t) is the nonlinear component of the ramp signal. The nonlinear component can be estimated using a set of basis function $F(t) = \sum a_j F_j(t)$. The INL measured by nonlinear ramp will be:

$$INL'_k = (N-2)t_k - k = INL_k - F(t_k) \qquad (2)$$

where INL'_k is the estimated INL using nonlinear ramp and INL_k is the corrected INL after removing the nonlinear component in the input signal.

$$INL_k = (N-2)t_k + F(t_k) - k \qquad (3)$$

For two ramp signal, $x_1(t)$ and $x_2(t)$ have a constant offset α.

$$x_2(t) = x_1(t) - \alpha \qquad (4)$$

Then, $x_1(t)$ and $x_2(t)$ can be expressed with transition points and the nonlinear components as:

$$x_1(t) = T_0 + (T_{N-2} - T_0)t + F(t) \qquad (5)$$

$$x_2(t) = T_0 + (T_{N-2} - T_0)t + F(t) - \alpha \qquad (6)$$

For the transition level from code k-1 to code k, we can get $T_k = x_1(t_{k,1}) = x_2(t_{k,2})$. Replace $x_1(t_{k,1})$ and $x_2(t_{k,2})$ with equation (1), equation (7) is obtained.

$$T_k = T_0 + (T_{N-2} - T_0)t_{k,1} + F(t_{k,1})$$
$$= T_0 + (T_{N-2} - T_0)t_{k,2} + F(t_{k,2}) - \alpha$$

$$(7)$$

Since the number of equations is much larger than the number of the unknowns, least square can be used to estimate the unknowns. The coefficients of F and the constant offset α can be obtained from equation (8).

$$\{\hat{a}_1, \hat{a}_2, \hat{a}_3, \cdots, \hat{a}_M, \hat{\alpha}\} = \arg\min \Big\{ \sum \Big((N-2)(t_{k,2} - t_{k,1}) -$$
$$\big[\sum a_j \big(F_j(t_{k,1}) - F_j(t_{k,2}) \big) + \alpha \big] \Big)^2 \Big\}$$

$$(8)$$

Then, the nonlinear component of the input ramp signal has been identified. The INL can be reconstructed by removing the nonlinear component from input signal. Either $INL_k^{(1)}$ or $INL_k^{(2)}$ can be used for the evaluated INL.

$$INL_k^{(1)} = (N-2)t_k^{(1)} + \sum_{j=1}^M \hat{a}_j F_j(t_k^{(1)}) - k \qquad (9)$$

$$INL_k^{(2)} = (N-2)t_k^{(2)} + \sum_{j=1}^M \hat{a}_j F_j(t_k^{(2)}) - k - \hat{\alpha} \qquad (10)$$

B. uSMILE

Different from the SEIR, the uSMILE algorithm was proposed to significantly reduce the test time as well as achieve better accuracy. By a system identification approach with a segmented non-parametric model, the algorithm is able to capture the nonlinearity of the ADC with much less test data.

The segmented non-parametric model in the INL curve is to break down the INL into different segments. For example, an INL curve can be broken into 64 segments if 6 MSB bits are used. For each MSB segment, this short INL curve can be further broken into smaller segments (for example, 5 ISB bits). Similarly, the ISB can be broken into LSB (for example, 5 LSB bits). For each segment, there is a corresponding error term. Define the MSB error term to be $e_M(C_{MSB})$, where C_{MSB} is the code of the MSB bits. Then, the errors for 64 segments are $e_M(0)$, $e_M(1)$, ..., $e_M(63)$ corresponding to the MSB code. Similarly, e_I and e_L are defined for ISB and LSB errors respectively and they are also called "segmented INL" in the following of this paper. The final INL value for code C will be:

$$INL(C) = e_M(C_{MSB}) + e_I(C_{ISB}) + e_L(C_{LSB}) \qquad (11)$$

For an input signal, there will be an ideal expected output C_{exp}. Due to the ADC nonlinearity, the actual output code becomes C. Then, the input output relationship can be created:

$$C_{exp} - C + q = e_M(C_{MSB}) + e_I(C_{ISB}) + e_L(C_{LSB}) \qquad (12)$$

where q is the noise.

In order to estimate the INL with the segmented non-parametric model, the linear input signal information is used. With a pure sine wave as the ADC's input signal, a linear ADC will get a linear sine wave in the output. However, due to the nonlinearity existing in the ADC, the output will have harmonics and other components. From the actual ADC's output code, the DC and fundamental components can be extracted in the frequency domain and an ideal ADC is constructed using the DC and fundamental only. In other words, after removing the DC and fundamental in the frequency domain, everything else are just noise and nonlinear components in the actual ADC. After identifying the MSB, ISB and LSB errors, the final full-code INL can be constructed.

III. USER-SMILE

This section proposes the USER-SMILE algorithm to identify the INL/DNL using nonlinear input signal and with much less test data, and achieve better test coverage and accuracy than the histogram test. Two identical input signals with constant offset between them are applied to the ADC. In the USER-SMILE, by subtracting the two sets of output data, the input signal information is no longer needed. Any error or nonlinearity in the stimulus is completely removed. At this point, there's no assumption on the input signal linearity. Some other restrictions on the input signal will be discussed later.

Apply two input signals $V_{in}^{(1)}$ and $V_{in}^{(2)}$ to the ADC with a constant offset α.

$$V_{in}^{(1)} = V_{in}^{(2)} + \alpha \quad (13)$$

The converted output codes from ADC are $C^{(1)}$ and $C^{(2)}$. Then, equation (14) and (15) are obtained:

$$V_{in}^{(1)} + w^{(1)} = T^{(1)} + q^{(1)} = C^{(1)} \cdot V_{LSB} + INL^{(1)} + q^{(1)} \quad (14)$$

$$V_{in}^{(2)} + w^{(2)} = T^{(2)} + q^{(2)} = C^{(2)} \cdot V_{LSB} + INL^{(2)} + q^{(2)} \quad (15)$$

where the noise $w^{(1)}$ and $w^{(2)}$ are the input-referred noise and T is the transition voltage for output code from C to C+1. And q in the equation is the quantization noise.

With the segmented non-parametric model, the INL can be broken into MSB segments, ISB segments and LSB segments. The total nonlinearity error for code C can be written into the same format in equation (11).

Then, the equation (14) and (15) can be expressed as:

$$
\begin{aligned}
V_{in}^{(1)} + w^{(1)} \\
= C^{(1)} \cdot V_{LSB} + e_M\big(C^{(1)}{}_{MSB}\big) \cdot V_{LSB} \\
+ e_I\big(C^{(1)}{}_{ISB}\big) \cdot V_{LSB} + e_L\big(C^{(1)}{}_{LSB}\big) \cdot V_{LSB} \\
+ q^{(1)}
\end{aligned}
\quad (16)
$$

$$
\begin{aligned}
V_{in}^{(2)} + w^{(2)} \\
= C^{(2)} \cdot V_{LSB} + e_M\big(C^{(2)}{}_{MSB}\big) \cdot V_{LSB} \\
+ e_I\big(C^{(2)}{}_{ISB}\big) \cdot V_{LSB} + e_L\big(C^{(2)}{}_{LSB}\big) \cdot V_{LSB} \\
+ q^{(2)}
\end{aligned}
\quad (17)
$$

By subtracting the two equations (16) and (17), we can get equation (18):

$$
\begin{aligned}
V_{in}^{(1)} - V_{in}^{(2)} + w^{(1)} - w^{(2)} \\
= V_{LSB} \cdot \{ C^{(1)} - C^{(2)} + e_M\big(C^{(1)}{}_{MSB}\big) + e_I\big(C^{(1)}{}_{ISB}\big) \\
+ e_L\big(C^{(1)}{}_{LSB}\big) - e_M\big(C^{(2)}{}_{MSB}\big) - e_I\big(C^{(2)}{}_{ISB}\big) \\
- e_L\big(C^{(2)}{}_{LSB}\big)\} + q^{(1)} - q^{(2)}
\end{aligned}
$$

$$(18)$$

Replace the $V_{in}^{(1)} - V_{in}^{(2)}$ with α and re-arrange the equation:

$$
\begin{aligned}
C^{(1)} - C^{(2)} - \frac{\alpha}{V_{LSB}} \\
= -\{e_M\big(C^{(1)}{}_{MSB}\big) + e_I\big(C^{(1)}{}_{ISB}\big) \\
+ e_L\big(C^{(1)}{}_{LSB}\big) - e_M\big(C^{(2)}{}_{MSB}\big) - e_I\big(C^{(2)}{}_{ISB}\big) \\
- e_L\big(C^{(2)}{}_{LSB}\big)\} + (q^{(2)} - q^{(1)} + w^{(1)} - w^{(2)})/V_{LSB}
\end{aligned}
$$

$$(19)$$

Assume that the input-referred noise is at a certain level and the quantization noise will be "whitened". So the term $(q^{(2)} - q^{(1)} + w^{(1)} - w^{(2)})$ can be considered as one random noise. For this overdetermined system, the least square algorithm can be used to find the unknowns e_M, e_I and e_L. With least square method, the noise term will be effectively averaged out. Then, the full code INL can be constructed.

Some crucial parts in the algorithm are discussed below:

A. Segmented non-parametric model

USER-SMILE leverages the segmented non-parametric model in the uSMILE algorithm. It treats the ADC itself as a black box and accurately models the actual INL curve. Any linear errors (mismatch and gain) and nonlinear error (voltage coefficients or code dependent parasitics) can be captured. Any advantages and restrictions in the segmented non-parametric model are also applied to the USER-SMILE algorithm. Therefore, with the segmented non-parametric model, the USER-SMILE algorithm can significantly reduce the test time. For a 16-bit ADC with 6-5-5 segmentation (6 MSB bits, 5 ISB bit and 5 LSB bits), only 128 unknowns need to be solved and the results can accurately reflect the actual INL. However, there are also some limitations. The segmented non-parametric model is intended for high resolution ADCs whose architecture facilitates a segmented structure of the INL curve. So, the USER-SMILE method is not intended for flash ADC or delta sigma ADC. For other types of ADCs such as SAR ADC, Cyclic ADC, Pipeline ADC, the USER-SMILE algorithm works well.

B. Stimulus Requirement

It is usually difficult to design a fast and highly linear signal generator on chip. If we can relax the requirement on the stimulus, the signal generator design complexity and cost can be significantly reduced. By subtracting the two equations (16) and (17), the information of input signal is no longer needed. And there is no assumption on the signal linearity or the signal shape.

However, there's still some constrains to the stimulus. Take an extreme case: if we use a fixed voltage to test the ADC, the ADC is always producing a similar code. There is no way to get the information of other codes. So, to achieve good estimation accuracy, the input signal should cover as most codes. Each segment (in MSB, ISB and LSB) needs to have sufficient coverage.

The signal generator design is simplified with the above consideration. It can be a very nonlinear signal generator but needs to cover most of the ADC input range. A low cost, nonlinear, ramp generator or sine wave generator can be easily built on chip as the stimulus with minimal area overhead.

C. Constant Offset

As showed in the previous derivation, a constant offset is required. Different methods have been proposed before for offset injections and good constancy can be achieved on chip.

The algorithm needs to know the exact value of the offset. Due to variation, the actual offset value may be different with the simulated value. In this case, the offset α can be simply estimated by the average difference between the output codes $C^{(1)}$ and $C^{(2)}$.

Usually, we want to minimize the amount of offset due to the offset generator design. Larger offset will increase the design cost or complexity. To make the USER-SMILE algorithm work, the offset cannot be too small. If we have a very small offset (for example, a few LSB), the MSB segments will hardly change after applying this offset. The e_M in this set of data are the same and they cancel each other so that there is no information from these data. Ideally we expect all segments to be changed after applying the offset and all data are fully used. If we use 6-5-5 segmentation (6 MSB bits, 5 ISB bit and 5 LSB bits), the ideal offset value is 1 MSB + 1 ISB + 1 LSB, which is 1057 LSB. Due to nonlinearity of ADC or the variation of the offset, for each set of data, the segments may not be all different. But the amount of such data is small and the effect on the estimation accuracy is neglectable. So, making the offset value to be slightly larger than 1 MSB is the best choice.

IV. SIMULATIONS

To verify the algorithm, extensive simulations have been done on different ADC architectures (SAR, Pipeline, and Cyclic) with various resolutions. The simulation results show that the algorithm works well with different architectures. SAR ADC is particularly studied due to its wide usage, high resolution and low power. A 16-bit SAR ADC is modeled with random capacitor mismatches. The true INL is constructed from the transition voltage. In all the following simulations, 0.5 LSB input-referred noise is added.

With two 1 hit/code nonlinear ramp signals, the INL (end point fitting) estimation is shown in Fig.1 (a). The true INL of the ADC is plotted with the red line. In this ADC, the INL is about 1 LSB. The estimated INL from the USER-SMILE is plotted in the blue line. From the plot, we can see that the blue line matches the red line very well. The estimation error is defined to be the difference between the estimated full-code INL with the true INL. Fig.1 (b) shows the estimation error in USER-SMILE. The maximum estimation error is about +/-0.15 LSB. It shows that the USER-SMILE method produces good estimation accuracy over all the codes.

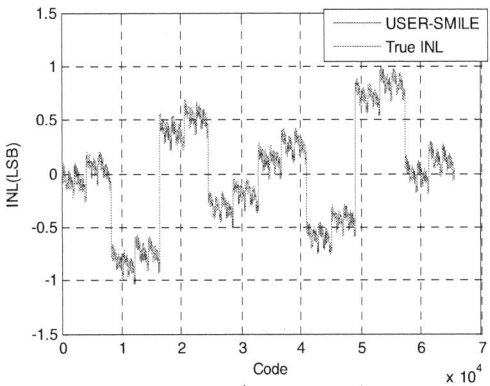

(a) INL Plot of USER-SMILE and True INL

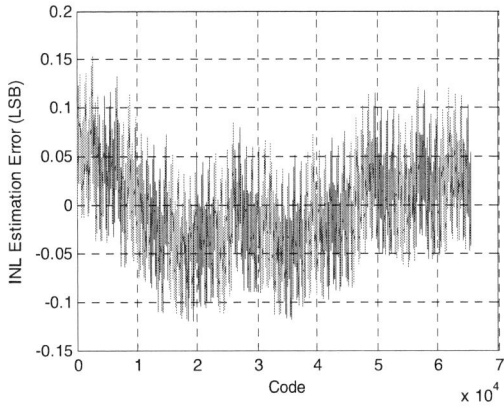

(b) INL Estimation Error of USER-SMILE

Fig. 1. INL Estimation for one ADC

To further verify the algorithm with different ADCs, a large number of simulations have been done. The test uses two 1 hit/code nonlinear ramp signals to test different ADCs and 100 test results are randomly selected. For each INL curve, the maximum estimation error and the minimum estimation error are recorded over all codes and shown in Fig.2. From the figure, the average maximum estimation error in USER-SMILE is around 0.2 LSB with few of them over 0.3 LSB. Fig.3 shows a different view of the estimation accuracy. The x-axis is the true INL and the y-axis is the estimated INL from the USER-SMILE. Ideally, if the estimated INL is the same as the true INL, this point will lie on the y=x line (the blue line). Due to the estimation error, these points will be away from this line. From the figure, the red points (estimated INL from the USER-SMILE) is very close to the y=x line, which means the USER-SMILE has very good accuracy. For ADCs with different performance (INL from 0.6LSB to 4.6LSB), the test accuracy stays the same. From the production test point of view, the USER-SMILE method will guarantee less yield loss.

With two 1 hit/code ramp signals for a 16-bit ADC, the overall test data are only 131k points. The statistical study shows that the USER-SMILE algorithm is robust over different ADCs (including good and bad ADCs). And from all tests, the maximum estimation error for INL is within +/-0.4LSB.

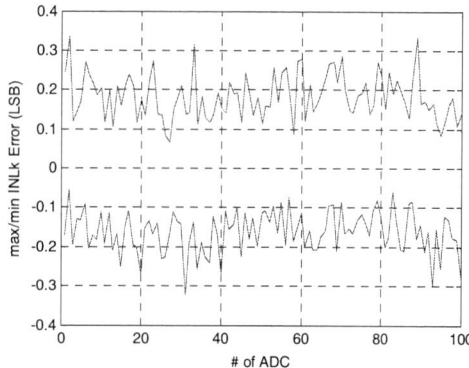

Fig. 2. Maximum/Minimum INLk Error

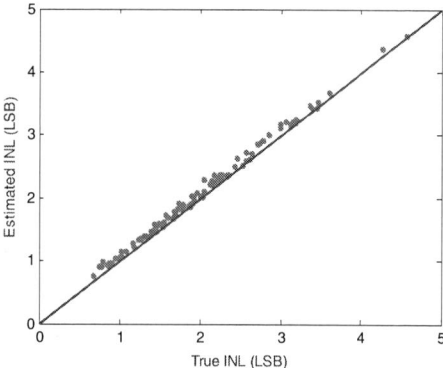

Fig. 3. INL Estimation in USER-SMILE over 100 ADCs

V. CONCLUSION

A fast and cost-effective method for ADC linearity test is presented in this paper. The USER-SMILE algorithm allows the stimulus signal's linearity requirement to be significantly relaxed and the test time to be reduced by orders of magnitude compared to the state-of-art histogram method, thus greatly reducing the test cost. The simulation demonstrates that the USER-SMILE can achieve superior test coverage and accuracy. With the USER-SMILE algorithm, a new BIST solution can be practical, which doesn't require highly accurate and expensive ATE as the signal generator. Furthermore, it simplifies the test board and interface design.

ACKNOWLEDGMENT

Materials presented in this paper are based upon the work supported in part by the Freescale Semiconductor and the Semiconductor Research Corporation. Any opinions, findings and conclusions or recommendations expressed in this paper are those of the authors and do not reflect opinions of the sponsors.

REFERENCES

[1] International Technology Roadmap for Semiconductors, 2011 edition, [Online]. Available: http://public.itrs.net

[2] T. Kuyel, "Linearity testing issues of analog to digital converters," in Proc. 1999 Int. Test Conf., 1999, pp. 747–756.

[3] J. Blair, "Histogram measurement of ADC nonlinearities using sine waves," IEEE Trans. Instrum. Meas., vol. 43, pp. 373–383, Jun. 1994.

[4] M. Burns and G. W. Roberts, An Introduction to Mixed-Signal IC Test and Measurement. New York: Oxford Univ. Press, 2000.

[5] J. Doernberg, H.-S. Lee, and D. A. Hodges, "Full-speed testing of A/D converters," IEEE J. Solid-State Circuits, vol. SC-19, pp. 820–827, Dec. 1984.

[6] IEEE Standard for Terminology and Test Methods for Analog-to-Digital Converters, IEEE Std. 1241-2010, Jan. 2011.

[7] J.L. Huang, C.X. Ong, K.T. Cheng, "A BIST scheme for on-chip ADC and DAC testing", Proc. of the Design, Automation and Test in Europe Conference and Exhibition, 2000, pp 216-220.

[8] L. Jin, et al, "Code-Density Test of Analog-to-Digital Converters Using Single Low-Linearity Stimulus Signal," IEEE Transactions on Instrumentation and Measurement, Vol. 58, No. 8, pp. 2679-2685, August 2009.

[9] L. Jin, K. Parthasarathy, T. Kuyel, D. Chen and R. L. Geiger, "Accurte Testing of Analog-to-Digital Converters Using Low Linearity Signals With Stimulus Error Identification and Removal," IEEE Trans. Instrum Meas., vol. 54, pp. 1188 – 1199, June 2005.

[10] L. Jin, et al, "SEIR Linearity Testing of Precision A/D Converters in Nonstationary Environments With Center Symmetric Interleaving," IEEE Transactions On Instrumentation And Measurement, Vol. 56, No. 5,pp. 1776-1785, October 2007.

[11] F. Adamo, et al, "FFT Test of A/D Converters to Determine the Integral Nonlinearity," IEEE Trans. on Instrumentation and Measurement, Vol. 51, No. 5, pp. 1050-1054, October 2002.

[12] F. Attivissimo, et al, "INL reconstruction of A/D converters via parametric spectral estimation," IEEE Trans. Instrum. Meas., vol. 53, no. 4, pp. 940–946, Aug. 2004.

[13] A. Cruz Serra , M. Fonseca da Silva , P. Ramos , R. Carneiro Martins , L. Michaeli and J. Saliga "Combined spectral and histogram analysis for fast ADC testing", *IEEE Trans. Instrum. Meas.*, vol. 54, no. 4, pp.1617 -1623 2005

[14] Z. Yu, D. Chen, R. Geiger, and Y. Papantonopoulos. "Pipeline ADC linearity testing with dramatically reduced data capture time," Proc. IEEE Int. Symposium on Circuits and Systems, pp. 792-795, 2005.

[15] S. Kook, et al, "Low-Resolution DAC-Driven Linearity Testing of Higher Resolution ADCs Using Polynomial Fitting Measurements," to appear in IEEE Transactions On Very Large Scale Integration (VLSI) Systems, 2012.

[16] S. Goyal, et al, "Test Time Reduction of Successive Approximation Register A/D Converter By Selective Code Measurement," International Test Conference, Nov, 2005.

[17] Z. Yu and D. Chen, "Algorithm for Dramatically Improved Efficiency in ADC Linearity Test," IEEE International Test Conference (ITC), pages 1-10, 2012.

Disturbance-free BIST for Loop Characterization of DC-DC Buck Converters

Navankur Beohar, Priyanka Bakliwal, Sidhanto Roy, Debashis Mandal, Philippe Adell[♦], Bert Vermeire[*],
Bertan Bakkaloglu, and Sule Ozev
School of Electrical, Computer, and Energy Engineering, Arizona State University, Tempe, AZ
[♦]NASA Jet Propulsion Laboratory, Pasadena, CA
[*]Space Micro Inc. San Diego, CA

Abstract—Complex electronic systems include multiple power domains and drastically varying dynamic power consumption patterns, requiring the use of multiple power conversion and regulation units. High frequency switching converters have been gaining prominence in the DC-DC converter market due to their high efficiency. Unfortunately, they are also subject to higher process variations jeopardizing stable operation of the power supply. This paper presents a technique to track changes in the dynamic loop characteristics of the DC-DC converters without disturbing the normal mode of operation using a white noise based excitation and correlation. White noise excitation is generated via pseudo random disturbance at reference and PWM input of the converter with the test signal energy being spread over a wide bandwidth, below the converter noise and ripple floor. Test signal analysis is achieved by correlating the pseudo random input sequence with the output response and thereby accumulating the desired behavior over time and pulling it above the noise floor of the measurement set-up. An off-the-shelf power converter, LM27402 is used as the DUT for the experimental verification. Experimental results show that the proposed technique can estimate converter's natural frequency and Q-factor within $\pm 2.5\%$ and $\pm 0.7\%$ error margin respectively, over changes in load inductance and capacitance.

I. INTRODUCTION

Switching mode DC-DC converters have been widely used as an integral part of power management integrated circuits (PMICs) and power management units (PMUs) in computer, communications, and consumer electronics. Quite often, electronic systems and SOCs contain many DC-DC converters to supply multiple voltage domains, current/voltage requirements of which may change dynamically. As an example, the Haswell processor has thirteen switching DC-DC converters [1]. In order to satisfy fast response requirements with a small form factor, the trend is to employ higher frequency switching. The switching frequency of the Haswell DC-DC converters is 140MHz [1]. Dynamic performance and stability of DC-DC converters greatly depend on the overall loop characteristics. Loop dynamics in turn are determined primarily by off-chip output filter inductance (L), inductor DC resistance (DCR), load capacitance (C), capacitor equivalent series resistance (ESR), and the on resistance of the power train transistors (Rds,on). In [1], converters with small output L-C filter are demonstrated using package trace inductors. Design of such high switching rate converters becomes a bigger challenge

as the smaller form-factor output filter (LCR filter) suffers from higher manufacturing variations. For an example, average DCR of Vishay IHLP-5050FD 4.7μH inductor is about 9.32mΩ with 1.3% manufacturing 3-sigma variation and the average DCR of Vitec 59P9022 100nH inductor is 0.3mΩ with 14.8% manufacturing 3-sigma variation. In addition to the manufacturing variations, temperature and aging also cause drift in the converter loop. In [2], it has been shown that for a typical DC-DC converter, open loop frequency response phase margin drops 13° caused by $\pm 25\%$ variation in the output filter and load current (see Fig.1).

In order to ensure efficient and stable operation of the DC-DC converter, dynamic loop characteristics need to be determined and the controller needs to be tuned with respect to these characteristics. Since the loop characteristics may shift, albeit incrementally and gradually, over time and environmental conditions, this monitoring and tuning cannot be done only after manufacturing, but it needs to be repeated in the field, while the device is actively working. Any in-field measurement technique for loop characteristics requires several aspects: (a) self-test needs to be transparent with respect to the normal operation of the converter, (b) the measurement needs to be conducted within the closed-loop and at operation point, (c) output

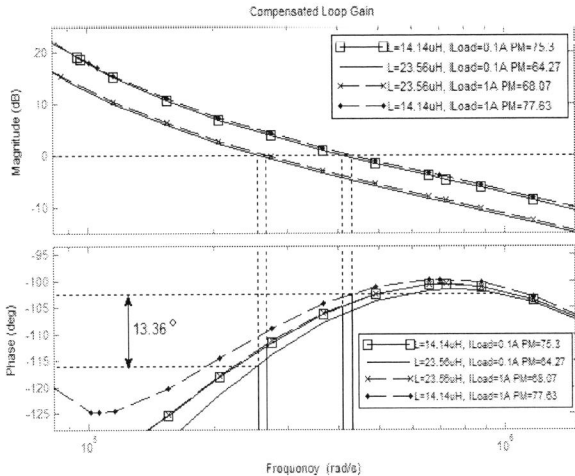

Fig. 1 Impact of output filter variation on the open loop AC response of a typical DC-DC converter [2]

response needs to be large enough to be immune to noise, and (d) the measurement needs to present with little to no computational overhead. These requirements are contradictory as measurement requires an observable response to a specific test input, and an observable response by the very definition disturbs the operation of the converter. Analysis of the observed response also typically requires extensive computational resources that are not available on the DC-DC converters.

In general, characterization of the transfer function of a given circuit is also known as the system identification problem. The methods for system identification can be categorized as parametric and non-parametric [3],[4]. Parametric identification methods start with a known structural model of the system and estimate the model parameters based on observed response. Non-parametric methods do not assume any particular model structure, but incrementally determine the time or frequency domain impulse response by exciting the system with known inputs and directly observing its response. Existing non-parametric methods are transient analysis [5], frequency response analysis [6], and correlation analysis [7],[8]. Two well-known transient methods are step-input injection and impulse injection methods. Injection of a sine-wave input and measuring the gain of the system for different input frequencies is a standard transfer function analysis method. Unfortunately, such system identification methods do not satisfy the above mentioned requirements since they disturb the operation of the converter.

In [9], the authors have proposed a safe test method to correlate the converter parameters under high and low current scenarios for PMUs. Similar statistical modeling approaches, as in [9], can also be used for the loop characterization. However, they too need an observable response at the output.

A number of methods have been proposed to self-test DC-DC converters at production time. An open-loop, control-to-output system identification technique using pseudo random input stimulus is proposed in [10]. Due to its open loop mode requirement, this technique is not suitable for on-line characterization of DC-DC converters. In [11], a Gaussian windowed closed-loop based transfer function measurement method is proposed using pseudo random input signals. The goal in [10],[11] is different than in-field self-test, it is to improve the high-frequency response of the system. As such, higher frequency observation is necessary, which complicates the self-test problem. A multi-period maximum length pseudo random binary sequence (MLBS) is used as the excitation signal in [12]. The transfer functions are identified from the measurement data with circular cross-correlation method. However, this technique uses digital low pass filtered MLBS, making it a multi-level analog excitation, therefore this technique requires a full D/A converter to generate the MLBS and a transformer to couple the signal to the regulator output.

In this work, we propose a method to determine the closed-loop transfer function of the DC-DC converters without significantly affecting its output noise beyond the existing ripple. We use a concept similar to spectrum spreading techniques (such as CDMA) in the communications domain to hide the test input/output signals within the existing noise floor of the DC-DC converter. We use a small perturbation pseudo random signal at an accessible input node of the converter (i.e. the reference and pulse width modulator, PWM, input). While the output response is not directly measurable, we use the correlation of the input and output signals to accumulate the response over time, thereby pulling it to above the noise floor. We aim to obtain the impulse response of the closed-loop operation. This impulse response (or the frequency domain equivalent, transfer function) can be analyzed to determine important stability characteristics, such as settling time, overshoot, and undershoot. In contrast to a previous similar technique [12], the proposed approach uses only a binary pseudo random bit sequence, does not require analog coupling, extracts stability parameters directly from measurements without resorting to computationally intensive procedures, and works within the closed-loop system, making it more suitable for fully integrated DC-DC converters.

We have used an off-the-shelf DC-DC converter to demonstrate our approach. Experimental results show that important parameters of the transfer function, such as natural frequency or quality factor, can be determined with high accuracy.

The organization of this paper is as follows. Section II presents the theoretical framework for the proposed method to characterize the loop transfer function. Section III presents proposed self-test method and its experimental verification, and conclusions are given in Section IV.

II. CORRELATION BASED DYNAMIC LOOP CHARACTERIZATION

In steady state operation, for small signal disturbances, a switching power converter can be approximated as a linear time-invariant discrete-time system [13]. A linear time-invariant sampled system can be described as

$$y[n] = \sum_{k=1}^{\infty} h[k]x[n-k] + v[n] \qquad (1)$$

where $y[n]$ is the sampled output signal, $x[n]$ is the sampled input signal, $h[n]$ is the discrete-time system impulse response and $v[n]$ represents unwanted disturbances, such as switching noise, quantization noise, etc. The cross-correlation of the input signal $x[n]$ and the output signal $y[n]$ is as follows:

$$R_{xy}[m] = \sum_{n=1}^{\infty} x[n]y[n+m]$$

$$= \sum_{n=1}^{\infty} h[n]R_{xx}[m-n] + R_{xv}[m] \qquad (2)$$

where $R_{xy}[m]$ is the cross-correlation of input and output signals, $R_{xx}[m]$ is the auto-correlation of input signal and $R_{xv}[m]$ is the cross-correlation of input signal with disturbances [3]. Now, if $x[n]$ is white noise, then correlation functions R_{xx} and R_{xv} have the following properties:

$$R_{xx}[m] = \delta[m] \qquad (3)$$

$$R_{xv}[m] = 0$$

where $\delta[m]$ is an ideal delta function. Auto-correlation of white-noise input is ideal delta function and cross-correlation of white-noise input with unwanted disturbances $v[n]$ is ideally zero. This simplifies Eq.(2) and the cross-correlation becomes the discrete-time system impulse response [3]. The discrete Fourier transform (DFT) of impulse response gives system frequency response.

$$R_{xy}[m] = h[m]$$

$$R_{xy}[m] \xrightarrow{DFT} H[j\omega] \qquad (4)$$

The properties presented in Eq.(3)-(4) need the injection signal to be white noise. In addition, it is desirable that the signal generation adds low overhead. In practical implementation, an approximate white noise is generated by pseudo random binary sequence (PRBS) generator consisting of shift registers and feedback taps [14]. The PRBS is periodic and deterministic, and the data length of the n-bit maximum length PRBS generator is given by $M=2^n-1$. Auto-correlation of white-noise is an ideal delta, but for PRBS, auto-correlation function is a mix of a delta function at $m=0$ and low amplitude components at $m\neq0$. Similarly, the cross-correlation of PRBS with system disturbances $v[n]$ is not zero. Hence, the cross-correlation (R_{xy}) of input PRBS with output signal has undesired noise terms in addition to the system impulse response due to the non-ideality in R_{xx} and R_{xv} of Eq.(3). In [11], this noise is reduced by windowing the measured cross-correlation function. It improves the measured response but still suffers from poor precision especially at high frequencies. Significant improvement in measured response is reported in [12] by using circular correlation, where the effect of zero padded ends and heads of the linear correlation procedure is reduced by circulating the two data sequences and multiplying the corresponding bits.

III. PROPOSED SELF-TEST METHOD FOR IN-FIELD MONITORING

As mentioned in the Introduction Section, we need a low-overhead method to extract the impulse response. Instead of a full system-ID of the transfer function, we aim to determine the parameters that affect the stability of the converter. These parameters are the natural frequency and the quality factor which are determined by the dominant poles and zeros. Hence, we can limit our observation to within a small region of the entire spectrum. This limitation will simplify the process in two ways: (a) the PRBS frequency can be much lower than the switching rate and (b) multi period PRBS sequence can be used along with time-domain averaging to suppress noise [10].

Fig. 2 shows the proposed dynamic loop characterization methodology. A PRBS generator drives the disturbance at multiple accessible nodes in the DC-DC converter (e.g. reference input node, PWM input node, etc.), the output of

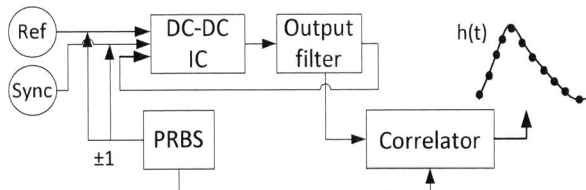

Fig. 2 Proposed dynamic loop characterization methodology

the loop filter is correlated with the binary PRBS data using circular correlation technique. This process is repeated multiple times and the results averaged in the time domain to suppress the effects of noise. It should be noted that since the samples of the impulse function are obtained at the peak SNR point after the correlator, this approach has the highest immunity to noise and will provide comparable results to oversampling and post-processing. Additionally, the shape of the impulse will be preserved in presence of process variations but will only result in a DC shift.

The extracted impulse response function can be processed in two ways. Measure the natural frequency and quality factor directly from the impulse response. Alternatively, DFT can be used to convert the time domain information into the frequency domain and measure these two parameters. The results from these two techniques are theoretically identical [15].

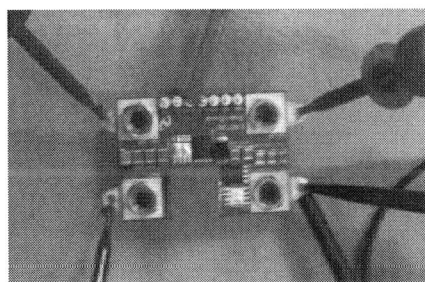

Fig. 3(a) Evaluation board LM27402 and experimental set-up

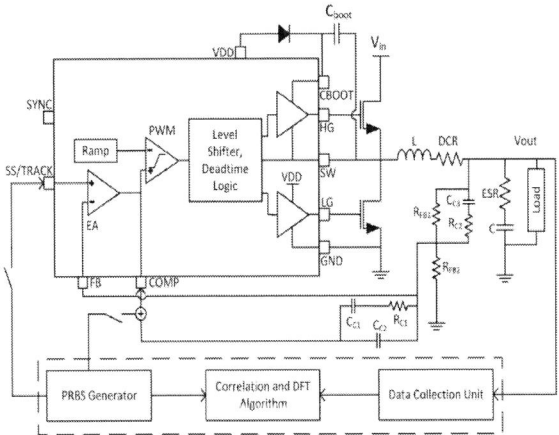

Fig. 3(b) Simplified block diagram of DC-DC converter on LM27402 evaluation board with external system identification blocks.

TABLE I   LOAD INDUCTANCES AND CAPACITANCES USED IN THE EXPERIMENT

L (nH)	% ΔL	DCR(mΩ)	C (μF)	% ΔC
560	-18	2.3	328	0
600	-12	2.3	375	14
680	0	2.34	406	24
700	3	2.5	-	-

A. Experimental Verification and Results

A synchronous single phase PWM DC-DC converter by Texas Instrument, LM27402 evaluation board has been used to experimentally verify the use of the PRBS based identification for the detection of load filter variations. Fig. 3(a) and 3(b) show the evaluation board setup and the converter block diagram with external system identification setup, respectively. Evaluation board consists of a PWM controlled DC-DC buck converter IC LM27402, N-type power FETs, L-C output filter, feedback resistive divider (R_{FB1} and R_{FB2}), compensation filter (R_{C1}, R_{C2}, C_{C1}, C_{C2} and C_{C3}) and bootstrap capacitor C_{boot}. Default component values of the converter on the evaluation board are: output filter inductance L=0.68μH with DCR=2.34mΩ, output filter capacitance C=240μF with ESR=0.75mΩ, R_{FB1}=20kΩ, R_{FB2}=13.3kΩ, R_{C1}=8.01kΩ, R_{C2}=261Ω, C_{C1}=3.9nF, C_{C2}=150pF, C_{C3}=820pF, C_{boot}=220nF. The transfer function is evaluated at the operating point where the input voltage, V_{in}=6V, output voltage, V_{out}=1.5V, load current, I_L=0A and the switching frequency, f_s=300kHz. The PRBS frequency is set at 50kHz, providing an observation bandwidth of 25kHz. From the above nominal values, we can confirm that the dominant poles and zeros of the loop are well within this bandwidth (BW). Although there are higher order poles and zeros, these do not affect the stability of the converter. Moreover, aging will result in degradation of circuit parameters (changing pole/zero locations). Hence, for the purposes of production and in-field testing for stability, the 25kHz BW is adequate.

For this experiment, the perturbation signal has been injected through two access nodes: reference input (SS/TRACK pin), to the error amplifier, and compensated signal input (COMP pin), to the PWM input, of LM27402 controller IC as shown in Fig. 3(b). The response of the converter has been observed at the output node (V_{out}). The impulse response obtained from the proposed method is compared with the traditional swept sine-wave method. Output filter variation has been introduced by using four different inductors and three different capacitances that have been removed from the board and replaced. TABLE I shows the parameters of inductors and capacitors.

PRBS noise has been injected using a function generator. Five periods of maximum length 10-bit PRBS data with 50kHz data rate has been applied. Total PRBS data length is M=5.(2^{10}-1) =5115. Converter output response at V_{out}, has been captured through a data acquisition block, setup at 50kHz sampling clock with 14-bit analog-to-digital converter (ADC) precision controlled by LabVIEW. The PRBS data and captured output data have been transmitted to a PC (personal computer) for the computation of cross-correlation, averaging and DFT based estimation of

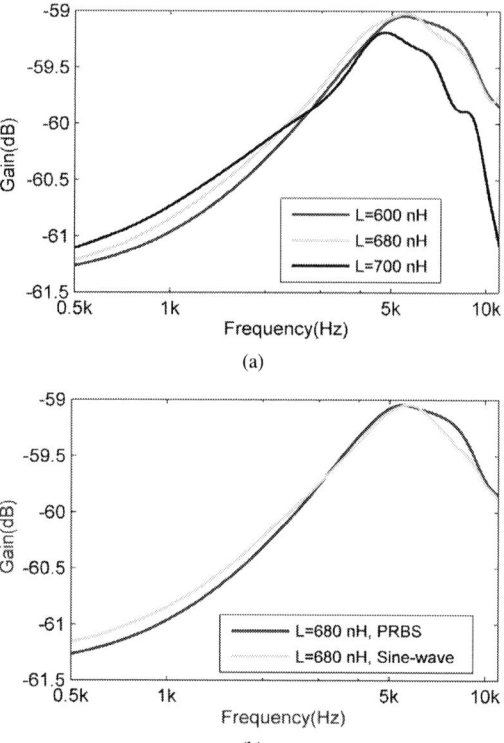

Fig. 4 Converter reference input to output frequency response measured: (a) for different inductance values using PRBS, and (b) comparison between PRBS and sine-wave method.

converter quality factor and natural frequency using MATLAB.

B. Transfer Function from Reference Input to Output

The transfer function from the reference input to the output node includes the output filter, controller poles and zeros, and the ON resistance of the power train transistors. Out of these, the poles and zeros due to the output filter and the ON resistance of the power train fall in-band for the proposed method. These parameters are also subject to the highest levels of process variation and aging and affect the stability of the system. For the remaining parameters, only catastrophic defects will shift them to within the BW of the proposed measurement, which will be detected. Other parametric variations may affect efficiency, but not stability of the system. This transfer function is effectively a second order transfer function which is slightly under-damped by design. We have used the proposed technique to measure the natural frequency (ω_0) and quality factor (Q) of this transfer function. For the comparison, we have used the traditional swept sine-wave method where the input node is excited by a single tone sine wave, the gain is measured, and the process is repeated for multiple frequencies. Input sine-wave has been generated by a function generator and output response has been measured on a high resolution oscilloscope.

Fig. 4(a) shows the measured transfer function for different inductance values using PRBS method while the comparison between PRBS and sine-wave methods is

978-1-4799-7598-3/15 $31.00 © 2015 IEEE

TABLE II ω_0 AND Q, MEASURED FOR DIFFERENT OUTPUT FILTER INDUCTANCES

L (nH)	PRBS		Sine-wave		Error (%)	
	ω_0 (Hz)	Q	ω_0 (Hz)	Q	$\varepsilon_{\omega 0}$	ε_Q
600	5713	1.311	5700	1.312	0.23	-0.08
680	5469	1.315	5400	1.306	1.28	0.69
700	4883	1.289	5000	1.294	-2.34	-0.39

TABLE III ω_0 AND Q, MEASURED FOR DIFFERENT OUTPUT FILTER CAPACITANCES

C (μF)	PRBS		Sine-wave		Error (%)	
	ω_0 (Hz)	Q	ω_0 (Hz)	Q	$\varepsilon_{\omega 0}$	ε_Q
328	5469	1.315	5400	1.306	1.28	0.69
375	5176	1.289	5300	1.294	-2.34	-0.39
406	4834	1.274	4800	1.273	0.71	0.08

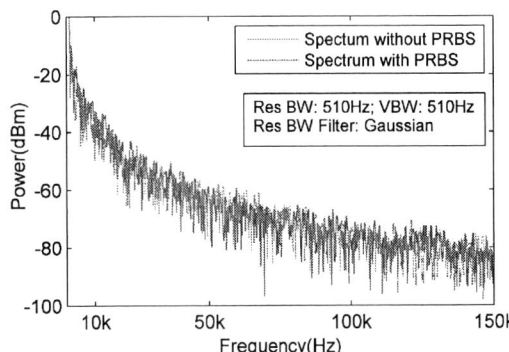

Fig. 5 Spectrum of output voltage (V_{out}): with and without PRBS injection at converter reference input (SS/TRACK) for correlation based system identification.

depicted in Fig. 4(b). The PRBS based response accurately tracks the sine-wave based measured response. For both PRBS and sine-wave methods, obtained ω_0 and Q values have been tabulated in TABLE II and III. TABLE II captures output filter inductance variation results, while TABLE III highlights output filter capacitance variation results. The comparison of the PRBS based method with the swept sine-wave method gives less than $\pm 2.5\%$ ω_0 error ($\varepsilon_{\omega 0}$) for inductance and capacitance variations. In comparison with the swept sine-wave measurement, the PRBS based Q measurement error (ε_Q), for inductance and capacitance variations, is below $\pm 0.7\%$.

It is clear that the proposed method achieves high accuracy in accordance with an industry standard method. As mentioned earlier, the advantage of the proposed method is the ability to spread the test signal energy over the 25kHz BW and hence not disturb the system operation. Fig. 5 shows the spectrum of the converter output voltage without and with the PRBS input signal. The comparison shows hardly any change in the output spectrum without and with the PRBS noise except the small increase in the noise floor. As the PRBS based method provides negligible disturbance in the output voltage, system identification based on this correlation method is the best suitable technique for in-field assessment of the DC-DC converter.

C. Transfer Function from PWM Input to Output

The proposed technique is versatile and low-impact in terms of disturbance. Hence, the same circuit can be used for multiple excitation points to extract more information from the DUT without adding more hardware overhead. We have also used the proposed technique to characterize the transfer function from the PWM input to the output. This transfer function has pass-band characteristics and is affected by the output filter, power train transistors, and the compensator poles and zeros. Deviations in this transfer function from its expected response indicate that one or more of the above parameters have drifted. From the earlier characterization, putting the two transfer functions together, it is possible to diagnose the location of the problems in terms of loop components or compensator. Fig. 6(a) shows

the transfer function with varying inductor values while Fig. 6(b) shows the comparison of the transfer function obtained using the PRBS method and with the traditional swept sine-wave based method. This result also shows a very good match between the two methods. Since this is a pass-band behavior, we compare the center frequency (ω_c) of the two transfer functions. The difference in the absolute gain is irrelevant since it does not affect the pole/zero configurations. Figure shows a good matching between the PRBS and sine-wave based measurements. Corresponding ω_c values obtained by the PRBS and swept sine-wave measurements have been summarized in TABLE IV for different output filter inductances and in TABLE V for different output filter capacitances. Inductances and

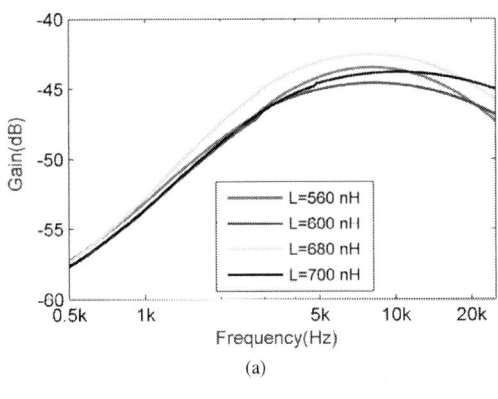

(a)

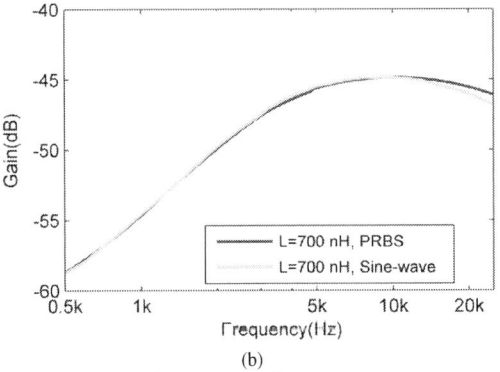

(b)

Fig. 6 Converter PWM input to output frequency response measured: (a) for different inductance values using PRBS, and (b) comparison between PRBS and sine-wave method.

978-1-4799-7598-3/15 $31.00 © 2015 IEEE

TABLE IV ω_c MEASURED FOR DIFFERENT OUTPUT FILTER INDUCTANCES

L (nH)	PRBS	Sine-wave	Error (%)
	ω_c (Hz)	ω_c (Hz)	$\varepsilon_{\omega c}$
560	8008	8000	0.10
600	8252	8200	0.63
680	8545	8600	-0.64
700	9570	9500	0.74

TABLE V ω_c MEASURED FOR DIFFERENT OUTPUT FILTER CAPACITANCES

C (µF)	PRBS	Sine-wave	Error (%)
	ω_c (Hz)	ω_c (Hz)	$\varepsilon_{\omega c}$
328	8545	8600	-0.64
375	7959	7900	0.75

capacitances are of the same values as noted in TABLE I. With respect to swept sine-wave, in PRBS method the measurement errors in ω_c ($\varepsilon_{\omega c}$) over L and C variations, are lower than $\pm 0.8\%$.

D. Test Time and Silicon Overhead

The overall time for transfer function extraction is determined by the PRBS frequency, the number of patterns in the PRBS signal, and the number of points evaluated on the time domain transfer function. While these numbers will depend on the individual DUT, the overall test time for the experimental circuit is about 100ms. It should be noted that the proposed technique works in the background, without disturbing the converter operation. Hence, the test time is not of high importance.

The technique requires IC implementation of (a) PRBS generator, and (b) analog correlator. Here, the proposed method has numerous advantages compared to techniques proposed in the literature. The proposed technique requires only binary excitation and binary correlation. This greatly reduces the hardware overhead of the entire measurement system and leads to easier, switched-capacitor based implementation. We have implemented these building blocks at the transistor level. Integrated CMOS implementation of the proposed stimulus and measurement blocks is made up of a linear feedback shift register (LFSR) based random number generator, a switched capacitor correlator and a low frequency ADC. In a typical DC-DC converter occupying 3.5mmx3.5mm die area, proposed BIST circuitry will occupy less than 4% die area and 150uA quiescent current.

IV. CONCLUSION

High frequency switching converters have been gaining prominence in the DC-DC converter market due to their high efficiency. Unfortunately, they are also subject to higher process variations jeopardizing stable operation of the power supply. This paper presents a technique to track changes in the dynamic loop characteristics of the DC-DC converters without disturbing the normal mode of operation using a white noise based excitation and correlation. White noise excitation is generated via pseudo random disturbance at reference and control input of the converter with the test signal energy being spread over a wide bandwidth, below the converter noise and ripple floor. The variation in loop characteristics is determined using a PRBS-based small perturbation white noise. The results are compared with those obtained using frequency swept sine-wave method. The technique is independent of the converter type and can be used without impacting the normal operation of the converter, during the closed-loop operation. The proposed technique is demonstrated on a TI LM27402 switch mode buck converter. The results obtained by PRBS method with the injection point at error amplifier input are within $\pm 2.5\%$ and $\pm 0.7\%$ for ω_0 and Q, respectively, over variation in inductance and capacitance. The error is less than $\pm 0.8\%$ for ω_c over inductance and capacitance variation when the injection point is at the PWM input. Single bit PRBS injection enables digital friendly implementation on silicon.

REFERENCES

[1] N. Kurd et al., "Haswell: A Family of IA 22nm processors," in *IEEE Int. Solid State Circuits Conf. Dig. Tech. Papers*, 2014, pp. 112-113.

[2] T. Liu et al., "A built-in self-test technique for load inductance and lossless current sensing of DC-DC converters," in *Proc. IEEE VLSI Test Symp.*, 2014, pp. 1-6.

[3] L. Ljung, *System Identification: Theory for the User*, 2nd ed. Englewood Cliffs, NJ: Prentice-Hall, 1999.

[4] G. F. Franklin, J. D. Powell, and M. L. Workman, *Digital Control of Dynamic Systems*, 3rd ed. New York: Addison-Wesley, 1997.

[5] D. Maksimovic, "Computer-aided small-signal analysis based on impulse response of DC/DC switching power converters," *IEEE Trans. Power Electron.*, vol. 15, no. 6, pp. 1183-1191, Nov. 2000.

[6] B. Johansson and M. Lenells, "Possibilities of obtaining small-signal models of DC-to-DC power converters by means of system identification," in *Proc. Telecommun. Energy Conf.*, 2000, pp. 65-75.

[7] P. Huynh and B. H. Cho, "Empirical small-signal modeling of switching converters using PSpice," in *Proc. IEEE Power Electron. Specialists Conf.*, vol. 2, 1995, pp. 809-815.

[8] A. Costabeber et al., "Digital autotuning of DC-DC converters based on a model reference impulse response," *IEEE Trans. Power Electron.*, vol. 26, no. 10, pp. 2915-2924, Oct. 2011.

[9] X. Wang et al., "Alternative "safe" test of hysteretic power converters," in *Proc. IEEE VLSI Test Symp.*, 2014, pp. 1-6.

[10] B. Miao, R. Zane, and D. Maksimovic, "System identification of power converters with digital control through cross-correlation methods," *IEEE Trans. Power Electron.*, vol. 20, no. 5, pp. 1093-1099, Sep. 2005.

[11] A. Barkley and E. Santi, "Improved online identification of a DC-DC converter and its control loop gain using cross-correlation methods," *IEEE Trans. Power Electron.*, vol. 24, no. 8, pp. 2021-2031, Aug. 2009.

[12] T. Roinila et al., "Circular correlation based identification of switching power converter with uncertainty analysis using fuzzy density approach," *Simulation Modelling Practice and Theory*, vol. 17, no. 6, pp. 1043-1058, Jul. 2009.

[13] R. W. Erickson and D. Maksimovic, *Fundamentals of Power Electronics*, 2nd ed. Boston, MA: Kluwer, 2001.

[14] K. R. Godfrey, "Introduction to binary signals used in system identification," in *Proc. IEEE Int. Conf. Control*, vol. 1, 1991, pp. 161-166.

[15] B. H. Karnopp and F. E. Fisher, "Determination of vibration parameters in moderately damped systems," *J. Franklin Institute*, vol. 327, no. 4, pp. 611-620, 1990.

978-1-4799-7598-3/15 $31.00 © 2015 IEEE

Innovative Practices Session 1C:
New Technologies, New Challenges – 1

Organizer: TM Mak (Globalfoundries)
Moderator: Paul Tracey (Altera)

Presenters & Abstracts:

Presenter #1: *Witek Maszara (Globalfoundries)*

Title: Designing with FinFETs

Abstract: FinFET, a fully depleted, multi-gate transistor entered the market with high performance microprocessor product at 22nm node in 2012 and became standard device for scaled technologies. FinFETs offer superior performance over incumbent planar devices due to their significantly improved electrostatics. FinFET technology faced two key barriers to their implementation in products: demanding process integration and its significant impact on layout and circuit design methodology. In this presentation, we focus on key features and challenges in designing with FinFETs. fin formation, metrology, device parasitics. Performance as well as design challenges for logic and SRAM circuits will be discussed.

Presenter #2: *Greg Yeric (ARM)*

Title: Test Implications of FinFETs

Abstract: FinFETs have recently overtaken bulk CMOS transistors as the device of choice for systems-on-chip. This paper provides some background on FinFETs together with their associated manufacturing processes and shows how they influence design and test of synthesized designs, including variability and margining.

Presenter #3: *Yervant Zorian (Synopsys)*

Title: Design, Test & Repair Methodology for FinFET-based Memories

Abstract: Due to their spatial structures, FinFETs have several advantages including controlled Fin body thickness, low threshold voltage variation, reduced variability, and lower operating voltage. With significantly higher wafer prices, maximizing yield is even more important. Modern FinFET-based memories can be impacted by unique defects that require new test and repair algorithms. The approaches used for planar-based memories cannot provide the appropriate level of defect coverage for FinFET-based memories. This presentation will describe new FinFET-specific defects and novel test algorithms to detect them. It will also describe built-in self-test (BIST) infrastructure with high-efficiency test and repair capabilities and showcase how this methodology has been validated on silicon across multiple foundries.

2015 IEEE 33rd VLSI Test Symposium (VTS)

A Multi-Layered Methodology for Defect-Tolerance of Datapath Modules in Processors

Hsunwei Hsiung and Sandeep K. Gupta

Ming Hsieh Department of Electrical Engineering, University of Southern California, Los Angeles, CA 90089-2562
{hsunweih, sandeep}@usc.edu

Abstract—Technology scaling increases circuits' susceptibility to manufacturing imperfections and dramatically decreases processor yields. Traditional defect-tolerance approaches add explicit redundant circuitry to improve yield and hence are very expensive for datapath modules in processors. We propose a multi-layered methodology to develop new and efficient defect-tolerance approaches for processors. Specifically, we develop a microarchitecture layer approach for arithmetic logic units (ALU), a circuit layer approach for multipliers, and an ISA layer approach for floating-point units (FPU). We demonstrate that our three approaches improve performance-per-fabricated-die-area of a modern processor core by 3.5%, 2.4%, and at least 9%, and hence collectively provide significant gains.

1 INTRODUCTION

Technology scaling has increased transistor density and reduced fabrication cost per transistor. To efficiently utilize these, processor designers pack more functionality and more performance enhancing features into each new processor. However, as technology advances deep into nano-scale, the improvements in cost, power, and delay, provided by each technology generation have started to slow down, or even reverse. One reason is increase in circuits' susceptibility to manufacturing imperfections. We use the term defect to refer to two types of imperfections, namely, process variations and random defects that affect circuit operation. Defect rates are now increasing with each scaling generation and hence reducing yield, especially at the top-levels of performance. Reduction in yield diminishes or may even negate reduction in die fabrication cost provided by scaling.

Traditionally, use of *explicit redundancy* has been explored in memory and logic circuits to improve yield. Spare rows and columns are commonly used in memories [1] [2]. However, cost of such explicit circuit redundancy is prohibitively high for logic circuits, such as most datapath modules. Recent advanced defect-tolerance (DT) approaches for processors use *implicit redundancy*, i.e., utilize existing features in processors and add a minimal level of reconfiguration to achieve DT with low overheads. For instance, many have exploited microarchitecture property of caches to tolerate defects [3] [4] [5] [6]. The fact that branch predictors are speculative has also been studied for the purpose of DT in [7]. However, there is no systematic methodology to guide the development of such advanced approaches and there is no study on using implicit redundancy for datapath modules.

In this work, we propose a methodology to develop implicit-redundancy DT approaches for processors. The methodology systematically develops approaches to maximize fabrication efficiency, which is measured as *performance-per-fabricated-die-area* (or performance-per-area). In particular, we propose new and efficient approaches for datapath modules used in processors, namely, ALU, fixed-point multiplier, and FPU. In this work, we use a single core processor to demonstrate that our three approaches collectively provide significant improvements in performance-per-fabricated-die-area. These improvements are expected to compound over several cores in multicore processors. In addition, we assume an artificial defect density so that the single core processor has 30% yield. In our on-going study for multicore processors using projected defect densities, the improvements are comparable or even higher than the ones we report in this work.

2 DEFECT-TOLERANCE METHODOLOGY FOR PROCESSORS

We consider the following factors when we develop our approach to improve fabrication efficiency. (1) *Yield benefit*: additional yield provided by defective processors salvaged by our approach. This also provides *performance benefit*, i.e., the additional performance provided by defective processors we salvage as these would have been otherwise discarded. In addition to the percentage of chips it can salvage, our approach also focuses on maximizing architecture-level performance for salvaged chips. (2) *Yield and area cost*: the reduction in yield and increase in area caused by area overhead of our approach. (3) *Performance cost*: the reduction in defect-free processors' performance caused by circuit delay overhead of an approach.

We have developed the following principles to maximize the benefits of our approaches and minimize their costs.

- **Multi-layered system view.** We view a processor core as a part of a multi-layered system and exploit the existing implicit redundancy in the layers, namely, software/OS layer, ISA layer, microarchitecture layer, and circuit layer. These layers provide many opportunities for low-overhead approaches, and approaches from different layers can be combined to achieve defect-tolerance in uniquely-efficient ways.

- **Yield-aware module selection.** Modules targeted for development of our approach are selected to maximize yield benefits. Having previously targeted cache modules, here we target ALUs, multipliers, and FPUs as they have appreciable impact on yield due to their large areas.

- **Utilization-aware module selection and approach implementation.** To maximize performance benefits, the selection of target modules and our approach implementations are significantly driven by how modules are utilized by typical applications. For example, a processor with a rarely-used defective module can potentially achieve the same performance as a defect-free processor, provided the defective processor can still guarantee correct results. This is also the case for a defect in a part of a module that is rarely used.

Next we describe our new approaches for datapath modules which have appreciable areas but are largely ignored by previous advanced defect-tolerance approaches for processors.

3 CIRCUIT LAYER APPROACH FOR MULTIPLIER

In this section, we demonstrate how our methodology facilitates the development of this circuit layer approach. We use a 4×4 carry-save adder (CSA) based multiplier to illustrate the approach. Then we estimate the overheads for a 32×32 CSA multiplier used in a processor. We will show that this enhancement incurs low area overheads yet significantly improves the overall performance-per-area of the processor core.

Utilization-aware module selection: We studied the instruction mix for benchmarks and found that fixed-point multiplication constitutes a small percentage of all instruction executed for a variety of benchmarks. Hence, we expect low performance penalty for processors with defective multipliers enhanced using our approach, provided that our approach guarantees correct execution of every multiplication operation and does not slowdown operations other than multiplication.

Multi-layered view from the circuit layer: Similar to memory modules, many datapath modules are designed using multiple copies of cells. For example, a single-bit full-adder cell that produces *sum* and *carry-out* can be used repeatedly to design a multi-bit adder module. Each such circuit-cell is functionally identical and each cell can act as redundancy for others. This is also the case for the CSA multiplier.

3.1 CSA multiplier design and enhancement

Figure 1 depicts a 4×4 CSA multiplier and its state machine controller. The array has sixteen CSA cells and four carry-propagate adder (CPA) cells. When signaled with the assertion of *rdy* (input ready), the multiplier produces an 8-bit product ($p_7 \sim p_0$) by multiplying a 4-bit X operand ($x_3 \sim x_0$) with a 4-bit Y operand ($y_3 \sim y_0$) in one clock cycle. At the end of the cycle it asserts the *done* signal. In the cell array circuit of the multiplier, each diagonal series of four CSA cells and one CPA cell along a y_i input constitute a *bit-slice*.

Every bit-slice performs the same functionality. Bit-slice i generates a partial product by multiplying X with y_i, adding the product generated by bit-slice $i + 1$, and adding the carry-bit propagated from bit-slice $i - 1$. If only one bit-slice is designed in a multiplier implementation, with a modified controller design and additional multiplexors, a 4×4 multiplication can be performed by rotating through y_i each cycle and storing the partial product generated so that it can be added to the product computed in the next cycle. Hence, 4-bit multiplication can be completed by one bit-slice in 4 cycles.

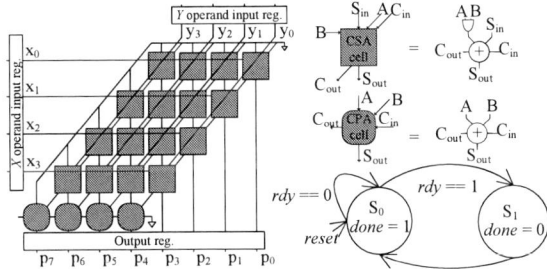

Figure 1. 4×4 CSA multiplier and its state machine controller

Figure 2 depicts the enhanced defect-tolerant CSA multiplier design which exploits the above circuit layer property. In this design, two neighboring bit-slices are grouped together to form a *bit-slice group (bg)*. Specifically, bit-slices along y_0 and y_1 form one group (bg_0) and those along y_2 and y_3 form the other (bg_1). Hence, the entire multiplier is divided into two halves, and each half contains one bit-slice group.

Depending on how a multiplier is affected by defects on a fabricated chip, the enhanced multiplier can be defect-free, defective 1st-half only, defective 2nd-half only, or defective both halves. The enhanced controller in Figure 2 shows that, depending on the *dfct* signal, set based on the result of testing, the multiplier can operate respectively in *mode 0* when both halves are defect-free, and in *mode 1* or *mode 2* when one half is defect-free.

Each bit-slice group is enhanced with multiplexors to select the proper connections based on the operation mode. When operating in one of the "defectively-functional" modes (*mode 1* and *mode 2*), the multiplication is segmented into two shorter

multiplications which are executed on a defect-free enhanced bit-slice group in two separate states. Three types of multiplexors are required: (1) *Input operand muxes* select which Y operand bits are to be computed, i.e., the bits that go to A inputs of CSA cells. (2) *Output product muxes* select which S_{out} outputs from CSA or CPA cells should be selected to be stored in output registers as the intermediate partial product or the final product. (3) *Partial product muxes* select if partial products, 0s, or the original paths are connected to S_{in} inputs in CSA cells for computation, or to A inputs of CPA cells at the boundary of two groups. The select signals of all muxes are controlled by the state machine.

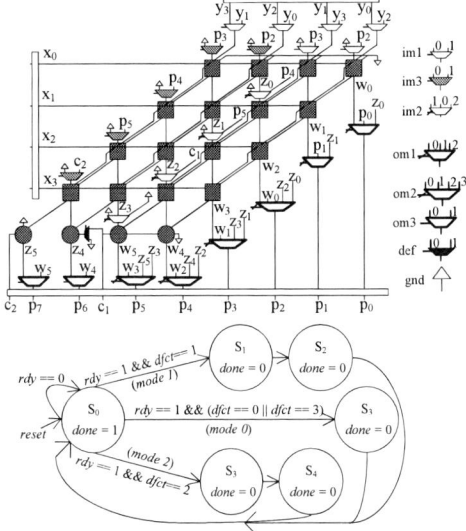

Figure 2. Enhanced 4×4 CSA multiplier and modified controller

It is clear that there are many more options for implementing such enhancements for the CSA. In our approach, an enhanced CSA multiplier design is defined by two parameters: the number of bit-slices per group, NS_{bg}, and the number of enhanced group, N_{bgEn}. The example in Figure 2 corresponds to the option ($NS_{bg} = 2, N_{bgEn} = 2$). Design options with different parameter values result in different overheads and effectiveness. The smaller the number of bit-slices per group, the higher the probability that an enhanced group will be defect-free when fabricated. However, a smaller bit-slice number also requires (1) multiplexors with higher complexity, and (2) more cycles to complete a multiplication. The more the number of enhanced groups, the higher the probability that one of them is defect-free. However, it also requires more multiplexors and area overhead is high.

3.2 Enhancement options for the 32×32 CSA multiplier

To maximize the overall efficiency, we empirically investigate the circuit layer overheads, i.e., area overhead and delay overhead, for various enhancement options. Based on these estimated circuit layer overheads, we can further characterize their overheads at microarchitecture layer.

An unenhanced 32×32 CSA multiplier is first implemented using a standard cell library. Area and delay are then measured. The three types of multiplexors required for different design options are implemented in RTL and synthesized for area and delay information. Area overhead of each enhancement option is estimated by accumulating the area of different types of multiplexors required. Delay overhead is estimated by accumulating multiplexor delays along the original critical paths of the unenhanced design. Table I shows area and delay ratios of the enhanced multiplier for each option with respect to those of the unenhanced multiplier. We have implemented option (16, 2) in detail and the overheads for our actual implementation and our estimation are negligible ($<1\%$).

Table I. Enhancement options: performance and overhead estimation

Options (NS_{bg}, N_{bgEn})	Circuit layer		Microarchitecture layer	
	Area ratio	Delay ratio	Number of processor cycles	
			Defect-free	Defectively- functional
Unenhanced*	1	1	10	N/A
(16,1)	1.028	1.011	11	22
(16,2)	1.051	1.016		22
(16,2)*	1.045	1.022		22
(8,1)	1.042	1.022		44
(8,2)	1.071	1.024		44
(8,4)	1.146	1.032		44
(4, 1)	1.032	1.028		88
(4, 2)	1.116	1.031		88
(4, 4)	1.237	1.038		88
(4, 8)	1.470	1.045		88

*These design options were implemented

Microarchitecture layer performance for each option is captured by the number of cycles required to execute a fixed-point multiplication. The number of processor cycles required for execution of complete 32×32 multiplication in the unenhanced multiplier can be derived by rounding up the ratio of the measured circuit layer delay to the given processor clock cycle. The number of cycles required for other options can be derived in the same way. However, when an enhanced multiplier operates in one of the defectively-functional modes, the number of cycles increases according to the number of segmented multiplications that need to be executed. We will later show that architecture-level performance-per-area of the enhanced processors can be significantly improved at extremely small overheads, i.e., low area overheads and just one extra cycle for defect-free processors to execute a multiplication instruction.

Note that our approach can be implemented with any positive integer NS_{bg} value < 32. However, many options, e.g., ($NS_{bg} = 17$, $N_{bgEn} = 1$), do not provide more performance or yield benefits compared to an option we consider, namely, ($NS_{bg} = 16$, $N_{bgEn} = 1$) for the above example, since it still requires 22 cycles to complete a full multiplication in defectively-functional mode, and requires larger area to operate error-free in a defectively-functional mode. It is also possible that the approach can be implemented such that the boundary of the first enhanced group is not aligned with the original boundary of the multiplier. However, such implementations result in higher multiplexor area overheads.

3.3 Processor under study

Throughout this paper, we study a single-core processor that is capable of out-of-order execution with instruction issue-width of four and is equipped with L1/L2 caches. The processor is equipped with two 64-bit ALU modules, one 32×32 integer multiplier, and a double-precision FPU which occupy 5%, 3.6%, and 18.4% of the chip area, respectively. The floating-point multiplier in the FPU occupies 51% of the FPU area. Hence, our three approaches collectively target defects in 27% of the area of the processor core. In our numerical calculations, we consider the case where the processor is subject to high defect density and an unenhanced processor has 30% yield.

3.4 Processor tiering

Fabricated processors with multiplier enhancement can be graded into three tiers based on how the defects affect the multiplier: tier 1 (*T1*), defect-free processors, tier 2 (*T2*), defectively-functional processors, and tier 3 (*T3*), non-functional defective processors, i.e., processors to be discarded. The

parameters associated with each enhancement option are shown in Table II.

Table II. Multiplier enhancement parameters

Parameter	With enhancement	No enhancement
Area of a processor chip	$A_{\mu pME}$	$A_{\mu p}$
Area of multiplier	A_{ME}	A_{mul}
Yield	Y_{Ti} (for tier i)	Y_o
Instructions per cycle	IPC_{Ti} (for tier i)	IPC_o

The yield for each tier i (Y_{Ti}) can be computed for each design option as follows. $Y_{T1} = PS(d, A_{\mu pME}, 0)$, where $PS(d, A, x)$ denotes the probability that there are x defects in an area A given that defect density is d and defects are uniformly distributed.

$$Y_{T2} = (1 - PS(d, A_{ME}, 0) - P_{mulFail}) \times PS(d, A_{\mu pME} - A_{ME}, 0)$$

where $P_{mulFail}$ is the probability that all enhanced bit-slice groups are defective, i.e., the probability that a defective multiplier does not work even under any of the defectively-functional modes. Finally, $Y_{T3} = 1 - Y_{T1} - Y_{T2}$.

3.5 Tier yields for enhanced CSA multiplier

To measure the unit-area yield benefits provided by each option, the metric $(Y_{T1} + Y_{T2})/A_{\mu pME}$ is used to capture the sellable-chip-per-die-area. Figure 3 shows the metric normalized to that of an unenhanced processor when its yield (Y_o) is 30%. It can be observed that the option (16, 2) yields the most sellable processors per area because of its lower area overhead and its slightly higher Y_{T1}. In the next section, the performance aspect of each option will be considered by performing microarchitecture simulations.

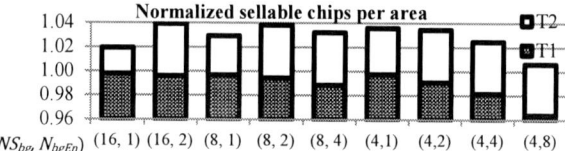

Figure 3. Normalized sellable chips per area

3.6 Evaluation of efficiency of options

Metric *performance-per-area* is used to measure the efficiency of each option. The *normalized efficiency* ($nEFF$) of each option can be evaluated as the following

$$nEFF = \left[\sum_{i=1}^{2} Y_{Ti} \times IPC_{Ti}/A_{\mu pME}\right] / \left[Y_o \times IPC_o/A_{\mu p}\right].$$ IPC for each tier is derived from microarchitecture simulation. SPEC2000 [8] benchmarks are simulated using a microarchitecture simulator [9]. The multiplication latency of each tier is set according to processor cycle estimation derived from circuit layer nano-second delay as shown in Table I.

Figure 4 shows the $nEFF$ across benchmarks simulated for option (16, 2) when the unenhanced processor has 30% yield. For most benchmarks, the enhancement achieves higher efficiency by using a small area overhead by making functional more processors fabricated with defective multipliers. Although both *T1* and *T2* processors have lower multiplication performance than the unenhanced processors, the architecture performance of *T1* and *T2* processors for most benchmarks does not degrade much, since multiplication instructions account for very small percentages of total instructions in the benchmarks. Also shown in the figure, the two benchmarks which result in lower efficiency have substantially higher multiplication percentages (shown by the curve labeled "mult %") than the other benchmarks.

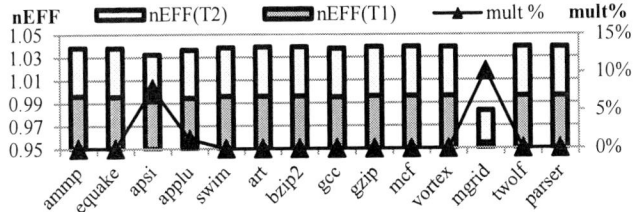

Figure 4. nEFF for option (16, 2) and multiplication instruction %

Overall *nEFF* for all options is derived by averaging the *nEFF* for all benchmarks and the (16, 2) option achieves the highest nEFF. The option is even more attractive when microarchitecture performance is considered. Since not only the sellable-chip-per-area of the option is larger, the performance provided by tier 2 is also higher than other options with comparable sellable-chip-per-area measures, i.e., options (8, 2) and (4, 1).

In summary, our methodology effectively maximizes the improvement provided by our defect-tolerance approach. The multi-layered system view enables the use of circuit layer implicit redundancy. The utilization-aware module selection targets the less frequently utilized multiplier module to minimize the performance penalty and to maximize the performance benefits.

4 MICROARCHITECTURE LAYER APPROACH FOR ALU

In this section, microarchitecture layer is explored for the ALU modules. Approaches based on coarse-grained and fine-grained redundancy are described. Our yield and utilization analysis further indicates a fine-grained approach can further improve the performance benefit.

4.1 Coarse-grained approach: ALU-disabling

Fixed-point instructions are among the most commonly used instructions. Hence multiple ALUs are commonly implemented in each processor core to boost performance by supporting parallel execution of multiple ALU instructions per cycle. For example, microarchitecture in recent processors [10] features two or more ALUs respectively in their out-of-order cores.

Figure 5 depicts a generic organization of dispatching, issuing, and execution of ALU instructions. Decoded instructions are dispatched into a unified *reservation station* waiting for their input operands to become available. The en-queued instructions have their ready flags in the reservation station set when all input operands are ready. An *instruction scheduler* delivers the operands of a fixed-point instruction to one of the ALUs for computation in the *execution unit*, which contains 2 ALUs in the processor core we study. Instructions are allowed to execute out-of-order and the instruction scheduler controls the instructions' routing between the reservation station and the ALUs.

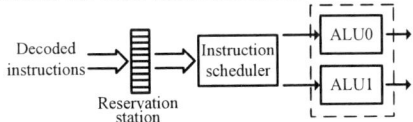

Figure 5. Generic organization for ALU instruction execution

By using test information regarding which ALU is defective and slightly modifying the instruction scheduler, the use of defective ALU can be avoided. Hence, a fabricated processor with defective ALUs can be used if at least one defect-free ALU is available in the processor. This can be achieved at negligible overhead by implementing a mechanism to deactivate the cycle counter [11] of each defective ALU in the instruction scheduler. The defective ALU will never appear available to the instruction

scheduler during operation. Therefore, the defective ALU will be disabled and no data error will be produced due to the defective ALU. This approach is referred as *ALU-disabling (AD)*.

4.2 Fine-grained approach: Adder error-masking (EM)

Each ALU consists of different types of functional units (i.e., adder, logic units, shifter, etc.). Since every ALU in a processor is identical, it is possible that a disabling approach can be implemented at finer granularity, namely at the level of individual functional units.

Yield benefit and utilization analysis: Finer disabling granularity indicates that most defective ALUs can correctly execute the instructions processed by defect-free functional units. Hence, we can expect smaller performance degradation than the *AD* approach. Our study on a synthesized 2-ALU instruction execution unit shows that the *adder* takes up the majority of the area of an ALU. It indicates that an ALU is most likely defective due to a defective adder. Hence, we target the adder functional unit in ALU to study the fine-grained approach to maximize the yield benefit. The straightforward way is to enhance the instruction scheduler so it issues *add-type* instructions (i.e., add, *subtract*, and so on) only to the ALU with defect-free adder.

However, the straightforward way does not improve performance-per-area much over our *AD* approach for the following reason. Our benchmarks profiling shows that in 100M instructions profiled, more than half of the instructions are add-type instructions. Hence, we carry out a finer-gran utilization analysis and find that, among the add-type instructions, 92% require additions limited to the 33-LSB positions. In other words, the 31-MSBs of an adder's output can be directly masked by *0s* for 92% of the add-type instructions (Note an adder in ALU produces 64-bit outputs). In the next section, we exploit this property to develop an error-masking approach to utilize defective adders and maximize performance benefit.

4.2.1 Adder error-masking mechanism

By checking the number of LSBs required for an add-type instruction and comparing this number with the number of guaranteed functional LSBs of a defective adder, it can be determined whether the defective adder is guaranteed to produce an error-free result for the instruction. Figure 6 shows the microarchitecture level mechanism we propose. The instruction scheduler issues instructions to ALUs as their operands become ready. In a processor with an adder with defects in the 31-MSB bit-slices, an operand width inspector dynamically checks if the number of LSB bit-slices required for an add-type instruction is larger than 33. The operands are checked in parallel with their computation in adders. If the add-type instruction executed on the defective adder does not require more than 33-LSBs computation, the 31-MSBs of the output are masked by *0s*. On the other hand, if the instruction requires more than 33-LSB computation, the output from the ALU with the defective adder is invalidated, and subsequently the instruction is re-issued and re-executed on a defect-free adder in the defect-free ALU.

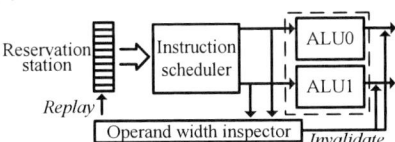

Figure 6. Adder EM microarchitecture support

To minimize area overhead of this approach, we exploit an existing microarchitecture mechanism used in most modern processors to serve the purpose of re-executing add-type instructions with possibly erroneous results. Modern processor designs speculate on the cycles required for *load* instructions to

978-1-4799-7598-3/15 $31.00 © 2015 IEEE

complete [12]. They allow the instructions dependent on a *load* instruction to be issued speculatively assuming that the *load* instruction will have a hit in the level 1 cache. In case of a level 1 cache miss, a correcting mechanism invalidates the speculatively issued instructions and re-executes by *replaying* the dependent instructions. Our approach uses this mechanism to re-execute add-type instructions whose outputs are possibly erroneous. Hence, the area overhead of the re-execution mechanism in our adder error-masking approach is negligible.

Our simulation result shows that, averaged across various benchmarks, the defective processors saved by *AD* approach can achieve 0.72 of the IPC performance of a defect-free processor, and the processors saved by *EM* approach achieve 0.99 of the IPC performance of a defect-free processor. We have computed the overheads by implementing the operand width inspector mechanism and measuring the overheads. These overheads will be used to evaluate the effectiveness of our multi-layered composite approaches in the next section. Note that we pick the specific masking width (31-bit) of an adder based on qualitative judgments. Decreasing the masking width decreases the number of add-type instructions to be re-executed, hence more performance benefit is provided. However, this improvement is very small since the 31-bit masking width implementation already achieves 99% of the IPC performance and provides high yield.

Table III summarizes the key parameters of the unenhanced processors and the processors enhanced with *AD* and *EM*. Without our approach, unenhanced processors yield is Y_{unenc}. Hence, $1 - Y_{unenc}$ of all processors fabricated are discarded. With our approaches, enhanced processors have area A_{enc} and yield of Y_{enc}. The area overhead of the approach ($A_{unenc} - A_{enc}$) is derived by synthesizing a modified fixed-point execution unit [13] with 2 ALUs. The Y_{enc} portion of all processors fabricated, which we call *T1* processors. *T2* processors have one defective adder which can be used by error-masking, and *T3* processors have one disabled defective ALU. *T2* and *T3* processors account for PF_{T2} and PF_{T3} fraction of the $1 - Y_{enc}$ portion of the processors fabricated respectively (described ahead).

Table III. Tier yield formulation for ALU enhancement

	Tier	Yield
Unenhanced area: A_{unenc}	Defect-free	Y_{unenc}
	Defective	$1 - Y_{unenc}$
Enhanced area: A_{enc}	T1	$Y_{tier1} = Y_{enc}$
	T2	$Y_{T2} = PF_{T2} \times (1 - Y_{enc})$
	T3	$Y_{T3} = PF_{T3} \times (1 - Y_{enc})$
	Discard	$(1 - PF_{T2} - PF_{T3}) \times (1 - Y_{enc})$

Note that *T2* and *T3* processors are defective processors that would have been thrown away if our approaches were not deployed. Let n represent the maximum number of defects considered, and let $P_{alu}(P_{adder})$ be the area percentage of an ALU (adder) in the processor (which can be derived from synthesis). Let $P(k)$ be the probability that there are k defects on the processor chip, and by assuming Poisson distribution of defects, it can be calculated as $e^{-dA_{chip}}(dA_{chip})^k/k!$, where d is defect density and A_{chip} is the area of the entire processor. Hence, PF_{T2} is calculated as $\sum_{k=1}^{n} 2(P_{adder})^k P(k)$ and PF_{T3} is calculated as $\sum_{k=1}^{n} 2(P_{alu})^k P(k) - PF_{T2}$.

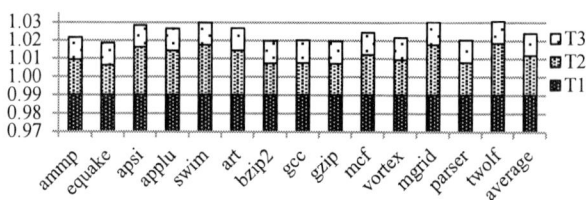

Figure 7. nEFF for AD+EM

Figure 7 shows the *nEFF* of our approaches for the benchmarks. The unenhanced processor has 30% yield and the yields of T1, T2, and T3 processors are 29.97%, 0.87%, and 0.37% respectively. Since the enhanced processors have slightly larger area, the efficiency of T1 processors, i.e., defect-free processors, is always slightly less than 1. However, the approach allows us to use additional T2 and T3 processors, where each of these has a slightly low performance than a defect-free processor.

5 ISA LAYER APPROACH FOR FPU

Multi-layered view and yield benefit analysis: At ISA layer, floating-point instructions can be emulated using fixed-point instructions. For a processor with a defective FPU, its FPU can be effectively disabled if all the floating-point instructions are emulated using fixed-point instructions during compilation. We will refer this ISA layer approach as FP-DIS.

However, instead of disabling all types of floating-point instructions if the FPU is defective, a fine-grained disabling approach can be developed. It is possible to disable only the functions provided by the individual defective floating-point functional units. For example, if an FPU has defects in its multiplier only, disabling floating-point multiplications (FPMUL-DIS) is sufficient to maintain the correctness and other floating-point functions can still execute on the FPU. Hence, the fine-grained disabling will have significantly lower penalty than disabling all floating-point instructions.

According to our synthesis study of a double precision FPU [14], its multiplier occupies more than 50% of the entire FPU area. Next we perform utilization analysis for FP-DIS and FPMUL-DIS, respectively. Then we demonstrate the efficiency that can be achieved when the two approaches are combined.

Utilization analysis for FP-DIS: Table IV shows the distribution of floating-point instructions and the instruction counts with and without emulation for the benchmarks. First, we observe the instruction distribution within a 100M-instruction window selected using SimPoint [15]. In general, an application that has high floating-point instruction percentage will have low performance if a defective FPU is used under the FP-DIS mode to execute the application.

To account the performance degradation more accurately, we count the total number of instructions to completion when no emulation is used (original) and the total number of instructions to completion when emulation is used (FP-DIS and FPMUL-DIS). We then estimate the performance for the ISA layer approaches by calculating the relative performance as $\frac{Number\ of\ original\ instructions}{Number\ of\ emulation\ instuctions} \times 100\%$.

Table IV. Floating-point utilization analysis

Bench.	Instruction %		Relative performance estimate	
	All FP	FPMUL	FP-DIS	FPMUL-DIS
ammp	32.9%	7.38%	5.95%	62.62%
equake	25.7%	11.23%	3.37%	15.17%
apsi	17.7%	3.49%	4.78%	24.88%
swim	42.5%	0.00%	3.30%	10.90%
art	15.2%	5.83%	5.43%	59.61%
applu	30%	4.66%	4.29%	64.10%
mgrid	6%	0.70%	14.49%	68.49%
Integer Benchmarks gzip, mcf, gcc, parser, vortex, bzip2			100%	100%
twolf	0.64%	0%	61.3%	100%

As shown in Table IV, the relative performance of FP-DIS ranges from 3.3% to 14.4% of the original applications. This implies that the processors with defective FPU will have approximately 3%~14.4% of the performance of defect-free processors. Nevertheless, for most integer benchmarks, there will be no degradation at all for processors with defective FPU.

Three tiers of processors will be fabricated if FP-DIS and FPMUL-DIS are applied. Table V summarizes the tiers and their yields. The yield for each tier is calculated as the following.

$$Y_{T1} = PS(d, A_{\mu p}, 0)$$

$$Y_{T2} = \left(PS(d, A_{\mu P} - A_{FPU}, 0)\right) \times \left(1 - PS(d, A_{FPU} - A_{FPMUL}, 0)\right)$$

$$Y_{T3} = \left(PS(d, A_{\mu P} - A_{FPMUL}, 0)\right) \times \left(1 - PS(d, A_{FPMUL}, 0)\right)$$

Note that $A_{\mu P}$, A_{FPU}, and A_{FPMUL} represents the area of the processor, FPU and FP-multiplier respectively.

Table V. FP-DIS and FPMUL-DIS tiers and yields

Tiers	T1	T2	T3
Description	Defect-free	Any of the FP functional units other than FP-multiplier is defective	Defective FP-multiplier (other FP functional units are defect-free)
Yield	30%	3.84%	3.59%

Figure 8 shows the overall efficiency provided by the three tiers. The IPC performance of each tier is obtained via simulation and multiplied by the relative performance in Table IV. The overall efficiency of FP-DIS only is shown in hollow bars for comparison. As expected for integer benchmarks, the efficiency is identical for FP-DIS only and FP-DIS with FPMUL-DIS. However, the combined approach shows significant improvement for several floating-point benchmarks over FP-DIS only.

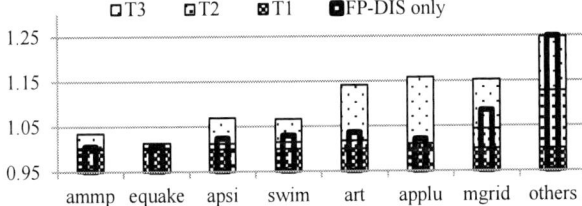

Figure 8. nEFF for FP-DIS+FP-MUL

6 CONCLUSION

We propose a defect tolerance methodology based on exploiting implicit redundancy in processors and use it to derive defect-tolerance approaches for datapath modules. Our multi-layered methodology identifies DT approaches that maximize the performance-per-area for processors by minimizing area and performance overheads of defect-free processors and maximizing

the yield and performance benefits provided by the defective processors. We have demonstrated that our approaches provide substantial performance-per-area improvement. Averaging across the benchmarks, our first two approaches increase the performance-per-area 3.6% and 2.4% respectively for the multiplier and the ALUs. Our ISA layer approach for FPU increases performance-per-area by 9% for floating-point benchmarks and 25% for integer benchmarks. When our three approaches are deployed together, the overall performance-per-area is roughly equal to the product of the improvements provided by individual approaches. Hence, collectively our approaches provide significant improvements.

We are extending this study to multicore processors and future generations of CMOS by projecting defect densities and the number of cores. Initial results show higher improvements than reported here, since the single-core improvements we report here compound over multiple cores.

Acknowledgement

This research was supported by National Science Foundation and the Semiconductor Research Corporation.

Reference

[1] M. Horiguchi, "Redundancy techniques for high-density DRAMs," in *2nd Annual IEEE International Conference Innovative Systems Silicon*, 1997.

[2] S. E. Schuster, "Multiple word/bit line redundancy for semiconductor memories," *IEEE J. Solid-state Circuits*, vol. 13, no. 5, pp. 698-703, October 1978.

[3] Hsunwei Hsuing, Byeongju Cha, Sandeep K. Gupta, "Salvaging chips with caches beyond repair," in *DATE*, 2012.

[4] A. Agarwal, B. C. Paul, H. Mahmoodi, A. Datta, and K. Roy, "A Process-Tolerant Cache Architecture for Improved Yield in Nanoscale Technologies," *IEEE TRANSACTIONS ON VERY LARGE SCALE INTEGRATION (VLSI) SYSTEMS*, vol. 13, no. 1, 2005.

[5] G. S. Sohi, "Cache memory organization to enhance the yield of high performance VLSI processors," *IEEE Trans. Computers*, vol. 38, pp. 484-492, April 1989.

[6] P. P. Shirvani and E. J. McCluskey, "PADded cache: a new fault-tolerance technique for cache memories," *17th IEEE VLSI Test Symposium*, pp. 440-445, April 1999.

[7] T.-Y. Hsieh, M. A. Breuer, M. Annavaram, S. K. Gupta, and K.-J. Lee, "Tolerance of Performance Degrading Faults for Effective Yield Improvement," in *Proceeding of International Test Conference*, 2009.

[8] J. L. Henning, "SPEC CPU2000: measuring cpu performance in the new millennium," *IEEE Computer*, vol. 33, no. 7, pp. 28-35, July 2000.

[9] Eric Larson , Saugata Chatterjee, and Todd Austin, "MASE: A Novel Infrastructure for Detailed Microarchitectural Modeling," in *ISPASS*, 2001.

[10] D. Kanter, "Inside Nehalem: Intel's Future Processor and System," 2008.

[11] E. Morancho, J. M. Llaberia and A.Olive, "Recovery Mechanism for Latency Misprediction," in *Proc. of International Conference on Parallel Architectures and Compilation Techniques*, 2001.

[12] Ilhyun Kim, Mikko H. Lipasti, "Understanding Scheduling Replay Schemes," in *Proceedings of the 10th International Symposium on High Performance Computer Architecture*, 2004.

[13] N. Choudhary, S. Wadhavkar, T. Shah, H. Mayukh, J. Gandhi, B. Dwiel, S. Navada, H. Najaf-abadi, and E. Rotenberg, "FabScalar: composing synthesizable RTL designs of arbitrary cores within a canonical superscalar template," in *ISCA*, 2011.

[14] D. Lundgren, "Dobule precision FPU," OpenCores, 2009.

[15] Erez Perelman, Greg Hamerly, Michael Van Biesbrouck, Timothy Sherwood, and Brad Calder, "Using SimPoint for Accurate and Efficient Simulation," in *ACM SIGMETRICS the International Conference on Measurement and Modeling of Computer Systems*, 2003.

PPB: Partially-working Processors Binning for Maximizing Wafer Utilization

Da Cheng
Xilinx, inc.[1]
2100 Logic Drive, San Jose, CA
dache@xilinx.com

Sandeep K. Gupta
Department of Electrical Engineering
University of Southern California, Los Angeles, CA
sandeep@usc.edu

Abstract— Hardware redundancy, such as spare processors and cores, has been added to chip multi-processors (CMPs) to improve yield while sustaining all functionalities of CMPs. During post-silicon testing, spares processors and cores are used for repair. Even after repair, some CMPs may have processors with insufficient number of cores; in such CMPs some processors are disabled and such chips are sold at lower prices to improve yield per area. Despite binning on the number of processors, substantial functional resources are wasted in disabled components.

In this work, we propose a new utility function and a new repair algorithm which enable utilization of every working core on a CMP. We demonstrate the benefits of the proposed approach for benchmarks from ISPASS and Nvidia CUDA SDK using GPGPU-sim to compute the instructions per cycle (IPC). Results show that our design and repair approaches provide above 50% IPC per wafer area even with 10x the current defect density.

Keywords—hardware redundancy; defect; yield; CMP

I. INTRODUCTION

Transistor count on chip multi-processors has increased dramatically in the last decade due to development of new manufacturing technologies and new micro-architectures, such as many-core processors and chip-multiprocessors, for sustaining Moore's law. Low yield has become a severe problem for digital systems, especially for CMPs since they usually comprise billions of transistors [1][2].

Adding hardware redundancy has been used as an effective approach to improve silicon *yield*, typically defined in terms of chips that provide all functionalities specified in the *nominal* specifications [3][4][7][8]. Generally speaking, these approaches can be viewed as a combination of two main components: a repair algorithm developed so that spare copies are used in an efficient manner for each defective chip, and an approach for identifying optimal spare configuration by systematically evaluating different spare configurations.

The optimal level of granularity at which to add redundancy has been explored in theoretical and practical studies. Theoretical: In [7], a CMP design is partitioned into modules of arbitrary sizes, each with an equal amount of functional logic. Spare modules are then incorporated with a *global scope*, i.e., in such a manner that each spare module can replace any module that is defective. This approach assumes that a spare module with a global scope incurs a fixed area overhead and ignores the overheads of interconnects required to use the spare modules in a global manner. Hence this approach provides an upper bound on the yield benefits of adding spares. Finally, the approach identifies the size of a module, which is also the *granularity* of each spare module, to maximize yield per area. Practical: In contrast, the practical approaches proposed to date start with models of the chip's floorplan and typically view a chip in terms of processors and sub-processor level modules. In case of CMPs, sub-processor modules are cores, caches shared by cores, and interconnects, or parts thereof. The approaches study yield trends across technology generations for various processor- and sub-processor-level redundancies [1][4]. These approaches take into account the area overhead of interconnects and the optimal designs they derive are more realistic.

In particular, a systematic spare cores sharing (SCS) approach was developed to add spare processors and cores to CMPs [3][4]. In this approach, all spare configurations are enumerated in terms of: (a) scope of sharing of spare cores, i.e., the number of original cores which share the same spare cores, and (b) the number of spare cores per scope. The notation $[\dots, n_{sc,k}, \dots, n_{sc,4}, n_{sc,2}, n_{sc,1}]$ is used to represent a spare configuration, where $n_{sc,k}$ is the number of spare cores that are each shared by k original cores. We use n_p to denote the number of nominal (original) processors per chip, and use n_c to denote the number of nominal (original) cores per processor. Hence the maximum value of k is equal to $n_p \cdot n_c$. Experiments show that this approach provides performance per wafer that is 65% of the performance per wafer for the *ideal scenario*, i.e., the scenario where defect density is zero and no redundancy is required. Substantial functional resources are still wasted in this approach since the latter requires that (i) a processor is considered working if the number of working cores within is equal to or greater than the nominal number of cores (n_c, which is 32 for their example CMP), and (ii) a chip is considered working if the number of working processor it has is equal to or greater than the nominal number of processors (n_p, which is 16 for their example CMP).

Number of processors binning (NPB) has been proposed in [4] to reduce such wastage of working cores and processors by allowing defective chips with fewer than the nominal number of working processors. In this approach, non-working streaming multiprocessors (SM, the Nvidia terminology for processors) are disabled, where an SM is defined as working if it has the nominal number or more working cores. (In the remainder of this paper, our target CMP is a state-of-the-art Nvidia GPU standard, where the nominal number of processors (SMs) in a chip is $n_p = 16$ and the nominal number of cores in each processor is $n_c = 32$.) This approach was motivated by three GPUs from Nvidia, namely, GTX 480, GTX 470, and GTX 465, all of which are the same Fermi design, sold with different numbers of SMs disabled. Once it considers NPB, the aforementioned SCS approach provides

---

[1] This work was finished by Da Cheng as a Ph.D candidate at University of Southern California.

performance per wafer that is 82.5% of the performance for the ideal scenario.

However, NPB still wastes (disables) a large number of working resources on defective CMPs. Specifically, this approach results in wastage in two ways: (1) SMs with fewer than 32 cores are disabled, where a disabled SM may have up to 31 working cores. Such wastage is significant especially when random defects are the main concern and hence the number of failing cores in each SM is typically small. (2) SMs with more than 32 working cores (including the available non-defective spares) will disable the additional cores, i.e., the 33rd core, the 34th core, and so on. To clearly contrast with the approach we present in this work, we adopt a more descriptive term for NPB, namely fully-working processors and partially-working chip binning (FPPCB).

In this work, *we present a new utility function to minimize wastage of working resources by removing the requirements on a working processor.* Specifically, a processor with any number of working cores is deemed acceptable and the value assigned to a chip with specific numbers of cores in each enabled processor is proportional to the performance the chip can provide for a desired set of benchmark programs. In comparison to FPPCB, we use the term *partially-working processors binning (PPB)* to describe the proposed approach.

We develop PPB as a procedure with three main steps. First, via simulations we estimate performance of target benchmarks, for the nominal chip configuration as well as the configurations of chips that are likely to be obtained after fabrication and repair. Second, we develop repair algorithms that ensure that spare copies are used in an efficient manner for each defective chip. Third, we develop an approach to identify the optimal spare configuration by systematically evaluating different spare configurations.

II. GPU BACKGROUND

We use state-of-the-art GPU models as case studies for CMPs in this work. This section presents the background related to this research. Here we describe some background information about how application programs are executed on GPUs.

A GPU benchmark usually consists of multiple kernels, where each kernel is a grid of blocks of threads. SMs receive *thread blocks* in a round robin fashion. The maximum number of threads that can be assigned to each SM is determined by the size of register file per SM.

During execution within each SM, threads are further separated into *warps*, where each warp has a certain number of threads, e.g., warp size w is equal to 32 for Nvidia GTX 480. Threads inside a warp are executed in a lock-step manner. Consider number of cores per SM to be n. Then the time in execution stage of one instruction in a warp is proportional to $\lceil \frac{w}{n} \rceil$. For example, execution of an integer instruction for each warp takes 2 cycles in GTX 480, since in this case $w = 32$ and $n_c = 16$. Both warp size and number of cores is identical for all SMs as a common practice [12][13].

In this work, we use $w = n$ due to GPGPU-sim constraints, where the simulator only supports $w = n$ [11].

A warp is issued to cores if all hazards and dependency are resolved for all threads, e.g., bank conflicts, memory dependency, branch divergence, and register dependency. Enabling *many* warps for concurrent execution in a single SM can hide the impact of latencies, which are due to either hazard or dependency, on overall GPU performance.

III. PROBLEM STATEMENT

We are looking for an optimal order in which defective SMs should be repaired, so that the evaluated metric, e.g., performance per wafer, is maximized after a defective CMP is repaired for a given spare configuration. We first explore the optimal strategy to use spare cores to repair any defective CMP. Once we have such a repair strategy, we develop approaches to identify an optimal spare configuration.

PPB does not specify on the numbers of working cores required for a SM to be working, i.e., it allows SMs to have various numbers of cores instead of a fixed identical value of 32 for each SM. As a result, SMs can have different speeds. It is easy to imagine that all SMs should be granted equal priorities on being repaired under PPB if the performance of an SM is proportional to the number of working cores in the SM. Through experiments, however, we found that performance of a CMP doesn't always increase proportionally with the number of cores per SM for certain benchmarks, e.g., NQU, Dct8x8, VectorAdd, and FastWalshTransform, due to diminishing return effect and compiler limitations, which is true in common sense. In other words, it might be better to repair SMs with x working cores other than SMs with y working cores since the former adds to more performance when we assign it a single spare core. Absolute values of x, y, and their relationship will be presented in Section IV.

For each spare configuration, we use Monte Carlo approach to create a large number of CMPs that have different numbers of defective cores in various SMs. Effectiveness of a spare configuration can be evaluated based on the average performance of all the CMPs after we apply our repair strategy. Approaches which explore all spare configurations to identify the optimal one have been proposed in previous research [4]. In this work, we integrate the newly proposed repair strategy into the exploring process.

We break our problem into the following two sub-problems.
(1) What is the most effective repair strategy given a specific spare configuration?
(2) What is the optimal spare configuration?

IV. PROPOSED APPROACH

We separate benchmarks into different categories. Then we show an approach to estimate performance of a defective CMP. We also present new repair heuristics to efficiently apply spare cores to repair defective SMs for different categories of benchmarks. In the end, we present our approach to identify the optimal spare configuration for newly-proposed utility functions.

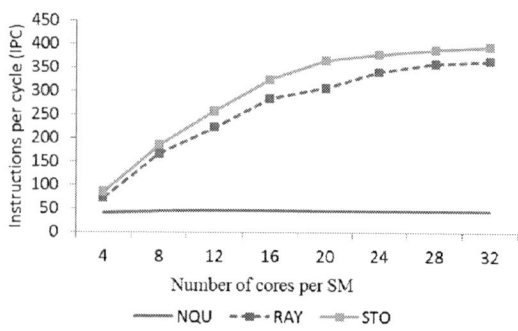
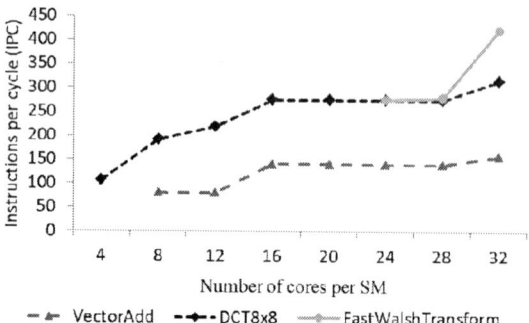

Figure 1. Performance for benchmarks: (a) ISPASS and (b) Nvidia CUDA SDK.

A. Benchmark categories

We have studied both ISPASS benchmarks, e.g., NQU, RAY, and STO, and those from Nvidia's CUDA SDK [9], e.g., VectorAdd, DCT8x8, and FastWalshTransform, using GPGPU-sim [11]. Instructions per cycle (IPC) are shown for different number of cores per SM for selected benchmarks in Figure 1. Based on performance (IPC), these benchmarks can be separated into following three categories.

Category I: Benchmarks without parallelism, e.g., N-Queens Solver (NQU), which solves a classic puzzle of placing N queens on an NxN chess board. Most computation in such benchmarks is performed by a single thread. As a result, such benchmarks have low and constant IPCs, which does not increase with increasing number of cores per SM. In other words, IPCs for such benchmarks are bottlenecked by inherent lack of parallelism. Hence, *for applications in this category, there is no benefit of obtaining an additional working core from the repair process for any SM if it already has at least one working core.*

Category II: Some benchmarks have high parallelism, e.g., Store-GPU (STO) and Ray-tracing (RAY). For such benchmarks, more cores per SM result in higher IPCs since these allow more threads to be executed in parallel. At the same time, however, the number of bank conflicts in the shared memory within each SM, which occur when more than one thread accesses the same memory bank, increases with the number of memory access threads that are executed concurrently. Also the fraction of the L1 cache that is assigned to each thread decreases as total number of threads increases, which results in higher L1-cache miss rate and consequently more global memory writes/reads if one assumes that the number of warps is constant. Lastly, the probability of branch divergence increases accordingly [12]. As a result, *we can observe diminishing returns on IPC curves as the number of working cores increases.* If we primarily focus on *applications in this category, SMs with more defective cores should be granted higher priority of being repaired since the obtained benefit of one additional working core decreases with number of cores per SM.*

Category III: Some benchmarks have high parallelism but only benefit from certain values of number of cores per SM, e.g., powers of 2. Such benchmarks include DCT8x8, VectorAdd, FastWalshTransform, and so on. For such

applications, the performance increases in step functions at specific numbers of cores. *For applications in this category, it is not as straightforward to decide the right defective SM to repair due to distinctive characteristics of such step functions.*

Due to the discreet nature of values of warp sizes, we consider IPCs of benchmarks as an arbitrary step function. Our proposed repair algorithm should be capable of dealing with such arbitrariness, i.e., improvements due to additional cores being arbitrary functions of the number of existing working cores. Note that such an arbitrary step function subsumes as a special case the diminishing marginal improvements obtained for category 2.

B. Estimating performance of defective CMPs

Performance of a CMP where each SM has n working cores can be obtained by setting $w = n$ in the simulations on GPGPU-sim. Similarly, we can simulate benchmarks on GPGPU-sim with decreased warp sizes to obtain the performance of each defective CMP, i.e., IPC. Note that warp size (w) being an identical value for all SMs in GPGPU-sim is a common feature of GPU compilers. But SMs can have different numbers of working cores (n_i) in defective CMPs, i.e., $n_1, n_2, ..., n_{16}$ may not be identical for a CMP with 16 SMs. Hence performance of a defective CMP can't be obtained directly from simulations since such heterogeneity is not supported by available versions of GPGPU-sim. For simplicity, we estimate performance of a defective CMP by computing the average performance of 16 CMPs with $w_i = n_i$, where $i \in [1,16]$. Figure 2 further illustrates the way that performance of a defective CMP is estimated, where the CMP has 4 SMs with 2, 4, 4, and 8 working cores, respectively.

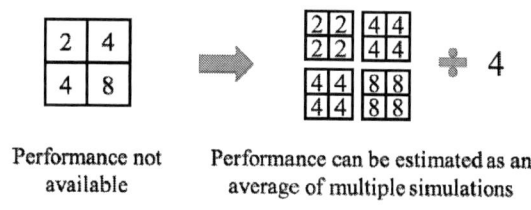

| Performance not available | Performance can be estimated as an average of multiple simulations |

Figure 2. Estimation of defective CMPs.

Then we can obtain IPCs for all warp sizes through simulations and create a lookup table. For each defective CMP, we decide its IPC by indexing the lookup table. By doing so, we don't need to simulate each defective CMP, which greatly reduces the simulation time.

C. Repair heuristic

A greedy repair algorithm has been proven to be optimal [4] for fully functional CMPs, i.e., with no flexibility. However, it does not guarantee the optimal spare configuration for our other different utility functions. For example, the optimal spare configuration that maximizes the evaluation metric in one FPPCB model doesn't maximize that in another FPPCB model. Hence, here we proposed Heuristic 1 in [4] to repair defective CMPs and compute the yield.

For an arbitrary IPC function and a specific spare configuration, we propose a general Greedy Repair Heuristic (GRH-x), which is shown in Heuristic 2. We use $f(n_i)$ for the IPC of a CMP with all SMs having n_i working cores. We define the first-level benefit α_i^1 as the benefit obtained by repairing one defective core for SM i, i.e., $\alpha_i^1 = f(n_i + 1) - f(n_i)$. Similarly, we define the second-level benefit α_i^2 as the benefit obtained by repairing two defective cores for SM i, where $\alpha_i^2 = f(n_i + 2) - f(n_i)$, and so on. This heuristic always picks the SM which generates or will generate the maximum benefit in terms of IPC with best effort, by comparing different α_i^j for all j and i.

For benchmarks in category I, we have α_i^j of zero for all i and j values. In this situation, all defective SMs should be granted with equal priority of being repaired. In other words, any order of SMs being repaired works equally well. For benchmarks in category II, $\alpha_i^j - \alpha_i^{j-1}$ decreases as j increases for the same i. As a result, our heuristic should repair SMs with the most defective cores in a greedy manner, according to our best-effort philosophy. Hence we propose GRH-r, which is shown in Heuristic 3, as a simplified version of GRH-x. Note that "r" in GRH-r denotes the fact that defective SMs are picked in the *reverse order* compared to previously developed GRH, where we repair the SM with fewest defective cores. As can be seen, the key difference among the three repair heuristics lies in *the order in which defective SMs are repaired.*

Different impacts of these three GRHs can be observed if both of the following two conditions are satisfied.

(1) Spare cores are required, i.e., designs with spare cores added have better performance per wafer than the nominal design. Note that the optimal spare configuration can be adding no spares due to substantial area overheads of spare cores in certain scenarios, especially when performance per wafer is already high or close to 1.

(2) Spare cores are shared across different SMs, where the order in which SMs are repaired becomes critical. To the contrary, if spare cores are only added within SMs, each spare core is dedicated to only one SM or certain cores in one SM. Then the order in which SMs are repaired is inconsequential, hence GRH, GRH-r, GRH-x will provide identical results.

To show the differences among the three repair heuristics, we use an arbitrary step function of IPC for different numbers of cores per SM, which is shown in Figure **3**. Since a nominal SM consists of 32 cores, we are only interested in spare cores that can be shared by more than 32 cores. We apply three repair heuristics to identify the optimal $n_{sc,x}$ for the scope of sharing x=32, 64, 128, 256, and 512, respectively, when no cores with other scopes of sharing exist. Figure 4 (a) shows that the proposed GRH-x provides obvious improvements compared to GRH and GRH-r in the shaded area, where two above-mentioned conditions are satisfied. Note that to the left of the shaded area, area overhead of adding spare cores is so large that no spares should be added. And to the right of the shaded area, spare cores are added within each SM and hence all three GRHs are equivalent. In this experiment, we use a high defect density of 15,000/ mm^2 for the purpose of demonstration of effectiveness of GRH-x. This defect density gives a mean value of 5.3 defective cores for a 32-core SM

Heuristic 1: GRH	
Step 1:	Sort processors in *ascending* order with respect to the number of defective cores within the processor.
Step 2:	Repair the processor that has the minimum number of defective cores, using the *greedy repair algorithm* [4].
Step 3:	Repeat until either of following conditions is met: (a) No more defective processors can be repaired due to lack of spare cores, or (b) All processors have required number of working cores to be fully functional.
Step 4:	Repair the processor that has the minimum number of defective cores, using the *greedy repair algorithm* [4].

Heuristic 2: GRH-x	

Step 1:	$k = 1$;
	$f(n_i)$: the IPC of a CMP with all SMs having n_i working cores;
	While (1)
	 Foreach SM_i
	  Compute $\alpha_i^k = f(n_i + k) - f(n_i)$;
	 End foreach
	 If equal α_i^k's are found
	  $k = k + 1$;
	  Continue;
	 Else
	   Break;
	 End if
	End while

Step 2:	Sort SMs in *descending order* using $\alpha_i^1, \alpha_i^2, \ldots, \alpha_i^j$, etc., where significance of α_i^j decreases with j.
Step 3:	Repair the first SM.

Heuristic 3: GRH-r	
Step 1:	Sort SMs in *descending order* using number of defective cores in each.
Step 2:	Repair the first SM.

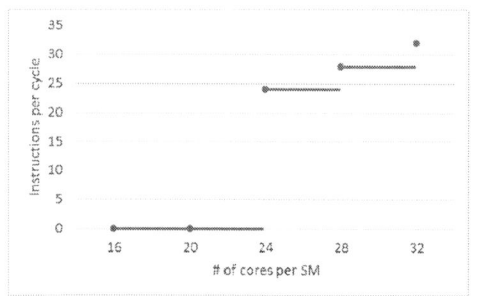

Figure 3. IPC vs. # of cores per SM for an arbitrary benchmark.

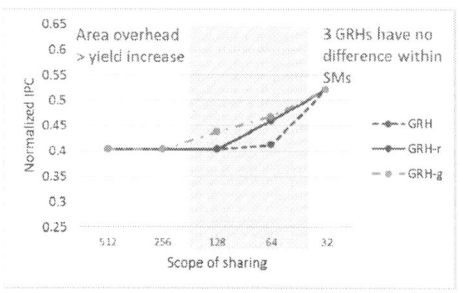

(a) Defect density of of 10,000/cm^2

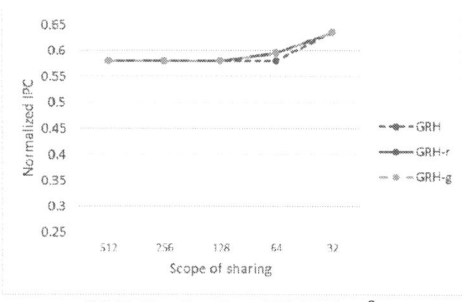

(b) Defect density of 15,000/cm^2

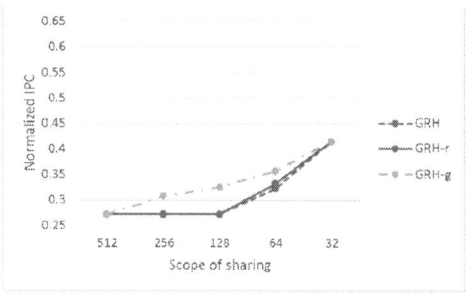

(c) Defect density of 20,000/cm^2

Figure 4. Optimal numbers of spares at various scopes of sharing.

design. For comparison, Figure 4 (b) and (c) show the result obtained using defect density of 10,000/ mm^2 and 20,000/mm^2, which provide mean values of defective cores of 3.8 and 6.4.

D. Proposed redundancy approach

We enumerate all possible values of $n_{sc,x}$ for different x for different spare configurations. Then we design the

floorplan for each configuration to estimate its area overhead as well as delay penalty. In the end, we compute performance per wafer for each configuration using our proposed approach described above: (a) generate a large number of CMP copies with different numbers of defective cores, where number of defective cores in each SM is a Poisson variable, (b) apply the repair heuristic to replace defective cores with spare cores, and (c) compute performance for the defective CMP through previously obtained CMP performances with various warps sizes.

Figure 4 (a), (b), and (c) show that GRH, GRH-r, GRH-x provide the same result if the aforementioned two conditions are not satisfied. *Hence GRH-x should be replaced with GRH or GRH-r once applicable since it has higher complexity than the other two.* In our redundancy approach, we will check the applicability of each spare configuration, especially in two scenarios: (a) no spares are required since yield is already high or close to 1 and/or area overhead is substantial, and (b) only spare cores inside SMs are available in the configuration. When applicable, we should simply apply GRH or GRH-r for those spare configurations as a practice to reduce program run time.

V. EXPERIMENTAL RESULTS

We have obtained the optimal performance per wafer and corresponding spare configuration designs for three benchmarks, namely, NQU, STO, and FastWalshTransform (FWT), which respectively represent Categories I, II, and III. Since NQU is mainly executed in a single thread, the presence of defective cores won't affect the performance because there are always available working cores for the single thread. The IPC for NQU is 0.029 for defect density values considered in this work. Table 1 shows optimal IPCs obtained for different benchmarks for defect densities of 2,000/mm^2, 10,000/mm^2, 15,000/mm^2, and 20,000/mm^2, where 2,000/mm^2 is current random defect density as reported by ITRS [10]. IPC values are normalized to the IPC value obtained for each benchmark for the nominal design. Table 2 shows the corresponding spare configurations, where $n_{sc,k}$ are not shown for $k > 32$ since they are all zeros. We also show yield of a core for every defect density. We have following observations.

(1) For benchmarks with limited number of threads, such as NQU, the optimal spare configuration for PPB is adding no spares. The performance per wafer for such benchmarks is limited by the fact that they must be executed in single thread. Improving CMP configuration doesn't add to any benefit even for 10x current defect density, when each 32-core SM still has an average of 19 working cores.

(2) For benchmarks with sufficient parallelism, such as STO, the optimal spare configuration for PPB is also adding no spares. The reason is that adding additional spare cores doesn't increase performance per wafer by much, especially when it is already high (if not close to 1), and adding spare cores incurs area overhead.

(3) For benchmarks with distinct step functions, such as FWT, adding no spares is the optimal spare configuration for

Table 1. Optimal IPC for PPB and FPPCB.

Defect density	Core yield	No utility function		PPB		FPPCB	
		STO	FWT	STO	FWT	STO	FWT
2,000	95.1%	0.650	0.695	0.892	0.893	0.825	0.868
10,000	82.4%	0.539	0.576	0.772	0.720	0.704	0.676
15,000	77.2%	0.468	0.500	0.724	0.659	0.617	0.588
20,000	73.3%	0.395	0.422	0.687	0.594	0.548	0.551

Table 2. Optimal spare configurations $[n_{sc,32}, n_{sc,16}, n_{sc,8}, n_{sc,4}, n_{sc,2}, n_{sc,1}]$.

Defect density	No utility functions		PPB		FPPCB	
	STO	FWT	STO	FWT	STO	FWT
2,000	[16,0,48,0,0,0]	[16,0,48,0,0,0]	No spares	No spares	[32, 32, 0, 0 ,0, 0]	[32, 32, 0, 0 ,0, 0]
10,000	[16,16,64,0,0,0]	[16,16,48,0,0,0]	No spares	[64, 64, 0, 0, 0, 0]	[32, 64, 0, 0 ,0, 0]	[64, 64, 0, 0 ,0, 0]
15,000	[64,48,32,0,0,0]	[64,48,0,0,0,0]	No spares	[0, 64, 64, 0, 0, 0]	[0, 64, 64, 0 ,0, 0]	[0, 64, 128, 0 ,0, 0]
20,000	[64,112,0,0,0,0]	[64,96,32,0,0,0]	No spares	[0, 64, 64, 0, 0, 0]	[0, 32, 128, 0 ,0, 0]	[0, 64, 128, 0 ,0, 0]

low defect density of $2,000/mm^2$. When defect density increases, yield of a core becomes low. At the same point, the benefits of adding spare cores become more significant compared to the area overheads incurred due to these spare cores. In such scenarios, spare cores should be added. It is also observed that number of spare cores required for the optimal design in PPB is less than that in FPPCB for both STO and FWT.

(4) We also show the result for defect density of $2,000/mm^2$, which is the current defect density reported by ITRS. For STO, compared to the optimal designs obtained from an approach without any utility functions, PPB can utilize additional 24.2% of cores on a nominal design and provide 37.2% improvement on performance per wafer, i.e., 89.2%, versus 65%. Also comparing the optimal designs obtained for PPB and our previously-developed FPPCB, we found that PPB utilizes additional 6.7% working cores and provides an 8% improvement on performance per wafer, i.e., 89.2%, versus 82.5%. [4]. Theoretically, the proposed PPB can utilize all working cores with best effort.

(5) For FWT, for defect density of $2,000/mm^2$, improvement of PPB over FPPCB is about 3%. This is lower compared to the improvement for STO because processors with fewer than 24 cores do not provide much lower performance (see Figure 1).

(6) As defect density increases, the improvement of PPB over FPPCB also increases (in almost all cases). For STO, the improvements are 8.1%, 9.6%, 17.3%, and 25.3% respectively for the four defect densities shown in above tables. For FWT, the corresponding improvements are 2.9%, 6.5%, 12.2%, and 7.7%. The reason is that the number of fully-working processors decreases as defect density increases. As a result, wastage in FPPCB increases. (The reason that the increase is lower for FWT for the highest defect density is that few fabricated processors have 24 or more cores, even after repair.)

VI. CONCLUSIONS

This work presents a new utility function called Partially-working Processors Binning (PPB) to remove the requirement on the number of working cores per processor on a CMP. A repair algorithm has also been developed for different benchmarks for PPB. Also we have presented an approach to estimate performance for defective CMPs. Results show that the proposed approach provides the capability of utilizing all working cores on each CMP. For current defect density, PPB shows significant improvement over FPPCB. Such improvement will increase as defect density increases.

VII. REFERENCES

[1] P. Sivakumar, et al., "Exploiting microarchitectural redundancy for defect tolerance," *Proc. ICCD, 2003.*

[2] C. Demerjian, "Nvidia's Fermi GTX480 is broken and unfixable," *http://semiaccurate.com/*

[3] D. Cheng and S. K. Gupta, "A systematic methodology to improve yield per area of highly-parallel CMPs," *Proc.* DFT, 2012.

[4] D. Cheng and S. K. Gupta, "Maximizing yield per area of highly parallel CMPs using hardware redundancy," *TCAD, 2014.*

[5] D. Cheng and S. Gupta, "Optimizing Redundancy Design for Chip Multiprocessors for Flexible Utility Functions," *Proc. Int'l Test Conf., 2013.*

[6] D. Cheng and Sandeep Gupta, "A novel software-based defect-tolerance approach for application-specific embedded systems," *Proc. ICCD,* 2011.

[7] Y. Markovsky and J. Wawrzynek, "On the opportunity to improve system yield with multi-core architectures," *Int'l Workshop Design for Manufacturability and Yield,* 2007.

[8] R. Kumar, et al., "Interconnections in multi-core architectures: understanding mechanisms, overheads and scaling", *ISCA,* 2005.

[9] NVIDIA CUDA SDK, *https://developer.nvidia.com/*

[10] ITRS 2011, *http://www.itrs.net/*

[11] GPGPU-Sim 3.x, *http://gpgpu-sim.ece.ubc.ca/*

[12] L. Zhang, et al., "Defect tolerance in homogeneous many-core processors using core-level redundancy with unified topology," *Proc. DATE, 2008.*

[13] A. Lashgar, B. Amirali, and K. Ahmad, "Investigating Warp Size Impact in GPUs," 2012.

978-1-4799-7598-3/15 $31.00 © 2015 IEEE

2015 IEEE 33rd VLSI Test Symposium (VTS)

In-depth Soft Error Vulnerability Analysis using Synthetic Benchmarks

Shahrzad Mirkhani, Balavinayagam Samynathan, and Jacob A. Abraham

Computer Engineering Research Center, The University of Texas at Austin
Email: {shahrzad, balavins, jaa}@cerc.utexas.edu

Abstract—Statistical fault injection is widely used for analyzing hardware in the presence of soft errors. Although this method can give accurate results for averaged erroneous outcomes with a fairly small sample size, it will not be accurate for vulnerability analysis of each sequential element in the design with small sample sizes. This paper describes a novel and highly efficient technique which is suitable for detailed vulnerability analysis of a processor. The technique involves specific sets of assembly language routines, and is shown to be much more efficient and comprehensive compared with traditional statistical error injection on a predetermined set of benchmarks. We have shown the effectiveness of the method using error injection in an ARM Amber25 processor model. Our analysis is based on more than 330,000 simulation runs with single bit-flips on the sequential elements of this processor running our synthetic benchmarks and 40,000 FPGA-based error injections for 4 conventional benchmarks.

I. INTRODUCTION

The impact of soft errors has been increasing with shrinking feature size in CMOS technology. This becomes a concern not only in mission- and life-critical systems, but also in general systems such as servers and routers[1]. Therefore, analyzing a system in the presence of soft errors has (or should) become one of the stages in the chip design process. If, based on the customer requirements, the designers find out that their design is vulnerable to soft errors, they need to incorporate resilience techniques in their design to make it more dependable. In general, adding full redundancy is expensive and therefore, designers need to apply partial redundancy and cross-layer techniques to obtain an acceptable resilient system with reasonable cost. In order to apply partial redundancy at the circuit level (e.g., using BISER [24] or DICE [5]), designers need to know which parts of the system are more vulnerable to soft errors. Such an insight into soft error vulnerability requires a good model of soft errors. Several error models have been proposed for soft errors (e.g., [6, 15, 23]). However, experiments in [3] show that the outcomes of injecting single errors in flip-flops (FF) of a system are very close to those when using radiation. On the other hand, it is shown in [7] that the high-level error models that can be simulated fast have a significant outcome difference from the error model that injects single bit-flips into the FFs. In this paper, we use the error model of single bit-flips into FFs as our golden model[2].

In order to apply partial redundancy to a design, we need to have a means of determining which FFs in the system are most vulnerable to soft errors. This calculation should ideally be done by injecting all possible error candidates and observing their outcomes under each error while running all possible applications for that system. Due to the enormous number of error candidates, which is the number of FFs in the design multiplied by the number of clock cycles a program takes to run on that design, this option is not feasible. Another option for this calculation is *Statistical Fault (Error) Injection* or SFI [18]. In SFI, a sample of the error candidates is randomly chosen and the average outcome results for this sample can represent the average outcomes for all error candidates, with a margin of error (MOE) and a confidence level. Error injection with a sample of errors has been

---

[1] Although multi-gate transistors such as Tri-gate FETs and FinFETs are less susceptible to soft errors [19], they can be affected by soft errors and are still an issue for mission-critical systems.

[2] We do not consider soft errors in combinational logic, since the effects of such errors would propagate to the flip-flops, and soft errors in gates are decreasing in new technologies, such as multi-gate transistors [19].

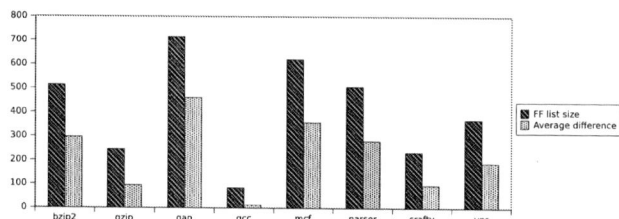

Fig. 1: Differences in the set of most vulnerable FFs for each SPEC benchmark in an out-of-order processor (IVM)

utilized to calculate the vulnerability of each FF as well as the average vulnerability of the design in several papers [9, 16, 20].

Using SFI, with a reasonable confidence level and MOE, we can perform error injection on a design with a sample small enough to be practical for large designs and applications. This sample size is suitable if we want to calculate average outcomes of error injection for that system. However, we have to consider each FF individually, if we are to determine highly vulnerable ones. This can result in millions of injections if we have a few thousand FFs in our design. The lack of sufficient sampling can cause a large MOE in calculating vulnerable FFs. Figure 1 shows an example of vulnerability analysis in an out-of-order processor based on the DEC alpha architecture (IVM [21]) with approximately 14,000 FFs. In this experiment which was done on 8 SPECINT2000 benchmarks, 160,000 soft errors were injected for each benchmark. This figure shows the number of highly vulnerable FFs[3] for each benchmark (dark gray bars). Also, the average number of FFs in each list that cannot be found in the vulnerable FFs for other benchmarks is displayed as light gray bars in this figure. For example, 500 FFs are highly vulnerable when running bzip2 program on IVM and, on average, 300 FFs out of these 500 FFs are not found in the most vulnerable FF lists for the other 7 applications. In this experiment, the MOE of the average outcomes [7] is around 0.25%, while the MOE of the FF vulnerabilities is around 30% (these MOEs are absolute errors).

In this paper, we develop a set of assembly language programs for a given processor that, along with error injection, can be used to efficiently analyze the vulnerability of a given FF in the design. These benchmarks can be simulated very fast, even when running the Register Transfer Level (RTL) model of the processor. Therefore, we can easily inject a sufficient number of errors, increasing the sample size. These **synthetic benchmarks** have been designed in a way to be less biased to data values or instruction sequences than typical benchmark programs. Figure 2 shows an overall view of sampling problems and the applications of the proposed synthetic benchmarks in this problem. SFI using synthetic benchmarks can increase the quality and decrease the injection time for soft error analysis when compared to SFI using application level benchmarks.

There are alternative solutions to SFI for calculating the most vulnerable FFs or parts of a design. Some analytical methods calculate the probabilities of soft error detection [2, 11]. These methods are fast, but their accuracy can be low in complex and/or large designs. There

---

[3] In this case, highly vulnerable FFs cause an erroneous outcome at least 70% of the times that an error is injected into them. Changing the cutoff point of 70% does not change the fact that different sets of vulnerable FFs are identified for the applications.

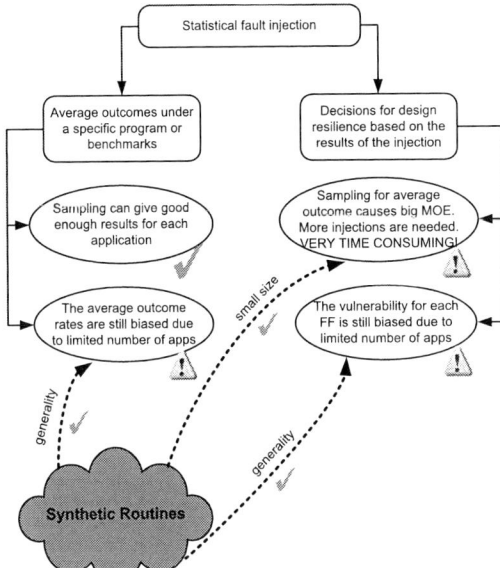

Fig. 2: Potential strength of synthetic benchmarks in error injection

are other approaches to vulnerability analysis such as architectural-level analysis [13, 14]. These methods calculate soft error vulnerability for specific structures using instruction set simulators. Applying such methods to a general structure can be very inefficient. There are several approaches for reducing the number of error candidates (e.g., [10]) or to decrease the injection time (e.g., [21]). Using our proposed synthetic benchmarks along with such methodologies can lead to faster injections.

The highlights of this paper are the following.

- Developing a set of synthetic benchmarks for vulnerability error analysis with better confidence for pipelined processors. To our knowledge, such synthetic benchmarks are novel in this area.
- Showing the effectiveness of the proposed synthetic benchmarks by injecting more than 330,000 errors into the ARM Amber25 processor model [1] running these benchmarks. We have also injected 40,000 errors into this processor model running four small programs and compared the list of most vulnerable FFs resulting from them to the list we extracted from our synthetic benchmarks. Our results show that these synthetic benchmarks are able to identify 9 FFs as highly vulnerable FFs, while the four programs identify only 1 FF in this list and this FF exists in the list of FFs from the synthetic benchmarks. Also, the injections in synthetic benchmarks result in a MOE around 8.6%, while the MOE for injections in the programs is around 31%.

The rest of this paper is organized as follows. After a brief discussion on sampling MOE for vulnerability analysis in Section II, we introduce the structure of our synthetic benchmarks in Section III. Section IV shows our observations we have made from error injection on the ARM Amber25 processor model while running the synthetic benchmarks and other existing benchmarks. Section V has additional discussion and Section VI concludes the paper.

II. SFI AND DESIGN RESILIENCE

As stated in the introduction, in SFI we need to find an acceptable sample size based on the goal of error injection (which can be an average outcome calculation for all FFs, or calculating highly vulnerable FFs for adding partial resilience to the design). In this section, we show the amount of error (MOE) introduced by SFI in FF vulnerability analysis if we want to use the same sample as we used for average outcome calculation.

Soft error candidates are distributed in two dimensions: FFs (spatial) and clock cycles (temporal). If we want to calculate the

average outcome of error injection while a set of benchmarks are running on our design, we can take a sample of candidates distributed in both spatial and temporal dimensions. If the number of FFs in a design is d, the total number of necessary clock cycles for running all the applications is c, which makes the population size equal to $d \times c$, and the sample size is n, due to [12], we can calculate the sampling MOE as in Eq. 1. In this equation, t is the cut-off point due to confidence level (e.g., $t = 1.96$ for 95% confidence level), and p is the estimated rate of injections that have erroneous outcomes. We can replace p with 0.5 to calculate the maximum possible sample size, or replace the calculated failure rate as p after the injection.

$$e = t \times \sqrt{\frac{p \times (1-p)}{n} \times \frac{c \times d - n}{c \times d - 1}} \quad (1)$$

On the other hand, if we want to use the sample with size n for finding the most vulnerable FFs in the design, we can say that an error is injected on each FF around $\frac{n}{d}$ times, on average. Therefore, if we want to calculate the MOE for the injection, based on each FF, our sample size is $\frac{n}{d}$, and our population size is c. Therefore,

$$e' = t' \times \sqrt{\frac{p' \times (1-p')}{\frac{n}{d}} \times \frac{c - \frac{n}{d}}{c - 1}} \quad (2)$$

If we want to calculate both e and e' with the same confidence level, i.e., $t = t'$, then by combining Equations 1 and 2, we will have e' as a function of e, the number of FFs in the design, and the number of clock cycles needed for running the applications. Note that in this equation, we assume $p = p'$. However, p' should be usually set as 0.5 since the vulnerability factors for each FF can vary between 0% and 100%, while p is usually less than 30%. Even if we consider $p \neq p'$, it does not affect our conclusion.

$$e' = e \times \sqrt{\frac{c \times d - 1}{c - 1}} \quad \therefore \quad \lim_{c \to \infty} e' = e \times \sqrt{d} \quad (3)$$

As an example, if:
d = 3000
c = 200,000
e = 1%
confidence level = 95%
Then, e' with 95% confidence will be:

$$e' = 0.01 \times \sqrt{\frac{200000 \times 3000 - 1}{199999}} \approx 0.55 \quad (4)$$

Equation 4 shows that an absolute MOE of 1% in calculating average outcomes results in an absolute error of 55% if we calculate the vulnerability of each FF with the same sample that we calculated the average outcomes. This error will be even higher if we have more FFs in our design! In this example, if we want each FF vulnerability to have $\hat{e} = 1\%$ of absolute MOE with 95% confidence level, we need n' samples for each FF. n' can be calculated by Eq. 5.

$$n' = \frac{c}{1 + \hat{e}^2 \times \frac{c-1}{t^2 \times p \times (1-p)}} = \frac{200000}{1 + 0.01^2 \times \frac{199999}{1.96^2 \times 0.5 \times (1-0.5)}} \approx 9164 \quad (5)$$

Therefore, we need to perform a total of $9164 \times 3000 = 27,492,000$ injections for a detailed vulnerability analysis! This number of injections cannot be done in a reasonable time although it is still much smaller than all the possible error candidates (which is $200000 \times 3000 = 600,000,000$). Since, in vulnerability analysis, we basically categorize FFs, e' can be set to a higher number like 10%. In this case, we still need to perform 288,000 injections which is still a large number due to the time for running a long application. In the next section, we discuss our synthetic benchmarks and how we can decrease the MOE for vulnerability analysis of a design.

III. SYNTHETIC BENCHMARKS

In this section we describe our synthetic benchmarks and explain how they can guide design resilience. These benchmarks consist of small assembly routines, and each of them analyzes one instruction in the instruction set. We follow the approach of Burch and Dill [4] who used a similar strategy to verify a pipelined processor. They match the implementation and specification by flushing the pipeline so that each instruction can be retired and finished independently. Also, Rahimi et al. [17] exploited the concept of individual instruction tests to analyze the effects of process variations (such as voltage and temperature) on a given design.

The set of assembly codes, when running on a pipelined processor model during a period of time, will have only one instruction surrounded by NOP[4] instructions in the pipeline. We refer to this instruction as the instruction under test or IUT. We use NOP before and after each IUT to minimize the effect of other instructions on the processor state. This way, we will have a more precise analysis based on the IUT. These synthetic benchmarks are specified for pipelined processors. However, we believe that the concept of testing individual instructions can be used for non-pipelined processors as well. During the period that an IUT is entering the pipeline, until it gets retired, a soft error is injected into a random FF (at a random cycle) in the design and the output of the benchmark execution (register file, status bits, IRQ registers, etc.) is compared to the golden output after the run is finished (or timed out).

Depending on the number of cycles that the IUT stays in the pipeline, we can calculate the sample size using [12]. For the IUTs that stay in the pipeline for no more than 13 cycles, we inject 7800 errors, while for the IUTs that stay more than 14 cycles, we inject 9600 errors[5]. The general structure of these benchmarks is shown in Fig. 3. As shown in this figure, we pre-load all registers with some random data, choosing them to have a different value from each other. This way, the erroneous and golden values of the destination register of an instruction has a very low probability to be the same if an error causes register r_j to be used instead of r_i.

Note that for a few IUTs (e.g., branch), we might need more instructions before/after the NOPs. Also, for cache-related instructions, we need to have a larger program (around 100 instructions) to take care of cache hits and misses. In general, the benchmarks like the one shown in Fig. 3 test the core logic. For analyzing system-level effects, we can develop a few special benchmarks which can test a specific system-level effect. We have already developed two benchmarks which inject an error under cache miss/hit conditions.

A. Advantages

In this section, we will discuss the advantages of these benchmarks over using an application benchmark (e.g., SPEC benchmarks).

- The synthetic benchmarks are **general** in contrast with application benchmarks, which might use special data values and instructions to implement a specific functionality. These assembly codes test each and every instruction, independently from other instructions.
- The freedom of choosing values of registers and choosing which registers to be used in every instruction gives us the opportunity to have **more general and pessimistic** erroneous results. Also, we introduce a high observability for errors in our synthetic benchmarks. If an error affects an instruction in the pipeline, it will show up in the output of the program immediately and will not be masked by the instructions running after injection, since there is only one IUT in the pipeline. Using NOPs after IUT, in our benchmarks, decreases the masking probability. By

[4]No operation
[5]These numbers are calculated based on having Max. MOE=1% with confidence level=95% for average outcomes of each benchmark.

```
#include "amber_registers.h"
.section .text
.globl main
main:
ldr r0, r0_data
ldr r1, r1_data
...
ldr r14, r14_data

nop
...
nop
add r0, r11, r8
nop
...
nop
testpass:
str r3, [r6]
b testpass
/* Write a non-zero value to this address to generate a Test Passed message */
AdrTestStatus: .word ADR_AMBER_TEST_STATUS

Data: .word 0x12345678
r0_data: .word 0xff666f90
r1_data: .word 0xc3a8efff
...
r14_data: .word 0x5a78006f
```

Fig. 3: A sample benchmark for "add" instruction as the IUT

injecting a reasonable number of errors, these benchmarks test each instruction at each stage of the pipeline. Due to the variety of instructions in an application, such precise testing is only feasible with a very large number of error injections.

- Each assembly code is very **small** (in contrast to an entire application). Therefore, the RTL model of the processor for these codes can be simulated very **fast** (in a few seconds) compared to hours of simulation for a program such as gzip in the SPEC benchmark set. Even if we have a mixed simulation environment (like [21]), we usually simulate the model at the RT or gate-level for at least 10,000 cycles (or even more). In this case (10,000 cycles), running a mixed-mode simulation for SPECINT2000 benchmarks (12 programs) takes 35x longer than our synthetic benchmarks (average of 100 cycles), if our processor has 35 instructions (i.e., 35 benchmarks).

- The portion that we are considering for error injection (the time that IUT enters the pipeline until it is retired) is very small (9 to 114 clock cycles, 20 cycles on average). Therefore, the number of soft error candidates (i.e., the number of FFs × the number of cycles an error can be active) needed to analyze each IUT is manageable. This gives us the opportunity to have more precise vulnerability factors for each FF under each IUT, because the number of our samples per each FF is **closer to exhaustive** (complete) error injection. In other words, the results from these synthetic benchmarks are *statistically* more accurate than the results of a long application with the same number of injections.

B. Applications

In this section, we will discuss the use of our benchmarks in the analysis of design vulnerability under soft errors.

1) Vulnerability Analysis: By injecting a sample[6] of error candidates and simulating the processor model for each benchmark, we can observe the outcome of the benchmark for each injected error. In this case, the output of our program is all the core-related software visible components (such as register file and IRQ registers) and the outcome of the injection can be categorized as: *masked, SDC, timeout,* and *fatal.* When the effect of an error is masked and does not appear in any software visible component then we label it as a masked outcome. Silent data corruption (SDC) is the outcome when the program has completed successfully but there are some erroneous

[6]We are even able to inject the whole population of soft error candidates in cases that the IUT takes a small number of cycles.

values in registers. Timeout happens when the program is running for a long time (2x to 4x more than its error-free runtime) and does not exit. In our experiments on ARM Amber25, the usual reason for timeout is the change of program counter (r15) to an erroneous (but not invalid) value. When the program crashes before finishing, we categorize it as fatal. An invalid opcode or invalid code/data address range usually causes a fatal outcome.

By analyzing the outcomes of each error injection for each benchmark, we can calculate the vulnerability factor (VF) of each FF in the design, which is shown in Eq. 6, and categorize all FFs based on their VF value. The higher the VF of a FF, the more it is vulnerable to soft errors. Therefore, to have a resilient design (i.e., circuit-level resiliency), the designers need to apply a resilience technique to those FFs. In our experimental results, we will show that the list of FFs with high VF have a large number of common FFs in all of our benchmarks. Also, the list of high VF FFs (considering all synthetic benchmarks) is a superset of the list in the case of running an application on the processor.

$$VF_{FF_i} = \frac{number\ of\ times\ a\ soft\ error\ on\ FF_i\ causes\ not\ masked\ outcome}{number\ of\ times\ a\ soft\ error\ injected\ into\ FF_i}$$
(6)

Note that depending on the designer's goal, the definition in Eq. 6 can be specified for each outcome (e.g., SDC and fatal) separately.

2) Deducing a High-level Error Model: As discussed above, the structure of our benchmarks helps us observe the effects of single bit-flips for each instruction. Based on our observations of the error injection experiments, we can develop models of the effect of a single bit-flip at the architecture level. Such errors can be injected and simulated in a high-level simulator (e.g., an instruction-set simulator), but, unlike the high-level error models that we discussed earlier, the outcomes of error injection will be highly correlated with the outcomes of the single bit-flip on FF injection. The results of this high-level error injection can also provide the necessary information on the vulnerability factor for each FF at the architecture level. Due to space limitations, we will not consider this application here (it will be part of our future work).

IV. ERROR INJECTION RESULTS

As mentioned earlier, we have performed low-level error injections (bit-flips on FFs) on an ARM Amber25 processor core with 3012 FFs.

A. Configuration

We used the Xilinx iSim simulator as our injection environment for synthetic benchmarks. Every simulation for a synthetic benchmark takes around 18-20 seconds and we have performed 333,740 injections in total for the synthetic benchmarks. In order to show the effectiveness of our benchmarks, we have also injected around 9,900 errors in the core for four programs (AES or Advanced Encryption Standard, PID or Proportional Integral Derivative, Dhrystone [22], and CRC or Cyclic Redundancy Check). Since the simulation run times for these programs would be high, we synthesized the ARM Amber25 core and implemented it, along with peripherals, on Xilinx SPARTAN6 FPGA boards. The injection methodology in our implementation is similar to the one discussed in [8]. We have automated the injection process by creating an extra scan chain of FFs in the design. Using this extra scan chain, we can inject an error on a random FF during a random cycle, using *Pyserial*. This method gives us the ability to inject every error without the obligation of re-synthesizing the design for each (or a group) of injections. Our FPGA setup runs an application in a reduced version of Linux operating system (version 2.4.27).

TABLE I: List of FFs with vulnerability factors more than 0.6

Program	
Run time (sec)	
Synthetic benchmarks ($VF \geq 0.6$)	Four applications ($VF \geq 0.6$)
u_execute.status_bits_mode[0] (VF=1)	u_execute.status_bits_mode[0] (VF=0.84)
u_execute.status_bits_mode[1] (VF=1)	u_execute.status_bits_mode[1] (VF=0.72)
u_execute.status_bits_irq_mask (VF=1)	-
u_execute.status_bits_firq_mask (VF=0.99)	-
u_execute.status_bits_flags[1] (VF=0.97)	-
u_execute.status_bits_flags[3] (VF=0.92)	-
u_mem.uncached_wb_req_r (VF=0.91)	-
u_execute.status_bits_flags[2] (VF=0.89)	-
u_execute.status_bits_flags[0] (VF=0.84)	-
u_execute.o_daddress_valid (VF=0.80)	-
u_write_back.mem_read_data_valid_r (VF=0.73)	-
u_wishbone.o_wb_we (VF=0.69)	-
u_wishbone.u_a25_wishbone_buf_p1.wbuf_used_r[1] (VF=0.68)	-
u_wishbone.u_a25_wishbone_buf_p0.wbuf_used_r[0] (VF=0.67)	-
u_wishbone.o_wb_stb (VF=0.66)	-
u_wishbone.u_a25_wishbone_buf_p1.wbuf_used_r[0] (VF=0.61)	-

B. The Most Vulnerable Flip-flops

Our experiments show that among 3012 FFs, only 530 of them are involved in erroneous results in all the synthetic benchmarks. As can be seen in Fig. 4, only a small portion of them (e.g., 9 FFs) have a high vulnerability factor (e.g., more than 0.8). This shows that only a small portion of FFs in a design need to be considered for circuit-level design resilience, and a considerable amount of area and power can be saved when we prevent over-designing through this detailed vulnerability analysis. The error bars in Fig. 4 show the standard error of the VFs of all the synthetic benchmarks for each FF. Standard error for each FF is measured by dividing the standard deviation of the VFs of that FF for each benchmark by the number of benchmarks. It can be used as a mean for measuring MOE. We have prepared another diagram similar to Fig. 4 for the VFs of the same FFs by running four applications, shown in Fig. 5. As can be seen in this figure, the standard error of the VFs among these four applications is very large. We have sorted the VFs in descending order, and therefore the FF IDs in this figure are not necessarily the same as the FF IDs in Fig. 4. Figure 6 shows the sorted FFs based on their vulnerability factors calculated by injections while running four applications. The MOE based on the number of injections has been calculated (around 0.3) and shown as error bars on the curve. The dots in this figure show the vulnerability factors calculated by injections while running synthetic benchmarks. As can be seen in this figure, most of the vulnerability factors calculated with synthetic benchmarks are within the range of the factors calculated by four applications. The exceptions are the cases that applications do not calculate because of the insufficient number of injections, or they do not use the complete set of instructions. This will be discussed more in Section V. A list of FFs with vulnerability factor more than or equal to 0.6 is shown in Table I.

It is interesting to see that in the VFs calculated for the four applications, we have only one FF with VF more than 0.8 and this FF (u_execute.status_bits_mode[0]) is in the list of most vulnerable FFs calculated in our synthetic benchmarks (with VF=1.0). A brief discussion on the vulnerable FFs is found in Section V. We should mention that there are some FFs that have VF greater than 0 when we run our applications, but they have zero VF in our synthetic benchmarks. We have studied these FFs. They are related to the errors propagated to memory mapped IO (e.g., UART), and we did not consider them as one of our visible outputs currently. We will add memory mapped IOs as one of the visible components in the future to have a more complete list of vulnerable FFs. However, the vulnerability factors in this list are less than 0.3 and they are not included in most vulnerable list of FFs in the application runs. The rest of the FFs that have a vulnerability factor greater than 0 are the same in the synthetic benchmarks and the four applications.

C. MOE Analysis

By using Eq. 2, we can calculate the MOE for our synthetic benchmarks as well as the four applications that we ran on the FPGA. A summary of the results are shown in Table II. This table shows the

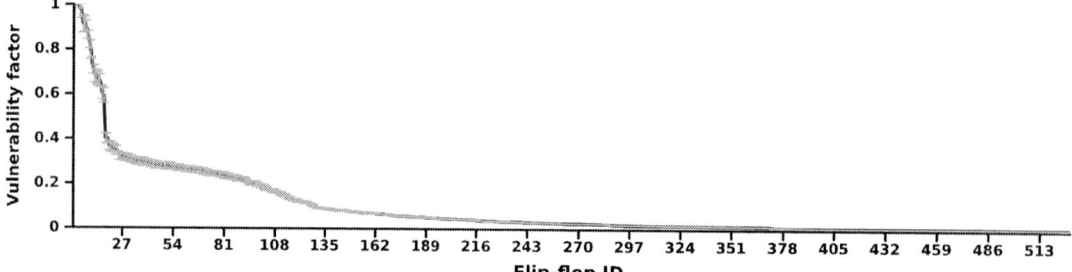

Fig. 4: Vulnerability factors of the FFs in ARM Amber25 core which cause erroneous results in synthetic benchmarks

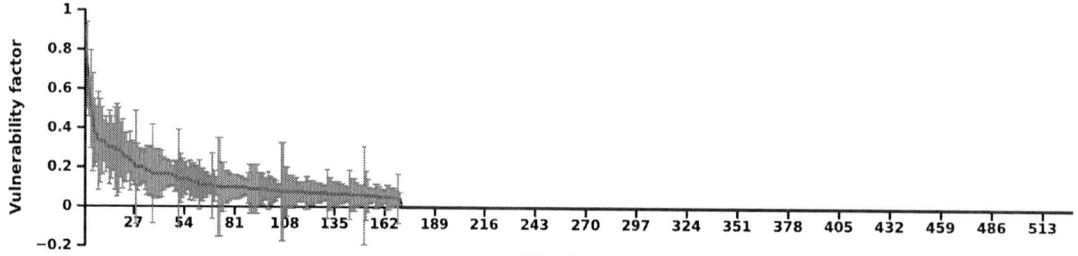

Fig. 5: Vulnerability factors of the FFs which cause erroneous results in four applications. Note the IDs are different from the ones in Fig. 4

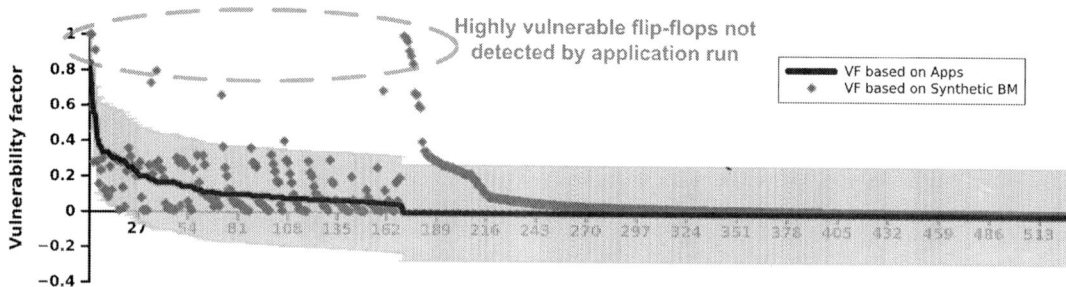

Fig. 6: Vulnerability factors of the FFs in ARM Amber25 core which cause erroneous results in synthetic benchmarks vs. applications

TABLE II: MOE in synthetic benchmarks vs. applications

Program	Avg. injection per FF	Number of cycles	MOE
Synthetic benchmarks	110	777	0.086
AES	3.35	98,897	0.53
CRC	3.36	88,392	0.53
Dhrystone	3.31	1,172,649	0.54
PID	3.34	253,218,253	0.54
Total (apps)	13.37	254,578,191	0.27

TABLE III: Program run times on ARM Amber25 for each injection

Program	Single run time (sec)	Total injection run time (days)
Synthetic benchmarks (14 servers)	18	5
AES (FPGA)	80	9
CRC (FPGA)	80	9
Dhrystone (FPGA)	80	9
PID (FPGA)	80	9
Total (1 board)		37
AES (simulation)	102	12
CRC (simulation)	88	10
Dhrystone (simulation)	952	40
PID (simulation)	4680	195
Total (14 servers)		18.5

average number of cycles and the total number of cycles for each case and based on these numbers and 95% confidence level, it shows the average MOE for FF vulnerability factors. As can be seen, using the synthetic benchmarks will result in vulnerability factors with around 8.6% MOE while vulnerability factors in all four applications have a large MOE (around 27% absolute error).

Now according to Eq. 5 with MOE, or $\hat{e}$, equal to 8.6% (the one in the synthetic benchmarks injections), we will need around 130 injections per FF. This means that we need to perform $130 \times 3012 = 391,560$ error injections while running our four applications in order to have results with same MOE as the results of the synthetic benchmarks.

D. Run Times

The synthetic benchmarks for ARM were simulated in a distributed system consisting of 14 servers. The error injections for the applications were done on the same ARM model in an FPGA system. Table III shows the time needed for a single run of each application or one synthetic benchmark, as well as run times for total number

of injections (330K injections for synthetic benchmarks and 10K injections for each application). We have extrapolated the time needed for running injections on ARM for both the synthetic benchmarks and the four applications. As can be seen, the run times for all four applications on the FPGA board are almost the same, since the main overhead in this case is programming the FPGA, loading the operating system, loading the program, and collecting the results.

If we wanted to inject 391,560 errors while running our four applications to gain an MOE around 8.6%, it would take around 362 days on an FPGA board and 471 days using simulation (on 14 servers) to perform the injections, as compared with 5 days using the synthetic benchmarks for the same MOE.

V. DISCUSSION

This section discusses the reasons that we can trust the results of the synthetic benchmarks and use them in a detailed vulnerability factor analysis instead of running long applications.

Since these static benchmarks are very short programs, the cases where an error remains in the design, without showing up in software-visible components can happen. Such cases are less likely to happen when running long programs. Previous work such as [21] refer to these cases as *gray area*. We call them *residuals*. We have done residual error analysis to see if errors reside in the system without showing up in any of the software-visible components. These residuals can reside in a single FF or multiple FFs. Our results show that the major portion of residual errors are already marked by synthetic benchmarks with high vulnerability factors. This shows that FFs with high error propagation probabilities are already captured by our benchmarks. Within the smaller subset that is not marked by our benchmarks as highly vulnerable, most of them reside on a single FF in the design. Since we are injecting single errors on FFs, their effect will be captured when error injection is done on that FF. In other words, the outcomes of such residuals are equivalent to other error injection outcomes. The only cases that are not marked by synthetic benchmarks and the residuals reside in multiple FFs are two cases: one is the read request FF of data cache, and the other is the control state FF of the instruction cache. We have manually traced these two errors and both of them affect only the performance of the design and not its functionality. Therefore, we can conclude that synthetic benchmarks are safe to use.

As shown in Table I, synthetic benchmarks find more FFs in their list of highly vulnerable FFs than the applications. If we take a look at the FFs that are not found by the application error injections, we can see that they are either related to the state flags or interrupts, or they are related to the wishbone interface. Wishbone is a critical part of Amber25 since all the memory operations go through it. The other FFs are related to interrupt handling which was not tested by the applications, but they are important bits since they can change the control flow of the instructions running on the pipeline. Also, the status bit flags are listed in synthetic benchmarks while they are not listed in the applications. These flags might not be checked frequently inside the applications and they might be simply overwritten before the error is observed. But they are also very important bits since they can change a program control flow.

An advantage of these synthetic benchmarks is that we have the vulnerability information based on each individual instruction. This way, if we want to consider some instructions more important than other instructions (which are not executed very frequently), we can calculate the vulnerability factors of FFs based on the importance of each instructions. In this paper, the weight of each instruction is 1. But, due to the sensitivity of the design or the types of applications running on it, we may assign different weights to each instruction.

VI. CONCLUSIONS

In this paper we developed synthetic benchmarks for efficiently determining the vulnerability of every flip-flop in a pipelined processor, and evaluated them on a model of the ARM Amber25 core. Our results, based on 333,740 error injections for the benchmarks and 40,000 for four applications show the effectiveness of the approach. Future work includes the use of these benchmarks to define an accurate high-level error models for processors.

VII. ACKNOWLEDGMENTS

This work is sponsored in part by a grant from Cisco through Silicon Valley Community Foundation. The authors also acknowledge Xilinx Inc. for providing the Spartan 6 FPGA boards.

REFERENCES

[1] ARM Amber25 Processor. http://opencores.org/project,amber.

[2] G. Asadi and M. B. Tahoori. An analytical approach for soft error rate estimation in digital circuits. In *IEEE International Symposium on Circuits and Systems*, 2005.

[3] C. Bottoni, M. Glorieux, J. Daveau, G. Gasiot, F. Abouzeid, S. Clerc, L. Naviner, and P. Roche. Heavy ions test result on a 65nm sparc-v8 radiation-hard microprocessor. In *Proceedings of IEEE International Reliability Physics Symposium*, 2014.

[4] J. R. Burch and D. L. Dill. Automatic verification of pipelined microprocessor control. In *Computer Aided Verification*, 1994.

[5] T. Calin, M. Nicolaidis, and R. Velazco. Upset hardened memory design for submicron CMOS technology. In *IEEE Transactions on Nuclear Science*, 1996.

[6] G. Chen, M. Kandemir, N. Vijaykrishnan, and M. J. Irwin. Object duplication for improving reliability. In *Proceedings of IEEE ASP-DAC Conference*, 2006.

[7] H. Cho, S. Mirkhani, C.-Y. Cher, J. A. Abraham, and S. Mitra. Quantitative evaluation of soft error injection techniques for robust system design. In *Proceedings of IEEE/ACM Design Automation Conference*, 2013.

[8] P. Civera, L. Macchiarulo, M. Rebaudengo, M. S. Reorda, and M. Violante. Exploiting circuit emulation for fast hardness evaluation. In *IEEE Transactions on Nuclear Science*, 2001.

[9] J.-M. Daveau, A. Blampey, G. Gasiot, J. Bulone, and P. Roche. An industrial fault injection platform for soft-error dependability analysis and hardening of complex system-on-a-chip. In *Proceedings of IEEE International Reliability Physics Symposium*, 2009.

[10] A. Evans, M. Nicolaidis, S.-J. Wen, and T. Asis. Clustering techniques and statistical fault injection for selective mitigation of seus in flip-flops. In *Proceedings of the International Symposium on Quality Electronic Design*, 2013.

[11] M. Fazeli, S. G. Miremadi, H. Asadi, and M. B. Tahoori. A fast analytical approach to multi-cycle soft error rate estimation of sequential circuits. In *Proceedings of Euromicro Conference on Digital System Design: Architectures, Methods and Tools*, 2010.

[12] R. Leveugle, A. Calvez, P. Maistri, and P. Vanhauwaert. Statistical fault injection: quantified error and confidence. In *Proceedings of IEEE Design, Automation & Test in Europe Conference*, 2009.

[13] S. S. Mukherjee, C. Weaver, J. Emer, S. K. Reinhardt, and T. Austin. A systematic methodology to compute the architectural vulnerability factors for a high-performance microprocessor. In *Proceedings of IEEE/ACM International Symposium on Microarchitecture*, 2003.

[14] A. A. Nair, L. K. John, and L. Eeckhout. Avf stressmark: Towards an automated methodology for bounding the worst-case vulnerability to soft errors. In *Proceedings of IEEE/ACM International Symposium on Microarchitecture*, 2010.

[15] K. Pattabiraman, G. P. Saggese, D. Chen, Z. Kalbarczyk, and R. Iyer. Automated derivation of application-specific error detectors using dynamic analysis. In *IEEE Transactions on Dependable and Secure Computing*, 2011.

[16] A. Pellegrini, K. Constantinides, D. Zhang, S. Sudhakar, V. Bertacco, and T. Austin. Crashtest: A fast high-fidelity fpga-based resiliency analysis framework. In *Proceedings of IEEE International Conference on Computer Design*, 2008.

[17] A. Rahimi, L. Benini, and R. K. Gupta. Analysis of instruction-level vulnerability to dynamic voltage and temperature variations. In *Proceedings of IEEE Design, Automation & Test in Europe Conference*, 2012.

[18] P. Ramachandran, P. Kudva, J. Kellington, J. Schumann, and P. Sanda. Statistical fault injection. In *Proceedings of IEEE International Conference on Dependable Systems and Networks*, 2008.

[19] N. Seifert, B. Gill, S. Jahinuzzaman, J. Basile, V. Ambrose, Q. Shi, R. Allmon, and A. Bramnik. Soft error susceptibilities of 22nm tri-gate devices. In *IEEE Transactions on Nuclear Science*, 2012.

[20] A. L. Silburt, A. Evans, I. Perryman, S.-J. Wen, and D. Alexandrescu. Design for soft error resiliency in internet core routers. In *IEEE Transactions on Nuclear Science*, 2009.

[21] N. J. Wang, J. Quek, T. M. Rafacz, and S. J. Patel. Characterizing the effects of transient faults on a high-performance processor pipeline. In *International Conference on Dependable Systems and Networks*, 2004.

[22] R. P. Weicker. Dhrystone: a synthetic systems programming benchmark. In *Communications of the ACM*, 1984.

[23] K. S. Yim, Z. Kalbarczyk, and R. K. Iyer. Measurement-based analysis of fault and error sensitivities of dynamic memory. In *Proceedings of IEEE/IFIP International Conference on Dependable Systems and Networks*, 2010.

[24] M. Zhang, S. Mitra, T. Mak, N. Seifert, N. J. Wang, Q. Shi, K. S. Kim, N. R. Shanbhag, and S. J. Patel. Sequential element design with built-in soft error resilience. In *IEEE Transactions on Very Large Scale Integration (VLSI) Systems*, 2006.

TMO: A New Class of Attack on Cipher Misusing Test Infrastructure

Sk Subidh Ali and Ozgur Sinanoglu
New York University Abu Dhabi (NYUAD)

Abstract—We present a new class of scan attack on hardware implementation of ciphers. The existing scan attacks on ciphers exploit the Design for Testability (DfT) infrastructure of the implementation, where an attacker applies cipher inputs in the functional mode and then by switching to the test mode retrieves the secret key in the form of test responses. These attacks can be thwarted by applying a reset operation when there is a switch of mode. However, the mode-reset countermeasure can be thwarted by using only the test mode of a secure chip. In this work we show how a Test-Mode-Only (TMO) attack can overcome the constraints imposed by a mode-reset countermeasure and demonstrate TMO attacks on private key as well as public key ciphers.

I. INTRODUCTION

Ciphers are among the most essential crypto primitives used in data security. There are two different types of ciphers; public key cipher and private key cipher. Public key ciphers are mainly authentication, signature schemes, session key generation etc., while private key ciphers provide encrypted data communication. To increase throughput, often these ciphers are implemented in hardware. These hardware implementations may be vulnerable to different side-channel attacks. Scan based attack is one such side-channel attack, where the attacker exploits the scan based DfT to leak the secret key of the cipher [1]. Scan based DfT improves the testability of manufacturing defects, i.e., enhance the controllability and the observability of the IC by providing direct access to the flip-flops of the chip. Scan aids debug, enabling in-field testing. However, it also provides access to the internal registers which may hold secure data such as the secret key of a cipher implementation.

Ciphers are implemented in an iterative fashion where certain operations of the cipher are repeated with different inputs. The security of any cipher lies in complete execution of the cipher. In presence of scan, an attacker can run the cipher for only one iteration in functional mode, and then by switching to the test mode, can shift out the result of the cipher iteration in the form of a test response. Thus, for a private key cipher implementation, an attacker can retrieve the individual round outputs [2]–[4], while in the public key ciphers such as RSA or ECC the internal states of the scalar multiplication can be retrieved [5], [6]. In either case the retrieved information leaks certain information of the cipher key.

These attacks rely on one condition that is while switching the mode from functional to test, the intermediate data in the registers (scan cells) will remain intact. Violating this condition will fail all the existing attacks that relies on mode switching. A mode-reset countermeasure was proposed in [7],

which resets the scan chain while switching the mode. The mode-reset countermeasure blocks the data flow from the functional mode to the test mode. To overcome this countermeasure the attacker has to use only the test mode. In test mode, boundary scan cells drive the primary inputs, and thus there is no direct access to the primary inputs. Therefore, in order to apply inputs to the cipher the attacker has to use scan chains. However, the main challenge in TMO attacks is to determine the mapping between cipher register and the scan cells.

In this work we will show how an attacker can misuse scan architecture to leak the secret key of a cipher by staying only in the test mode. We propose two different attack strategies, one in which the internal scan chain is used to apply inputs while in the other boundary scan chain is configured to apply input. To demonstrate our attack we have targeted AES as a representative private key cipher and ECC as a representative public key cipher. However, our attack strategy is straightforward and can be extended to any cipher.

II. PRELIMINARIES

A. Advanced Encryption Standard

AES is a 128-bit symmetric key block cipher available in three different key lengths: 128, 192 and 256 bits. The entire AES algorithm is divided into several identical round operations. The number of rounds in the three different versions of AES are 10 (128-bit key), 12 (192-bit key) and 14 (256-bit key) respectively. Each round comprises of following four basic transformations,

- *SubBytes* is a non-linear substitution operation.
- *ShiftRows* is the byte-wise permutation.
- *MixColumns* is the four-byte mixing operation.
- *AddRoundKeys* is the XORing the state with the round key.

We will refer to these operations as *SB*, *SR*, *MC* and *ARK* respectively. Fig. 1 shows the structure of the first round of AES which contains an extra key XORing at the beginning.

In iterative implementation of AES the first few cycles of the execution transfer the input plaintext to the round register; and then in each iteration the round register value is applied to the round input, the round operation is executed, and the results are stored back to the round register for the next round operation. At the completion of all the rounds the content of the round register is transferred to the primary outputs as a ciphertext. More details about the AES cipher can be found in [8].

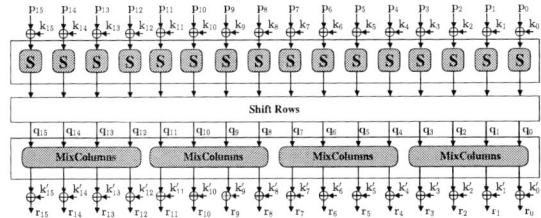

Fig. 1. First round of AES: p_i is the plaintext byte, k_i is the initial key byte, q_i is the SR output byte, k_i is the round key byte, and r_i is the round output byte.

B. Elliptic Curve Cryptography

An elliptic curve is a set of points $P = (x, y)$ on a curve given by the equation (1). The equation is known as Weierstraβ equation. The curve is defined over a large finite field F_q, where $a_1, a_2, a_3, a_4, a_6 \in F_q$ and the integer q defines the size of the field. $\mathcal{O}$ represents the point at infinity.

$$y^2 + a_1 xy + a_3 y = x^3 + a_2 x^2 + a_4 x + a_6 \qquad (1)$$

Each point on the curve is chosen from F_q. The set of points on the elliptic curve form a *group* under the addition rule, where the point $\mathcal{O}$ is the identity element. Given a point $P = (x, y)$ on the elliptic curve, determining k from the scalar multiplication kP, where $k \in F_q$, is a discrete logarithmic problem, which is computationally infeasible. In ECC, the scalar multiplication is the most computation intensive operation. The widely used algorithm to perform fast and secure side-channel attack resistant scalar multiplication is the Montgomery Ladder Multiplication (MLM) shown in Algorithm 1.

Algorithm 1: Montgomery Ladder Multiplication

Input: $P = (x, y)$; Scalar $k = (k_{m-1}, k_{m-2} \ldots k_0)_2$, where $k_{m-1} = 1$
Output: Point on the curve $Q = kP$

1 $Q_0 = P$
2 $Q_1 = 2P$ /*Point doubling*/
3 **for** $i = m - 2$ *to* 0 **do**
4      $Q_{1-k_i} = Q_0 + Q_1$ /*Point addition*/
5      $Q_{k_i} = 2Q_{k_i}$ /*Point doubling*/
6 **end**

Fig. 2 shows the ECC operation hierarchy, where MLM invokes the ECC point addition (line 4 of Algorithm 1) and doubling operations (lines 2 and 5 of Algorithm 1), which in turn use the finite field arithmetic operations: addition, square, multiplication and inverse. For more detailed description of

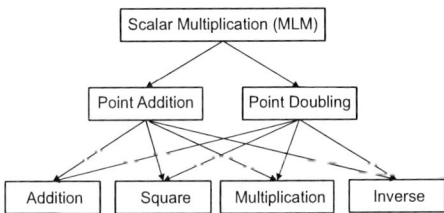

Fig. 2. ECC operation hierarchy

ECC one may refer to [9].

C. DFT Structure and Boundary Scan

The scan design converts the sequential circuit into combinational by making each flip-flop a fully accessible scan cell that can be controlled and observed easily through shift operations. The fact that tester equipments typically support a limited number of test channels presents a challenge for integrated circuits with a large number of input/output pins. IEEE Standard 1149.1 [10] solves the problem by defining the test access port and boundary scan architecture (Fig. 3). It consists of two major components, TAP (Test Access Port) and the Boundary Scan register. Each primary input and primary output port is supplemented with a special memory element, referred to as boundary scan cell. During the test mode, the test vectors are shifted in the boundary scan cells while in the capture operation, these boundary scan cells drive the primary inputs and sample the output logic that normally drives the primary outputs. The TAP controller is the control logic that can connect one of the five registers (boundary scan register, internal registers, instruction register, identification register, bypass register) between TDI and TDO.

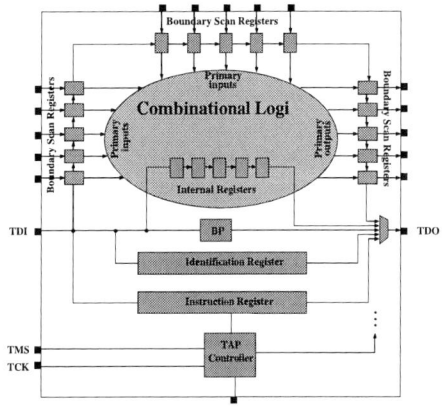

Fig. 3. Boundary scan architecture

D. Attack Assumptions

The following are the assumptions we make to perform the attack. We assume that the attacker:

1) knows the cipher implementation inside the security chip.
2) has access to the JTAG port, and thus, to the scan chains and test capabilities.
3) knows that the boundary scan cells corresponding to a particular input is consecutive but does not know which boundary scan cell corresponds to which input.
4) does not know the DfT scan architecture implementation details such as the number of scan cells, the number of scan chains, and the positions of the scan cells.
5) can apply arbitrary test vectors in the test mode.

III. TMO ATTACK ON AES

In this section we demonstrate two different attacks on AES, one using only the internal scan chain, while the other using both the boundary scan chain as well as the internal scan chain.

978-1-4799-7598-3/15 $31.00 © 2015 IEEE

A. TMO Attack on AES: Using Internal Scan Chains

In this attack we apply plaintexts through the scan cells corresponding to the round register. The challenge in this attack is that the round register flip-flops are randomly mapped to the scan cells by the physical placement tool [1]. If the mapping is not known, we can not precisely determine the round input corresponding to the value in the scan chain. Hence, we must identify the exact mapping between the round register and the scan cells. We determine the mapping by following steps.

1) Identify AES Input Words in the Scan Chain: In order to identify the scan cells corresponding to an input word we apply two test vector pairs (V, V_i) and (V, V_j), where V consists of 0-bits only, V_i and V_j are the one-hot values with 'one' in i-th and j-th bit, respectively. We apply these three test vectors and capture the responses. We observe the output differences corresponding to the input pairs (V, V_i) and (V, V_j). As per the AES algorithm if the two differences have any common bit-flip then the i-th and j-th bits correspond to the same input word of AES. By applying this technique we identify bits corresponding to a AES input word.

2) Itedentify AES Input Bytes in the Scan Chain: In this step we apply a fourth test vector V_{ij} with two-hot value, in addition to the three test vectors in the previous step. In AES S-box, two different pairs of inputs, each with the same input difference, will always produce two different output difference patterns. The input pairs (V, V_i) and (V_j, V_{ij}) have the differences in bit i. The bit-flip in j makes the two pairs different from each other. If i and j are in the same S-box, the two test vector pairs will produce two different input pairs to the S-box with one bit difference. Hence, the output differences corresponding to the two input pairs will be different. Using this technique all the scan cells in a word are divided into four bytes.

3) Recovering the key: In this step we apply selectively chosen eight test vector pairs (V, V_0) ... (V, V_7) to each of the S-boxes with bit-flips in the eight different bits of the S-box input. We apply the test vector pairs and calculate Hamming distances in the output differences. A precomputed Hamming distance signature table provides the eight Hamming distances for all possible key bytes, given the eight input test vector pairs. There are sixteen such Hamming distance tables corresponding to sixteen S-boxes. The captured eight Hamming distances are searched in the precomputed tables. The matched row will give the key byte value and the table will identify the S-box. Using the proposed attack we uniquely identify the AES key.

B. TMO Attack on AES: Using Boundary Scan Chain

In this attack we use the boundary scan chain as the input to the cipher and the round outputs are retrieved through the internal scan chain. As per the third assumption, the bits of the AES input are consecutive. Therefore, we need to identify either LSB or MSB of the input in the boundary scan chain.

1) Identify AES input in Boundary Scan Cells: We configure the test mode to apply selectively chosen test vectors to the boundary scan chain. In order to identify the AES inputs in the boundary scan chain we apply a pair of test vectors (V, V_i) similar to the previous attack. The test vector pair creates a one bit difference at the i-th scan cell of the boundary scan chain. If the i-th scan cell corresponds to any one of the AES inputs, we will find a Hamming distance in the captured response pair ranging from 5 to 24. We keep on applying the test vector pairs with different values of i until we get all the 128 input bits of AES in the scan chain. As per the third assumption if the LSB of x is the i-th boundary scan cell then the MSB could be either $i + 127$-th boundary scan cell or $i - 127$-th boundary scan cell. This implies that there are two possible orders of the input. We have to determine the key based on the two different orders.

2) Key Recovery: For the key recovery, we choose an input byte and apply all possible 127 input pairs with one bit difference in LSB of the byte. Among the output Hamming distances, only a few (9, 12, 23, and 24) appear exactly once. Once a unique Hamming distance is observed, the key byte can determined by XORing the precomputed S-box input with the AES input byte value. We apply the same technique across all the bytes we get 2^{16} possible key values. By considering two different mappings between the boundary scan cells and the AES inputs we get 2^{17} possible keys of AES.

IV. TMO ATTACK ON ECC

In this section we demonstrate how the TMO attack can be extended to public key cipher ECC. We consider an ECC on non-supersingular curve shown in equation (2), defined over a finite field F_2^{233}.

$$y^2 + xy = x^3 + a_2x^2 + a_6 \qquad (2)$$

This is a special instance of equation (1). The rest of the parameters of the ECC are taken from [11]. We consider Lopez-Dahab (LD) projective coordinate representation of points. In LD projective coordinate, a point is represented as $(X{:}Y{:}Z)$ and the corresponding affine point is $(X/Z, Y/Z^2)$. The point at infinity is represented as $(1{:}0{:}0)$. Our aim is to leak the secret scalar k of the MLM core. As a first step we identify the input x and y in the boundary scan chain.

A. Identify the Inputs x and y in the Boundary Scan Cells

The first operation of any ECC core is to check whether the input point is a point at infinity or the input point is a valid point of the elliptic curve. We target the infinity point checking mechanism to identify the LSB of x. We apply a one-hot test vector and perform capture(s). If the 'one' in the test vector corresponds to LSB of x, the circuit will activate the point at infinity signal. Once the location of LSB is obtained we will have two possible mappings of x as shown in Sec. III-B1. Next, we use one of the points where the ECC intersects the X axis ($y = 0$) and try two possible mappings of x with the help of point validation module. The test vector corresponding to the correct mapping will be validated by the point validation module. We apply a similar technique to identify y in the boundary scan chain.

B. Identify the Intermediate Registers in Scan Cells

Due to the presence of internal registers in the addition and doubling operations, we can not directly identify the Q_0 and Q_1 registers in MLM (Algorithm 1); there may be common data in these registers. We will use depth-first search to identify the internal registers. We first identify the internal registers used in the field multiplication, and subsequently determine the registers used by the higher level modules.

1) Identify the Internal Registers of Field Multiplication: In point validation module each input point is tested by equation (2). In order to do the testing two multiplication operations xy and a_2x^2 (equation (2)) need to be performed. Fig. 4 shows an example of five bit MSB first multiplier with reduction polynomial $f(z) = z^5 + z^2 + 1$. In each cycle, MSB of b is ANDed with each bit of a and XORed with the corresponding bit in c. We need to identify registers b and c, which vary in each iteration. We apply a test vector with a valid point and capture the response of each cycle of the multiplier. Let us assume that there are s scan cells; we get 233 values of the s scan cells corresponding to 233 iterations of the multiplication operation. We store these values in a table, where each row represents the 233 values of a scan cell. We also precompute a table with values corresponding to bits of c register only. We compare each row of the precomputed table with the other table. The matching row will identify a bit of c in the scan chain. We apply the same technique to identify the mapping for the register b.

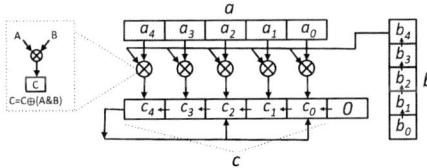

Fig. 4. Most significant bit first multiplier

2) Identify the Register in the Doubling Operation: In this step we determine the intermediate registers used in the point doubling operation in a step-by-step fashion. Algorithm 2 shows the sequence of operations executed in point doubling operation. There are five internal registers $(X_2, Y_2, Z_2, T_1, T_2)$

Algorithm 2: Point doubling in LD cordinate system

Input: $P=(X_1:Y_1:Z_1)$ /* Input point in LD coordinate */
Output: $2P=(X_2:Y_2:Z_2)$ /* Output point in LD coordinate*/

1   **if** $P = (1 : 0 : 0)$ **then**   9 $T_2 = T_1.b$
2   Return (P) 10 $X_2 = X_2 + T_2$
3   **end**   11 $T_1 = Y_1^2$
4   $T_1 = Z_1^2$ 12 $T_1 = T_1 + Z_2$
5   $T_2 = X_1^2$ 13 $T_1 = T_1 + T_2$
6   $Z_2 = T_1 \cdot T_2$ 14 $Y_2 = X_2 \cdot T_1$
7   $X_2 = T_2^2$ 15 $T_1 = T_2 \cdot Z_2$
8   $T_1 = T_1^2$ 16 $Y_2 = Y_2 + T_1$
     17   Return $(X_2:Y_2:Z_2)$

---

of the point doubling module. We target T_2 first, which varies

with a change in X_1. In order to identify the register, we apply n random input points. We run the MLM core until it executes $T_2 = X_1^2$ (line 5) and capture the corresponding response. We get n responses which we store in a table. We compare each row with a precomputed table and the matching row will identify the register bit in the scan chain. We experimentally observed that $n = 16$ is sufficient to uniquely identify T_2. Using this technique and the technique shown in Sec IV-B we determine the scan cells corresponding to registers X_2, Y_2, Z_2, T_1, and T_2. The registers X_2, Y_2, Z_2 represent Q_1. Once Q_1 is known, we apply the technique again on the initialization phase of MLM, when the input point P is assigned to register Q_0 (line 1 of Algorithm 1). Based on the same 16 input points we easily determine the register Q_0.

C. Recover the Scalar k

We apply a valid point to the MLM core and observe the values in registers Q_0 and Q_1 corresponding to 233 iterations. We compare each value with a precomputed value, where corresponding to each iteration we have two values of registers Q_0 and Q_1 which correspond to two possible values of the key bit k_i. The matching value will determine the value of the key bit. In this way we can determine all the bits of k except for the MSB, which can be determined by comparing the final output kP with the precomputed output. In total we need 678 test vectors to retrieve the 233-bit secret key.

V. CONCLUSIONS

In this paper we propose a new class of scan-based attack against cipher implementation using only the test mode. We demonstrate our attack on the public key as well as the private key cipher. The proposed attacks can overcome the constraints imposed by the mode-reset countermeasure.

REFERENCES

[1] B. Yang, K. Wu, and R. Karri, "Scan Based Side Channel Attack on Dedicated Hardware Implementations of Data Encryption Standard," in *ITC-2004*, pp. 339–344.

[2] B. Yang, K. Wu, and R. Karri, "Secure scan: a design-for-test architecture for crypto chips," in *DAC-2005*, pp. 135–140.

[3] A. Das, B. Ege, S. Ghosh, L. Batina, and I. Verbauwhede, "Security Analysis of Industrial Test Compression Schemes," *TCAD-2013*, vol. 32, pp. 1966–1977.

[4] S. S. Ali, O. Sinanoglu, and R. Karri, "AES design space exploration new line for scan attack resiliency," in *VLSI-SoC-2014*, 2014, pp. 1–6.

[5] R. Nara, N. Togawa, M. Yanagisawa, and T. Ohtsuki, "Scan-based attack against elliptic curve cryptosystems," in *ASP-DAC-2010*, pp. 407–412.

[6] J. DaRolt, A. Das, G. D. Natale, M.-L. Flottes, B. Rouzeyre, and I. Verbauwhede, "A scan-based attack on Elliptic Curve Cryptosystems in presence of industrial Design-for-Testability structures," in *DFT-2012*, pp. 43–48.

[7] D. Hely, F. Bancel, M.-L. Flottes, and B. Rouzeyre, "Test Control for Secure Scan Designs," in *ETS-2005*, pp. 190–195.

[8] " National Institute of Standards and Technology, Advanced Encryption Standard, NIST FIPS PUB 197, 2001."

[9] D. Hankerson, A. J. Menezes, and S. Vanstone, *Guide to Elliptic Curve Cryptography*. Secaucus, NJ, USA: Springer-Verlag New York, Inc., 2003.

[10] "IEEE Standard Test Access Port and Boundary-Scan Architecture," *IEEE Std 1149.1-2001*, pp. i–200, 2001.

[11] P. Gallagher, D. D. Foreword, and C. F. Director, "Fips pub 186-3 federal information processing standards publication digital signature standard (dss)," 2009.

A Call to Action: Securing IEEE 1687 and the Need for an IEEE Test Security Standard

Jennifer Dworak
Department of Computer Science & Engineering
Southern Methodist University
Dallas, Texas, USA

Al Crouch
ASSET InterTech, Inc.
Richardson, Texas, USA

Abstract— **Today's chips often contain a wealth of embedded instruments, including sensors, hardware monitors, built-in self-test (BIST) engines, etc. They may process sensitive data that requires encryption or obfuscation and may contain encryption keys and ChipIDs. Unfortunately, unauthorized access to internal registers or instruments through test and debug circuitry can turn design for testability (DFT) logic into a backdoor for data theft, reverse engineering, counterfeiting, and denial-of-service attacks. A compromised chip also poses a security threat to any board or system that includes that chip, and boards have their own security issues. We will provide an overview of some chip and board security concerns as they relate to DFT hardware and will briefly review several ways in which the new IEEE 1687 standard can be made more secure. We will then discuss the need for an IEEE Security Standard that can provide solutions and metrics for providing appropriate security matched to the needs of a real world environment. [1]**

Keywords—IEEE Standard; DFT; P1687; JTAG; IJTAG; security; scan; BIST; lock; trap; LSIB

I. INTRODUCTION

There are many security concerns associated with packaged semiconductor chips. Of particular concern is the fact that the test/debug ports and associated design for testability (DFT) logic can provide unauthorized access to the chip internals. Unfortunately, security and test/debug are polar opposites. Test and debug require controllability, observability, and traceability, whereas security demands the inverse (no unauthorized access, control, trace, or observation).

For example, test and debug ports on a microprocessor might allow attackers to investigate internal logic states through registers and scan test architectures in addition to obtaining sensitive data such as secret keys or ChipIDs. Test and debug ports also allow the operation of embedded internal instruments, such as environment monitors (temperature, current, voltage, frequency); BISTs (logic, memory, PLL, SerDes, etc.), configuration controllers (e.g. for PLLs, power modes, bus interfaces, etc.), and debug and trace logic (triggers, assertions, breakpoint counters, trace buffers, cycle counters, tags, etc.). These items can aid in reverse engineering, maliciously changing configurations, or operating destructively and disrupting the mission mode of the chip.

Different chip providers may apply different types of security schemes associated with the test and debug ports. This can create a chaotic security environment on a system board

---

[1] This paper was supported in part by NSF grant CCF-1061164.

when different chips from different providers are used on the same board. Some chip providers disable the ports completely after IC test—disabling all access to test and debug during board development, board test, and software development. Other chip providers disable some portions of the internal test and debug logic while leaving other portions intact for the use of board test and software development. Unfortunately, the very instruments that are left active may provide the most danger. Other organizations do not attempt to allocate any hardware for security, but restrict their efforts to the software level using data encryption and obfuscation.

Ultimately, a board test organization must manage, track, and preserve the different operating sequences, encryption schemes, and delivered secrets (codes, keys, ID numbers, etc.) for all the chips. This leads to several issues: Does the collection of chip security schemes at the chip level provide any security at the board level? Do the chip schemes open up security holes at the board level? Is additional board security required? How much additional information must be managed, tracked, and preserved to increase security?

Several IC provider and system provider companies, as well as universities, are asking if there is a need to look into a "compatible" security system that is more uniform across the many ICs created by chip providers. It may be time to investigate and possibly develop a security standard. This paper will investigate concerns involving test and debug security from the chip provider's point of view and from the board/system developer's point of view, and will discuss several possible methods for enhancing security in the presence of IEEE 1687 scan architectures. It will then consider some topics that should be explored in a test standard study group to evaluate the viability of such a standard.

II. WHICH HACKERS SHOULD WE CONSIDER?

Attackers of chips and boards may vary from the "curious individual with a JTAG box" to the "low dollar tamper and hacking shop" to the "well financed chip-delayering shop." Even today's mature chip-level data security can't easily make it past the deep pockets and fully destructive chip-delayering process that leads to reverse-engineering a device mask or netlist. In addition, stolen masks may lead to exact clones of a chip. Such attacks are difficult to thwart. However, countermeasures that attempt to identify such chips after the fact through PUF-based chip IDs or prevent their operation through hardware metering have been proposed (e.g. [1], [2]).

Here, we focus specifically on the security provided to test

and debug circuitry. Generally, it isn't necessary to protect against all possible attacks, but to provide reasonable security against attackers during a chip's or board's useable lifetime. The cost and level of the security should be commensurate with the value of the secrets being protected. Thus, we are interested in preventing system modification or data extraction by curious individuals that have an 1149.1 test setup and more dedicated attackers that have a bench setup and side-channel attack equipment such as temperature control, thermal imaging cameras, current monitors, or logic analyzers. In addition, many chip providers need to prevent the 3rd party individuals that touch the chip in the manufacturing, test and debug process from learning about the chip, and these individuals may have legal authorized tools to conduct their jobs.

III. CHIP-BASED TEST SECURITY CONCERNS

In many chips, the 1149.1 Test Access Port (TAP) and controller are being used as the de facto test and debug port/controller and can provide critical access to the chip internals. For a full-scan device, they can be used to produce a dump of the entire state of the chip. Many companies use the scan architecture to conduct silicon evaluation—switching from functional operation into a scan mode and mapping the state of key registers. Thus, techniques that totally disable such functionality are less than ideal. Direct Memory Access can also be used to snoop the structure and contents of embedded memories.

Another concern is the use of some of the chip's test and debug instruments to disable the chip after it is in mission mode on a board. For example, the chip could be placed in a scan or memory test mode during functional operation. The chip could even be damaged by running logic BIST constantly to the point of overheating. Alternatively, they could be used to reconfigure the system to operate in an unauthorized manner. Such attacks are especially problematic when a chip's test and debug hardware in a system can be accessed remotely through Internet access to the JTAG network.

Obtaining physical access to an example board or chip and monitoring side channels while investigating the JTAG port can help an attacker match JTAG instructions to embedded instruments and chip operations to enable reverse engineering, data theft, or sabotage. Test and debug actions are generally localized and may be susceptible to being mapped. For example, scanning in random instructions to an 1149.1 instruction register (IR) can invoke a memory BIST, and if the attacker has sprayed the chip with a coolant and is taking thermal images, he may see a hot spot at the memory array. Other side channels such as the current draw and pin activity level may also increase dramatically. At the very least, for modern chips based on 1149.1, 1500, and 1687, the indicator that an embedded instrument has been found may simply be that the length of the scan chain has changed [3].

Another concern of the chip provider and the whole supply chain, is the validation of the chip's trust value—whether or not the chip is real and genuine, or is a counterfeit. One option is to read public or hidden chip IDs through the test port. Unfortunately, if the chip ID is public and is easy to read, then even a PUF-based chip ID could potentially be spoofed with an on-chip memory holding sufficient challenge/response pairs.

When security techniques are applied to combat or counteract these chip concerns, then "goodness metrics" become important and temper the "size" or "complexity" of the solution. In general there are three main metrics that are used to evaluate DFT security when it is being designed into a chip:

1. Silicon impact (area, timing, routing)

2. Test time or test cost impact

3. Strength (how long and how difficult to break)

There may be other considerations as well. The complexity of the bureaucracy needed to maintain and propagate the "secrets" to authorized individuals during the lifetime of the chip must be considered. In addition, how the current security scheme mixes with other security schemes in the chip's end use environment (i.e. in the board or system) must be analyzed. Finally, the security scheme applied cannot be more costly and complex than the perceived value of the chip's secret.

IV. BOARD-BASED TEST SECURITY CONCERNS

When chips are placed on boards, they become a part of a bigger system that also has security concerns. The concerns are similar: reverse engineering, denial of service attacks, protection of on board codes and critical operating system data, and management of counterfeit issues (both chip and board). One of the main differences between dealing with chip security issues and board security issues is that there are more "probe-able" weak spots on a board, including connectors, test points, sockets, and in some cases, chip leads.

To make a counterfeit board, a counterfeiter must have the BOM (bill of materials) and the board netlist. Both items are easily available from contract manufacturers. What is missing is the "firmware" component—the part that defines the board's operating system—and the contents of memory during board operations that could identify the firmware actions of a legal board. There are active techniques to minimize access to the firmware, and generally on-board memory data is encrypted. However, when making a copy of a board, having memory data exactly as it is on the board (i.e. encrypted) may be what the attacker wants. Methods to prevent access are needed.

One common method for investigating memory chips on a board is to use the 1149.1 boundary scan circuitry in the interfacing chip to 'bit bang' the memory chip. Preventing the on-board memory from being illegally read requires restricting access to at least the data and control interfaces of the memory chip and the chip that provides the driving and receiving interface. Since most memory chips do not currently support boundary scan, and since most soldered down chips are ball grid arrays with no easy access to the board connections, it is the interfacing chip that brings the most risk. Thus, there is a need to restrict access to the common test features provided by boundary scan on such chips.

To obfuscate the actual test data applied and captured, it is often thought that decryption and encryption should be applied at the beginning and end of the scan paths (for both manufacturing and JTAG types of scan paths) so as to cover the whole scan path. However, on a board the scan path may traverse many chips, and some of those chips may have TDI or TDO pins that could be physically probed—allowing any decrypted data on the board to be observed. The encryption function needs to be within the chips and should be applied to internal scan paths only, not scan paths that pass through visible registers such as the chip's boundary scan ring.

Because the 1149.1 JTAG port may also be used to program FPGA and PLD devices, it may be difficult to hide programming firmware from attackers. The firmware, when stored prior to being fed to the test port for programming, should be in an encrypted state and should be decrypted either on the board near the target programmable device or inside the programmable device itself. Unfortunately, power analysis attacks have been shown to defeat some internal FPGA encryption [4]. After programming, the JTAG port on the programmable device should ideally be hidden or restricted from use so that the firmware can't be "dumped." Locking the JTAG port may be needed. Another option is to design the firmware itself so that it only works on valid boards [5], [6].

Board security analysis must consider the *interaction* of various chips on the board and their security protocols. Chips on the board should have security techniques that are compatible or at least don't interfere with each other. Even better would be chips whose security could be actively used together to provide a level of security that is greater than the sum of their parts. Ultimately, the security scheme must protect at least the high value targets on the board (e.g. the firmware and memory data on the board, including the chips that interface to them). Thus, metrics to consider include:

1. Security technique compatibility across chips

2. Security enhancement by combining chip techniques

3. Security strength (time and effort to break)

4. Security coverage of high-value targets

5. Overhead

V. GENERIC SECURITY FOR IEEE 1687

Many have proposed measures to protect internal scan chains and obfuscate data extracted from them (e.g. [7]–[11]). In addition, several recent papers have investigated the security of 1149.1-like test architectures, such as 1149.1, 1500, and the new IEEE standard 1687 (e.g. [3], [6], [12]–[16]). Figure 1 shows a hierarchical 1687 scan network that allows access to embedded instruments by opening and closing SIBs (Segment Insertion Bits). When a SIB is open, it allows access to a new segment of the scan network. When the SIB is closed, it bypasses that segment, making the overall scan path shorter. One advantage of 1687 is the ability to dynamically reconfigure the network with the data scanned through it instead of through the creation of new instructions in the JTAG instruction register. This provides easy access to hierarchically grouped sets of instruments, such as those in Figure 1.

Although pure IEEE 1687 is generally quite insecure due to the ability of an attacker to easily find and open all SIBs by walking 1's or 0's through the network, we have previously shown that security may be significantly enhanced through the use of "Locking SIBs" (LSIBs). Figure 2 shows a schematic of

a Pre-LSIB [3] (that inserts the new scan path before the shift bit). When a logic 1 is placed in the Update Register, the Select* line is asserted, which enables the scan chain elements in the new segment and allows the first mux to select the signal FROM_TDO2 as the shift data. An LSIB can be created from a normal SIB by gating the Update signal, UpDR, with a set of *key bits* (shown in purple). The key bits correspond to values in other predefined scan or update cells in the 1687 network. Opening an LSIB requires both clocking the correct value into the Update cell and scanning the correct data into the key bits in the chain. Trap bits may also be distributed in the chain so a hacker trying different scan values may accidentally set a bit that would make the Lock inoperable until system reset.

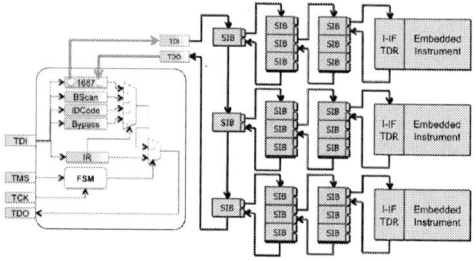

Fig. 1 Hierarchical 1687 network accessed with IEEE 1149.1 TAP. Accessing each instrument shown requires opening 3 SIBs.[3]

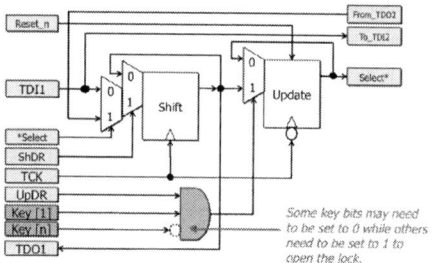

Fig. 2. Example schematic for a Locking SIB (LSIB) [3].

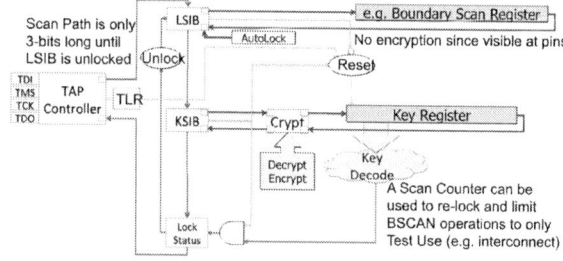

Fig. 3. Example boundary scan Lock with separate Key register

Figure 3 shows one generic technique for security, hiding registers and embedded instruments behind a Lock and providing Keys distributed in either the scan path or in hidden registers selected by other 1149.1 instructions, such as the KSIB. Data obfuscation may also be placed behind LSIBs to protect particular embedded instruments or key registers, as shown with the "Crypt" block behind the KSIB. Ideally, the data should be decrypted and then re-encrypted such that data coming into the chain is the same as the data going out when data is scanned straight through. This minimizes the attacker's

978-1-4799-7598-3/15 $31.00 © 2015 IEEE   60

knowledge of the actual data present in the chain internally. Depending on the instrument or chip designer's needs, encryption strength could vary from simple XOR's of scan data with multiple LFSRs to lightweight ciphers.

Such techniques cover many of the IC concerns: 1) the 1149.1 port is accessible and not disabled by a fuse; 2) instruments considered a security threat can be hidden behind a Lock or nested Locks, while items that are not security threats can be left publicly accessible; 3) the data in and out of the chip through TDI and TDO may be obfuscated with decryption-encryption and would only apply after a Lock has been opened; and 4) the known and public 1149.1 IR cannot directly invoke secure embedded instruments because all secure instruments would be behind Locks.

Locks and keys may be created from scan cells that would already be present in the scan chain—leading to negligible silicon overhead. Test time requires the time needed to scan-in and update a valid and known key to open each lock. Security strength is adjustable by increasing the number of Keys, by adding Traps, and by using multi-LSIB architectures [3], [16].

At the board level, similar techniques can be applied. For example, placing a gateway LSIB [6] at each chip's boundary scan register makes it more difficult for an attacker to use a JTAG test unit to bit-bang the interfaces and operate on-board chips that hold high-value data. Scan path linkers [17] can also be enhanced to provide security [18]. Traditionally, these are chips with a TAP controller that act like SIBs at the board level. Security functions, such as locks and encryption can be added to such a chip to help protect the JTAG ports of sensitive devices and restrict unauthorized access. By embedding the TAP traces between the scan path linker and the sensitive device under critical routes (such as the clock), the successful tampering of the board (such as by drilling to expose the buried traces), may be significantly reduced.

Anti-counterfeiting techniques at the board level can also benefit from locks—making it more difficult for an attacker to access electronic chip IDs or other identifying circuitry, making them more difficult to find and "spoof." Furthermore, on-chip monitors may be used to query other chips and reset the board when suspicious IDs or test behavior indicative of attacks are detected.

VI. AN IEEE 1687 TEST SECURITY STANDARD

The techniques summarized in the previous section provide an initial architectural toolbox for a system that uses IEEE 1149.1 and 1687 to provide access to embedded instruments within chips and to chips on boards. Such techniques could serve as some of the building blocks for a potential standard that aims to make security approaches compatible while still protecting internal security information and allowing implementation variability.

Standardization requires a toolbox and must consider:

- *What type of documentation should be used to specify security hardware so that software tools can be used to generate appropriate patterns for communication? Can a "secure" file format be created for encrypted file transfer?*
- *What kinds of databases are needed to store security*

configurations and keys and who should control access?

- *How should security be managed hierarchically when security techniques are present in 3rd party cores, instruments, chips, and boards? What information should be passed between levels? How much trust should be placed in users of licensed software at each level?*
- *What metrics should be used to quantify the security provided in different application environments? What types of encryption are really needed at different levels, for different applications, and for different circuit lifespans? Where does side channel analysis fit in?*

To help answer such questions, exploration of an IEEE standard for Test Security has been sanctioned by the TTSC at the 2014 meeting at ITC. Anyone interested in the joining a study group should contact Al Crouch at acrouch@asset-intertech.com.

VII. REFERENCES

[1] F. Koushanfar, "Integrated circuits metering for piracy protection and digital rights management: an overview," in *Proceedings of the 21st edition of the great lakes symposium on Great lakes symposium on VLSI*, Lausanne, Switzerland, 2011, pp. 449–454.

[2] G. E. Suh and S. Devadas, "Physical Unclonable Functions for Device Authentication and Secret Key Generation," in *Design Automation Conference, 2007. DAC '07. 44th ACM/IEEE*, 2007, pp. 9–14.

[3] Jennifer Dworak, Al Crouch, John Potter, Adam Zygmontowicz, and Micah Thornton, "Don't Forget to Lock your SIB: Hiding Instruments using P1687," in *Proceedings of the IEEE International Test Conference*, 2013.

[4] A. Moradi, A. Barenghi, T. Kasper, and C. Paar, "On the vulnerability of FPGA bitstream encryption against power analysis attacks: extracting keys from xilinx Virtex-II FPGAs," in *Proceedings of the 18th ACM conference on Computer and communications security*, Chicago, Illinois, USA, 2011, pp. 111–124.

[5] Catalin Baetoniu, "FPGA IFF Copy Protection Using Dallas Semiconductor/Maxim DS2432 Secure EEPROMs," Application Note XAPP780 (v1.1), May 2010.

[6] J Dworak, Z Conroy, A. Crouch, and J. Potter, "Board Security Enhancement using New Locking SIB-based Architectures," in *International Test Conference*, 2014.

[7] D. Mukhopadhyay, S. Banerjee, D. Roychowdhury, and B. B. Bhattacharya, "CryptoScan: A Secured Scan Chain Architecture," *Test Symp. 2005 Proc. 14th Asian*, pp. 348–353, 18.

[8] J. Lee, M. Tehranipoor, C. Patel, and J. Plusquellic, "Securing Scan Design Using Lock and Key Technique," *Defect Fault Toler. VLSI Syst. 2005 DFT 2005 20th IEEE Int. Symp. On*, pp. 51–62, 5.

[9] R. Nara, H. Atobe, Youhua Shi, N. Togawa, M. Yanagisawa, and T. Ohtsuki, "State-dependent changeable scan architecture against scan-based side channel attacks," *Circuits Syst. ISCAS Proc. 2010 IEEE Int. Symp. On*, pp. 1867–1870, May 2010.

[10] G. Sengar, D. Mukhopadhyay, and D. R. Chowdhury, "Secured Flipped Scan-Chain Model for Crypto-Architecture," *Comput.-Aided Des. Integr. Circuits Syst. IEEE Trans. On*, vol. 26, no. 11, pp. 2080–2084, Nov. 2007.

[11] Y. Atobe, Youhua Shi, M. Yanagisawa, and N. Togawa, "Dynamically changeable secure scan architecture against scan-based side channel attack," *SoC Des. Conf. ISOCC 2012 Int.*, pp. 155–158, 4.

[12] R. F. Buskey and B. B. Frosik, "Protected JTAG," *Parallel Process. Workshop 2006 ICPP 2006 Workshop 2006 Int. Conf. On*, p. 8 pp.–414, 2006.

[13] C. J. Clark, "Anti-tamper JTAG TAP design enables DRM to JTAG registers and P1687 on-chip instruments," *Hardw.-Oriented Secur. Trust HOST 2010 IEEE Int. Symp. On*, pp. 19–24, 2010.

[14] K. Rosenfeld and R. Karri, "Attacks and Defenses for JTAG," *Des. Test Comput. IEEE*, vol. 27, no. 1, pp. 36–47, Feb. 2010.

[15] L. Pierce and S. Tragoudas, "Enhanced Secure Architecture for Joint Action Test Group Systems," *Very Large Scale Integr. VLSI Syst. IEEE Trans. On*, vol. PP, no. 99, pp. 1–1, 2012.

[16] Adam Zygmontowicz, Jennifer Dworak, Al Crouch, and John Potter, "Making it Harder to Unlock an LSIB: Honeytraps and Misdirection in a P1687 Network," in *Design, Automation & Test in Europe Conference*, 2014.

[17] Lattice Semiconductor Corporation, "Using Multiple Boundary Scan Port Linker (BSCAN2)--Application Note AN8081." Jul-2009.

[18] Al Crouch and Jennifer Dworak, "JTAG | IJTAG Semiconductor and Board Test Security," http://www.asset-intertech.com/eResources/eBooks-Chip-Debug/JTAG-IJTAG-Semiconductor-and-Board-Test-Security, 2014.

Enabling Unauthorized RF Transmission below Noise Floor with no Detectable Impact on Primary Communication Performance

Doohwang Chang, Bertan Bakkaloglu, and Sule Ozev
School of Electrical, Computer, and Energy Engineering
Arizona State University

Abstract— **With increasing diversity of supply chains from design to delivery, there is an increasing risk of unauthorized changes within an IC. One of the motivations for this type change is to learn important information (such as encryption keys, spreading codes) from the hardware and pass this information to a malicious party through wireless means. In order to evade detection, such unauthorized communication can be hidden within legitimate bursts of transmit signal. In this paper, we present a stealth circuit for unauthorized transmissions which can be hidden within the legitimate signal. A CDMA-based spread spectrum with a CDMA encoder is implemented with a handful of transistors. We show that the unauthorized signal does not alter the circuit performance while being easily detectable by the malicious receiver.**

I. INTRODUCTION

Increasing diversity in the supply chain in the manufacturing of an integrated circuit also increases the risk of unwanted changes to the hardware structures. Of particular interest is the insertion of hardware Trojans that are small scale circuit structures that perform a malicious function not intended by the designer of the system [1]. These Trojans can be inserted at multiple points in the supply chain; for instance, by the production foundry or by an IP provider. In order to evade detection, these Trojans have a small footprint both in terms of die-size and in terms of performance impact. A secure supply chain therefore needs mechanisms to verify that the manufactured devices do not contain any unintended functionality designed for malicious purposes [1].

Unfortunately, detection of such carefully hidden Trojans is not straightforward as the intent is to hide their behavior during production testing [2]. For digital circuits, enhanced functional testing or additional parametric measurements can be used to detect anomalous behavior [3-5]. The challenge for digital circuits is the obfuscation that is introduced by process variations and the ability to hide the effects of the Trojan within these variations. Various techniques in variance reduction for parametric testing of digital circuits can be used for Trojan detection [5-8]. More sophisticated training tools can be used if Trojan-free samples of the circuit are available [2][7-8].

While Trojans can be inserted for multiple malicious reasons, one such motivation is to spoof data out of the legitimate functionality of the device and pass this information to an unauthorized party. For instance, stealing encryption keys has been one of the most researched Trojan intentions

[2][8-9]. The easiest form of communication between the Trojan and the attacker is through the use of an existing wireless transceiver on the chip since the medium of transmission is open to anyone. Once information is leaked out of the device, the attacker can simply listen on the designated frequency band to detect the transmitted signal, demodulate, and decode it [2]. It is rather difficult to monitor and detect such unauthorized transmissions as the performance of the original circuit is seldom affected and the effect of the Trojan can be hidden in the process variations. In this scenario, existing specification-based techniques will fail in detecting the Trojan even if it is active during the production test process.

In this work, we show that hardware Trojans can hide unauthorized transmit signals within the ambient noise floor of authorized transmissions while retaining the ability to detect and decode the information thanks to spread-spectrum techniques. We demonstrate a simple 0.18μm CMOS BPSK modulator circuit which generates Trojan transmit signals that can be added to the legitimate transmit signal using a simple low-gain directional coupler. We show that this signal does not alter the performance of the circuit, while being detectable by the malicious party.

II. SPREAD SPECTRUM COMMUNICATION TECHNIQUES

Spread Spectrum (SS) techniques were originally developed to achieve secure communications by spreading the modulated signal over a much wider frequency band. There are several benefits of this spreading, including multiplexing of high number of users, resistance to jamming and interferers,

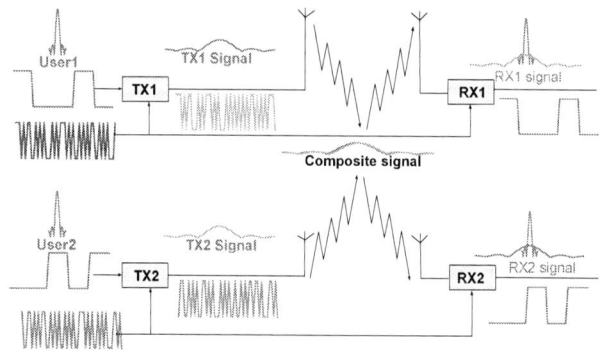

Figure 1: Spread-spectrum communications principle

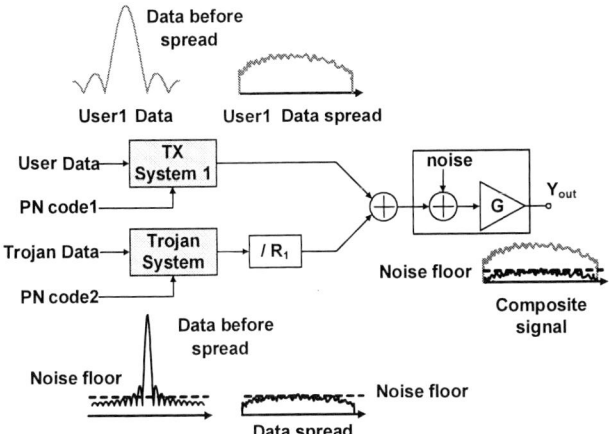

Figure 2: Generic system-level model for using spread spectrum techniques to evade detection for wireless hardware Trojans

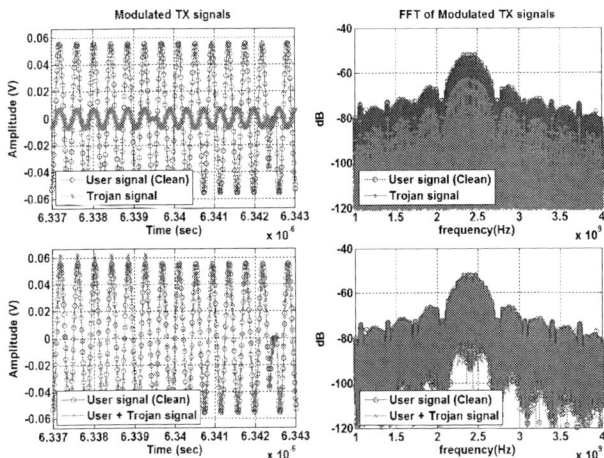

Figure 3: Time domain and Frequency domain representation of the authorized User transmission and the Trojan transmission

and multi-path fading.

Several state of the art communication and geo-location services use this principle for obtaining a robust communication link, including GPS, 4G mobile telecommunications, W-LAN, and Bluetooth [10].

To generate a spread-spectrum modulated signal, the baseband signal bandwidth is intentionally spread over a larger bandwidth by multiplying it with a higher-frequency spreading code or sequence. The ratio between the spread bandwidth and the original bandwidth is called processing gain, which represents the effective reduction in the necessary signal-to-noise ratio (SNR) to effectively decode the signal. In the receiver, the received signal is demodulated by using the same sequence, de-spreading the signal back into its original form.

Figure 1 shows the principle of a spread-spectrum transmitter/receiver chain. The lower-rate data is multiplied with a higher rate SS code to generate a higher rate sequence. This higher bitrate digital sequence is transmitted over the medium, which may include multiple interferers, multi-path fading, or multiple users in the same channel. At the receiver, both the intended signal and interferers are mixed with the same SS code, de-spreading the original information, and spreading the interferers instead. Spectrum spreading techniques are also particularly useful for hiding unauthorized transmissions. Since the bitrates of such malicious transitions may be relatively low, the occupied bandwidth is also very narrow. By spreading this ultra-narrow band signal over a much larger bandwidth, effective communication can be attained by extremely low signal power levels. In this way, even in the presence of a strong interferer in the same channel (i.e. the authorized transmission), the unauthorized transmission can be detected and decoded by the malicious attacker, unknown to the user.

This spreading has two advantages from the perspective of the malicious attacker: (a) the performance of the authorized

transmission is not affected in any adverse way, thereby evading any performance-based testing or monitoring, and (b) its effect on enhanced measurements, such as supply current, is negligible, thereby evading most proposed detection techniques for hardware Trojans.

III. GENERIC SYSTEM MODEL

We present a generic system model that uses spread spectrum techniques to evade detection of an unauthorized transmission. The underlying principles of this system are as follows:

(a)   The Trojan transmits a very low number of bits per transmit burst to achieve the highest processing gain thereby communicating with extremely low power levels.

(b)   The Trojan takes advantage of the existing training and channel equalization techniques to enable coherent demodulation.

(c)   The processing gain of the Trojan signal is limited by the maximum chip-rate and maximum baseband bandwidth of the original circuit due to multiple layers of filtering in the channel and in the band.

(d)   The resolution of the transmit digital-to-analog converter in the wireless transmitter is set by the original design.

Figure 2 shows the block diagram of such a system that uses the CDMA coding techniques. The Trojan bitrate is much lower compared to the legitimate user. The original user may use the same coding technique or another spectrum spreading technique, which is more suitable for higher data rates. The Trojan attenuates the signal by a ratio R_1 that can be determined by the processing gain. Hence, the signal power is pushed below the ambient noise level, but above the quantization noise level of the transmit DAC since this would hamper demodulation process for the Trojan receiver. The two signals are added in the analog domain. While the signals can

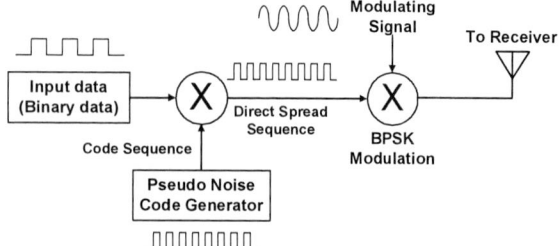

Figure 4: Architecture of CDMA BPSK transmitter system

also be added in the digital domain, the purpose of the work is not to optimize the Trojan circuit; rather to show that such stealth communication is feasible. The composite signal is transmitted with an unknown gain (due to process variations and path variations), where additional noise is added.

A. Performance Assessment

We demonstrate that the Trojan can modify the transmit signal marginally to communicate effectively with a malicious attacker who uses a receiver with the same spreading code. First, we set the Trojan bit rate to $1/40^{th}$ of the bit rate of the legitimate transmission to transmit 3 bits per transmit burst. Figure 3 shows the time domain information for the authorized User and the Trojan signals. The transmit signal for the Trojan is attenuated by a factor of 20 (20dB reduction) and added to the authorized transmit signal. Figure 3 shows the spectrum of the modulated TX signals with (red) and without (blue) the Trojan signal after white noise with an SNR of 20dB is added. These two spectra are practically identical, and the minute variations can easily be attributed to environment noise and process variations.

Performance evaluation of the transmitter shows that the Trojan signal results in no bit errors for the authorized transmission and the waveform quality factor is within acceptable limits for CDMA systems. This is due to the fact that the Trojan signal SNR is less than -15dB, hence the information is hidden in the noise level of acceptable transmission. Yet, when the Trojan receiver de-spreads the information using the same SS-code, it can recover the bit sequence correctly. Hence, this unauthorized transmission, which can be detected and demodulated by the malicious attacker, will not result in noticeable performance degradation for the transmitter and will evade performance testing during production testing.

IV. CDMA TRANSMITTER DESIGN

In this section, we use a CDMA-based spread spectrum system with a binary phase shift keying (BPSK) modulator to generate Trojan signal. Our goal here is to show that this can be achieved with a relatively small circuit. It should be noted that the Trojan circuit is not optimized for compactness. The block diagram of the Trojan circuit is given in Figure 4.

A Pseudo-random Noise (PN) code is a binary sequence that exhibits randomness properties but has a finite length, and

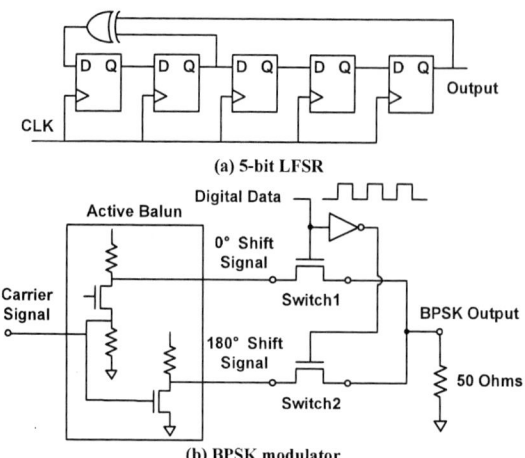

Figure 5: Block diagram of (a) LFSR and (b) BPSK modulator

can be used as SS code sequence in CDMA system. The PN code generator can be designed using N-bit linear feedback shift register (LFSR). LFSR consists of N-stages shift registers, normally D-flip flops and XOR gates. Spreader device consists of an XOR gate which simply performs logical operation between input data and PN code sequences. The signal spreader is generating PN spread sequences when the input data bit is '1' and the complement of the PN spread sequence if the input data bit is '0'. The spread bit sequences are used for modulation by BPSK modulator.

The phase shift keying is a digital modulation scheme that transmits data by changing or modulating the phase of a carrier signal. BPSK uses a finite (two) number of phases that the phase of carrier sinusoidal signal changes abruptly by 180° for every transition of modulating binary sequence (spread bit sequence). The block diagram of BPSK modulator is shown in Figure 5. It consists of an active balun circuit using common-gate and common-source pair that converts the carrier from single-ended to differential, and two complementary MOSFETs switches that are controlled by the baseband data sequence (spread bit sequence) [11]. The output from the common-source device provides the 180° phase shift and 0° phase shift can be achieved from the common-gate device. Finally, 0° phase signal is passing to the load when the voltage of the switch gate is high (a bit '1') and 180° phase signal is connected to the load during a bit '0', thus achieving BPSK modulation [12].

As mentioned before, Trojan signal generator circuit should not be detectable to the chip designer, and thus it has to be implemented with a very compact layout area. Furthermore, the Trojan signal has to be delivered to without physical connections at sensitive nodes to avoid detection. One such delivery mechanism is using a directional coupler. Directional coupler has the additional advantage of a natural attenuation in the signal and can be placed at a safe distance from the primary signal line to avoid detection.

V. EXPERIMENTAL RESULTS

The transceiver of a CDMA-based spread spectrum system is emulated using a transistor-level design kit and software (MATLAB). The Trojan signal is generated using transistor-level simulations, whereas the coding, processing, and detection are achieved by using a MATLAB/Simulink model of the receiver. 5-bit LFSR and CMOS BPSK modulator circuits are implemented using a 0.18μm CMOS process. The total circuit layout area for both LFSR and BPSK circuit is 400μm x 400μm which is less than 0.4% of the total area of a single transceiver system chip (41mm^2) [13]. The sub-blocks of the CDMA BPSK transmitter are designed to generate the transient signals of both 120 bits authorized and 3 bits Trojan data with pseudo random bit sequences as spread bit sequences. The User data begins with a rate of 10.75Mbps (268.8kbps for Trojan) and then it is spread to a rate of 333Mbps. These spread-spectrum modulated signals with 2.4GHz carrier frequency are applied to a CMOS BPSK modulator to provide the actual signal pattern of CDMA system. Random noise is added with 20dB SNR for the User transmit signal, which puts the Trojan transmit signal at a negative SNR.

As shown in Figure 6, the User data and Trojan data can be demodulated correctly at the receiver systems with 20dB suppression in Trojan signal. Finally, the Trojan signal SNR is less than -15dB, hence this signal can be hidden in the noise level of transmitted signal. When the Trojan receiver de-spreads the received signal using the same SS-code in the transmitter side, it can recover the baseband signal correctly. Furthermore, the authorized signal (User data) can be reconstructed without bit errors at the receiver. Table I shows the summary of Bit error rate (BER) of 100 random User and Trojan data at the receiver system with various random noise (SNR) at the transmitter system. Therefore, this unauthorized signal (Trojan signal), which cannot be undetected through

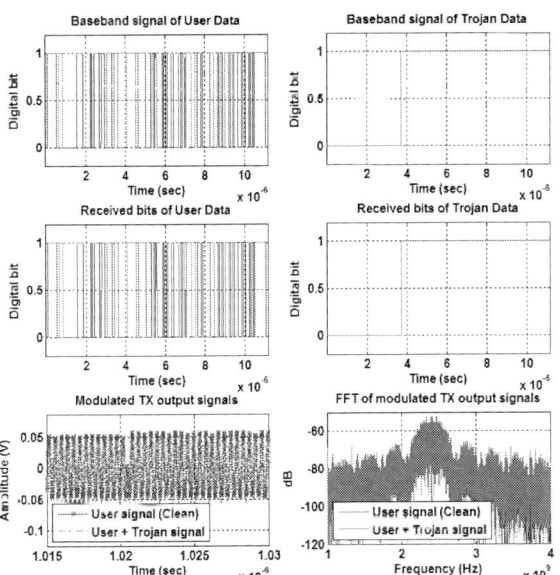

Figure 6: Transmitted and received baseband signals

TABLE I
BIT ERROR RATE OF USER AND TROJAN DATA AT RECEIVER

BER		SNR (dB)					
		-30	-20	-10	0	10	20
TX output without Trojan	User Data (120bits)	0.40	0.39	0.29	0.10	0	0
TX output with Trojan	User Data (120bits)	0.39	0.38	0.31	0.12	0	0
	Trojan Data (3 bits)	0.57	0.49	0.43	0.36	0	0

performance testing can be demodulated correctly by the Trojan user.

VI. CONCLUSIONS

In this paper, we show that with relaxed requirements on the bit rate and using spread-spectrum techniques, it would be possible to hide the Trojan transmission within the noise level and avoid altering the performance of the primary circuit. Furthermore, Trojan data can be successfully recovered when the Trojan receiver de-spreads the received signal using the same PN code sequence from the Trojan transmitter. The undetectable Trojan transmitter is implemented with a compact CMOS BPSK modulator and LFSR to provide CDMA-based spread spectrum code sequences. For each of the 100 random simulations, the Trojan signal results in no performance degradation for the primary transmission while being correctly decoded by the malicious receiver.

REFERENCES

[1] Rad, R.M. et al. "Power supply signal calibration techniques for improving detection resolution to hardware Trojans," *IEEE ICCAD*, 2008.

[2] Jin, Yier, and Yiorgos Makris. "Hardware Trojans in wireless cryptographic ICs," *Design & Test of Computers, IEEE*, 2010, pp.26-35.

[3] F. Wolff, C. Papachristou, S. Bhunia, and R. S. Chakraborty, "Towards Trojan-free trusted ICs: Problem analysis and detection scheme," *IEEE DATE*, 2008.

[4] R. S. Chakraborty and S. Bhunia, "Security against hardware Trojan through a novel application of design obfuscation," *IEEE/ICCAD*, 2009.

[5] R. Rad, J. Plusquellic, and M. Tehranipoor, "Sensitivity analysis to hardware Trojans using power supply transient signals," *IEEE HOST*, 2008.

[6] Baktir, S., et al. "Detection of Trojans in integrated circuits," *IEEE International Symposium on INISTA*, 2012.

[7] Gwon, Youngjune, et al. "Statistical screening for IC Trojan detection," *IEEE ISCAS*, 2012.

[8] Jin, Yier, Dzmitry Maliuk, and Yiorgos Makris. "Post-deployment trust evaluation in wireless cryptographic ICs," *IEEE DATE*, 2012.

[9] Hély, David, et al. "Malicious key emission via hardware Trojan against encryption system," *IEEE ICCD*, 2012.

[10] A. J. Viterbi, "Spread Spectrum Communications: Myths and Realities," *IEEE Communications Magazine*, vol.17, no.3, May 1979.

[11] Tokumitsu, T., et al, "A K-band bi-phase modulator MMIC for UWB application," *Microwave and Wireless Components Letters, IEEE*, vol.15, no.3, pp.159,161, March 2005.

[12] Jackson, B.R.; You Zheng; Saavedra, C.E., "A CMOS Direct-Digital BPSK Modulator Using an Active Balun and Common-Gate Switches," *IEEE ISCAS*, 2007.

[13] Mehta, S., et al, "An 802.11g WLAN SoC," *Solid-State Circuits,IEEE Journal of* , vol.40, no.12, pp.2483,2491, Dec. 2005.

Innovative Practices Session 2C:
New Technologies, New Challenges – 2

Organizer: TM Mak (Globalfoundries)
Moderator: Suraj Sindia (Intel)

Presenters & Abstracts:

Presenter #1: *Jon Colburn (NVIDIA)*

Title: Test Challenges for Complex 2.5D SOC Designs

Abstract: As the economics of traditional devices scaling changes, alternative solutions to increase transistor counts in semiconductor packages are being explored, including various multi-die integration techniques. These solutions include 3D die stacking, 2.5D with dies sitting side-by-side on substrate, Package-on-Package (PoP), System in Package (SiP), etc. In particular, 2.5D and 3D device integration may create new challenges and old challenges seen in multi-chip module (MCM) manufacturing also re-appear to affect new users. What are the specific test challenges and how can Design-for-Test (DfT) solve them?

Presenter #2: *Sandeep Goel (TSMC)*

Title: 3D IC design and test challenges: System architect perspective

Presenter #3: *Raghunandan Chaware (Xilinx)*

Title: Yield challenges in stacked silicon interconnects technology and methods to achieve high yields

Abstract: Stacked die technology enables high bandwidth connectivity between multiple dies by providing significantly large number of connections via microbumps. This interposer based die stacking approach provides low power and latency, but also adds significant manufacturing complexity. This presentation discusses key challenges observed during manufacturing of 28nm 3DIC products with chip-on-wafer-first process. During the initial product ramp stage, most of the failures observed were related to interposer level assembly process. Common failure modes were microbump opens, interposer metal line opens and shorts, interposer metal line shorts and TSV to C4 opens. This presentation will provide a brief summary of these failures and the interconnection yield improvement process that involves a close loop feedback process of failure analysis and inline inspection. Another unique failure mode is transistor damage caused by stacked die assembly. In order to identify the root cause and isolate this transistor degradation problem, different assembly process splits and process corner studies were performed. A C4 probe card was designed to provide an intermediate test point at a major process loop after wafer level die assembly and before flip chip assembly of the stacked die on the organic package.

978-1-4799-7598-3/15 $31.00 © 2015 IEEE

2015 IEEE 33rd VLSI Test Symposium (VTS)

Extracting Effective Functional Tests from Commercial Programs

Sreekumar Vadakke Kodakara, Mehul V. Sagar, Joel Yuen

Intel® Corporation

sreekumar.v.kodakara@intel.com, mehul.v.sagar@intel.com, joel.yuen@intel.com

Abstract—**We describe a tool and methodology for extracting short and effective functional tests from long running commercial programs and manufacturing system tests for testing microprocessors and SOCs. The tool combines fast Instruction Set Architecture (ISA) simulator and Design for Test (DFT) capabilities of the microprocessor to enable tracing of long running workloads. The trace is then converted into short functional test programs that can be replayed back in silicon. The tool can extract test programs from BIOS, operating systems, application programs and long running manufacturing system test programs. Using data from silicon experiments on recent microprocessor products, we show that the short tests extracted with our tool was able to screen the defective units as effectively as the original long running application with 6X to 15X reduction in test time.**

Keywords—hvm, software, architecture, microprocessor, functional tests

I. INTRODUCTION

Modern high performance microprocessors are fabricated in the latest process technologies. The small device sizes of these process technologies combined with very high integration densities can magnify the effect of defects. Tester patterns may not be able to screen all the different types of defects that can occur in the processor. System level functional test programs are known to be effective in screening defects and are used in addition to wafer level tests and tester patterns. System level functional test programs are assembly code that is designed to stress the different features of the microprocessor such as cache, out-of-order pipeline and branch predictors. BIOS, operating systems, commercial applications, benchmarks and custom assembly tests are examples of functional test programs that can be used to screen defects.

Although functional test programs are effective in screening defects, they can also significantly increase test time. These programs are not designed to run within the test time constraints of High Volume Manufacturing (HVM) environment. For example, operating systems can take anywhere between a few seconds to several minutes to boot up and complete the initialization process. Application programs such as benchmark programs and post-silicon tests can take several minutes to complete their execution. The long execution times of these programs can significantly increase the test cost. Another problem when working with commercial programs such as OS and application programs is the variability of their behavior from run-to-run. For example, modern operating systems implement Address Space Layout Randomizations (ASLR) for enhanced

security [23]. Due to ASLR the page table structures and virtual memory mappings are different from one boot to another and from one process invocation to another. This variability makes it difficult to debug and root cause the defect.

One way to shorten the runtime of system level functional tests and commercial programs is to extract multiple short sequences of instructions from different regions of the original program and convert them into short functional tests. If these short tests are as effective as the original test program to screen defective units, then we have a way to reduce test time while still maintaining the fidelity of the original test program to screen defects. Furthermore, when converting the original program into short functional tests, we can remove the variability caused by ASLR, by fixing the page table mappings, making the debug process predictable. It also paves a way to apply standard tools and methodologies to create test patterns from these test programs.

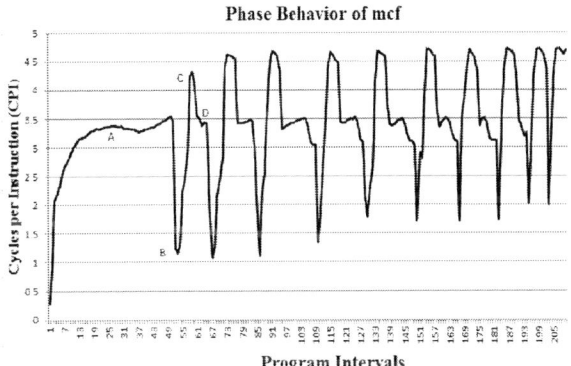

Figure 1: Phase behavior of mcf, a benchmark program from the SPEC CPU2000 benchmark suite. The graph shows how the performance of the program varies during the execution of the program. Each point in the graph represents the performance of the program in a 1billion instruction interval. The x-axis is the interval count and y-axis is the average number of cpu cycles required to execute one instruction (CPI). The label A, B, C and D identifies regions of stable performance behavior. These regions maps to different nested loops in the programs.

Short tests extracted from the original program can be effective in screening defective units because software programs are known to exhibit a phased behavior during their execution

978-1-4799-7598-3/15 $31.00 © 2015 IEEE   67

[11]. A phase in a software program is a region of execution where the program exhibits stable performance behavior [11]. Figure 1 shows the phased behavior of *mcf*, a benchmark program from the SPEC CPU2000 benchmark suite [16]. The y-axis is the Cycles-Per-Instruction (CPI), which is the average number of cpu clock cycles required to execute one instruction and x-axis plots the execution time of the program split into 1 billion instruction intervals. The larger the CPI number, the lower the performance of the microprocessor. *mcf* shows four distinct phases of execution, which maps to different loops in the source code of *mcf*. Each loop in the program exhibit different cache and pipeline behavior resulting in different performance numbers. Phase behavior of programs has been exploited for performance optimizations and to speedup microarchitecture simulation [12, 13].

We propose that functional test programs exhibit phased behavior in the context of screening defects. There are regions in the code that are effective in finding defects in the processor while other regions are not so effective. We believe that creating small tests by sampling different regions of a long running program that are effective in exposing defects in the processor would give us the best of both worlds, namely shorten the test time while still maintain their defect screening fidelity. Our results and observations from experiments conducted on silicon supports our theory.

We describe an elegant tool and a methodology that combines fast architecture simulator with the Design for Test (DFT) capabilities built into the microprocessor to trace long running commercial programs. We then convert this trace into valid system level functional test. We use this tool and methodology to automatically extract short functional tests from commercial and long running test programs to test microprocessors. Using data from silicon experiments on a recent microprocessor product we show that the test programs extracted using our method are as effective as the original long running test programs while providing a 6X-15X reduction in test time.

Prior work on system level functional tests was done on generating random or directed tests to target defects that can occur in the microprocessor. In [1] native mode random instruction-based testing from the cache was discussed and was applied to real microprocessors through the FRITS methodology [2]. In [3, 4, 5, 6] functional tests targeting stuck at faults were generated. Tests targeting delay faults were discussed in [7, 8, 9] but were restricted to simple processor cores. In [10] tests were generated to perform marginality validation of silicon prior to manufacturing. In all the above works, emphasis was on creating new test content targeting different types of faults that can occur in the processor. In [22] additional hardware was added to the processor to trace the execution of instruction for post-silicon debug and algorithms were introduced to localize bugs with the collected data. Our work is focused on creating short functional tests for test time reduction without any additional hardware support other than what is already present in the microprocessor. To the best of our knowledge this is the first work to propose a tool and methodology to generate short and effective system level functional tests from commercial and long running programs, and to provide results from silicon experiments on recent microprocessor products.

The rest of the paper is organized as follows. In section 2 we describe the background on ISA simulators and In-Target Probe [14] used in our tool; section 3 describes our tool and methodology for extracting tests; section 4 describes the results from silicon experiments; and section 5 concludes the paper.

II. BACKGROUND

We will give a brief description of two important components of our tool flow in this section, namely a fast simulator and In-Target Probe [14], a tool used in silicon debug.

A. Instruction Set Architecture Simulator

The Instruction Set Architecture (ISA) of the processor describes the details of the instructions and features in the processor that the system software programmer and compiler writer use to program the processor [21, 17]. It includes details about the opcode and operands of different instructions, the different modes of execution of the processor, the details of virtual memory and paging support, and other hardware feature that the processor implements such as virtualization.

An ISA simulator is a software program that implements all the features of the architecture in enough detail to execute assembly code. The simulator takes as input the assembled binary file and executes the code. The simulator, unlike silicon, gives complete control to the user to inspect and modify the state of the processor during the execution of the code, including useful debug capabilities such as reverse execution.

ISA simulators have many different uses. They are used in a virtual platform to simulate a complete computer including the motherboard [18]. The virtual platform is used to develop the software for the new processor before the first silicon is available. Architecture simulators are also used extensively in pre-silicon simulation based validation environment to check the correctness of the RTL of the microprocessor design [19]. In our tool we use an ISA simulator that is fast enough to run real software applications. We use the simulator in our tool to trace the memory addresses accessed by the program during its execution.

B. ITP – In-Target Probe

In-Target Probe (ITP) [14] is a hardware debugger unit developed at Intel®. ITP is connected to the JTAG port of the processor and is primarily used as a silicon debug tool. Some features of ITP include the ability to set breakpoints to halt the processor when different events are triggered and to inspect/change the state of the processor including the registers, memory and model specific registers (MSRs). We use ITP to halt the processor at interesting places during the execution of the program and read the register and memory state of the system. In the next section we describe how we combine an ISA simulator with ITP to create effective functional tests.

978-1-4799-7598-3/15 $31.00 © 2015 IEEE

III. TOOL AND METHODOLOGY

There are two parts to the problem of generating short and effective functional tests from long running programs. (a) Identify the region of execution of the program that is effective in detecting defects (b) Extract the region of execution into a functional test program. In this section we will describe in detail how our tool and methodology will address the two parts of the problem.

A. Identify program regions effective in finding defects:

We analyze the runtime profile of the program to identify regions in the program that are effective in finding defects. We profile the program using performance monitoring hardware present in the microprocessor. A performance monitoring hardware (*perfmon*) is a hardware unit in the microprocessor that has many programmable counters in it [20]. The counters can be programmed by the user to count interesting events that can happen in the processor during the execution of a program. For example, *perfmon* can be programmed to count the number of instructions retired, the number of cache misses at different cache levels, the number of pipeline stalls and branch mis-predictions. *perfmon* is present in most modern microprocessors and is used extensively in performance studies of software programs [15].

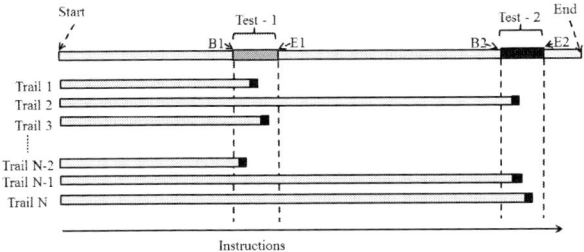

Figure 2: We illustrate the profiling step in this figure. The first line represents the complete run of a program in a good unit. The labels 'Start' and 'End' mark the beginning and end of the program respectively. The other lines represent different trial runs of the program in different bad units. The trial runs do not complete and stop at different points during the execution of the program marked with the symbol ■. We take the instruction counts at which the program stops and use that to find out the regions of the program from which to extract small tests. Regions marked 'Test-1' and 'Test-2' illustrates two small tests extracted from the program.

The profiling step is illustrated in Figure 2. We count the number of retired instructions from the beginning of execution of the program to profile the program. We boot the defective unit, and setup the counters in the *perfmon* to count the number of instructions retired by the processor. We also setup a breakpoint to halt the processor when a failure is detected. Next, we run the long running program on the system. The *perfmon* counters will count the number of instructions retired in the pipeline, and the processor will halt at the breakpoint when the failure is detected. The performance counter value tells us the instruction count at

which the defect was observed. We repeat the profiling step over multiple runs of the program and over multiple defective units. The instruction counts from these runs over multiple units give us enough information about the region in the program that exposes the defect. In our experience, the instruction count cluster together in groups, as illustrated in Figure 2, ie., there are few regions in the program's execution where defects are observed.

At the end of the profiling step, we know the instruction counts at which the defects were detected by the program. In Figure 2, this instruction count is marked with the symbol ■ at the end of each trial. We now need to determine the instruction counts at which to extract the tests. In Figure 2, the instruction counts to extract the tests are marked as B1, E1 and B2, E2 for Test-1 and Test-2 respectively. The start of the extracted tests cannot be very close to where the defects were found. If we start the test too close to the instruction where the defects were found, the pipeline and cache may not be sufficiently initialized like the original program run to cause the defect to occur. Furthermore, the start of the test cannot be too far from the point when the defect was observed as it will increase the test time.

We use engineering judgment to determine the instruction counts at which to start and end each extracted test. For example, we found that extracting 100 million instruction tests around the point at which the defect was observed was sufficient to create high quality short functional tests. These test lengths were sufficient to initialize the cache and pipeline structures to cause the defect to occur, while short enough to execute quickly. In Figure 2, the regions marked Test-1 and Test-2 will be extracted into two different test programs.

B. Extracting the test, an example:

```
Initialize:
  mov eax, 0x0     #Initialize eax to 0x0
  mov ebx,0xaaaa   #Initialize ebx with read address
  mov ecx, 0x100   #Initialize counter ecx to 0x100
begin:
  mov edx, [ebx]   #mov data from address [ebx] to edx
  add eax, edx     #Add edx to eax and store in eax
  add ebx, 0x4     #increment ebx to next address
  dec ecx          #decrement the counter
  jnz begin        #jump to begin if counter !0
end:
  mov ebx,0xbbbb   #Initialize ebx with write address
  mov [ebx], eax   #Store result eax to address [ebx]
```

Figure 3: An example assembly code. We need to capture the following to extract the code between labels begin and end (a) the memory region where the code is stored, (b) the values of registers ebx, eax and ecx (c) and the values stored in address 0xaaaa to 0xaaaa+0x100 (d) any additional register state such as control registers, flag registers and model specific register values (MSRs) (e) the page translation tables from memory for code and data.

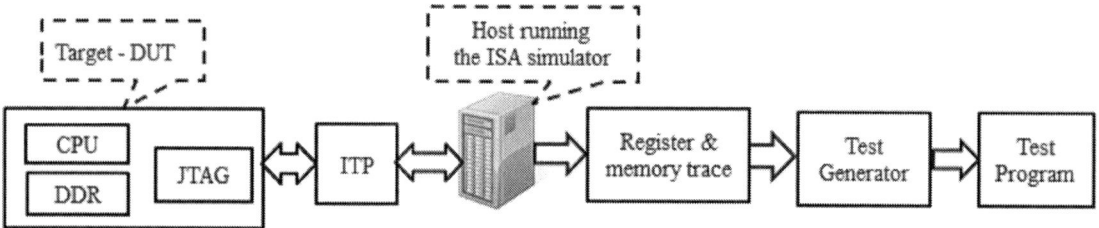

Figure 4: Block diagram depicting the architecture of the tool. There are three main components in our tool and methodology. The DUT is the microprocessor under test which runs the long running system test. The ITP is the In-Target Probe hardware unit. The host is a computer which runs the ISA simulator. The host is connected to the DUT via the ITP and has complete control of the DUT. The simulator will generate a register and memory trace which is fed to a test generator. The test generator will add a prolog and epilog to the trace and create a valid test program

In Figure 3 we show a sample assembly code. Registers eax, ebx and ecx are initialized in the block of code marked with label `Initialize`. The assembly codes between labels `begin` and `end` is a loop and it does the following. It loads the value present in address ebx to register edx, adds it to the value present in eax. The counter present in ecx is decremented and the address present in ebx is incremented. The code checks to see if the counter ecx has reached 0x0. If not it jumps back to the beginning of the loop. If ecx reaches 0x0, the address to store the result 0xbbbb is loaded to ebx and the value of eax is stored in that address.

Let us assume that we have identified the loop marked with labels `begin` and `end` to be the code that we want to extract and convert into a test. For the extracted test to work correctly, we need to setup the state of the registers and memory values that the snippet depends on. In this case, the assembly code in the loop depends on the following (a) the memory region where the code is stored, (b) the values of registers ebx, eax and ecx (c) and the values stored in address 0xaaaa to 0xaaaa+0x100 (d) any additional register state such as control registers, flag registers and model specific register values (MSRs) (e) the memory corresponding to the page translation tables for code and data. In real programs the state of the extracted code will depend on the values stored in general purpose registers, control registers, model specific registers, and the working memory set of the process. The working memory set of the process includes the entire memory region that the program depends on, such as the code, the data and the page tables stored in memory.

Determining the working memory set is non-trivial due to the different modes in which the processor can execute the program. The virtual memory management and address translation schemes are very different for each mode. For example, in real-mode, a simple segmented memory model is used for virtual memory implementation, while in protected mode a complicated 3 level paging implementation is used to perform memory management. There are many enhancements to the memory model in recent years to support extended page tables and larger memory sizes.

C. Our solution: Combine Architecture simulator with ITP

The goal of the trace capture tool is to capture the register state, MSR's and working memory set of the program. We combine an architecture simulator with ITP to extract these states. The high level architecture of the state capture tool is given in Figure 4. The setup consists of two systems, one is the target system and other is the host system. The target system is the device under test (DUT) and it will have the new microprocessor which runs the program that we want to extract the test from. The ITP hardware unit is connected to the JTAG port of the target system and to the USB port of the host system. The architecture simulator is running on the host system and is connected to the target system via the API's exposed by ITP.

Let us assume that we have halted the target at the beginning of the region where we want to extract test from. We start the simulator in the host. On startup, the simulator will copy the register state, namely the general purpose registers such as RIP, RAX, RBX etc, flag registers and segment registers from the target using the API's exposed by ITP. We do not copy the model specific registers or the memory to the simulator. We cannot capture the MSR when the tool is started as there are many model specific registers that should not be accessed in specific mode of execution of the processor. If accessed, it can result in the target system to hang. Similarly, we do not copy the memory from the target to the simulator because when we start tracing we do not know the working memory set of the test program. To summarize, we copy only the register state from silicon to the simulator on startup.

To understand how the simulator can capture the MSR and working memory set of the program, first we need to understand what happens when we execute one instruction in the simulator. When we execute one instruction in the simulator, the simulator will use the RIP and segment registers to calculate the addresses of the instruction bytes and fetch the bytes. This translates to a sequence of memory reads. We use the API's exposed by ITP to read the memory of the target system for the addresses requested by the simulator. In other words, we will lazily fetch the memory from the system as and when the memory value is requested by

978-1-4799-7598-3/15 $31.00 © 2015 IEEE

the simulator. A similar mechanism is also implemented for MSR. With this mechanism in place, the working memory set of the test program and MSR's values can be built during simulation. We simulate the program to the end of the test region. At the end of simulation, we will get two files namely the *initial-state-file* and *final-state-file*. The *initial-state-file* contains all the register and the memory state on which the test reads, including the memory region where the code resides. The *final-state-file* contains the register and memory state of the program at the end of the test.

D. Test Generation and Replay

These two files obtained from the previous step are given as input to a test generator. The test generator converts the *initial-state-file* file into a test program format and adds a preamble and epilog to the program. The preamble is an assembly code snippet which restores the register state and MSR state of the processor and jumps to the beginning of the test code. The epilog is another assembly code snippet which uses a signature checking mechanism to determine if the test passed or failed during execution. The signature is calculated using a Linear Feedback Shift Register (LSFR) implemented in assembly code. The LSFR code shifts the register and memory state that is modified by the program and calculate the signature. This signature is compared against the signature calculated from the *final-state-file* got from state capture. A mismatch in the state indicates that a defect was detected by the program.

E. Test Quality Assessment

The extracted tests were run repeatedly on multiple good and defective units to test their effectiveness in detecting defects. We ran the tests on good silicon to ensure that the generated tests are valid. We ran the tests on multiple defective units and collected the failure statistics. We observed the following in our experiments: (a) Not every run of the short test on a defective unit will result in a failure. (b) The failure rate on defective units can be increased to 100% by running the test back-to-back a few times. The first observation that the test did not fail on every run in the bad unit is expected. This is because the original long running program did not fail at the exact same point in their execution as illustrated in Figure 2. For example, the original program could fail either one second into the test or five seconds into the test. We attribute this behavior to the inherent variability in the state of the out-of-order pipeline during program execution. This inherent variability causes the defect to manifest at different locations in the program. In the silicon experiments that we conducted, we found that bad units can be reliably screened if we repeat running the test a small number of times. For example, we will run the extracted test about 10 times on bad units. If the unit was defective, one of the 10 runs will screen the unit. Running the tests multiple times did not increase the test time significantly since the extracted tests have very short runtimes.

F. Optimizations in State Capture

The simulator in our tool accesses the code and data for executing the program via ITP. ITP uses the JTAG port of the target system to access the registers and memory state. JTAG is a very slow interface and repeated calls to the JTAG interface significantly slows down the speed of simulation. Since we are simulating 10's of millions of instructions from large real-world applications, fast simulation is critical for the tool to be effective. To speed up the simulation performance, we need to reduce the calls to the JTAG interface. We do that by implementing a cache in our tool. The cache stores the values of memory and MSR's read from the system. When a memory address is requested by the simulator, instead of reading only the requested address, we read a complete 4K page of memory that contains the address and store it in the cache. Subsequent read/write access to an address in that 4K page will then be serviced from the cache rather than through ITP. This optimization speeds up the tracing performance by many orders of magnitude due to the high spatial and temporal locality of memory accesses in real programs. Similarly, we also implement a cache to store the values of MSR's read from the system.

G. Limitations and Future Enhancements

The architecture simulator used in our tool models the processor and the memory interface. It does not simulate other devices such as the graphics processor, audio and IO devices. The tool can be used to trace functional programs that do not interact with these devices. This limitation can be overcome if we simulate models of these devices during trace collection.

Our tool generates functionally correct assembly code which replicates the behavior of the original test at the architecture level of abstraction. We do not store the microarchitecture state such as the pipeline registers, the branch predictor tables and cache at the beginning of trace capture. Capturing some microarchitecture state at the start of trace collection can result in higher likelihood of failure reproduction with a shorter test. This is an enhancement that we are looking to try in the future.

IV. RESULTS

Table 1 shows the test time reductions achieved by our tool when applied in four different recent microprocessor products (P0-P3) and four different programs (TC0-TC3).

Table 1: Test Time Reduction

Product	Test Program	Number of Short tests	Test Time Reduction
P0	TC0	2	6X
P1	TC1	1	7X
P2	TC2	5	15X
P3	TC3	1	7X

The four programs TC0-TC3 consist of two functional test programs, one commercial OS and one application program running on top of a commercial OS. Each program can take anywhere between 30 seconds to several minutes to complete execution. These programs used different features of the ISA including different processor modes, page table structures and instruction sets. We were successful in extracting short tests in all cases and the extracted tests were as effective as the original program in screening defects. For P0, we profiled one defective unit, extracted two short tests based on the profile and ran the extracted tests on five bad units. The two tests were able to screen all the bad units correctly. Depending on the test program and processor on which our methodology was applied we extracted between 1 and 5 short tests from the original program and achieved test time reductions between 6X to 15X The test time used to calculate the results in Table 1 include the time it took to load the test program, to run the test program multiple times and to verify the results.

V. CONCLUSION

High performance microprocessors are manufactured using the latest process technologies. The higher integration densities of these process technologies can increase the impact of defects. Wafer level tests and tester patterns may not be enough to screen all defects. Running system level tests and commercial programs is one way to screen defective units. System level tests and commercial programs increase test time significantly, resulting in higher test cost. We proposed a tool and a methodology to extract short and effective functional tests from long running programs. Our tool combines a high speed ISA simulator and In-Target Probe to capture traces from long running programs. We profiled the long running program using the performance monitoring hardware to understand regions in the program that are effective in finding defects. We then traced these regions of execution using our tool and converted the trace into short and effective test programs. We applied the tool and methodology on recent microprocessor products and showed 6X-15X reduction in test time while still maintaining the screening quality. We plan to extend this tool to extract functional tests from commercial programs for the graphics processor and other IP units.

VI. ACKNOWLEDGEMENTS

We would like to thank our colleagues Paraag Vaishampayan and Vanessa F. Eusebio for their help in developing and testing the tool.

VII. REFERENCES

[1] Shen, J; Abraham, J.A, "Native mode functional test generation for processors with applications to self-test and design validation," IEEE ITC, 1998.

[2] Parvathala, P.; Maneparambil, K.; Lindsay, W., "FRITS - a microprocessor functional BIST method," IEEE ITC, 2002

[3] Guramurthy, S.; Vasudevan, S.; Abraham, J.A, "Automated mapping of pre-computed module-level test sequences to processor instructions," IEEE ITC 2005

[4] Gurumurthy, S.; Vasudevan, S.; Abraham, J.A, "Automatic generation of instruction sequences targeting hard-to-detect structural faults in a processor," IEEE ITC 2006

[5] Kranitis, N. et.al. "Effective software self-test methodology for processor cores," IEEE DATE 2002

[6] Corno, F. et.al. "Fully automatic test program generation for microprocessor cores," IEEE DATE, 2003.

[7] Lai WC; Krstic, A; Kwang-Ting C, "Test program synthesis for path delay faults in microprocessor cores," IEEE ITC, 2000.

[8] Singh, V. et.al. "Instruction-based delay fault self-testing of pipelined processor cores," IEEE ISCAS 2005.

[9] Gurumurthy, S. et.al. "Automatic Generation of Instructions to Robustly Test Delay Defects in Processors," ETS 2007.

[10] Natarajan, S. et.al., "Path coverage based functional test generation for processor marginality validation," IEEE ITC, 2010.

[11] Sherwood, T et.al. "Automatically characterizing large scale program behavior," ACM ASPLOS, 2002.

[12] Kim, J et.al. "Dynamic code region (DCR) based program phase tracking and prediction for dynamic optimizations," HiPEAC 2005.

[13] Sherwood, T; Perelman, E; and Calder, B, Basic Block "Distribution Analysis to Find Periodic Behavior and Simulation Points in Applications," IEEE PACT, 2001.

[14] In-Target Probe Manual, Intel® Corporation, http://www.intel.com/design/Xeon/guides/24967914.pdf.

[15] Ammons, G; Ball, T; Larus, J. Exploiting hardware performance counters with flow and context sensitive profiling. ACM PLDI, 2007.

[16] Henning, J.L. "SPEC CPU2000: Measuring CPU Performance in the New Millennium," IEEE Computer, July 2000.

[17] Intel® Architecture Programmers Reference Manual http://www.intel.com/content/www/us/en/processors/architectures-software-developer-manuals.html

[18] Rosenblum, Mendel. "VMware's virtual platform™." Hot Chips, 1999.

[19] Kodakara, S. V. et.al. "Model Based Test Generation for Microprocessor Architecture Validation," IEEE VLSI Design 2007.

[20] Performance Monitoring Unit Sharing Guide, Intel® Corporation, https://software.intel.com/file/30388\

[21] Hennessy, J. L; Patterson, D. A; 2 Computer Architecture: A Quantitative Approach (3 ed.). Morgan Kaufmann Publishers Inc., San Francisco, CA, USA, 2003.

[22] Park, T. Hong and S. Mitra, "IFRA: Instruction Footprint Recording and Analysis for Post-Silicon Bug Localization in Processors," IEEE Trans. CAD, 2009.

[23] Bhatkar, S; DuVarney,D; and Sekar, R, "Address Obfuscation: An Efficient Approach to Combat a Broad Range of Memory Error Exploits," USENIX Security Symposium (USENIX Security) August, 2003.

2015 IEEE 33rd VLSI Test Symposium (VTS)

Statistical Techniques for Predicting System-Level Failure using Stress-Test Data

Harry H. Chen[†], Shih-Hua Kuo[†], Jonathan Tung[†], Mango C.-T. Chao[‡]

[†]MediaTek Inc, Hsinchu, Taiwan
[‡]Dept. of Electronics Engineering & Institute of Electronics, National Chiao Tung University, Hsinchu, Taiwan
[harry-h.chen, shih-hua.kuo, jonathan.tung]@mediatek.com, mango@faculty.nctu.edu.tw

Abstract—**In this paper we describe a novel scheme for collecting and analyzing a chip's failure signature. Incorrect outputs of digital chips are forced by applying scan patterns under non-destructive stress conditions. From binary mismatch responses collected in continue-on-fail mode, numeric data features are formed by grouping and counting mismatches in each group, thus defining a chip's "analog" failure signature. We use machine learning to explore prediction models of system-level test (SLT) failures by comparing signatures of chip samples from known SLT pass/fail bins. Important features that clearly separate the SLT pass/fail chips are identified. Experimental results are presented for a 28-nm 1.2-GHz quad-core low-power processor.**

I. INTRODUCTION

Nanometer scaling with its attendant process variation has made it increasingly difficult to meet quality goals in complex system-on-chip (SoC) products. Across the die, operating conditions can show significant local variation in terms of power level and temperature due to electrical activity which can differ greatly between test mode and functional mode. The functional operating condition where a subtle defect can trigger failure may never be encountered under normal test mode conditions running production scan patterns. Chips passing production scan patterns with high structural fault coverage end up failing functional operation in the end-user system degrade the product quality as indicated by a rise in defective parts per million (DPPM).

To minimize defect escapes missed by scan testing, system-level testing (SLT) that mimics the end-user application can be inserted as an extra and final test step [1]. But SLT cannot guarantee complete functional coverage and it limits high-volume throughput. SLT can help in the early production stage to identify manufacturing test holes and ensure adequate initial product quality. But it should not be a permanent part of the production test flow; and effort should be made so its role reverts to that of occasional sample monitoring. Of course even with SLT, some defects can still end up as customer in-field failures, i.e., RMA (return merchandise authorization) parts.

Statistical outlier detection is a recognized and well-developed approach that complements deterministic testing to improve outgoing quality [2], [3]. In cases of defective devices, certain features may exhibit clear outlying values. Tests that measure analog parameters are well-suited for outlier analysis. Hence much work has been devoted to analog circuit applications [4], [5], [6].

For digital circuits, the availability of analog attributes is limited. On-chip embedded sensors such as ring oscillators, voltage droop detectors, and temperature monitors do provide more internal analog visibility [7]. But their relatively small numbers only cover the neighborhood of their placed locations. In contrast, the scheme we will shortly describe collects information from every scan flop, increasing observability by orders of magnitude. Indeed taken at face value, the previous statement offers no new revelation. The novelty lies in converting the collected digital data into amenable analog features for treatment by outlier analysis.

Subjecting a device to stress conditions can expose normally hidden defects. By stress, we mean operating at voltage and temperature extremes as well as running faster than the specified datasheet F_{max}, i.e., over-clocking. Burn-in and high-temperature operating life (HTOL) are prime examples of destructive stress testing. Examples of non-destructive stress testing include very-low-voltage [8], higher-than-at-speed [9], and substrate bias manipulation [10]. The combination of non-destructive stress testing and outlier analysis can further enhance defect exposure [11].

A. Overview of Work

The work described in this paper is motivated by the goal to make predictions, upstream in the production test flow, either during wafer chip probe (CP) or package final test (FT), about downstream SLT/RMA failures. If we attain our goal, it means less reliance on SLT, thus saving test cost, and faster time to low DPPM, i.e., better quality. To be cost effective, carrying out the prediction step should incur minimal overhead and preferably be done in-line. But we do not rule out post-processing.

Since our focus is on digital circuitry, we decided on an approach based on answering this key question: *Can production scan patterns be leveraged to make SLT/RMA failure predictions?* Under normal testing conditions, the answer is clearly negative as all SLT/RMA fails are CP/FT escapes from production scan patterns. Therefore, we must seek the answer under stress test conditions.

Assuming we obtain an affirmative answer to the first key question, then a second key question follows: *Can we make SLT/RMA failure predictions in an efficient manner using stressed scan patterns?* These two key questions frame our research into two phases. This paper presents results of phase-1 showing that to a first order, the answer is affirmative to question-1. As we are still exploring numerous facets of phase-1, we do not have a definitive answer to question-2 yet. Early clues though give us hope that it is also affirmative.

The scan patterns we use target transition delay faults. They are generated using on-chip clocks in launch-on-capture mode. Underlying this choice is the assumption that most defects, regardless of actual physical mechanism, can manifest as delay faults. Note that the traditional stuck-at-faults are also included since they are simply special cases of slow-to-rise/fall faults with infinite delay.

For stress test conditions, we collect data at multiple below-nominal voltage and above-maximum frequency steps. We also tried above-typical temperature early on, but decided that the contribution to learning was relatively minor. The stress conditions are aimed at reducing slack margins on all timing paths such that even small delay defects (SDD) have a chance to cause capture failures.

By running above F_{max}, we are in effect doing higher-than-at-speed testing (HTAST). But there is a key difference — we do not use capture flop masking. In standard HTAST, path endpoint flops that do not meet timing are masked to prevent test overkill. Masking decisions are based on approximate and static corner-derived timing models which do not reflect dynamic activity in actual silicon and stress test conditions. In fact, our phase-1 investigation relies on capture failures, the more the better. So much so that scan patterns are applied in continue-on-fail mode to record all capture failures in all patterns. Chips that pass provide little insight. Much more is revealed about a chip's true personality when stressed to the point of failure, and scan flops serve as internal probes to

978-1-4799-7598-3/15 $31.00 © 2015 IEEE

portray that personality in great detail. Performing HTAST without masking also forms the basis of the work in [12]. However, we collect failing data at a much finer granularity than the simple pass/fail status of a chip.

We call our scheme "Stressed **O**n-chip-clock test **M**ism**A**tch **C**ount", or **SOMAC** for an easy to say and remember acronym. SOMAC uses outlier analysis and machine learning to explore prediction models. In a sense, delay is the analog attribute of a digital circuit that is analyzed. We don't actually measure delays for that would be totally impractical [13]. But capture failures can be viewed as crude samplings of delays, in every flop input cone, and in every pattern with diversely sensitized paths. This information-rich content is what SOMAC exploits.

In the following, details of the SOMAC scheme are described in Section II. Experimental results applying SOMAC to a 28-nm 1.2-GHz quad-core processor are presented and discussed in Section III. Our conclusion in section IV will cover limitations, and future work, and potential applications.

II. THE SOMAC SCHEME

A. SOMAC Data Collection

The SOMAC data collection flow diagram is shown in Fig. 1. Transition delay fault patterns using on-chip-clock (OCC) design-for-testability (DFT) is applied to the DUT on the automated test equipment (ATE) under stress conditions. Compressed OCC scan patterns are allowed. Continue-on-fail test responses are shifted out of the DUT's scan output pins and all response mismatches are logged. Each failed response records scan output pin, pattern number, cycle number, and expected value.

For the purpose of explaining "analog" data feature extraction later, we represent the output responses by a two-dimensional grid. Each step on the horizontal axis corresponds to a single scan shift out cycle, and each step on the vertical axis corresponds to a single scan output pin. The horizontal steps are ordered in ascending order by pattern number and within each pattern by ascending cycle number. Order on the vertical axis is immaterial. Given N patterns, C cycles per pattern, and S scan outputs, the total number of grid positions would equal $N \times C \times S$.

B. Data Feature Extraction

SOMAC data collection produces a response grid for each DUT at each test condition corner. Comparative analysis of DUT response grids could be done to find similarities or dissimilarities. If SLT-pass/fail chips make up the set of DUT's, perhaps we can find certain grid positions whose values define a clear separation between SLT-pass and SLT-fail chips. Treating each grid position as an individual analysis feature appears daunting as the number of positions can be very large – over 100 million per DUT per corner in the case of the SoC processor block we experimented with. Instead we propose to group positions into a smaller set of data features for analysis.

A data feature comprising a group of positions has a value defined to be the count of mismatches seen at the positions. There can be many grouping schemes. The simplest one groups the entire response grid into a single feature, which we call "Total **M**is**M**atch **C**ount" (**TMMC**).

Shown in Fig. 2 is an example of another scheme which we call "**P**attern-**S**et **S**can-**O**utput **MMC**" (**PS-SO-MMC**). This scheme relates to the way scan patterns are organized for the ATE. Due to tester memory restriction, the complete set of scan patterns is split into multiple pattern sets each containing a smaller number of patterns. During test operation, patterns sets are loaded into ATE memory one at a time. In this scheme, response grid positions in each pattern set and scan output are grouped to form one PS-SO-MMC feature. Assuming R pattern sets and using the same symbolic variables as before, there would be $R \times S$ features. Each feature would contain $(N/R) \times C$ grid positions.

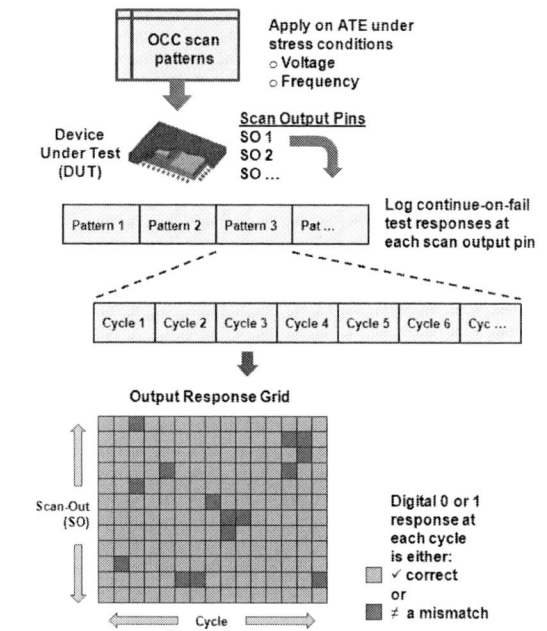

Figure 1. SOMAC data collection flow.

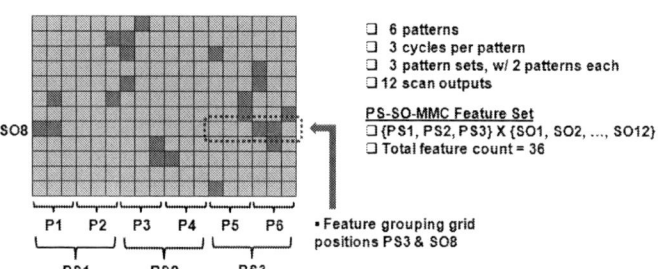

Figure 2. Example of PS-SO-MMC feature formation.

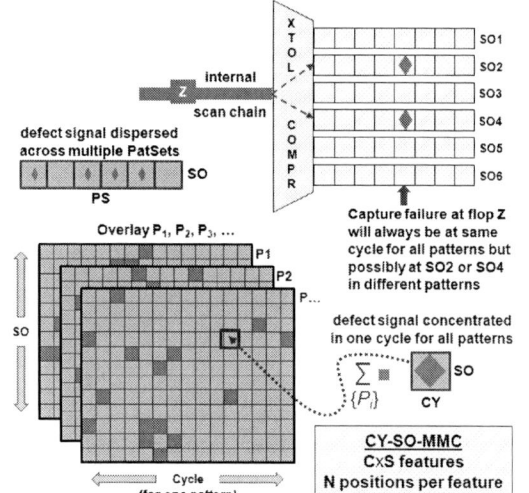

Figure 3. Forming CY-SO-MMC feature.

It's clear that the response grid can be carved up in any number of ways to form data features. It's impossible to enumerate them all. Here we describe one more scheme called "**CY**cle Scan-**O**utput **MMC**" (**CY-SO-MMC**) which is motivated by the desire to increase a defect's signal strength. Referring to Fig. 3, it shows a flop Z capturing failures when tested under stress due to a defect in an SLT-fail chip. Depending on conditions unique to each pattern, capture failure at Z may or may not occur with each pattern. In the example scenario, scan compression with X-tolerance is deployed. That means the internal scan chain containing flop Z may fan out to multiple external scan outputs. However, due to flop Z's internal chain position, it will always be observed at the same cycle position in all patterns.

With PS-SO-MMC, capture failures at flop Z could be dispersed over too many features. As data collected under SOMAC is inherently noisy, the defect's signal-to-noise ratio (SNR) under PS-SO-MMC may be too low for reliable detection. For CY-SO-MMC, each scan pattern is aligned on top of each other so that all response positions at the same cycle and scan output form a single feature. This will concentrate capture failures at flop Z into the same cycle and a few SO positions determined by X-tolerance routing.

C. Outlier Analysis Using Data Features

The objective of outlier analysis is to detect numerically outlying samples from the multivariate data. The detected outliers can be potential candidates of the abnormal samples. A successful outlier analyzer should be able to not only help the prediction of abnormal sample but also report the important features that contribute the most to the distinction between normal samples and outlying samples. Those important features can then help us to discover the root causes of the abnormal samples.

There are two approaches to analyze the data, supervised learning and unsupervised learning. Supervised learning builds the model from labeled training data. With the help of supervised learning, we can build prediction models of system-level test (SLT) failures by comparing the signatures of the chip samples from known SLT pass/fail bins. On the other hand, unsupervised learning finds the statistical structure of data from unlabeled samples.

In this paper, we use a supervised learning technique, random forest (RF), and an unsupervised learning technique, Density-Based Spatial Clustering of Applications with Noise (DBSCAN), to analyze the feature data.

1) Random Forest: Random forest (RF) [14] is a classification (or regression) algorithm using an ensemble of decision trees to make the final prediction to a sample's class. RF uses the bootstrap sampling to choose the designated training samples for building each decision tree. Each decision tree is constructed with different random subsets of features. An internal split node of a tree divides the training samples associated with the node into two subgroups.

We use RF to build the classification model to analyze the SOMAC data. RF is relatively robust to noise and can deal with high-dimensional data sets comparing to other classification methods. The confidence of a built RF model can be evaluated by examining the consistency of fraction votes among individual decision trees. The procedure of RF can also be used to rank the importance of feature variables. The white-box nature of RF operation makes it well-suited to gain insights about the data being analyzed.

2) DBSCAN: Density-based spatial clustering of applications with noise (DBSCAN) [15] is an unsupervised data clustering algorithm that categorizes samples based on finding density boundary. Besides grouping a set of samples into clusters, DBSCAN identifies isolated samples as noise based on density relationships.

We perform DBSCAN to detect outliers that are far away from the boundaries of clusters. DBSCAN, as an unsupervised pure outlier analysis method, can detect outliers based on the abnormal behavior of data. DBSCAN can deal with arbitrary shape of distributions and is suited to analyze the SOMAC data with unknown feature distributions.

D. Addressing Mislabeled Input Data

At the beginning of applying supervised machine learning, the actual relation between the features and the prediction target (label) has not been determined. If the provided training set contains samples with labels unrelated to the features, the learning algorithm may be confused and produce an unreliable model. Fig.4a shows an example of building a classification model with mislabeled samples.

In our case, a 2-class classification problem, the label of a sample depends on whether it passed or failed SLT. However, SOMAC data is collected only from the CPU core. Some SLT fail samples may be directly caused by errors in the CPU core, but some may not. In other words, those SLT fail samples that are not directly caused by errors in the CPU core have no bearing on SOMAC data and hence should be relabeled in our learning process.

The issue of mislabeled inputs is well-known in machine learning and some research works focused on correcting minor mislabels. Ours, however, is a case of severe mislabeling. Under such a scenario, even the most powerful machine learning algorithm will be misled and produce trained models that fail to account for the truly important distinguishing features.

The proposed re-labeling flow utilizes the concept of unsupervised learning methods which detect the outliers by recognizing abnormal behavior of the data. We try to find the proper labels that incur least confusion to classification and make best separation between the two-class samples. The samples will be re-labeled to the other class if the data behaviors of the same class are statistically similar. Fig. 4b shows how the re-labeling flow affects the classification model. After re-assigning labels to the possible mislabels samples, a better classification model can be built.

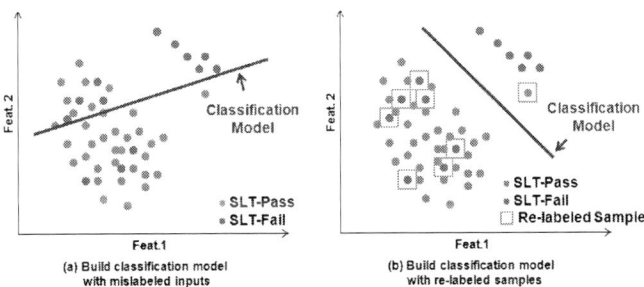

Figure 4. **Illustration of mislabeled samples.**

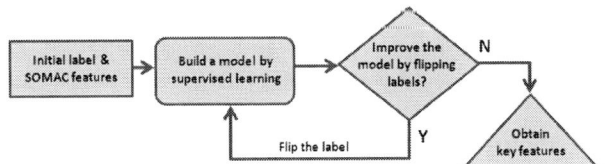

Figure 5. **Iterative re-labeling flow.**

Fig. 5 illustrates the iterative re-labeling flow, which improves the models by flipping the labels of suspecting mislabeled samples, starting from the data with initial labels. The labels that can build the best-separating model are kept as the final labels of the samples. The proposed re-labeling flow is a wrapper method that can co-work with any supervised learning. The experimental results of re-labeling will be later discussed in Section III using RF as the supervised learning technique.

III. SILICON EXPERIMENTAL RESULTS

The product chosen for SOMAC experimental study is a fully integrated SoC that powers Android-based smartphones. It is fabricated in a 28-nm low-power process node and produced in high volume up towards 100M units. At the heart of the SoC is a quad-core application processor designed to run at 1.2 GHz. OCC DFT structure was implemented only for the processor block (referred to as CPU below). Since our scheme requires the ability to apply OCC patterns, SOMAC data was collected for the CPU only, not the full SoC.

The CPU core contains ~148K scan flops. The scan compression architecture implements ~1200 short internal chains. The short chains feed an X-tolerant combinational scan compressor merging by XOR-tree into 16 compressor outputs. The 16 outputs are further halved by a two-stage pipeline serializer to end in 8 SoC top-level primary scan outputs. In each scan pattern, responses in 240 external shift cycles are measured.

We initiated SOMAC work as the product entered volume production. At the time, 60K transition delay fault OCC patterns were generated for the CPU using timing slack data to enhance SDD coverage. Test coverage for the block was slightly below 85%. The 60K patterns were further split into 60 sets with each containing 1K patterns to fit in ATE memory. So for this product, SOMAC variables from Section II are:

$$N = 60K, \ R = 60, \ C = 240, \ S = 8$$

A full lot was pulled from production containing ~40K packaged chips dedicated to SOMAC experimentation. SLT was performed on chips that passed production FT, resulting in 92 SLT failures. Not all of the 92 failures are due to the CPU. Others can be attributed to graphics processor, video decoder, and composite causes. Based on previous production experience, we can expect CPU failure to be roughly 20~30% of total SLT failures. Our original plan had called for collecting SOMAC data for a few thousand SLT-pass chips. That was pared back to 200 due to high ATE time consumption and large data volume. Thus SOMAC data was collected for a total of 292 chips – 200 SLT-pass (SLTP) and 92 SLT-fail (SLTF).

The data sheet specifies nominal supply at 1.2 V and F_{max} at 1200 MHz. Without a clear idea of what might show up under stress, we chose to collect SOMAC data at five 50 mV steppings below nominal supply and four frequency steps above F_{max} for a total of 20 stress condition corners:

5 V_{DD} supplies = { **1.15, 1.10, 1.05, 1.00, 0.95** } V
4 Frequencies = { **1482, 1508, 1600, 1700** } MHz

A. CY-SO-MMC Feature Results

Extracting the CY-SO-MMC feature set requires re-mapping the ATE's 1K pattern set cycle numbers to the equivalent single-pattern cycle numbers. The number of features is 240[C] x 8[S] = 1920 with each feature containing 60K[N] grid positions.

At the (1.00V, 1700MHz) corner, we observe that many SLTF chips show up as outliers in most of the features which is well-suited for prediction using multivariate machine learning techniques such as decision tree classification and clustering. In supervised learning, an empirical rule states that the ratio #features/#samples should be less than $1/10 \sim 1/15$ to avoid data over-fitting. In our case, the ratio is 1920/292. We chose to use the Random Forest (RF) algorithm which is more robust and avoids over-fitting. In addition, we applied leave-one-out cross validation (LOOCV) to further enhance decision robustness.

TABLE I.   RANDOM FOREST WITH LOOCV.

	Predict Fail	Predict Pass
True Fail	22	70
True Pas	18	182

The result is shown in Table I achieving prediction accuracy of 22/92 = 23.9% with overkill of 18/200 = 9%.

B. The Issue of Mislabeled Inputs

Fig. 6 shows the 92 SLTF chips being split among three bins based on the failing block. Bin #2 combines CPU and GPU; bin #3 is for MM/VDEC; and bin #4 is a catch-all where the SLT program hangs and times out without cause indication. Any of the aforementioned blocks could contribute to failures in bin #4. Since SOMAC data is only obtained from the CPU, it is unlikely to reflect failures in non-CPU blocks. Specifically, we can exclude the 13 from bin #3; and only a subset of the 79 chips from bins #2 and #4 should be attributed to the CPU. Unfortunately, in the normal procedure to prevent counterfeiting, the 92 chips were physically destroyed by the SLT sub-contractor after we collected SOMAC data and returned the chips. Thus we do not know the SLTF bins of the 92 chips.

The SOMAC data we have on hand can only be reasonably expected to predict CPU SLTF which by historical record is only 20~30% of the 92 total SLTF. Another way to state this is that as input to supervised learning, 70~80% of the chips labeled SLTF are wrong when the label is taken to mean CPU SLTF.

The confusion caused by severe mislabeling is clearly seen in Fig. 7 where histograms of RF class votes for SLTF and feature importance are shown for the previous CY-SO-MMC data. In the left histogram, the two classes are hardly separable; and in the right histogram, all 1920 features are bundled close together towards non-importance with relatively few stand-outs.

The scheme to iteratively apply RF and class re-labeling to simultaneously identify outliers and the key features that differentiates two-class samples is described next.

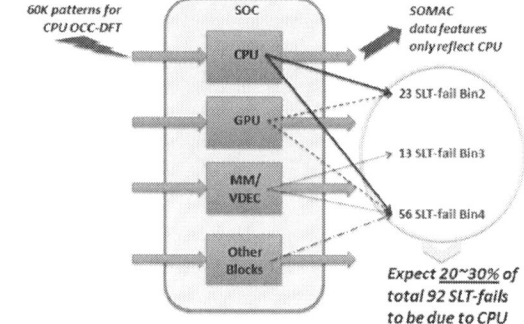

Figure 6.   SLT-fail bin partitions and failed block assignments. MM/VDEC is multimedia video decoder.

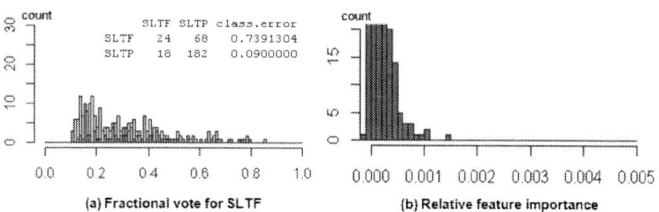

(a) Fractional vote for SLTF   (b) Relative feature importance

Figure 7.   In the left histogram of RF votes for SLTF, 200 SLTP chips are in green and 92 SLTF chips are in red. In the bottom histogram of importance for 1920 features, the upper portion is cut off to save display space. In the resulting 2x2 confusion matrix at the top of the figure, off-diagonal entries are incorrect prediction counts.

C. Re-labeling Using Random Forest

The iterative relabeling flow shown in Fig. 8 makes use of RF voting results because the voting percentage is a measure of confidence about the chip's class prediction.

When a chip ends up on the wrong side of the vote, i.e., SLTP chip receives majority SLTF votes, or SLTF chip gets majority SLTP votes, it is perhaps an indication that the input label is wrong. Respecting the collective wisdom of thousands of decision trees, the chip's label is flipped to match the majority vote. Then the re-labeled chips are run through RF again. The new labels guide the decision trees to tease out the more relevant features and increase vote separation (or vote of confidence) between the classes. The loop terminates when perfect prediction is achieved.

The effect of re-labeling on voting confidence and feature importance can be seen in Fig. 9. First pass re-labeling (pass-1 in the figure) is applied to the results of Fig. 7 and causes the biggest change. The intervening passes are mostly moving the few uncertain chips that have marginal majorities on either side of 50%. Perfect prediction is achieved by pass-4. We found that the few uncertain chips may switch labels multiple times in this process. Re-labeling can be controlled by a parameter λ which has a value range from 0.5 to 1.0. It determines the majority threshold that flips a label. Fig. 9 used the most aggressive setting of $\lambda = 0.5$. Setting λ closer to 1.0 is more conservative, i.e., require higher vote of confidence to flip a label.

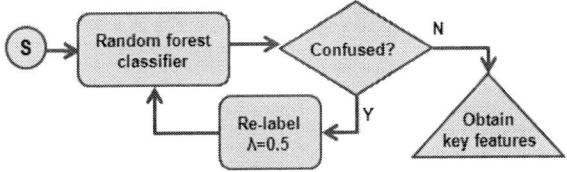

λ = Re-label control parameter

Figure 8. Iterative RF re-labeling flow. Confusion exists when off-diagonal entries of the confusion matrix are non-zero (see Fig. 7 caption).

Also shown in Fig. 9 are the counts of chips that ended up with labels flipped from the original. Of the 92 original SLTF, 71 changed to SLTP while 21 are true SLTF. Note that 21/92 = 23% falls within the expected percentage range of CPU-SLTF. Of the original 200 SLTP, 10 or 5% became SLTF. The (1.00V, 1700MHz) corner is 17% below nominal supply and 42% above F_{max}. Under such deep stress conditions, it is not unreasonable to expect some marginal CPU-SLTP chips to degrade faster to resemble CPU-SLTF chips. Less stressful corners can be analyzed to see if these chips could be recovered. We use r-SLTP and r-SLTF to denote final chip labels at the end of the RF re-labeling process. From the distribution of feature importance, 27 key features which are 4σ above mean are flagged as key features.

To gain more confidence that re-labeling isn't just "cooking the data", we did a further experiment by removing the 31 r-SLTF chips, reverting to the original 190/71 SLTP/SLTF labels for the remaining 261 r-SLTP chips, and re-ran the RF re-labeling flow. This time, instead of separation, all 261 chips ended up as r-SLTP and all features became equally unimportant after three iterations. In other words, the 261 chips are truly indistinguishable despite distinct starting labels. Re-labeling has an amplifying effect that forces convergence to one of two clear end-states: (1) strong distinction with important features, or (2) complete merging into one class with no important features. There might exist a third possibility of non-convergence where relabeling continues to occur near the 50% voting threshold for some samples. In one sense, RF re-labeling can be thought of as a kind of hybrid learning - unsupervised with an initial suggestion of labels which guides but need not be strictly obeyed as the process works to expose outliers.

To further validate the re-labeling results, we apply unsupervised outlier detection methods to find the outliers. We perform outlier detection by DBSCAN. A simple procedure is used to identify samples which easily locate as outliers. The procedure exhaustively performs 2D-DBSCAN to mark outliers in all possible feature pairs, and identify the most frequently marked samples as outliers. Note that not all 1920

features are used in the analysis, features with narrow-ranged values are avoided in order to form the 2D spatial distribution. Features are sorted by mean and standard deviation and only top-k features are used.

Table II shows the result of detecting outliers by DBSCAN. The first row shows the features number k we used to explore the 2D outliers. The second/third row shows the number of chips detected as outlier in the r-SLTP/r-SLTF class. The fourth row shows the threshold of identifying outliers. Result shows consistently 28/31 r-SLTF sample are picked from greater than 5.84 sigmas which tend to locate as outliers in various feature spaces. This result validates that re-label separates classes by their different data distributions.

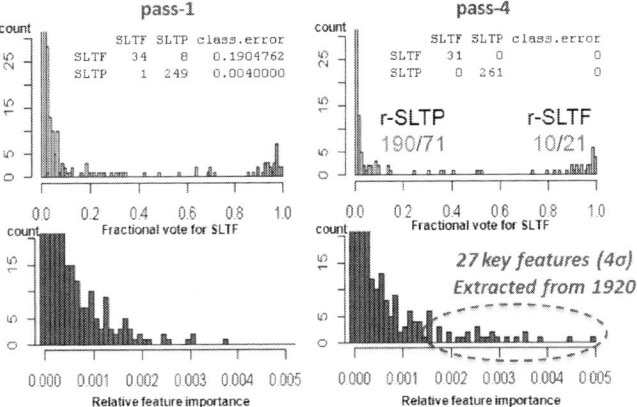

Figure 9. Results of interative RF re-labeling showing first and final passes. As in Fig. 7, x-axis is fractional vote for SLTF in top histograms, and feature importance for bottom histograms.

TABLE II.   OUTLIERS DETECTED BY DBSCAN.

Used feat.#	384	576	768	960
r-SLTP#	11/261	9/261	8/261	11/261
r-SLTF#	28/31	28/31	28/31	28/31
Threshold(sigma)	11.53	7.47	7.40	5.84

D. Analysis of Key CY-SO-MMC features

With 27 important features, the #features/#samples ratio is now 27/292 or $\approx$ 1/11, falling within the empirical rule of thumb. A simple decision tree classifier was applied to the re-labeled chips using these 27 features. In the resulting decision tree shown in Fig. 10c, only one feature, CY58_SO2, was needed to complete classification, although Fig. 10b shows the distribution of feature values with no margin of tolerance. Three other primary features with slightly lower splitting power were also identified by the classifier: CY38_SO2, CY46_SO4, and CY47_SO6 which are indicated on the feature importance histogram in Fig. 10a.

We describe a possible prediction scheme that utilizes all four features. First define a tolerance margin around each feature's decision threshold. For a given chip, evaluate each feature to determine one of three possible outcomes:

(1) Value is inside tolerance margin.
(2) Value is outside tolerance margin on r-SLTP side.
(3) Value is outside tolerance margin on r-SLTF side.

If all four features have outcome #2 (#3), predict SLTP (SLTF). Since no SLT is run for firm predictions, cost saving is realized. All other cases result in "unable to predict" and the associated chips need to be run through SLT. Using data for the 292 chips, Table III illustrates three tolerance margins with percentage of chips requiring SLT. Wider margin should reduce test escapes and yield loss, but with less SLT cost-savings.

The prediction scheme just described is still not practical for production test however, since each CY-SO-MMC feature value is derived from 60K patterns. Considering the most important feature, CY58_SO2, only 128 patterns are needed to cover MM contributions from all 292 chips. Upon further examination of the 128 patterns, more than 60% contribute little to distinction between r-SLTP and r-SLTF chips. Contribution is assessed by considering the difference in percentage of r-SLTF and r-SLTP chips that show MM in each pattern.

We are developing prediction schemes around the truly useful patterns found at the few key CY-SO-MMC features and possibly at multiple stress conditions. If we can limit the total number of patterns to the low hundreds for continue-on-fail operation, then we may have a practical solution for production test.

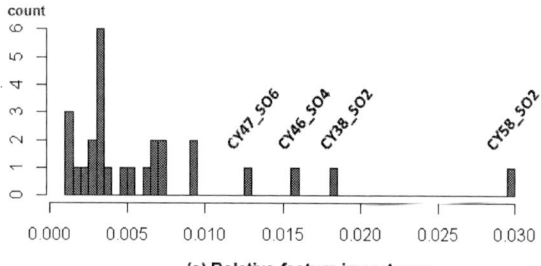

(a) Relative feature importance

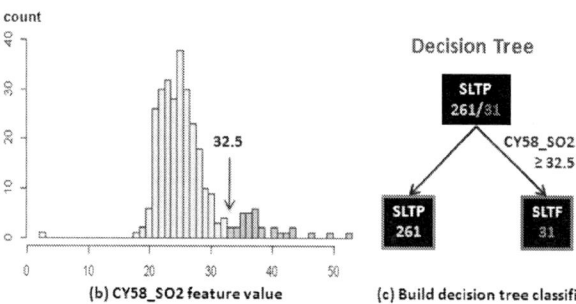

(b) CY58_SO2 feature value   (c) Build decision tree classifier

Figure 10.   Decision tree built with 27 important features.

TABLE III.   SOFT PREDICTION USING 4 PRIMARY SPILT FEATURES.

Feature	Decision Threshold	Tolerance Margin (inclusive)		
		5% SLT	9% SLT	15 %SLT
CY58_SO2	32.5	[32, 33]	[31, 34]	[30, 35]
CY38_SO2	19.5	[19, 20]	[18, 21]	[17, 22]
CY46_SO4	38.5	[38, 39]	[37, 40]	[36, 41]
CY47_SO6	25.5	[25, 26]	[24, 27]	[23, 28]

IV. CONCLUSIONS

To re-cap, we started by defining a humongous set of fine-grain data features collected over many low-voltage and over-clocked conditions. Given the high degree of input noise, i.e., mislabeling, we had to follow a script described in [16] to adopt multiple perspectives via feature aggregation, and remove noise by domain knowledge injection. With the aid of machine learning tools, we made iterative discoveries of relevant aspects, pushing forward until we ended up with a much smaller and valuable set of fine-grain features again.

Results presented here show much promise. We will explore more feature extraction and analysis algorithms to improve accuracy. The fact that only a few features are needed for differentiation in this particular case seems to indicate that SLT failures are due mostly to systematic variations. We need to increase sample size and validate outside of the 292 chips used in the current study. This experiment is being repeated on

a new octa-core smartphone SoC which have OCC DFT implemented in most major blocks. And we will do a better job of device tracking this time around.

Other longer-term possibilities include linking SOMAC data to diagnostics which may identify systematic design issues that are yield-limiting or pose reliability risks [17]. Are there new DFT approaches that could help reduce noise and improve accuracy in SOMAC data? Is it possible to reduce scan pattern size and use SOMAC to augment? We can view SOMAC as a new kind of deeply probing instrument providing enhanced visibility into chip internals. Hopefully, our work can stimulate interest in others to try and develop it further, and extend it to other applications.

REFERENCES

[1] S. Biswas and B. Cory, "An Industrial Study of System-Level Test," IEEE Design & Test of Computers, vol. 29, issue 1, pp. 19-27, 2012.

[2] W. R. Daasch and R. Madge, "Data-Driven Models for Statistical Testing: Measurements, Estimates and Residuals," IEEE International Test Conference, pp. 322-331, 2005.

[3] P. M. ONeill, "Statistical Test: A New Paradigm to Improve Test Effectiveness & Efficiency," IEEE International Test Conference, pp. 1-10, 2007.

[4] H.-G. Stratigopoulos, S. Mir, and Y. Makris, "Enrichment of Limited Training Sets in Machine-Learning-Based Analog/RF Test," IEEE Design Automation & Test in Europe Conference, pp. 1668-1673, 2009.

[5] E. Yilmaz, S. Ozev, and K. M. Butler, "Adaptive Multidimensional Outlier Analysis for Analog and Mixed Signal Circuits," IEEE International Test Conference, pp. 1-8, 2011.

[6] H. H. Chen, R. Hsu, P. Yang, and J. J. Shyr, "Predicting system-level test and in-field customer failures using data mining," IEEE International Test Conference, pp. 1-10, 2013.

[7] A. Gattiker, "Yin and Yang of Embedded Sensors for Post-Scaling Era," IEEE VLSI Test Symposium, pp. 324-327, 2011.

[8] J. T.-Y. Chang and E. J. McCluskey, "Detecting Delay Flaws by Very-Low-Voltage Testing," IEEE International Test Conference, pp. 367-376, 1996.

[9] T. Yoneda, K. Hori, M. Inoue, and H. Fujiwara, "Faster-Than-At-Speed Test for Increased Test Quality and In-Field Reliability," IEEE International Test Conference, pp. 1-9, 2011.

[10] A. Gattiker and P. Nigh, "Using Well/Substrate Bias Manipulation to Enhance Voltage-Test-Based Defect Detection," IEEE International Test Conference, pp. 1-6, 2011.

[11] R. P.. Turakhia, W. R. Daasch, J. Lurkins and B. Benware, "Changing Test and Data Modeling Requirements for Screening Latent Defects as Statistical Outliers," IEEE Design & Test of Computers, pp. 100-109, 2006.

[12] S. H. Wu, B. N. Lee, Li-C. Wang and M. S. Abadir, "Statistical Analysis and Optimization of Parametric Delay Test," IEEE International Test Conference, pp. 1-10, 2007.

[13] J. Lee and E. J. McCluskey, "Failing Frequency Signature Analysis," IEEE International Test Conference, pp. 1-8, 2008.

[14] L. Breiman, "Random Forest," Machine learning, vol. 45, pp. 5-32, 2001.

[15] M. Ester, H.-P. Kriegel, J. Sander, and X. Xu, "A Density-Based Algorithm for Discovering Clusters in Large Spatial Databases with Noise," International Conference on Knowledge Discovery and Data Mining, vol. 96, pp. 226-231, 1996.

[16] N. Sumikawa, J. Tikkanen, Li-C. Wang and M. S. Abadir, "Iterative Knowledge Discovery for Screening Customer Returns," IEEE International Workshop on Digital and Analog Test and Data Analysis (DATA), September 2013, Paper 4.2.s

[17] L. M. Huisman, M. Kassab, and L. Pastel, "Data Mining Integrated Circuit Fails with Fail Commonalities," IEEE International Test Conference, pp. 661-668, 2004.

[18] H. H. Chen, S.-H. Kuo, J. Tung, M. C.-T. Chao, "Learning from Chips Behaving Badly," IEEE International Workshop on Defects, Adaptive Test and Data Analysis (DATA), 2014.

2015 IEEE 33rd VLSI Test Symposium (VTS)

Yield Prognosis for Fab-to-Fab Product Migration

Ali Ahmadi*, Ke Huang†, Amit Nahar‡, Bob Orr‡, Michael Pas‡, John M. Carulli Jr.§ and Yiorgos Makris*

*Department of Electrical Engineering, The University of Texas at Dallas, Richardson, TX 75080
†Department of Electrical and Computer Engineering, San Diego State University, San Diego, CA 92115
‡Texas Instruments Inc., 12500 TI Boulevard, MS 8741, Dallas, TX 75243
§GlobalFoundries, 400 Stone Break Road Extension, Malta, NY 12020

Abstract—We investigate the utility of correlations between e-test and probe test measurements in predicting yield. Specifically, we first examine whether statistical methods can accurately predict parametric probe test yield as a function of e-test measurements within the same fab. Then, we investigate whether the e-test profile of a destination fab, in conjunction with the e-test and probe test profiles of a source fab, suffice for accurate yield prognosis during fab-to-fab product migration. Results using an industrial dataset of ~3.5M devices from a 65nm Texas Instruments RF transceiver design fabricated in two different fabs reveal that (i) within-fab yield prediction error is in the range of a few tenths of a percentile point, and (ii) fab-to-fab yield prediction error is in the range of half a percentile point.

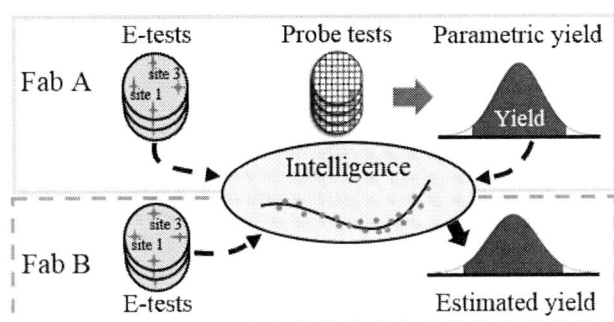

Fig. 1: Yield Prediction during fab-to-fab migration.

I. INTRODUCTION

The rapidly growing and dynamically changing consumer electronics market introduces interesting challenges to production planning of semiconductor manufacturing companies, calling for agility and flexibility in order to efficiently respond to fluctuating demand. Contingency plans for dealing with catastrophic events, such as earthquakes, floods, and hurricanes, which have severely hampered the market in the past, as well as political or sheer financial reasons, often place similar constraints in production planning as well. Migrating a product from one fab to another, however, is not a trivial endeavor. Different fabs, implementing the same technology node and even employing identical equipment and software suites, are bound to exhibit variations in the parametric profile of the silicon they produce, and by extension, the yield of a device fabricated therein. Accurate yield prognosis, however, is an indispensable piece of information during device migration and production planning.

Predicting parametric yield for a design produced by a specific fab is not a new problem, with solutions varying from pure simulation-based to silicon measurement-driven. Traditionally, Monte Carlo based approaches have been used to generate a large number of random samples based on expected process variations, in order to estimate the distribution of each performance of interest [1]. Alternatively, modeling techniques, which approximate a performance of interest as a linear or non-linear function of device-level parameters have also been employed. Subsequently, the distribution of a desired performance is estimated by numerical methods [2], [3]. Such simulation-based methods, however, are of limited accuracy. Along another direction, the authors of [4] introduced the use of Bayesian model fusion for yield estimation, wherein pre-silicon simulations are reinforced with a small set of post-silicon measurements to enhance model accuracy. Similarly, in [5], high volume manufacturing yield of a product is estimated

through spatio-temporal wafer correlation models learned from early silicon wafers. Such silicon measurement-based methods, however, assume access to probe measurements from a tangible number of wafers of the device, produced in the fab of interest, in order to estimate parametric production yield. Yet in the context of fab-to-fab product migration, this information is not available since the device has yet to be produced in the target fab.

To address this problem, in this work we develop a yield prognosis method which does not require target fab measurements from the device to be migrated. Instead, as shown in Figure 1, it relies on e-test and probe test measurements from the source fab, where the device is currently produced, as well as the e-test profile of the target fab, which can be obtained from other devices produced therein, as e-tests are typically common across devices in the same technology node. Toward this end, in Section II we first discuss a regression-based solution to the problem of correlating e-test measurements to parametric probe test yield within a single fab. Then, in Section III, we introduce three methods for extending this capability to the fab-to-fab product migration scenario, namely *model migration, importance sampling* and *predictor calibration*. Experimental results demonstrating the effectiveness of parametric yield prediction based on e-test measurements on actual production data for both the within-fab and the fab-to-fab migration scenarios are presented in Section IV and conclusions are drawn in Section V.

II. WITHIN-FAB CORRELATION

Before we address the problem of predicting parametric probe test yield from e-tests during fab-to-fab product migration, we discuss the simpler version of doing so within the same fab. Meaningful correlations among measurements from various stages of semiconductor manufacturing are known to

978-1-4799-7598-3/15 $31.00 © 2015 IEEE

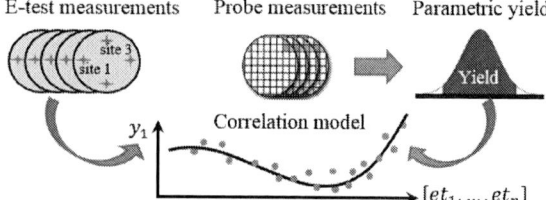

Fig. 2: E-test/yield correlation model.

exist and have been utilized in several tasks, such as test cost/time reduction and yield improvement in the past. In one such approach, described in [6], the authors proposed a statistical method for predicting probe test outcomes from e-test data of a wafer. This method employed a genetic algorithm-based approach to select a key subset of all e-test parameters and, then, build a multivariate nonlinear correlation model between selected e-tests and probe test outcomes. The identified correlations were iteratively refined through designer feedback, with the main objective of providing useful information for process health monitoring, rather than reducing test cost/time. Indeed, while the accuracy of these e-test to probe test prediction models is high, it would not suffice for omitting probe test of *individual dies* with acceptable test error. Nevertheless, if one is interested in predicting parametric probe test yield across an *entire wafer* from its e-tests, as is the case in our problem, these models are very accurate.

Figure 2 depicts how parametric probe test yield can be predicted through e-test measurements of wafers in a fab. First, for a training set of wafers, both e-tests and parametric probe test measurements are obtained. The device specifications are then used to compute parametric yield for each probe test across the wafer. Using the e-tests and the parametric probe test yield, a correlation model is trained. Finally, for a new wafer produced by this fab, its e-test measurements can be provided to the trained correlation model in order to predict parametric yield. The key component in this scheme is the construction of the correlation model, whereby the dependent variables (parametric probe test yield) are expressed as functions of predictors (e-test measurements).

Several methods exist in the literature for multivariate regression such as *Multivariate Adaptive Regression Splines (MARS)*, *Least-Angle Regression Splines (LARS)*, *Projection Pursuit Regression*, *Multi-Layer Perceptrons*, and *Radial Basis Function Networks* [7]. Among them, in this work, we use MARS [8], which was also used in [6] and several other test cost reduction methods in the past [9].

III. YIELD PROGNOSIS IN FAB-TO-FAB MIGRATION

We now turn our attention back to the fab-to-fab migration problem, wherein we seek a prognosis of the parametric yield of a product migrating from a source fab to a target fab. This prognosis may be based on the e-test and the parametric probe test of the source fab, where the device is currently produced so ample data is available. In addition, it may also be based on the e-test profile of the target fab, which can be obtained from other devices produced therein, as most e-tests

are typically shared across designs on the same technology node. The probe test profile for the target fab, however, is not available, since the device has yet to be produced therein. Our objective is to statistically predict parametric probe test yield in the target fab based on the above data. After introducing notation and formulating the problem, we describe three such methods, namely *model migration*, *importance sampling*, and *predictor calibration*.

A. Notation and Problem Formulation

Given a set of e-test measurements, $\mathbf{eT_S}$, and probe test measurements, $\mathbf{PT_S}$, from the source fab, we can use the device specification limits to compute the parametric probe test yield vector for every wafer in our dataset. Each wafer can, then, be represented by:

$$\mathbf{wafer_S^i} = (\mathbf{eT_S^i}, \mathbf{y_S^i}) \qquad (1)$$

where $\mathbf{eT_S^i} = [et_S^{i1}, ct_S^{i2}, ..., et_S^{im}]$ is the $\mathbf{m}$−dimensional vector of e-test measurements for wafer $\mathbf{i}$ and $\mathbf{y_S^i} = [y_S^{i1}, y_S^{i2}, ..., y_S^{ik}]$ is the $\mathbf{k}$−dimensional parametric yield vector for the probe test measurements of wafer $\mathbf{i}$. Let us also denote by $\mathbf{p_S(eT_S)}$ the density function of e-tests over $\mathbf{n_S}$ wafers of the source fab. Similarly, given a set of e-test measurements, $\mathbf{eT_T}$, from the target fab, with their density function over $\mathbf{n_T}$ wafers denoted by $\mathbf{p_T(eT_T)}$, a wafer can be represented by:

$$\mathbf{wafer_T^j} = [\mathbf{eT_T^j}] = [et_T^{j1}, et_T^{j2}, ..., et_T^{jm}] \qquad (2)$$

Our objective is to predict the $\mathbf{k}$−dimensional parametric yield vector for the probe tests of wafer $\mathbf{j}$, $\mathbf{\hat{y}_T^j} = [\hat{y}_T^{j1}, \hat{y}_T^{j2}, ..., \hat{y}_T^{jk}]$, for each of the $\mathbf{n_T}$ wafers of the target fab.

B. Model Migration

A straightforward approach is to use the method discussed in Section II to express parametric yield in the source fab as a function of its e-tests, $\mathbf{Y_S} = \mathbf{f_S(eT_S)}$. Then, the trained regression function can be applied directly to the e-tests of the target fab, in order to predict its parametric yield, $\mathbf{Y_T} = \mathbf{f_S(eT_T)}$. Model migration success relies on two assumptions:

1. Homogeneous distribution between training and testing data sets, i.e. e-tests in the source and target fabs must come from the same distribution, $\mathbf{p_S(eT_S)} = \mathbf{p_T(eT_T)}$.
2. Identical conditional distribution of yield values for training and testing data sets, $\mathbf{p_S(Y_S \mid eT_S^i)} = \mathbf{p_T(Y_T \mid eT_T^j)} \Rightarrow \mathbf{eT_S^i} \approx \mathbf{eT_T^j}$. In other words, if a wafer from the source fab and a wafer from the target fab have the same yield, they must also have similar e-test vectors.

As these assumptions do not necessarily hold true in a semiconductor manufacturing context, the accuracy of model migration is expected to be limited.

C. Importance Sampling

Another approach, which revokes the homogeneity assumption but retains the identical conditional distribution assumption discussed above, is importance sampling. In order to build a model using the training data, which will retain its accuracy

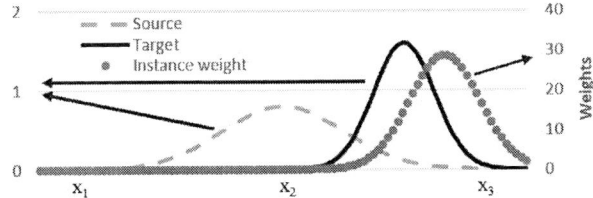

Fig. 3: Instance weighting example.

when used on the testing data, importance sampling biases the selection of the instances from which the model is built. Specifically, preference is given to the most relevant region of the training distribution (source fab e-tests), i.e. the region which overlaps the most with the testing distribution (target fab e-tests). To achieve this, higher weights are assigned to samples from the relevant region, hence the name importance sampling [10]. This method has been successful in various real-world applications [11]–[14].

In our context, we seek to assign weights (importances) to the instances of the source fab e-test distribution, favoring those which are frequently encountered in the target fab e-test distribution with higher weights and penalizing those that are rarely encountered with lower weights. To achieve this, we compute the weight of each instance in the source distribution as the ratio of the target density over the source density. Therefore, we first estimate the density of e-tests for the source and target fabs separately, and then we compute the weight vector through the following equation:

$$\mathbf{W} = \frac{\mathbf{p_T}(\mathbf{eT_S})}{\mathbf{p_S}(\mathbf{eT_S})} \qquad (3)$$

Figure 3 shows an example of source and target fab densities for one e-test, along with the instance weights calculated using Equation 3. The dashed and solid lines represent the density of the training and testing data, while the dotted graph is the weight corresponding to each training instance.

To apply importance sampling to the fab-to-fab migration problem, we first allocate the instance weights for the source fab. Next, we train a regression model to express parametric yield in the source fab as a function of its e-tests, $\mathbf{Y_S}=\mathbf{f_{wS}}(\mathbf{eT_S}, \mathbf{W})$, but elements from the training set are selected with probability commensurate with their assigned weights, instead of equal probability, as in model migration. The trained regression function is, then, applied directly to the e-tests of the target fab to predict its parametric yield, $\mathbf{Y_T}=\mathbf{f_{wS}}(\mathbf{eT_T})$.

Importance sampling is expected to perform better than model migration, as it does not rely on the homogeneity assumption. Nevertheless, it still assumes identical conditional distributions, hence there remains room for improvement.

D. Predictor Calibration

A third approach, which does not rely on any of these two assumptions, is predictor calibration. In this method, the distribution of e-tests in the target fab (i.e. predictors) is calibrated based on the distribution of e-tests in the source

fab, $\widehat{\mathbf{eT_T}} = \mathbf{h}(\mathbf{eT_T}, \mathbf{eT_S})$, prior to being used for predicting parametric probe test yield in the target fab. A simple way of achieving this would be mean calibration, which removes the mean shift, $\mathbf{\Delta}(\mu)$, from each instance of target distribution:

$$\widehat{\mathbf{eT}}_\mathbf{T}^\mathbf{i} = \mathbf{eT}_\mathbf{T}^\mathbf{i} - \mathbf{\Delta}(\mu), \quad \Delta(\mu) = \mu(eT_T) - \mu(eT_S) \qquad (4)$$

However, in order to achieve better precision, other parameters of the distribution, such as variance, skewness and kurtosis, also need to be calibrated. To accomplish this, we employ a two-step procedure. First, in the cumulative distribution function (CDF) of the target fab, we find the cumulative probability associated with each sample in the target distribution, $\mathbf{x_i} = \mathbf{F_T}(\mathbf{eT}_\mathbf{T}^\mathbf{i})$. Then, using the inverse CDF of the source fab, we determine the e-test value associated with cumulative probability $\mathbf{x_i}$, $\widehat{\mathbf{eT}}_\mathbf{T}^\mathbf{i} = \mathbf{F_S}^{-1}(\mathbf{x_i})$, where $\mathbf{F_S}^{-1}$ is the inverse CDF of the source fab distribution. This procedure is applied to all $\mathbf{n_T}$ instances (i.e. wafers) of the target fab distribution. The predictor calibration algorithm is summarized below:

$\widehat{\mathbf{eT_T}} = \oslash$
for $eT_T^i \in eT_T$ **do**
    $\mathbf{x_i} \leftarrow \mathbf{F_T}(\mathbf{eT}_\mathbf{T}^\mathbf{i})$
    $\widehat{\mathbf{eT}}_\mathbf{T}^\mathbf{i} \leftarrow \mathbf{F_S}^{-1}(\mathbf{x_i})$
    $\widehat{\mathbf{eT_T}} \leftarrow \widehat{\mathbf{eT_T}} \cup \widehat{\mathbf{eT}}_\mathbf{T}^\mathbf{i}$
end for

Using this method, the mapping function is defined as:

$$\widehat{\mathbf{eT_T}} = \mathbf{h}(\mathbf{eT_T}) = \mathbf{F_S}^{-1}(\mathbf{F_T}(\mathbf{eT_T})) \qquad (5)$$

In order to utilize predictor calibration in fab-to-fab migration, a regression function is first trained to express parametric yield in the source fab as a function of its e-tests, $\mathbf{Y_S}=\mathbf{f_S}(\mathbf{eT_S})$. Then, the prediction calibration algorithm maps the distribution of e-tests in the target fab into the distribution of e-tests in the source fab, $\widehat{\mathbf{eT_T}} = \mathbf{h}(\mathbf{eT_T})$. Eventually, the trained regression model is applied to the calibrated e-tests of the target fab, in order to predict parametric yield, $\mathbf{Y_T} = \mathbf{f_S}(\widehat{\mathbf{eT_T}})$.

Since predictor calibration does not make any of the two assumptions stated earlier, it is expected to outperform both model migration and importance sampling.

IV. EXPERIMENTAL RESULTS

In order to experimentally evaluate the effectiveness of the proposed yield prognosis methods, we use actual production data from a 65nm analog/RF device currently in high volume manufacturing (HVM) production by Texas Instruments[1]. This data, which is depicted in Figure 4, comprises devices from two geographically dispersed fabs wherein this device is fabricated, which we will refer to as fab A and fab B. The dataset for fab A includes 54 e-test and 168 parametric probe test measurements from a total of 1800 wafers, each of which has 9 e-test measurement sites and approximately 1500 die per wafer. The dataset for fab B includes the same e-test and parametric probe test measurements from a total of 500 wafers,

[1]Details regarding the device cannot be released due to an NDA under which this data has been provided to us.

978-1-4799-7598-3/15 $31.00 © 2015 IEEE

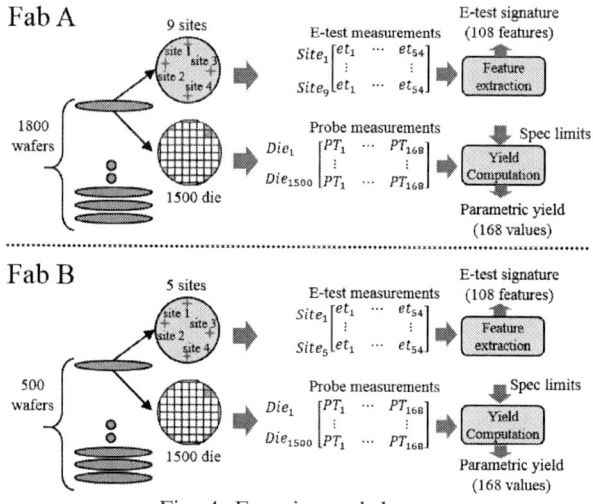

Fig. 4: Experimental dataset.

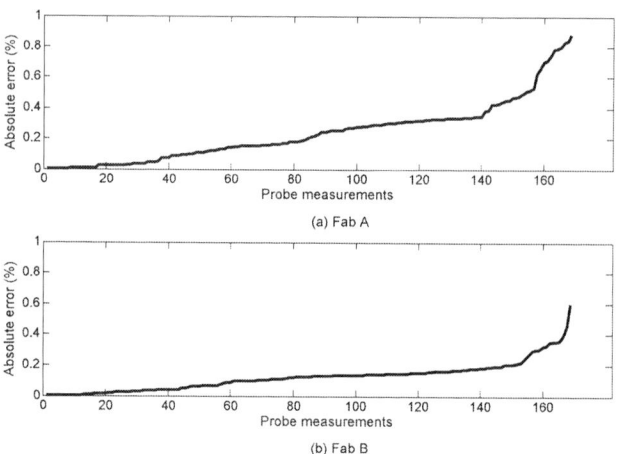

Fig. 5: Within-fab yield prediction error.

with the only difference being that e-tests are obtained on only 5 instead of 9 sites. These two datasets were obtained from the two fabs at approximately the same time period. Along with the data, we are also provided with the specification limits for each of the 168 parametric probe tests, hence we can compute the yield of each performance on every wafer for each of the two fabs. Additionally, for each of the 54 e-test measurements, we compute the mean and the standard deviation across the 9 sites on wafers produced in fab A (5 sites on wafers produced in fab B), hence the e-test signature of each wafer consists of 108 parameters. Using this dataset, we seek to:

- Quantify the accuracy of statistically predicting parametric yield from e-test measurements within a single fab.
- Quantify the accuracy of the described prognosis methods in statistically predicting yield during fab-to-fab product migration based on e-test and probe test profiles of the source fab and only an e-test profile of the target fab[2].

In both cases, we use two metrics to quantify prediction accuracy. The first metric is the average absolute difference, δ_i, between predicted and actual yield for the i-th probe test:

$$\delta_i = \frac{1}{n_T} \sum_{j=1}^{n_T} |\hat{y}_{ij} - y_{ij}| \qquad (6)$$

where n_T is the number of wafers for which the prediction is applied, while $\hat{y}_{ij}$ and y_{ij} are the predicted and the actual yield of the i-th probe test on the j-th wafer, respectively.

The second metric, ϵ_i, normalizes the average absolute difference to the yield range:

$$\epsilon_i = \frac{1}{n_T} \sum_{j=1}^{n_T} \frac{|\hat{y}_{ij} - y_{ij}|}{\max(y_i) - \min(y_i)} \qquad (7)$$

where $\max(y_i)$ and $\min(y_i)$ are the highest and lowest yield values, respectively, of the i-th probe test across all wafers. Expressing prediction error as a percentage of this range is important towards gaging its significance.

[2]We note that the e-test profile of the target fab should be obtained from a different product fabricated therein. Since we only have data from one device, however, we use its e-test profile as a proxy.

A. Within-Fab Yield Prediction

In order to quantify the accuracy of statistically predicting parametric yield from e-test measurements within a single fab, we use 90% of the available wafers from a fab as the training set and the remaining 10% as the test set. Using the 108 e-test features and the 168 yield values reflecting each wafer in our training set, we train a separate regression model (i.e. MARS) for each of the 168 probe tests. The trained regression models are then applied to the 108 e-test features of each wafer in the test set, in order to predict the yield of each of the 168 probe tests across this wafer. The predicted results are, then, compared to the actual values, which are available in the dataset, in order to estimate prediction accuracy. To establish statistical significance, we apply a 10-fold cross validation approach where results are averaged over 10 repetitions, each time randomly splitting the dataset into training and test sets.

Figures 5(a) and 5(b) present the results for the datasets of fab A and fab B, respectively, using the first metric, δ_i, defined in Equation 6. The horizontal axis shows the 168 probe tests, sorted in increasing prediction error, while the vertical axis shows the corresponding average absolute difference between the predicted and actual yield[3]. As may be observed, this difference is in the order of a few tenths of a percentage point, corroborating that parametric probe test yield can be predicted very accurately from the e-test measurements of a wafer.

Figures 6(a) and 6(b) demonstrate the same results, this time using the second metric, ϵ_i, defined in Equation 7. In each histogram, the horizontal axis is the prediction error, while the vertical axis shows the percentage of probe tests that are predicted within a given error range. For example, the first bar shows the percentage of probe test measurements whose normalized average prediction error is below 2%, with the corresponding value being 60% and 21% for fab A and fab B, respectively. As may be observed, the yield of the vast majority of probe tests can be predicted using e-test measurements with an error which is well below 10% of their yield range.

[3]We note that since our test data is Continue on Fail (COF) and a device might fail multiple probe tests, the sum of the yield prediction errors over the 168 probe tests does not reflect the overall yield prediction error.

978-1-4799-7598-3/15 $31.00 © 2015 IEEE   82

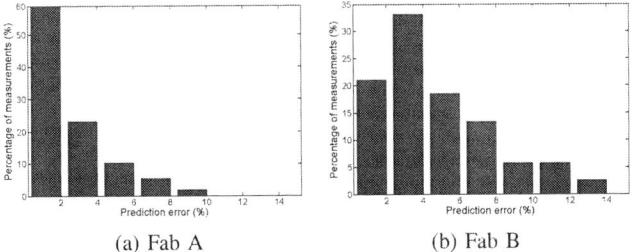

(a) Fab A   (b) Fab B

Fig. 6: Normalized within-fab yield prediction error.

TABLE I: Average within-fab yield prediction error.

Metric	Fab A	Fab B
δ_i	0.23%	0.16%
ϵ_i	3.2%	5.6%

These results are summarized in Table I, where the average absolute prediction error is calculated at 0.23% for fab A and at 0.16% for fab B. Normalized to the yield range of each probe test, the results are 3.2% for fab A and 5.6% for fab B.

B. Yield Prognosis for Fab-to-Fab Migration

In order to quantify the accuracy of the described prognosis methods in predicting yield during fab-to-fab product migration, we performed the following experiment, first using fab A as the source and fab B as the target, and then reversing the roles: using the 108 e-test features and the 168 yield values from every wafer in the dataset of the source fab, as well as the 108 e-test features from every wafer in the dataset of the target fab, we apply the three methods described in Section III to predict the yield for the 168 probe tests in the wafers of the target fab. The predicted values are, then, compared to the actual yield values, which are available in our dataset, in order to estimate prediction accuracy. As a baseline for prediction accuracy, we use the within-fab yield prediction results for the target fab, which were presented in the previous subsection.

Figures 7(a) and 7(b) present the results for product migration from fab A to fab B and vice-versa, respectively, for each of the three methods of Section III, using the first metric, δ_i, defined in Equation 6. The within-fab baseline results are also shown as a point of reference. The horizontal axis shows the 168 probe tests, sorted in increasing prediction error, while the vertical axis shows the corresponding average absolute difference between the predicted and actual yield for each method. We note that, in this case, the vertical axis is in logarithmic scale and in the *model migration* plot connecting lines are omitted in order to enhance figure readability.

As may be observed, *model migration*, wherein the correlation models learned on the source fab are directly applied to the e-tests of the target fab, results in prediction error in the range of 10% and 5% for the two experiments, respectively. This is expected, since this approach assumes homogeneous e-test distributions and identical conditional yield distributions in the two fabs, something that is typically not the case. *Importance sampling*, on the other hand, reduces the yield prediction error to within a couple of percentage points. Evidently, the weighting policy used therein is effective in modeling the

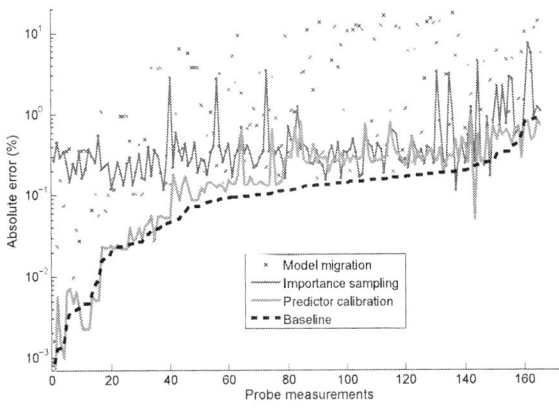

(a) Migrating from fab A to fab B

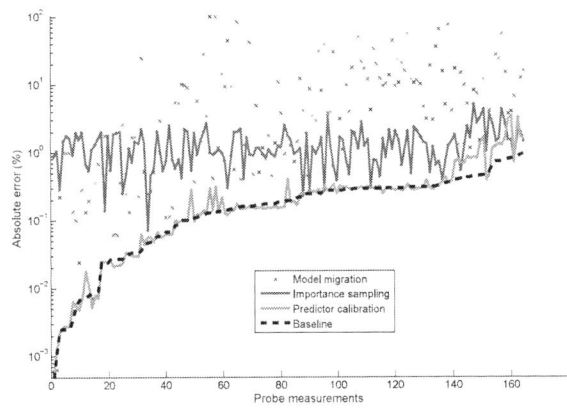

(b) Migrating from fab B to fab A

Fig. 7: Absolute fab-to-fab yield prediction error.

difference in the marginal distribution of e-tests in the source and target fabs. Finally, *predictor calibration* outperforms the other two methods due to the accurate mapping of the target into the source distribution. Indeed, the yield prediction error drops to below half of a percentage point and is very close to the baseline achieved by within-fab correlation models.

Figures 8(a) and 8(b) demonstrate the results of the same experiments, this time using the second metric, ϵ_i, defined in Equation 7. In each histogram, the horizontal axis is the prediction error, while the vertical axis shows the percentage of probe tests that are predicted within a given error range. Separate histograms are shown for each of the three fab-to-fab yield prognosis methods, as well as the within-fab baseline method. As may be observed, the results corroborate our previous observation that the *predictor calibration* method is almost as efficient as the baseline within-fab yield prediction method, achieving accuracy which is within a single-digit percentage of the yield range. In contrast, *model migration* and *importance sampling* are far less accurate, resulting in a normalized prediction error of more than 16% of the yield range for the majority of the probe test measurements.

These results are summarized in Tables II(a) and II(b) for the two experiments, respectively. When migrating from fab A to fab B, the average absolute prediction error over all probe tests and wafers is calculated at 5.52%, 0.98%, and 0.54% for

978-1-4799-7598-3/15 $31.00 © 2015 IEEE

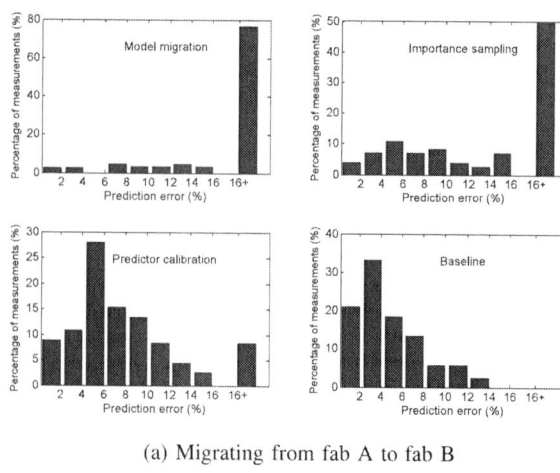

(a) Migrating from fab A to fab B

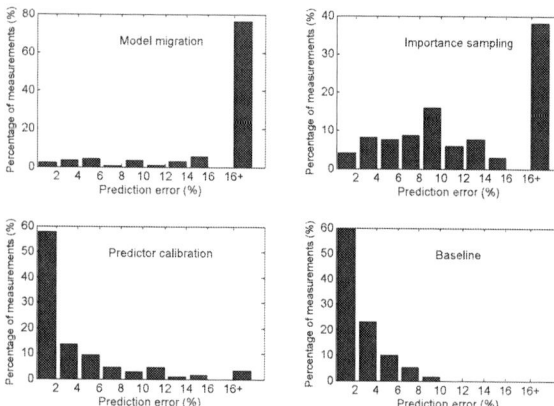

(b) Migrating from fab B to fab A

Fig. 8: Normalized fab-to-fab yield prediction error.

the three fab-to-fab migration methods, respectively, compared to 0.16% for the within-fab baseline yield prediction. Normalized to the yield range of each probe test, the results are 87.1%, 31.4%, 10.3%, and 5.6%, respectively. When migrating from fab B to fab A, the average absolute prediction error over all probe tests and wafers is 9.84%, 1.37%, and 0.35% for the three fab-to-fab migration methods, respectively, compared to 0.23% for the within-fab baseline yield prediction. Normalized to the yield range of each probe test, the results are 82.6%, 28.2%, 4.9%, and 3.2%, respectively.

V. CONCLUSION

E-test and probe test measurements exhibit strong correlation which can be statistically harnessed for yield learning purposes. As we demonstrated using a large dataset from a 65nm Texas Instruments RF transceiver produced in two different fabs, these correlations enable very accurate prediction of parametric probe test yield from e-test measurements within the same fab, with error ranging in the order of a few tenths of a percentage. Moreover, using the e-test and probe test profiles of a source fab and only the e-test profile of a target fab, which can be obtained from prior devices fabricated therein, these correlations facilitate highly accurate yield prognosis when migrating a product across these fabs, with error ranging in the order of half of a percentage.

TABLE II: Average error for fab-to-fab migration.

(a) Migrating from fab A to fab B

Metric	Model migration	Importance sampling	Predictor calibration	Baseline
δ_i	5.52%	0.98%	0.54%	0.16%
ϵ_i	87.1%	31.47%	10.29%	5.6%

(b) Migrating from fab B to fab A

Metric	Model migration	Importance sampling	Predictor calibration	Baseline
δ_i	9.84%	1.37%	0.35%	0.23%
ϵ_i	82.6%	28.2%	4.9%	3.2%

VI. ACKNOWLEDGEMENT

This research has been partially supported by the Semiconductor Research Corporation (SRC) Task 1836.131.

REFERENCES

[1] F. Gong, H. Yu, and L. He, "Stochastic analog circuit behavior modeling by point estimation method," in *ACM International Symposium on Physical Design*, 2011, pp. 175–182.

[2] X. Li, J. Le, P. Gopalakrishnan, and L. Pileggi, "Asymptotic probability extraction for non-normal distributions of circuit performance," in *IEEE/ACM International Conference on Computer-Aided Design*, 2004, pp. 2–9.

[3] R. H. Myers, D. Montgomery, and C. Anderson-Cook, "Response surface methodology: process and product optimization using designed experiment," *John Wiley and Sons, New York*, pp. 343–350, 2002.

[4] C. Fang, F. Yang, X. Zeng, and X. Li, "BMF-BD: Bayesian model fusion on Bernoulli distribution for efficient yield estimation of integrated circuits," in *ACM Design Automation Conference*, 2014, pp. 1–6.

[5] A. Ahmadi, K. Huang, S. Natarajan, J. Carulli, and Y. Makris, "Spatio-temporal wafer-level correlation modeling with progressive sampling: A pathway to HVM yield estimation," in *IEEE International Test Conference*, 2014, pp. 1–10.

[6] N. Kupp, M. Slamani, and Y. Makris, "Correlating inline data with final test outcomes in analog/RF devices," in *IEEE Design, Automation & Test in Europe Conference & Exhibition*, 2011, pp. 1–6.

[7] V. Cherkassky and F. Mulier, *Learning from data: concepts, theory, and methods*, John Wiley & Sons, 2007.

[8] J. H. Friedman, "Multivariate adaptive regression splines," *The annals of statistics*, pp. 1–67, 1991.

[9] P. Variyam, S. Cherubal, and A. Chatterjee, "Prediction of analog performance parameters using fast transient testing," *IEEE Transactions on Computer-Aided Design of Integrated Circuits and Systems*, vol. 21, no. 3, pp. 349–361, 2002.

[10] T. C. Hesterberg, *Advances in importance sampling*, Ph.D. thesis, Stanford University, 2003.

[11] P. Baldi and S. Brunak, *Bioinformatics: the machine learning approach*, MIT press, 2001.

[12] J. J. Heckman, "Sample selection bias as a specification error," *Journal of the Econometric Society*, pp. 153–161, 1979.

[13] M. Sugiyama, M. Krauledat, and K. R. Müller, "Covariate shift adaptation by importance weighted cross validation," *Journal of Machine Learning Research*, vol. 8, pp. 985–1005, 2007.

[14] H. Shimodaira, "Improving predictive inference under covariate shift by weighting the log-likelihood function," *Journal of Statistical Planning and Inference*, vol. 90, no. 2 pp. 227–244, 2000.

3D microelectronic with BEOL compatible devices

D Drouin[1,2], M A-Bounouar[1,2], G Droulers[1,2], M Labalette[1,2], M Pioro-Ladriere[3], A Souifi[4], S Ecoffey[1,2]

[1]Laboratoire Nanotechnologies Nanosystemes (LN2) - CNRS UMI-3463, Universite de Sherbrooke

[2]Institut Interdisciplinaire d'Innovation Technologique (3IT), Universite de Sherbrooke

3000 Boul. Universite, Sherbrooke, J1K OA5, Quebec, Canada

[3]Departement de physique, Universite de Sherbrooke, 2500 Boul. Universite, Sherbrooke, J1K 2R1, Quebec, Canada

[4]Institut des Nanotechnologies de Lyon - UMR CNRS 5270, 7 av. Jean Capelle, 69621 Villeurbanne cedex

e-mail: Dominique.Drouin@usherbrooke.ca

Abstract—This presentation will address the potential of nanoelectronic devices 3D monolithic integration in the CMOS back-end-of-line (BEOL) to add functionality and enhance integrated circuits (ICs) performances.

Keywords—nanoelectronics; 3D integration; single electron transistor; tunneling transistor; resistive RAM

The microelectronic industry has been able to drastically increase the computing performance while reducing the cost-per-chip over the years. In the last years, to keep this performance pace, system innovations regarding architectures, interconnects and packaging were required in addition to traditional CMOS scaling[1]. Furthermore, the power density tendency of assembled modules is reaching level where standard air cooling is not adequate anymore[2]. Due to the ever increasing demand for mobility, this energy consumption issue is becoming the predominant factor for introducing new devices or systems. Tremendous efforts have been deployed to increase the number of devices by mm^2 through 3D integration at the chip or package level. Such approach, can drastically increase the performance by reducing data the latency due to the reduction of interconnects lengths. This will also reduce the power consumption by reducing the interconnects RC losses. To achieve 3D integration heterogeneous and monolithic approaches can be pursued. Among those, the most common heterogeneous technology has been developed at the packaging level where different chips can be stacked on each other and/or integrated on an interposer to adapt the I/O pitch from the chip to the printed circuit board[3]. Recently, a monolithic integration technology has been proposed by the CEA Leti. A Si wafer is bonded on a processed and functional CMOS substrate at low temperature and the devices are fabricated on this second substrates giving rise to two levels of CMOS ICs. Keeping the thermal budget below 450°C when producing the second transistors layer is very challenging but mandatory to avoid deteriorating the underlying CMOS devices[4].

With our approach, we propose to integrate low-power nanoelectronic devices above CMOS ICs using a 3D monolithic technology. While nanoelectronic devices will be used to integrate low-power logic, memory and/or new functionalities such as sensors, CMOS devices will be used for high-performance logic and memory, and for I/O signals restoration. We have developed the nanodamascene process that can produce tunnel junctions with extraordinary small capacitances while maintaining an accurate control on the junction surface. Chemical mechanical planarization (CMP) is used to minimize and control the devices height with a nanometric resolution. This process is compatible with CMOS BEOL technology, materials, and thermal budget. We will present how this process can be used to integrate nanoelectronic devices such as single electron transistors (SETs), tunneling transistors, resistive random-access memories (RRAMs), and complementary resistive switching (CRS) memories. CRS memories do not require 1T1R or 1D1R configuration as in bipolar resistive switching (BRS). The proposed planar CRS architecture yields a high reliability by solving all crosstalk issues of a common BRS passive crossbar array and an increased storage density due to 3D scalability.

We will also introduce the developed SET CAD tools and discuss the circuit design at the nanoscale for a double gate-SET (DG-SET) compact model for hybrid SET-CMOS circuit design. Afterwards, a focus on circuit design methodology based on DG-SET will be reported, wherein the second gate makes the polarity of the device controllable, inspiring new opportunities in low-power logic design. A static versus dynamic biasing of the polarity gate will be presented aiming at a new computational paradigm with low-voltage operation. We will discuss how the static biasing can be used to design CMOS-like DG-SET architectures enabling thereby to reuse existing knowledge and design tools. Thereafter, A circuit design approach based on transmission gates logic style will be addressed This design exhibit extended functionalities offered by conventional MOSFETs in terms of device count, design flexibility, regularity, and reconfigurability.

REFERENCES

[1] ITRS Organisation. Executive summary - international technology roadmap for semiconductors. International Technology Roadmap for Semiconductors Web site. . Updated 2012.

[2] Meyerson B. Echoes of DACs past: From prediction to realization, and watss next? Design Automation Conference (DAC), 2010 47th ACM/IEEE. 2010

[3] Iyer, S. S., T. Kirihata, and J. E. Barth. "Three Dimensional integration-Considerations for memory applications." Custom Integrated Circuits Conference (CICC), 2011 IEEE. IEEE, 2011

[4] Vinet, M., et al. "Monolithic 3D integration: A powerful alternative to classical 2D scaling." SOI-3D-Subthreshold Microelectronics Technology Unified Conference (S3S), 2014

2015 IEEE 33rd VLSI Test Symposium (VTS)

Innovative Practices Session 3C:
Advances in Silicon Debug & Diagnosis

Organizer & Moderator:
Mike Ricchetti (Synopsys)

Presenters & Abstracts:

Presenter #1: *CJ Clark (Intellitech)*

Title: P1149.10 High Speed JTAG – Debug using a fire hose rather than a straw

Abstract: Debug and diagnosis using the IEEE 1149.1 TAP has been a useful tool for engineers for some twenty years. The TAP however is limited as it only can provide a single full duplex data stream of 50 to 100mb/s. IEEE 1500 provides higher bandwidth via parallel acccss to multiple scan-channels however providing physical access to hundreds of pins has become more challenging. This parallel access has little benefit in debug when the SoC is in the system. This presentation focuses on the solution proposed by IEEE P1149.10 which uses a packet protocol over SERDES to access on-chip DFT (instruments) like the TAP but with multi-gigabit SERDES. With the standardization of the IEEE 1149.1-2013 PDL language (Procedural Description Language) which abstracts the TAP, PDL can be used with a higher speed interface as proposed by P1149.10. Use cases of how the proposed standard are shown with benefits for silicon debug.

Presenter #2: *Eric Rentschler (Mentor Graphics)*

Title: The New Era of EDA Supported Post-Silicon Validation

Abstract: In the history of chip design, we have witnessed great efficiency improvements through IP re-use and EDA support. Today, relatively small teams can tackle complex designs by working more efficiently at higher levels of abstraction. EDA has off-loaded much of the tedium of the past, especially in the areas of test, physical design and pre-silicon verification. We now have industry standards and off-the-shelf EDA solutions. Design-for-test has gone from a black-art to a science, complete with academic classes. However, one of the most challenging and costly areas of development that continues to grow has been post-silicon validation. Yet EDA has yet to step-up in this space. This talk will address how Mentor Graphics is stepping-up in order to provide powerful off-the-shelf solutions for post-silicon validation and debug.

Presenter #3: *Eric Thorne (Xilinx)*

Title: FPGA as Programmable Diagnosis Platform

Abstract: A FPGAs inherent structure lets us easily test and provide accurate, precise, targeted diagnostics to specific regions and layers of the units under test. Aggregating this test and diagnostic data helps us significantly improve both defect driven and parametric driven yield in short time frames. We'll show how memory cell diagnostics can find defects in logic circuits, how to create tests that primarily target specific back end metal layers, and how to find process uniformity issues.

978-1-4799-7598-3/15 $31.00 © 2015 IEEE

PANEL:

When Will the Cost of Dependability End Innovation in Computer Design?

Organizer: Valeria Bertacco - University of Michigan

Abstract

As silicon feature sizes approach atomic scales, device reliability is waning and the cost of dependability is on the rise. Post silicon devices, such as CNTs or TFETs, promise better performance but at the cost of even worse reliability. Will we reach the point where the cost of reliability for future silicon substrates is too expensive to justify their existence? Or will we discover new ways to contain the cost of dependability? If we do discover low-cost reliability mechanisms, how much time do we have before we must deploy them? If not, how much life does silicon have left?

Moderator

Tim Cheng – Professor, University of California at Santa Barbara

Panelists

Andrew Kahng – Professor, University of California at San Diego

Ritesh Parikh – Power and Performance Architect, PDG Architecture Group, Intel Corp.

Siva Kumar Sastri Hari – Research Scientist, Architecture Research Group, NVIDIA

Todd Austin – Professor, University of Michigan

Special Session: Hot Topics: Statistical Test Methods

Organizers: Manuel J. Barragan, *TIMA, CNRS-Univeristé Grenoble-Alpes*, France
Gildas Leger, *IMSE-CNM, CSIC-Universidad de Sevilla*, Spain
Speakers: Florence Azais, *LIRMM, CNRS-Université Montpellier 2*, France
R. D. (Shawn) Blanton, *Carnegie Mellon University*, USA
Adit D. Singh, *Auburn University*, USA
Stephen Sunter, *Mentor Graphics*, Canada

The process of testing Integrated Circuits involves a huge amount of data: electrical circuit measurements, information from wafer process monitors, spatial location of the dies, wafer lot numbers, etc. In addition, the relationships between faults, process variations and circuit performance are likely to be very complex and non-linear. Test (and its extension to diagnosis) should be considered as a challenging highly dimensional multivariate problem.

Advanced statistical data processing offers a powerful set of tools, borrowed from the fields of data mining, machine learning or artificial intelligence, to get the most out of this data. Indeed, these mathematical tools have opened a number of novel and interesting research lines within the field of IC testing.

In this special session, prominent researchers in this field will share their views on this topic and present some of their last findings. The first talk will discuss the interest of likelihood prevalence in random fault simulation. The second talk will show how statistical data analysis can help diagnosing test efficiency. The third talk will deal with the reliability of Alternate Test of AMS-RF circuits. The fourth and last talk will address the idea of mining the test data for improving design manufacturing and even test itself.

A. *Random sampling for fault simulation: intuition vs. theory and reality*, by Stephen Sunter

Most engineers understand the theory of random sampling, but often their intuition is inconsistent with the theory or reality. Theory for a new likelihood-weighted random sampling technique was presented in our ITC 2014 paper [1]. This presentation provides real examples of inconsistencies based on experiences with a commercial analog fault simulator at multiple companies. The examples: intuition says that simulating more defects might produce a higher estimated coverage, but theory shows this is unlikely; intuition says that pre-simulation analysis of a circuit could reveal defects that do not need to be simulated, but reality shows this is impractical; intuition says you need to simulate at least a few percent of all potential defects, but theory shows otherwise; intuition says that coverage of portions of a circuit can be gleaned from results for faults randomly injected into the

whole circuit, but theory shows otherwise; intuition says you must simulate coverage of every defect type, but reality shows otherwise; intuition says that improving a test to detect the most-likely defects that were undetected will have the greatest impact on coverage, but theory shows otherwise. We conclude that likelihood-weighted random sampling is more general and practical than other approaches for reducing fault simulation time, but its efficiency can be counter-intuitive.

B. *Targeting Opens versus TDF in Two-Pattern Scan Testing: What Defect Statistics May Be Telling Us*, by Adit D. Singh

Industry has now had over a decades worth of experience with scan based TDF timing tests, yet many questions still remain regarding the effectiveness of this methodology, and even how it is best applied. Many companies strongly believe that launch-on-capture (LOC) tests alone are sufficient for screening manufacturing defects since they can test all transitions between functional states that are encountered in normal functional operation; this view also often holds that launch-on-shift (LOS) tests can potentially lead to 'overtesting' and yield loss. Others consider LOS timing tests essential for high TDF coverage and are willing to make the design-for-test (DFT) investment in timing closed scan enable control signals needed to support at-speed LOS testing. So what physical defects are we really catching with each type of test? Are we 'overtesting'? Is small delay testing worth the extra test cost and potential yield loss? These questions can only be reliably answered with detailed statistics from volume production tests on a range of manufactured parts.

In the absence of such comprehensive data, at least in the public domain, we piece together the best available evidence and show that it appears to challenge conventional wisdom and current test practice. We make the case that TDF tests, even when applied with aggressive timing, appear to mostly detect open defects, the majority of which can be detected at somewhat slower test speeds without the risk of unnecessary yield loss from test noise. Meanwhile, many other open defects that can cause operational failures remain undetected by current LOC, and even LOS, TDF tests, as has been shown by recently published studies with Cell-Aware tests. These test

escapes can significantly compromise product defect levels. However, even Cell-Aware tests currently do not target a significant class of open defects that appear to be redundant but can in fact frequently cause functional failure due to circuit hazards. We therefore suggest that it may be better for two-pattern tests to explicitly target all open faults in the circuit, with the tests being applied at the highest possible speed that avoids yield loss from test noise. TDF faults will implicitly be covered by such an approach.

C. *Statistical techniques and metrics for alternate testing of analog/RF integrated circuits*, by Florence Azaïs

The concept of alternate testing emerged in the late 90s with the objective to reduce testing costs of analog integrated circuits by replacing the conventional specification measurements with a single transient acquisition using a carefully optimized test stimulus [2], [3]. This concept has then been extended to RF circuit testing with the objective to replace the costly RF performance measurements by simple low-cost indirect measurements. In both cases, the fundamental idea of the technique is to exploit the underlying relationships that exist between indirect measurements and conventional measurements in order to build prediction models that permit to evaluate the device performances using only the low-cost indirect measurements. Because these relationships are non-trivial, statistical methods have to be used to build prediction models.

This approach has been widely explored and demonstrated in the literature on various case studies over the past twenty years. However alternate testing is still not widely used in industry mainly because of a lack of confidence in the achieved test efficiency. Many factors influence this efficiency such as the choice of adequate indirect parameters, the choice of the prediction model, the order of mapping between indirect measurements and device specifications, or the size and composition of the training set. The objective of this talk is to discuss various statistical methods for the choice of adequate indirect measurements [4]–[7], pertaining to both filter and wrapper categories in the field of feature selection. Their impact on alternate test efficiency will be evaluated in terms of model and prediction accuracy by using classical metrics such as average and maximal errors, but also in term of prediction reliability by introducing a new metric called Failing Prediction Rate (FPR). Results are illustrated on two case studies for which we have experimental test data, i.e. a power amplifier and a RF transceiver.

D. *What Gold can be Mined from Test Data?*, by Shawn Blanton

Test data can take on many forms, ranging from pass-fail information for both binary and analog tests, to various parametric measurements that include min VDD and IDDQ to full-blown quantification of various specifications. Undoubtedly these measurements are taken to determine if the circuit under test is functioning as desired. It is becoming more and more evident however that there is valuable information,

beyond go/no-go, that is hidden within the test data. The challenge however is developing sound methodologies that de-convolute the test data in order to make statistically-significant conclusions about the information derived.

For over ten years now, researchers in the Advanced Chip Testing Laboratory (www.ece.cmu.edu/ actl) have and continue to develop various methodologies for mining test data to uncover actionable information for improving design, manufacturing and even test itself. A key enabling technology is software-based diagnosis, a topic of significant interest as of late because of its role in test-data mining. In this talk, we will make a case for all the gold that can be mined from test data and the challenges involved that range from having a precise and accurate chip-level diagnosis methodology to obtaining significant levels of data for making sound conclusions using actual in-production chips. Finally, we will make the case for increasing the fidelity of test-data mining through the design and fabrication of product-like test chips which are actual ICs that are designed to be highly testable and diagnosable while at the same time reflecting characteristics of actual customer designs.

References

[1] S. Sunter, K. Jurga, P. Dingenen, and R. Vanhooren, "Practical random sampling of potential defects for analog fault simulation," in *International Test Conference*, 2014.

[2] P. N. Variyam and A. Chatterjee, "Enhancing test effectiveness for analog circuits using synthesized measurements," in *16th IEEE VLSI Test Symposium, 1998. Proceedings*, 1998, pp. 132–137.

[3] P. N. Variyam, S. Cherubal, A. Chatterjee, T. I. Inc, and T. X. Dallas, "Prediction of analog performance parameters using fast transient testing," *IEEE Transactions on Computer-Aided Design of Integrated Circuits and Systems*, vol. 21, no. 3, pp. 349–361, 2002.

[4] H. Ayari, F. Azais, S. Bernard, M. Comte, M. Renovell, V. Kerzerho, O. Potin, and C. Kelma, "Smart selection of indirect parameters for DC-based alternate RF IC testing," *VLSI Test Symposium (VTS), 2012 IEEE 30th*, pp. 19–24, 23-25 April 2012.

[5] M. Barragan and G. Leger, "Efficient selection of signatures for analog/RF alternate test," in *Test Symposium (ETS), 2013 18th IEEE European*, May 2013, pp. 1–6.

[6] J. Liaperdos, A. Arapoyanni, and Y. Tsiatouhas, "Adjustable RF Mixers' Alternate Test Efficiency Optimization by the Reduction of Test Observables," *Computer-Aided Design of Integrated Circuits and Systems, IEEE Transactions on*, vol. 32, no. 9, pp. 1383–1394, Sept. 2013.

[7] S. Larguech, F. Azais, S. Bernard, V. Kerzerho, M. Comte, and M. Renovell, "Evaluation of indirect measurement selection strategies in the context of analog/RF alternate testing," *Test Workshop - LATW, 2014 15th Latin American*, pp. 1–6, 12-15 March 2014.

2015 IEEE 33rd VLSI Test Symposium (VTS)

ExTest Scheduling for 2.5D System-on-Chip Integrated Circuits*

Ran Wang[†], Guoliang Li[‡], Rui Li[‡], Jun Qian[‡], and Krishnendu Chakrabarty[†]

[†]ECE Dept., Duke University, Durham, NC, USA   [‡]AMD Inc. Shanghai, China

{rw118, krish}@duke.edu {guoliang.li, peter.li, jun.qian}@amd.com

Abstract—**Interposer-based 2.5D integrated circuits (ICs) enable high-density interconnects, but introduce new challenges for the testing of a system-on-chip (SoC) die on an interposer. This paper presents an efficient ExTest scheduling strategy that implements interconnect testing between tiles inside an SoC die while satisfying the practical constraint that the number of required test pins cannot exceed the number of available pins at the chip level. The tiles in the SoC are divided into groups based on the manner in which they are interconnected. In order to minimize the test time, two optimization solutions are introduced. The first solution minimizes the number of input test pins, and the second solution minimizes the number of output test pins. We present scheduling and optimization results for a "monster" die with 50 million flip-flops in a 2.5D IC, which is currently in production, to highlight the effectiveness of the proposed test strategy.**

I. Introduction

In keeping with Moore's law, integrated circuits (ICs) are being aggressively scaled in order to achieve increased functionality and higher performance. However, continued scaling results in increasing interconnect delay, which is a key limiter for chip performance. A potential solution to this problem lies in the exploration of new types of interconnects. As technology advances, chip-scale wires and through-silicon vias (TSVs) are emerging as promising solutions to reduce the interconnect length [1]. These solutions are being incorporated in 2.5D ICs [2], which are viewed as a precursor to 3D integration, but with lower fabrication cost and design complexity [3], [4].

A 2.5D IC relies on the use of a passive silicon interposer that is placed between the package and the dies [5]. The microbumps between the dies and the interposer can be as small as 10 μm in diameter and 40 μm in pitch [6]. Therefore, a large number of input and output (I/O) ports are available for the dies in a 2.5D IC. However, the majority of the I/O ports are connected to other dies through horizontal interconnects inside the interposer. External I/O ports for a large die are connected to TSVs, but they are much fewer in count than the total number package pins available for the same die in a 2D IC [7]. As a result, the number of test pins available for testing a die in a 2.5D IC is much smaller than that in 2D package.

Consider the following example based on an actual design. Die 1 is packaged as a 2D IC and it has a total of 1087 I/O ports. These I/Os can be probed for testing. Next suppose that Die 2 is integrated in a 2.5D IC, but it has similar functionality and size as Die 1. Because Die 2 is mounted on the interposer

and it has a new I/O interface, it has a total of 7055 I/O ports. However, 6576 I/O ports are connected to other dies on the interposer, and only 479 I/O ports are connected to external I/Os through TSVs. Even if these 479 I/O ports are available as dedicated test pins, Die 2 cannot be fully tested using only these I/Os.

The typical structure of a die in a 2.5D IC for a real application is shown in Fig. 1. The die consists of multiple tiles, and tiles are located in the four regions of the die: top-left (TL), top-right (TR), bottom-left (BL), and bottom-right (BR). Each tile can only be accessed by the test pins that are in the same region. Two types of testing are involved when the dies are tested: InTest and ExTest. InTest refers to the testing of the internal logic of all of the tiles. ExTest refers to the testing of the interconnects between different tiles. InTest can be carried out easily because the internal logic of each tile is independent from the other tiles. Thus, each tile can be tested independently. However, ExTest cannot be carried out in this way because multiple tiles must be enabled simultaneously to test the interconnects between them. If all of the tiles are enabled for interconnect testing at the same time, the number of required test pins exceeds the available test pins, hence ExTest cannot be carried out as desired.

In addition, ExTest solutions for core-based SoCs are not applicable for testing the interconnect between the tiles in a die in a 2.5D IC. In core-based SoCs, the hierarchy is SoC → cores and ExTest targets the interconnects among cores. When solutions for core-based SoCs are applied to a 2.5D IC, the chip can be viewed as the SoC and dies on the interposer can be viewed as cores. In this scenario, interposer interconnects are tested because they can be considered as interconnects between the cores in the SoC. However, in the realistic scenario being considered in this paper, the hierarchy is 2.5D IC → dies → tiles, and the objective of testing is to target the interconnects among tiles, which is one level deeper.

In the ExTest method currently used for the industry design, the CPU time for test generation is as high as 30 days for a large die while the interconnect fault coverage is less than 86%. These are serious concerns that necessitate a rethinking of the ExTest scheduling problem.

In this paper, we present an efficient ExTest scheduling strategy that reduces CPU run time and increases fault coverage while satisfying the constraint that the number of test pins required does not exceed the number of available test pins at the chip level. In the proposed strategy, tiles are placed in groups based on the interconnect relationship between them. The tiles in a group are enabled simultaneously, and testing of one group is referred to as a single test round. In this way, we ensure that the groups of tiles are mutually independent

---

*This research was supported in part by the National Science Foundation under grant no. CCF-1017391 and by the Semiconductor Research Corporation under contract no. 2470.

978-1-4799-7598-3/15 $31.00 © 2015 IEEE

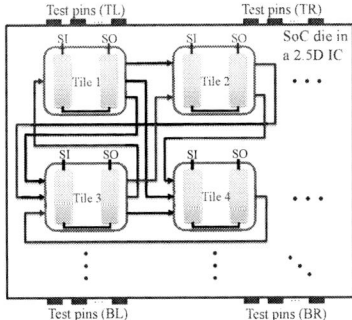

Fig. 1. Structure of a die in 2.5D ICs.

from each other; all transition and stuck-at faults for the interconnects can be detected. In addition, in order to minimize the number of test rounds, we introduce two optimization solutions. The first solution minimizes the number of input test pins required, and the second solution minimizes the number of output test pins. We present scheduling and optimization results for a "monster" die with 50 million flip-flops in a 2.5D IC in volume production to highlight the effectiveness of the proposed strategy.

The rest of this paper is organized as follows. Section II describes the drawbacks of the ExTest method currently in use for production test; the ExTest architecture is also introduced. Section III presents the proposed ExTest scheduling strategy that can detect all tile-internal interconnect faults. In Section IV, two optimization methods are described. Section V presents experimental results on ExTest scheduling and optimization for a large die. Section VI concludes the paper.

II. TEST ARCHITECTURE AND CURRENT SOLUTION

In the ExTest architecture in use at our industry collaborator for 2.5D ICs, each tile is wrapped by several wrappers. The structures and locations of these wrappers for a single tile are illustrated in Fig. 2. These wrappers are not based on the IEEE 1500 Std; instead, they are based on traditional scan cells [8]. Because the number of interconnects between each pair of tiles is extremely large (e.g., 13652 in the case of a real design), wrappers cannot be added to all primary inputs and outputs of a tile.

The wrappers are therefore classified by their locations, which are shown in Fig. 2. The first type of wrapper is one that is connected directly to a primary input or output; these wrappers are referred to as dedicated wrappers. The second type of wrapper is one that reuses the internal scan cell; these wrappers are referred to as shared wrappers. Because the majority of wrappers are shared wrappers, the area overhead is dramatically reduced compared to a design with dedicated wrappers for each primary input and output. In our design, parts of a tile's internal (combinational) logic are between primary inputs (outputs) and the shared wrappers; transition and stuck-at faults for these combinational logic blocks need to be targeted in ExTest. As a result, ExTest is needed not only for interconnect testing but also for the testing parts of combinational logic internal to the tiles.

The scan chains for ExTest are shown as the vertical dashed lines in Fig. 2. Embedded deterministic test (EDT)

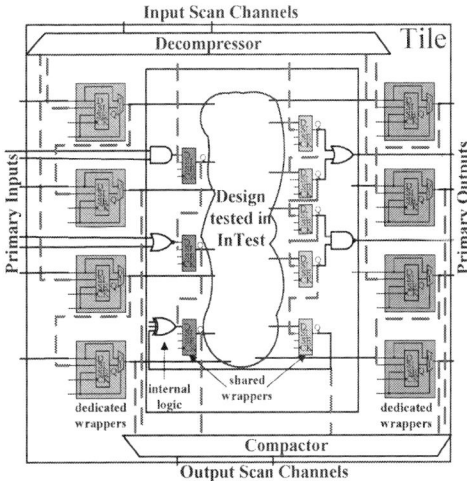

Fig. 2. Tile design wrapped with dedicated/shared wrappers and interfaced with EDT scheme.

[9] is used to reduce scan test-data volume for production test. The EDT design consists of a decompressor on the scan input side and a compactor on the scan output side. In Fig. 2, there are two input scan channels and six scan chains; this design corresponds to a compression ratio of three. In practice, to ensure nearly equal scan-chain length for all of the tiles, the ExTest compression ratio for different tiles varies from 20 to 50. Because the compression ratio is related to the linear-feedback-shift register (LFSR) configuration time during decompression, it has only a limited impact on the test time, as discussed in Section V. The compression ratio is varied by changing the number of input or output scan channels because the number of internal scan chains for a tile is fixed. Therefore, the number of scan channels does not affect the ExTest test time.

The ExTest procedure can be summarized as follows. First, the compressed test patterns are shifted to the decompressor through the input scan channels. Then, the test patterns are decompressed and shifted to the scan chains. Second, the test patterns are launched to the interconnects, either through dedicated wrappers or through shared wrappers with combinational logic circuits. These patterns pass through the interconnects and are captured either by dedicated wrappers or by shared wrappers. Next, the test responses are shifted to and compressed by the compactor. Finally, the compressed responses are shifted out through the output scan channels.

In the test method used thus far for the industry design, the tiles are grouped randomly, without considering the interconnect relationship between the different tiles. However, the tile designs are compiled together for design-rule checking (DRC) and automatic test-pattern generation (ATPG). In contrast to traditional ATPG for interconnect testing [10], ExTest ATPG for dies in 2.5D ICs is much more complex and can take a very long time because it not only conducts open/short and at-speed test for interconnects but also targets combinational logic for stuck-at faults and transition faults. For the die used in our industry collaborator's 2.5D IC with a total of 531 tiles, the entire design was loaded to a server with 512 GB memory. The generation of 64 test patterns (the DRC and ATPG steps) took 5-7 days, and the generation of 512 test patterns took

978-1-4799-7598-3/15 $31.00 © 2015 IEEE

as much as 30 days. In addition, test patterns generated by ATPG are required for debug based on the test data for first silicon. As a result, the time-to-market (TTM) is affected by the inefficiencies of the currently used test-generation method.

A total of 512 test patterns is sufficient to detect all faults in the interconnects and the associated combinational logic circuits. However, the fault coverage obtained using the current solution is much less than the desired value of 100% because the tiles are randomly grouped: the interconnects and combinational logic circuits connecting different test groups remain untested. For the large die with a total of 531 tiles that we consider, when its design is divided into four groups, 14% of the faults cannot be detected. If the design is divided into more groups to reduce CPU time, the fault coverage drops below 86%.

III. PROPOSED SCHEDULING STRATEGY

The proposed strategy places tiles into groups on the basis of the functional interconnect between them. In this way, no interconnects exist between groups and each group of tiles is independent from others, thus test patterns can be generated separately and in parallel for each group. Because the design within each group is much smaller than the entire design, it can be easily compiled. The DRC and ATPG run time for a single group can be dramatically reduced.

The goal of forming independent test groups is to minimize the total test time. In each test round, test patterns are applied in parallel to all the tiles in the same test group. Thus, the test time for a single test round is determined by the tile with the longest test chain. Because the tiles in each group have similar scan chain lengths, the test time the different test rounds are nearly equal. Hence, the objective of minimizing the total test time can be viewed as minimizing the number of test rounds. Because the total number of interconnects for a die is fixed, the number of test rounds can be minimized by testing as many interconnects as possible in one test round.

The optimization problem can be defined as follows. We are given a die with a set of M tiles. As shown in Fig. 1, we divide the die into four regions (TR, TL, BR, BL) and use pins available at the boundary of a region to test tiles in that region. The parameters considered for optimization are defined in Fig. 3. Note that some test pins are bidirectional. The sum of TR_{input} and TR_{output} is larger than TR. The test pins on the other three regions have similar parameters, namely TL, TL_{input}, TL_{output}, BR, BR_{input}, BR_{output}, BL, BL_{input}, and BL_{output}. If Tile i is located in the top-right region, tr_i is 1 and the other three parameters are 0. The goal is to find a test group such that the number of tested interconnects is maximized while enforcing the constraint that the number of test pins required cannot exceed the total number of test pins available in the four regions.

We use integer linear programming (ILP) to solve the above problem. Although ILP models are computationally intractable and often not feasible for large problem instances, for up to a few hundred tiles per die, the problem instance is small enough to be amenable to ILP.

During the testing of a test group, the tiles that belong to the group are referred to as "enabled tiles", and tiles that are not in the group are referred to as "disabled tiles". Each enabled tile can be classified as an "assistive tile" or a "tested tile" based on its functionality. The assistive tiles can only be used to launch patterns to the interconnects under test, and the interconnects under test are only connected to primary outputs of assistive tiles. The tested tiles are used to capture responses from the interconnects under test, and these interconnects are connected to the primary inputs of tested tiles. An illustration of the different types of tiles is shown in Fig. 4. Tiles can be reused as assistive tiles for different test groups if their primary outputs are connected to tiles in different test groups; e.g., see Tile 4 in Fig. 4. Some tested tiles can be used to launch patterns if their primary outputs are connected to the primary inputs of other tested tiles, e.g., Tile 5 in Fig. 4.

---

1. TR: the total number of test pins in top-right region of the die;
2. TR_{input}: the number of test pins that can serve as inputs in TR;
3. TR_{output}: the number of test pins that can serve as outputs in TR;
4. I_i: the number of input scan channels for Tile i;
5. O_i: the number of output scan channels for Tile i;
6. tr_i, tl_i, br_i, and bl_i: binary values, indicate the location of Tile i;
7. w_{ij}: the number of interconnects from Tile i to Tile j.

Fig. 3.   Definition of parameters in the optimization problem.

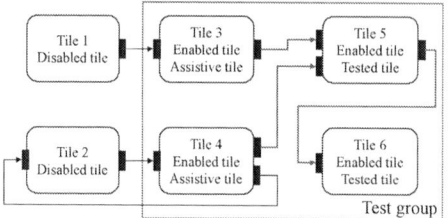

Fig. 4.   Illustration of different types of tiles.

Based on the definition of enabled tiles and tested tiles, two binary variables x_i and y_i, $1 \le i \le M$, are defined. The variable x_i is equal to 1 if tile i is included in the test group and utilized as a tested tile. Similarly, y_i is equal to 1 if tile i is included in the test group and utilized as an enabled tile. Two constraints on variables x_i and y_i are first defined as follows:

$$x_i \le y_i, \forall i, \text{ and } (y_i - x_j) \cdot w_{ij} \ge 0, \forall i, j, \ \forall w_{ij} \ge 1.$$

The first constraint defines the relationship between x and y for the same tile: a tested tile must be an enabled tile. This is based on the definitions of an enabled tile and a tested tile. In the second constraint, if x_j is equal to 1 and w_{ij} is greater than 1, then variable y_i must be equal to 1. This indicates the relationship between x and y for different tiles: if Tile j is a tested tile, all tiles whose primary outputs connect to Tile j must be enabled tiles.

As discussed in Section II, after the test patterns are launched to the interconnects, most patterns go through combinational logic circuits and are captured by shared wrappers. As a result, each test response must be determined based on several test patterns from different tiles. Therefore, to avoid the capture of unknown responses, all interconnects feeding a tested tile must be fully controlled in each test round. In other words, all interconnects feeding a tested tile must be tested simultaneously; no interconnects can be tested in other test groups. Hence, the tested tiles in one test group cannot be reused as tested tiles in other test groups. This leads to two other constraints on x_i and y_i that are defined as follows:

$$\sum_{i=1}^{M} w_{ij} \geq x_j, \forall j, \text{ and } x_i + \sum_{j=1}^{M} w_{ij} \geq y_i, \forall i.$$

The first constraint addresses the situation where x_i is equal to 0. If there are no interconnects feeding Tile j, Tile j cannot be a tested tile. After one test round, the tested interconnects are not considered in subsequent test rounds. Therefore, no interconnects will be allowed to feed the already tested tiles. This constraint guarantees that tested tiles in one test group cannot be reused as tested tiles in other test groups. The second constraint considers the case when y_i is equal to 0. If Tile i is not a tested tile and no interconnects connect to its primary outputs, then Tile i cannot be an enabled tile. The constraints on the number of test pins in the top-right region are defined as: (1) $\sum_{i=1}^{M} I_i \cdot tr_i \cdot y_i \leq TR_{input}$; (2) $\sum_{i=1}^{M} O_i \cdot tr_i \cdot y_i \leq TR_{output}$; (3) $\sum_{i=1}^{M} (I_i + O_i) \cdot tr_i \cdot y_i \leq TR$.

Constraint (1) indicates that the sum of the input scan channels of all of the enabled tiles in the top-right region cannot exceed the number of available input test pins; Constraint (2) ensures that this requirement is satisfied for the output scan channels and test pins. Constraint (3) denotes the fact that the sum of the scan channels of all of the enabled tiles in the top-right region cannot exceed the total number of test pins available in this region. Because the sum of TR_{input} and TR_{output} is larger than TR, these constraints provide the flexibility for arranging the input and output test pins inside a region. The constraints on the number of test pins for the other three regions are similar to Constraints (1), (2), and (3). With the variables defined above, our objective is to maximize the total number of interconnects tested in one test round for a die with a set of M tiles, which is given by:

$$\sum_{j=1}^{M} \left[\left(\sum_{i=1}^{M} w_{ij} \right) \cdot x_j \right]$$

The quantity $\sum_{i=1}^{M} w_{ij}$ represents the total number of interconnects feeding Tile j. The product $\left(\sum_{i=1}^{M} w_{ij} \right) \cdot x_j$ indicates whether interconnects must be added to the objective function based on whether Tile j is a tested tile. Finally, the total number of tested interconnects is the total number of interconnects feeding all the tested tiles in one test round.

Note that the ILP model is used to generate one test group. Hence the ILP model must be invoked several times in order to generate all the test groups for ExTest. First, the initial interconnection matrix is loaded into memory and the ILP model is invoked to generate the first test group. After one test round, the tested interconnects are eliminated from the interconnection matrix and all interconnects feeding the tested tiles are eliminated. The ILP model is then used to generate the next test group based on the updated interconnection matrix. This process is repeated until that all interconnects are tested or the updated interconnection matrix cannot satisfy the constraints listed above.

IV. SCHEDULE OPTIMIZATION

Because the test time is proportional to the number of test rounds and the test times for the different rounds are similar,

---

If Tile i does not share the same input scan channels with other tiles
    $s_{ii} = 1$;
    $s_{ij} = 0, \forall i \neq j$;
If S_i exists and Tile i is the representative Tile of S_i
    $s_{ij} = 1, \forall j \in S_i$;
    $s_{ij} = 0, \forall j \notin S_i$;
    $s_{jk} = 0, \forall j \neq i \ \& \ j \in S_i, \forall k$;

Fig. 5. Definition of the binary parameter s_{ij} in different cases.

---

the test time can be minimized by decreasing the number of test rounds. This can be accomplished by including more tiles in a single test round. However, the number of tiles in a test round is limited by the number of available test pins. As a result, the number of scan channels for each tile must be reduced such that the available test pins can be assigned to more tiles. In this section, two optimization methods are introduced: (i) a sharing of inputs method to decrease the number of input scan channels; (ii) an output removal method to decrease the number of output scan channels.

A. Sharing of inputs

Some of the tiles in a die are instances of the same design, and therefore they have similar functions even though they are connected to different tiles. These tiles can thus share the same input scan channels for ExTest. As a result, the number of input scan channels needed for these tiles can be reduced, and the test pins that are originally connected to these input scan channels can be assigned to other tiles. For example, if 10 tiles are instances of the same design and each of them requires two input scan channels, a total of 20 input scan channels are required. However, if they share the input scan channels, only two input scan channels are required and the other 18 input scan channels can be utilized elsewhere. Therefore, sharing of inputs decreases the number of test rounds.

The sharing of inputs can be easily implemented in a die. To update the ILP model for optimization, several new parameters are introduced based on the interrelationship between the tiles. The set S is defined as a set of tiles that share the same input scan channels. One tile is chosen as the representative tile for each such set. If Tile i is the representative tile for a set, this set is denoted as S_i. A new binary variable s_{ij} is defined in Fig. 5.

When inputs are shared, not all enabled tiles are considered in the input constraint (Constraint (1)); only the representative tiles and the tiles that do not share input scan channels are considered. Thus, a new binary variable, z_i, is defined to update the ILP model. The variable z_i is equal to 1 if tile i is considered in the input constraint, else it is 0. There are two constraints on variable z_i, one for each value of z_i, that are defined as follows:

$$z_i \geq y_i \cdot s_{ij}, \forall i, j, \text{ and } z_i \leq \sum_{j=1}^{M} (y_i \cdot s_{ij}), \forall i.$$

The first constraint defines the situation where z_i is equal to 1: if any of the tiles in S_i are enabled tiles, the representative tile (Tile i) will be considered in the input constraint. The second constraint defines the situation where z_i is equal to 0: tile i will not be considered either if Tile i is a representative tile but no tiles are enabled tiles in S_i or if tile i is not a representative tile but is in a shared set. Based on the above constraints, it can be seen that z_i is equal to y_i for a tile that does not share input scan chains with others. The constraints on the number of test pins in the top-right region are updated as follows:

$$\sum_{i=1}^{M} I_i \cdot tr_i \cdot z_i \leq TR_{input}, \quad \sum_{i=1}^{M} (I_i \cdot z_i + O_i \cdot y_i) \cdot tr_i \leq TR$$

The constraints on the number of test pins in the other three regions can be similarly updated.

B. Output removal

During an ExTest test round, after the test patterns are shifted into the enabled tiles and launched to the interconnects, the test responses are captured by the tested tiles. The test responses are later shifted out from the output scan channels of the tested tiles. Therefore, the output scan channels of the tested tiles need to be connected with the output test pins to observe the responses. For purely assistive tiles during a test round, no responses are recorded from their output scan channels. As a result, the output scan channels of assistive tiles need not be connected to the output test pins. If the output scan channels of the assistive tiles are not connected to anything for the ExTest round, the saved output test pins can be assigned to more tested tiles. Therefore, output removal is an effective method for decreasing the number of test rounds.

The method of output removal can be easily implemented in a die. In the ILP model, not all enabled tiles are now considered in the output constraint (2); only the tested tiles are considered. Thus, no additional parameters are required to update the model. The constraints on the number of test pins in the top-right region are updated as follows:

$$\sum_{i=1}^{M} O_i \cdot tr_i \cdot x_i \leq TR_{output}, \quad \sum_{i=1}^{M} (I_i \cdot z_i + O_i \cdot x_i) \cdot tr_i \leq TR$$

In these constraints, y_i is replaced by x_i. Because one tested tile may receive signals from several assistive tiles, the total number of tested tiles is much smaller than the total number of enabled tiles. Therefore, the number of required output test pins is significantly reduced. The constraints on the number of test pins in the other three regions can be updated similarly.

V. EXPERIMENTAL RESULTS

The proposed ExTest scheduling method is applied to the largest SoC die used in our industry collaborator's 2.5D IC production. We refer to it as A531. It is a large design with 50 million flip-flops and 35 asynchronous clock domains.

A. Compression ratio analysis

We first present results to show that the overall ExTest time is determined by the number of test rounds, and not by the number of scan channels. In EDT, the compression ratio is related to the configuration time of the LSFR inside the decompressor. Therefore, different compression ratios result in differences in the number of shift cycles per pattern and the total number of test patterns for a design. In order to analyze the impact of compression ratio on test time, EDT

structures with different compression ratios were added to several industry designs of different sizes. Table I shows the number of shift cycles per pattern, the number of test patterns, and the test time reduction (T. reduc.) as a function of the compression ratio (CR) for these designs.

Note that the compression ratio is equal to the number of scan chains divided by the number of scan channels. The number of shift cycles decreases with the compression ratio because a smaller compression ratio increases the number of scan channels. An increase in the number of scan channels reduces the configuration time for the LSFR in the decompressor. When the compression ratio is changed from a value larger than 50 to a value smaller than 50, the percentage change of the test time is as large as 13%. This is in contrast to when the compression ratio is changed but remains below 50; in these cases, the percentage change in test time is less than 4% for all designs. Because the ExTest compression ratio for the different tiles in our design varies from 20 to 50, it has only a limited impact on the test time. As a result, we conclude that the test time is not affected by the number of scan channels; it depends mainly on the number of test rounds.

B. Scheduling results

The A531 die has 531 tiles and 479 test pins that are in the following locations: 131 tiles and 120 test pins in the top-right region, 132 tiles and 120 test pins in the top-left region, 131 tiles and 119 test pins in the bottom-right region, and 137 tiles and 120 test pins in the bottom-left region.

The number of interconnects from Tile i to Tile j is a non-negative integer w_{ij}. Consequently, A531 is fully characterized by the number of test pins in the regions and its interconnection matrix $W = [w_{ij}]$. Note that the w_{ij} values can be zero, indicating that a connection from Tile i to Tile j does not exist in A531. A 531×531 interconnection matrix is generated based on the design netlists. Among all the 281,961 possible elements in the interconnection matrix, only 3% of the elements are nonzero. Therefore, a tile in A531 is connected to only a small number of tiles.

The scheduling problem is solved using the advanced ILP solver Xpress-MP [11], using the given parameters (interconnection matrix, test-pin-count numbers, and locations of tiles). The numbers of input and output scan channels are obtained from the A531 design data[1]. In addition, due to the results obtained using EDT, the number of output scan channels is larger than the number of input scan channels; the width of the input scan channels is one or two for most of the tiles in the die. The average width of the input scan channels is 1.7 for all of the tiles. Therefore, each tile is also assumed to have one or two input scan channels.

The scheduling results based on grouping only are shown in Table II. If each tile has one input scan channel, the proposed strategy targets 522 tested tiles in 11 test rounds. In other words, all the interconnects that are connected to the primary inputs of these 522 tested tiles are successfully tested while the interconnects feeding the remaining 9 tiles are untested. These 9 tiles are not tested because each of them receives

---

[1]Details not disclosed due to confidentiality reasons.

TABLE I
COMPRESSION RATIOS, AND THE ASSOCIATED NUMBER OF SHIFT CYCLES AND TEST PATTERN NUMBERS FOR DIFFERENT DESIGNS

	CR	Shift cycles	No. of patterns	CR	Shift cycles	No. of patterns	T. reduc.	CR	Shift cycles	No. of patterns	T. reduc.
Design I	94	343	10310	47	325	9495	13%	23	316	9443	3%
Design II	93	340	17208	46	323	17163	5%	23	315	17175	2%
Design III	90	336	8058	45	322	7573	10%	23	314	7489	4%
Design IV	62	373	6958	31	340	6605	13%	15	324	6660	4%
Design V	52	363	8334	26	335	8327	8%	13	321	8300	4%
Design VI	44	355	30116	22	331	29804	8%	11	319	29742	4%

TABLE II
SCHEDULING AND OPTIMIZATION RESULTS

I. Scheduling results (grouping only)						
II. Method based on sharing of inputs						
III. Method based on sharing of inputs & output removal						
No. of input scan channels in each tile	Total number of test rounds			Test coverage (tested tiles/total tiles)		
	I	II	III	I	II	III
1	11	9	6	522/531	522/531	531/531
2	17	12	8	521/531	521/531	531/531
Based on design data for A531	15	12	6	521/531	521/531	531/531

interconnects from a large number of tiles. Therefore, to test even one of these 9 tiles, we need more test pins for the assistive tiles than the total number of available test pins. By increasing the number of input scan channels, the number of test rounds increases whereas the number of tested tiles decreases because each enabled tile requires more test pins.

Table II also shows the optimization results for the methods based on the sharing of inputs and output removal. When the input scan channels are shared by tiles with the same design, test pins can be assigned to more tiles in a single test round. For example, in the scheduling results, the test pins are assigned to 105 tiles in the first test round when the A531 design data is used. When inputs are shared, test pins can be assigned to 130 tiles in the first round. Therefore, the number of test rounds decreases. After the output removal method is added, the number of test rounds further decreases to only six. In addition, because the output scan channels of all of the assistive tiles need not be connected to test pins, the required number of test pins does not exceed the number of available test pins. As a result, previously untestable tiles can now be targeted, and all the interconnects are successfully tested.

C. Run-time and fault coverage analysis

In the non-optimized method in use until recently, all tiles in A531 were compiled together to run DRC and ATPG. When the entire design was analyzed on a server with 512 GB memory, it took 30 days, 11 hours, and 20 minutes to generate 512 test patterns. In the proposed method, the largest test group has 374 enabled tiles and the smallest test group has 221 enabled tiles. When the design is analyzed in the same environment, the generation of 512 test patterns for the largest test group takes 11 days, 2 hours, and 40 minutes; the generation of 512 test patterns for the smallest test group takes 6 days, 2 hours, and 20 minutes. Since DRC and ATPG can be run in parallel for each group, the total run-time for the proposed method is 11 days, 2 hours, and 40 minutes, which is only one-third of the runtime of the previous method. This run-time is acceptable in industry for such a large design. Compared to the 2D design with similar functionalities, the run-time is reduced 65%. In addition, using the previous ExTest scheduling, 10 days of CPU time is required for a design with 28 million flip-flops. Hence for the same time budget, a much larger design can now be handled.

In addition, the fault coverage is only 86% for the method in use until recently because many interconnects and a lot of combinational logic connecting different test groups remain untested. With the proposed method, since each test group is independent and no interconnects are considered between different test groups, all the interconnects and combinational logic are successfully tested. The fault coverage increases to 100%. Therefore, the proposed method can both reduce run-time and increase fault coverage.

VI. CONCLUSION

Although interposer-based 2.5D ICs are being advocated as the next-generation ICs, efficient ExTest for the tiles within the large dies on interposers remains a major bottleneck. We have introduced a new scheduling strategy for ExTest involving the tiles within dies in 2.5D ICs. The proposed strategy can implement interconnect testing inside a die while enforcing the constraint that the required number of test pins cannot exceed the number of available test pins of the die. We have presented comprehensive scheduling and optimization results for a large SoC design in actual production to demonstrate the effectiveness of the proposed strategy.

REFERENCES

[1] K. Banerjee et al., "3-D ICs: A Novel Chip Design for Improving Deep-Submicrometer Interconnect Performance and Systems-on-Chip Integration," Proceedings of the IEEE, vol. 89, no. 5, pp. 602–633, 2001.

[2] M. Jackson, "A Silicon Interposer-based 2.5D-IC Design Flow, Going 3D by Evolution Rather than by Revolution," 3D Architecture for Semiconductor Integration and Packaging Conference, 2011.

[3] J. H. Lau, Y. Chan, and R. Lee, "3D IC Integration with TSV Interposers for High-Performance Applications," Chip Scale Review, vol. 14, pp. 26–29, 2010.

[4] Y.-K. Ho and Y.-W. Chang, "Multiple Chip Planning for Chip-Interposer Codesign," in IEEE DAC, pp. 1–6, May 2013.

[5] M. Sunohara et al., "Silicon Interposer with TSVs (Through Silicon Vias) and Fine Multilayer Wiring," IEEE Electronic Components and Technology Conference, pp. 847–852, 2008.

[6] B. Banijamali et al., "Advanced Reliability Study of TSV Interposers and Interconnects for the 28nm Technology FPGA," IEEE Electronic Components and Technology Conference, pp. 285–290, 2011.

[7] K. Kumagai et al., "A Silicon Interposer BGA Package with Cu-filled TSV and Multi-layer Cu-plating Interconnect," IEEE Electronic Components and Technology Conference, 2008.

[8] IEEE Computer Society, IEEE, New York, NY, USA, IEEE Std 1500™-2005, IEEE Standard Testability Method for Embedded Core-based Integrated Circuits, August 2005.

[9] J. Rajski et al., "Embedded Deterministic Test," IEEE Trans. CAD, vol. 23, pp. 776–792, 2004.

[10] P. T. Wagner, "Interconnect Testing with Boundary Scan," IEEE Int. Test Conf., pp. 52–57, 1987.

[11] Xpress-MP, http://www.fico.com/en/Products/DMTools/xpress-overview/Pages/Xpress-Mosel.aspx, 2012.

978-1-4799-7598-3/15 $31.00 © 2015 IEEE

Pulse Shrinkage Based Pre-bond Through Silicon Vias Test in 3D IC

Chang Hao
School of Computer and Information
Hefei University of Technology
No.193, Tunxi Road, Hefei, Anhui, China, 230009
Email: 007changhao@163.com

Liang Huaguo
School of Electronic Science & Applied Physics
Hefei University of Technology
No.193, Tunxi Road, Hefei, Anhui, China, 230009
Email:huagulg@hfut.edu.cn

Abstract—**Defects in TSV not only lead to variation in the propagation delay but also in the transition delay of the net connected to the TSV. A non-invasive approach for pre-bond TSV test based on pulse shrinkage is proposed to detect resistive open and leakage fault. TSVs are used as capacitive loads of their driving gates, then the pulse visiting the cyclic shrinkage cells will be shrunk until it vanishes completely. The shrinkage amount is digitized into a digital code to compare with an expected value of fault free. Experiments on fault detection are presented through HSPICE simulations using realistic models for a 45 nm CMOS technology. The results show the effectiveness in the detection of resistive open defects $0.2k\Omega$ above and equivalent leakage resistance less than $40M\Omega$. The estimated design for testability area cost of our method is negligible for realistic dies.**

I. INTRODUCTION

Three dimensional integrated circuits (3D-ICs) are invented to address the scaling challenge by stacking 2D dies and connecting them with through silicon vias (TSVs). 3D-ICs offer many significant benefits over traditional stacking with wire-bonds including small footprint, high bandwidth, lower power and heterogeneous integration[1]. It has been estimated that the switch to vertical interconnects may reduce power consumption in half, increase bandwidth by a factor of eight, and shrink memory stacks by some 35 percent[2].

A TSV is a vertical via formed between tiers through silicon of oxide layers. TSVs support higher density and generate low-capacity interconnects compared to traditional wire-bonds. Testing for manufacturing defects is inherently important to satisfy the required product quality since one defect in a TSV may damage the whole stack. Hence, TSVs need to be thoroughly tested for reasons ranging from fault detection, fault diagnosis, performance characterization, and built-in self-repair (BISR) for yield and reliability enhancement, etc[3].

The test methods under development currently either rely on probe/contactless communication with TSVs, or on built-in self-test (BIST) schemes. The former ones require development of dedicated probe cards. Before wafer thinning, one TSV end is buried in silicon and thus not accessible. After wafer thinning, contact based probing of TSVs arrays of some *um* at array pitches of some tens of *um* without damaging TSVs tips and thinned wafers is still a challenge. On the other hand, contactless probing techniques require the implementation of extra logic and possible antennas leading to considerable area overhead.

Limitations of probe/contactless solutions have led to several BIST approaches as alternative solutions. Classical challenges for BIST approaches are to provide sufficient coverage of possible defects, while requiring a wide detection range, a high resolution and a small area overhead.

This paper presents a novel test method for TSVs, using a technique named pulse shrinkage test (or PS test for short). For a given TSV under test, the PS test applies a pulse signal at the driving end. Due to the loading effect of the TSV, the signal arriving at the receiving end tends to have a constant rise/fall times. If the TSV is fault free, the pulse signal visiting the shrinkage cell will be shrunk by a definite amount of width per cycle until it vanishes completely. On the other hand, if the TSV is faulty with excessive resistances or larger equivalent capacitance, then the voltage waveform may be distorted and thus will lead to a different rise/fall times. As a result, the output counter can capture the amount of circulations and generate a corresponding digital code to compare with an expected value of fault free.

In summary, the proposed PS test measuring the transition delay with picosecond-level high resolution has several technical merits compared to previous works, making it not only suitable for BIST but also supportive of timing-aware BISR. To the best of our knowledge, little research has focused on the TSV test with the pulse shrinkage scheme. The technical merits include the following.

- Unlike the ring oscillator based on the measurement of propagation delay deviation with nanosecond level resolution[4][5], our PS test is based on the measurement of the transition delay deviation and has the characteristic of a higher picosecond level resolution with a larger range of fault detection. Simulations show the effectiveness of the method in the detection of resistive open defects $0.2k\Omega$ above and equivalent leakage resistance less than $40M\Omega$. Meanwhile, PS test can support on-the-spot pass/fail fault detection, avoiding sophisticated post-processing.

- Unlike the pulse vanishing test in [3] sensitive to the pulse width launched, there is no strict requirement on our PS test, thus it can effectively deal with various TSVs of different manufacturing process or size. It will effectively alleviate the burden of the ATE or test controller in the early stage of manufacture without any knowledge on the fault type and the degree of

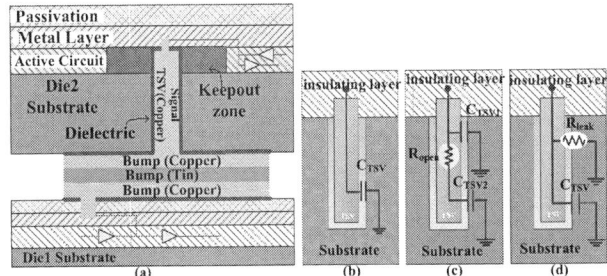

Fig. 1. TSV models (a) TSV bonding, (b) fault free, (c) resistive open, (d) pinhole

failure.

- Unlike the method in [6] that is characteristic of leakage binning with 8 levels, our PS test supports not only test but also digitizes the degree of resistive open and leakage fault into a specific digital code with 132 levels. This feature is especially important when it is applied to diagnose and built-in self-repair.

- Last, but not the least, the test cost of the proposed PS test is low since we do not require TSV probing and the design for testability (DfT) area overhead is negligible. Furthermore, the pulse-shrinking capability of the element is controlled by the relative dimension ratio of the adjacent gates. There is no special requirement on the circuit customization.

The remainder of the paper is organized as follows. Section II presents the electrical model of TSVs and related preliminary works. The pulse shrinkage test scheme is detailed in section III. The corresponding experiments will be elaborated in section IV. Finally, we conclude this work in section V.

II. PRELIMINARY WORKS

A. Electrical models of TSVs

When two dies are bonded together, the goal is to successfully connect topside micro-bumps on the lower die with associated bottom side TSV micro-bumps on the upper die as shown in Fig. 1(a)[7]. The tin material is the solder that makes a low resistance connection between the copper surfaces. TSV protrudes from the bottom of a die and travels up through a substrate where it passes through the active circuit region of the top of the die and then into the metal layer region. There, a connection to the metal layer may be made. The top layer of the die is a passivated, insulating layer. The dielectric layer between the TSV copper and the substrate is very thin and does not materially increase the diameter of a TSV body. The "Keepout Zone" (KOZ) is an area of the circuitry layer that should not contain active circuits. This zone is determined by the mechanical tolerances that exist when drilling/etching and filling the TSV.

TSV can be considered as a cylindrical metal bar insolated to the silicon body by a SiO_2 layer. Its resistance and capacitance can be expressed as[8]:

$$R_{TSV} \approx \frac{4\rho l}{\pi d^2} \qquad (1)$$

TABLE I. TEST METHODS FOR INTERPOSER OR TSV

Work	Basic scheme	Fault model	Resolution	Area overhead	Extra feature
Chen[11]	L2VCC	Open	8.9%	$10 \ um^2$	Analog
Chen[8]	CAF-WAS	Open,leakage	5%	NULL	Capacitance test
Natale[10]	CAF-WAS	Open,leakage	2%	NULL	Delay circuit
Huang[6]	CAF-WAS	Open,leakage	ns level	$54.43 um^2$	Leakage binning
Krish[4]	RO	Open,leakage	ns level	$0.01 mm^2$	NULL
Montanes[12]	DC	Open	ns level	NULL	$\geq 1k\Omega$
Huang[3]	PV	Bridging,open	ns level	54.9%	diagnosis
This work	PS	Open,leakage	ps level	$75.54 um^2$	Flexible

$$C_{TSV} = \frac{2\pi \varepsilon h}{\ln[(d + 2t_{ox})/d]} + \frac{\pi \varepsilon d^2}{4t_{ox}} \qquad (2)$$

where d is the diameter, h is the height of a TSV, t_{ox} is the thickness, and ε is the dielectric constant of the isolation layer, respectively. Resistance of TSVs is around $20m\Omega$ with limited variation, and typical value of capacitance is $50fF$.

Electrical model of TSVs using R, L, and C elements have been presented in many publications (e.g. [5][9][10]). L and R components are generally neglected in the pre-bond phase and a simplified model based on the predominant parasitic capacitance C between the TSV and the substrate is considered (see Fig.1(b)). We focus on two types of TSV faults: resistive open and leakage faults. Several TSV defects can be modeled by these faults. Micro-voids due to insufficient TSV filling increase the TSV resistance R_{open} at the defect location and thus can be modeled as a resistive open fault as shown in Fig.1(c). Pinholes due to silicon side wall imperfection create a conduction path between the TSV and the substrate and can be modeled as a leakage fault R_{leak} as shown in Fig.1(d).

For simplicity without losing generality, we use a lumped RC model to represent a TSV. The method to be proposed applies to TSVs in the pre-bond condition as well as the post bond condition. The only difference between these two test conditions is the existence of an extra microbump resistance and an extra gate input capacitance at the ending point of the TSV, as it is physically connected to the receiving logic at the other end in the post bond stage. In the following discussion we will assume the pre bond condition.

B. Related Prior Works

Several BIST methods have been proposed in the literature for detecting full-open, micro-voids, or pin-hole defects. TSVs are used as capacitive loads and the deviation of their expected RC parameters are detected by indirect measures, namely the delays required for charging or discharging the nets connected to their front ends - the only ends accessible before bonding[10].Table I compares the previous works on interposer or TSV test in terms of five characteristics, including the basic scheme, fault model targeted, resolution, area overhead and extra features.

1) The L2VCC, CAF-WAS scheme and variants: The first method referred to as L2VCC (Leakage to Voltage Conversion and then Comparison), initially presented in [11] converted the leakage amount into a voltage by imposing a pull-up device on the TSV node. The pull-up device may be a resistor, or an always-on transistor. Along with the leakage path to ground, this will form a voltage divider. After the circuit has stabilized, the voltage level of the TSV node reflects the amount of leakage. One major drawback existing in the L2VCC method is low resolution. Furthermore, analog or custom circuit might be required to perform analog voltage detection.

The second approach called CAF-WAS (Charge-and-Float, Wait-and-Sample), initially described in [8], targeted pinhole defects and was based on a leakage current sensor. First, turn on the tri-state buffer to charge the TSV and then turn off the tri-state buffer to float the TSV and wait for a certain time. Second, sample the value at the output and perform the pass/fail fault detection based on the binary result. G. Di. Natale improved the CAF-WAS method by designing a dedicated delay circuit[10] and Shi-Yu Huang enhanced the CAF-WAS method with 8 leakage binning levels[6]. However, one issue may hinder the CAF-WAS method from being applicable to the TSV leakage test. The wait time generation circuit needs to be designed carefully as it is sensitive to the threshold of fault targeted. Furthermore, there is little flexibility as it is difficult to fix an exact wait time for various TSVs of different sizes. In practical, the CAF-WAS method needs to be modified to be flexible enough to accommodate TSVs from different manufacturing process.

2) Ring oscillators based scheme and variants: In [4][5][6], based on the measurement of the propagation delay deviation, ring oscillators were created with TSVs drivers, receivers and extra inverters. The oscillation period is captured by extra binary counters that use the oscillating signal as clock. As TSV manufacturing defects change the propagation delay of the ring oscillator nets, TSV parameter deviations are detected through the deviation of the expected ring's oscillation periods. In [12], variations in the Duty Cycle of transmitted signals after balanced logic gates were used to detect weak open defects in TSVs. However, the oscillation period lies in the nanosecond level, while the propagation delay caused by TSV RC parameter deviations spreads in the picosecond level, hence some subtle variations in TSV may be masked by the DfT circuitry. Furthermore, defects in TSV not only lead to variation in the propagation delay but also in the transition delay of the net connected to the TSV. In this paper, we will exploit the transition delay based on the pulse shrinkage scheme to detect the resistive open and leakage fault with a higher resolution of picosecond level.

3) Pulse vanishing scheme: In [3], the author proposed a test method for interposer wires, using a technique called pulse-vanishing test (or referred to as PV test). For a given interposer wire under test, a short-duration pulse is launched at its driving end. If there is excessive resistance along the interposer wire, the pulse signal may vanish along its way. A pulse signal detected at the receiving end indicates "fault-free," while no pulse signal indicates "faulty." However, we can not apply the pulse vanishing scheme to TSV test directly. The capacitance of TSV is usually much smaller than that of interposer wire, hence to detect a small defects in TSV, the pulse width should be much smaller, such as 300ps or less, which will put a strict requirement on the pulse generator on-chip or ATE. On the other hand, it is notable that the test results strongly depend on the pulse width, the type and degree of fault. At the early stage of manufacture, it is a difficult task to determine a right pulse width without any knowledge on

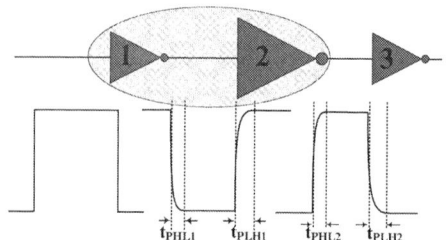

Fig. 2. One stage pulse-shrinking delay element

the fault type and the degree of failure. Furthermore, there are various types of TSVs such as signal TSV, power TSV, test TSV and thermal TSV of different manufacturing processes or sizes. In pulse vanishing scheme, it may need to launch multiple test pulses with different width, however, which will burden the clock generator and test controller.

III. PULSE SHRINKAGE TEST SCHEME

A. The principle of Pulse Shrinkage

Sizing the aspect ratio (W/L) of the transistors is the most powerful and effective performance optimization tool in the hands of the designer. The rise and fall times, t_r and t_f, are metrics that apply to signal waveforms and express how fast a signal transits between the different levels.

Fig. 2 shows the conceptual block diagram of the proposed pulse shrinkage cell. The dimensions of the gates 1 and 3 are the same, and only that of the gate 2 is different. The inhomogeneous dimension of the gates makes the input pulse undergo different rising and falling time at the interface boundaries among the gates. This mechanism can be used to accurately control the pulse shrinking time.

Due to the asymmetrical transfer characteristics, when the pulse signal pass through the first inverter, the pulse width will be shrunk for $t_{PHL1} - t_{PLH1}$. The t_{PHL1} defines the response time of the first inverter for a high to low output transition, while t_{PLH1} refers to a low to high transition. For the same reason, the shrunken pulse from former inverter will be shrunk for $t_{PHL2} - t_{PLH2}$ again when passing through the latter inverter. Thus, for one shrinkage cell the pulse width will be shrunk for:

$$\Delta W = (t_{PHL1} - t_{PLH1}) + (t_{PHL2} - t_{PLH2}) \quad (3)$$

To simplify the derivation, the input pulse is supposed to be stepwise at each stage for the first-order approximation. When the pulse visits from gate 1 to gate 2, the transition time was given in[13]:

$$t_{PHL1} = \frac{2C_2 V_{TN}}{k_{N1}(V_{DD}-V_{TN})^2} + \\ \frac{C_2}{k_{N1}(V_{DD}-V_{TN})} \cdot \ln\left(\frac{1.5V_{DD}-2V_{TN}}{0.5V_{DD}}\right) \quad (4)$$

$$t_{PLH1} = \frac{-2C_2 V_{TP}}{k_{P1}(V_{DD}+V_{TP})^2} + \\ \frac{C_2}{k_{P1}(V_{DD}+V_{TP})} \cdot \ln\left(\frac{1.5V_{DD}+2V_{TN}}{0.5V_{DD}}\right) \quad (5)$$

where k_{N1}, k_{P1} are the transconductance parameters of the gate 1 and C_2 is the effective input capacitance of the gate 2.

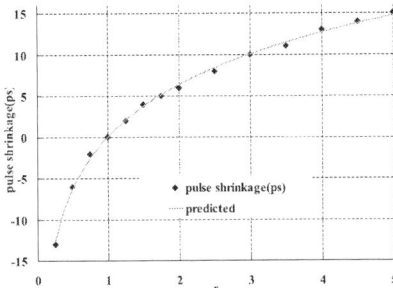

Fig. 3. Influence of the dimension ratio on the pulse shrinkage

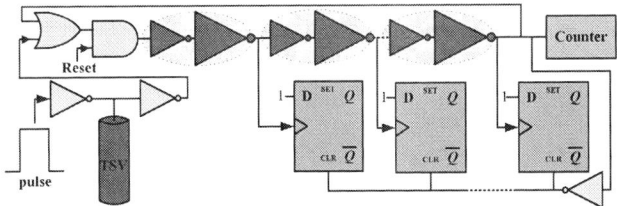

Fig. 4. TSV test architecture based on cyclic pulse shrinkage

B. Pulse Shrinkage Test Scheme

Assume $-V_{TP} = V_{TN}$, then the pulse-shrinking time from gate 1 to gate 2 can be analyzed as $\Delta W_1 = t_{PLH1} - t_{PHL1}$.

$$\Delta W_1 = C_2 \left(\frac{1}{k_{P1}} - \frac{1}{k_{N1}} \right) \left[\frac{2V_{TN}}{(V_{DD}-V_{TN})^2} + \frac{1}{V_{DD}-V_{TN}} \right].$$
$$\ln \left(\frac{1.5V_{DD}-2V_{TN}}{0.5V_{DD}} \right) \tag{6}$$

Similarly, the pulse-shrinking time from gate 2 to gate 3 can be analyzed as $\Delta W_2 = t_{PHL2} - t_{PLH2}$.

$$\Delta W_2 = -C_3 \left(\frac{1}{k_{P2}} - \frac{1}{k_{N2}} \right) \left[\frac{2V_{TN}}{(V_{DD}-V_{TN})^2} + \frac{1}{V_{DD}-V_{TN}} \right].$$
$$\ln \left(\frac{1.5V_{DD}-2V_{TN}}{0.5V_{DD}} \right) \tag{7}$$

where $C_3 = C_1$. The total pulse-shrinking time can be found as

$$\Delta W = \Delta W_1 + \Delta W_2$$
$$= \partial_i \left[C_2 \left(\frac{1}{k_{P1}} - \frac{1}{k_{N1}} \right) - C_1 \left(\frac{1}{k_{P2}} - \frac{1}{k_{N2}} \right) \right] \tag{8}$$

where

$$\partial_i = \frac{2V_{TN}}{(V_{DD}-V_{TN})^2} + \frac{1}{(V_{DD}-V_{TN})} . \ln \left(\frac{1.5V_{DD}-2V_{TN}}{0.5V_{DD}} \right) \tag{9}$$

is a constant factor which is approximately layout independent. By varying the dimension ratio (r) of the inverters, the pulse-shrinking time can be easily controlled. For example, let the length and width of the gate are $L_1 = L_2 = L_3$ and $W_2 = rW_1 = rW_3$. Then we have $k_{N2} = rk_{N1}$, $k_{P2} = rk_{P1}$, $C_2 = rC_1 = rC_3$, and the total pulse-shrinking time can be further simplified to be

$$\Delta W = \left(r - \frac{1}{r} \right) C_1 \left(\frac{1}{k_{P1}} - \frac{1}{k_{N1}} \right) \partial_i \tag{10}$$

For $r = 1$, this means that all inverters have the same size and the amount of pulse shrinking is zero. The input pulse will be shrunk for $r > 1$ or expanded for $r < 1$. To verify the derivation, the simulation result for different r value is shown in Fig. 3. The simulation result matches the prediction curve well.

From formula (10), the pulse signal will be shrunk for ΔW when passing through one stage pulse shrinkage cell. With multiple shrinkage cells connected forming a delay line, the pulse with width T will vanish after $T/\Delta W$ cells. The resolution is dependent on ΔW that is much smaller than the single cycle of the delay line, thus it can reach relatively higher resolution compared with the ring oscillator scheme. The effective resolution will be elaborated in IV.A.

Defects in TSV not only lead to variation in the propagation delay but also in the transition delay of the net connected to the TSV and thus variations in the shrinkage of pulse width. These variations can be measured by pulse shrinkage cells. In this paper, borrowing the wisdom of time digitization, we propose a new BIST scheme based on pulse shrinkage.

We create a pulse shrinkage net containing two balanced inverters used as TSV drivers and receivers in the front side of the TSV as shown in Fig. 4. The pulse shrinkage delay line is modified from linear to cyclic. The AND and OR gates are used as a coupling unit, and pulse shrinkage unit is composed of an even number of inverters. The input pulse from TSV will visit every shrinkage element in the cyclic delay line once per cycle. The pulse shrinking amount is determined by the aspect ratio between the inhomogeneity (AND an OR gates) and homogeneity gates (inverters in the shrinkage cells). The effective resolution is determined by one shrinkage element $T_{LSB} \approx \Delta W$.

The output of each shrinkage element is connected to the clock terminal of the D flip flop. The D terminal is constantly tied to "1", thus the D flip flop will latch "1" if a pulse exists. The D flip flop will be reset each cycle until the pulse vanishes. Before measurement starts, the reset signal is set to "0", and then keeps a high level. The input pulse from TSV circulates in the cyclic shrinkage cells and will be shrunk by a definite amount of width per cycle until it vanishes completely. The output counter manages to plus 1 per cycle and the signal return to the OR gate. Meanwhile, the signal is shared to reset all D flip flops by an inverter. With the aid of an output counter and D flip flops to generate the corresponding digital code, the number of the circulations can be captured and compared with an expected value. The digital code m can be calculated by formula (11)

$$m = N_c * N_s + n_D \tag{11}$$

where N_c is the number of circulations captured by the output counter, N_s is the number of shrinkage cells, and n_D is the number of "1" recorded by D flip flops.

For the last few cycles, the input pulse will become too narrow to make the counter toggle its states. A big offset error occurs when the pulse-shrinking time per cycle is relatively small. However, the offset error can be eliminated by a careful calibration process. First, feed two reference periods T_{ref1} and T_{ref2} to the shrinkage net. Assuming the digital code of these two reference pulses are N and N' respectively. Then we have $T_{ref1} = \alpha N + T_{offset}$ and $T_{ref2} = \alpha N' + T_{offset}$ where α is the effective resolution, and T_{offset} is the measurement offset. By resolving α and T_{offset} from the above equations,

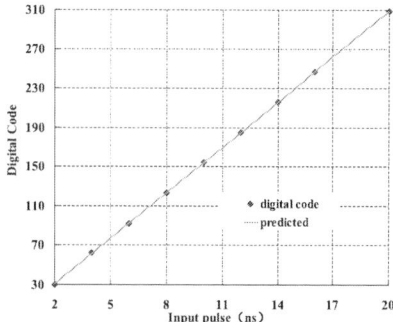

Fig. 5. Digital code measured of input pulse

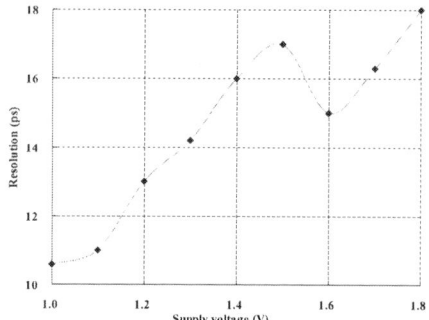

Fig. 6. Digital code measured of input pulse

the measured width of an input pulse T_{in} with output code M can be calculated as

$$T_{in} = \alpha M + T_{offset} = \frac{(M-N)T_{ref2} + (N'-M)T_{ref1}}{N'-N} \tag{12}$$

If T_{ref2} is realized as twice of T_{ref1} by dividing the stable reference frequency of T_{ref1} by 2, we will have $T_{ref2} = 2T_{ref1} = 2T_{ref}$ and $T_{in} = \frac{M+N'-2N}{N'-N}.T_{ref}$. When necessary, the calibration can be done just before measurement to get the best accuracy.

IV. EXPERIMENTAL RESULTS

To evaluate the performance of the proposed pulse-shrinking delay element, we perform the simulation with HSPICE and 45 nm predictive technology model (PTM)[14]. The cyclic delay line is composed of 5 INV X1 and 5 INV X4 used as shrinking cells, extra 3 INV X1 used as TSV driver, receiver and reset respectively, 1 AND X1 gate, 1 OR X1 gate, 5 DFFR X1 and an 7 bit counter composed by 7 DFFR X1. All the gates come from the Nangate 45 nm open cell library[15]. The aspect ratios of shrinkage cells is $r = 4$.

A. The effective resolution and dependence of supply voltage

To verify the effective resolution of the cyclic pulse shrinkage element, a series of pulse width with different width are sent for coding. The measured results with the theoretical prediction line are depicted in Fig. 5. The valid output code range can be further increased by lengthening the delay line. The experimental data agrees with the linear prediction very well. The effective resolution was estimated as the ratio of the pulse width difference over the output code difference and it is calculated to be 13ps per cell.

To find out the supply voltage dependence, another series of experiments were conducted for supply voltage ranging from 1.0 to 1.8V with 0.1V increments. For each supply voltage, only two different single-shot pulses were coded. The corresponding results without calibration are shown in Fig. 6. The effective resolution only varies less than 8 ps for a typical supply voltage range from 1.0 to 1.8V.

B. Detection of resistive open and leakage fault

We simulate a resistive open fault at the location $x = 0.5h$ in TSV with typical value of $50fF$ as depicted in Fig. 1(c),

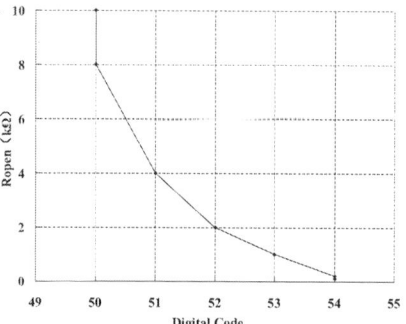

Fig. 7. Digital code versus the resistive open fault (R_{open})

where h is the height of TSV. The input pulse width is set to $W = 3ns$ and we sweep R_{open} from $0.1k\Omega$ (no fault) to $10k\Omega$ (strong resistive open) at the typical supply voltage $V_{DD} = 1.2V$. With this model, we perform transient analysis and record the digital code. Fig. 7 shows the obtained digital code in case of a resistive open fault. As expected, an increase in the resistance R_{open} leads to a reduction of digital code from 54 to 50. This indicates that we can detect resistive opens of a sufficient size by measuring the digital code directly. When the resistance R_{open} increases from $8k\Omega$ to $10k\Omega$, the digital code does not change and stay at 50, which indicates the TSV is completely broken at $h/2$ position. When the resistance R_{open} reduces from $0.2k\Omega$ to $0.1k\Omega$, the digital code keeps constant and stays at 54, thus open resistances higher than $0.2k\Omega$ can be detected by means of the proposed technique and it is assumed to be no resistive open fault if R_{open} is less than $0.2k\Omega$.

Generally, it is difficult to distinguish a TSV with small resistive open fault (small resistance R_{open}) from a normal one; however, the minimum resistive open fault identified in our scheme can reach $0.2k\Omega$ above with the fault position assumed to be $h/2$.

Leakage faults exhibit a different behavior from resistive open fault. To show this, we use the same simulation approach as described above. Fig. 8 shows the dependence of digital code on the leakage resistance R_{leak}. First, we observe that the digital code reduces from 132 to 56 as the leakage resistance increases, which makes them distinguishable from the fault free case as well as resistive open faults. Second, strong leakage faults below a certain threshold, e.g. $R_{leak} \approx 0.5k\Omega$, leads to the counter overflow (132 is the largest number

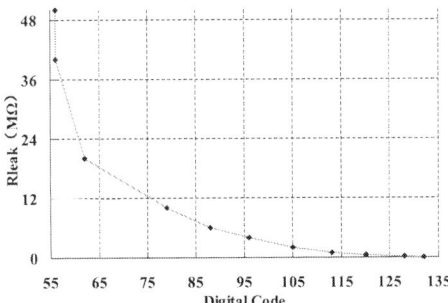

Fig. 8. Digital code versus leakage fault (R_{leak})

recorded by a 7 bit counter plus another 5 D flip flops). Simply, the overflow can be alleviated by a 8 bit counter or more at the expense of larger area overhead. Normally, it is also difficult to distinguish a TSV with small leakage fault (large leakage resistance R_{leak}) from a fault free one. When the leakage resistance R_{leak} increases from $40M\Omega$ to $50M\Omega$, the digital code stays at 56 unchanged, which indicates leakage current is too small to be detected. Hence the minimum leakage fault identified in our scheme can reach $40M\Omega$.

It is worth noting that there exists a small possibility of alias (about $1/(132 - 50) \approx 0.76\%$) when the digital code equals to 55. It is reasonable to believe that the errors, corresponding to such deep sub-nanosecond resolution, may be mostly induced by the jitter effect of the pulse generator and the inherent measurement error of the universal counter.

C. Area overhead

One advantage of our scheme is that cyclic delay line can save the area overhead compared with delay chains. As described above, the pulse with width T will completely vanish after passing through T/T_{LSB} shrinkage cells, which indicates the number of shrinkage cell and D flip flop is T/T_{LSB} and will increase with T. However, in our cyclic scheme, only N_s shrinkage cells and D flip flops are needed. In general case, T_{LSB} is relatively small and $T/T_{LSB} \gg N_s$, thus our scheme saves $T/T_{LSB}.(N_s - 1)X$ area overhead compared with shrinkage chains. Meanwhile, due to a small number of shrinkage cells, errors brought by process variation can also be reduced.

In total, the DfT circuitry occupies an area of $0.532 * 13 + 0.798 + 0.798 + 5.586 * 5 + 5.586 * 7 = 75.54um^2$. Generally, taken the keepout zone and the pitch into account, one TSV approximately occupies a square area of up to $1600 \ \mu m^2$[1]. Therefore, the DfT area overhead can be negligible compared with one single TSV area overhead. Compared with oscillator scheme, the area increases by one inverter and D flip flops, which is relatively small and it is worthy from the resolution point of view. Furthermore, we can remove the D flip flops to save the area overhead at the expense of a lower resolution (from 13ps to 65ps). Even so, we still have better resolution (picosecond level) than ring oscillator scheme (nanosecond level).

V. CONCLUSION

A versatile test method should be able to adapt to different test thresholds, TSV size and fault degree. Existing methods fail to achieve such a one-method-fits-all ideal. We have demonstrated that it can be easily achieved by a cyclic pulse shrinkage element with the pulse-shrinking capability controlled by the relative dimension ratio of the adjacent gates. With its accurate resolution, the method can detect the resistive open defects $0.2k\Omega$ above and equivalent leakage resistance less than $40M\Omega$. With its high flexibility and no special requirement on pulse generator, it can effectively deal with various types of TSVs of different manufacturing process or size. This feature is extremely import in the early stage of manufacture without any knowledge on the fault type and the degree of failure.

VI. ACKNOWLEDGEMENT

This research is supported by the National Nature Science Foundation of China under Grant No. (61274036, 61371025, 61474036, 61204046).

REFERENCES

[1] Chang Hao, Liang Huaguo, Li Yang, et al, "Optimized stacking order for 3D-stacked ICs considering the probability and cost of failed bonding," in 2014 International Symposium on VLSI Design, Automation and Test (VLSI-DAT), Hsinchu, Taiwan, 2014, pp. 283-286.

[2] J. Rajski and J. Tyszer, "Fault diagnosis of TSV-based interconnects in 3-D stacked designs," in 2013 IEEE International Test Conference (ITC), 2013, pp. 1-9.

[3] S. Y. Huang, J. Y. Lee, K. H. Tsai, et al., "Pulse-Vanishing Test for Interposers Wires in 2.5-D IC," IEEE Transactions on Computer-Aided Design of Integrated Circuits and Systems, vol. 33, pp. 1258-1268, 2014.

[4] S. Deutsch and K. Chakrabarty, "Contactless Pre-Bond TSV Test and Diagnosis Using Ring Oscillators and Multiple Voltage Levels," IEEE Transactions on Computer-Aided Design of Integrated Circuits and Systems, vol. 33, pp. 774-785, 2014.

[5] L. R. Huang, S. Y. Huang, S. Sunter, et al., "Oscillation-Based Prebond TSV Test," IEEE Transactions on Computer-Aided Design of Integrated Circuits and Systems, vol. 32, pp. 1440-1444, 2013.

[6] S. Y. Huang, Y. H. Lin, L. R. Huang, et al., "Programmable Leakage Test and Binning for TSVs With Self-Timed Timing Control," IEEE Transactions on Computer-Aided Design of Integrated Circuits and Systems, vol. 32, pp. 1265-1273, 2013.

[7] K. P. Parker, 3D-IC Defect Investigation, in Provisional Report of the IEEE P1838 Defect Tiger Team, 2012. http://grouper.ieee.org/groups/3Dtest/statusReports/.

[8] P. Y. Chen, C. W. Wu and D. M. Kwai, "On-chip testing of blind and open-sleeve TSVs for 3D IC before bonding," in 2010 28th VLSI Test Symposium (VTS), 2010, pp. 263-268.

[9] B. Noia and K. Chakrabarty, "Pre-bond probing of TSVs in 3D stacked ICs," in 2011 IEEE International Test Conference (ITC), 2011, pp. 1-10.

[10] G. Di Natale, M. L. Flottes, B. Rouzeyre, et al., "Built-in self-test for manufacturing TSV defects before bonding," in 2014 IEEE 32nd VLSI Test Symposium (VTS), 2014, pp. 1-6.

[11] P. Y. Chen, C. W. Wu and D. M. Kwai, et al., "On-Chip TSV Testing for 3D IC before Bonding Using Sense Amplification," in Asian Test Symposium, 2009, pp. 450-455.

[12] R. Rodriguez-Montanes, D. Arumi and J. Figueras, "Post-bond test of Through-Silicon Vias with open defects," in 2014 19th IEEE European Test Symposium (ETS), 2014, pp. 1-6.

[13] T. A. Demassa and Z. Ciccone, Digital Integrated Circuits. New York: Wiley, 1996.

[14] PTM. 45nm Predictive Technology Model, http://ptm.asu.edu.

[15] Nangate. Nangate 45 nm Open Cell Library, http://ptm.asu.edu.

2015 IEEE 33rd VLSI Test Symposium (VTS)

Testing of 3D-Stacked ICs With Hard- and Soft-Dies – A Particle Swarm Optimization Based Approach

Rajit Karmakar, Aditya Agarwal and Santanu Chattopadhyay
Dept. of Electronics & Electrical Comm. Engineering
Indian Institute of Technology Kharagpur, India, Kharagpur, 721302
Email: {rajit, adityaagarwal, santanu}@ece.iitkgp.ernet.in

Abstract—This paper presents a test architecture optimization and test scheduling strategy for TSV based 3D-Stacked ICs (SICs). A test scheduling heuristic, that can fit in both session-based and session-less test environments, has been used to select the test concurrency between the dies of the stack. The proposed method minimizes the overall test time of the stack, without violating the system level resource and TSV limits. Particle Swarm Optimization (PSO) based meta search technique has been used to select the resource allocation of individual dies and also their internal test schedules. Incorporation of PSO in two stages of optimization produces a notable reduction in the overall test time of SIC. Experimental results show that upto 51% reduction in test time can be achieved using our strategy, over the existing techniques.

Keywords—3D-SIC, TSV, Test scheduling, PSO, Optimization.

I. INTRODUCTION

[1] With increasing demand for high performance and low-power chips, present day's semiconductor industry is heading towards smaller feature size with reduced chip area. Interconnects, which cannot be scaled down with transistors, are becoming main stumbling block in IC design. Long interconnects in 2D-ICs hamper circuit performance with its high delay and power consumption. Recently, 3D-IC has emerged to be a potential solution to this problem. Instead of designing 2D-IC with long global interconnects, interconnect lengths can be reduced significantly by designing circuit components into several layers and bonding them together. This helps to achieve high bandwidth, low latency circuit with higher packaging density and low footprint. Based on different stacking methodologies, 3D stacking can be categorized as wafer-to-wafer, die-to-wafer, and die-to-die stacking [1]. In 3D-SICs different dies are stacked and interconnected using through-silicon vias (TSVs) bonding. These TSVs are vertical metal interconnects that can be integrated into a substrate during manufacturing. TSVs are very important in 3D integration as they are used to provide functional signals, power/ground, clock, as well as test access to logic blocks of different layers of the device [2]. Figure 1 shows a typical example of 3D-SIC with dies at different layers of the stack.

Although 3D-SIC provides several advantages over 2D-IC, testing of 3D-SIC has become more challenging because of its high complexity. Individual dies need to be tested before stacking (pre-bond testing [3]) to ensure stacking of defect-free dies. Post-bond testing [3] is required after completion of stacking of all the dies, to ensure defect-free thinning, alignment, and bonding during stacking. Mid-bond testing [3]

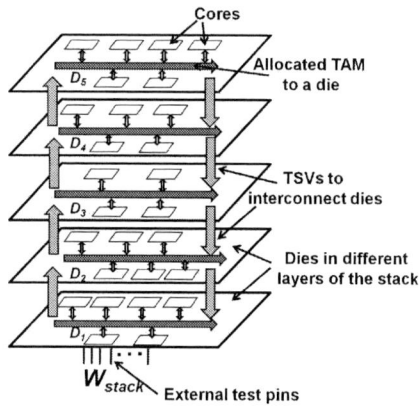

Figure 1. A typical example of 3D-SIC

or partial-stack testing is also carried out optionally in the intermediate process of stacking of pre-bond tested dies. In post-bond testing, test data can only be fed through the pins of the lowest die of the stack. A 3D test access mechanism (3D-TAM) must be designed to transport the test data to the cores of the dies at different layers of the stack to facilitate both pre-bond and post-bond testing [4]. TSVs are used to route the 3-D TAM to different layers of the stack. However, total number of TSVs must be within a certain limit to reduce its area overhead as well as TSV defects and TSV-induced defects in devices [5].

Some initial works in 3D testing have proposed wrapper design for TSV based 3D-SICs [2] while some other works [6], [7] have considered test architecture design. However, test optimization method has not been discussed in these papers. Although, the paper [1] has proposed test optimization techniques for 3D SoC consisting of cores distributed over multiple layers, the claim of no maximum limit on TSVs may not be a valid assumption. Moreover, the proposed 3D test architecture may not be feasible in practical cases. An Integer Linear Programming (ILP) based optimization method has been proposed in [8] for complete stack testing as well as for multiple test insertion during bonding. Another work [3] has considered a matrix partitioning based cost model to account for various test costs incurred during 3D integration and a also test flow selection to optimize the test cost.

Test architecture optimization for 3D-SIC can be divided into two parts, *Die level optimization* to be followed by *stack level optimization*. In *stack level optimization*, a 3D test architecture and test schedule is developed for all the dies of the stack, under the resource and TSV constraints. Given an allocated TAM to a die, *die level optimization*

---

[1]This work is partially supported by the research project No. 9(5)/ 2010-MDD dt. 23/1/2011, sponsored by the DeitY, Govt. of India

978-1-4799-7598-3/15 $31.00 © 2015 IEEE

deals with resource allocation to individual cores and selection of core ordering during testing to minimize the test time of the die. The authors in [4] have classified 3D-SIC into three categories, "*hard die*", "*soft die*" and "*firm die*". 2D test architectures of the dies are fixed for 3D-SIC with *hard die*, while for *soft die*, test engineers have the flexibility to decide how much test resources should be allocated to each die, *die level optimization* is also possible. In *firm dies*, test resources and TSV requirements of the dies can be reduced at the expense of an added serial/parallel conversion hardware. The authors in [4] have proposed an ILP based test architecture optimization, where they have considered session-based testing for all three categories of 3D-SICs. However, the schedules generated using this method, fail to satisfy the maximum TSV limit for hard dies in some cases (as discussed in Section IV).

In this paper, we have presented test architecture optimization and test schedule generation strategies for 3D-SIC with hard dies as well as soft dies. A TSV constrained test scheduling heuristic has been used to select the test concurrency between the dies without violating the resource and TSV constraints. Both session-based and session-less testings have been considered. Particle Swarm Optimization (PSO) based search procedure is used to take the decision about the test resource allocation for individual dies for 3D-SIC with soft die. This helps to evolve towards better test schedule of the stack. Although, *Die level optimization*, which is similar to test architecture optimization of 2D-SoC, has been well researched in the past few years [9], [10], none of the reported approaches could produce better solution for all the test cases. This justifies the search for newer techniques. So, unlike other 3D test architecture optimization approaches [4], which consider one of the previously reported 2D test architecture optimization methods [10], we have developed a new PSO guided 2D test architecture optimization technique to generate test schedule of individual dies. It gives us the flexibility of both *die level* and *stack level optimization*. Experimental results show that our strategy can achieve upto 51% improvement in test time over the works reported in the literature.

The rest of the paper is organized as follows. Test architecture optimization strategy for 3D-SIC with hard die has been described in Section II. Section III presents the PSO guided test architecture optimization technique for 3D-SIC with soft die. Results of our experimentation and related discussions have been presented in Section IV. Section V draws the conclusion of the paper.

II. TEST-ARCHITECTURE OPTIMIZATION FOR 3D-SIC WITH HARD DIE

For 3D-SIC with hard dies, the vendors provide fabricated dies with fixed test bandwidth and test time to the 3D-integrator. Hard dies offer less flexibility for optimization, as the test engineers are limited to selection of test concurrency between dies. The test architecture and test schedule problem of 3D-SIC with hard dies can be described as follows.

A. *Problem Formulation*

Suppose a stack consisting of M dies D_i $(1 \leq i \leq M)$ is to be tested with a maximum available test pins, W_{stack} and maximum number of allowable TSVs, TSV_{stack} to route the 3D-TAM. Each die D_i has a predefined test pin value of WD_i $(WD_i \leq W_{stack})$ and associated test time, TD_i. Determine an optimal TAM design and corresponding test schedule for the stack such that the total test time TT for the stack is minimized without violating resource and TSV constraints.

B. *Scheduling Strategy*

As all the dies have to be tested using predefined number of test resources and test architecture of each die is fixed, *die level optimization* is not applicable for 3D-SIC with hard die. We can only decide the test concurrency and ordering of the dies in the schedule. In this section, we present a test scheduling heuristic considering both session-based testing and session-less testing.

1) Session-based testing: For session-based testing, maximum test time of all the dies tested concurrently in k^{th} test session (TS_k), is the test time (TST_k) of that session. It may be noted that, if all the dies are tested serially, a maximum number of M test sessions are possible. Total test time (TT) of the stack is the summation of all the test session's time. Our test scheduling heuristic $3D\_Test\_Schedule$ (Algorithm 1) starts with sorting the dies D_i $(1 \leq i \leq M)$ in descending order of test time TD_i. We choose a die D_i in sorted order and check its schedulability in a test session without violating resource (W_{stack}) and TSV (TSV_{stack}) constraints. A $TSV\_Checker$ (Procedure 2) checks TSV requirement between different layers of the stack to facilitate concurrent testing of D_i with other dies tested in the same session TS_k. The number of TSVs required between layer j and $j-1$ is determined by two factors- (i) the maximum number of test pins required by a layer at or above j and (ii) the sum of test pins for parallel tested dies at or above layer j in the same test session. TSVs between layers j and $j-1$ must be equal to the maximum of these two quantities. If D_i does not satisfy resource and TSV constraints, we move to the next die or try to schedule D_i in next test session TS_{k+1}. We repeat this process until all the dies get scheduled. However, it must be kept in mind that, TSVs cannot be allocated dynamically. It may happen that the TSV requirement between layers j and $j-1$ (TSV_j) in test session TS_{k+1} is less than that in TS_k. Still we have to use TSV_{j_k} number of TSVs for TS_{k+1}, otherwise it will hamper the test schedule of TS_k. So, we only update the TSV_j if some later test session requires more TSVs between layers j and $j-1$. In that case, earlier test session will have some unused TSVs. However, total TSV requirement at any test session must not violate the limit of TSV_{stack}.

2) Session-less testing: In a session-based testing, the dies tested in a test session, may not have the same test times. Session time depends on the test time of the die with the largest test time, scheduled in that session. As no new die can be scheduled in the middle of a test session, the resources occupied by the die, which finishes its testing earlier in the session, have to remain unutilized till the session ends. As a result, total test time increases unnecessarily. This shortcoming of session-based testing can be overcome using session-less testing, which allows scheduling of a die at any point in the schedule depending upon resource availability. In session-less testing, instead of considering different test sessions, we consider two important data-structures throughout the schedule. $Break\_Point\_List (BP_{stack})$ note the points in the schedule,

Algorithm 1: *3D_Test_Schedule*

Input : List of dies $D_i (1 \leq i \leq M)$ to be scheduled
with assigned test pins WD_i and test time
TD_i; W_{stack}: maximum test pins of the stack;
TSV_{stack}: maximum TSV limit;

Var : TS_k: k^{th} test session ($k \leq M$); TST: Test
session time; TSW: Test session width; TT:
Total time; TSV_j: number of TSVs between
layer j and $j - 1$ ($2 \leq j \leq M$);

begin
  $TT \longleftarrow 0; k \longleftarrow 1$;
  for $j \leftarrow 2$ **to** M **do**
    $TSV_j \longleftarrow 0$;

  Sort the dies in descending order of TD_i;
  Mark all dies as unscheduled;
  while there exists unscheduled die **do**
    $TSW_k \longleftarrow W_{stack}$;
    $TST_k \longleftarrow 0$;
    while there exists unscheduled and unchecked
    die with $WD_i \leq TSW_k$ **do**
      Select an unscheduled die D_i in sorted
      order;
      if $(WD_i \leq TSW_k)$ **then**
        $TSV\_Checker()$;
        if yes **then**
          Schedule D_i; Mark D_i as scheduled;
          $TSW_k = TSW_k - WD_i$;
          if $(TD_i > TST_k)$ **then**
            $TST_k = TD_i$;

      else
        Check next die of the sorted list;

    $TT = TT + TST_k$; k++;
  Return TT as total schedule generation time;

Procedure 2: *TSV_Checker*

$TSV_{total} \longleftarrow 0$;
for $j \leftarrow 2$ **to** M **do**
  $TSV_{parallel} \longleftarrow 0$;
  $WD_{max} \longleftarrow 0$;
  for $l \leftarrow j$ **to** M **do**
    if $(WD_l > WD_{max})$ **then**
      $WD_{max} = WD_l$;
    if D_l is being tested **then**
      $TSV_{parallel} = TSV_{parallel} + WD_l$;

  $TSV_{j_k} = max\ (WD_{max}, TSV_{parallel}, TSV_{j_{k-1}})$;
  $TSV_{total} = TSV_{total} + TSV_{j_k}$;
if $(TSV_{total} < TSV_{stack})$ **then**
  Return yes;
else
  Return no;

---

where a die finishes its test and releases occupied test pins. An unscheduled die can be scheduled at any of the break-points, $bp_{stack_k} \in BP$. $Available\_Test\_Pins(ATP_{stack})$ keeps track of the available test resources at each bp_{stack_k}. We choose a die D_i in descending order of test time and check its schedulability in the minimum break-point bp_{stack_k}. The same $TSV\_Checker$ mentioned in session-based testing, is used to check TSV constraint at each bp_{stack_k}, where an unscheduled die can be scheduled. BP_{stack} and ATP_{stack} get updated with the schedule of each die. If sufficient resources are not available to schedule any unscheduled die at any break-point, we shift to the next break-point. This process continues till all the dies get scheduled. Finally maximum test finish time among all the dies is reported as the test time TT of the stack.

III. TEST-ARCHITECTURE OPTIMIZATION FOR 3D-SIC WITH SOFT DIE

3D-SIC with soft dies provide better opportunity of optimization of test time than 3D-SICs with hard dies. Unlike 3D-SIC with hard dies, the test architecture of individual dies is not fixed for soft dies. Any number of test resources within W_{stack} can be allocated to any die and 2D test architecture

of individual dies can also be developed. The test architecture optimization and test schedule generation problem for 3D-SIC with soft dies can be formulated as follows.

A. Problem Formulation

Suppose a stack consisting of M dies D_i ($1 \leq i \leq M$) is to be tested with a maximum available test pins W_{stack}. Assume that the maximum number of allowable TSVs to route the 3D-TAM is TSV_{stack}. Each die D_i consists of N_i number of cores $C_1, C_2 \ldots C_{N_i}$. The number of test patterns required to test core C_j ($1 \leq j \leq N_i$) is p_j. Determine an optimal TAM design and corresponding test schedule for the stack, as well as for each die, such that the total test time TT for the stack is minimized without violating resource and TSV constraints.

B. Scheduling Strategy

Although the flexibility of choosing TAM width for individual dies and *die level optimization* for each die offer better opportunity to optimize overall test time of the stack, test architecture design and scheduling of 3D-SIC with soft dies becomes more complex than hard die cases. Both the problems of (i) 3D-TAM distribution among dies of the different layers and (ii) determining an optimized test schedule for cores in a die, are NP-hard [4]. We have approached these problems in two steps. First, we have carried out *die level optimization*. The results of *die level optimization* are used in later part of *stack level optimization*, where we have carried out the 3D-TAM design. Particle Swarm Optimization (PSO) guided heuristics have been used in both levels of optimization.

1) Die level optimization: Suppose a die with N cores $C_1, C_2 \ldots C_N$ is to be tested with a maximum of W_{die} test pins. The die level test scheduling problem is to allocate test resources and test times to the cores so that, the test time (TAT) of the die is minimized.

We have used PSO guided Rectangular 2D-bin packing approach to generate optimized test schedule. Each core C_i ($1 \leq i \leq N$) is represented by a set of wrapper configurations R_i. The test resource of core C_i with j^{th} wrapper configuration can be represented by a rectangle whose height and width represent allocated test pins (w_{ij}) and the corresponding test time ($T(w_{ij})$) respectively. To get a schedule for the full die, the rectangles are to be packed into a bin of fixed height (W_{die}) so that TAT (width of the bin) is minimized. Selection of one test rectangle per core has been performed using PSO.

- *Particle Swarm Optimization Formulation:*

PSO is a population based evolutionary technique designed by Eberhart and Kennedy [11]. It starts with an initial population of particles. Each particle corresponds to a solution to the optimization problem being solved. Each particle has its fitness value. Particles evolve over generations guided by three factors – its own intelligence (*pbest*), global (swarm) intelligence (*gbest*), and the inertia factor.

Each core has a set of test rectangles and the maximum number of rectangles for any core be R. Let $B = \lceil \log_2^R \rceil$. A particle consists of $N \times B$ number of bits. First B bits identify the test rectangle selected for the first core, second B bits for the second core, and so on. Figure 2 shows a sample particle with $N = 4$ and $B = 4$. Fitness of a particle is equal to the total test time (TAT) of the SoC after scheduling the test rectangles using the *2D_Test_Schedule* procedure (Algorithm 3). For the initial generation, particles are generated randomly. In the successive generations, new particles are created using a *replace* operator, which attempts to align a particle with its *pbest* and the *gbest* particles, with some probability. For the sake of this alignment, the *replace* operator is applied at each bit position of a particle. For bit position i of a particle, the bit is replaced by the corresponding bit of *pbest* particle with probability α. After the operator has been applied for *pbest*, the same is done with respect to *gbest* with probability of replacement, β. In our experimentation, we have kept both α and β values at 0.1.

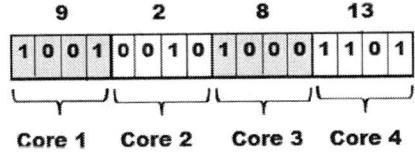

Core 1   Core 2   Core 3   Core 4

Figure 2.   Sample particle structure of 4 cores with W_{die} =16 ($B = 4$)

- *Scheduling of cores in a die:*

The algorithm takes as input the rectangle set corresponding to a particle, the maximum test pins W_{die}. BP_{die} and ATP_{die} keep track of the scheduling points and corresponding resource availability at those points. As the still unscheduled cores get scheduled, the list BP_{die} and ATP_{die} get updated. The rectangles are sorted on their area values (test pins (w) × test time (T)) in a descending order. The break-point list BP_{die} is scanned from the minimum to the maximum value. For the break-point bp_{die_k}, the algorithm scans the unscheduled rectangle list to check for the largest rectangle that can be scheduled at bp_{die_k}. If none are feasible, the algorithm advances to the next break-point. When rectangles corresponding to all cores have been scheduled, the maximum end time of testing of all cores gives the total test application time for the SoC. The *2D_Test_Schedule* algorithm to produce the schedule is presented next.

Algorithm 3: *2D_Test_Schedule*

Input : List of rectangles to be scheduled; W_{die}, the maximum test pins;

Var : BP_{die}: A list of break points; ATP_{die}: List of available test pins at each break point
$bp_{die} \in BP_{die}$;

begin
  Sort list of rectangles on decreasing area;
  Mark all rectangles as unscheduled;
  while there exists unscheduled rectangles **do**
    Check if any rectangle picked up in sorted order can be scheduled at next break point bp_{die_k} with available TAM resource atp_{die_k};
    if yes **then**
      Update BP_{die}, ATP_{die} and Rectangle List;
      Mark corresponding rectangle scheduled;
    else
      Continue with next $bp_{die_k} \in BP_{die}$;
  Return the maximum test end time of all rectangles;

---

2) Stack level optimization: In *stack level optimization*, total test resources W_{stack} are optimally allocated to all the dies and a test schedule of the dies is generated in such a way, so that the total test time TT to test the stack is minimized. The problem is similar to test architecture optimization of 3D-SIC with hard dies with an added complexity of selection of TAM allocation of each die. Again, it is an NP-hard problem [4], which we have solved by incorporating meta search technique like PSO with the *3D_Test_Schedule* heuristic mentioned in Section II. Similar kind of PSO formulation used for *die level optimization*, is adopted here. Each die D_i is represented by a set of assigned test pin value WD_{i_k} ($1 \leq k \leq W_{stack}$) and associated test time TD_{i_k}. These values are obtained from *die level optimization* by varying the W_{die} value from 1 to W_{stack} and noting the corresponding test times. Each particle selects a tuple of assigned test pins and associated test time (WD_{i_k}, TD_{i_k}) for each die. Fitness of a particle is evaluated by calculating the test time of the stack by using *3D_Test_Schedule* heuristic. The *replace* operator, guided by *pbest* and *gbest*, evolves particles towards better solution. A similar *TSV_Checker* (as mentioned in Section II) is used to check TSV constraint. Both session-based and session-less testing have been performed. Figure 3 describes the total flow of the test architecture design and test scheduling procedure for 3D-SIC with soft dies.

IV.   EXPERIMENTAL RESULTS

In this section we present the results of our experimentation for both *hard die* and *soft die* cases. For the sake of comparison, we have considered the same 3D-SIC benchmarks presented in [4]. Figure 4 presents the SICs, which are formed using ITC'02 benchmarks as dies. For 3D-SIC with hard die, we have considered the same test architectures of individual dies reported in [4]. Table I reports the details of test architectures of each die.

Table II shows the comparison of our test scheduling approaches (both session-based and session-less) with the

Table I. Test Lengths and Test Pins for Hard Dies [4]

Die	$d695$	$f2126$	$p22810$	$p34392$	$p93791$
Test length	96927	669329	651281	1384949	1947063
Test pin	15	20	25	25	30

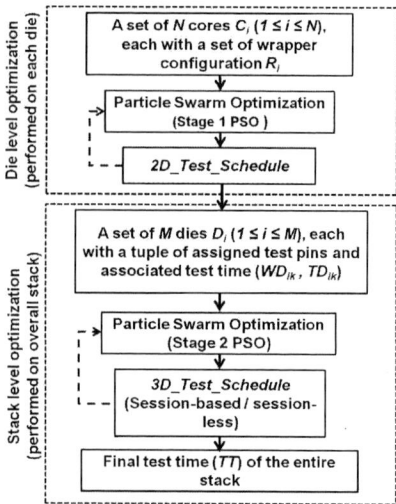

Figure 3. Test Flow for 3D-SIC with soft dies.

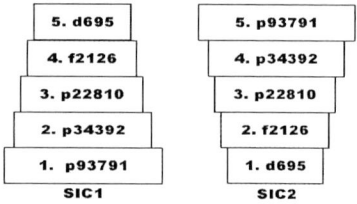

Figure 4. 3D-SIC benchmarks [4]

approach presented in [4] for 3D-SIC with hard dies. The first two columns of Table II describe the maximum allowable TSV limit and number of allocated test pins to test the SIC. Columns 3, 4 and 8, 9 present the test time and corresponding test schedule reported in [4] for SIC1 and SIC2 respectively. The test times and schedules obtained from our $3D\_Test\_Schedule$ heuristic considering session-based (SB) testing for both SIC1 and SIC2 are reported in columns 5, 6 and 10, 11 respectively while columns 7 and 12 report our session-less (SL) test times for SIC1 and SIC2 respectively. It may be noted from Table II that the schedules reported in [4] violate maximum TSV limit TSV_{stack} in several cases. For example, the number of test pins of dies 3, 4 and 5 of SIC1 are 25, 20 and 15 respectively. To facilitate parallel testing of these three dies, 60 TSVs (25 + 20 + 15) are required between each of the layers 2 and 1 and layers 3 and 2. The TSV requirement between dies 4 and 3 is 35 (20 + 15) and finally 15 TSVs are required between layers 5 and 4. So, the schedule reported in [4], for TSV_{stack} = 160 and W_{stack} = 60, which shows parallel scheduling of dies 3, 4 and 5 requires a total of 170 (60 + 60 + 5 + 15 = 170) TSVs, which clearly violates the maximum TSV limit. Similarly, for SIC2, all the schedules (W_{stack} = 60, 70, 80, 90 and 100), which report parallel testing of dies 4 and 5, require a total of 195 TSVs. It again violates the maximum TSV limit. In contrast, no TSV limit violation can be noted in the schedules generated by our $3D\_Test\_Schedule$ heuristic. Moreover, our session-based test time results are same with the results reported in [4], for all the cases where [4] does not violate TSV_{stack}. Session-less testing further improves test time over session-based testing.

Next, we present the results of *stack level optimization*

for soft dies. The *die level optimization* results of individual dies are fed to PSO guided $3D\_Test\_Schedule$ heuristic, which evolve over generations to explore a large search space of solution to find a near optimal test schedule of the stack. Table III presents the comparison between our scheduling strategy and the approach reported in [4] for different values of W_{stack} for SIC1. Column 3 of Table III reports the test time reported in [4] while columns 4 and 9 report our session-based and session-less test time results respectively. Column 6 reports the test schedule obtained from our session-based testing. Corresponding test pin allocation to each die and associated TSV requirements between all the layers are reported in columns 7 and 8 respectively. For example, to test SIC1 with 40 test pins, 26 test pins are allocated to die 1. Dies 2, 3, 4 and 5 are tested using 14, 40, 34 and 26 test pins respectively. Similarly 40, 40, 34 and 26 TSVs are required between dies 2 and 1, 3 and 2, 4 and 3, 5 and 4 respectively. It may be noted that, both of our session-based and session-less test architecture and scheduling strategies can reduce test time of SIC1 upto 51% compared to the results reported in [4]. Figure 5 presents a pictorial illustration of the session-based test schedule of SIC1 for W_{stack} = 70 and TSV_{stack} = 140, obtained from our PSO guided $3D\_Test\_Schedule$ heuristic. All the dies are tested in three test sessions in descending order of their individual test times and satisfying the resource and TSV constraints. The figure describes how $TSV_{Checker}$ updates the TSV requirements between different layers of the stack in successive test sessions. The number of TSVs used and the actual number of TSVs assigned between each two dies in each session is mentioned in the figure. It may be noted that, although 14 TSVs are required to send test data to die 3 in session 1, we have to assign 48 TSVs between die 2 and 1, as die 2 is allocated 48 test pins. The TSV value between dies 2 and 1 gets updated to 66 in session 2, to facilitate parallel testing of dies 2 and 4. In the final session, only 24 out of 66 TSVs are used between die 2 and 1. Rest of the TSVs remain unused. Table IV reports the test time results of our techniques for SIC2. It may be noted from Table III and Table IV that session-less testing can reduce test time over session-based testing for both the SIC1 and SIC2.

V. Conclusion

In this paper we have presented a PSO guided heuristic for test architecture optimization and test scheduling for 3D-SIC with hard dies and soft dies. Both session-based and session-less testing have been considered. Our heuristic can produce optimized results without violating resources and TSV constrains. Incorporation of PSO in both 2D and 3D optimizations has given us the flexibility of two stage optimization, which has been instrumental in minimizing the overall test time of the stack to a large extent.

References

[1] L. Jiang, L. Huang, and Q. Xu, "Test architecture design and optimization for three-dimensional socs," in *Proc. Conf. Design, Automation and Test in Europe*, 2009, pp. 220–225.

Table II. Comparison of Our Test Scheduling Approach (Session-Based and Session-Less) With [4] For 3D-SIC With Hard Die

3D-SIC with hard die		SIC1					SIC2				
TSV_{stack}	W_{stack}	Test time [4]	Schedule [4]	Test time (SB)	Schedule (SB)	Test time (SL)	Test time [4]	Schedule [4]	Test time (SB)	Schedule (SB)	Test time (SL)
160	30	4748920	1, 2, 3, 4, 5	4748920	1, 2, 4, 3, 5	4748920	4748920	1, 2, 3, 4, 5	4748920	5, 4, 2, 3, 1	4748920
160	40	4652620	1, 2, 3, 4 ∥ 5	4652620	1, 2 ∥ 5, 4, 3	4652620	4652620	1 ∥ 3, 2, 4, 5	4652620	5, 4 ∥ 1, 2, 3	4652620
160	50	3428310	1 ∥ 4, 2 ∥ 3, 5	3428310	1 ∥ 4, 2 ∥ 3, 5	3332012	3428310	1, 2 ∥ 5, 3 ∥ 4	3428310	5 ∥ 2, 4 ∥ 3, 1	3332012
160	60	2616390	1 ∥ 2, 3 ∥ 4 ∥ 5	2712690	1 ∥ 2, 4 ∥ 3, 5	2598340	2616390	1 ∥ 2 ∥ 3, 4 ∥ 5	3428310	5 ∥ 2, 4 ∥ 3, 1	3332012
160	70	2616390	1 ∥ 2 ∥ 5, 3 ∥ 4	2616390	1 ∥ 2 ∥ 5, 4 ∥ 3	2598340	2616390	1 ∥ 2 ∥ 3, 4 ∥ 5	3332012	5 ∥ 2 ∥ 1, 4 ∥ 3	3332012
160	80	2598340	1 ∥ 2 ∥ 4, 3 ∥ 5	2598340	1 ∥ 2 ∥ 4, 3 ∥ 5	1947063	2616390	1 ∥ 2 ∥ 3, 4 ∥ 5	3332012	5 ∥ 2 ∥ 1, 4 ∥ 3	3332012
160	90	2598340	1 ∥ 2 ∥ 4, 3 ∥ 5	2598340	1 ∥ 2 ∥ 4 ∥ 5, 3	1947063	2616390	1 ∥ 2 ∥ 3, 4 ∥ 5	3332012	5 ∥ 2 ∥ 1, 4 ∥ 3	3332012
160	100	2043360	1 ∥ 2 ∥ 3 ∥ 4, 5	2043360	1 ∥ 2 ∥ 4 ∥ 3, 5	1947063	2616390	1 ∥ 2 ∥ 3, 4 ∥ 5	3332012	5 ∥ 2 ∥ 1, 4 ∥ 3	3332012

Table III. Comparison of Test Scheduling Results Between [4] and Our Approaches (Session-Based and Session-Less) For SIC1 (3D-SIC With Soft Die)

TSV_{stack}	W_{stack}	Test time [4]	Test time SB	Impv. over [4] %	Schedule (SB)	Test pins (SB)	TSV (SB)	Test time (SL)	Impv. over [4] %
140	30	4795930	3755885	**21.69**	1, 2, 3, 4, 5	30, 30, 30, 30, 30	30, 30, 30, 30	3724309	**22.34**
140	40	3841360	2881060	**25.0**	2 ∥ 1, 3, 4, 5	26, 14, 40, 34, 26	40, 40, 34, 26	2802280	**27.05**
140	50	3090720	2335908	**24.42**	1 ∥ 2, 4 ∥ 3 ∥ 5	32, 18, 22, 18, 10	50, 50, 28, 10	2265193	**26.71**
140	60	2873290	1915438	**33.34**	1 ∥ 3, 2 ∥ 4, 5	48, 44, 12, 16, 26	60, 26, 26, 26	1875655	**34.72**
140	70	2743320	1701271	**37.98**	1 ∥ 3, 2 ∥ 4, 5	56, 48, 14, 18, 24	66, 24, 24, 24	1623521	**40.82**
140	80	2439760	1481323	**39.28**	1 ∥ 4, 2 ∥ 5, 3	66, 56, 24, 14, 18	80, 24, 18, 18	1448635	**40.62**
140	90	2395760	1396718	**41.70**	1 ∥ 4, 2 ∥ 3, 5	74, 56, 24, 16, 18	80, 24, 18, 18	1267313	**47.10**
140	100	2369680	1153410	**51.32**	4 ∥ 1 ∥ 3 ∥ 2, 5	50, 28, 12, 10, 30	50, 30, 30, 30	1153169	**51.33**

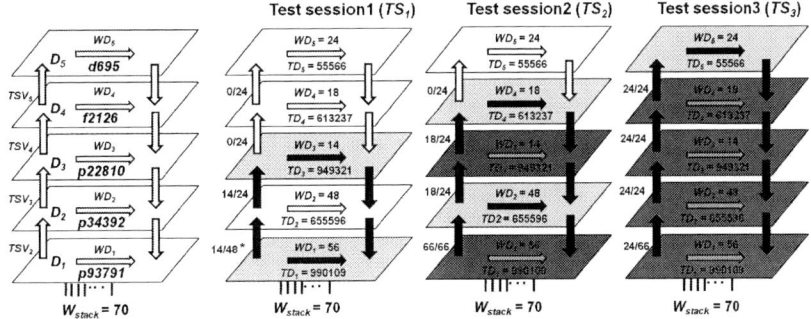

Figure 5. Session-based test schedule of SIC1 for $W_{stack} = 70$ and $TSV_{stack} = 140$ (considering 3D-SIC with soft dies)

Table IV. Test Scheduling Results Of Our Approaches (Session-Based and Session-Less) For SIC2 (3D-SIC With Soft Die)

TSV_{stack}	W_{stack}	Test time (SB)	Schedule (SB)	Test pins (SB)	TSVs (SB)	Test Time (SL)
140	30	3731707	5, 4, 3 ∥ 2, 1	30, 14, 16, 30, 30	30, 30, 30, 30	3695406
140	40	2960916	5 ∥ 3, 4 ∥ 2, 1	40, 12, 10, 28, 30	40, 40, 30, 30	2874935
140	50	2846974	4 ∥ 5 ∥ 2, 3 ∥ 1	10, 10, 40, 12, 22	44, 40, 34, 22	2735816
140	60	2846974	4 ∥ 5 ∥ 2 ∥ 1, 3	10, 10, 40, 12, 22	44, 40, 34, 22	2735816
140	70	2846974	4 ∥ 5 ∥ 2 ∥ 1, 3	10, 10, 40, 12, 22	44, 40, 34, 22	2735816
140	80	2846974	4 ∥ 5 ∥ 2 ∥ 1, 3	28, 10, 40, 12, 22	44, 40, 34, 22	2735816
140	90	2846974	4 ∥ 5 ∥ 2 ∥ 1, 3	28, 10, 40, 12, 22	44, 40, 34, 22	2735816
140	100	2846974	4 ∥ 5 ∥ 2 ∥ 1, 3	46, 10, 40, 12, 22	22, 20, 17, 11	2735816

[2] B. Noia, K. Chakrabarty, and Y. Xie, "Test-wrapper optimization for embedded cores in tsv-based three-dimensional socs," in *Comput. Design, 2009. ICCD 2009. IEEE Int. Conf.* IEEE, 2009, pp. 70–77.

[3] M. Agrawal and K. Chakrabarty, "Test-cost optimization and test-flow selection for 3d-stacked ics," in *VLSI Test Symposium (VTS), 2013 IEEE 31st*, April 2013, pp. 1–6.

[4] B. Noia, K. Chakrabarty, S. K. Goel, E. J. Marinissen, and J. Verbree, "Test-architecture optimization and test scheduling for tsv-based 3-d stacked ics," *IEEE Trans. Comput.-Aided Des. Integr. Circuits Syst.*, vol. 30, no. 11, pp. 1705–1718, 2011.

[5] K. Chakrabarty, S. Deutsch, H. Thapliyal, and F. Ye, "Tsv defects and tsv-induced circuit failures: The third dimension in test and design-for-test," in *Rel. Physics Symp. (IRPS), 2012 IEEE Int.*, 2012, pp. 5F–1.

[6] C.-Y. Lo, Y.-T. Hsing, L.-M. Denq, and C.-W. Wu, "Soc test architecture and method for 3-d ics," *IEEE Trans. Comput.-Aided Des. Integr. Circuits Syst.*, vol. 29, no. 10, pp. 1645–1649, 2010.

[7] E. J. Marinissen, J. Verbree, and M. Konijnenburg, "A structured and scalable test access architecture for tsv-based 3d stacked ics," in *VLSI Test Symp. (VTS), 2010 28th.* IEEE, 2010, pp. 269–274.

[8] B. Noia, K. Chakrabarty, and E. J. Marinissen, "Optimization methods for post-bond die-internal/external testing in 3d stacked ics," in *Test Conf. (ITC), 2010 IEEE Int.* IEEE, 2010, pp. 1–9.

[9] C. Giri, S. Sarkar, and S. Chattopadhyay, "Test scheduling for core-based socs using genetic algorithm based heuristic approach," in *Advanced Intelligent Computing Theories and Applications. With Aspects of Artificial Intelligence.* Springer, 2007, pp. 1032–1041.

[10] S. K. Goel and E. J. Marinissen, "Control-aware test architecture design for modular soc testing," in *Test Workshop, 2003. Proc. The Eighth IEEE European.* IEEE, 2003, pp. 57–62.

[11] J. Kennedy and R. Eberhart, "Particle swarm optimization," in *Neural Networks, 1995. Proceedings., IEEE International Conference on*, vol. 4, Nov 1995, pp. 1942–1948 vol.4.

2015 IEEE 33rd VLSI Test Symposium (VTS)

Improving Diagnosis Resolution of a Fault Detection Test Set

Andreas Riefert *   Matthias Sauer *   Sudhakar Reddy †   Bernd Becker *

* Albert-Ludwigs-Universität Freiburg
Georges-Köhler-Allee 051
79110 Freiburg, Germany
{ riefert | sauerm | becker }@informatik.uni-freiburg.de

† University of Iowa
5324 Seamans Center
Iowa City, United States
reddy@engineering.uiowa.edu

Abstract— Manufactured VLSI circuits using a new technology typically suffer from systematic defects that are process-dependent and at sub-nanometer feature sizes such defects may be even design-dependent. The root causes for systematic defects must be determined to ramp up yields. Volume diagnosis is becoming popular to identify root causes for systematic defects. Volume diagnosis uses logic diagnosis based on failing circuit responses to production tests of a large number of failing devices, followed by statistical analysis methods to determine the root cause(s) for yield limiters. Typically production tests use fault detection tests and hence may have limited diagnosis resolution. To improve diagnosis resolution diagnostic ATPGs can be used to generate test sets to distinguish all pairs of distinguishable faults in one or more fault models. The sizes of such tests tend to be considerably higher than fault detection test sets used as production tests. For this reason, generation of test sets that detect faults and also possess a high diagnosis resolution is important.

In this work we present a method to improve the diagnosis resolution of a compact fault detection test set without increasing pattern count or decreasing fault coverage. The basic idea of the approach is to generate a SAT formula which enforces diagnosis and is solved by a MAX-SAT solver which is a SAT-based maximization tool. We believe this is the first time a method to improve diagnosis resolution of a test set of given size has been reported. Experimental results on ISCAS 89 circuits demonstrate the effectiveness of the proposed method.

I. INTRODUCTION

Logic diagnosis of failing VLSI circuits is used to identify the location(s) and the nature of defects [1]–[10] causing the circuit to fail test. The data on defective circuit responses, gathered from production tests, is analyzed to perform logic diagnosis of the devices. Logic diagnosis results can be used to help guide physical failure analysis (PFA) of failing devices to pinpoint the defects [11]. Because of the cost and time required for PFA, volume diagnosis is becoming popular, especially for the current deep sub-micron devices, to identify root causes for systematic defects and to take steps to eliminate the root causes. Volume diagnosis uses logic diagnosis on failing circuit responses, of a large number of failing devices, to production tests followed by statistical analysis methods to determine the root cause(s) for yield limiting defects [12]–[14]. Production tests use fault detection tests and hence may have limited diagnosis resolution. To improve diagnosis resolution of tests used one can use diagnostic ATPGs [15]–[26] to generate test sets to distinguish all pairs of distinguishable faults in one or more fault models. The sizes of such test sets tend to be considerably larger compared to fault detection test sets used as production tests. Thus, test sets generated by diagnostic ATPGs may be used in offline diagnosis, often using incremental test generation [6], [21], [22] flows, but are not considered for use in volume diagnosis flows. For this reason, generation of fault detection test sets with higher diagnosis resolution without inflating pattern counts is important [27].

In this work we present a method to generate a test set with improved diagnosis resolution starting from a compacted fault detection test set without increasing pattern count. We believe this is the first time a method to maximize diagnosis resolution of a test set of given size has been reported. The method utilizes two algorithms. The *Implicit Diagnosis* distinguishes faults by requiring a unique detection signature for each fault and is able to effectively partition a large number of easy-to-distinguish faults at once. The *Explicit Diagnosis* considers fault pairs and can effectively differentiate hard-to-distinguish faults. The proposed algorithms utilize a MAX-SAT solver which is a SAT-based maximization tool [28]. To illustrate the effectiveness of the procedure presented in this work we consider increasing the diagnosis resolution of compacted transition fault detecting test sets that were generated by a commercial ATPG tool using launch off capture test mode. However, the method we propose is applicable to any test sets using any test generation procedures. We use a commonly used measure of diagnostic resolution of a given test set called *diagnostic coverage* [23]–[25] to evaluate the effectiveness of test sets. Experimental results for ISCAS 89 circuits show that on average diagnostic coverage of a compacted launch off capture transition fault detecting test set is improved by 5.85 % using the proposed procedure.

As an overview, the procedure we present in this paper works as follows: it will generate a test set T' with improved diagnostic coverage starting from a given fault detection test set T. Additionally, the size of T' will be no more than the size of T and the fault detection coverage of T' will be no less than that of T. As shown in Figure 1, T' is initially empty. Then, each pattern t_i of T is processed. First, faults detected by t_i and not detected by tests currently in T' are determined. Next, a MAX-SAT formula is encoded and solved which requires the detection of the determined faults and maximizes diagnostic coverage. The returned pattern

978-1-4799-7598-3/15 $31.00 © 2015 IEEE

108

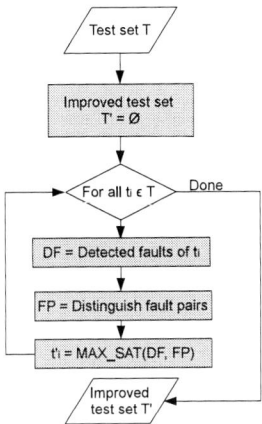

Figure 1. Diagnostic ATPG flow

t'_i is added to T'.

The remainder of the paper is organized as follows. Section II discusses the diagnosis resolution of test sets as well as the underlying maximization engine used in this work. Section III describes the proposed algorithms in more detail. Finally, in Section IV the experimental results are shown and conclusions are drawn in Section V.

II. Preliminaries

In this section we first review a commonly used metric to measure diagnosis resolution of tests. Next we give an overview of the utilized MAX-SAT solver.

A. Diagnosis resolution

We say that a set of tests S distinguishes a pair of modeled faults f and g in a circuit C if the tests in S produce different responses when fault f is present in C compared to when fault g is present in C. A pair of faults f and g are said to be (functionally) *equivalent* if faults f and g cannot be distinguished by any set of tests. The set of faults in a fault model can be partitioned into disjoint subsets of equivalent faults called *equivalence classes*. Each equivalence class contains all faults that are pairwise equivalent to each other. Given a fault model, responses to a given set of tests T partition the set of faults into disjoint subsets of faults. The faults in each subset are those whose responses to all the tests in T are the same. Such subsets are called *T-equivalent* faults. For brevity we call T-equivalent subsets of faults *clusters* in this work. A commonly used measure of the diagnosis resolution achieved by a given test set is called *diagnostic coverage* (*DC*) [24]. We define it as follows. Let C be a circuit with N_{eq} equivalence classes of faults and let T be a test that divides the set of faults into N_{T-eq} T-equivalent clusters. Then the diagnostic resolution DC of a test set T is the ratio N_{T-eq} / N_{eq}. Notice that $DC <= 1$, and when $N_{T-eq} = N_{eq}$, $DC = 1$ and T has distinguished all pairs of distinguishable faults.

B. SAT-based maximization

The underlying engine of our algorithm is an efficient SAT-based maximization tool which solves the so-called maximum satisfiability problem (MAX-SAT). An ordinary SAT problem consists of a number of clauses. A SAT solver has to find a solution which satisfies each single clause and thus satisfies the SAT formula. If no such solution exists, the formula is unsatisfiable. A MAX-SAT problem is a generalization of SAT and its target is to determine the maximum number of clauses that can be satisfied simultaneously.

In the following, we provide a brief overview of the employed MAX-SAT solver [28]. The solver distinguishes between two types of clauses, namely hard clauses and soft clauses. A valid solution has to satisfy all hard clauses and the maximum number of soft clauses. This problem is solved by incrementally calling a SAT solver. In order to transform the original problem, consisting of hard and soft clauses, into a standard SAT problem which only consists of hard clauses, the formula has to be modified. For this purpose a bitonic sorting network is employed and encoded into the formula. This network can be viewed as a circuit with n inputs (corresponding to the soft clauses) and n outputs. Its function is to sort all '0' and '1' (corresponding to a satisfied soft clause) applied at the inputs and output a non-decreasing sequence. Alternatively, the network can be understood as a counter which counts all applied '1's and outputs the result in a unary representation. This enables us to count the number of satisfied soft clauses and incrementally adjust the bounds for this number until the optimal solution is found.

To prevent excessive runtimes, the employed MAX-SAT solver implements a timeout. If the timeout is reached, the solver returns the best solution that it has found so far. Furthermore, the solver comprises a partial mode which does not return the optimal solution but incrementally optimizes blocks of soft clauses. The block size is a user-defined parameter. Small block sizes yield increased performance but in general will produce worse results.

III. Diagnostic ATPG framework

In this section we describe the proposed diagnostic ATPG framework based on the SAT-based maximization engine described above. For convenience, we refer to the procedure we use to improve the diagnostic coverage of a fault detecting test set as diagnostic ATPG procedure. We implemented two basic algorithms, namely the *Implicit Diagnosis* and the *Explicit Diagnosis*. The implicit approach enforces diagnosis by requiring a pre-computed unique detection signature for each fault. This allows to handle a large number of faults as the partitioning of the faults is enforced by the unique signature and fault pairs do not have to be explicitly considered. However, requiring a specific detection signature may be too restrictive and prevent valid solutions. The explicit approach explicitly distinguishes fault pairs and requires a difference in their output signatures under a test pattern. This allows to effectively differentiate between hard-to-distinguish faults. However, the number of faults that can be considered at once is limited as each possible fault pair has to be considered. Our framework exploits the advantages of both algorithms by first executing the implicit algorithm which partitions the initially large fault clusters. Then the

978-1-4799-7598-3/15 $31.00 © 2015 IEEE

explicit algorithm processes the remaining small clusters which usually contain several hard-to-distinguish faults.

The proposed diagnostic ATPG framework requires a detection test set T as input which we obtained with a commercial ATPG tool. The improved test set T' will have the same pattern count and at least the same fault coverage as T. Our framework uses a fault list which contains all slow-to-rise and slow-to-fall transition delay faults at each input and output of each node of the circuit under test (CUT). We then collapse all trivially equivalent faults (e.g. a slow-to-rise fault at the input of an inverter and a slow-to-fall fault at its output). Next, we identify all untestable faults, by solving an accordingly generated SAT formula, and remove them from the fault list as they cannot be distinguished from each other.

After these preprocessing steps the *Implicit Diagnosis* is executed for N_I iterations, where N_I is a user-defined parameter. Then the *Explicit Diagnosis* is executed for $N - N_I$ iterations, where N is the number of patterns in test set T. Each iteration will process a pattern t_i from T and generate a pattern t'_i for T' with improved diagnostic capabilities. For $N_I = N$ the implicit approach will generate all patterns in T'; and when $N_I = 0$ the explicit approach is used for all patterns. We maintain a data structure which contains all fault clusters which were distinguished so far. Initially only one cluster is contained which consists of all faults from the preprocessed fault list. The fault clusters are updated after each generated pattern. If a cluster contains only one fault, this fault has been uniquely diagnosed and is not considered any further.

The fault coverage of T is maintained in each iteration i by first determining the faults which are detected by t_i and are not detected by the already generated patterns in T'. For each of these detected faults we extract the signal values which enable its detection. These values are denoted as *detection requirements* and are added to the generated MAX-SAT formula as constraints, i.e. as unit clauses. This guarantees that the fault coverage of the improved test T' will be no less than that of T.

Following the diagnostic ATPG procedure we evaluate each fault pair which is still not distinguished by the generated test set T'. By encoding and solving a corresponding SAT formula we can determine whether this fault pair can be distinguished at all. This final analysis enables us to give the number of (functionally) equivalent fault classes N_{eq} and compute the diagnostic coverage.

A. Implicit Diagnosis

The basic idea of the *Implicit Diagnosis* is to generate a pattern which detects a large number of faults with a unique output signature for each fault which enables the pairwise distinction of these faults. In Figure 2, four faults f_k with their structural output cones $OC(f_k)$ are shown: $OC(f_1)$ = $\{o_1\}$, $OC(f_2)$ = $\{o_3\}$, $OC(f_3)$ = $\{o_1, o_2\}$, $OC(f_4)$ = $\{o_2, o_3\}$. Assume a pattern which detects these faults at the outputs which are indicated by the colored lines. Then this pattern will also pairwise distinguish all of the faults as it produces a unique signature for each of the faults. A pseudocode description of the approach is given in Algorithm

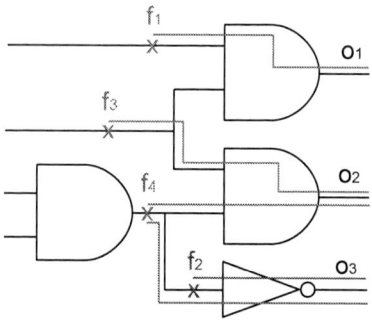

Figure 2. Implicit Diagnosis

1. The user can define the maximum number of faults F which are considered for diagnosis within one MAX-SAT formula.

Algorithm 1 Implicit Diagnosis

1: dr = Compute_detection_requirements(t_i);
2: **for all** $fault\_cluster_j$ **do**
3:      Create_fault_groups();
4:      Select_target_faults(F);
5:      Compute_detection_signatures();
6: Encode_formula(dr);
7: t'_i = MAX_SAT_solve();
8: Update_fault_clusters(t'_i);

Before the encoding of the MAX-SAT formula each fault cluster $fault\_cluster_j$ is processed. In a first step all faults f_k are grouped with respect to the outputs o_l in $OC(f_k)$. Thus for all faults f_k within one group $OC(f_k)$ is equal. The grouping assures that two faults from different groups will have differing structural output cones. Then one fault is selected from each group. The selection process is guided by a heuristic which is explained in the following. As the intention is to detect as many of the selected faults as possible (and thus implicitly distinguish them) we compute for each fault a set of *necessary assignments*. This set contains all signal values which are necessary for the sensitization of the fault. During the selection process a list is maintained which contains an entry for each signal line to which a value has been assigned. Initially this list contains the values from the *detection requirements*. Each fault is checked for compliance with the values contained in the list. If it is compatible its *necessary assignments* are added to the list and the next fault is evaluated. If no fault from a fault group is compatible then this fault group is skipped. This process avoids adding faults which cannot be sensitized simultaneously. After the target faults have been selected, a unique detection signature has to be computed for each fault. If a fault is detected with this signature, it is guaranteed to be distinguished from each of the other targeted faults. The computation starts with the faults which fan out to the smallest number of outputs as they permit the smallest number of possible detection signatures. By evaluating all possible permutations the algorithm tries to find a detection signature which does not conflict with any of the already computed signatures. This will always succeed as each target fault has a differing output cone. For an example look again at Figure 2. First the faults f_1 and f_2 would be required to be detected at o_1 and o_3. Next f_3 would be processed and output o_2 would be chosen as unique detection signature as it does not conflict with the yet chosen signatures for f_1 and f_2. Finally for f_4 it is not enough to choose only one output as both of them would conflict with already computed signatures. Therefore this fault has to be detected at both outputs o_2 and o_3. The colored lines in Figure 2 show the computed signatures.

When the preprocessing steps are finished the actual encoding of the MAX-SAT formula commences. The circuit is encoded for two unrollings which constitutes the two time frames of a launch off capture pattern. Additionally, for each target fault the part of its output cone is encoded which is required to determine the precomputed detection signature (i.e. the colored lines in Figure 2). An additional AND-gate is encoded which outputs a '1' if the fault is detected at each output that was specified in the detection signature. The output of this AND-gate is added for each fault as a soft clause to the formula. The remaining encoding is added as hard clauses. Also the computed *detection requirements* are added as hard unit clauses. By solving the resulting formula the MAX-SAT solver will find a solution which detects the maximum number of targeted faults with the required detection signature and detects all faults from the *detection requirements*. The computed pattern t_i' is added to the diagnostic test set and the fault clusters are updated accordingly.

B. Explicit Diagnosis

Algorithm 2 Explicit Diagnosis

1: $dr = $ Compute_detection_requirements(t_i);
2: $t_i' = NULL$;
3: **for** $j = 0$ to L **do**
4:     Select_target_faults(F, FP);
5:     Encode_formula(dr);
6:     $tmp_pat_j = $ MAX_SAT_solve();
7:     **if** $DC(tmp\_pat_j) > DC(t_i')$ **then**
8:         $t_i' = tmp\_pat_j$;
9:     Update_target_faults();
10: Update_fault_clusters(t_i');

A pseudocode description of the *Explicit Diagnosis* is given in Algorithm 2. The user can define the maximum number of faults F and the maximum number of fault pairs FP which are considered for diagnosis within one MAX-SAT formula. Furthermore an iteration count L has to be given which determines how many times the algorithm tries to improve a pattern t_i.

In order to construct the MAX-SAT formula we first select target faults out of all fault clusters which shall be considered for diagnosis. The selection is done randomly, but faults which have not yet been detected are prefered. The number of selected target faults is bounded by F and FP; depending on whether a lot of faults from one cluster are

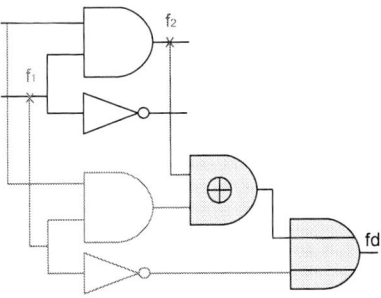

Figure 3. Explicit Diagnosis Encoding

chosen (which induces a lot of target fault pairs) or faults from a lot of distinct clusters are chosen (which induces a lot of target faults). Next the circuit is encoded for two unrollings which constitutes the two time frames of a launch off capture pattern. Furthermore for each selected fault its structural output cone is encoded in order to determine whether a fault effect is visible at an output. Then the logic has to be added which evaluates whether two faults are distinguished by the applied pattern. To do so, each pair *(f1, f2)* of the selected faults, which belongs to the same fault cluster and consequently has not yet been distinguished, has to be considered.

We have to differentiate between 3 cases: (1) A circuit output which lies neither in the output cone of *f1* nor in the output cone of *f2* can be ignored. (2) An output which is only contained in the cone of one fault distinguishes the fault pair if the corresponding fault is detected at this output. (3) An output which is contained in both output cones distinguishes the faults if only one of them is detected at this output. In this case an additional XOR-gate has to be encoded whose output is '1' if case (3) holds. Finally an OR-gate is encoded whose inputs consist of the signals for cases (2) and (3). Thus the output of the OR-gate is '1' when either case (2) or (3) holds for at least one output in the output cone of *(f1, f2)* and thus the fault pair is distinguished.

Figure 3 shows an example with a CUT (black) and two faults *f1*, *f2*. The colored output cone (red, blue) of the faults determines whether a fault effect is visible at the corresponding logic nodes. As both faults can be detected at the output of the AND-gate an additional XOR-gate (filled grey) is required to determine whether case (3) holds. Only *f1* can be detected at the output of the inverter which corresponds to case (2). Thus the added OR-gate (filled grey) is '1' when either only *f1* or *f2* are detected at the output of the AND-gate (case (3)) or *f1* is detected at the output of the inverter (case (2)). The outputs *fd* of all accordingly created OR-gates are added as soft clauses to the formula. All the remaining encoding is added as hard clauses. Also the computed *detection requirements* are added as hard unit clauses. This way the MAX-SAT solver finds a pattern tmp_pat_j which distinguishes the maximum number of encoded fault pairs and detects all faults from the *detection requirements*.

For large fault clusters it is not feasible to consider and encode all possible fault pairs at once. Therefore the inner loop iterates several times to incrementally find a selection of target faults which yields the best diagnostic pattern. For each generated pattern tmp_pat_j we determine its diagnostic coverage. If it is better than the coverage of the best found pattern t'_i, t'_i is updated accordingly. Next, the target faults for the following iteration are updated. This is done by identifying distinguished fault pairs. These faults are added as target faults for the following iteration. The remaining, undistinguished faults are removed and marked such that they will not be selected anymore in further iterations. The intuition is to stepwise find a set of target faults which yields a pattern tmp_pat_j with the highest diagnostic coverage. After L iterations, the best found pattern t'_i is added to the diagnostic test set and the fault clusters are updated accordingly. For our experiments we used $L = 10$.

IV. EXPERIMENTAL RESULTS

For our experiments we utilized a commercial ATPG tool which generated a compacted test set for fault detection for each benchmark circuit. We considered slow-to-rise and slow-to-fall transition delay faults at each input and output of each gate. All experiments were run on one core of an Intel Xeon processor running at 3.3 GHz. The results are reported in Table I. After the circuit name the number of patterns in the compacted transition fault detecting test set is given, followed by N_{eq}, the number of fault equivalence classes.

The column *Com ATPG DC %* gives the percentage diagnostic coverage of the detection test set generated by the commercial ATPG. The columns *Implicit Diag*, *Explicit Diag* and *Combined Diag* give the results for executing only the implicit approach ($N_I = N$), for executing only the explicit approach ($N_I = 0$), and for executing the implicit approach on the first 50 % of the patterns and the explicit approach on the remaining patterns ($N_I = N/2$), respectively. $N_I = N/2$ was chosen as the best trade-off based on experiments with different values. The corresponding sub-columns show the diagnostic coverage and the required runtime in seconds. The last row provides the average percentage diagnostic coverage and the average runtime normalized to the runtime of *Explicit Diag*. Thus on average, *Implicit Diag* requires 0.0493 times and *Combined Diag* requires 0.4123 times the runtime of *Explicit Diag*, respectively. The values in bold give the best result for the corresponding circuit. For the explicit approach we set the maximum number of faults to 200 and the maximum number of fault pairs to 3,000. Thus one MAX-SAT formula considered a maximum of 200 faults and 3,000 fault pairs for diagnosis at once. The implicit approach considered up to 1,000 faults at once.

The results show that the initial detection test sets produced by the commercial ATPG yield a low diagnostic coverage for several circuits. Our proposed procedure is able to significantly increase the diagnostic capabilities of the test set at no additional costs w.r.t. fault coverage and pattern count. On average, *Implicit Diag* increases the diagnostic coverage by 4.89 percentage points, *Explicit Diag* by 5.74 and *Combined Diag* by 5.85 percentage points, respectively. In the following, we compare our results to other diagnostic

ATPG tools which also considered transition delay faults and used procedures to append additional patterns to a fault detection test set to achieve 100 % diagnostic coverage. We also generated additional patterns using the proposed explicit approach until 100 % diagnostic coverage was reached. For *Combined Diag* on average, over all ISCAS 89 circuit, 34.35 % of additional diagnostic patterns have to be added to achieve 100 % diagnostic coverage. In [25], only the larger ISCAS 89 circuits are considered but a gate netlist with complex gates is used which yields a much lower number of faults. This approach requires on average 60 % additional diagnostic patterns to achieve 100 % diagnostic coverage. When considering only the circuits evaluated in [25], we require on average 25.85 % additional patterns. In [24], also a subset of ISCAS 89 circuits is considered. The number of faults given here is very similiar to ours, thus the considered gate netlists are probably comparable to our netlists. The approach in [24] requires on average 274.57 % additional patterns. When considering the same circuits, we require on average 30.71 % additional patterns to achieve 100 % diagnostic coverage. This demonstrates the effectiveness of the proposed procedure to enhance the diagnostic coverage of fault detection test sets. Finally, our approach is able to achieve diagnostic coverages close to 100 % for several circuits (see e.g. s15850, s38417) without adding any patterns.

On average, *Combined Diag* yields the best trade-off between diagnostic coverage and runtime. The reason is that this procedure combines the advantages of the *Implicit Diagnosis* and the *Explicit Diagnosis*. First, the initially large fault clusters are partitioned by the implicit approach. This can be done efficiently as large clusters usually contain several easy-to-distinguish faults. After half of the patterns have been generated and the fault clusters have been updated accordingly, the majority of the cluster will be small and contain hard-to-distinguish faults. The explicit approach is now able to effectively partition these faults.

V. CONCLUSIONS

We presented a diagnostic ATPG framework for increasing the diagnostic coverage of production test sets without increasing the pattern count and whilst retaining their fault coverage. The experimental results show that the diagnostic coverage can be significantly improved at reasonable runtimes. Furthermore, the number of additionally required diagnostic patterns for 100 % diagnostic coverage can be considerably reduced compared to earlier works.

In the future, we will work on heuristics which dynamically determine the best point to switch between the *Implicit Diagnosis* and the *Explicit Diagnosis*. Furthermore we will investigate heuristics for the selection of target faults and more efficient procedures for the computation of unique detection signatures.

ACKNOWLEDGEMENTS

Parts of this work were supported by the German Research Foundation (DFG) under grant GRK 1103.

978-1-4799-7598-3/15 $31.00 © 2015 IEEE

Table I
EXPERIMENTAL RESULTS

Circuit	# TP	N_{eq}	Com ATPG DC %	Implicit Diag DC %	Runtime	Explicit Diag DC %	Runtime	Combined Diag DC %	Runtime
s00027	8	49	57.14	69.39	0.02	**73.46**	2.40	71.43	0.83
s00208	40	215	80.93	83.26	0.47	87.44	45.79	**87.91**	11.30
s00298	33	348	86.21	91.38	0.56	92.53	40.09	**93.68**	7.89
s00344	40	458	92.14	97.60	0.66	100.00	43.54	99.78	7.02
s00349	38	459	91.94	97.39	0.69	98.91	46.35	**99.13**	10.33
s00382	43	395	91.90	97.97	0.76	**99.24**	68.95	98.99	26.34
s00386	58	478	78.45	86.40	0.36	**86.61**	76.60	**86.61**	22.77
s00400	41	407	91.40	98.53	0.74	98.28	67.33	**98.77**	26.84
s00420	89	419	84.01	81.86	1.21	**85.44**	140.27	84.49	52.40
s00444	43	418	93.54	97.61	0.84	**99.04**	63.41	98.56	23.21
s00510	84	716	89.53	90.64	1.20	**91.06**	129.26	90.78	48.92
s00526	66	452	89.82	93.36	1.00	93.36	108.51	**95.35**	37.16
s00641	71	663	89.14	97.89	1.16	100.00	135.73	100.00	37.21
s00713	72	658	87.54	99.39	1.12	100.00	685.10	100.00	436.24
s00820	130	1147	81.43	**87.18**	2.25	85.70	199.19	85.61	72.84
s00832	131	1149	80.33	**87.21**	2.55	83.11	203.97	85.81	93.12
s00838	198	827	85.01	84.40	11.76	84.89	388.87	**85.25**	177.53
s00953	117	1530	86.99	91.44	3.21	**93.53**	211.50	93.07	86.11
s01196	187	2102	85.01	92.72	7.09	**94.24**	394.32	93.29	168.07
s01238	208	2215	84.51	93.14	7.23	93.54	432.35	**93.63**	184.06
s01423	93	1867	93.04	96.57	15.65	100.00	1169.96	99.89	842.95
s01488	146	2191	88.91	**93.06**	6.47	92.70	253.35	92.42	115.59
s01494	150	2201	90.37	**93.68**	7.73	92.23	256.56	92.46	115.21
s05378	204	6303	91.75	98.46	25.47	98.83	796.95	**99.03**	384.85
s09234	377	8487	91.67	98.32	329.30	99.34	2265.09	**99.39**	1135.16
s13207	433	11247	93.80	96.53	860.85	96.76	4502.37	**97.03**	2644.40
s15850	306	13407	96.09	97.40	825.34	**99.55**	12926.60	99.46	8253.49
s35932	74	48025	97.74	**99.04**	876.04	97.69	3371.99	98.00	1336.88
s38417	228	45117	96.23	98.64	3019.74	99.30	11077.00	**99.41**	6006.00
s38584	364	53151	95.76	97.36	2492.90	97.95	12343.20	**98.37**	5372.94
Average			88.08	92.97	0.0493	93.82	1.0000	**93.93**	0.4123

REFERENCES

[1] I. Pomeranz and S. Reddy, "Location of stuck-at faults and bridging faults based on circuit partitioning," in *IEEE TC*, p. 1124, 1998.

[2] S. Venkataraman and S. B. Drummonds, "Poirot: A logic fault diagnosis tool and its applications," in *ITC*, pp. 253–262, 2000.

[3] T. Bartenstein, D. Heaberlin, L. Huisman, and D. Sliwinski, "Diagnosing combinational logic designs using the single location at a time (slat) paradigm," in *ITC*, pp. 287–296, 2001.

[4] Z. Wang, M. Marek-Sadowska, K.-H. Tsai, and J. Rajski, "Multiple fault diagnosis using n-detection tests," in *ICCD*, pp. 198–201, 2003.

[5] X. Wen, T Miyoshi, S. Kajihara, L. Wang, K. Saluja, and K. Kinoshita, "On per-test fault diagnosis using the x-fault model," in *ICCD*, pp. 633–640, 2004.

[6] S. Holst and H.-J.Wunderlich, "Adaptive debug and diagnosis without fault dictionaries," in *ETS*, pp. 7–12, 2007.

[7] X. Yu and R. D. Blanton, "Multiple defect diagnosis using no assumptions on failing pattern characteristics," in *DAC*, pp. 361–366, 2007.

[8] W. C. Tarn, O. Poku, and R. D. Blanton, "Precise failure localization using automated layout analysis of diagnosis candidates," in *DAC*, pp. 367–372, 2008.

[9] X. Yu and R. D. Blanton, "An effective and flexible multiple defect diagnosis methodology using error propagation analysis," in *ITC*, pp. 1–9, 2008.

[10] J. Ye, Y. Hu, and X. Li, "Diagnosis of multiple arbitrary faults with mask and reinforcement effect," in *DATE*, pp. 885–890, 2009.

[11] T. W. Bartenstein, "Panel 9-diagnostics vs failure analysis," in *ITC*, p. 1439, 2004.

[12] H. Tang, S. Manish, J. Rajski, M. Keim, and B. Benware, "Analyzing volume diagnosis results with statistical learning for yield improvement," in *ETS*, 2007.

[13] M. Sharma, C. Schuermyer, and B. Benware, "Determination of dominant-yield-loss mechanism with volume diagnosis," in *IEEE Design & Test of Computers*, pp. 54–61, 2010.

[14] W. C. Tam, O. Poku, and R. Blanton, "Systematic defect identification through layout snipet clustering," in *ITC*, 2010.

[15] P. Camurati, D. Medina, P. Prinetto, and M. S. Reorda, "A diagnostic test pattern generation algorithm," in *ITC*, pp. 52–58, 1990.

[16] T. Gruning, U. Mahlstedt, and H. Koopmeiners, "Diatest: A fast diagnostic test pattern generator for combinational circuits," in *ICCAD*, pp. 194–197, 1991.

[17] F. Corno, P. Prinetto, M. Rebaudengo, and M. S. Reorda, "Garda: A diagnostic atpg for large synchronous sequential circuits," in *Eur. Des. Test*, pp. 267–271, 1995.

[18] I. Pomeranz and S. M. Reddy, "A diagnostic test generation procedure for synchronous sequential circuits based on test elimination," in *ITC*, pp. 1074–1083, 1998.

[19] A. Veneris, R. Chang, M. S. Abadir, and S. Seyedi, "Functional fault equivalence and diagnostic test generation in combinational logic circuits using conventional atpg," in *J. Electron. Test.: Theory Appl. (JETTA)*, vol. 21, no. 5, pp. 495–502, 2005.

[20] I. Pomeranz and S. M. Reddy, "Output-dependent diagnostic test generation," in *VLSI Des. Conf.*, pp. 3–8, 2010.

[21] I. Pomeranz and S. M. Reddy, "Gradual diagnostic test generation based on the structural distance between indistinguished fault pairs," in *DFTS*, pp. 349–357, 2010.

[22] I. Pomeranz, "Gradual diagnostic test generation and observation point insertion based on the structural distance between indistinguished fault pairs," in *IEEE TVLSI*, pp. 1026–1035, 2012.

[23] Y. H. Y. Kurose, "Diagnostic test generation for transition faults using a stuck-at ATPG tool," in *ITC*, 2009.

[24] Y. Zhang and V. Agarwal, "Reduced complexity of test generation algorithms for transition fault diagnosis," in *ICCD*, pp. 96–101, 2011.

[25] K. Lee and C. Wu, "An efficient diagnosis-aware pattern generation procedure for transition faults," in *ITC*, 2014.

[26] S. Prabhu, M. S. Hsiao, L. Lingappan, and V. Gangaram, "A smt-based diagnostic test generation method for combinational circuits," in *VTS*, pp. 215–220, 2012.

[27] Semiconductor Research Corporation, "Research challenges in test and testability," 2013.

[28] S. Reimer, M. Sauer, T. Schubert, and B. Becker, "Incremental encoding and solving of cardinality constraints," in *Automated Technology for Verification and Analysis*, pp. 297–313, 2014.

Improving the Accuracy of Defect Diagnosis by Considering Reduced Diagnostic Information

Irith Pomeranz
School of Electrical & Computer Eng.
Purdue University
W. Lafayette, IN 47907, U.S.A.

Abstract - It was noted earlier that the accuracy of defect diagnosis may be improved if certain tests are removed from consideration by the defect diagnosis procedure. This paper observes that the effects, which support the removal of tests, also support the removal of observable outputs from consideration during defect diagnosis. Specifically, a test may create an output response that a defect diagnosis procedure will not be able to interpret correctly. This may affect some observable outputs more strongly than others. Therefore, the removal of observable outputs from consideration can improve the accuracy of diagnosis. This paper describes a generalized augmented defect diagnosis procedure that removes tests and observable outputs from consideration. It presents experimental results to demonstrate the effects of removing observable outputs on the accuracy of diagnosis.

I. INTRODUCTION

A chip that produces a faulty output response may undergo defect diagnosis in order to identify the defects that caused it to fail [1]-[20]. Based on the observed faulty output response of the chip, a defect diagnosis procedure computes a set of candidate faults that is expected to point to the locations of the defects in the chip. A small set of candidate faults is important for facilitating failure analysis, which needs to consider fewer locations in the chip when the set of candidate faults is smaller. In addition, it is important for the set of candidate faults to include at least some of the defect locations in the chip. The number of candidate faults, and the overlap between the set of candidate faults and the defect locations in the chip, determine the accuracy of diagnosis.

A defect diagnosis procedure will produce accurate diagnosis results even if it is given only part of the observed output response of the chip. This is the basis for the computation of small fault dictionaries [21], for the reduction of test data volume for defect diagnosis [22],

This work was supported in part by SRC Grant No. 2013-TJ-2469.

and for increasing the throughput of defect diagnosis [23]. In [22], fail data collection for diagnosis is stopped before all the tests are considered based on the expectation that additional fail data, for additional tests, will not contribute to the accuracy of diagnosis. In [23], defect diagnosis is performed based on a partition of the circuit that includes a subset of the circuit outputs. The goal in [23] is to identify the smallest partition whose use is sufficient for producing the same defect diagnosis results as when the complete circuit is considered. This increases the speed of defect diagnosis and reduces its memory requirements.

Moreover, it was noted earlier that the addition of diagnostic tests to a fault detection test set does not always improve the accuracy of diagnosis. In fact, the accuracy of diagnosis may decrease with the addition of diagnostic tests. As a generalization of this observation, it is possible that if certain tests are removed from consideration by the defect diagnosis procedure, the accuracy of diagnosis will improve. This issue was considered in [24]. An augmented defect diagnosis procedure is described in [24] to address this issue. The procedure considers tests for removal one at a time. It decides to remove a test from consideration during defect diagnosis if this results in a reduced set of candidate faults. Removing a test from consideration implies only that the defect diagnosis procedure ignores the output response to this test when it computes a set of candidate faults. The chip does not need to be tested again. The effect of removing tests from consideration on the accuracy of diagnosis is captured in [24] by a parameter called the improvement factor. The improvement factor captures the reduction in the number of candidate faults and the change in the overlap between the set of candidate faults and the defect locations.

Only tests are considered for removal in [24]. However, an output response of a chip can be considered conceptually as a two-dimensional array with a row for every test and a column for every observable output (a primary output or a next-state variable whose value is scanned out). Thus, in addition to removing tests (or rows of the array), it is also possible to remove from consideration observable outputs (or columns of the array). This paper

explores the possibility of removing observable outputs from consideration by the defect diagnosis procedure. As in the case of tests, the removal of an observable output implies only that the defect diagnosis procedure will ignore the output response for this output when it computes a set of candidate faults. The chip does not need to be tested again.

To generalize the procedure from [24], a test or an observable output is referred to as an entry of the output response. The motivation for removing tests comes from the observation related to the accuracy of diagnosis when diagnostic tests are added to a fault detection test set. The effectiveness of removing entries from consideration can be explained as follows.

A test may create an output response that a defect diagnosis procedure will not be able to interpret correctly. For example, with a defect diagnosis procedure that is based on specific fault models, as in [1], [2], [4], [14] or [20], the presence of modeled faults at the sites of the defect present in the chip may not explain the observed response for a test when it is substantially different from the responses of the modeled faults. In this case, the defect diagnosis procedure may not be able to find the correct modeled faults to match the observed output response.

This effect may be stronger for some observable outputs, and weaker for others. Therefore, the removal of observable outputs from consideration by the defect diagnosis procedure may contribute to the accuracy of diagnosis similar to the removal of tests.

To study this issue, the paper generalizes the augmented defect diagnosis procedure from [24] to address the removal of tests and observable outputs. The main contribution of the paper is in pointing out the possibility of increasing the accuracy of diagnosis by removing observable outputs from consideration, and studying the effects on the accuracy of diagnosis.

The paper is organized as follows. Section II describes the generalized augmented defect diagnosis procedure. Section III describes the experiment used for demonstrating the effectiveness of the generalized procedure, and presents experimental results. Section IV discusses the effects on the accuracy of diagnosis.

II. GENERALIZED PROCEDURE

This section describes the generalized augmented defect diagnosis procedure. The generalization allows the procedure to consider the removal of tests and observable outputs.

Let R_{obs} be an observed response of a circuit with a set of observable outputs Z to a test set T. The set $E = T \cup Z$ is the set of entries of the observed response R_{obs}. The generalized augmented defect diagnosis procedure may remove from consideration any entry of E.

The generalized augmented defect diagnosis procedure assumes the existence of a defect diagnosis procedure, which is denoted by $Defect\_Diag$ (). For the discussion in this paper, $Defect\_Diag$ () is based on specific modeled faults. The set of modeled faults is denoted by F. The defect diagnosis procedure accepts the observed response R_{obs}, a set of entries $E_d \subseteq E$ that it should consider, and a set of modeled faults $F_d \subseteq F$. The procedure returns a set of candidate faults $CND \subseteq F_d$. For the computation of CND, the defect diagnosis procedure uses only the entries of R_{obs} that are indicated by E_d. It ignores all the other entries. In addition, it uses the set of faults F_d to define the set of candidate faults.

Initially, procedure $Defect\_Diag$ () is called with R_{obs}, the complete set of entries $E = T \cup Z$, and the complete set of modeled faults F. It returns a set of candidate faults CND, which is stored in CND_{best} for reference.

The generalized augmented defect diagnosis procedure removes entries from E one at a time. It calls procedure $Defect\_Diag$ () to check the effects of the removal of each entry on the set of candidate faults. When the decision is to remove an entry $e_i \in E$, the removal of additional entries is considered with e_i excluded from E. This proceeds as follows.

For every $e_i \in E$, the procedure removes the entry e_i from E, and calls procedure $Defect\_diag$ (R_{obs}, E, CND_{best}). The procedure returns a set of candidate faults $CND \subseteq CND_{best}$. Using CND_{best} as the set of modeled faults implies that the defect diagnosis procedure considers significantly fewer faults than the number of faults in F. This reduces its fault simulation effort significantly. It is important for making the iterative generalized augmented defect diagnosis procedure feasible.

The generalized augmented defect diagnosis procedure decides whether or not to accept the removal of e_i from E based on the following considerations.

If $CND = CND_{best}$, the removal of e_i from E does not reduce the set of candidate faults. In this case, the procedure does not accept its removal, and it reintroduces e_i into E.

If $CND \subset CND_{best}$, the removal of e_i from E reduces the set of candidate faults. In this case, the procedure accepts its removal. It assigns $CND_{best} = CND$ such that the reduced set of candidate faults becomes the reference for additional removals.

The generalized augmented defect diagnosis procedure is given next.

Procedure 1: Generalized augmented defect diagnosis procedure

(1) Include in E all the tests and observable outputs. Assign $E_{curr} = E$.

(2) Call procedure $Defect\_Diag(R_{obs}, E, F)$ to find a set of candidate faults CND. Assign $CND_{best} = CND$.

(3) Select an entry $e_i \in E_{curr}$. Remove e_i from E and from E_{curr}.

(4) Call procedure $Defect\_Diag(R_{obs}, E, CND_{best})$ to find a set of candidate faults CND.

(5) If $CND \subset CND_{best}$, assign $CND_{best} = CND$. Else, add e_i back to E.

(6) If $E_{curr} \neq \phi$, go to Step 3.

III. EXPERIMENTAL RESULTS

The generalized augmented defect diagnosis procedure was applied to benchmark circuits as described next.

Observed responses are obtained by simulating multiple transition faults with multiplicities in the range between five and 20. This creates output responses that are difficult to diagnose because of the interactions among multiple faults. Such responses are interesting for the study in this paper.

The number of observed responses is determined as follows. As observed responses are considered, the maximum number of candidate faults obtained for any observed response is stored in a variable denoted by max_cnd. Additional observed responses are considered until the last 10 observed responses do not increase max_cnd.

A multiple transition fault that is used for creating an observed response is denoted by f_{obs}. The members of f_{obs} are single transition faults that together form a multiple transition fault.

The defect diagnosis procedure used for $Defect\_Diag()$ is the one from [20]. This procedure is based on a scoring algorithm that ranks modeled faults. The faults with the very highest ranks are included in the set of candidate faults. The procedure is applied using single transition faults. This implies that a set of candidate faults CND that the procedure produces consists of single transition faults. It is thus possible to define the overlap between CND and f_{obs} as $OVRLP = CND \cap f_{obs}$. The members of $OVRLP$ are the single transition faults that the defect diagnosis procedure identifies correctly.

For comparison with [24], the generalized augmented defect diagnosis procedure is applied in two modes. In mode 1, it attempts to remove tests, and then observable outputs. In mode 2, it only removes observable outputs. The initial set of candidate faults for a given observed response with $E = T \cup Z$ is denoted by CND_0. In mode 1, the set of candidate faults after the procedure removes tests from consideration is denoted by CND_T. In each one of the modes, the set of candidate faults after the procedure removes r observable outputs from consideration is denoted by $CND_{Z,r}$, for $r = 1, 2, \cdots$. The set of candidate faults obtained for the last value of r is also denoted by CND_Z.

Corresponding to CND_0, CND_T and $CND_{Z,r}$, for $r = 1, 2, \cdots$, there are overlaps that are denoted by $OVRLP_0$, $OVRLP_T$ and $OVRLP_{Z,r}$. For example, $OVRLP_0 = CND_0 \cap f_{obs}$.

The accuracy of diagnosis depends on the number of candidate faults and on the size of the overlap. In general, a smaller number of candidate faults, or a larger overlap, imply improved accuracy of diagnosis. However, when the number of candidate faults is decreased, the overlap may decrease together with it. One of two approaches can be taken to assess the change in the accuracy of diagnosis depending on the relative importance of the two parameters.

For the case where the parameters are equally important, an improvement factor is defined in [24] to capture the combined effect as follows.

Before the removal of entries from E, failure analysis needs to consider $|CND_0|$ locations, and $|OVRLP_0|$ of them point correctly to the defect locations in the chip. The fraction of correct locations is $|OVRLP_0|/|CND_0|$. After the removal of tests from E, the fraction of correct locations is $|OVRLP_T|/|CND_T|$. The improvement factor captures the change through the formula
$$\rho_T = \frac{|OVRLP_T|/|CND_T|}{|OVRLP_0|/|CND_0|}.$$

In a similar way, after removing r observable outputs from E, the improvement factor is
$$\rho_{Z,r} = \frac{|OVRLP_{Z,r}|/|CND_{Z,r}|}{|OVRLP_0|/|CND_0|}.$$

With an improvement factor that is larger than one, the fraction of correct locations increases when entries are removed from E. This facilitates failure analysis.

For the case where the number of candidate faults is more important than the size of the overlap, it is possible to require only that the overlap would be non-empty. In this case, reducing the number of candidate faults

TABLE I. Largest Numbers of Candidates

circuit	obs	CND_O	mode 1 CND_T	CND_Z	mode 2 CND_Z
s1423	3	28	28	6	6
s1423	9	34	27	10	13
s5378	10	56	50	37	37
s5378	9	50	47	24	24
s9234	20	65	65	28	28
s9234	2	64	61	25	27
s13207	3	97	83	20	26
s13207	9	84	83	19	19
s15850	7	74	74	24	24
s15850	6	71	71	6	6
s35932	23	37	37	11	11
s35932	20	35	34	25	25
s38584	3	45	45	14	14
s38584	9	40	39	25	25
b04	6	24	19	3	3
b04	15	22	18	4	5
b07	6	33	22	9	10
b07	3	28	18	2	4
b14	3	40	23	12	15
b14	15	24	23	6	6
b15	4	34	20	10	14
b15	9	26	19	6	9
b20	19	26	24	11	12
b20	11	130	22	10	10
aes_core	6	31	25	18	22
aes_core	7	26	25	23	24
des_area	4	43	27	27	40
des_area	9	30	24	24	30
i2c	15	29	24	3	3
i2c	11	37	23	5	8
pci_spoci_ctrl	3	34	25	6	6
pci_spoci_ctrl	4	27	23	15	15
sasc	11	30	29	2	2
sasc	9	29	26	1	1
simple_spi	17	35	29	6	6
simple_spi	11	31	26	3	3
spi	3	37	22	7	10
spi	6	29	19	5	5
systemcaes	11	31	30	7	7
systemcaes	20	24	24	5	5
systemcdes	11	35	26	15	14
systemcdes	3	36	22	15	13
tv80	11	30	25	17	17
tv80	9	64	23	15	20
usb_phy	6	31	27	4	4
usb_phy	7	34	24	6	6
wb_dma	3	26	24	6	6
wb_dma	11	23	21	6	6

improves the accuracy of diagnosis by allowing failure analysis to consider fewer locations in the chip. The non-empty overlap ensures that at least one of these locations will point correctly to a defect location.

The results of the generalized augmented defect diagnosis procedure are reported in Tables I-IV. In all the cases, a non-empty overlap was obtained before and after removing entries from consideration by the defect diagnosis procedure.

Table I lists all the benchmark circuits that were considered for this experiment. For every circuit, the two observed responses for which CND_T is the largest in

mode 1 are reported.

Table I shows the index of an observed response under column *obs* (this is its identifier among the observed responses that are considered for the circuit). Next, it shows the numbers of candidate faults in CND_0. For mode 1, it shows the number of candidate faults in CND_T and CND_Z. For mode 2, Table I shows the number of candidate faults in CND_Z.

Detailed results for mode 1 are shown in Tables II-IV for several circuits. The observed response with the largest set CND_T in mode 1 is reported for every one of these circuits. A set of rows that is separated by horizontal lines corresponds to one observed response.

For each observed response, the first row corresponds to defect diagnosis with all the tests and observable outputs. The second row corresponds to the removal of tests. Additional rows correspond to the removal of observable outputs one at a time.

In every case, column *tests* shows the number of tests that are used for diagnosis. Column *po* shows the number of primary outputs that are used as observable outputs for diagnosis. Column *sv* shows the number of next-state variables that are used as observable outputs for diagnosis.

Column *cand* shows the number of candidate faults. Column *ovlp* shows the number of faults in the overlap. Column *imprv* shows the value of the improvement factor.

Column *ntime* shows the increase in run time because of the removal of tests or observable outputs. Let the run time for defect diagnosis using all the tests and observable outputs be rt_0. The run time of the generalized augmented defect diagnosis procedure is divided by rt_0 to show the increase in run time.

The results are discussed in the next section.

IV. ACCURACY OF DIAGNOSIS

Tables I-IV demonstrate the following points.

Removing observable outputs from consideration by the defect diagnosis procedure reduces the number of candidate faults below the number obtained by removing only tests. In many cases, the final set of candidate faults contains a small number of faults. Overall, the removal of observable outputs from consideration has a more significant effect on the number of candidate faults than the removal of tests. This is related to the fact that multiple transition faults are used for producing observed responses. The interactions between the faults are stronger on some observable outputs, and weaker on other observable outputs. This can be true across all the tests.

TABLE II. Mode 1 for ISCAS-89 Benchmarks

circuit	obs	tests	po	sv	cand	ovrlp	imprv	ntime
s5378	10	229	49	179	56	16	1.00	1.00
s5378	10	224	49	179	50	15	1.05	2.13
s5378	10	224	48	179	43	14	1.14	2.15
s5378	10	224	48	178	38	14	1.29	2.46
s5378	10	224	48	177	37	13	1.23	2.72
s9234	20	344	22	228	65	16	1.00	1.00
s9234	20	344	22	228	65	16	1.00	2.22
s9234	20	344	21	228	62	15	0.98	2.29
s9234	20	344	21	227	57	14	1.00	2.54
s9234	20	344	21	226	54	13	0.98	2.59
s9234	20	344	21	225	49	12	0.99	2.62
s9234	20	344	21	224	44	11	1.02	2.69
s9234	20	344	21	223	37	10	1.10	2.72
s9234	20	344	21	222	35	9	1.04	2.73
s9234	20	344	21	221	34	8	0.96	2.78
s9234	20	344	21	220	33	7	0.86	2.84
s9234	20	344	21	219	28	6	0.87	2.89
s13207	3	468	121	669	97	17	1.00	1.00
s13207	3	464	121	669	83	14	0.96	2.55
s13207	3	464	120	669	76	13	0.98	2.56
s13207	3	464	119	669	73	12	0.94	2.70
s13207	3	464	118	669	69	11	0.91	2.75
s13207	3	464	117	669	56	10	1.02	2.83
s13207	3	464	116	669	53	9	0.97	2.85
s13207	3	464	116	668	40	8	1.14	2.91
s13207	3	464	116	667	39	7	1.02	3.00
s13207	3	464	116	666	37	6	0.93	3.40
s13207	3	464	116	665	31	5	0.92	3.55
s13207	3	464	116	664	26	4	0.88	3.63
s13207	3	464	116	663	21	3	0.82	3.74
s13207	3	464	116	662	20	2	0.57	3.75

TABLE III. Mode 1 for ITC-99 Benchmarks

circuit	obs	tests	po	sv	cand	ovrlp	imprv	ntime
b14	3	183	54	247	40	16	1.00	1.00
b14	3	175	54	247	23	12	1.30	1.27
b14	3	175	54	246	22	11	1.25	1.36
b14	3	175	54	245	20	10	1.25	1.37
b14	3	175	54	244	18	9	1.25	1.38
b14	3	175	54	243	16	8	1.25	1.47
b14	3	175	54	242	15	7	1.17	1.48
b14	3	175	54	241	14	6	1.07	1.54
b14	3	175	54	240	12	5	1.04	1.56
b15	4	320	72	447	34	11	1.00	1.00
b15	4	313	72	447	20	10	1.55	1.23
b15	4	313	72	446	19	9	1.46	1.36
b15	4	313	72	445	18	9	1.55	1.37
b15	4	313	72	444	15	9	1.85	1.40
b15	4	313	72	443	13	8	1.90	1.42
b15	4	313	72	442	12	7	1.80	1.43
b15	4	313	72	441	11	6	1.69	1.43
b15	4	313	72	440	10	5	1.55	1.44
b20	19	266	22	494	26	16	1.00	1.00
b20	19	264	22	494	24	14	0.95	1.20
b20	19	264	22	493	23	13	0.92	1.50
b20	19	264	22	492	18	12	1.08	1.51
b20	19	264	22	491	17	11	1.05	1.52
b20	19	264	22	490	16	10	1.02	1.53
b20	19	264	22	489	15	9	0.98	1.54
b20	19	264	22	488	14	8	0.93	1.55
b20	19	264	22	487	13	7	0.88	1.57
b20	19	264	22	486	11	6	0.89	1.59

Therefore, removing observable outputs is more effective than removing tests at preventing these interactions from reducing the accuracy of diagnosis.

This observation applies in general to the case where multiple defects are present, and explains why the removal of observable outputs may be more effective than the removal of tests. It is important since multiple defects are likely to occur.

As the number of candidate faults is reduced, the overlap is also reduced. However, the overlap remains non-empty in all the cases considered. Thus, the accuracy of diagnosis is improved by providing fewer candidate faults.

There are typically steps in the process of removing observable outputs that produce higher values of the improvement factor than the removal of tests. However, the improvement factor does not increase monotonically as observable outputs are removed. In addition, there are steps where it is lower than one. Overall, keeping the last several observable outputs that can be removed results in a reduced set of candidate faults and a high improvement factor.

Finally, removing only observable outputs, without removing tests, provides similar results to the removal of both tests and observable outputs.

V. CONCLUDING REMARKS

An augmented defect diagnosis procedure that was described earlier removed tests from consideration by a defect diagnosis procedure in order to improve the accuracy of diagnosis. This paper generalized the procedure to remove from consideration both tests and observable outputs. This was based on the following observations. A test may create an output response that a defect diagnosis procedure will not be able to interpret correctly. Removing a test that is affected by this phenomenon from consideration allows the defect diagnosis procedure to produce more accurate results. Certain observable outputs may be affected by this phenomenon more than others. Therefore, the removal of observable outputs from consideration can also improve the accuracy of diagnosis. Experimental results supported the possibility of improving the accuracy of diagnosis by removing from consideration observable outputs in addition to tests.

REFERENCES

[1] M. Abramovici, M. A. Breuer and A. D. Friedman, *Digital Systems Testing* and *Testable Design*, IEEE Press, 1995.

[2] D. B. Lavo, B. Chess, T. Larrabee and I. Hartanto, "Probabilistic Mixed-Model Fault Diagnosis", In Proc. Intl. Test Conf., 1998, pp. 1084-1093.

[3] J. Ghosh-Dastidar and N. A. Touba, "Adaptive Techniques for Improving Delay Fault Diagnosis", in Proc. VLSI Test Symp., 1999, pp. 168-172.

TABLE IV. Mode 1 for IWLS-05 Benchmarks

circuit	obs	tests	po	sv	cand	ovrlp	imprv	ntime
sasc	11	36	12	117	30	19	1.00	1.00
sasc	11	35	12	117	29	18	0.98	1.98
sasc	11	35	12	116	26	17	1.03	2.68
sasc	11	35	12	115	24	16	1.05	2.87
sasc	11	35	12	114	23	15	1.03	3.18
sasc	11	35	12	113	22	14	1.00	3.23
sasc	11	35	12	112	20	13	1.03	3.31
sasc	11	35	12	111	18	12	1.05	3.35
sasc	11	35	12	110	17	11	1.02	3.45
sasc	11	35	12	109	16	10	0.99	3.48
sasc	11	35	12	108	14	9	1.02	3.95
sasc	11	35	12	107	9	7	1.23	4.55
sasc	11	35	12	106	7	6	1.35	4.56
sasc	11	35	12	105	6	5	1.32	4.90
sasc	11	35	12	104	5	4	1.26	4.91
sasc	11	35	12	103	4	3	1.18	5.04
sasc	11	35	12	102	2	2	1.58	5.08
systemcaes	11	166	129	670	31	19	1.00	1.00
systemcaes	11	165	129	670	30	19	1.03	1.12
systemcaes	11	165	128	670	28	18	1.05	1.18
systemcaes	11	165	128	669	27	17	1.03	1.28
systemcaes	11	165	128	668	25	16	1.04	1.29
systemcaes	11	165	128	667	23	15	1.06	1.30
systemcaes	11	165	128	666	21	14	1.09	1.31
systemcaes	11	165	128	665	19	13	1.12	1.38
systemcaes	11	165	128	664	15	12	1.31	1.39
systemcaes	11	165	128	663	14	11	1.28	1.40
systemcaes	11	165	128	662	13	10	1.26	1.40
systemcaes	11	165	128	661	12	9	1.22	1.40
systemcaes	11	165	128	660	11	8	1.19	1.41
systemcaes	11	165	128	659	10	7	1.14	1.42
systemcaes	11	165	128	658	9	6	1.09	1.46
systemcaes	11	165	128	657	8	5	1.02	1.46
systemcaes	11	165	128	656	7	4	0.93	1.47
tv80	11	706	32	359	30	18	1.00	1.00
tv80	11	703	32	359	25	17	1.13	1.67
tv80	11	703	32	358	24	16	1.11	1.74
tv80	11	703	32	357	23	15	1.09	1.83
tv80	11	703	32	356	22	14	1.06	1.83
tv80	11	703	32	355	21	13	1.03	1.86
tv80	11	703	32	354	18	12	1.11	2.00
tv80	11	703	32	353	17	11	1.08	2.01

[4] S. Venkataraman and S. B. Drummonds, "POIROT: A Logic Fault Diagnosis Tool and its Applications", in Proc. Intl. Test Conf., 2000, pp. 253-262.

[5] S.-Y. Huang, "On Improving the Accuracy of Multiple Defect Diagnosis", in Proc. VLSI Test Symp., 2001, pp. 34-39.

[6] T. Bartenstein, D. Heaberlin, L. Huisman and D. Sliwinski, "Diagnosing Combinational Logic Designs Using the Single Location At-A-Time (SLAT) Paradigm", in Proc. Intl. Test Conf., 2001, pp. 287-296.

[7] D. B. Lavo, I. Hartanto and T. Larrabee, "Multiplets, Models, and the Search for Meaning: Improving Per-Test Fault Diagnosis", in Proc. Intl. Test Conf., 2002, pp. 250-259.

[8] Z. Wang, M. Marek-Sadowska, K.-H. Tsai and J. Rajski, "Multiple Fault Diagnosis Using n-Detection Tests" in Proc. Intl. Conf. on Computer Design, Oct. 2003, pp. 198-201.

[9] I. Pomeranz, S. Venkataraman, S. M. Reddy and E. Amyeen, "Defect Diagnosis Based on Pattern-Dependent Stuck-At Faults", in Proc. VLSI Design Conf., 2004, pp. 475-480.

[10] J. B. Liu and A. Veneris, "Incremental Fault Diagnosis", IEEE Trans. on Computer-Aided Design, Feb. 2005, pp. 240-251.

[11] R. Desineni and R. D. Blanton, "Diagnosis of Arbitrary Defects Using Neighborhood Function Extraction", in Proc. VLSI Test Symp., 2005, 366-373.

[12] C. Liu, "Improve the Quality of Per-Test Fault Diagnosis Using Output Information", Journal of Electronic Testing - Theory and Application, Feb. 2007, Vol. 23, Issue 1, pp 11-24.

[13] R. Adapa, S. Tragoudas and M. K. Michael, "Accelerating Diagnosis Via Dominance Relations Between Sets of Faults", in Proc. VLSI Test Symp., 2007, pp. 219-224.

[14] S. Holst and H.-J. Wunderlich, "Adaptive Debug and Diagnosis without Fault Dictionaries", in Proc. European Test Symp., 2007, pp. 7-12.

[15] W. C. Tarn, O. Poku and R. D. Blanton, "Precise Failure Localization Using Automated Layout Analysis of Diagnosis Candidates", in Proc. Design Autom. Conf., 2008, pp. 367-372.

[16] X. Yu and R. D. Blanton, "An Effective and Flexible Multiple Defect Diagnosis Methodology Using Error Propagation Analysis", in Proc. Intl. Test Conf., 2008, pp. 1-9.

[17] W.-T. Cheng, B. Benware, R. Guo, K.-H. Tsai, T. Kobayashi, K. Maruo, M. Nakao, Y. Fukui and H. Otake, "Enhancing Transition Fault Model for Delay Defect Diagnosis", in Proc. Asian Test Symp., 2008, pp. 179-184.

[18] V. J. Mehta, M. Marek-Sadowska, K.-H. Tsai and J. Rajski, "Timing-Aware Multiple-Delay-Fault Diagnosis", IEEE Trans. on Computer-Aided Design, Feb. 2009, pp. 245-258.

[19] X. Tang, W.-T. Cheng, R. Guo and S. M. Reddy, "Diagnosis of Multiple Physical Defects Using Logic Fault Models", in Proc. Asian Test Symp., 2010, pp. 94-99.

[20] I. Pomeranz, "OBO: An Output-By-Output Scoring Algorithm for Fault Diagnosis", in Proc. IEEE Computer Society Annual Symp. on VLSI, 2014.

[21] P. G. Ryan, W. K. Fuchs and I. Pomeranz, "Fault Dictionary Compression and Equivalence Class Computation for Sequential Circuits", in Proc. Intl. Conf. on Computer-Aided Design, 1993, pp. 508-511.

[22] H. Wang, O. Poku, X. Yu, S. Liu, I. Komara and R. D. Blanton, "Test-Data Volume Optimization for Diagnosis" in Proc. Design Autom. Conf., 2012, pp. 567-572.

[23] X. Fan, H. Tang, Y. Huang, W.-T. Cheng, S. M. Reddy and B. Benware, "Improved Volume Diagnosis Throughput Using Dynamic Design Partitioning", in Proc. Intl. Test Conf., 2012, pp. 1-10.

[24] I. Pomeranz, "Improving the Accuracy of Defect Diagnosis by Considering Fewer Tests", IEEE Trans. on Computer-Aided Design, Dec. 2014, pp. 2010-2014.

Signature Oriented Model Pruning
to Facilitate Multi-Threaded Processors Debugging

Fatemeh Refan, Bijan Alizadeh, and Zainalabedin Navabi

School of Electrical and Computer Engineering, College of Engineering, University of Tehran, Tehran, Iran

{fatemeh_refan, b.alizadeh, navabi}@ut.ac.ir

Abstract—In this paper, we propose a signature based pruning technique to facilitate the debugging of multi-threaded processors. To accomplish this, a pipelined implementation of the multi-threaded processor model is checked for correspondence against the specification model based on flushing proof. Then, a two-stage signature oriented pruning method is proposed to avoid the space explosion problem caused by inserting debugging facilities in the model. The results show an average improvement of 47%, and 71% in the size of decision formula and CPU time for the DLX processor, respectively.

Keywords— Multi-threaded Processors; Formal Debugging; Model Pruning; Correspondence Checking; State Space Explosion;

I. INTRODUCTION

Formal verification methods, although leading to absolute result, still suffer from poor scalability. This limits the use of such methods for complex systems such as multithreaded (MT) processors. Furthermore, the need to insert additional logic for debugging purposes worsens the problem size and scalability issues.

In [1], the method of modifying a single-threaded processor to support multi-threading and its verification is discussed. It proposes two approaches for modeling: (1) duplicate storage elements for each thread; (2) abstract the threading and have one storage element for all threads, while checking the thread identifier (TID) through an un-interpreted predicate (UP). The first approach is more precise, but the size of the model grows exponentially, as the number of threads increases. In the second approach, the size is constant for any arbitrary thread number, but the model is not precise and the chance of encountering spurious counterexamples (CEX) increases. The authors of [1], however, discuss neither a method of debugging, nor a solution to reduce the size of final formula caused by adding debugging facilities, which are the main focus of this paper.

Well-known methods to formally debug different design models make use of satisfiablity (SAT) and satisfiability modulo theories (SMT) solvers. The authors of [2] propose the MUX-based debugging approach which is based on the insertion of debugging multiplexers (MUXes) over suspicious lines of the model, and find the appropriate controlling and alternative value for each multiplexer (MUX), in a way that model behaves correctly using a SAT solver. The method is then extended to register transfer level (RTL) designs in [3], and SMT solvers are used. Authors of [4] also focus on RTL debugging by computing the dominance relationships between RTL blocks and leveraging them to reduce the problem to be solved by SAT solver. Another approach in [5] defines input constraints as hard and other parts of design as soft clauses and solves the resulting conjunctive normal form (CNF) formula by using maximum satisfiability (Max-SAT) solvers to find the candidate fault locations.

The automatic processor debugging is discussed in [6], using the idea of [2] to insert debugging MUXes in the model. The authors of [6] have applied this method to a higher level of abstraction and have automated it using a modeling and verification tool called UCLID [7]. In this method, a CEX is achieved from this automatic tool by comparing the buggy model with the model enriched with debugging MUXes. This CEX determines which debugging MUXes should be activated and which alternative value can correct the design. Although the method proposed in [6] effectively localizes bugs in an abstract model, the final formula size may become too big for complex designs, making it inapplicable for debugging of multi-threaded processors. Another similar work in the field of automatic debugging is presented in [8], where after receiving CEX, the debugging MUXes are inserted into the model, and the CEX is applied to the CNF formula of the model enriched with MUXes. Finally, a SAT solver determines the possible bug correction. The same method is used in [9] for verification and debugging of cycle-accurate out-of-order processors in the event of exceptions and external interrupts. Note that the approach of [8] is more scalable than [6] and [9], but still suffers from producing large CNF due to added debugging facilities.

In this paper, the scalability problem of formal debugging is addressed by proposing a two-stage signature oriented pruning approach. The goal is to have an effective debugging method for multi-threaded processors. The main idea of our proposed method and its difference with the conventional MUX-based debugging approach [6] is shown in Fig. 1. In MUX-based debugging approach as shown in Fig. 1 (a), when the verification fails and a CEX is generated, first a list of candidate bug locations is prepared then the repair MUXes are inserted in the specified locations, and finally the related decision formula is solved using a SAT-solver. If this formula is satisfiable, the repair phase can be started while having the candidate locations and determined MUX values as the starting clues; otherwise, the bug locations should be re-chosen. As Fig. 1 (b) demonstrates, in the proposed approach before inserting repair MUXes a set of primary and easy-to-check invariants which are extracted according to the main attributes of the design class, referred to as the rule set, are checked. If any of these rules does not hold, an early, simple repair phase is applied to correct them. Then based on the rule set, the signature of model and CEX are extracted to prune the decision formula obtained from the debugging phase in the final step. The model pruning stage

is composed of several steps, including extraction of model and CEX signature, disabling signals, CEX constraints and finally pruning the implementation model.

The rest of the paper is organized as follows. The necessary background is discussed in Section II. Section III presents the details of the proposed two-stage signature oriented pruning approach. The experimental results are presented in Section IV. Finally, Section V concludes the paper.

II. BACKGROUND

A. UCLID

The UCLID tool proposes a language based on the logic of Counter Arithmetic with Lambda Expressions and Uninterpreted Functions (CLU) to specify a state machine, and a verification engine based on symbolic simulation supporting decision procedure for CLU logic. The CLU modeling language supports uninterpreted function (UF) and uninterpreted predicate (UP) symbols, arithmetic of counters, and restricted lambda expressions. Un-interpreted symbols are mainly used to abstract the details of the model. A function and a predicate of arity 0 are called TERM (integer value) and TRUTH (Boolean value) respectively. UFs and UPs of larger arity, satisfy only functional consistency. Counter arithmetic provide a mechanism to define precedence between TERMs, and lambda expressions are used to describe interpreted functions and predicates, and behavior of UFs and UPs. Lambda expressions are especially applicable for defining memory and storage elements [7].

B. DLX Processor

The storage elements of DLX processor are program counter (PC), Register File (RF), and Data Memory (DM). The DLX processor supports five types of instructions, including Register-Register (RR), Register-Immediate (RI), Load (LD), Store (ST), and conditional branch (BR). The pipelined model of DLX processor has five stages of Fetch (F), Decode (D), Execute (E), Memory (M), and Write Back (W). Therefore, the corresponding pipeline register sets are named Fetch/Decode (FD), Decode/Execute (DE), Execute/Memory (EM), and Memory/Write Back (MW). This processor supports stall and forwarding mechanisms to avoid data hazards, and squash signal to handle control hazard.

C. Multithreaded Processor Modeling and Verification

The modeling and verification approach of this paper is based on the method proposed in [1], but the implementation, here is done based on the CLU logic which is different from that of [1]. First of all, for each thread, a separate storage element should be defined. Each thread reads from its own storage element and writes on its own storage element. To accomplish this, as many instances of each storage element as the number of threads are defined. The next issue to be considered is that at each pipeline stage, at most one thread should be active. A thread identifier (ID) is used when checking conditions of squash, stall, and forwarding signals, to prevent control interference among active threads executing in different pipeline stages. Furthermore, an individual ID is considered in the case where no thread is active in the pipeline stage. To verify this model, an extension of the flush-based correspondence checking to a multi-threaded processor is used. The basic idea of flushing proof [10] is that if both implementation and specification models are let to execute for one step starting from the same initial state, they should lead to the same final state. In a MT processor, this final state includes the status of all storage elements. When checking the model using flushing proof in the verification tool, if the model is buggy the verification tool outputs a CEX, including a set of value assignments to the state variables and storage elements of the model in different steps of simulation.

III. TWO-STAGE SIGNATURE ORIENTED PRUNING TO DEBUG MULTI-THREADED PROCESSORS

A. Stage One: Rule Checking

In the first stage, the rule set, including valid, instruction type, and thread ID are checked in the implementation model. If they do not hold an early repair is done, otherwise they are used to prune the model.

Since the processor to be checked has a pipelined implementation, it has sets of pipeline registers. The valid register at each pipeline stage determines whether the registers in that pipeline stage are valid or not. Therefore, valid rule is based on two observations in processors: (1) An invalid instruction never gets valid, while the vice versa is possible; and (2) Only a valid instruction can modify the value of storage elements. These two observations can be checked in the implementation model by deciding about a number of invariants. To do this, in each step of simulation, the values of valid registers and storage elements are stored, then in the next step, both conditions are checked using the current and previous values of registers.

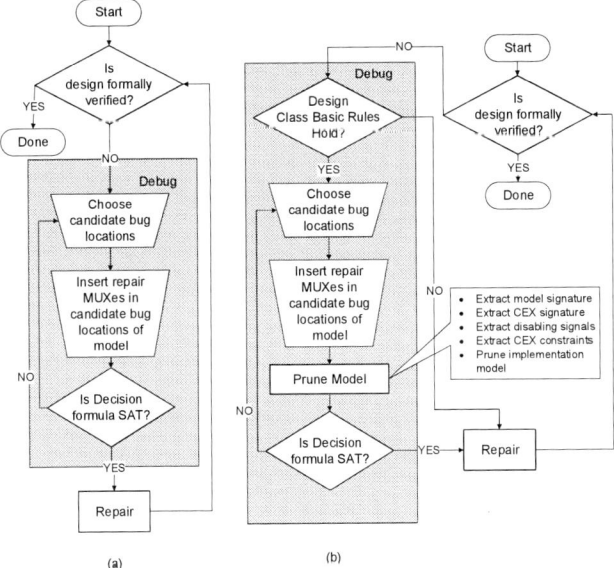

Fig. 1. (a) Conventional MUX-based debugging approach, (b) proposed two-stage signature oriented pruning approach

Instruction type rule is based on the observation that the type of an instruction never changes after being determined in the decode stage. This observation can be checked in the implementation model using a number of invariants. To do this, in each step of simulation, the values of type registers are stored, then in the next step the stored value is compared to the current value of type registers considering that each instruction has executed for one step.

The last rule, Thread ID rule, is based on two observations in processors: (1) Thread ID of an instruction never changes after being determined in the fetch stage; and (2) Each instruction can only modify the storage elements with the same thread ID. These two observations can be checked in the implementation model by deciding about a number of invariants. To do this, in each step of simulation, the values of TID registers and storage elements are stored, then in the next step, both conditions are checked using the current and previously stored values.

If these invariants are not satisfied, an early repair phase is required. Considering that each statement in the implementation can be buggy with the same probability and a single bug is present, as the number of threads increases the probability of rule violation approaches 10.5%. I.e. for about one out of ten buggy designs, one is detected in the first stage by rule checking.

B. Stage Two: Debugging based on Signature Oriented Pruning

The second stage of proposed debugging approach aims at pruning the final decision formula according to the signature extracted based on the rule set as summarized in Fig. 2. As it can be seen, the inputs to the method are buggy model from modeling phase, and the CEX generated during the verification phase. The goal of the debugging process that consists of six steps shown in Fig. 2 is to determine candidate bug locations.

The first step includes extraction of a set of information from CEX and buggy model according to the rule set checked in the first stage. In the second step, the rule set is used in conjunction with the CEX signature in an automatic process to achieve model disabling signals. Disabling signals are used to disable some parts of the model in special simulation steps. In step three, the model is pruned using the generated disabling signals from step two. A MUX enriched model is developed in step four, based on the candidate bug locations. A set of constraints (Cc) forcing the verification tool to develop the same CEX is extracted from CEX signature in step five. The Cc constraints in addition to the correspondence checking constraints achieved from the pruned buggy model of step three (Ci) and MUX enriched model of step four (Ce) are then used as the final decision formula to the verification tool in step six. If the tool generates a CEX, the MUXes are placed in correct locations, and can be used as the starting clue for finding the root cause of a bug. Otherwise, other candidate bug locations should be considered, and the MUX-enriched model is modified accordingly. The details of each step are discussed in the following subsections.

1) Model and CEX Signature Extraction

Considering the rule set checked in the first stage, not all data used for modeling processor and reported in CEX are useful. Therefore, these two sets of data are analyzed carefully to extract the rule-related information, called model signature and CEX signature. A sample model signature required for applying the rule set to enable pruning extracted from the processor model is summarized in TABLE I, which reports the name of pipeline registers and their usage in different instructions. The first two columns determine the pipeline stage and name of the state variable, respectively. The major column "Instruction Type" indicates which state variables are used by a given instruction. An 'X' in the related minor columns means that the specified state variable is used during execution of this instruction.

The same type of signature is extracted from the buggy model for storage elements. Considering RF, while all instructions read from it in FD stage, only instructions of types RI, RR, and LD write to it in MW stage. DM, on the other hand is only used by LD and ST instruction types for reading and writing in EM stage respectively. All copies of storage elements in the multi-threaded processor model, behave in the same way. In contrast to pipeline registers which change value in each simulation step, and their value is only checked exactly in the

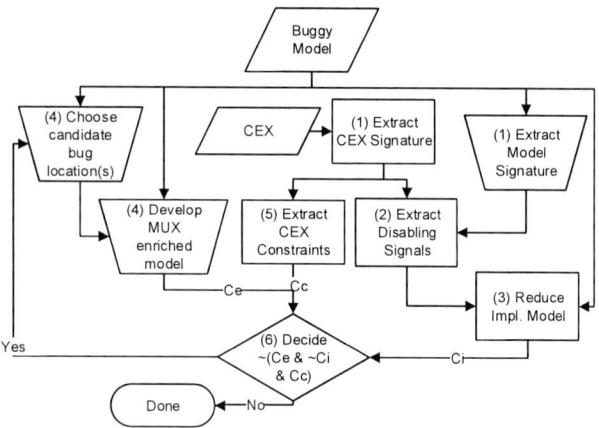

Fig. 2 Proposed signature oriented pruning approach

TABLE I

MODEL SIGNATURE: STATE VARIABLES

Pipeline Stage	State Variable	Instruction Type				
		RI	RR	LD	ST	BR
FD	fdValid, fdInstr, fdType, fdTID	X	X	X	X	X
	fdPC					X
DE	deValid, deInstr, deArg1, deType, deTID	X	X	X	X	X
	dePC					X
	deArg2		X		X	X
EM	emValid, emTID, emType	X	X	X	X	X
	emBr, emTar					X
	emArg2				X	
	emVal	X	X	X	X	
	emInstr	X	X	X		
MW	mwData, mwDest, mwTID	X	X	X		
	mwValid	X	X	X	X	X

978-1-4799-7598-3/15 $31.00 © 2015 IEEE

next step, storage elements have different reading and writing pipeline stages. These set of information are referred to as the model signature in this paper. In the process of pruning, the first part of model signature is used by valid and instruction type rules, while the second is required by all three rules. Note that since the model signature is extracted from the buggy model, its contents may change according to the type of bug, and the reported signature is only a sample one.

Now, considering a bug in the processor model, a CEX is reported from the verification tool. This CEX is a lengthy file, including values assigned to all state variables and storage elements in different simulation steps. However, just a summary of the rule-related signature should be extracted from this raw data to be used for the process of pruning. A part of this is reported in TABLE II for a bug in taking a branch. The first column indicates the simulation step among the total of 15 steps. The next four columns, determine which type of instruction is executing in each pipeline stage, where shaded cells indicate invalid instructions. These cells are called simulation slot in the rest of this paper. Since after memory stage the type of instruction does not matter, and any valid instruction writes back in RF, 'X' is used in MW column. The number in parenthesis determines the thread ID of the executing instruction. CEX signature is extracted automatically from any generated CEX. Note that CEX signature is selected according to the rule set. I.e. validity, instruction type, and thread ID of each instruction are pieces of information required by valid, instruction type, and thread ID rules, respectively.

2) Extraction and Insertion of Disabling Signals
The pruning of the processor model is done through a set of disabling signals. Each state variable has an associated disabling signal, which disables reading from it. Each storage element has two disabling signals, to disable reading and writing, called read-disable and write-disable signals respectively. Although the number of these signals is fixed, their value may change at each simulation step. The goal is to determine disabling steps for each disabling signal; i.e. the simulation steps in which each disabling signal should be activated to disable the state variable or storage element. The result of applying valid, instruction type, and thread ID rules to the sample model signature and CEX signature is partly reported in TABLE III and TABLE IV. The columns of TABLE III determine the pipeline register set and name of state variables, and the disabling steps according to valid and instruction type rules, respectively. TABLE IV reports the read-disabling and write-disabling steps for RF1 and DM2 according to valid, instruction type, and thread ID rules. The union of all disabling steps obtained from the rule set for each state variable and storage element determines the total disabling steps, as shown in the last columns of TABLE III and TABLE IV.

3) State Variables and Storage Elements Disabling Method
Next, the obtained disabling steps should be applied to the model. The method of disabling a special state variable or storage element with the aim of model size reduction is discussed in this subsection. In the proposed method, all changes to the model are done directly on the model description code. There are two main issues when applying the disabling signals: how to control the value of each disabling signal in different simulation steps, and how to disable a state variable or a storage element using a disabling signal. To handle the first issue, the value of each disabling signal is controlled in the control part of the verification tool by assigning value TRUE. E.g., the value of *fdInstr* disabling signal is set in steps two to six, and eight to twelve. For the second issue, in order to disable a state variable according to its disabling steps, its disabling signal is used to disable the cases of reading it. For storage elements the disabling signal should also be utilized for disabling write.

4) MUX Enriched Model and Final Decision Formula Construction
In this subsection, the last three steps of the signature oriented pruning method are discussed in detail. The method used throughout this paper for developing the final decision formula is an improvement to the method proposed in [6] and [9] where MUX-based debugging approach is used in the UCLID model when using flush based verification method. To do this, two implementation models are described: A MUX enriched and an original. The MUXes are inserted as additional case statements

TABLE III

THE SAMPLE DISABLING STEPS FOR DISABLING SIGNALS OF STATE VARIABLES

Pipeline Register Set	State Variable	Applied Rule		All Rules
		Valid	*Instruction*	
Fetch-Decode	fdInstr	2-6,8-12	-	2-6,8-12
	fdPC	2-6,8-12	2-6,10-14	2-6,8-14
Decode-Execute	deValid	-	-	-
	deArg1	3-6,9-13	-	3-6,9-13
Execute-Memory	emType, emTID	4-6,10-14	-	4-6,10-14
	emTar	4-6,10-14	0-1,4-8,12-14	0-1,4-8,10-14
	emBr	4-6,10-14	0-1,4-8,12-14	0-1,4-8,10-14
	emArg2	4-6,10-14	0-14	0-14
Memory-Writeback	mwData	0,3-7,10-14	-	0,3-7,10-14
	mwDest	0,3-7,10-14	-	0,3-7,10-14

TABLE IV

THE SAMPLE DISABLING STEPS FOR DISABLING SIGNALS OF STORAGE ELEMENTS

Storage Element		Action	Rule Type			All Rules
Type	*Thread ID*		*Valid*	*Instruction*	*TID*	
Register File	1	Read	2-6,8-12	-	1-6,8-13	1-6,8-13
		Write	0,3-7,10-14	-	1-2,4-6,8-9,11-14	0-14
Data Memory	2	Read	4-6,10-14	0,2-3,6-7,9-14	2,9	0,2-7,9-14
		Write	4-6,10-14	0-14	2,9	0-14

TABLE II

SAMPLE CEX SIGNATURE

Sim. Step	Pipeline Register Set			
	FD	*DE*	*EM*	*MW*
0	BR(1)	LD(2)	RI(2)	X(1)
1	BR(2)	BR(1)	LD(2)	X(2)
8	BR(2)	BR(1)	LD(2)	X(2)
14	RI(1)	RI(2)	RI(2)	X(2)

978-1-4799-7598-3/15 $31.00 © 2015 IEEE   123

for state variable assignments, as shown in step four of Fig.2. Each MUX has an additional Boolean constant as controlling signal and an alternative Term/Boolean constant as the correcting value. The decision about the place of MUX insertions is done based on a list of candidate bug locations. Note that, the pruning is applied on the original model, and the enriched model remains unchanged.

In the control part of UCLID description in [6] and [9], the decision is made in the case where the enriched model is correct but the original one is incorrect, with the constraint of having the same constants. Calling correction of enriched and implementation models Ce and Ci respectively, the decision condition would be defined as ¬ (Ce & ¬Ci) as shown in the final step of Fig.2. This leads the tool to assign appropriate values to data and control lines of MUX(es) to correct the model errors, if it is possible. In this paper, however, to enable the correct pruning, a CEX constraint (Cc) to determine the CEX is added to the final decision formula, as shown in step five of Fig.2. The set of constraints is added based on the CEX signature, including the type, thread ID, and validity of instructions. Therefore, the final decision formula changes to ¬(Ce & ¬Ci & Cc). This will force the tool to produce the same sequence of CEX, while assigning appropriate values to debugging MUX(es). Finally, the verification tool is executed to achieve the appropriate controlling and alternative values for debugging MUX(es). If no CEX is found, the inserted MUX(es) cannot correct the design, and other candidate fault locations should be considered for inserting debugging MUX(es).

IV. EXPERIMENTAL SETUP AND RESULTS

In order to demonstrate the effectiveness of the proposed approach on processors, three sets of experiments have been performed, reporting the effectiveness of applying each rule and impact of increasing the number of threads on the formula size and CPU time, reported in seconds and number of parameters of propositional formula respectively. The test case in these experiments is the multi-threaded DLX processor explained in section II. To show the effectiveness of the proposed approach regardless of bug type, the model with three different bug types is considered, where the first bug (B1) is the same as the sample bug reported in section III.B, the second buggy model (B2) has bug in the forwarding logic, and the third one (B3) is a threading-related bug. The results are compared with the results obtained through debugging method of [6] and [9], where no pruning is applied. Note that M.E. and S.E. denote memory and size error respectively. Memory error is caused by incomplete process of verification after construction of final CNF formula, while size error happens in the step of propositional formula construction. I.e. in the case of M.E. the size of formula is reported, while in S.E., no report is available; but none of them produces CEX.

TABLE V shows the results of first experiments, where the columns determine the applied rule, the sizes of the four buggy models, and the evaluation parameters, including average size reduction and average effectiveness, respectively. The average size reduction equals to the average percentage of size reduction obtained using each rule, while the effectiveness of each rule is defined as the ratio of reduction caused by the rule to total reduction achieved when applying the rule set. The results

show that applying the rule set results in average 41% reduction in the size of final propositional formula. The most effective rule is the valid rule which is 81% effective in average, due to the fact that from the total of 60 instruction slots, most of them are occupied by invalid instructions. Instruction type rule, as the next most effective rule, reduces the model size for about 11%. Although instructions are of five different types, since most state variables and storage elements are shared among them, the percentage of reduction is far less than the ideal 80% value. The reduction caused by this rule may increases by the increment to the number of instruction types. The least effective rule is TID, which only applies to RF and DM storage elements and only about 18% of overall reduction is caused by it. This is because the results from this experiment are obtained from two-threaded DLX processor, and in most steps, only one storage element is free of read/write operations.

TABLE VI demonstrates the effect of applying the rule set when the number of threads increases, where the rows are dedicated to the number of supported threads in the processor model, the buggy model name, and the formula size and CPU time for original buggy model, model obtained by the debugging approach of [6], pruned model, and the percentage of achieved individual and average reduction as the evaluation parameters, respectively. As the number of threads increases, the reduction becomes more effective up to 61% at most for buggy model B2 supporting 16 threads, and 53% in average for all buggy models supporting 16 threads. This improvement is mostly due to the TID rule, because while the number of supported threads increases, the number of CEX slots remains fixed. This lowers the chance of each thread to participate in the CEX. Considering CPU time, the results show an improvement of 47% to 88% for different cases. In the same way as formula size, more the number of supported threads more the reduction in CPU time is achieved in average. However, CPU time is not an accurate effectiveness measure, because it is highly dependent on the SAT solver efficiency and methodology. E.g. in all cases, the debug process when using the proposed approach is even faster than the original model, because in the pruned case the CEX is guided toward a specific case, while in the original model, the generation of CEX is not restricted and the searching space is bigger.

The results of the final experiment dedicate to the models supporting more than 20 threads are reported in TABLE VII where

TABLE V

EFFECT OF APPLYING THE RULE SET ON THE SIZE OF MODEL (N.A. : NOT APPLICABLE)

Applied Rule	Size for buggy designs			Evaluation	
	B1	B2	B3	Avg. Size Reduction	Avg. Effectiveness
no pruning ([6])	30392	35517	39535	N.A.	N.A.
Valid	24247	21573	23279	33.53%	81.42%
Instruction	28151	28054	37588	11.10%	26.96%
TID	29070	32969	35342	7.38%	17.91%
All	20480	19071	21889	41.18%	100%

TABLE VI

EFFECT OF INCREASING THE NUMBER OF SUPPORTED THREADS ON THE
EFFECTIVENESS OF PROPOSED METHOD (M.E. : MEMORY ERROR)

Model		Formula Size					CPU Time				
		Size (No. of Param.)			Eval.		Time (sec.)			Eval.	
No. of Threads	Bug Type	Original	[6]	Proposed	Reduction (%)	Average Reduction (%)	Original	[6]	Proposed	Reduction (%)	Average Reduction (%)
2	B1	15437	30392	20480	32.61	41.18	2	5	1.2	76.2	65.76
	B2	15418	35517	19071	46.3		1.7	2.8	1.1	60.71	
	B4	15444	39535	21889	44.63		1.2	2.7	1.1	60.37	
4	B1	32759	77489	46392	40.13	46.47	11	37	5	86.49	66.81
	B2	32740	96306	46348	51.87		4.4	8.6	2.9	66.28	
	B4	32755	94801	49857	47.41		3.5	6.4	3.4	47.67	
8	B1	98485	267297	145787	45.46	52.13	54	93	16	82.8	78.24
	B2	98466	344410	141561	58.9		22.3	98	10.8	88.98	
	B4	98478	310670	149020	52.03		10.2	32.5	12	62.94	
16	B1	370264	1049644	559876	46.66	53.66	132	395	57	85.57	76.28
	B2	370245	1388461 (M.E.)	530828	61.77		86	N.A.	51	N.A.	
	B4	370271	1194324	566777	52.54		48.7	206	68	67	
20	B1	530232	1622667 (M.E.)	851670	47.51		261	144	126	N.A.	

TABLE VII

EFFECT OF INCREASING THE NUMBER OF SUPPORTED THREADS ON THE
EFFECTIVENESS OF PROPOSED METHOD : THE CASES WHERE [6] IS NOT
APPLICABLE

Model		Formula Size (No. of Param.)		CPU Time (sec.)	
No. of Threads	Bug Type	Original	Proposed	Original	Proposed
20	B2	530213	806945	120	98
	B3	530245	818915	82	113
23	B1	748394	1124726	1341	177
	B2	748375	1045440	287	152
	B3	748401	1052187	156	211

determine the number of supported threads in the processor model, the buggy model name, and the formula size and CPU time for original buggy model, and pruned model, respectively. Note that since in these cases using [6] results in S.E., the effectiveness is not reported in this table. The size of model increases by an increment in the total number of supporting threads, e.g. for B2, therefore [6] is not able to produce CEX for any number of threads between 16 and 23 either.

According to the results, the proposed method affects the size of different buggy models, even when supporting the same number of threads, e.g. ranges from 46% to 61% for the case of 16 threads. As another example, although both B1 and B3 have nearly the same original model size, their effectiveness results show up to 8% difference for different number of threads. This is due to the differences in the type of bugs, the place of de-

bugging MUXes, and the produced CEXes. In fact, the produced CEX can affect the effectiveness of each rule thoroughly. E.g. according to TABLE I, instruction types RI, LD, and BR do not use six state variables, while the number changes to five and eight for RR and ST instruction types respectively. Therefore, the reduction caused by instruction type rule for a case with a CEX containing only ST instructions could be more effective. In the same way if the CEX includes ID0 (i.e. no active thread) or a few valid instructions, the TID and Valid rules could be more effective respectively. Note that, despite the impact of CEX on the effectiveness of proposed method, the average effectiveness reported in TABLE VI, shows that by increasing the model size, all buggy models show average size improvement of 12%.

V. CONCLUSIONS

In this paper, a two-stage signature oriented pruning method is proposed to avoid the space explosion problem caused from inserting debugging facilities in the multi-threaded processors, which is implemented on an automatic verification tool. The pruning method utilizes the facilities of an automatic verification tool effectively, without interfering the internal steps. According to the results, the two-stage signature oriented pruning method leads to 32% to 61% improvement in the final size of propositional formula. For smaller designs the most effective rule is the valid rule, but by increasing the number of supported threads, the effectiveness of thread ID rule and the overall effectiveness improve. The results are compared to that of [6], which uses the same approach, without performing any pruning. As future work, the entire process including model signature extraction should become automatic. Furthermore, it is recommended to use the rule set to prune candidate bug locations, extend pruning to enriched model, and utilize the obtained results in post-silicon debugging phase.

REFERENCES

[1] M. N. Velev and G. Ping, "Automatic formal verification of multithreaded pipelined microprocessors," in *Proc. of ICCAD'11*, 2011, pp. 679-686.

[2] A. Smith, A. Veneris, M. Fahim Ali, and A. Viglas, "Fault diagnosis and logic debugging using Boolean satisfiability," *IEEE TCAD*, vol. 24, pp. 1606-1620, 2005.

[3] S. Mirzaeian, Z. Feijun, and K. T. T. Cheng, "RTL Error Diagnosis Using a Word-Level SAT-Solver," in *Proc. of ITC'08*, 2008, pp. 1-8.

[4] H. Mangassarian, L. Bao, and A. Veneris, "Debugging RTL Using Structural Dominance," *IEEE TCAD*, , vol. 33, pp. 153-166, 2014.

[5] Y. Chen, S. Safarpour, J. Marques-Silva, and A. Veneris, "Automated design debugging with maximum satisfiability," *IEEE TCAD*, vol. 29, pp. 1804-1817, 2010.

[6] B. Alizadeh and M. Fujita, "Debugging and optimizing high performance superscalar out-of-order processors using formal verification techniques," in *Proc. of ISQED'11*, 2011, pp. 1-6.

[7] R. E. Bryant, S. K. Lahiri, and S. A. Seshia, "Modeling and verifying systems using a logic of counter arithmetic with lambda expressions and uninterpreted functions," in *Proc. of CAV'02*, 2002, pp. 78-92.

[8] M. N. Velev and P. Gao, "Automated debugging of counterexamples in formal verification of pipelined microprocessors," in *Proc. ASP-DAC'12*, 2012, pp. 689 - 694.

[9] B. Alizadeh, "Formal verification and debugging of precise interrupts on high performance microprocessors," *ACM Trans. on Design Automation of Electronic Systems*, vol. 17, pp. 37:1-37:8, 2012.

[10] J. R. Burch and D. L. Dill, "Automatic verification of pipelined microprocessor control," in *Proc. ICCAD'94*, 1994, pp. 68-80.

Innovative Practices Session 5C:
Advancements in Test –Keeping Moore Moving!

Organizer & Moderator:
Enamul Amyeen (Intel)

Presenters & Abstracts:

Presenter #1: *John Bowling (Intel)*

Title: Challenges with manufacturing huge chips –an interplay between test and debug

Abstract: This talk focuses on test and debug challenges that are unique to large die products. A technical review of problems being faced today specific to test quality, test time, test cost, electrical and speed content, and debug will be outlined along with some current solutions being pursued to drive large die products to production quality. The talk concludes with a discussion about new solutions and areas of innovation that will be necessary to keep future generations of these products manufacturable, and on a cadence that meets the stringent time to market requirements.

Presenter #2: *Greg Yeric (ARM)*

Title: Technology and the pursuit of Moore's Law

Abstract: Basic Moore's Law scaling is hitting some fundamental barriers. Various technology advancements attempt to continue the chip-level progress, be the new transistors, lithography scaling work-arounds, etc., but almost all advancements in these areas come at the cost of added complexity, which filters forward into design and test. This talk will cover many of the anticipated technology scaling trends and relate them in terms of their impact on testing concerns.

Presenter #3: *T.M. Mak (GlobalFoundries)*

Title: Test Technology at a standstill with "More than Moore"

Abstract: More than Moore" is a term to mean heterogeneous integration above and beyond the exponential scaling of transistor scaling - the Moore's Law. This form of integration actually has existed long before this term was invented, namely: MCM, MCP, SiP, and now 2.5D/3D packaging and IoT. While it is not new, it has been presenting challenges to the test community with a poorly defined requirement: KGD, (Known Good Die). The term is misleading as it implies that there is a way to test/predict a die to be absolutely good before it is to be packaged. Test Technology in the past 30-40 years has been evolving in standalone silos: scan for logic, algorithmic self-test engine for large embedded memories, IO (loopback) self-test, boundary scan (board test) with analog test (MEMS, opto included) largely remained instrument based, Will this silo approach solves the challenges presented by More-than-Moore, when all these heterogeneous dies all come into the package one way or the other? Will we resort to system test as the only solution? What does system test look like when we have billions of transistors, sensors/actuators of motion, pressure, magnetism, light and laser all on board? Does system test even provide the quality level that are expected for our products? It is time to move out of our silos and the comfort zone.

2015 IEEE 33rd VLSI Test Symposium (VTS)

At-Product-Test Dedicated Adaptive Supply-Resonance Suppression

Kohki Taniguchi, Noriyuki Miura, Taisuke Hayashi, and Makoto Nagata

Graduate School of System Informatics, Kobe University, Japan
taniguchi@cs26.scitec.kobe-u.ac.jp, miura@cs.kobe-u.ac.jp

Abstract—This paper presents an adaptive supply-resonance (SR) suppression scheme at a product testing stage. Dedicated to each product in different assembly forms, an on-chip power-delivery-network analyzer identifies SR frequency and auto-tunes notch filter for SR noise suppression. The feasibility has been silicon-proven by a prototype demonstration in 0.18μm CMOS successfully.

I. INTRODUCTION

Power integrity management is one of the key issues in a design of highly-reliable electrical controller systems. Especially in automotive applications, an electrical Engine Control Unit (ECU) is a fatal component to human life and must reliably operate even under a stringent condition on large-power noise injection directly or due to Electro-Magnetic (EM) interference. There are many strict noise and EM Compatibility (EMC) requirements for automotive components such as CISPR12, CISPR25, ISO11451-X, and ISO11452-X series [1]. A product that is failed in this standard EMC test can not be shipped worldwide.

The critical device in the ECU is a semiconductor Micro Controller Unit (MCU) chip. Although the supply voltage is as low as 1V, the MCU chip has to properly operate with surrounded noisy motors and other electro-mechanical components operating under >1,000V supply. The situation is even more adverse. The noise and EM immunity strongly depend on all the chip, package, and board assembly implementations. At a wafer sort test stage, the chip functionality can be only checked for know-good-die decision. The actual immunity test has to be done at a final product testing stage (Fig.1) including other package and board assembly. Conventionally, the chip, package, and board are separately designed based on a guideline of individually-predefined Power-Delivery Network (PDN) specification on typical supply impedance characteristics. However, since the same-design chips are used in multiple different products, the PDN characteristics are largely fluctuated. Moreover, the process and assembly variations increase this fluctuation further broad wide. This results in a huge design margin and/or modification at each chip, package, and board design, causing cost increase and/or time loss. The similar problem can also be found in any other electronic systems such as mobile, server, M2M, and sensor network.

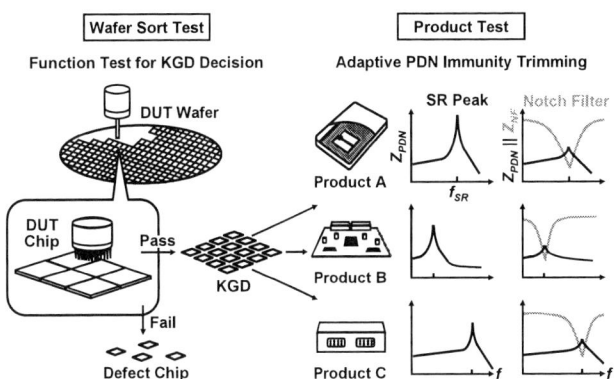

Fig.1 Conceptual sketch of at-product-test adaptive PDN trimming.

In this work, a concept of an on-chip one-time automatic PDN trimming scheme is introduced at a product testing stage (Fig.1). A product-level complete PDN characteristics are auto-adjusted adaptively to each different product (assembly) over broad manufacturing variations. The product yield can be improved by securely providing highly-reliable PDN with proper EMC performance even under the wide variety of PDN characteristics fluctuations.

For trimming the PDN characteristics, one of the main problems is Supply Resonance (SR) in PDN. Due to parasitic inductance of the supply interconnect and decoupling capacitance, a resonant tank is unfortunately always formed in PDN which causes impedance peaking at the self-resonant frequency f_{SR}. Due to this impedance peaking at f_{SR}, large persistent supply noise is produced as a result of the periodic supply current drawn near at f_{SR} of typically around several hundreds of MHz in electrical systems, which often becomes a dominant cause of PDN dependability degradation [2]. On-chip active SR suppression and supply-noise reduction techniques were proposed [3-6]. In all these active approaches, there is a risk of even increasing the noise due to excess noise-suppression current flow. To solve this problem, a precise current control is necessary however this is not fully discussed in [3-6]. In addition, since all the noise-suppression current becomes power loss, the active approach consumes large additional active power and also static power for monitoring the supply noise. A passive SR suppression approach on the

978-1-4799-7598-3/15 $31.00 © 2015 IEEE   127

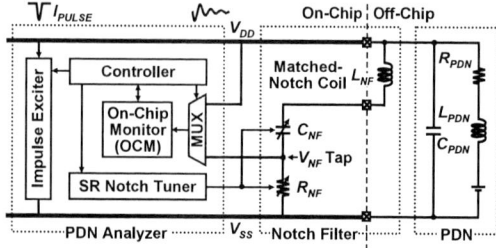

Fig.2 Block diagram of adaptive SR suppression system.

other hand can solve this problem [7] where a passive notch filter composed of a bonding-wire coil and an on-chip capacitor bank is utilized for suppressing the PDN impedance peaking. In the passive filtering approach, there is no excess suppression issues. Also the static power consumption is effectively zero and active power consumption is only caused by the small resistive loss in the filter. For the SR suppression, the notch frequency f_{NF} must be tuned at f_{SR}. However an effective tuning scheme is not yet well-established. In this paper, an on-chip adaptive SR suppression scheme with closed-loop filter auto-tuning is proposed. An on-chip PDN analyzer with an on-chip waveform monitor [8] precisely identifies f_{SR} and auto-tunes f_{NF} instantaneously by monitoring an internal node of the notch filter. Introducing the proposed scheme at the product testing stage enables adaptive noise immunity enhancement in each individual product including broad manufacturing variations. The waveform monitoring for f_{SR} identification and the f_{NF} auto tuning is done once at the product testing stage. All the PDN analyzer circuits are switched off during the filter operation and no static power is consumed in the circuits.

The rest of the paper is organized as follows. Section II will describe the detail of the proposed adaptive SR suppression scheme, architecture, and circuit. The PDN impedance characteristics will be theoretically analyzed and the closed-loop tuning mechanism will be derived. Section III will presents prototype test-chip measurement results for feasibility demonstration. The test-chip is designed and fabricated in 0.18μm standard digital CMOS for the proof of concept. Finally Section IV will summarize this paper with some conclusive remarks.

II. ADAPTIVE SUPPLY-RESONANCE SUPPRESSION

Figure 2 depicts the block diagram of the adaptive SR suppression system at the product testing stage. It is composed of the on-chip PDN analyzer and the tunable notch filter. The analyzer identifies f_{SR} in PDN and adaptively tunes f_{NF} to f_{SR} for SR suppression. A simplified equivalent circuit model of

PDN is also shown in Fig.2. The PDN impedance Z_{PDN} can be seen as a parallel LC resonant tank from the chip side and is expressed as

$$Z_{PDN} = \frac{2\pi j f L_{PDN} + R_{PDN}}{1 - 4\pi^2 f^2 L_{PDN} C_{PDN} + 2\pi j f C_{PDN} R_{PDN}}, \quad (1)$$

where L_{PDN}, C_{PDN}, R_{PDN} is self-inductance, capacitance, resistance of PDN respectively. The PDN impedance peaking appears at the SR frequency f_{SR} as illustrated in Fig.1. f_{SR} is given by

$$f_{SR} = \frac{1}{2\pi\sqrt{L_{PDN} C_{PDN}}}. \quad (2)$$

The notch filter acts as shunt impedance to suppress the impedance peaking at the block (notch) frequency f_{NF} as the impedance of the filter Z_{NF} is given by

$$Z_{NF} = \frac{1}{2\pi j f C_{NF}} + 2\pi j f L_{NF} + R_{NF}, \quad (3)$$

where L_{NF}, C_{NF}, R_{NF} is self-inductance, capacitance, resistance of the notch filter respectively. Z_{NF} is minimized at f_{NF} which is given by

$$f_{NF} = \frac{1}{2\pi\sqrt{L_{NF} C_{NF}}}. \quad (4)$$

The total impedance seen from the chip side Z_{TOT} is given by $Z_{PDN}\|Z_{NF}$. By tuning f_{NF} to f_{SR}, the PDN peaking due to SR can be shunted by the notch filter as shown in Fig.1. When f_{NF} is tuned to f_{SR} ($f_{NF}=f_{SR}$), Z_{TOT} at f_{SR} is given by

$$Z_{TOT}(f_{SR}) = Z_{PDN}(f_{SR}) \| Z_{NF}(f_{SR}) = \frac{Z_{PDN}(f_{SR})R_{NF}}{Z_{PDN}(f_{SR}) + R_{NF}}. \quad (5)$$

The ratio of the noise amplitude without and with the filter $V_{N,WOF}/V_{N,WF}$ is therefore calculated as follows (derivation is given in APPENDIX)

$$\frac{V_{N,WOF}}{V_{N,WF}} = \frac{Z_{PDN}(f_{SR})}{Z_{TOT}(f_{SR})} = 1 + Q_{PDN}^2 \frac{R_{PDN}}{R_{NF}} - j Q_{PDN} \frac{R_{PDN}}{R_{NF}}. \quad (6)$$

Eq.(6) is approximated as

$$\frac{V_{N,WOF}}{V_{N,WF}} \approx 1 + Q_{PDN}^2 \frac{R_{PDN}}{R_{NF}}, \quad (7)$$

where Q_{PDN} is a quality factor of PDN ($=L_{PDN}^{0.5}/C_{PDN}^{0.5}R_{PDN}$). Note that R_{NF} cannot be zero because too sharp notch filter can

978-1-4799-7598-3/15 $31.00 © 2015 IEEE   128

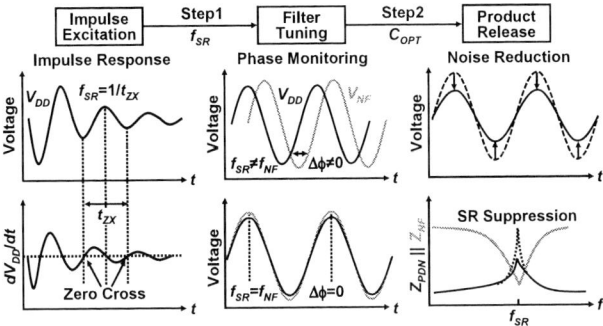

Fig.3 Operation flow of adaptive SR suppression.

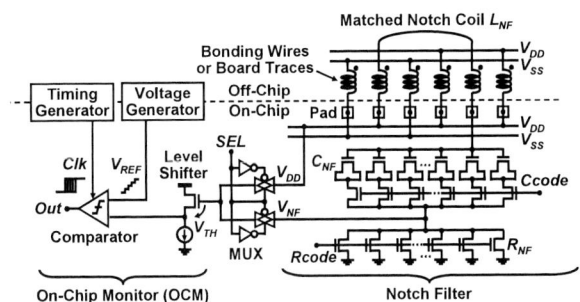

Fig.4 Detailed circuit schematic of key building blocks.

only suppress the impedance peaking exactly at f_{SR} and it just makes two split impedance peaks in PDN. In the carefully-designed PDN, Q_{PDN} is reduced to be around 2~3 by inserting large enough decoupling capacitance both on-chip and on-board. Practically, $V_{N,WOF}/V_{N,WF}$ is close to 2 and hence the noise reduction effect by the filter is expected to be around 50%.

Based on the above theoretical analysis, a closed-loop filter tuning scheme is derived. Figure 3 describes the detail operation flow of the proposed SR suppression system with the tuning scheme. The tuning process consists of two steps.

(Step1) The first step is f_{SR} identification based on PDN impulse response monitoring. An impulse exciter applies a narrow impulse current I_{PULSE} into PDN to generate the impulse response in the supply voltage V_{DD}. The impulse response becomes a damped oscillation waveform which is given by

$$V_{DD}(t) = V_{DD} + V_0 \exp(-\frac{t}{\tau})\sin(2\pi f_{SR}t + \phi), \quad (8)$$

where V_{DD} is the DC supply voltage, V_0 is an amplitude of the damped oscillation waveforms, τ is a decay constant equal to $2L_{PDN}/R_{PDN}$ and the fundamental frequency of the response corresponds to f_{SR}. f_{SR} can be therefore identified by measuring the oscillation cycle of the response. An On-Chip Monitor (OCM) [8] is integrated on-chip to in-situ capture accurately the response waveform. The f_{SR} can be easily calculated in a small controller by taking the first-order derivative form of the captured waveform and measuring the interval of the zero-cross timings t_{ZX}. In most of the cases, existing processor hardware resources can be utilized for this calculation. The impulse exciter can also be substituted with an existing clock tree on the chip. Only the OCM frontend circuit is needed to be on-chip and no additional dedicated hardware resources are required for this calculation. The hardware overhead of the OCM itself is also very small. The detail of the hardware implementation will be explained later.

(Step2) The second step is f_{NF} tuning based on phase monitoring between V_{DD} and the internal node tap voltage of the filter V_{NF}. Unlike the first step, the impulse exciter in the

second step is continuously driven at the f_{SR} frequency to generate a continuous sinusoidal resonant waveform in V_{DD}. By on-chip capturing both V_{DD} and V_{NF} waveforms and comparing their relative phase difference $\Delta\phi$, f_{NF} can be precisely tuned at f_{SR} under the closed-loop control. The tap voltage V_{NF} excited at f_{SR} is given by

$$V_{NF}(t) = \frac{2\pi jfC_{NF}R_{NF}}{1 - 4\pi^2 f^2 L_{NF}C_{NF} + 2\pi jfC_{NF}R_{NF}}V_{DD}(t).$$
$$(9)$$

It denotes that, when f_{NF} is tuned at f_{SR}, $V_{NF}(t)=V_{DD}(t)$ and their phase difference $\Delta\phi$ is zero. The V_{NF} phase is delayed when $f_{NF}<f_{SR}$ and hence C_{NF} is larger than the optimal C_{OPT}, and vice versa. A binary search for C_{OPT} can be thus performed for fast trimming to save test time and cost.

Figure 4 describes transistor-level circuit schematics of key building blocks. The notch filter consists of matched coils L_{NF}, an on-chip capacitor bank C_{NF}, and damping resistor R_{NF}. The matched coils are made by the same interconnects used for the last PDN paths, such as bonding wires or on-board power traces. Since the last paths are always narrow to fit for the dense pads location on the chip, the self-inductance of the last path becomes large and therefore produces the lowest dominant f_{SR} by the combination with the large on-chip decoupling capacitance. By using the matched same interconnects for the coils, f_{NF} can be easily matched (tuned) to f_{SR} even with small-size of on-chip capacitor bank for on-chip layout area saving. In addition, by exploiting mutual inductive coupling between the power connections and the coils, the L_{NF} inductance can be enhanced for further layout area saving [7]. The hardware overhead due to the matched coil is very small because two out of several tens of power and ground connections are only replaced for the matched coil. The capacitor bank C_{NF} is implemented by the binary-weighted MOS capacitors. The binary search for f_{NF} tuning is digitally performed by changing a binary capacitance code $Ccode$. The damping resistor R_{NF} is made by the unit-size MOS resistors. It can be roughly scaled with the size of C_{NF} to keep the proper quality factor of the filter. Only several-steps

978-1-4799-7598-3/15 $31.00 © 2015 IEEE   129

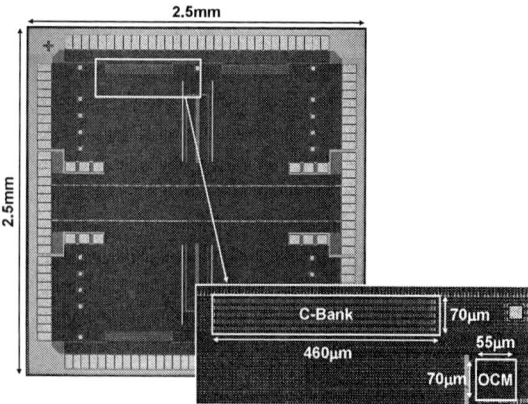

Fig.5 Microphotograph of test chip.

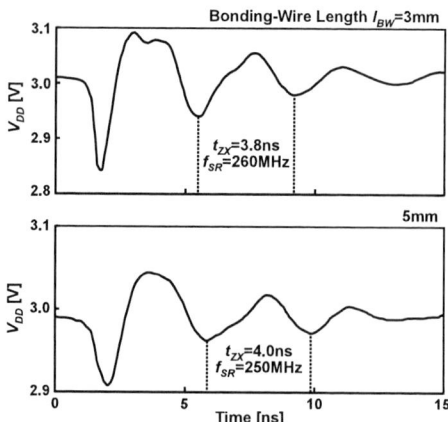

Fig.7 Measured impulse response.

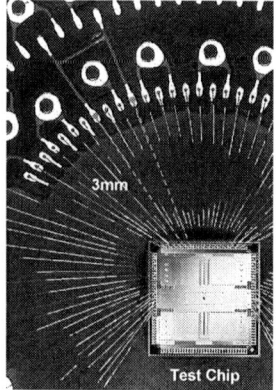 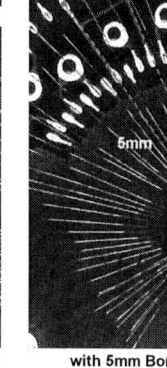

Prototype with 3mm Bonding Wires   **with 5mm Bonding Wires**

Fig.6 Test setup.

transition in the comparator digital output, the waveform voltage level can be indentified. The waveform can be captured by shifting the Clk timing and performing the V_{REF} comparison repeatedly. The Clk timing generator and the V_{REF} voltage generator can be implemented in either on-chip or off-chip depending on total cost considerations. By integrating the timing and the voltage generator on-chip, the required test time can be reduced while the chip layout area is increased. When off-chip, the sampling clock timing and the reference voltage must be provided from a test equipment through two additional input pads and connectors. A fully-integrated on-chip implementation such as in [8] maybe a better choice for future scaled CMOS technologies.

III. TEST-CHIP MEASUREMENT

A test chip has been designed and fabricated in a 0.18μm 1-poly 6-metal layer standard digital CMOS process for the proof of concept of the proposed adaptive SR suppression scheme. Figure 5 shows the microphotograph of the test chip. The chip size is 2.5mm x 2.5mm. It integrates the clock-tree-like impulse exciter which later also acts as a intentional noise generator for evaluating the effect of the filter. The OCM frontend (source follower and comparator) is only integrated which occupies only 55μm x 70μm. The capacitor bank occupies 460μm x 70μm for the SR suppression notch filter covering the several hundreds MHz frequency range.

The test chip is mounted on an FR4 evaluation board and bonding wires are used for power, ground, and signal connections in this implementation. For the matched notch coils, the bonding wires are therefore utilized. Note that, in case of area-bump packaging, the on-board power traces would be used for the power and ground connections and hence same for coils in the filter for the matching purpose. It is also possible to apply the proposed scheme to the area-bump flip-chip packaging. In order to demonstrate the wide application feasibility of the proposed scheme, the different assembly forms are emulated by making evaluation boards with two different bonding wire length l_{BW}: one is 3mm and the other 5mm as shown in Fig.6. The controller in the PDN

coarse linear adjustment is practical enough for digital R_{NF} tuning. The On-Chip Monitor (OCM) precisely captures waveforms of either V_{DD} or V_{NF} through transfer-gated analog multiplexer (MUX). The main frontend consists of only two small circuits, a source follower and a clocked comparator. The source follower acts as a buffer to reduce input loading and also acts as a level shifter to provide an adequate input bias voltage to the following comparator, respectively. The level shift is required otherwise the comparator can not operate correctly since its input is biased at the supply-voltage V_{DD} level which is the highest voltage in the supply rail, making the input transistor of the comparator to be a low-gain triode region and degrading the comparator accuracy. The source follower is implemented by using NMOS which level shifts the supply-voltage waveforms down by the NMOS threshold voltage V_{TH}. The following comparator operates like a similar manner as in a sampling oscilloscope. Both a reference voltage V_{REF} and a sampling timing of Clk are scanned two-dimensionally in voltage and time domains, and the level-shifted supply-voltage waveform are repeatedly compared with V_{REF} at each sampling point. At the fixed Clk timing, V_{REF} is swept and the comparator output is changed from 0 to 1 when the measured waveform voltage at the sampling timing is equal to the V_{REF} level. By detecting this

978-1-4799-7598-3/15 $31.00 © 2015 IEEE   130

analyzer and the timing/voltage generators are implemented by using an off-chip FPGA, an external data timing generator, and a PC for this testing purpose. All these peripheral circuit components can be implemented in an actual test equipment. The measured resolutions of OCM in this test chip are measured to be 10ps and 1mV in time and voltage respectively in this test setup.

First, f_{SR} identification process is evaluated. Figure 7 presents the measured impulse response in the supply voltage waveform V_{DD}. The response waveforms are successfully measured by OCM in both the evaluation board implementations of 3mm and 5mm l_{BW}. The measured zero cross interval t_{ZX} in the derivative of the response is measured to be 3.8ns and 4.0ns for 3mm and 5mm l_{BW} respectively. The SR frequency f_{SR} is therefore calculated to be around 260MHz and 250MHz, respectively. The continuous sinusoidal excitation measurement is conducted by sweeping the excitation frequency from 200MHz to 300MHz. The maximum-amplitude response waveform was successfully observed at the measured f_{SR}. This denotes that the PDN impedance peaking is also appeared at the measured f_{SR}. This measured fact supports good agreement between the identified f_{SR} and the actual PDN characteristics in both board implementations.

Next, f_{NF} tuning process is evaluated. Figures 8 and 9 present measured V_{DD} and V_{NF} tap voltage response waveforms by the continuous sinusoidal excitation at f_{SR}. It can be clearly seen in the measured waveforms that the phase difference becomes zero when the capacitor bank is properly adjusted to the optimal capacitance code C_{OPT} and the f_{NF} is successfully tuned to f_{SR}. It is also confirmed in the measured waveforms that the noise amplitude is minimized when $f_{NF}=f_{SR}$. The SR noise suppression ratio is measured to be 38% and 39% for 3mm and 5mm l_{BW} respectively. The measured noise reduction ratio is well matched with the expected value theoretically discussed in Section II.

IV. CONCLUSION

Supply Resonance (SR) is one of the critical characteristics to degrade noise and EM immunity of PDN in electronic controller systems. The SR frequency f_{SR} fluctuation due to product and manufacturing variations further degrades the immunity of the systems or causes cost increase because of excess design margin requirements. This paper presents a concept of an adaptive SR suppression scheme at a product testing stage. An on-chip PDN analyzer precisely identifies f_{SR} and tunes a block frequency f_{NF} of a built-in notch filter for SR suppression dedicated to each product including the manufacturing variations. A prototype demonstration by using a 0.18μm CMOS test chip successfully exhibits the feasibility of the proposed scheme for highly-reliable electronic controller systems.

REFERENCES

[1] P. Andersen, "The Present Status of the International Automotive EMC Standards," *Proc. of IEEE Electro Magnetic Compativility (EMC)*, pp.98-102, Aug. 2009.

[2] M. Swaminathan and E. Engin, "Power Integrity Modeling and Design for Semiconductors and Systems," *Prentice Hall*, Nov. 2007.

[3] J. Xu, P. Hazucha, M. Huang, P. Aseron, F. Paillet, G. Schrom, J. Tschanz, C. Zhao, V. De, T. Karmik, and G. Taylor, "On-Die Supply-Resonance Suppression Using Band-Limited Active Damping," *IEEE Intenational Solid-Stete Circuits Conference (ISSCC) Dig. Tech. Papers*, pp.286-287, Feb. 2007.

[4] T. Nakura, M. Ikeda, and K. Asada, "Feedforward Active Substrate Noise Cancelling Technique using Power Supply di/dt Detector," *IEEE Symp. on VLSI Cir. Dig. Tech. Papers*, pp.284-287, June 2005.

[5] T. Tsukada, Y. Hashimoto, K. Sakata, H. Okada, and K. Ishibashi, "An On-Chip Active Decoupling Circuit to Suppress Crosstalk in Deep Sub-Micron CMOS Mixed-Signal SoCs," *IEEE International Solid-State Circuits Conference (ISSCC) Dig. Tech. Papers*, pp.160-161, Feb. 2004.

[6] K. Fukuda, S. Maeda, T. Tsukada, and T. Matsuura, "Substrate Noise Reduction Using Active Guard Band Filters in Mixed-Signal Integrated Circuits," *IEICE Trans. Fundamentals*, pp.313-320, Feb. 1997.

[7] T. Hayashi, N. Miura, K. Yoshikawa, and M. Nagata, "A Passive Supply-Resonance Suppression Filter Utilizing Inductance-Enhanced Coupled Bonding-Wire Coils," *Proc. of VLSI Design Automation and Test (VLSI-DAT)*, pp.1-4, Apr. 2014.

[8] T. Hashida and M. Nagata, "An On-Chip Waveform Capturer and Application to Diagnosis of Power Delivery in SoC Integration," *IEEE Journal of Solid-State Circuits (JSSC)*, vol.46, no.4, pp.789-796, Apr. 2011.

APPENDIX

The ratio of the noise amplitude without and with the filter $V_{N,WOF}/V_{N,WF}$ is given by

$$\frac{V_{N,WOF}}{V_{N,WF}} = \frac{Z_{PDN}(f_{SR})I_{EXT}}{Z_{TOT}(f_{SR})I_{EXT}} = \frac{Z_{PDN}(f_{SR})}{Z_{TOT}(f_{SR})}, \quad (10)$$

where I_{EXT} is drawn current due to the circuit operation. Based on Eqs.(1),(2), and (5), Eq.(10) can be expressed as

$$\frac{V_{N,WOF}}{V_{N,WF}} = 1 + \frac{1}{R_{NF}} \cdot \frac{2\pi j f_{SR} L_{PDN} + R_{PDN}}{2\pi j f_{SR} C_{PDN} R_{PDN}}$$
$$= 1 + \frac{1}{R_{NF}} \left(\frac{L_{PDN}}{C_{PDN} R_{PDN}} - j \frac{1}{2\pi f_{SR} C_{PDN}} \right)$$
$$= 1 + \frac{1}{R_{NF}} \left(\frac{L_{PDN}}{C_{PDN} R_{PDN}} - j \sqrt{\frac{L_{PDN}}{C_{PDN}}} \right) \quad (11)$$
$$= 1 + Q_{PDN}^2 \frac{R_{PDN}}{R_{NF}} - j Q_{PDN} \frac{R_{PDN}}{R_{NF}}.$$

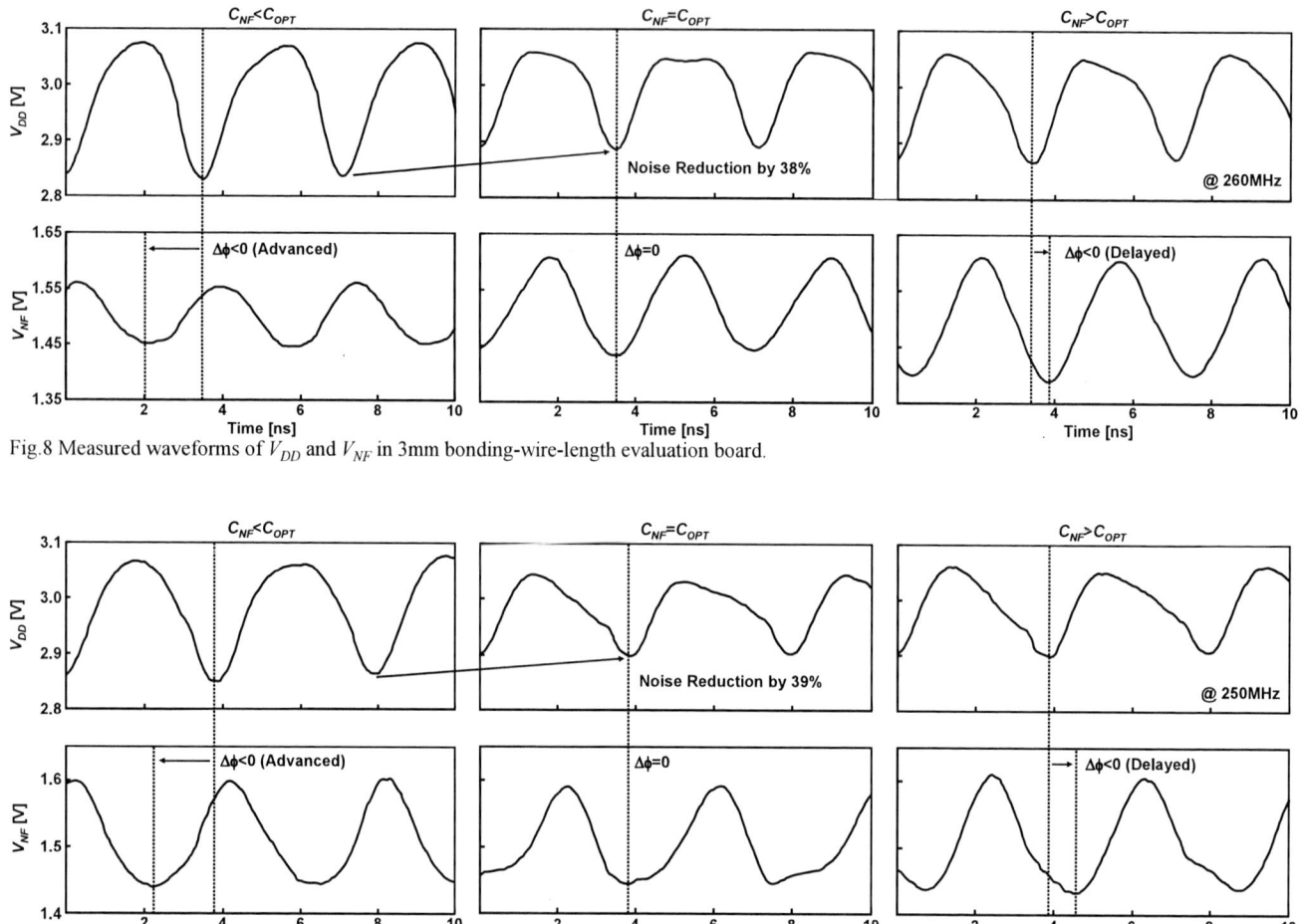

Fig.8 Measured waveforms of V_{DD} and V_{NF} in 3mm bonding-wire-length evaluation board.

Fig.9 Measured waveforms of V_{DD} and V_{NF} in 5mm bonding-wire-length evaluation board.

Low Cost High Frequency Signal Synthesis: Application to RF Channel Interference Testing

Xian Wang, Debashis Banerjee, and Abhijit Chatterjee
School of Electrical and Computer Engineering
Georgia Institute of Technology
Atlanta, USA

Abstract—The quality of a communication link is commonly indicated by signal to noise/interference ratio and affected by noise and interference variance. To ensure a power efficient adaptive OFDM system's performance, the system needs to be tested against various channel conditions. Conventional algorithms and experiments assume that the noise and interference density stays constant over the OFDM frequency band but in reality the communication link often populated with both white noise and interference that results an uneven impact on OFDM spectrum. To create such channel condition, an up-conversion is commonly used. However, due to the non-linearity of the mixer, the up-converted interference is always distorted. The proposed method uses a modified higher-than-Nyquist-rate RF signal generation algorithm to minimize the distortion of generated interference within a pre-defined output bandwidth. Both concept validation and experimental measurement have been conducted to prove the effect of proposed method.

Keywords—Adaptive system testing; OFDM; RF Signal synthesis

I. INTRODUCTION

Due to its efficient usage of available frequencies and robustness against multi-path fading, orthogonal frequency division multiplexing (OFDM) is used extensively in the field of LAN/MAN communication systems [1-2]. After 7500MHz of bandwidth (3.1GHz to 10.6GHz) was opened up for UWB devices by the FCC, much research has focused on reliable UWB communication with the expectation that UWB devices will provide low cost solutions that can satisfy the increasing demand for high data rates [3]. With wide communication bandwidths available, it is necessary to test modern OFDM communication systems for performance in the presence of interferers with different spectral characteristics [4-11]. This drives the need for generating high frequency signals with specified spectral content to function as "channel interferers" during communication systems performance testing and requires the use of expensive test instrumentation.

To create channel interferers with specified spectral characteristic (shapes), the conventional approach is to up-convert a baseband signal with the desired spectral characteristics to the frequency range of interest using an up-conversion mixer. However, this can cause signal (interferer) distortion due to mixer nonlinearity and is particularly troublesome when the interferer bandwidth is large. To solve this problem, higher-than-Nyquist-rate RF signal synthesis [12] can be used. This method takes advantage of a specific type of digital-to-analog converter (DAC) reconstruction filter and generates an optimized input for two interleaved DACs in such a way that higher order image replicas in the frequency domain conform to the desired spectrum. This image replica is filtered and used as the primary output of the system, achieving higher-than-Nyquist rate signal generation.

In this paper, a modified higher-than-Nyquist-rate direct RF signal synthesis method is proposed to generate high fidelity signals with specified spectral characteristics. Such signals can function as interferers during testing of OFDM based communications systems. The method optimizes the input to a DAC with a specified output reconstruction filter to minimize the difference between the first image of the output and the target signal across a pre-defined signal bandwidth. The first image (*beyond the Nyquist frequency of the DAC*) is then passed through a bandpass filter and used as the primary output of the system. This allows the proposed method to shape higher-than-Nyquest-rate signal, which cannot be achieved by the Direct Digital Synthesis (DDS) method [13]. Compared to the method of [12], the proposed technique introduces a scalable and efficient way to calculate the optimized input data string for a desired image with specified bandwidth. The method also differs from that of [12] in that it simplifies signal generation by using only one DAC instead of an interleaved system of DACs. The concept is validated through both simulation and experiments. The quality of the generated channel interference is also compared against that generated by the traditional mixer based approach.

This paper is organized as below. Section II gives an overview of the basics of DAC reconstruction filters and higher-than-Nyquist-rate signal generation. Section III describes the proposed signal generation approach for a signal with specified spectral characteristics and discusses the signal generation architecture. Sections IV and V discuss simulation and experimental results pertaining to the proposed method. Concluding notes are given in Section VI.

II. BACKGROUND

The process of signal generation by a DAC can be mathematically represented as convolution of the input discrete-time signal with an output reconstruction filter [14]. This process results in a continuous time waveform at the output of the DAC. A typical reconstruction filter used by DACs is called the zero-order-hold (ZOH) filter, which holds the value of the output waveform according to the DAC's

978-1-4799-7598-3/15 $31.00 © 2015 IEEE

133

input for an entire clock cycle. The impulse response of such a ZOH reconstruction filter is given by

$$h_{ZOH}(t) = \frac{1}{T_s} rect(\frac{t - T_s/2}{T_s}),\qquad(1)$$

where T_s denotes the sampling time interval, and $rect(.)$ is the rectangular function[15]. The continuous-time signal reconstructed by using this filter is given by

$$x_{ZOH}(t) = \sum_{k=0}^{n-1} x[k] \cdot rect(\frac{t - T_s/2 - kT_s}{T_s}),\qquad(2)$$

where $x[k]$ denotes the discrete-time samples of the signal being generated, and n is the number of samples. As the ZOH data conversion is observed in the frequency domain (Equation (3)), the reconstructed continuous-time signal contains multiple spectral components even beyond the Nyquist frequency, which are known as spectral image replicas.

$$X_{ZOH}(j\omega) = \frac{1}{T_s} \sum_{n=-\infty}^{+\infty} X_B(j(\omega - \frac{2\pi n}{T_s})) \cdot \frac{2\sin(\omega T_s/2)}{\omega} e^{-j\omega T_s},\qquad(3)$$

Equation (3) also indicates that the higher order spectral components appear over multiple frequency locations (image replicas) with a *sinc* frequency response as shown in Figure 1(a).

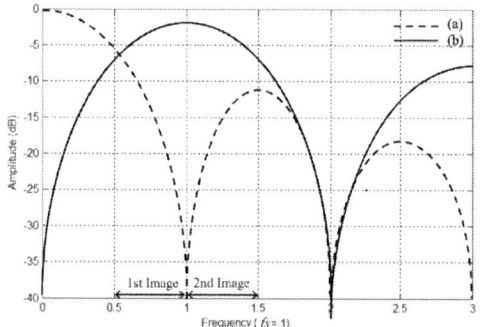

Fig. 1. Spectral response of various reconstruction filters for data converter: (a) ZOH filter (b) mix filter.

There exist other types of reconstruction filters that are integrated with commercially available DACs. The frequency domain representation of a mix reconstruction filter is presented below [16].
Mix filter:

$$H(j\omega) = \frac{\sin(\omega T_s/4)}{\omega} e^{-j\omega T_s/4} + \frac{\sin(\omega T_s/4)}{\omega} e^{-j\omega 3T_s/4}\qquad(4)$$

The spectral responses corresponding to the two different filters can therefore be calculated and are shown in Figure 1.

Since the mix reconstruction filter offers the least attenuation of the 1st and 2nd image replicas, it is selected as the reconstruction filter for the DAC used in the proposed method. For ease of understanding, Figure 2 shows how a time domain signal is generated using a DAC with ZOH vs. a mix output reconstruction filter.

III. PROPOSED INTERFERENCE GENERATION ARCHITECTURE

There are two limitations of the direct RF signal synthesis method proposed in [12]. The first involves the underlying computing complexity. Since a pseudo inversion needs to be performed on a matrix whose size is proportional to the number of sample points of the reference waveform, a reference waveform with a large number of sample points, which is essential for channel interference generation, leads to extensive computation. Second, the interleaved system is very sensitive to the frequency response of the power combiner used to combine the outputs of the interleaved DACs and requires careful calibration.

The proposed waveform generation method reduces the required computational cost by allowing input vector generation via analysis of matrices of smaller size than that proposed in [12] and simplifies the system to use of a single DAC. The proposed method is shown in Figure 3 below. Assuming that the discrete reference waveform y has M sample points obtained at a sampling frequency of $4f_{clk}$, it is divided into M/N sub-groups where N is the length of the matrix model used in the waveform synthesis algorithm and is discussed in the later section. Optimized $N/4$-dimensional vectors are then obtained through this process and re-grouped to form the input data sequence x for the DAC. The targeted band-limited signal (interferer) is then selected by a bandpass filter and used as the primary output of the system. To perform such input data generation, two matrix models, which mimic the waveform generation process and the output signal band-limiting process respectively, are required.

(a) Zero order hold (ZOH) reconstruction filter.

(b) Mix reconstruction filter

Fig. 2. Convolution of a discrete-time signal with various reconstruction filters.

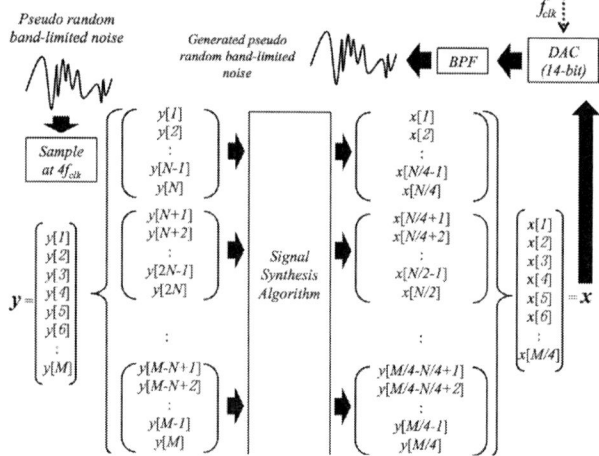

Fig. 3. Block diagram of proposed RF band-limited interference generation with a single DAC

1) System Matrix Model

The system matrix model is a matrix that maps the digital input data sampled at f_s to an output waveform sampled at $4f_s$. The matrix model for a DAC using the *mix* reconstruction filter is given by Equation (5). Note that 1 and -1 in the reconstruction filter matrix represent the normal and inverse sampling modes of the mix filter respectively (refer to Figure 2(b)). Each input data point $x[i]$, corresponding to a sampling frequency of f_s is mapped to four output values corresponding to a sampling frequency of $4f_s$.

$$H_{4n \times n} = \begin{bmatrix} 1 & 0 & \dots & 0 \\ 1 & 0 & \dots & 0 \\ -1 & 0 & \dots & 0 \\ -1 & 0 & \dots & 0 \\ 0 & 1 & \dots & 0 \\ 0 & 1 & \dots & 0 \\ 0 & -1 & \dots & 0 \\ 0 & -1 & \dots & 0 \\ \vdots & \vdots & \vdots & \vdots \\ 0 & 0 & \dots & 1 \\ 0 & 0 & \dots & 1 \\ 0 & 0 & \dots & -1 \\ 0 & 0 & \dots & -1 \end{bmatrix} \quad (5)$$

The input data sequence for a DAC is given by $x^T=[x[1]\ x[2]\ x[3]\ \dots\ x[n]]$. The output waveform ($y^T=[y[1]\ y[2]\ y[3]\ \dots\ y[4n]]$) sampled at 4 times the clock rate of DAC can be calculated by Equation (6) below.

$$y_{4n \times 1} = H_{DAC(4n \times n)} \cdot x_{(n \times 1)} \quad (6)$$

2) Band-limiting Matrix Model

The band-limiting matrix model is a matrix that removes unwanted frequency components in the digital sample set sampled at $4f_s$ and extracts the desired image replica at the higher than Nyquist desired frequencies. Due to the existence of unwanted image tones that are inherently present in the synthesized waveform, it is not feasible for the synthesized waveform to have a frequency spectrum that is properly matched with that of the reference waveform across a wide frequency band. For this reason, we apply a filtering matrix to the existing system matrix model to define the frequency range of interest (or in-band frequency range) within which the objective is to obtain accurate output signal reconstruction. This limitation is very critical for the proposed algorithm to operate efficiently.

Assume that the waveform obtained after application of the system matrix model is given by $y^T_{in} = [y_{in}[1]\ y_{in}[2]\ y_{in}[3]\ \dots\ y_{in}[4n]]$ (n is an integer and power of 2), The band-limiting matrix filter consists of three parts and is presented in Equation (7)

$$H_{filter(4n \times 4n)} = W^{-1}_{(4n \times 4n)} F_{(4n \times 4n)} W_{(4n \times 4n)}, \quad (7)$$

where W is the discrete Fourier transform (DFT) matrix [17], which converts the discrete-time domain signal y^T_{in} in the time domain to the frequency domain and W^{-1} is the inverse DFT matrix which translates the frequency-domain signal back to the time domain. F is a diagonal matrix that defines the frequency bands of interest to attenuate any unwanted signals that are present outside the band. If the band of interest is defined as $[f_1, f_2]$ where f_2 is larger than f_1, F is given as

$$F_{i,j} = \begin{cases} 1 & if \quad \dfrac{f_1}{f_s}n < i < \dfrac{f_2}{f_s}n \quad and \quad i=j \\ 0 & otherwise \end{cases} \quad (8)$$

Therefore, by performing the operation of Equation (9) below

$$y_{out(4n \times 1)} = H_{filter}y_{in(4n \times 1)}, \quad (9)$$

the discrete-time domain waveform y^T_{in} is converted into the discrete frequency domain and the signal power outside the pre-defined frequency bands (or in-bands) is attenuated by the matrix F and the signal is converted back to the time domain given by y^T_{out}. By multiplying the two matrix models, a complete matrix model for the waveform generation process within a limited bandwidth is obtained as in Equation (10).

$$y_c = H_{filter(4n \times 4n)} H_{DAC(4n \times n)} x_c = H_{total(4n \times n)} x_c, \quad (10)$$

Note that H_{DAC} is not square and its rank is not equal to $2n$. For this reason, the inverse of the matrix does not exist and the correct input value x_c can only be estimated through Equation (11) below:

$$x_c = pinv(H_{total(4nxn)})y_c, \quad (11)$$

where *pinv(.)* denotes the Moore–Penrose pseudoinverse [18].

IV. CONCEPT VALIDATION

The proposed wide-band spectrum shaping and channel interference synthesis technique is validated using numerical simulation for a band-limited channel interferer. In this section, the waveform synthesis process is first demonstrated for sinewave generation. Then, wide-band channel interferer generation is discussed. All the simulations are performed under the assumption that a 14-bit DAC with a mix reconstruction filter (as shown in Figure 2(b)) clocked at 2.5GHz is used.

A. Sinewave Generation

To demonstrate the proposed method, a time domain comparison between waveforms at different stages of the generation process is shown in Figure 3 and performed with a 3.2GHz sinewave as the reference signal.

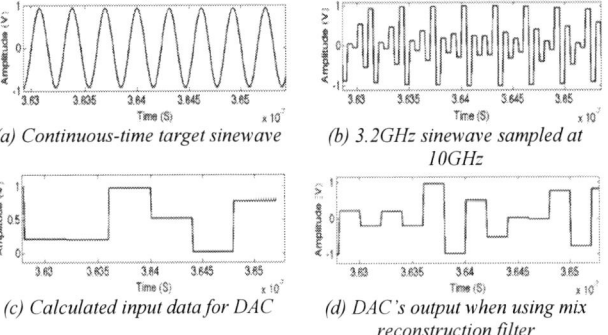

(a) Continuous-time target sinewave   (b) 3.2GHz sinewave sampled at 10GHz

(c) Calculated input data for DAC   (d) DAC's output when using mix reconstruction filter

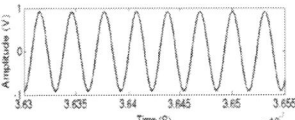

(e) DAC's output wave after filtering and normalization

Fig. 4. Waveforms at difference stages of the proposed method during sinewave generation.

In Figure 4, the original targeted continuous sinewave is shown in Figure 4(a). A sampling process is performed on the continuous-time target waveform at *10GHz* (4 times the clock rate of DAC) and all the sampling points are listed according to their sampling times in y (Figure 3). The optimized data subsequence x_c for the DAC is obtained through the transformation of y (as y_c) as shown in Equation 11. Combining all the relevant subsequences as shown in Figure 3, an optimized input data sequence x is obtained. The time domain representation of each data point in x is shown in Figure 4(c), where a ZOH reconstruction filter is used to convert the discrete data points into the corresponding time domain waveform. With this input data sequence x into the DAC, the output waveform generated with a mix reconstruction filter is shown in Figure 4(d). As shown in the figure, the output value for each data point changes at the last half sampling period. By passing this generated waveform through a bandpass filter, the unwanted fundamental and image replicas are removed and the obtained waveform is shown in Figure 4(e), and is identical to the target waveform. Note that the amplitude of the generated waveform is normalized and does not reflect its real value.

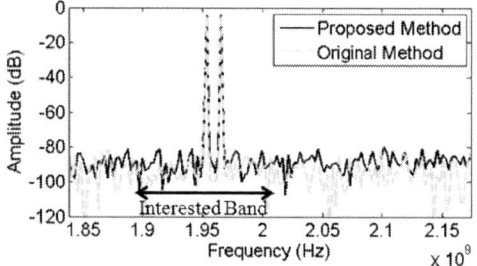

Fg. 5. Multi-tone Signal Generation

Figure 5 shows a multi-tone signal generated by the method proposed. The two tones are at 1.953GHz and 1.965GHz and are generated by a single 10-bit DAC clocked at 2.5GHz. Since the image tones are directly used as the primary output, there is no nonlinear distortion observed. Compared to the signal generated by the method proposed in [12], the in-band noise floor is slightly higher but the complexity of the signal generation system has been largely reduced.

B. *Wide-band Channel Interference Generation*

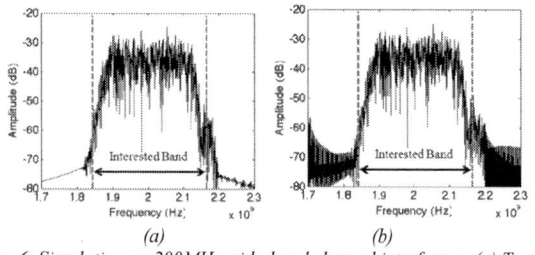

(a)   (b)

Fig. 6. Simulation on 200MHz wide-band channel interference(a) Target spectrum (b) Generated spectrum

To demonstrate the proposed method's ability to generate complex channel interferers, a reference waveform (interference with specific characteristics) with 20480 sample points is used. The size of the matrix model of the system is 4096x1024. To fit the long reference waveform, it is further divided into 5 sub-groups with 4096 sample points each, as explained previously.

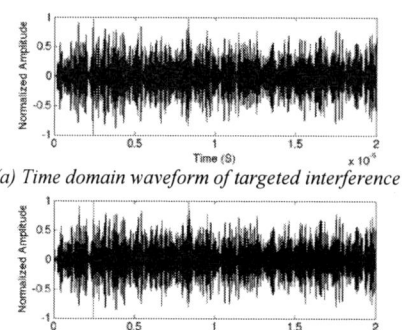

(a) Time domain waveform of targeted interference

(b) Time domain waveform of generated interference

Fig. 7. Time domain comparison between generated waveform and targeted one

Figure 6 compares the frequency domain representation of the generated band-limited channel interferer (right) with the target interferer (left). The attenuation and distortion within the pre-defined bandwidth (1.9GHz to 2.1GHz) has been minimized by the proposed method. Also, the time domain comparison is presented in Figure 7. Note that this high frequency interferer is generated by a single DAC clocked at 2.5GHz and the first image replica is used as the primary output signal.

V. EXPERIMENTAL RESULTS

Two experiments have been conducted as proof of concept of the proposed signal/interferer generation method. In the first experiment, a band-limited signal is generated directly using the proposed method and compared against one that is generated using an up-conversion mixer. The second experiment demonstrates how this method can be used in adaptive RF system testing.

A. *"BATMAN" Signal Generation*

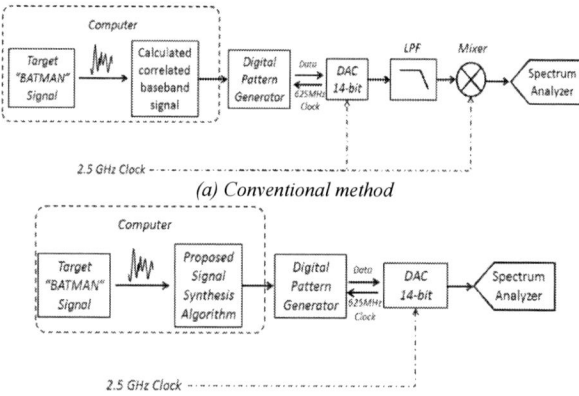

(a) Conventional method

(b) Proposed method

Fig. 8. Experimental setups for (a) Conventional method (b) Proposed method.

Since using a mixer to up-convert a baseband signal to radio frequency is commonly used for high-speed signal generation, this experiment was set up to demonstrate that the RF waveform generated using the proposed method suffers significantly less distortion as compared to one generated using an up-conversion mixer. The test setups for both methods are presented in Figure 8.

978-1-4799-7598-3/15 $31.00 © 2015 IEEE   136

To demonstrate the benefits of the proposed approach on a wideband RF waveform, a signal with sharp band-limited spectrum is used as the reference waveform. As shown in Figure 9, the signal has a bandwidth of 200MHz and is centered at 2GHz. Due to the shape of the spectrum, this target waveform is referred as the "BATMAN" signal for future reference.

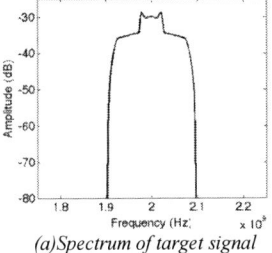

(a)Spectrum of target signal

In Figure 10(a), the baseband of the target "BATMAN" signal was calculated and used as the digital input to the DAC (AD9739a, operated with ZOH reconstruction filter). This baseband signal was mixed with a 2.5GHz carrier wave and the spectrum of the up-converted waveform is shown in Figure 10(b) while the spectrum of the baseband signal is shown in Figure 10(a).

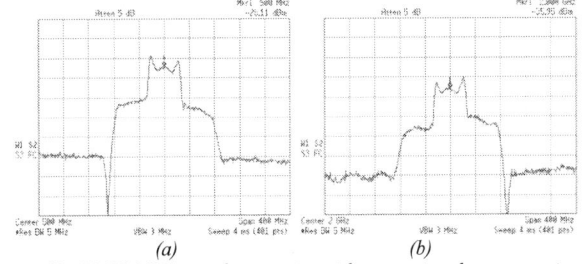

(a)   *(b)*

Fig. 10. "BATMAN" signal generation with conventional up-converting method: (a) Spectrum of baseband signal (b) Up-converted signal after mixer

In the proposed method, the target waveform is converted into digital samples at a sampling rate of 10GHz. These samples are then used as the reference waveform y_c in Equation (11). Note that during the experiment for the proposed method, the DAC was operated with a mix reconstruction filter. The spectrum of the generated waveform is shown in Figure 11.

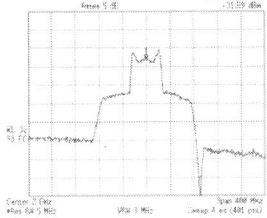

Fig. 11. Spectrum of signal generated through proposed approach

By comparing the up-converted signal in Figure 10(b) and the directly synthesized one in Figure 11, distortions were clearly observed in the up-converted signal using a mixer. This experiment demonstrates that the proposed signal generation approach produces signals with less distortion than existing schemes that use mixer-based signal up-conversion.

B. Application in Channel Interference Generation for Adaptive System

To demonstrate the proposed channel interferer generation technique we observe its effect on the performance of a SISO front-end as shown in Figure 12. The OFDM base-band signal generated by the PC and PXI 5412 DAC is upconverted to a center frequency of 2 GHz by MAX2039 upconversion mixer passed the channel with interferer added, downconverted by HMC687LP4 downconversion mixer and captured by PXI5105 digitizer. The interferer generation is performed as per the block diagram shown in Figure 13. By using different simulated channel interferers as target waveforms, different channel conditions were emulated.

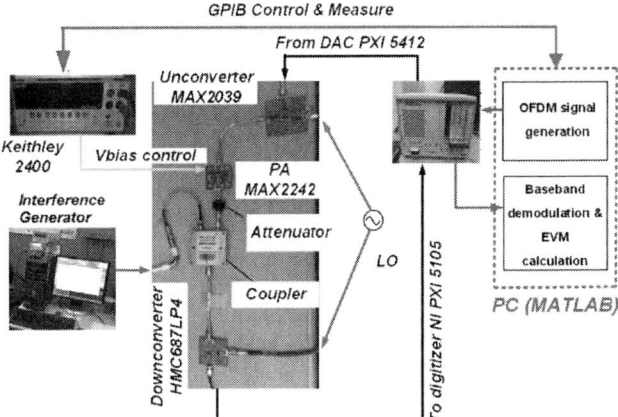

Fig. 12: Transmitter adaptation hardware setup

Figure 14 shows the spectrum of the generated band-limited interferer.

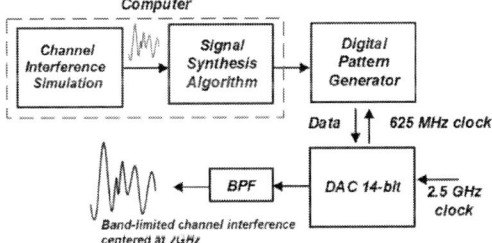

Fig. 13: Band-limited interference generation setup

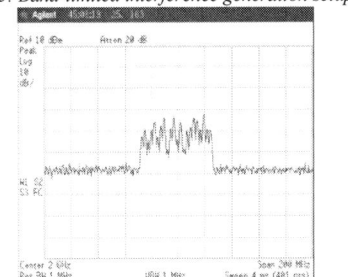

Fig. 14: Band-limited interference spectrum

The interferer generation setup injects variable amounts of interference into the system to create different channel conditions. Depending upon the channel condition (or interference level encountered) the performance of the system changes. This is observed as an increase in EVM (and correspondingly BER) for the system. The degradation in EVM with injected interference strength is shown in Figure 15 for the nominal transmitter and also the degraded transmitter (consumes less power but worse performance). As clearly shown the performance varies with the degraded emulated channel (more interference) allowing the interference generation algorithm to be suitable for the testing of the performance of such systems.

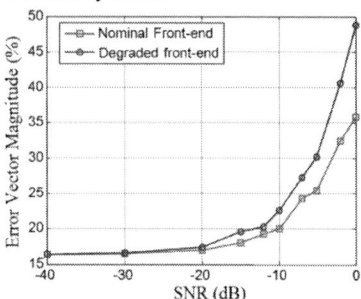

Figure 15: Variation of EVM with increasing amount of interference added to the channel

VI. CONCLUSION

This paper proposes a DAC based wide-band signal/interferer generation method. This method optimizes the input data sequence to the DAC to minimize the attenuation and distortion of a specific image replica which is used as the primary output signal. It is possible therefore, to generate higher than Nyquist signals using a single DAC with a "mix" output reconstruction filter. It is shown that the signal generation approach has less distortion as compared to methods which use up-conversion mixers for generating RF signals. Also, an application to interference testing of adaptive RF systems is demonstrated.

ACKNOWLEDGMENT

This research was supported in part by Samsung and by NSF Grant No. ECCS 1407542.

REFERENCES

[1] Nee, Richard van, et al. "OFDM for wireless multimedia communications". Artech House, Inc., 2000.

[2] Sun, Jian, et al. "Frame Detection of OFDM System with periodic pattern preamble." Communications and Mobile Computing (CMC), 2010 International Conference on. Vol. 2. IEEE, 2010.

[3] Balakrishnan, Jaiganesh, et al. "A multi-band OFDM system for UWB communication." Ultra Wideband Systems and Technologies, 2003 IEEE Conference on. IEEE, 2003.

[4] Yucek, Tevfik, et al. "Interference plus interference power estimation in adaptive OFDM systems." Vehicular Technology Conference, 2005. VTC 2005-Spring. 2005 IEEE 61st. Vol. 2. IEEE, 2005.

[5] Debaillie, B. Bougard et. al., "Energy-scalable OFDM transmitter design and control", 43rd IEEE Design Automation Conference, July 24-28, pp. 536-541.

[6] Sen S., et al. "A Power-Scalable Channel-Adaptive Wireless Receiver Based on Built-In Orthogonally Tunable LNA," in IEEE Transactions on Circuits and Systems 1, vol. 59, no. 5, 2012.

[7] Debashis Banerjee et. al., "Real-Time Use-Aware Adaptive MIMO RF Receiver Systems for Energy Efficiency Under BER Constraints", 50th Design Automation Conference, 2013.

[8] Debashis Banerjee et. al., "Low-Power Adaptive RF Systems using Real-time Fuzzy Interference-Distortion Control," in IEEE International Symposium on Low Power Electronics and Design (ISLPED 2012).

[9] Meghdadi, M, et al. "Two-Dimensional Multi-Parameter Adaptation of Interference, Linearity, and Power Consumption in Wireless Receivers," Circuits and Systems I: Regular Papers, IEEE Transactions on , vol.PP, no 99, pp.1,11.

[10] Shreyas Sen et. al., "Pro-VIZOR: Process Tunable Virtually Zero Margin Low Power Adaptive RF for Wireless Systems," in ACM/IEEE 45th Design Automation Conference 2008

[11] Sen, S, et al. "Environment-Adaptive Concurrent Companding and Bias Control for Efficient Power-Amplifier Operation," Circuits and Systems I: Regular Papers, IEEE Transactions on , vol.58, no.3, pp.607,618, March 2011

[12] Wang, Xian, et al. "Higher than Nyquist test waveform synthesis and digital phase interference injection using time-interleaved mixed-mode data converters." Test Conference (ITC), 2012 IEEE International. IEEE, 2012.

[13] Brigati, Simona, et al. "An 0.8-µm CMOS mixed analog-digital integrated audiometric system." Solid-State Circuits, IEEE Journal of 34.8 (1999): 1160-1166.

[14] Taleie, S. Mehdizad, et al. "A Bandpass/spl Delta//spl Sigma/RF-DAC with Embedded FIR Reconstruction Filter." Solid-State Circuits Conference, 2006. ISSCC 2006. Digest of Technical Papers. IEEE International. IEEE, 2006.

[15] Bracewell, R. "Rectangle Function of Unit Height and Base, ." In The Fourier Transform and Its Applications. New York: McGraw-Hill, pp. 52-53, 1965.

[16] Chen, S. S, et al. (2008, October). Multi-mode sub-Nyquist rate digital-to-analog conversion for direct waveform synthesis. In Signal Processing Systems, 2008. SiPS 2008. IEEE Workshop on (pp. 112-117). IEEE.

[17] Winograd, Shmuel. "On computing the discrete Fourier transform." Mathematics of computation 32.141 (1978): 175-199.

[18] G. Golub and C. van Van Loan, "Matrix Computations" (Johns Hopkins Studies in Mathematical Sciences), 3rd Edition, The Johns Hopkins University Press, 1996

Automated Testing of Mixed-Signal Integrated Circuits by Topology Modification

Anthony Coyette[1], Baris Esen[1], Ronny Vanhooren[2], Wim Dobbelaere[2] and Georges Gielen[1]

[1] Department of Electrical Engineering, KU Leuven,
Kasteelpark Arenberg 10, 3001 Leuven, Belgium
{anthony.coyette, georges.gielen, baris.esen}@esat.kuleuven.be

[2] ON Semiconductor Belgium,
{ronny.vanhooren, wim.dobbelaere}@onsemi.com

Abstract—A general method is proposed to automatically generate a DfT solution aiming at the detection of catastrophic faults in analog and mixed-signal integrated circuits. The approach consists in modifying the topology of the circuit by pulling up (down) nodes and then probing differentiating node voltages. The method generates a set of optimal hardware implementations addressing the multi-objective problem such that the fault coverage is maximized and the silicon overhead is minimized. The new method was applied to a real-case industrial circuit, demonstrating a nearly 100 percent coverage at the expense of an area increase of about 5 percent.

Keywords—Design-for-Testability, controllability, observability, low-overhead, co-optimization.

I. INTRODUCTION

The electronic world is evolving towards systems with a higher complexity and with increasing quantities of integrated circuits (ICs). For example, automotive chips are built into systems that contain hundreds of other electronic components and chips. As a consequence, the probability of a failing system increases due to the multiplication of the different defectivities. Also, applications such as the bio-medical or aerospacial ones require critical quality of the chips. In both cases a demand emerges and becomes stronger to acquire test techniques allowing to reduce and assess the defect level in the shipped products.

Research in digital ICs testing has led to defect level reaching the part per billion (ppb) range thanks to different advancements. First of all, the functional tests aiming at verifying the circuit performances and functionality were replaced by structural tests. These techniques focus on alternative features helping to match structurally the circuit under test with its design without considering the functionality. Next, progress was made by developing the concepts of controllability and observability. Algorithms such as PODEM [1] combining the two aspects form the basis of the efficient way to test digital ICs in the present industry.

In parallel, research on the testing of analog and mixed-signal ICs has tried to reiterate the digital success. Controllability has been studied in many perspectives. In [2] circuits are considered as black-boxes. The stimuli set is composed of sine waves resulting as the solution of a search problem. By including a generic insight about analog circuits [3] proposes to ramp up the power supply of the circuits under test to make some faults observable. In [4], specific information about the circuit under test is exploited thanks to the development of Testability Transfer Factors (TTF) to find stimuli maximizing the fault coverage. Finally, as for the digital case, controllability can be improved by Design-for-Testability (DfT) circuitry. In [5] a Built-in-Self-Test (BIST) technique creates a feedback system in order to make the circuit oscillate.

Similarly, observability offered by the outputs of the circuits has first been exploited [6]. Thanks to DfT, access to the internal nodes was gained and optimized. The choice of the optimal node voltage set, referred as test point selection, has been studied for electronic circuits in works such as [7] [8]. Non-intrusive techniques have been developed to enhance the observability as in [9].

However, a generic solution should result from a co-optimization combining observability and controllability in a flexible way. This has been presented in [10] [11]. In the same way as for digital testing, scan-chains have been added to observe node voltage and inject signals at specific nodes. But these solutions suffer from the need to open the signal path which is a practice considered as a bad practice among analog designers. Furthermore the problem of the signal to be injected in the circuit stays unsolved. [12] uses current branches and does not require the opening of signal paths. However, like the other scan-chain techniques previously cited, it suffers from a relatively high hardware overhead.

In this paper, a new generic method is introduced to enhance the controllability and the observability of mixed-signal circuits. Simple DfT building blocks with small hardware overhead are added and combined to improve the fault coverage. The main idea is to reconfigure the architecture of the circuit instead of injecting signal on nodes. This technique does not require the opening of the signal path. First, an overview of the fault-oriented methodology used to assess the efficiency of the method is addressed in Section II. Then, a general method enhancing the controllability and the observability of mixed-signal circuits is outlined in Section III. Afterwards, a possible hardware implementation is given in Section IV. Results for

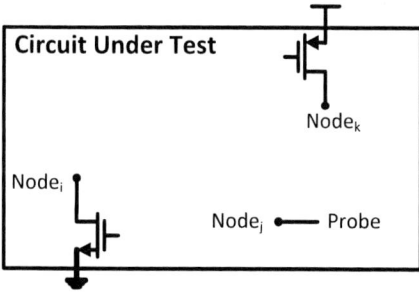

Fig. 1. Building blocks of the method.

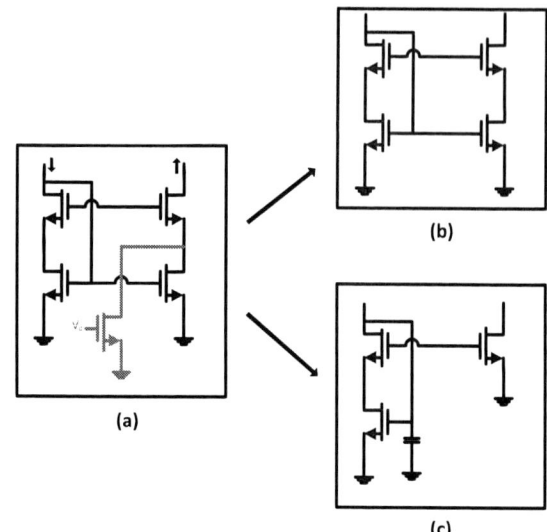

Fig. 2. Illustration of a topology modification:
(a) Current mirror with a Pull Down transistor.
(b) Topology in normal mode (V_c=0) : C_0.
(c) Topology in test mode (V_c=VDD) : C_1.

an industrial case study demonstrate the effectiveness of the method in Section V. Finally, conclusions are drawn in Section VI.

II. FAULT-ORIENTED TESTING

Given an analog or mixed-signal circuit C_0, the defect-oriented methodology proposes to model the possible defects occurring at the transistor level and simulate them [13]. The flow starts with the generation of a list of the physical defects likely to appear in the manufactured chips. In the scope of this work, the faults modelling the defects are based on the schematic, but it could also start from the layout. In general, defects are categorized as catastrophic or parametric. The former emerges from a problem in the manufacturing process such as a dust particle or an over-etching. This causes a definitive change in the circuit i.e. shorts and opens. The latter results from the imperfect control of the process-voltage-temperature (PVT) conditions which may lead to tolerances causing certain devices to shift outside their specification range.

The present work focuses on applications using well mastered technologies (above 100nm) where parametric defects can be neglected to first order. Therefore, only catastrophic defects are studied. Furthermore, since the test circuit illustrating the presented method in Section V is designed in a technology using BJTs and MOSFETs, two different fault models often encountered in literature [14] are implemented. The fault model for the BJTs is a 6-fault model assuming that a short can happen between any pair of its three terminals and each terminal can be open. The model for the MOSFETs is a 5-fault model making the same assumptions as the 6-fault one. However the open gate fault is excluded because of the absence of an appropriate model for the DC simulations on which the presented technique is based. In both cases, the values of 100Ω and 1TΩ are used for modelling the short and open circuits respectively.

Based on the list of faults $L = \{F_1, ... F_D\}$, a list of possible faulty circuits is created. The assumption is made that two defects cannot occur in the same circuit. Hence, each fault is injected separately in the original circuit to create a faulty circuit. A circuit possessing D possible defects leads then to a set of D+1 circuits, i.e. the original circuit and the D faulty ones. Each of these D+1 circuits is simulated in SPICE in the presence of process variations. Signatures are extracted to distinguish the faulty cases from the good case. These signatures can be as simple as the measurement of the

current consumption or of a node voltage. It is also possible to use signatures such as a Fast-Fourier-Transform or more sophisticated scheme based on measured transient signals [15].

In the scope of this work, the exploited signatures are node voltages coming from DC simulations in the presence of process variations. Therefore, each of the D+1 circuits delivers a distribution which is assumed to be gaussian and represents the span of possible values for the voltage on the considered node. To assess the fault coverage, the signature of each of the D faulty circuits is compared to the signature of the original circuit. A fault F_i is considered as covered if its distribution $\mathcal{N}(\mu_{F_i}, \sigma_{F_i})$ is distant from the fault-free distribution $\mathcal{N}(\mu_G, \sigma_G)$ by at least 10mV and 3σ, where $\sigma = max(\sigma_G, \sigma_{F_i})$.

III. TEST SCHEME

As in the case of digital testing, and already applied in [10], the main idea is to enhance the controllability and observability simultaneously in a co-optimization. DC voltages are probed from internal nodes and primary output while the circuit is forced into different topologies. This technique distinguishes itself from existing schemes using the control offered by the circuit inputs or the internal injection of a signal by opening the signal path. The stimuli come from the modification of the circuit topology thanks to small and generic building blocks.

Instead of having a set of waveform as search space to optimize the fault coverage, the problem is transformed into a search for new topologies activating faults which were unobservable in the original circuit. Nevertheless, it is worth noting that the inclusion of the circuit inputs in the optimization system would lead to even more efficient solutions in terms of hardware overhead. This point will be illustrated in the study case of Section V.

The concept is first explained in details and a test procedure is proposed. Then, the method is expressed as an optimization

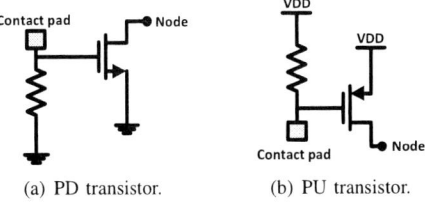

(a) PD transistor. (b) PU transistor.

Fig. 3. Control of PXs at the wafer level.

system to solve. Finally, possible improvements on the method are presented.

A. Topology Modification Approach

As introduced before, topology modifications are introduced into the circuit. These modifications aim at transforming the original circuit C_0 into new circuits having different behaviors. By operating this reconfiguration faults which were undetectable are made observable. This observability is based on the probing of voltages. While generally, for integrated circuits, only the input and output signals are assumed to be accessible [16], the internal nodes are supposed to be measurable through extra probes in the scope of this work. This hypothesis is supported by the hardware implementation proposed in Section IV and other works such as [17].

In summary, the test procedure can be summarized to a set of topologies $C_1, ..., C_k$ for which sets of circuits nodes are assigned $S_1, .., S_k$. Testing the circuit consists in applying each topology modification C_i, probe the corresponding set of nodes S_i and test these measurements against the corresponding decision threshold Th_i.

In this paper, topology modifications are realized by connecting nodes to the ground or the power supply. This is made possible by adding pull-down (PD) or pull-up (PU) transistors as illustrated in Figure 1. In the following, when the distinction between PD and PU is not essential, the discussion will be generalized by the use of the terminology PX to designate either a PD or a PU.

Figure 2 gives an illustration of the topology modification mechanism on a current mirror. The circuit originally consists of 4 transistors which are represented in black in Figure 2 (a). The gray nMOS transistor is a PD transistor that will operate the topology modification during the test mode. In normal mode the voltage imposed on the gate of this transistor is grounded. With a V_{GS} of 0V the transistor is in its cut-off region and the circuit C_0 is a normal current mirror as illustrated in Figure 2(b). In order to test the circuit, a set of node voltages S_0 is first measured in the original circuit. These values are compared to the ones expected from the simulations. Then, the PD transistor is activated such that the circuit topology changes to become the circuit C_1 seen in Figure 2(c). Another set of node voltages S_1 is measured and compared to the values expected from the simulations.

B. Optimization problem

In order to apply the proposed test procedure, the set of nodes to control (i.e. pull up or down) and the sets of node voltages to measure for each topology has to be calculated.

These sets come as the solution of the optimization system developed in the following.

Given a circuit C_0, its set of internal nodes is labeled by T. The selection of nodes to control is operated on the sub-set $N \subset T$. This pre-selection is done due to the large number of nodes present in industrial designs. In this work, the pre-selected nodes are the ones surrounding the transistors of the tested circuit with the exception of the digital gates. Other criteria can be added to refine or extend this pre-selection step. For instance, nodes that are extremely sensitive to parasitics can be excluded from the search set.

Each node contained in N leads to two possible topology modifications : one where the node is pulled up and one where the node is pulled down. This makes that $2\|T\| + 1$ circuits are finally considered i.e. the $2\|T\|$ topology variations and the original circuit, where $\|T\|$ designates the cardinality of the set T. For each of these, the fault-free and the D faulty circuits are simulated in the presence of process variations.

The second step after the simulations is to identify for each of the $2\|T\| + 1$ topologies which nodes allow to discriminate the good circuit from the faulty circuits. As said in Section II, for a topology C_i and a node $n \in N$, the fault F_i is considered detected if the distribution of simulated voltage for the good circuit and the faulty circuit are separated by at least 10mV or 3σ. The final results of these simulations is summarized in $2\|T\| + 1$ fault coverage vectors. Each vector contains D boolean values indicating for each fault if it is detected or not.

The third step consists in selecting which topologies should be used and which nodes should be probed. The problem of test points selection has already been addressed and solved in many works such as [7] [8]. More specifically, a co-optimization of the input stimuli and the selection of test points is formulated in [18]. However, instead of posing the optimization problem as a maximization of the fault coverage or a minimization of the tests set as it is usually done, the problem is set as a multi-objective optimization :

$$\max_{\substack{m \in \mathcal{P}(N) \\ p \in \mathcal{P}(T)}} [FC_{co}(m,p), -\|p\|, -\|m\|]$$

where $FC_{co}(m,p)$ is a function computing the fault coverage when the nodes contained in m are controlled and the ones in p are probed with the output signals and $\mathcal{P}(X)$ designates the power set of X i.e. the set containing all the subsets of X. The present formulation is expressed in terms of size of node sets in its mathematical terms. However, once the building blocks are defined and designed as it is done in the next section, the optimization systems can be expressed in terms of silicon area.

This two-objective function aiming at maximizing the fault coverage and minimizing the silicon overhead is solved for the study case in the next section using a genetic algorithm described in literature as NSGA-II [19]. This tool is well suited for managing the search of Pareto optimal solutions in big search spaces.

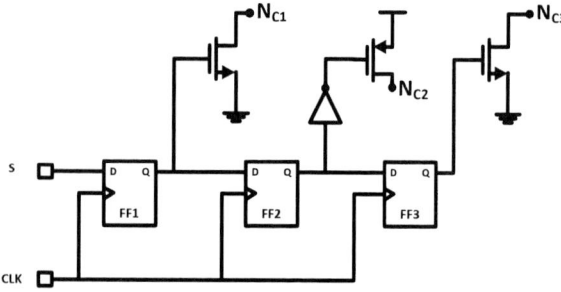

Fig. 4. A daisy-chain of flip-flops controlling the PXs.

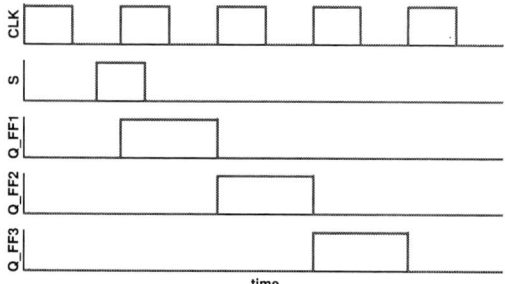

Fig. 5. Transient signals in the daisy-chain of Figure 4.

C. Extension

This technique forms a basis which can be further improved in controllability and observability. First, for the controllability, it should be noted that the PD and PU transistors are actually a particular case of topology modification. A more general approach consisting in connecting two nodes together could also be used at the expense of a rising computational complexity. This has not been studied in the scope of this paper, but it should be highlighted that this idea can also be seen as a generalization of the oscillation-based methods such as [5]. Indeed, these methods induce a feedback in a circuit in order to make them oscillate and a feedback connection is a particular case of the connection of two nodes. Furthermore, it is also worth noticing that the PXs are individually used. A finer optimization consisting in using several PXs in combination could lead to better results. However the complexity of such a method increases as the factorial of the number of nodes.

For the observability, this paper focuses on the probing of node voltages but other approaches can be used to improve the performances. Current monitoring systems were already studied in [20] and applied in control and observation structure (COS) in [12].

Finally, the method is presented for a set of DC measurements, but it should be noted that the same method can be exploited for transient signals. If a ramp is applied on the gate of a PX transistor the transition between the two topologies can be measured on an internal node. However, even if the question of dealing with transient signals can be addressed in a theoretical form, the hardware implementation rises technical issues.

IV. PROPOSED VLSI IMPLEMENTATIONS

In the previous section the basic building blocks and optimization method were explained. In this section practical solutions are presented to control the PX transistors and observe the selected node voltages. The basic hardware solutions which are proposed intend to demonstrate the feasibility and rely on existing work [10] [17]. These blocks can be enhanced to fulfill the resolution required by the applied detection mechanism.

It is worth noting that even though the case study developed in the next section is designed for a BCD technology, the proposed building blocks consists only of MOSFETs. This choice was taken due to the omnipresence of the CMOS technology and the aim of obtaining a technique tackling a large panel of different technologies.

A. Controlling

Industrial systems have a rapidly increasing complexity and are commonly composed of multiple sub-circuits organized hierarchically. Results have shown that a few PXs are required for each sub-circuit. The issue of controlling these elements independently rises rapidly since one pin cannot be assigned to each of them.

1) Wafer-Level Test: At wafer level, the routing of the signals controlling the gates of the PXs can be kept very small by introducing contact pads inside the circuit perimeter as illustrated in Figure 3. In the case of a PD transistor, the gate can be connected to GND through a resistor. Thanks to the contact pad connected to the gate an automated test equipment (ATE) can control directly the transistor. In normal mode, the voltage applied on the gate is the ground. With a V_{GS} equal to 0V the PD transistor is in its cut-off region and does not affect the circuit. When the wafer is under test, the ATE can impose a voltage on the gate. This allows the transistor to go into its active region and the desired topology modification takes place.

2) Packaged Device Test: In the case of a production test taking place on packaged dies, no direct access can be granted by contact pads. The controlling signal has to be routed from outside the ICs through a pin. In order to minimize the routing needed, a solution already used in digital and analog scan-chain [10] is proposed and illustrated in Figure 4.

In this approach only two interconnect lines are needed : one for the clock signal CLK and one for the control signal S. A daisy-chain of flip-flops can scan the control signal and the PX transistors can be activated successively. This mechanism is illustrated in Figure 5.

B. Observing

As the proposed technique not only relies on output signals from the circuits but also exploits internal node voltages, circuitry has to be added to obtain the access.

1) Wafer-Level Test: At wafer level, since basic DC measurements are required, the ATE can directly probe the node voltages through a contact pad added during the design step.

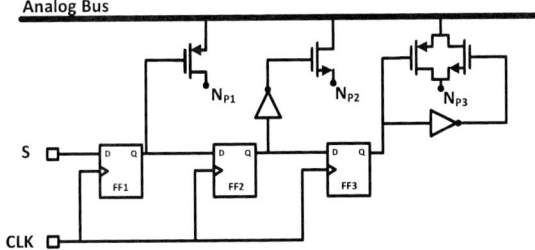

Fig. 6. A daisy-chain of flips-flops controls the pass gates.

Fault Coverage (percent)	Number of probes	Number of PXs
91.4	8	4
90.7	7	3
90	6	2
89.2	5	2
87.9	4	2
...	...	...
82.9	2	2
...	...	...
47.9	4	0
46.4	3	0
42.1	2	0
35	1	0
26.4	0	0

TABLE I.   FIRST AND LAST 5 OPTIMAL SOLUTIONS IN THE PARETO FRONT.

2) Packaged Device Test: In the case of the packaged dies, the proposed solution for the probing of the internal nodes makes also use of daisy-chains scanning configuration bits in the DfT circuits. Figure 6 illustrates an example of a chain which involves the three possible configurations. A combination of these three different cases is needed because of the limited driving capacity of pMOS transistors when used as pass gates. A voltage in the range $[\text{VDD} - V_{th}; \text{VDD}]$ can not be transmitted since V_{GS} forces the transistor in its subthreshold region, where V_{th} is the threshold voltage of the transistor and VDD is the supply voltage of the circuit. The same limitation is encountered with nMOS transistors in the band $[0; V_{th}]$.

Therefore, if the set of DC voltages that have to be measured from a node of the circuit stays in the band $[0; \text{VDD} - V_{th}]$ (resp. $[V_{th};\text{VDD}]$) a pMOS (resp. nMOS) is enough. In contrast, if the full range $[0; \text{VDD}]$ is required for a node, a full transmission gate is required to connect the node to the analog bus.

V. EXPERIMENTAL RESULTS

The proposed method has been applied to an industrial mixed-signal circuit designed using the 0.35μm I3T50 BCD technology provided by ON Semiconductor. Figure 7 shows the schematic of the Power-On-Reset (POR) circuit which will serve as case study. This circuit consists of an analog part and a digital part, containing in total 4 BJTs and 30 MOSFETs. This circuit possess one primary output and no primary input. It is

an essential building block which keeps the chip in power-down mode as long as the voltage supply is too low. It is recognized as a hard-to-test circuit because of the Schmitt trigger it contains. Furthermore, faults on transistors like the 3 pMOS connected to the Power Saving Mode signal (PSM) are difficult to detect because they do not participate to the function of the POR.

First, the fault models introduced in Section II were applied on the schematic to generate a full list of the possible faults. The redundant cases (i.e. two different faults leading to the same simulation) were then removed and the test scheme proposed in Section III was carried out for the circuit with a list of 140 faults.

Table I shows the 5 first and the 5 last entries from the set of Pareto optimal solutions for the problem. An example of a hardware implementation is illustrated in Figure 7. The proposed hardware DfT is added in gray and allows to detect 116 faults with 2 PXs, 2 internal probes and 1 output probe.

Among the set of optimal solutions, two extremes can be found and show the importance of the controllability/observability co-optimization. The first extreme is the case where no PX is used. This observe-only solution covers 69 faults at the price of 5 probes, which is twice less than the 128 faults covered by the highest-coverage solution offered in the set of optimal solutions. This improvement demonstrates the limitations of an observability-only and the added value brought by the topology modification. The second extreme is the case where only PXs are used while the output signal is observed. In this case, 37 faults are covered proving the importance of enhancing the observability. To conclude, the three cases presented above show the importance to simultaneously enhance and co-optimize the controllability and observability of the tested circuits.

The analysis of the 12 remaining uncovered faults reveals that 10 of them are located in the digital gates at the end of the circuit. The lack of coverage for these gates emerges from the insufficient control on the digital input signals A and B. This results from the exclusion of the digital nodes and can be fixed by including the nodes of the digital gates in the set of usable nodes. This has not been done in this work since it is believed that the optimal solution should come from the combination of this method and a path sensitization technique such as typically used in digital test techniques.

Finally, the overhead created in terms of silicon area can be estimated. If the case where 128 faults are detected is considered, Table I indicates that 4 PXs are required. If it is chosen to implement this at package level, the proposed solution is largely dominated by the area of the flip-flop since the transistors for the pass gates and PXs are minimal size. Based on industrial designs, it can be estimated that the implementation of this solution would require an increase in silicon area of about 5 percent for the example POR circuit.

VI. CONCLUSION

A structured and automated method has been presented to address the problem of detecting faults in analog and mixed-signal integrated circuits. This was done in a generic way by replacing the traditional problem of finding a set of input

978-1-4799-7598-3/15 $31.00 © 2015 IEEE

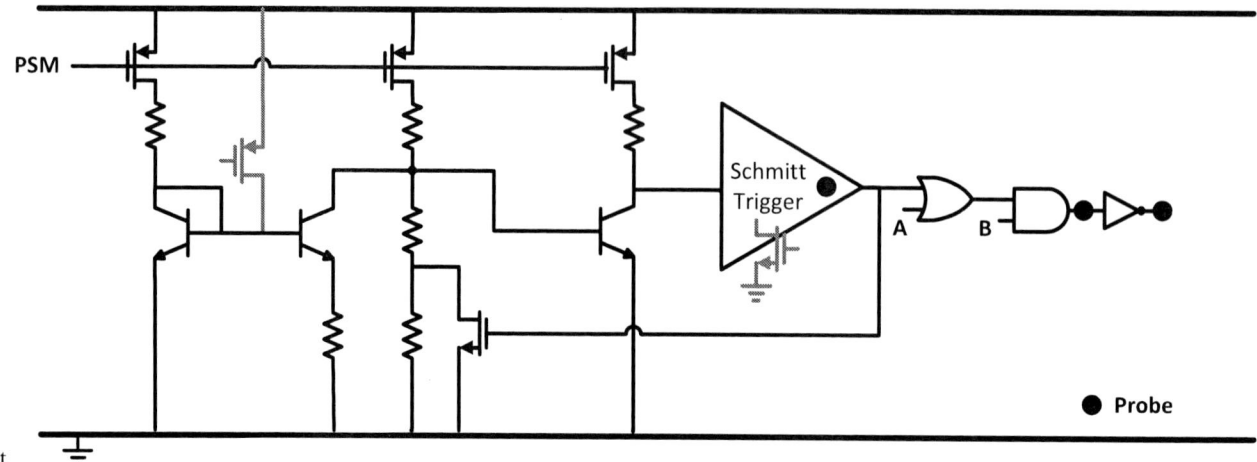

Fig. 7. Schematic of the Power-On-Reset with extra DfT blocks indicated.

signals allowing to control a circuit and make its faults detectable. Instead, a simple set of DC measurements combined to different circuit topology modifications are required. With different topologies, different faults are made observable.

Low-overhead hardware implementations were presented at wafer and package-levels. These allow to make the circuit under test adopt successively different topologies and probe DC node voltages during the test mode. Based on these building blocks, it was showed for an industrial Power-On-Reset circuit that nearly all the faults can be detected at the expense of around 5 percent of area overhead.

REFERENCES

[1] P. Goel, "An implicit enumeration algorithm to generate tests for combinational logic circuits," *Computers, IEEE Transactions on*, vol. 100, no. 3, pp. 215–222, 1981.

[2] N. Nagi, A. Chatterjee, A. Balivada, and J. A. Abraham, "Fault-based automatic test generator for linear analog circuits," in *Proceedings of the 1993 IEEE/ACM International Conference on Computer-aided Design*, ser. ICCAD '93. Los Alamitos, CA, USA: IEEE Computer Society Press, pp. 88–91.

[3] A. Zjajo, H. Bergveld, R. Schuttert, and J. de Gyvez, "Power-scan chain: design for analog testability," in *Test Conference, 2005. Proceedings. ITC 2005. IEEE International*, Nov 2005, pp. 8 pp.–83.

[4] M. Soma, S. Huynh, J. Zhang, S. Kim, and G. Devarayanadurg, "Hierarchical atpg for analog circuits and systems," *Design Test of Computers, IEEE*, vol. 18, no. 1, pp. 72–81, Jan 2001.

[5] K. Arabi and B. Kaminska, "Oscillation built-in self test (obist) scheme for functional and structural testing of analog and mixed-signal integrated circuits," in *Test Conference, 1997. Proceedings., International*, Nov 1997, pp. 786–795.

[6] N. Hamida and B. Kaminska, "Analog circuit testing based on sensitivity computation and new circuit modeling," in *Test Conference, 1993. Proceedings., International*, Oct 1993, pp. 652–661.

[7] J. A. Starzyk, D. Liu, Z.-H. Liu, D. E. Nelson, and J. O. Rutkowski, "Entropy-based optimum test points selection for analog fault dictionary techniques," *Instrumentation and Measurement, IEEE Transactions on*, vol. 53, no. 3, pp. 754–761, 2004.

[8] H. Luo, Y. Wang, H. Lin, and Y. Jiang, "A new optimal test node selection method for analog circuit," *Journal of Electronic Testing*, vol. 28, no. 3, pp. 279–290, 2012.

[9] L. Abdallah, H. Stratigopoulos, S. Mir, and J. Altet, "Defect-oriented non-intrusive rf test using on-chip temperature sensors," in *VLSI Test Symposium (VTS), 2013 IEEE 31st*, April 2013, pp. 1–6.

[10] L. T. Wurtz, "Built-in self-test structure for mixed-mode circuits," *IEEE transactions on instrumentation and measurement*, vol. 42, no. 1, pp. 25–29, 1993.

[11] H.-W. Ting and C.-W. Yang, "An infrastructure for analog circuits testing," in *Mixed-Signals, Sensors and Systems Test Workshop (IMS3TW), 2012 18th International*, May 2012, pp. 108–112.

[12] C.-L. Hsu, "Control and observation structure for analog circuits with current test data," *Journal of Electronic Testing*, vol. 20, no. 1, pp. 39–44, 2004.

[13] B. Kruseman, B. Tasic, C. Hora, J. Dohmen, H. Hashempour, M. van Beurden, and Y. Xing, "Defect oriented testing for analog/mixed-signal devices," in *Test Conference (ITC), 2011 IEEE International*, Sept 2011, pp. 1–10.

[14] R. Reis, M. Lubaszewski, and J. Jess, *Design of Systems on a Chip: Design and Test.* Springer, 2006.

[15] A. Coyette, G. Gielen, R. Vanhooren, and W. Dobbelaere, "Optimization of analog fault coverage by exploiting defect-specific masking," in *Test Symposium (ETS), 2014 19th IEEE European*, May 2014, pp. 1–6.

[16] L. Milor, "A tutorial introduction to research on analog and mixed-signal circuit testing," *Circuits and Systems II: Analog and Digital Signal Processing, IEEE Transactions on*, vol. 45, no. 10, pp. 1389–1407, Oct 1998.

[17] Y.-R. Shieh and C.-W. Wu, "Dc control and observation structures for analog circuits," in *Test Symposium, 1995., Proceedings of the Fourth Asian*, Nov 1995, pp. 120–126.

[18] A. Halder and A. Chatterjee, "Automated test generation and test point selection for specification test of analog circuits," in *Quality Electronic Design, 2004. Proceedings. 5th International Symposium on*, 2004, pp. 401–406.

[19] K. Deb, A. Pratap, S. Agarwal, and T. Meyarivan, "A fast and elitist multiobjective genetic algorithm: Nsga-ii," *Evolutionary Computation, IEEE Transactions on*, vol. 6, no. 2, pp. 182–197, 2002.

[20] J. Beasley, H. Ramamurthy, J. Ramírez-Angulo, and M. DeYong, "Idd pulse response testing on analog and digital cmos circuits," in *Test Conference, 1993. Proceedings., International*. IEEE, 1993, pp. 626–634.

Impact of Parameter Variations on FinFET Faults

G. Harutyunyan, G. Tshagharyan, Y. Zorian

Synopsys

gharutyu@synopsys.com, grigort@synopsys.com, zorian@synopsys.com

Abstract—The technology shrinking strategy below 20nm feature sizes adopted by the giants of the nowadays semiconductor industry has boosted the research on FinFET which is considered as an alternative to the conventional planar technology. This paper presents a comprehensive study carried out for FinFET-based memories using an advanced flow for fault modeling and test algorithm generation. Using this flow it has been shown that parameter variation (of process, voltage, temperature, frequency) has a significant impact on the fault coverage when dealing with FinFET-specific faults.

Keywords— FinFET, fault modeling, test algorithm, parameter variation, embedded memory

I. INTRODUCTION

It is commonly acknowledged fact that MOSFET or generally said planar transistor technology has already consumed all its resources as a semiconductor industry's number one choice and is no more capable of developing at the speed of industry advancement. Shrinking of transistor sizes down to 20, 16nm or even further reveals all the negative characteristics and shortcomings of conventional transistors starting from the growing amount of leakage up to various short-channel effects. Having this in mind the semiconductor industry was in a search for the long-term replacement of MOSFET technology during the last decade and it seems that among all the alternatives 3D or FinFET transistors are acknowledged as having all the necessary characteristics to succeed MOSFET and further shrink the technology (see [1]-[4]).

FinFET transistors are also referred as a tied-gate (TG) or shorted-gate (SG) memories since there is a single gate surrounding the Fins from 3 sides (see Fig. 1). Sometimes in literature, independent-gate (IG) memories having two gates in front and back sides of the Fins are also referred as FinFETs (e.g. [5]-[6]).

In contrast to the MOSFET, FinFET transistors conducting channel consist of one or more vertical thin silicon Fins, which are surrounded by gate electrodes. The channel's and the gate's spatial structures essentially improve the control of the channel thus providing better electrostatic properties at the same time diminishing the leakage current.

Yet another important feature of FinFET transistors is that the effective width of a channel is $W_{eff} = 2H_{Fin} + W_{Fin}$, where H_{Fin} is the height of the Fin and T_{Fin} is the width of the Fin or body thickness. Therefore channel's current drive capability can be increased by the means of using more Fins. As a consequence short-channel effects are drastically reduced for

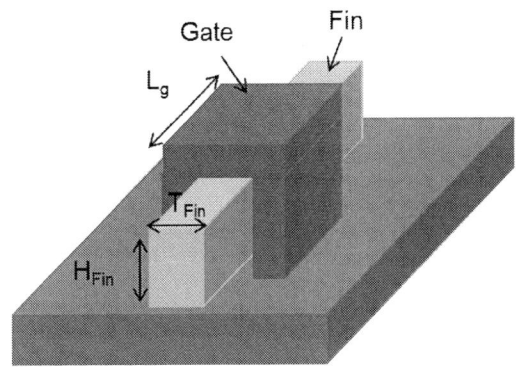

Fig. 1. FinFET structure

FinFET transistors. Besides that FinFETs have several more advantages, including controlled Fin body thickness, low threshold-voltage variation, reduced variability and lower operating voltage.

Despite having all these off-the-shelf benefits enabling construction of a new generation of memories, nevertheless design and manufacturing of such memories cause significant challenges due to the unique FinFET structure. With production of FinFET-based memories (memories using FinFET transistors), the problem of embedded memory test and repair has been considered as critical since the fault models and test algorithms used for conventional memories may not cover the whole aspect of possible defects in FinFET-based memories. The novelty of FinFET technology urges the researchers to take a closer look at the underlying defects, which are specific to this technology. This would enable to provide the comprehensive test mechanisms for their detection, diagnosis and repair therefore preventing it from affecting product quality and leading to yield loss at the same time keeping the test cost low.

In our previous work (see [7]), we have already mentioned that despite the importance of the problem, a relatively small number of research studies have been conducted in this area during the recent years. For example, in [8], the authors examined stuck-open, stuck-on and gate oxide short defects on different number of Fins within one FinFET transistor, in [9] the authors investigate Gate Oxide Shorts (GOS) in FinFETs while in [10] the authors examined stuck-open faults (SOF) for small nanometer technologies, including FinFET. Taking all this into account in [7], we have proposed a new strategy for investigation of FinFET-specific faults. According to it, an

automated flow is developed for SPICE simulation of FinFET-specific defects that are injected into memory layout or in memory spice net-list. Based on simulation results, fault models and their corresponding test sequences are identified. Afterwards on the basis of the obtained test sequences for the target fault set an optimal test algorithm is generated. One of the main advantages of the developed flow is that it is generic and has absolutely no dependency on input fault types. As one of the results of the study our experiments showed that FinFET-based memories compared with planar-based memories are more prone to dynamic faults and are more stable to process variation faults.

In the frame of this paper, we have continued the research on FinFET transistors and enhanced the scope of the conducted study to investigate also the influence of parameter variations (of process, voltage, temperature, frequency) on FinFET-specific fault coverage. For the ease of reference, we split the obtained results into two categories, first is the set of various test conditions (frequency, voltage, temperature) influence on FinFET fault coverage and second, the dependency of FinFET-specific faults on the process (used technology). The main points of the study are summarized in the next sections.

This paper is organized as follows. Section II provides the short summary of the defect models considered for FinFETs. Section III gives an overview of the obtained investigation results for the influence of different test conditions on fault coverage for FinFET-based memories. In Section IV, the summary of comparative study on FinFET-specific faults is presented in case when different technology nodes are used. Finally, Section V draws the main conclusions.

II. DEFECT MODELS FOR FINFETS

This section provides the overall understanding of the discussed defect models in FinFETs, which are later referenced along the paper. As it can be seen from Fig. 2 the following defect types are considered for FinFETs:

(a) Fin Open – Full and resistive open defects on Fin;

(b) Gate Open – Full and resistive open defects on Gate;

(c) Fin Stuck-On – Full and resistive short defects between Source and Drain;

(d) Gate-Fin Short – Full and resistive short defects between Gate and Fin;

(e) Fin-VDD/VSS Short – Full and resistive short defects between Fin and VDD or Fin and VSS.

(f) Process Variation – Variations in FinFET parameter values.

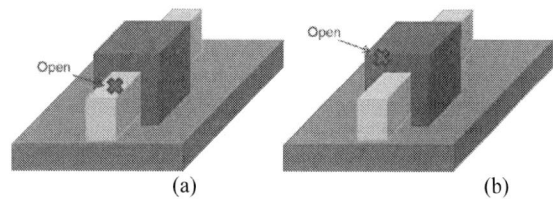

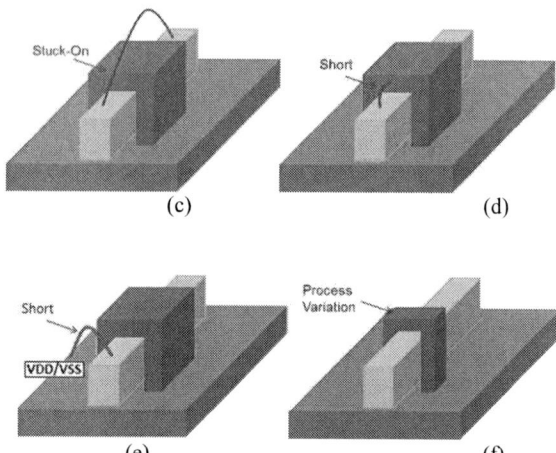

Fig. 2. Defect models considered for FinFETs

The injection of these defects in FinFET transistors leads to various fault models. Among these fault models there are those which are specific to FinFETs. For example injection of resistive Fin Open defect into a pull-down transistor results in seven-operation dynamic Deceptive Read Destructive Fault dDRDF0-7 (see Fig. 3), which is considered as FinFET-specific fault. FinFET-specific faults had been extensively discussed in the scope of our previous study (see [7]).

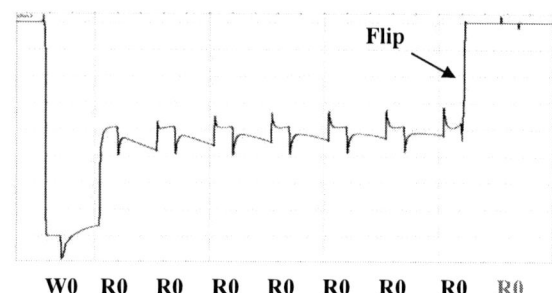

Fig. 3. Resistive Fin Open defect in PD transistor results in dDRDF fault

III. TEST CONDITIONS IMPACT ON FAULT COVERAGE

Fault coverage for a defect can be different depending on the selected test condition, i.e., it can directly affect the efficiency of the selected test algorithm used in a memory test. The fault coverage here is considered in terms of the range of resistance values, for which the defect is detected. This is especially true with regard to the family of special dynamic faults, which were discussed above as being FinFET-specific faults. In the scope of this study several test conditions, such as voltage, temperature and frequency were chosen and the level of their influence was investigated. The influence of voltage, temperature and frequency was examined for different corner values (min, max and nominal).

Tables I-III present the experimental results, which were obtained for Fin Open defect. As it can be seen from Table I in nominal case (Temperature=25°C, VDD=0.8V) the received fault model is dDRDF0-3, which implies that three consecutive read operations are needed to sensitize the fault. With temperature increase the fault model changes towards RDF (Read Destructive Fault) meaning that number of read operations needed to detect the fault is decreased to one.

On the other hand, with a temperature decrease number of consecutive read operations needed to detect the fault increases up to the point when the fault is not detected anymore. Similar inverse correlation was observed also in case of voltage change. Table II shows that with the voltage increase number of consecutive read operations decreases and when decreasing the voltage the fault is not detected anymore. Finally, Table III shows the results of frequency analysis, where Fin Open defect was injected in Pull Down and Pull Up transistors. It can be seen from the table that in both cases again inverse correlation was observed between the frequency and number of consecutive read operations needed to detect the fault.

Our experiments showed that the best corner value of test conditions varies for different defects. For example, as considered above for Fin Open defect the best fault coverage (maximum covered resistance range) was observed at maximum temperature, maximum voltage and maximum frequency while for Fin Stuck-On defect the best one is minimal temperature, minimal voltage and maximal frequency.

Based on the experimental results, the following observations can be highlighted for dRDF/dDRDF faults:

- In case of open defects: Temperature increase (decrease) leads to decrease (increase) in a number of consecutive read operations needed to sensitize dRDF/dDRDF faults.

- In case of short defects: Temperature increase (decrease) leads to increase (decrease) in a number of consecutive read operations needed to sensitize dRDF/dDRDF faults.

- In case of open defects: Voltage increase (decrease) leads to decrease (increase) in a number of consecutive read operations needed to sensitize dRDF/dDRDF faults.

- In case of short defects: Voltage increase (decrease) leads to increase (decrease) in a number of consecutive read operations needed to sensitize dRDF/dDRDF faults.

- In case of open/shorts defects: Frequency increase (decrease) leads to decrease (increase) in a number of consecutive read operations needed to sensitize dRDF/dDRDF faults.

Summarizing the section our observations showed that all corner values of test conditions (minimum and maximum values) are equally important in order to make sure that for each individual defect the highest possible coverage is obtained.

TABLE I.   TEMPERATURE IMPACT ON DYNAMIC FAULTS

Temperature	Voltage	Fault
T = 125	V = 0.8	RDF0
T = 29	V = 0.8	DRDF0
T = 28	V = 0.8	DRDF0
T = 27	V = 0.8	dDRDF0
T = 26	V = 0.8	dDRDF0
T = 25	**V = 0.8**	**dDRDF0-3**
T = 24	V = 0.8	dDRDF0-3
T = 23	V = 0.8	dDRDF0-4
T = 22	V = 0.8	dDRDF0-9
T = 21	V = 0.8	dDRDF0-16
T = 20	V = 0.8	No Fault

TABLE II.   VOLTAGE IMPACT ON DYNAMIC FAULTS

Voltage	Temperature	Fault
V = 0.85	T = 25	RDF0
V = 0.84	T = 25	DRDF0
V = 0.83	T = 25	DRDF0
V = 0.82	T = 25	DRDF0
V = 0.81	T = 25	DRDF0
V = 0.8	**T = 25**	**dDRDF0-3**
V = 0.75	T = 25	No Fault
V = 0.7	T = 25	No Fault

TABLE III.   FREQUENCY IMPACT ON DYNAMIC FAULTS

Defective Transistor	Frequency	Fault
Pull Down	400 MHz	dDRDF0-7
	800 MHz	dDRDF0-5
	1200 MHz	dDRDF0-3
Pull Up	400 MHz	dDRDF0-17
	800 MHz	dDRDF0-4
	1200 MHz	dDRDF0-3

IV. COMPARISON OF FAULTS FOR DIFFERENT TECHNOLOGY NODES

Besides examining the different test conditions, a comparative study of faults for different technology nodes is conducted. For currying this study three FinFET-based memory instances were selected, which were received from three different foundries. To preserve the confidentiality the foundry names are not disclosed, they are hereafter referred as Foundry 1, 2 and 3. Then we selected two FinFET-specific defects, namely Fin Open and Gate-Fin Short and injected these defects into all three memory instances. In order to get the full picture both these defects were injected in memory cell's Pass-Gate, Pull-Down and Pull-Up transistors.

TABLE IV.   FinFET-Specific Fault Comparison Among Different Foundries

Defect type	Technology node	Pass-Gate Transistor	Pull-Down Transistor	Pull-Up Transistor
Fin Open	Foundry 1	TF[1], SAF[2]	RDF[4], DRDF[5]	-
	Foundry 2	TF, SAF	RDF, DRDF, dRDF[6], dDRDF[7]	RDF, DRDF, dRDF
	Foundry 3	TF, SAF	RDF, DRDF, dRDF, dDRDF	TF
Gate-Fin Short	Foundry 1	TF, SAF, LF1[3]	TF, RDF, DRDF, dRDF, dDRDF	TF, RDF, DRDF, dRDF, dDRDF
	Foundry 2	TF, SAF, LF1	TF, RDF, DRDF, dRDF, dDRDF	TF, RDF, DRDF, dRDF, dDRDF
	Foundry 3	TF, SAF, LF1	TF, RDF, DRDF	TF, RDF, DRDF

[1] TF – Transition Fault, [2] SAF – Stuck-At Fault, [3] LF1 – Static Linked Fault, [4] RDF – Read Destructive Fault, [5] DRDF – Deceptive Read Destructive Fault,

[6] dRDF – dynamic Read Destructive Fault, [7] dDRDF – dynamic Deceptive Read Destructive Fault

Table IV summarizes the obtained results after the simulation of injected defects. Several important observations can be derived from the table.

First of all, one of the most important observation is that table proves the point which was already highlighted in [7], i.e., in FinFET-based memories the probability of occurring dynamic faults has significantly increased compared to planar technology. Thus it induces the necessity of the memory test mechanisms to be adjusted in a proper way for detecting dynamic faults and therefore securing high fault coverage.

Secondly, looking at the table it can be seen that the injection of both defect types in Pass-Gate (or Access) transistor does not lead to any FinFET-specific faults but mostly to transition or stuck-at faults which is probably expected behavior since Pass-Gate transistor, as the name suggests, controls the access to the cell's content. Instead Pull-Down transistor is the main contributor to FinFET-specific faults.

Last but not least, the situation seems to be a little bit unclear concerning the Pull-Up transistor simulation results since especially for Fin Open defect, its behavior significantly differs in case of memories from different foundries. This can be explained either by memory structural differences among different foundries or by having different levels of the process maturity. It requires additional research in this direction to investigate the observed phenomenon better.

The selected FinFET instances from different foundries provide enough confidence for making the above listed above observations. However taking into account the novelty of FinFET technology and constantly growing amount of foundries announcing their own line of FinFET-based memories production, the study is still in the progress and fresh results can be used in order to further confirm the made observations.

V. Conclusions

In this paper the results from the conducted comprehensive study were presented, which was conducted in the scope of FinFET-based memories. This stage of the study was devoted to investigation of the parameter variation impact on FinFET-specific fault coverage and overall test efficiency.

In the scope of the extended study first the influence of various test conditions on FinFET fault coverage was investigated and as the result several recommendations were worked out to select the best test condition for running memory test in order to achieve maximum possible fault coverage.

Next, the results of the comparative study were presented where the aim was to investigate the behavior of injected FinFET-specific defects in several FinFET-based memory instances from different foundries. This study once more confirmed the point that FinFET-based memories are becoming more prone to dynamic faults while in planar-based memories the vast majority of faults were of static nature.

References

[1] X. Huang, W.-C. Lee, C. Kuo, et al, "Sub 50-nm FinFET: PMOS", International Electron Devices Meeting Technical Digest, 1999, pp. 67-70.

[2] T.-J. King, "FinFETs for nanoscale CMOS digital integrated circuits", IEEE/ACM International Conference on Computer-Aided Design, 2005, pp. 207-210.

[3] M. Jurczak, N. Collaert, A. Veloso, T. Hoffmann, S. Biesemans, "Review of FinFET technology", IEEE International SOI Conference, 2009, pp. 1-4.

[4] H.-W. Cheng, Y. Li, "16-nm multigate and multifin MOSFET device and SRAM circuits", International Symposium on Next-Generation Electronics, 2010, pp. 32-35.

[5] Y. Liu, Q. Xu. "On modeling faults in FinFET logic circuits", IEEE International Test Conference, 2012, pp. 1-9.

[6] C.-W. Lin, M. C.-T. Chao, C.-C. Hsu, "Investigation of gate oxide short in FinFETs and the test methods for FinFET SRAMs", VLSI Test Symposium, 2013, pp. 1-6.

[7] G. Harutyunyan, G. Tshagharyan, V. Vardanian, Y. Zorian, "Fault modeling and test algorithm creation strategy for FinFET-based memories", VLSI Test Symposium, 2014, pp. 1-6.

[8] Y. Liu, Q. Xu. "On modeling faults in FinFET logic circuits", IEEE International Test Conference, 2012, pp. 1-9.

[9] C.-W. Lin, M. C.-T. Chao, C.-C. Hsu, "Investigation of gate oxide short in FinFETs and the test methods for FinFET SRAMs", VLSI Test Symposium, 2013, pp. 1-6.

[10] J. Vazquez, V. Champac, C. Hawkins, J. Segura, "Stuck-open fault leakage and testing in nanometer technologies", IEEE VLSI Test Symposium, 2009, pp. 315–320.

Memory Repair for High Defect Densities

Michael Nicolaidis, Panagiota Papavramidou
TIMA (CNRS, Grenoble INP, UJF)

Abstract : *We illustrate that memory repair for high defect densities allows improving yield, extending circuit life, reducing power, and improving reliability, and can be used to push aggressively the limits of technology scaling. Then we present several developments enabling low-cost memory repair for high defect densities, which alllow realising this promise.*

I. MOTIVATION

Aggressive process scaling has dramatic impact on reliability, yield, power density, and temperatures. Thus, Design for Reliability, Yield, and Low Power become mandatory.

As we scale deeper in the nanometric domain, defect densities induced by process, voltage and temperature variations; aging induced failures like NBTI, PBTI, HCI; EMI (like cross talk and ground bounce); steadily increase and affect dramatically fabrication yield, circuit life duration, and reliability. Fast increasing power densities and temperatures make mandatory stringent low-power requirements. In this context, disposing memory repair approaches for as high fault rates as possible becomes desirable as, sooner or later, they will allow pushing further the technology limits. In particular, in future ultimate CMOS and post-CMOS technologies affected by high defect densities, disposing memory self-repair able to cope with high fault rates will allow improving *manufacturing yield*, and subsequently *extend the life of the circuit* by employing periodic self-test and repair sessions in the field for fixing aging-induced faults. Furthermore, disposing self-repair for fault rates much higher than those required for the above faults, will also allow *drastic power reduction* by aggressively reducing the operating voltage and repairing the memory cells that exhibit faulty behavior under the reduced voltage. As a last concern, aging-induced faults may also impact the *reliability* by affecting the correct execution of the application during the time that elapses between their occurrence and the next test and repair session. Disposing a self-repair technique for fault rates even higher than those required for fixing the yield, extending circuit life, and reducing power, will also allow *solving this reliability issue* by testing and repairing the memory under more stringent conditions (lower voltage and/or higher speed) than the worst conditions used during application executions. Indeed, such tests can detect proactively and repair: memory cells that do not yet exhibit faulty behavior but are degraded due to aging and have increased chances to become faulty in the near future. Thus, thanks to this strategy, memory self-repair for high fault rates may *fix issues related with yield, life duration, low power, and reliability*, and allow pushing aggressively the technology limits.

II. ECC-BASED MEMORY REPAIR

Unfortunately, under high defect densities, conventional memory repair induces non-negligible area and excessive power penalties, which are not acceptable. To reduce area and power cost, the ECC-based repair scheme was proposed [3], which combines memory repair with the ECC codes. In fact, this repair approach reuses the ECC present in memories for fixing soft errors; in order to also fix memory words comprising just one faulty cell. Thus, we need to use repair only for replacing memory words comprising two or more faulty cells, reducing drastically the area and power of the repair circuitry. Indeed, as stated in [1] and confirmed in [2] "the number of repairable faults dramatically increases by combining the ECC and redundancy techniques together". Furthermore, as shown in [3], ECC-based repair is the only cost-effective solution (in terms of area) for achieving acceptable fabrication yield under high fault rates. Unfortunately, as shown later [4], ECC-based repair suffers by an important drawback: the size of diagnosis CAM required in ECC-based repair for discovering memory words comprising two or more faulty cells, is as large as the repair CAM required in conventional memory self-repair. Thus, the low-area advantage of ECC-based repair is lost, due to diagnosis issues. Furthermore, though ECC-based repair reduces drastically the size of the repair CAM, with respect to conventional repair, and thus reduces also runtime power dissipation, this dissipation remains non-negligible as CAMs are power hungry. Important developments, allowing to cope with these issues and to enable repairing memories affected by high fault rates at low cost are presented next.

III. SRDF TEST ALGORITHMS

To completely eliminate the diagnosis hardware for ECC based memory repair, a new family of memory test algorithms, coined as SRDF (single-read double-fault detection), was introduced [4]. These algorithms have the property to detect in a single read operation at least two faulty cells in any memory word comprising two or more faulty cells. This way, each memory word comprising two or more faulty cells is recognised immediately and is stored in the repair CAM, avoiding using a large diagnosis CAM for storing all memory words and analysing them at the end of the test algorithms to extract the ones comprising more than one faulty cells. However, the SRDF constraint is very strong and makes very challenging the development of SRDF test algorithms. This challenge was addressed in [4], by means of a formal theory comprising 16 lemmas and 8 propositions. Based on this theory, SRDF march tests targeting a comprehensive fault model (including all static unlinked single-cell and two-cell FFMs [5]) were developed [4]. Thanks to these algorithms, the area penalty induced by the diagnosis CAM was completely eliminated, achieving very low area penalty in ECC-based repair, as shown in

table 1. The results given in this table concern two SoCs having each a total memory capacity of 9,75 Gbit (i.e. a total of 250M words x 39 bits per word, for words of 32 data bits and 7 Hamming code bits). In the one SoC the 9,75 Gbit memory capacity is distributed over 300 memories, and in the other SoC it is distributed over 3000 embedded memories (as reported in column 2 of table 1). Also, in table 1, column 1 gives the considered fault rate P_f (expressed as the probability of a memory cell to be faulty). The columns labeled as %A and %P give the area and power penalties for conventional repair and for ECC based repair. The column labeled as "Test-time Increase" gives the factor of increase of test time required by the SRDF algorithms, in comparison with the test time required by the conventional test algorithms used in conventional repair. This increase takes into account both, the test length (expressed as the total number of read and write operations), as well as the power dissipated by each system during test (since the number of memories that can be tested in parallel is reduced when the power dissipations increases).

Table 1. Area and power cost comparison

P_f	Embed. Mem.	Conventional Repair		ECC-based Repair		
		%A	%P	%A	%P	Test-time Increase
10^{-4}	300	1.32	185.3	0.008	1.267	16.35
	3000	1.27	67.90	0.028	1.297	27.7
3×10^{-4}	300	3.93	532.9	0.036	5.337	7.65
	3000	3.46	177.9	0.078	3.676	17.15
10^{-3}	300	12.75	1629	0.249	39.56	3.71
	3000	13.49	581.5	0.344	17.56	7.94

In table 1, we observe that even in the worst case scenario for the proposed approach, i.e. for $P_f = 10^{-4}$ and for 3000 embedded memories - where the test-time increase takes its largest value (27.7) and the power penalty of the conventional repair takes its smallest value (67.9%) - the proposed approach is clearly more attractive, as this power penalty is totally inacceptable. The situation turns in the decisive advantage of the proposed approach as the **fault rate** increases. Thus, for $P_f = 10^{-3}$ and 300 embedded memories the test time is increased by a factor of 3.71, which is absolutely preferable than the huge 1629% power increase induced by the conventional repair approach. Also, the 12.75% and 13.49% area penalties induced by the conventional approach in two case in table 1, is totally undesirable. Indeed, as in most modern SoCs memories occupy more than 90% of the SoC area, a 12.75% area penalty induced by the conventional repair approach represents more than 11.5% of the total SoC area.

IV. ITERATIVE DIAGNOSIS APPROACH

By eliminating completely the diagnosis CAM, SRDF test algorithms reduce drastically the area cost of ECC-based repair. However, these algorithms are more complex than conventional test algorithms and increase test time. Thus, to provide a wider range of tradeoffs, in terms of test time and area penalty, we also developed an approach alternative to the SRDF test algorithms. This approach,

rather than using more complex algorithms to completely eliminate the diagnosis CAM, it uses conventional test algorithms and reduces the size of this CAM by means of an iterative diagnosis approach [6]:

- The conventional test algorithm is executed iteratively several times; at each iteration we diagnose a subset of faulty memory words (those stored in the diagnosis CAM until the instant in which all its locations are occupied).
- Then, we continue executing the test algorithm until its end, but we do not store new faulty words in the CAM. Instead, we only update the CAM contents for the already stored faulty words. Thus, at the end of the test algorithm, all these words are completely diagnosed.
- At the end of the current test iteration we delete all words stored in the CAM, which comprise a single faulty cell. Thus, we liberate CAM space for diagnosing new faulty words in the next test iteration.
- In the next iteration, we start storing faulty words in the CAM only from the instant of the previous iteration in which all CAM locations have been occupied.
- To reduce test time, in each new iteration we do not execute the test algorithm from its beginning. But this may mask certain faults that are sensitized by the non-executed part of the test algorithm. To address this issue, we show that the sensitization of faults detected during a sequence of the march test algorithm can be realized only during this sequence itself, or during the sequence preceding it. Thus, at each iteration, we start the test algorithm from the sequence preceding the sequence at which we start updating the CAM. This way we reduce the test length without inducing fault masking.

A major challenge concerns the evaluation of this approach. This evaluation requires: performing a large number of fault injections, in order to obtain statistically significant results; and simulating, for each of these injections, the faulty memory and the diagnosis circuit, by executing the test and diagnosis algorithm.

As fault simulation is very consuming in computation time (and this is also the case for the algorithmic simulation of CAMs), we developed a new approach that we called "fault pseudo-simulation", which reduces drastically simulation time while providing identical results as the conventional fault simulation [6]. In this approach, instead of injecting faults in the memory, we inject what we called "detection profiles" (which consist in the sequences of the test algorithm in which the injected fault is detected). This approach consists in:

- Identifying the sequences of the test algorithm in which each fault is detected (detection profiles).
- Performing in the memory cells probabilistic injection of the detection profiles (instead of the faults themselves), and create a table containing the results of this injection.
- Developing a fault pseudo-simulation algorithm, which simulates the diagnosis process by using this table. To avoid developing a new pseudo-simulation algorithm each time we use a different memory test algorithm, *we developed a generic algorithm that takes as input the*

978-1-4799-7598-3/15 $31.00 © 2015 IEEE

test algorithm and generates as output a "pseudo-simulation" algorithm dedicated to this test algorithm.

Thanks to this approach, we reduce drastically the simulation time. Thus, we were able to conduct intensive campaigns of statistical fault injections and simulations, allowing to obtain statistically significant results. In particular, for each considered case (SRAM capacity, CAM size, and fault rate) we conducted 1000 fault injection and simulation campaigns. The results are shown in Table 2.

Table 2. Area and power cost, and test length

P_f	#Embed. Mem	Non-ECC Repair		All ECC Rep.	ECC-Rep. CAM/2		ECC-Rep CAM/4		ECC-Rep CAM/6	
		%A	%P	%P	%A	T.T	%A	T.T	%A	T.T
10^{-4}	300	1.31	185.4	1.26	0.69	2.01	0.31	3.09	0.21	4.20
	1000	1.19	96.06	1.18	0.62	2.31	0.33	3.92	0.23	5.47
	3000	1.23	67.90	1.29	0.67	2.34	0.38	3.69	0.28	5.20
$3x$ 10^{-4}	300	3.93	533.0	5.33	2.16	1.86	1.02	2.30	0.70	2.83
	1000	3.87	283.3	4.43	2.05	1.98	0.93	2.72	0.65	3.57
	3000	3.45	177.9	3.67	1.82	2.01	0.96	3.16	0.68	4.01
10^{-3}	300	12.7	1629	39.5	6.83	1.69	3.69	2.30	2.65	2.50
	1000	13.0	913.9	24.2	7.33	1.78	3.58	2.33	2.50	2.65
	3000	13.5	581.5	17.5	6.50	1.79	3.14	2.18	2.24	2.70

In table 2, column 1 gives the fault rates. All considered cases concern a total SRAM capacity of 9,75 Gbit corresponding to a total number of 250M words x 39 bits pert (32 data bits et 7 Hamming code bits). The total capacity of 9,75 Gbit is partitioned into several memories embedded in a SoC. We considered 3 cases for this distribution: 300, 1000, and 3000 embedded memories, as reported in column 2. Columns 3 and 4 show the area and power costs of conventional repair (as a percentage of the area and power of the memory under repair).
In all ECC-based repair cases we use separates CAMs for repair and for diagnosis. Thus, as during application execution only the repair CAM is used (which is identical in all ECC-repair cases), runtime power is the same in all ECC-repair cases and is given in column 5.
Columns 6, 8 and 10 give the area cost for the approaches using a diagnosis CAM having respectively size equal to 1/2, 1/4, and 1/6 of the size of the CAM required in the case of conventional repair. Columns 7, 9, and 11 give the increase of test time, with respect to conventional repair. This increase takes into account both, the increase of the number of operations caused by the iterative approach, and the power dissipated by each approach during the test phase.

In table 2, we observe that the area and power penalties of conventional repair (columns 3 and 4) increase linearly with the fault rate. However, the area penalty is significant but not excessive, while the power penalty is excessive in all cases. We also observe that, all the ECC-based approaches reduce drastically runtime power. They also allow significant reduction of the area penalty, which is valuable as memories usually occupy more that 90% of a SoC area. These gains come at the expense of test time, as observed in

columns 7, 9, and 11. However, this increase is fully justified with respect to the large area and power gains. Considering the results of tables 1 and 2, we note also that the two approaches (the SRDF test algorithms, and the iterative diagnosis process), are complementary and offer to the designers a wide range of tradeoffs for satisfying the area, power, and test time constraints of each particular design.

V. LOW-POWER REPAIR ARCHITECTURES

The reduction of power penalty achieved by ECC-based repair is impressive (e.g. reduced from 1629% down to 39.56%, in the case of 300 embedded memories and $P_f = 10^{-3}$ shown in table 1). However, as in modern designs low-power is one of the most stringent constraints, further power reduction is highly desirable. In particularly this is necessary if we have to achieve all four goals stated in the motivation of this work: improve yield, extend circuit life, improve reliability, and reduce power.

Existing word-repair schemes store all faulty words in a CAM. Then, at each memory access the current address is compared against the addresses of the faulty words stored in the tag fields of the CAM. ECC-based repair reduces drastically the number of faulty words stored in the CAM. Thus, it reduces drastically the number of comparisons and the associated power dissipation. However, for very high defect densities, the number of memory words that have to be stored in the CAM becomes non-negligible even for ECC-based repair, resulting in non-negligible number of comparisons and power dissipation.

To reduce power, our first approach [7] partitions the memory address space into several subspaces and associates to each subspace a CAM for storing its faulty words. Thus, at each memory access, the current address is compared against the addresses of the faulty words stored in one of these CAMs, reducing the number of comparisons and the associated power dissipation. This way, by using a fine partitioning, power dissipation can be reduced drastically. It comes out that, this architecture consists in replacing the repair CAM by a set associative Cache, and is referred hereafter as Cache repair-architecture.

A drawback of this approach is that: while in the CAM repair-architecture any CAM word can repair any faulty memory word, in the Cache repair-architecture the words of each set of the set-associative cache can repair only faulty words belonging to the corresponding set of memory words. As a consequence, a memory in which the number of faulty words is lower than the total number of good locations of the cache, may not be repaired if any of the partitions of the memory comprises more faulty words than the *good* locations of the corresponding cache-set. Thus, the cache repair scheme requires larger number of cache locations for achieving a given repair efficiency. The increase in cache locations becomes higher if, in order to reduce the number of comparisons (for reducing power dissipation), we reduce the number of ways of each set of the repair cache and we increase the number of its sets.

978-1-4799-7598-3/15 $31.00 © 2015 IEEE

To cope with this issue, our second solution [7] adds in the repair architecture an Overflow CAM, which repairs the faulty memory words left unrepaired by some sets of the cache. Though the overflow CAM will usually be small, in some situations its power can be significant. Thus, in these cases we replace the Overflow CAM by an Overflow Set-Associative Cache.

The evaluation results of the new word-repair architectures are given in tables 3 and 4. Table 3 presents the evaluation of these architectures in the context of conventional (i.e. non-ECC) repair, while table 4 presents the evaluation of these architectures in the context of ECC-based repair. In both tables, column 1 gives the fault rate. All considered cases concern a total SRAM capacity of 9,75 Gbit corresponding to a total number of 250M words x 39 bits pert (32 data bits et 7 Hamming code bits). The total capacity of 9,75 Gbit is partitioned into several memories embedded in a SoC. We considered 3 cases for this distribution: 300 and 3000 embedded memories, as reported in column 2 of the tables. Also, the results presented in these tables are obtained for 97% yield after repair, for the total memory capacity of 9,75 Gbit.

In table 3, columns 3, 4 and 5 give the results (number of words of the repair CAM Ncw, area penalty %A, and power penalty %P) for the non-ECC repair using the conventional (i.e. CAM-based) word-repair architecture. In the same table, columns 6 to 11 give the results for the non-ECC repair using: a set-associative cache (CACHE 1), for repairing the faulty memory words; and an overflow set-associative cache (CACHE 2), for repairing the words left unrepaired by CACHE 1. In particular, columns 6 and 7 give the number of *sets* and the number of *ways* of CACHE 1; columns 8 and 9 give the same parameters for CACHE 2; columns 10 and 11 give the area and power penalties.

In table 3, we observe that for the conventional (i.e. non-ECC) repair approach, the new word-repair architecture allows drastic reduction of the power penalty at the expense of a slight increase of the area penalty.

Table 3. Evaluation of the new word-repair architectures in the context of conventional (i.e. non-ECC) repair

P_f	#Emb Mem	Non-ECC Repair CAM			Non-ECC Repair CACHE-1 / CACHE-2					
		N_{CW}	%A	%P	N_{S1}	N_{W1}	N_{S2}	N_{W2}	%A	%P
10^{-4}	300	3466	1.32	185.3	64	63	2	39	1.879	22.73
	3000	402	1.27	67.90	32	18	1	13	2.376	23.94
$3x10^{-4}$	300	10285	3.93	532.9	128	99	2	30	4.548	33.23
	3000	1121	3.46	177.9	64	24	2	20	6.251	42.33
10^{-3}	300	35325	12.75	1629	512	85	64	32	15.20	65.16
	3000	3693	13.49	581.5	128	39	2	27	15.19	70.33

As table 4 concerns ECC-based repair, the number of memory words that require repair is much smaller than in table 3. Thus, in some cases the traditional (i.e. CAM-based) word-repair, presented in columns 3, 4, and 5 of table 4, induces very low power penalty, as can be seen in column 5 of table 4. Thus, in these cases the new word-repair architectures provide marginal improvements. Then, in these cases, columns 6 to 11 (presenting the results for the

new repair architectures) are empty (-). For the remaining cases, in table 4 we also observe that: the Cache-based repair without overflow CAM/Cache, provides the best results in the case where the total capacity of 9,75 Gbit is partitioned into 300 embedded SRAMS, and Pf is equal to 10^{-4}; while for the other cases the best results are obtained by the approach using the Overflow Set-Associative Cache (Cache-1/Cache-2). On the overall, we observe that for high fault rates (Pf = 10^{-4}), the power penalty is less than 1.3%, for even higher fault rates (Pf = $3x10^{-4}$), the power penalty is less than 3.8%, and for very high fault rates (Pf = 10^{-3}), the power penalty is less than 10%. In more recent developments, not presented in this paper, an improved architecture brings this penalty down to 8.5%.

Table 4. Evaluation of the new word-repair architectures in the context of ECC-based repair

P_f	#Emb Mem	ECC Repair CAM			ECC Repair CACHE-1 / CACHE-2					
		N_{CW}	%A	%P	N_{S1}	N_{W1}	N_{S2}	N_{W2}	%A	%P
10^{-4}	300	16	0.008	1.267	4	8	-	-	1.201	0.017
	3000	6	0.028	1.297	-	-	-	-	-	-
$3x10^{-4}$	300	83	0.036	5.337	16	6	1	12	0.069	3.723
	3000	17	0.078	3.676	-	-	-	-	-	-
10^{-3}	300	720	0.249	39.56	64	14	2	30	0.544	9.68
	3000	98	0.344	17.56	16	8	1	10	0.646	9.93

VI. CONCLUSIONS

We presented recent developments allowing memory repair at low area and power penalties, even in the case of very high defect densities. Thanks to these capabilities, the proposed repair framework enables improving yield, extending circuit life, reducing power, and improving reliability, in advanced technologies affected by high defect densities and impacted by high power densities and temperatures. Thus, this framework can be used to cope with these issues and push aggressively the limits of technology scaling.

REFERENCES

[1] Horiguchi M., Itoh K., "Nanoscale Memory Repair", Springer, Series: Integrated Circuits and Systems, 1st Edition., 2011.

[2] Itoh K., "Adaptive Circuits for the 0.5-V Nanoscale CMOS", Keynote ISSCC 2009.

[3] M. Nicolaidis, N. Achouri, L. Anghel, "A Diversified Memory Built In Self Repair Approach for Nanotechnologies", IEEE VLSI Test Symposium/Best Paper Award, April-May 2004.

[4] P. Papavramidou, M. Nicolaidis, "Test Algorithms for ECC-based Memory Repair in Nanotechnologies", IEEE VLSI Test Symposium, April 2012.

[5] A.J. van de Goor and Z. Al-Ars, "Functional Fault Models: A Formal Notation and Taxonomy", In Proc. of IEEE VLSI Test Symposium, pp. 281-289, 2000.

[6] P. Papavramidou, M. Nicolaidis, "An iterative diagnosis approach for ECC-based memory repair", 2013 IEEE VLSI Test Symposiums (VTS).

[7] P. Papavramidou, M. Nicolaidis, "Reducing Power Dissipation in Memory Repair for High Defect ensities", 2013 IEEE European Test Symposiums (ETS).

Horizontal-FPN fault coverage improvement in production test of CMOS imagers

R. Fei[1,2], J. Moreau[1], S. Mir[2,3], A. Marcellin[1], C. Mandier[1], E. Huss[1], G. Palmigiani[1], P. Vitrou[1], T. Droniou[1]

[1] STMicroelectronics, 12 rue Jules Horowitz, F-38000 Grenoble, France
[2] Université Grenoble Alpes, TIMA, F-38000 Grenoble, France
[3] CNRS, TIMA, F-38000 Grenoble, France
richun.fei@imag.fr, jocelyn.moreau@st.com, salvador.mir@imag.fr

Abstract - Current production testing of CMOS imager sensors is mainly based on capturing images and detecting failures by image processing with special algorithms. The fault coverage of this costly optical test is not sufficient given the quality requirements. Studies on devices produced at large volume have shown that Horizontal Fixed Pattern Noise (HFPN) is one of the common image failures encountered on products that present fault coverage problems, and this is the main cause of customer returns for many products. A detailed analysis of failed devices has demonstrated that HFPN failures arise from changes of electronic circuit topology in pixel addressing decoders or the metal lines required for pixel powering and control. These changes are usually due to the presence of spot defects, causing some pixels in a row to operate incorrectly, leading to an HFPN failure. Moreover, defects resulting in partially degraded metal lines may not induce image failure in limited industrial test conditions, passing the optical tests. Later, these defects may produce an image failure in the field, either because the capture conditions would be more stringent, or because the defects would evolve into catastrophic faults due to electromigration. In this paper, we have first enhanced the HFPN detection algorithm in order to improve the fault coverage of the optical test. Next, a built-in self-test structure is presented for the on-chip detection of catastrophic and non-catastrophic defects in the pixel power and control lines.

Keywords - CMOS image sensor, industrial test, HFPN.

I. INTRODUCTION

CMOS Image Sensors are widely used in many domains such as consumer and industrial electronics, military, science, etc. Thanks to their continuous performance improvement, they are today replacing traditional Charge Coupled Devices (CCD) in many areas where CCD has been predominant such as security, automotive and medical applications [1] [2]. CMOS imagers benefit from low power consumption, on-chip functionality and low cost, key advantages for the consumer electronics market, especially for mobile imaging that has been a major driving force for the development of CMOS imaging technologies.

Like all other IC products, the fabrication of CMOS sensors is not free from manufacturing defects. The products must pass through a variety of tests after manufacturing, and only devices meeting quality criteria can be delivered to customers. However, the fault coverage of the actual optical tests methods

for the sensor array and its analog readout is not sufficient to meet quality levels. These tests are based on capturing images with the sensor and processing them with special algorithms to evaluate the performances. There is a large number of performance parameters that must be measured, due to the sensor complexity that includes several analog blocks and the array of pixels that can present pixel response non-uniformity. The selection of the parameters to be tested and the setting of the test limits are complex decisions, with a trade-off between fault coverage and yield.

Our studies on devices produced at large volume have shown that Fixed Pattern Noise (FPN) is one of the key parameters that face a fault coverage problem. In the literature [3] [4] [5], FPN is commonly defined as the constant non-uniformity of the image (forming a constant noise pattern) under the same uniform illumination conditions. However, during optical production testing, FPN errors are usually extracted from row and column signatures: the Horizontal-FPN (HFPN) presents a row to row FPN while Vertical-FPN (VFPN) presents a column to column FPN. Random FPN errors usually refer to Dark Signal Non Uniformity (DSNU) and Photo Response Non Uniformity (PRNU). The random FPN is mainly caused by process parameter variations, like the threshold of the pixel transistors. The VFPN arises from the column-parallel read out which is used today in most high speed imagers [6]. The HFPN typically arises from the offset variations when the pixels are read, e.g. during pixel reset and selection. The FPN phenomena in CMOS sensors have been largely reduced by the use of many FPN reduction techniques during data readout, such as analog Correlated Double Sampling (CDS) [6] [7], dual digital CDS [8], or other methods via optical flow [9] [10].

FPN due to process variations can be evaluated by corner simulation during design, reduced by FPN reduction techniques and corrected by digital processing in the field. Thus, devices that fail during production testing due to FPN faults present mainly spot defects in electronic components that produce open or short circuits. These defects in column amplifiers and bit lines lead to VFPN faults, while these defects in pixel addressing decoders [11] and horizontal lines [12] lead to HFPN faults. These faults produce image defects with residual stripes and bands, often hard to correct by digital processing. In addition, the trend towards increasing CMOS

sensor resolution requires smaller pixel sizes and higher density. As a result, the space to route interconnection lines becomes narrower and the number of pixels connected to them increases. This leads to higher defect probabilities of H/V FPN.

During production optical testing, the H/V FPN values can be extracted by comparing each row/column with adjacent rows/columns. Devices can then be sorted according to this. However, this optical test procedure is insufficient to reveal all defects due to a lack of time for capturing enough images in varied illumination conditions, or due to the lack of precision for image processing. Furthermore, weak open and short defects in interconnection lines, which do not produce image failure during production testing, may cause serious failure in the field. Some alternative techniques for VFPN test improvement have been applied in some products such as Bit Line Test and test of Analog to Digital Converter (ADC), but no solutions have been applied for HFPN test. In this paper, we will first present in Section II an improvement of the current HFPN optical test. Next, Section III will present an on-chip test structure for the detection of weak open/shorts in horizontal lines. Finally, Section IV will provide some conclusions.

II. HFPN DETECTION BASED ON OPTICAL TEST

A. Review of the algorithm for HFPN computation

In the current optical test program, an HFPN failure in a CMOS imager is detected by processing the images that are captured both under dark and light conditions. The HFPN value for a row is calculated as the absolute difference between the average values of the pixels of the row and the average values of the pixels in the neighboring rows (typically, 5 rows above and 5 rows below are considered, except for the border rows). This computation is performed for each color Bayer frame independently, and the maximum HFPN value of all rows is used for wafer sorting. A device will fail the HFPN test if its max-HFPN is above the limit specified in the test program.

However, some devices with passing HFPN values close to the limit have resulted in customer returns. The test limit is hard to adjust due to the trade-off between fault coverage and yield. This can be illustrated by the below analysis.

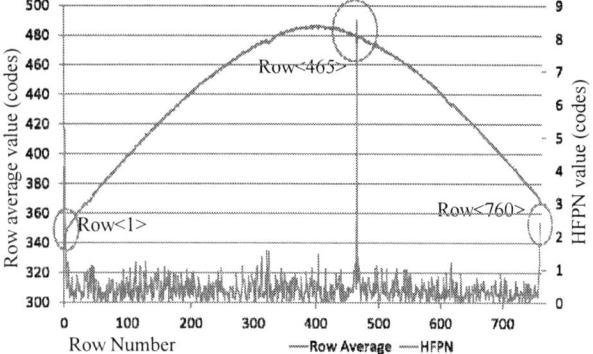

Figure 1. HFPN distribution of a failed sample.

Figure 1 illustrates the HFPN distribution of a failed sample for the case of a Green-Red color frame obtained under light conditions with a sensor that has 760 rows of Green-Red pixels. Rows are numbered starting from bottom to top. The red curve corresponds to the average value of the pixels in each row. Notice that, in production testing, the light diffusion is not uniform, being brighter at the center of the imager than at the borders. As a result, the row average values are larger at the center than at the top and the bottom of the image. The blue curve is the HFPN distribution. The mean of the HFPN of all rows is 0.425. There are three abnormal rows with their HFPN value much larger than the mean. The HFPN value of row <465> is 8.54 which is the maximum. This defective row is visible in test images. The first and last rows also have HFPN values much larger than the average, with 5.25 for row <1> and 2.42 for row <760>, but this is due to computation effects. As mentioned before, the HFPN is calculated considering 5 rows above and 5 below, except for the rows at the top and the bottom of the imager. The HFPN of the top (bottom) row is calculated considering as reference only the 5 rows below (above). Due to the illumination conditions the photo response is gradually weakened from the center to the borders, and the reference values have an offset for the top and bottom rows, as shown in Figure 2 for the case of the bottom rows.

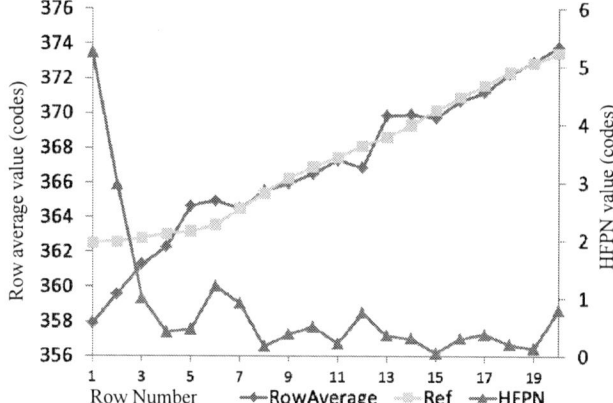

Figure 2. Pixel average value of the bottom rows (red) and its reference for HFPN calculation (yellow).

The HFPN of each row (the blue curve) is the difference between the average value of the row (red) and its reference for HFPN calculation (yellow). The reference for the first two rows has a significant offset with respect to the row average.

B. Improvement of the HFPN test

The analysis of the above failed sample shows that under light conditions the HFPN computation needs to be improved for top and bottom rows. It is also important to detect weaker HFPN faults that may be hidden when only the max-HFPN value is provided.

A first improvement is to use different Regions of Interest (ROI) for the top and bottom rows, with test limits that can be specific to these regions, while the other rows will be able to have tighter limits without impacting yield.

A second improvement is the use of column ROIs for the detection of weaker HFPN faults caused by the degradation of some of the pixels in a row. For example, Figure 3 shows the HFPN computation that has been performed for 6 ROIs of 160 columns each. The HFPN distribution of each ROI shows that

978-1-4799-7598-3/15 $31.00 © 2015 IEEE   154

row <465> is gradually degraded from left to right and that there may be a spot defect located in ROI-L2.

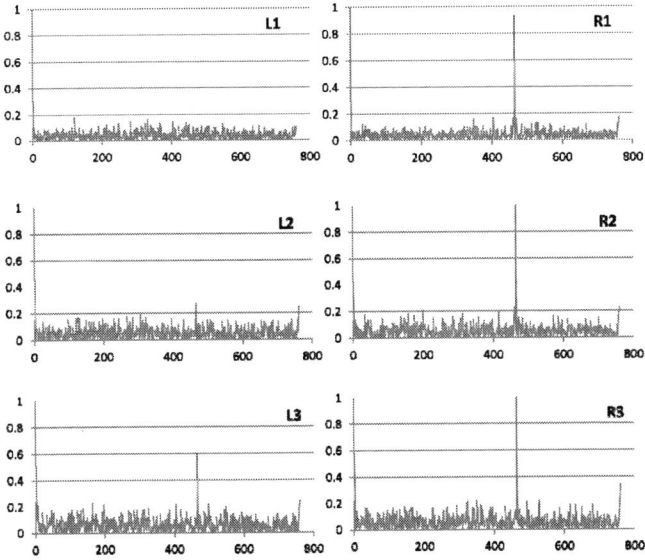

Figure 3. Partial HFPN fault detection using column ROIs (vertical axis correspond to the HFPN value and the horizontal one to the column number).

C. Experimental results

The test program has been modified to store the max-HFPN of different ROIs in order to verify the analysis of border effects on HFPN calculation. The results of a wafer test are shown in Figure 4.

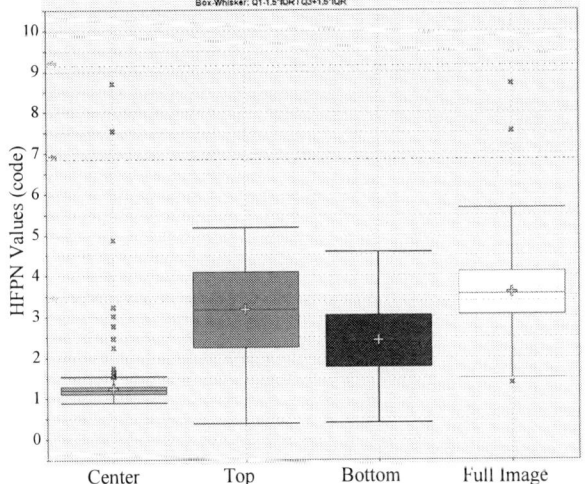

Figure 4. Wafer-level HFPN computation with improved test program.

The green box is for the central ROI; the purple and blue ones are for top and bottom ROIs, respectively; the yellow one is the max-HFPN of the full image that is the same as the max-HFPN extracted by the former test program. This figure illustrates that the max-HFPN of central ROI is much smaller than border ROIs. Therefore, the test limits for the central ROI can be tightened to improve HFPN fault coverage without impacting yield.

III. ON-CHIP TEST STRUCTURE FOR HFPN DETECTION

As mentioned before, HFPN faults are mainly caused by spot defects in pixel control and power lines. Weak open or short defects cannot be detected by optical tests. Therefore, a built-in self-test (BIST) technique has been developed to further improve the HFPN fault coverage. The technique targets the detection of Resistive Open (RO) or Resistive Short (RS) defects in pixel horizontal interconnection lines.

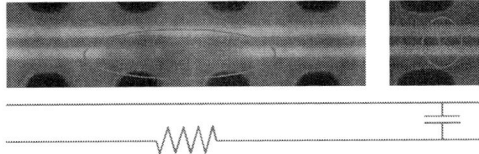

Figure 5. Non-catastrophic defects in horizontal lines are modeled as an additional impedance.

As shown in Figure 5, defects in an interconnection line can be modeled as an additional resistor and/or capacitor in the line, typically increasing the line impedance and the signal propagation delay. By measuring this delay for each Line Under Test (LUT), defective lines with high impedance can be distinguished from functional ones.

Many techniques have been studied in the past for path delay detection. A simple delay measurement scheme was proposed in [13]. This technique is based on the application of a critical width pulse to the path under test and on the detection of its successful propagation to the output. This simple technique is not applicable in the case of an image sensor that has thousands of pixel rows. This is because it would be necessary to modify the addressing decoders which are critically analog designed, and also because of the need to generate a special video timing for testing. Thus, a novel delay measurement structure that is well suited for CMOS image sensors has been developed.

A. Principle of the new test structure

The test scheme is based on the observation that a defect in a line will result in a change of its RC behavior. If we assume that a rising (falling) edge is sent at one end of the line, the measurement of the rise (fall) time of the signal at the other end of the line can be used to test it. We will consider in this paper the measurement of the rise time. The generation of a rising edge at one end of a horizontal wire is straightforward using the address decoders, with a rising time several orders of magnitude lower than at the other end of the LUT. We can then neglect the variability of the rising time at the input. For the measurement of the rise time of this edge at the other end of the line, we use a two-stage structure. The first stage, called Line Propagation Delay Detector (LPDD), generates two signals with a time interval between their rising edges that is

978-1-4799-7598-3/15 $31.00 © 2015 IEEE   155

proportional to the rise time of the signal in the LUT. The second stage compares this time interval with a reference time for LUT sorting.

For the design of the LPDD, both analog differential comparators and simple inverter-based comparators can be considered. A voltage-crossing detector based on an analog differential comparator is proposed in [14] showing high measurement precision and process independence. However, it is not convenient for CMOS imager test due to large silicon area and too high power consumption. The precision of the inverter-based comparators is sufficient for our purposes, requiring small silicon area and power consumption.

The schematics of the LPDD is shown in Figure 6. A Low Threshold Inverter (LTI) is used for low voltage level triggering and a half Schmitt trigger (SMT) for high level triggering. The LPDD circuits is designed with thick oxide transistors, except INV0 uses digital standard transistors. An analog power is supplied for the first stage of the SMT, and all other parts are supplied with digital power. In the first stage of the LTI, the W/L ratio is much larger for the NMOS transistor than for the PMOS one. The trigger threshold of the SMT (V_{HI}) mainly depends on MN1 and MN3. It can be roughly calculated as follows. When V_{in} increases from GND level, $V_{gs2} < V_{th2}$ and MN2 is cutoff. When $V_{in} = V_{HI}$, we get $V_{gs2} > V_{th2}$ and MN2 becomes active. Since MN1 and MN3 are in the saturation region, we have:

$$I_{MN1} = \frac{1}{2}K_1(V_{Hi} - V_{th1})^2 = I_{MN3} = \frac{1}{2}K_3(V_{DD} - V_{Hi})^2 .$$

$$\frac{K_1}{K_3} = \frac{(W/L)_1}{(W/L)_3} = \frac{(V_{DD} - V_{HI})^2}{(V_{HI} - V_{th1})^2}$$

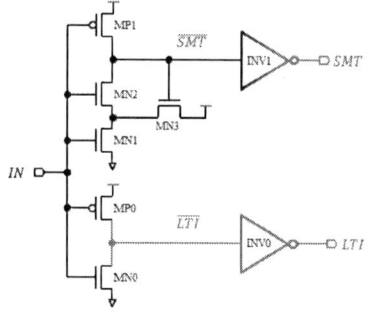

Figure 6. LPDD designed with inverter-based comparators.

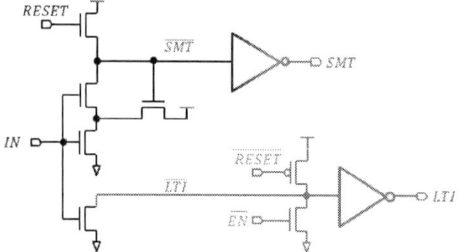

Figure 7. Schematics of the LPDD with the input stage made of NMOS transistors only.

For further saving wafer area and facilitating the layout, the first stages of the SMT and LTI can be designed with only NMOS transistors as shown in Figure 7. An additional test control signal, *RESET*, is added to reset the LPDD.

A timing diagram of the LPDD operation is shown in Figure 8. The LPDD is first reset by a RESET pulse, and both the SMT and LTI are set to '0'. When a rising edge arrives at the LPDD input, the LTI circuit is triggered at a low voltage level and the SMT circuit is triggered at a high level. The delay between the outputs of both circuits is proportional to the propagation delay of the input signal, and this will be used in the next stage to detect defective lines.

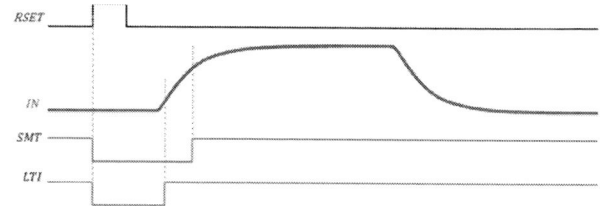

Figure 8. Timing diagram of the LPDD with reset.

Simulation results in Figure 9 show the delay measurement with respect to the line impedance that is typically of a few kΩ. The sensitivity of the LPDD is approximately 0.75 ns/kΩ. This is sufficient for the detection of the impedance increase caused by weak defects.

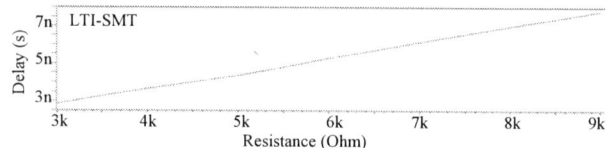

Figure 9. Delay between LTI and SMT signals as a function of the LUT resistance.

B. Operation

The time interval between the LTI and SMT output signals is compared to a reference time. The comparison result is stored in a memory that can later be read out.

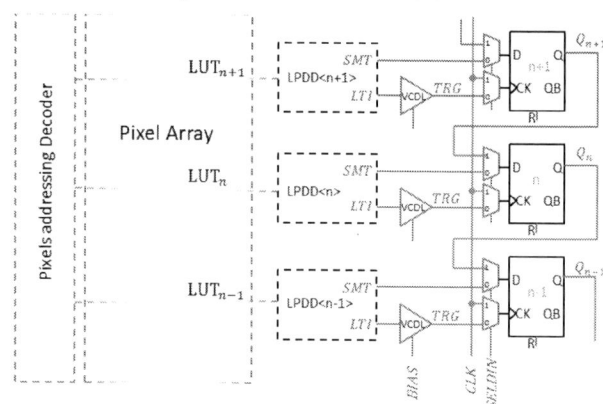

Figure 10. Comparison with a reference time implemented with a Voltage Controlled Delay Line.

978-1-4799-7598-3/15 $31.00 © 2015 IEEE   156

As shown in Figure 10, a Voltage Controlled Delay Line (VCDL) is added on the LTI output to implement the reference time. A positive-edge-triggered D flip-flop (DFF) can be used to store the test result. The mux added on the D-input and D-clock is used to switch the DFF between a test result acquisition mode and a read out mode.

Figure 11 illustrates the operation of the LPDD. When a rising edge arrives at the LPDD input, the two outputs of the LPDD successively change from '0' to '1'. The reference time added via the VCDL at the LTI output produdes a trigger signal (TRG), and the rising edge of this signal is used as the clock of the DFF. The signal at the output of the SMT circuit is used as the D-input of the DFF.

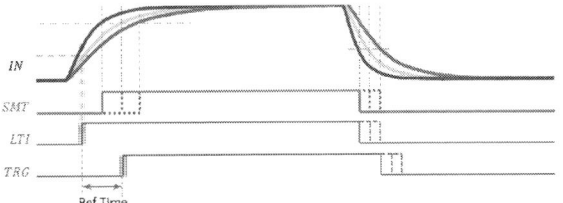

Figure 11. Operation of the LPDD.

In the case of a fault-free LUT, the input of the LPDD has a fast rising edge, and the SMT circuit switches to '1' before a rising edge of the TRG signal arrives. A value '1' will be stored into the DFF. If the LUT has a high impedance due to the presence of a defect, the SMT circuit will switch after the TRG signal, and a value '0' will be stored into the DFF indicating a defective LUT. In case of catastrophic defects present on the LUT, neither the TRG nor the SMT signals will show an edge, and the DFF will hold its initial value.

A third debug mode can also be added, with a global clock signal used as D-clock and the DFF taking as D-input the signal SMT. This way, LUT stuck-at-'0' and stuck-at-'1' faults can be distinguished.

C. An alternative read out operation

The read out of the test results may also use a pair of bit lines instead of storing the results in a memory for each line.

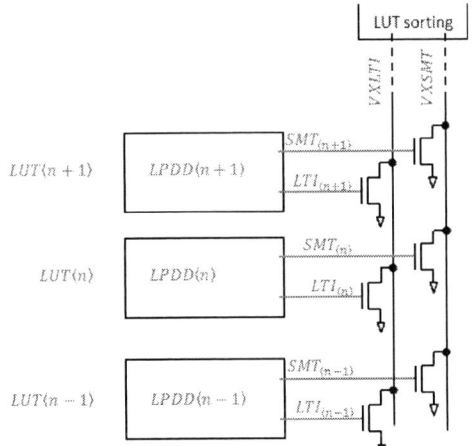

Figure 12. Column read out by using bit lines.

As shown in Figure 12, each output of the LPDD drives an NMOS switch transistor with the drain connected to the bit line and the source to ground. The time interval between $LTI\langle n \rangle$ and $SMT\langle n \rangle$ is transmitted via bit lines $VXLTI$ and $VXSMT$ to a ADC circuit that sorts out the LUT defective lines.

This scheme saves silicon area, but parallel testing of multiple lines is not possible, since only one LUT can be tested at a time. Notice also that the bit lines must first be reset to a high logic level before the test of an LUT.

D. Design of a prototype

We have designed and implemented a prototype CMOS imager using the schemes of Figure 7 and Figure 12. The time interval between the pair of bit lines $VXLTI$ and $VXSMT$ is converted to a voltage by means of an on-chip Time-to-Voltage Converter (TVC), and this voltage is read out by an on board ADC as an additional pixel value.

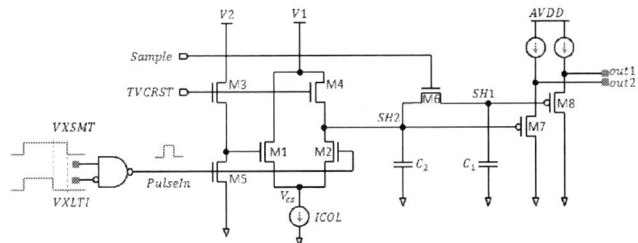

Figure 13. TVC for test results readout.

As shown in Figure 13, a simple TVC is designed with a layout that is compatible with the Correlated Double Sampling (CDS) read out circuitry of the CMOS sensor. The measurement is composed of three steps: the TVC is first reset by a TVCRST pulse through transistors M3 and M4; next, M6 is turned on before the signal PulseIn is generated, so that capacitor C_1 maintains the voltage of referenceV_{SH1}; finally, capacitor C_2 is discharged when transistor M2 is switched on by PulseIn. The voltage difference between V_{SH1} and V_{SH2} is read out for LUTs sorting.

The measurement range of the TVC is defined by the initial and minimum voltage levels of SH2 as follows:

$$T_{max} = \frac{(V_{SH2ini} - V_{SH2min})C_2}{I_{col}}$$

In our prototype, the hold capacitor C_2 is designed around 0.9 pF. The initial voltage at SH2 is reset to $V_1 = 1.6V$. To keep the current source of ICOL in the saturated region, the minimum voltage level of V_{SH2} is 0.2 V. Therefore, the measurement range is defined by the discharging current I_{col}. which may be adjusted between 0 to 7.4 uA by the external voltage V_{ICOL} in our prototype. The measurement ranges and transfer ratios are listed in Table 1.

V_{ICOL}	0.85	1.41	2.5	V
ICOL	2.41	4.01	7.45	uA
Linear Measurement Range	170	315	520	ns
Output Swing	1.3	1.3	1.3	V
TVC Transfer Ratio	8.2	4.4	2.6	mv/ns

Table 1. Measurement ranges and transfer ratio of the TVC.

The performance of the TVC circuit is evaluated through Monte-Carlo (MC) simulation, with $V_{ICOL} = 1.41V$. The mean value and standard deviation of the circuit main parameters is shown in Table 2.

V_{ICOL} = 1.41 V	Mean	σ
ICOL	4 uA	1.1 nA
TVC transfer ratio	4.4 mV/ns	7.9 uV/ns

Table 2. Mean and standard deviation of the TVC parameters.

Figure 14 shows the MC simulation for the measurement of a fault free line with R=2 kΩ and C=800 fF considering an input square wave amplitude of 2.75 V, where V_{sf} is the output of the TVC, and V_{buf} is the output provided by an analog buffer connected at the TVC output. The results are summarized in Table 3.

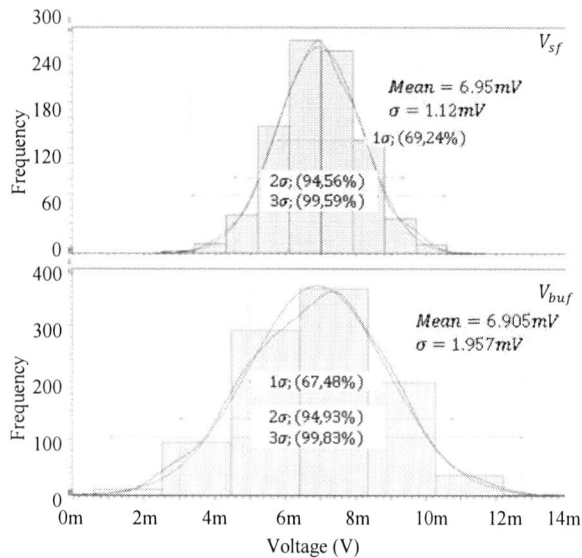

Figure 14. Output of the TVC for the measurement of a line with R=2 kΩ and C=800 fF.

Output	Mean	σ
V_{sf}	6.95 mV	1.21 mV
V_{buf}	6.91 mV	1.957 mV

Table 3. Measurement of a fault free wire.

Considering an LPDD sensitivity of 0.75 ns/kΩ and a TVC transfer ratio of 4.4 mV/ns (V_{ICOL}=1.41 V), the measurement sensitivity is 3.3 mV/kΩ, and the measurement resolution is much better than 1 kΩ. This is sufficient for the detection of weak spot defects that most often lead to an impedance increase much more important. Notice also that an LPDD with external inputs and an internal delay generation block is also considered on-chip for the calibration of the LPDD and TVC blocks.

IV. CONCLUSIONS

In this work, we have discussed a major source of failure in CMOS imagers and presented techniques to facilitate the detection of the actual defects. An improvement of the current optical test program has been applied to the testing of new products, and validated by statistical test results. In order to further increase defect coverage, a BIST technique for the CMOS sensor has been developed. The technique presents a very small silicon overhead equivalent to five CMOS imager columns. The test time corresponds to the time to read one imager column. Simulation results have demonstrated the interest of the technique. The BIST technique has been embedded in a CMOS imager prototype that has been sent for fabrication. Silicon results will be presented during the Conference.

ACKNOWLEDGEMENT

The authors would like to thank the STMicroelectronics Product Engineering team for providing production test results, and the pixel design team for their help on the design of the test chip.

References

[1] E. Fossum, "CMOS image sensors: electronic camera-on-a-chip," *IEEE Trans. Electron Devices*, vol. 44, no. 10, pp. 1689 - 1698, Oct. 1997.

[2] M. Bigas, E. Cabruja, J. Forest and J. Salvi, "Review of CMOS image sensors," *Microelectronics Journal*, pp. 433-451, May 2006.

[3] K. Matou and Y. Ni, "Precise FPN compensation circuit for CMOS APS," *Electronics Letters*, vol. 38, no. 19, pp. 1078 - 1079, Sept. 2002.

[4] E. Labonne, G. Sicard and M. Renaudin, "An on-pixel FPN reduction method for a high dynamic range CMO S imager," in *Solid State Circuits Conference*, Munich, pp. 332 - 335, Sept. 2007.

[5] C. de Moraes Cruz, D. de Lima Monteiro, E. Cotta, V. de Lucena and A. Souza, "FPN Attenuation by Reset-Drain Actuation in the Linear-Logarithmic Active Pixel Sensor," *IEEE Trans. Circuits and Systems*, vol. 61, no. 10, pp. 2825-2833, July 2014 .

[6] K. Yonemoto and H. Sumi, "A CMOS image sensor with a simple fixed-pattern-noise-reduction technology and a hole accumulation diode," *IEEE J. Solid-State Circuits*, vol. 35, no. 12, pp. 2038 - 2043, Dec. 2000.

[7] I. Fujimori, C.-C. Wang and C. Sodini, "A 256/spl times/256 CMOS differential passive pixel imager with FPN reduction techniques," *IEEE J. Solid-State Circuits*, vol. 35, no. 12, pp. 2031 - 2037, Dec. 2000.

[8] S. Yoshihara, Y. Nitta, M. Kikuchi and K. Koseki, "A 1/1.8-inch 6.4 MPixel 60 frames/s CMOS Image Sensor With Seamless Mode Change," *IEEE J. Solid-State Circuits*, vol. 41, no. 12, pp. 2998 - 3006, Dec. 2006.

[9] S. Lim and A. El Gamal, "Gain fixed pattern noise correction via optical flow," *IEEE Trans. Circuits Circuits and Systems I, Reg. Papers*, vol. 51, no. 4, pp. 779 - 786, Apr. 2004.

[10] S. E. Bohndiek, A. Blue, J. Cabello and A. Clark, "Characterization and Testing of LAS: A Prototype ´Large Area Sensor' With Performance Characteristics Suitable for Medical Imaging Applications," *IEEE Trans. Nucl. Sc*, vol. 56, no. 5, pp. 2938 - 2946, Oct. 2009.

[11] M. Purcell, G. Storm, R. Elliott, D. Rasaratnam, M. Qu, M. E. Hachimi, K. Moore, P. Dmochowsky, D. Tolmie, J. Hart, S. Pellegrini, L. Grant and A. Laflaquiere, "Image Artifacts Caused by Pixel Bias Cells in CMOS Imagers targeted for mobile applications," in *International Image Sensor Workshop*, Bergen, Norway, June 2009.

[12] R. Fei, S. Mir and J. Moreau, "Défauts catastrophiques dans les capteurs optiques CMOS 1T75 PIN photodiode," in *Journées GDR ondes*, Grenoble, France, 2013.

[13] C. L. Chen, C. L. Lee and M.-S. Wu, "A new path delay test scheme based on path delay inertia," in *IEEE 13th ATS*, Kenting, pp. 140 - 144, Nov. 2004.

[14] M. Safi-Harb and G. Roberts, "A CMOS circuit for embedded GHz measurement of digital signal rise time degradation," *Proc. IEEE ISCAS*, pp.3446 -3449, Island of Kos, 2006.

2015 IEEE 33rd VLSI Test Symposium (VTS)

Capacitive Coupling Mitigation for TSV-based 3D ICs

Ashkan Eghbal, Pooria M.Yaghini, and Nader Bagherzadeh
Center for Pervasive Communications and Computing
Department of Electrical Engineering and Computer Science, University of California, Irvine
Email:{aeghbal, pooriam, nader}@uci.edu

Abstract—TSV-to-TSV capacitive coupling has large disruptive effects on timing requirements of the circuit. The latency effect of TSV-to-TSV capacitive coupling for different characteristics of a TSV using circuit-level model is presented in this article. Two coding approaches are proposed to mitigate capacitive parasitic effects by adjusting the current flow pattern for any given $n \times n$ mesh of TSV arrangement to reduce the number of 8C/7C parasitic capacitance. The experimental results proves the efficacy of the proposed coding methods.

I. INTRODUCTION

IC transistors have reached the fundamental limits of miniaturization at the atomic levels; Three-Dimensional (3D) IC has been proposed as an emerging technology to keep Moore's Law ticking by packing a great deal of functionality into small die area. A 3D IC is actually composed of multiple tiers of thinned-active Two-Dimensional (2D) ICs which are stacked, bonded, and electrically connected with vertical vias formed called Through-Silicon Vias (TSVs) [1]. TSVs support higher bandwidth and core integration with lower power consumption and latency [1]. However, the impact of sub-micron TSVs on future 3D ICs is still under study [2]. TSVs are known as noticeable sources of coupling noise that deteriorate the Signal Integrity (SI) of 3D IC layouts in literature. This is an effect of fine pitch integration on conductive silicon substrate in smaller form factor of 3D ICs [3], introducing TSV coupling as a 3D IC design challenge. The term TSV coupling refers to capacitive and inductive couplings among neighboring TSVs. Electric field results in capacitance coupling and magnetic field is the source of inductive coupling. The capacitance coupling between TSVs depends on the permittivity of the oxide, TSV geometry, the arrangement of surrounding TSVs, and body contacts places. Inductive coupling is disruptive in higher operation frequencies (>5GHz) while capacitive coupling is considerable with application in lower range of frequencies (<5GHz). TSV-to-TSV Capacitive Coupling (TTCC) is considered in this article as one of the major issues of 3D IC design. We have previously proposed solely TSV-to-TSV inductive coupling aware coding for both low and high bandwidth application [4], [5], but they are not proper for TTCC. Based on our analysis, the TSV configuration pattern which is the most interference-free in inductive and capacitive coupling analyses are different.

Many crosstalk-aware methods have been suggested for 2D designs [6]–[8], they are not expandable to 3D designs due to different physical characteristics of 2D wires and TSVs. Furthermore, the number of neighbor TSVs in 3D architectures is not the same as the number of neighbor wires in 2D designs. Many research groups have addressed the effect of capacitive TSV coupling on delay and SI of circuits and interconnects in 3D ICs [9]–[11], but limited systematic solutions have been proposed to reduce TTCC effects in 3D ICs. Five solutions are suggested

in [9] to reduce the coupling including: increasing TSV distances, shielding the victim TSVs, inserting buffers at the victim net, decreasing the driver size at the aggressor net, and increasing the load at both victim and aggressor net. The last two approaches have negative implications for timing performance, and others need high effort at post-design time. A crosstalk avoidance coding for 3D VLSI has been proposed in [12]. This coding is suitable for a 2D array of $3 \times n$ by limiting some specific data bit patterns for transmission, but this method is not scalable for larger mesh of TSVs. This is because the computation process of their algorithm is increased exponentially for larger number of n, as they need to enumerate all the patterns and count the valid ones. 3DLAT [13] has been proposed with the goal of capacitive crosstalk reduction and power consumption overhead minimization in the TSV array. In this method they suggest to encode the input data to a codeword which contains limited number of 1's in every 3×3 TSV array. However they have not considered the vertical and horizontal overlap among 3×3 TSV arrays in their proposed technique. Furthermore the overhead of their design for larger mesh of TSVs is not negligible. In this paper, two TTCC mitigation coding for small and large 3D IC bandwidths with affordable overhead are proposed. Our proposed methods are scalable to support any $n \times n$ number of TSVs without limiting any specific data patterns. The main contributions of this work are:

- To introduce the worst class of TTCC by a circuit-level analysis.
- To devise a baseline and an enhanced system-level method to mitigate the TTCC effect for smaller and larger bandwidths.
- To evaluate the efficiency and overhead of both proposed methods.

II. TTCC CHARACTERIZATION

In this section the current flow of TSVs and parasitic capacitance coupling is first discussed and then the presented classes of parasitic capacitanceare compared using circuit-level model.

A. Current flow in TSVs

In order to characterize the effects of capacitive coupling between TSVs used in a CMOS digital circuit, it is first necessary to characterize the direction of current in a given TSV, based on the transmission direction and the the sequential data bit values. Fig. 1 illustrates six possible cases in which a TSV has three possible current directions including: downward, upward, and no-current. For the cases where the data is transmitted from an upper to a lower layer, Fig. 1(a) shows that the TSV current is conducted downward if its voltage makes a high-to-low transition; Fig. 1(b) shows that the TSV current is conducted upward if its voltage makes a low-to-high transition. For the cases where the data is transmitted from an upper to a lower layer, the currents are in the

978-1-4799-7598-3/15 $31.00 © 2015 IEEE
159

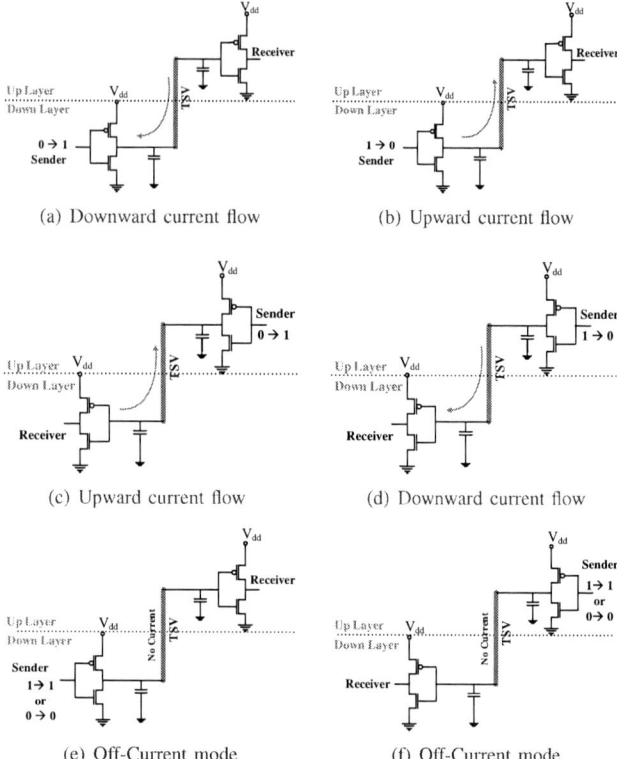

(a) Downward current flow

(b) Upward current flow

(c) Upward current flow

(d) Downward current flow

(e) Off-Current mode

(f) Off-Current mode

Fig. 1. Current flow direction in TSV.

opposite direction of those indicated in Fig. 1(a) and Fig. 1(b), as shown in Fig. 1(c) and Fig. 1(d), respectively. If there is no output data transition on the TSV, then no current will conduct, as shown in Fig. 1(e) and Fig. 1(f). In the rest of this article, a TSV which does not have any current flow is called an inactive TSV. The $\odot$, $\otimes$, and $\bigcirc$ symbols represent active TSV with upward, downward current flow directions, and inactive TSV, respectively.

The total capacitive coupling voltage on the victim TSV is equal to the sum of voltages coupled by each aggressor on the victim TSV [12]. Furthermore, the value of TTCC depends on the distance of each pair of TSVs. It is proven that the value of TTCC between a victim TSV and its diagonal neighbors is roughly 1/5 of the value of TTCC between a victim TSV and its adjacent TSVs [12]. So only adjacent neighbor TSVs are considered in this experiment for the sake complexity of the proposed algorithms. We use the same 9 presented classification of TTCC for each element of mesh of TSV as discussed in [12] to discuss our proposed methods. In this classification the severity of capacitive coupling voltage between each pair of TSVs is represented by:

- **0C** if they both have the same direction or they are both inactive TSVs (like $\otimes\otimes$ or $\bigcirc\bigcirc$).

- **1C** if one of them is inactive and the other is active (like $\bigcirc\otimes$ or $\odot\bigcirc$).
- **2C** if they have reverse current flow (like $\odot\otimes$).

With this definition the maximum capacitive coupling voltage on a victim TSV in this representation is 8C. Neglecting the diagonal neighbor TSVs, there are $3^5 = 243$ possible TSV configurations of active (upward or downward) and inactive TSVs, while many of them are similar as long as the capacitive coupling value is concerned. Table I summarizes all possible TSV configuration patterns with their occurrence frequency for each capacitive coupling class from the range of 0C to 8C. According to the table, the 3C parasitic capacitance has the highest probability of occurrence while 8C parasitic capacitance has the lowest one.

B. Circuit-level model

As discussed earlier, TSV coupling deteriorates the delay and SI of neighboring TSVs. To evaluate the effect of TTCC on timing of 3D IC, a victim and its four adjacent neighbors are modeled in HSPICE. In simulations, the top and bottom end of each TSV is connected to a D flip-flop to monitor the effect of TTCC on their sampling time. Predictive Technology Model (PTM) [14] FinFET transistor models are employed to implement D flip-flops in this experiment. The severity of each class of TTCC from 0C to 8C are reported for different range of process technologies (20nm to 7nm) in Fig. 2. The radius, length, pitch, and t_{ox} parameter values of TSVs in this experiment are $5\mu m$, 15μ, $21\mu m$, and $2\mu m$, respectively extracted from reported values in ITRS reports [15].

Timing Violation (TV) is defined as the additional delay, relative to the clock period, caused by the parasitic capacitive coupling which is given by:

$$TV = \frac{APD - NPD}{T_{clk}} = f_{clk}(APD - NPD) \quad (1)$$

where APD refers to actual path delay (when there is CTTC), NPD refers to nominal path delay (when there is no CTTC), T_{clk} is the clock period, and f_{clk} is the clock frequency. According to the experimental results, the effects of 8C and 7C parasitic capacitance are more critical than the other types, specially in higher frequencies, which are targeted in the proposed TSV-to-TSV Capacitive Coupling Mitigation Algorithm (TCMA).

III. PROPOSED CODING APPROACHES

The main goal of our proposed TCMA , is to reduce the probability of 7C and 8C parasitic capacitance emergence by adjusting the transmitting data bits. Mitigation is chosen in this experiment since eliminating all 7C and 8C parasitic capacitance imposes a complex architecture which is not scalable for any size of TSV meshes [12]. In this Section, our baseline TCMA is discussed for small interconnections and then issues of the baseline method for large interconnections are highlighted. Finally, the enhanced TCMA is presented which supports large mesh of TSVs.

Table I
TSV-TO-TSV CAPACITIVE COUPLING CATEGORIZATION

Types	0C	1C	2C	3C	4C	5C	6C	7C	8C
Sample pattern	$\odot$ $\odot\ \odot\ \odot$ $\odot$	$\odot$ $\odot\ \odot\ \odot$ $\bigcirc$	$\odot$ $\odot\ \odot\ \odot$ $\otimes$	$\odot$ $\odot\ \odot\ \bigcirc$ $\otimes$	$\odot$ $\odot\ \odot\ \otimes$ $\otimes$	$\bigcirc$ $\odot\ \odot\ \otimes$ $\otimes$	$\otimes$ $\odot\ \odot\ \otimes$ $\otimes$	$\otimes$ $\bigcirc\ \odot\ \otimes$ $\otimes$	$\otimes$ $\otimes\ \odot\ \otimes$ $\otimes$
Occurrence frequency	3	16	44	64	54	32	20	8	2
Occurrence probability	0.01	0.07	0.18	0.26	0.22	0.13	0.08	0.03	0.01

978-1-4799-7598-3/15 $31.00 © 2015 IEEE

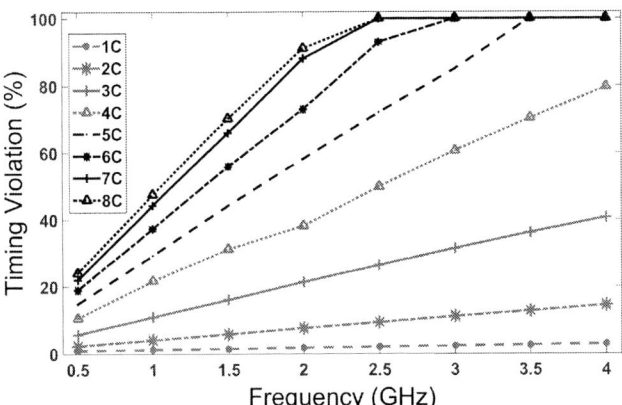

Fig. 2. Severity of TTCC of each classes

A. Baseline TCMA

The TTCC is data-dependent as described in Section II. The basic idea of the baseline TCMA is to encode, if necessary, the consecutive data bits transmitting over the TSVs in order to mitigate the frequency of 7C and 8C parasitic capacitance. This method does not limit any pattern of data transmission bits by encoding them before transmission and decoding them in receiver side, if needed. The inversion operation is chosen as a simple but light and efficient practical coding method in TCMA in order to keep the overhead low, while mitigating TTCC noise. In a mesh of TSVs, a single bit per row is needed in TCMA to determine whether the inversion process is needed or not at the receiver side. TCMA stores the last transmitted data bit of each TSV and compares it with the available data bit which has not been transmitted yet. The current direction matrix of all TSVs is generated by comparing these successive data bits as described in Section II. Then the parasitic capacitance for each of TSVs are calculated based on the the current flow of its neighbor TSVs. Each row of 2D array of TSVs including 8C or 7C parasitic capacitance values is nominated for the data encoding process. By encoding the ready to transmit data bits, 8C parasitic capacitance will be 4C and 7C parasitic capacitance will be 1C or 2C in this method.

B. Enhanced TCMA

Although the baseline TCMA reduces the quantity of 8C and 7C parasitic capacitance values, but it may have some undesirable side effects by converting a row of data bits. For some special data patterns, converting a single row of data bits may generate unexpected 8C or 7C parasitic capacitance values, which happens

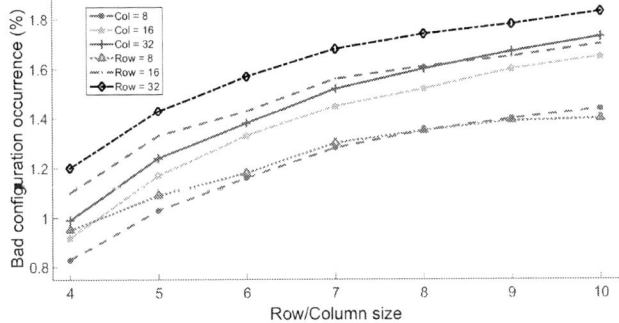

Fig. 3. Probability of bad configuration occurrence

Table II
CURRENT FLOW OF TSVS BEFORE AND AFTER ENCODING

Sent data	Ready to send data	CF_{bi}	CF_{ai}
0	0	○	⊙
0	1	⊙	○
1	0	⊗	○
1	1	○	⊗

in a mesh of TSV with more than 3 rows or 6 columns. We refer to these special cases as bad configuration in the rest of this article.

A bad configuration is a subset of TSV mesh which potentially generates unexpected 8C or 7C parasitic capacitance values by converting a single row of data bits. In more details, the row encoding affects the other data bits of the same row or the data bits in predecessor or successor rows in 2D matrix of TSVs. However, since the probability of bad configuration occurrence is low, specially for smaller matrices of TSVs, the baseline coding is still efficient for smaller data buses (less than 64 bits) which are considered in 3D Network-on-Chip (3D NoC applications). Fig. 3 shows the probability of bad configuration occurrences in different mesh size of TSVs. This experiment is done by running the Monte Carlo simulations for 10000 iterations for different row/column dimensions. According to experimental results the reported percentage of bad configuration for all of the experimented dimensions is less than 2%.

However, the baseline coding is not scalable for larger data buses (more than 64 bits) which are applied in 3D memory applications according to the increasing trend in Fig. 3. The enhanced version of TCMA is devised for these sorts of application to make sure the encoding process of a selected data bit of TSVs does not worsen the total capacitive coupling. First, we explore the bad configuration in detail and then present our solution.

Table II summarizes the TSV current flow direction before and after encoding its ready to send data bit. CF_{bi} shows the current flow of TSV before inverting the ready to send data, while CF_{ai} represents the current flow of TSV after inversion. Based on this table an inactive TSV current flow (○) may convert to active TSV (either ⊙ or ⊗), while an active TSV (either of the ⊙ or ⊗) is converted into an inactive one (○) after inverting the ready to send data bits. Based on our analysis, a bad configuration occurs in five cases, while two of them are potential to generate unwanted 8C parasitic capacitance and the other three may generate unwanted 7C parasitic capacitance. They are called bad_config$_{8\_1}$, bad_config$_{8\_2}$, bad_config$_{7\_1}$, bad_config$_{7\_2}$, and bad_config$_{7\_3}$. Fig. 4 illustrates these five cases in top view of 2D array of TSVs in a 3x3 mesh of TSVs. The candidate row for inversion is recognized by dashed lines in this figure. It also shows the parasitic capacitance value of middle TSV in the recognized row by dashed lines before and after encoding process. Each of these bad configurations affects the result of baseline coding with some conditions which are discussed in the following.

In the baseline method and in case of encoding, the 3C parasitic capacitance, if any, is converted into 7C (see Fig. 4(a))by encoding the second row of 2D array of TSVs with following four conditions:

- There are exactly two inactive TSV next to each other in potential row for encoding process as in TSV5 and TSV6.
- TSV2 and TSV8 are active with the same current direction.
- The current direction of TSV6 after encoding should be the

978-1-4799-7598-3/15 $31.00 © 2015 IEEE   161

same as the current direction of TSV2 and TSV8.

- The current direction of TSV5 should be reverse of the current direction in TSV2, TSV6, and TSV8 after encoding.

The 1C parasitic capacitance is converted into 7C (see Fig. 4(b)) by encoding the second row of 2D array of TSVs with following four conditions:

- There are at least three inactive TSV next to each other in potential row for encoding process as in TSV4, TSV5, and TSV6.

- Either of TSV2 or TSV8 is inactive and the other should be active.

- The current direction of TSV4 and TSV6 after encoding should be the same as the current direction of either TSV2 or TSV8 which was active.

- The current direction of TSV5 after encoding should be reverse of the current direction of TSV4, TSV6, and either TSV2 or TSV8 which was active.

The 6C parasitic capacitance is converted into 7C (see Fig. 4(c)) by encoding the third row of the 2D array of TSVs with following four conditions:

- In capacitive matrix there is a 6C parasitic capacitance in predecessor row which is selected for encoding in a way that TSV5 has reverse current direction of TSV2 and either of TSV4 or TSV6.

- TSV8 which is in the nominated row for encoding is inactive.

- One of TSV4 or TSV6 is inactive and the other should should be active with reverse current direction of TSV5.

- The current direction of TSV8 after encoding should be same as current direction of TSV2 and either of TSV4 or TSV6 which was active.

The 2C parasitic capacitance is converted into 8C (see Fig. 4(d)) by encoding the second row of 2D array of TSVs with following four conditions:

Algorithm 1 Enhanced TCMA coding algorithm

1: AMAT ← Sent data bits
2: BMAT ← To be sent data bits
3: CMAT ← Current direction of each TSV generated by AMAT & BMAT
4: CAPMAT ← Capacitive parasitic noise of each TSV generated by CMAT
5: INV ← Redundant vector for inversion process decision at receiver side
6: **for each** $R \in Rows$ **do**
7:      **for each** $C \in Columns$ **do**
8:          **if** CAPMAT[R][C] == 8 or CAPMAT[R][C] == 7 **then**
9:              $78C\_counter + +$
10:          **end if**
11:          **if** (there is a bad configuration $bad\_config_{7\_1}$ or $bad\_config_{7\_2}$ or $bad\_config_{7\_3}$) **then**
12:              $bad\_config_7\_counter + +$
13:          **end if**
14:          **if** (there is a $bad\_config_{8\_1}$ or $bad\_config_{8\_2}$) **then**
15:              $bad\_config_8\_counter + +$
16:          **end if**
17:      **end for**
18:      **if** ($78C\_counter > bad\_config_7\_counter + bad\_config_8\_counter$) **then**
19:          Encode the BMAT[R]
20:          INV[R]=1
21:      **end if**
22: **end for**

- There are at least three inactive TSVs beside each other in potential row for encoding process like TSV4, TSV5, and TSV6.

- TSV2 and TSV8 are active with same current direction.

- The current direction of TSV4 and TSV6 after encoding should be the same as the current direction of TSV2 and TSV8.

- The current direction of TSV5 should be reverse of the current direction in TSV2, TSV4, TSV6, and TSV8 after encoding.

The 7C parasitic capacitance is converted into 8C (see Fig. 4(e)) by encoding the third row of 2D array TSV, if the following conditions are satisfied:

- In capacitive matrix there is a 7C parasitic capacitance in predecessor row which is selected for encoding. The inactive TSV should be also in the selected row for encoding.

- TSV8 has the reverse current direction of TSV5 after encoding process.

The probability of bad configuration presence in a mesh of TSVs is very low since all the discussed conditions should be satisfied simultaneously. However, the goal of the enhanced TCMA, which is summarized in Algorithm 1 is to guarantee the encoding process will not worsen the total number of 7C and 8C parasitic capacitance in a 2D array of TSVs. In the enhanced version of TCMA the encoding process will be done if the total number of 7C and 8C parasitic capacitance in capacitive matrix is higher than the total number of bad configuration in each row.

IV. TCMA ELABORATION AND EVALUATION

Fig. 5(a) illustrates an example of the baseline and enhanced algorithm for 7×10 given AMAT and BMAT matrices. These matrices and the ones which are used in following sentences are defined in Algorithm 1. This dimension has been chosen to show the advantages of the enhanced approach over the baseline technique for higher bandwidth data buses. First, CMAT and then CAPMAT matrices are generated form the sent (AMAT) and not sent yet (BMAT) data lines. The current flow of each TSV is presented with the same method as discussed in Section II. Then, CAPMAT is generated from CMAT by counting the total mutual capacitive parasitic difference between each TSV and its adjacent neighbors. The INV matrix is evaluated in the receiver side to extract the original data values if they are encoded. $\text{INV}_{baseline}$

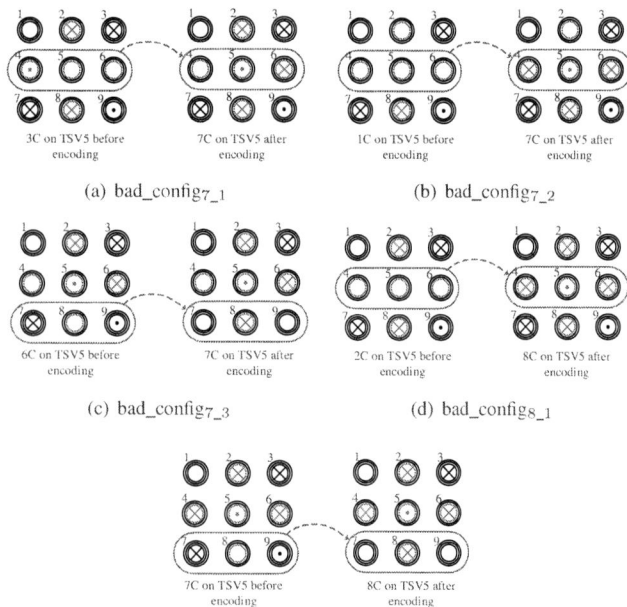

(a) bad_config$_{7\_1}$ (b) bad_config$_{7\_2}$

(c) bad_config$_{7\_3}$ (d) bad_config$_{8\_1}$

(e) bad_config$_{8\_2}$

Fig. 4. Potential configurations to generate 7C and 8C parasitic capacitance

978-1-4799-7598-3/15 $31.00 © 2015 IEEE

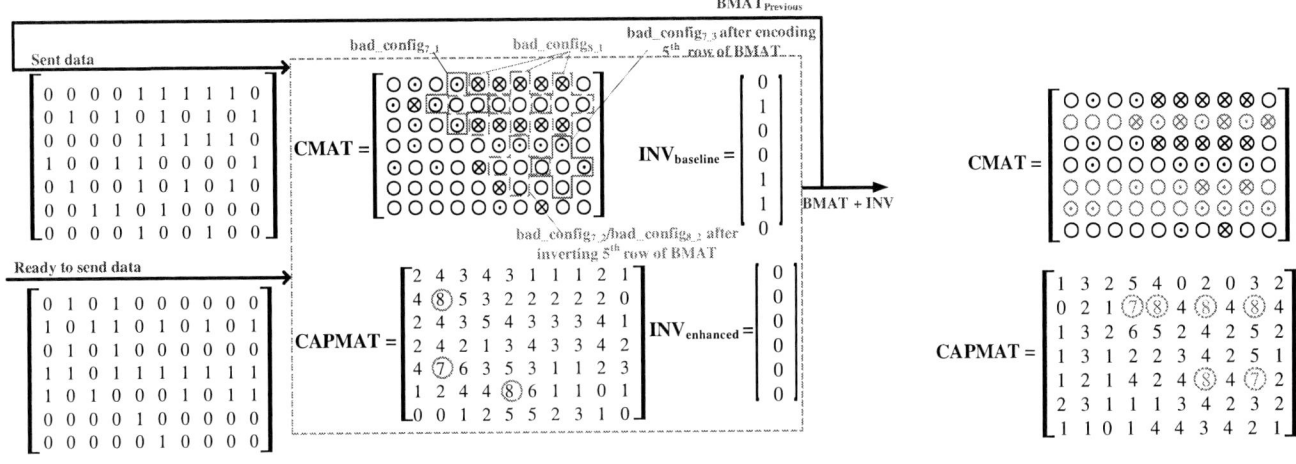

(a) Example of baseline and enhanced TCMA

(b) CMAT and PMAT after baseline TCMA

Fig. 5. An example that shows baseline algorithm issues

of this example shows that the second, fifth, and sixth rows of the BMAT matrix have been encoded since there are 8C or 7C parasitic capacitance values in these rows of CAPMAT matrix. Since the number of 7C and 8C parasitic capacitance are not higher than the number of bad configuration in enhanced method, the $INV_{enhanced}$ shows none of the rows the BMAT has been encoded. Fig. 5(b) represents the updated CMAT and CAPMAT matrices in the baseline approach after encoding the second, fifth, and sixth rows of BMAT matrix in which the total number of 7C and 8C parasitic capacitance increases from 3 to 6. This example illustrates all 5 possible bad configurations. The bad_config$_{7\_1}$ and bad_configs$_{8\_1}$ are depicted in second row of CMAT in Fig. 5(a), resulting in three 8C and one 7C after encoding second row of BMAT. The bad_config$_{7\_2}$ of fifth row is highlighted in CMAT matrix of Fig. 5(a). After encoding fifth row of BMAT, the undesirable 7C will be generated in CAPMAT[5][7], which is also bad_config$_{8\_2}$. Furthermore, encoding fifth row of BMAT generates a bad_config$_{7\_3}$ in CAMPMAT[5][9]. Since the encoding decision is supposed to be done row by row in one direction (from top to bottom in this example) or reverse, the unwanted generated 7C and 6C in fifth row of CAPMAT are potential to generate 8C and 7C, respectively by encoding the sixth row of BMAT. Due to the presence of 8C in sixth row of CAPMAT, it is selected for encoding process and both of bad_config$_{7\_3}$ and bad_configs$_{8\_2}$ generate undesirable 8C and 7C in fifth row of CAPMAT which is shown in Fig. 5(b). However, the enhanced algorithm prevents all of these bad effects by predicting them.

To evaluate the advantages of the baseline TCMA for smaller mesh size, Monte Carlo simulations for 10000 iterations on different sizes of TSV mesh are examined. The total number of 7C and 8C parasitic capacitance before and after applying the baseline TCMA for different mesh size of TSVs is shown in Fig. 6. It is depicted that the mitigation rate of 7C and 8C parasitic capacitance after applying the baseline TCMA are almost 98%, 94%, and 90% for 4×4, 6×6, and 8×8 mesh of TSVs. The information redundancy of the baseline TCMA method for these sizes of mesh of TSVs are 25%, 16%, and 12%. However, the mitigation rate of the baseline TCMA is increased for large mesh of TSVs, as expected. This is because of the probability

of bad configuration occurrence rises by increasing the sizes of TSV meshes. The Monte Carlo simulations for 10000 iterations for larger mesh of TSVs are also examined for both baseline and enhanced TCMA to show the advantages of enhanced TCMA. Although the mitigation rate of total number of 7C and 8C parasitic capacitance values is increasing by using larger mesh of TSVs, enhanced TCMA prevents encoding process if the result is worsen. This is shown in Fig. 7(a), in which the mitigation rate of 7C and 8C parasitic capacitance occurrence by applying enhanced TCMA are always higher than baseline approach.

PARSEC benchmark [16] as a realistic data traffic for large size of mesh of TSVs are also applied to check the performance of the baseline and enhanced TCMA. Memory traces of PARSEC applications have been employed in this experiment, which are extracted by the PIN tool [17], a dynamic binary instrumentation framework for the IA-32 and x86-64 instruction-set architectures. The total number of 7C and 8C parasitic capacitance values for memory traces of PARSEC application workloads through the TSVs are reported for a 8×32 mesh of TSVs in Fig. 7(b). The migration rate of TCMA for Blackscholes, Facesim, Vips, and Raytraces are between 80% to 90% and for the rest of them is almost 70%. Although the differences between the mitigation rates of baseline and enhanced TCMA are not very much, but the result of enhanced method is always better than baseline as it is expected. In other words, it is always guaranteed that by

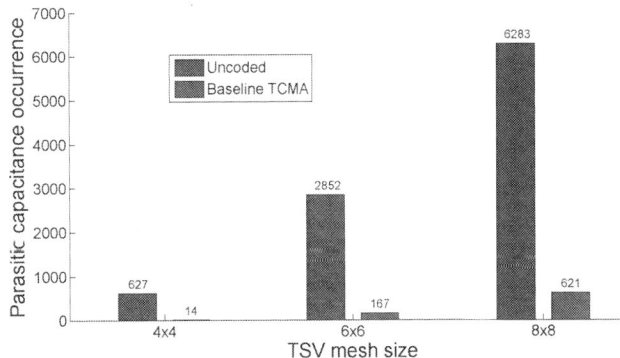

Fig. 6. Number of 7C/8C for random data bit patterns in small mesh of TSVs

978-1-4799-7598-3/15 $31.00 © 2015 IEEE

163

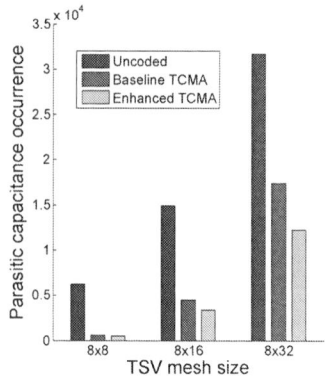
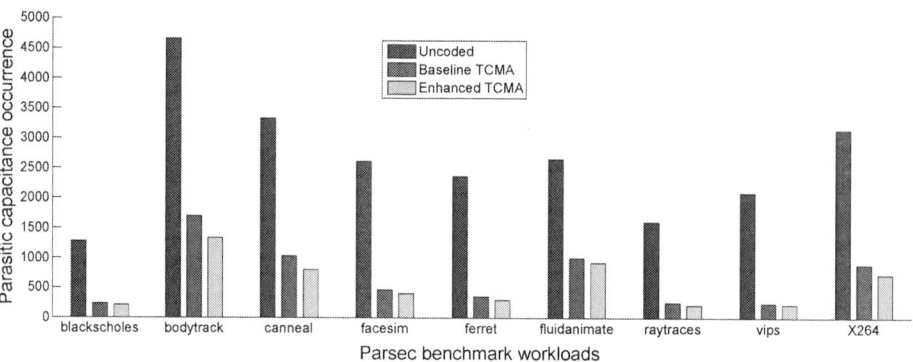

(a) Random data bit patterns in larger mesh of TSVs

(b) PARSEC application data bit patterns in 8×32 mesh of TSVs

Fig. 7. 7C and 8C parasitic capacitance for random and PARSEC applications data with/without TCMA

applying the enhanced TCMA the total number of 7C and 8C parasitic capacitance will never be worse off because of the bad configuration presence.

In order to evaluate the proposed coding methods, the baseline and enhanced TCMA encoders are implemented in Verilog and synthesized by Synopsys Design Compiler using 28nm TSMC library (1.05V, 25 °C). Table III reports the synthesis results as power consumption and occupied area. The latency of the enhanced method is reported by the critical path including: registers latching the adjusted output data bits toward the feedback input for subsequent CMAT computation. In other words, it does not depend on the dimension of TSV arrays. According to the experimental results the latency of the baseline and enhanced TCMA are reported 69.5ps and 74.9ps for all given TSV dimensions in Table III. The feasibility of both proposed coding algorithms are confirmed by considering the gained coupled parasitic capacitance mitigation and its tangible footprint and power consumption. Decoder units are not implemented in this experiment since they are only composed of a comparator and a mix of inverter gates. They are much lighter than encoder components in terms of area, power consumption, and latency.

V. CONCLUSION

Two baseline and enhanced algorithms have been proposed in order to minimize the TSV-to-TSV capacitive coupling issue. Baseline algorithm is proposed for small mesh of TSVs which are considered in 3D NoC applications, while enhanced method is suggested for large mesh of TSVs which are more applied in 3D memory applications. The enhanced method guarantees that the encoding process prevents generating undesirable parasitic capacitance values by recognizing all susceptible configurations. According to experimental results, the baseline method's mitigation rate is more than 90% for TSV meshes smaller than 10×10.

Table III
HARDWARE SYNTHESIZE RESULTS

Mesh size	Baseline		Enhanced	
	Area (μm^2)	Power (μW)	Area (μm^2)	Power (μW)
8×8	918	2340	1096	3000
8×16	1818	4520	2173	5900
16×8	2094	5260	2165	5880
8×32	3321	8840	4331	11700
32×8	4086	11000	4323	11700

The enhanced algorithm mitigates the TSV-to-TSV capacitive coupling more than 70% for 8×32 mesh of TSVs.

REFERENCES

[1] J. Burns, "Tsv-based 3d integration," in *Three Dimensional System Integration*, A. Papanikolaou, D. Soudris, and R. Radojcic, Eds. Springer US, 2011, pp. 13–32.

[2] D. H. Kim and S. K. Lim, "Design quality trade-off studies for 3-d ics built with sub-micron tsvs and future devices," *Emerging and Selected Topics in Circuits and Systems, IEEE Journal on*, vol. 2, no. 2, pp. 240–248, 2012.

[3] C. Liu and S. K. Lim, "A study of signal integrity issues in through-silicon-via-based 3d ics," in *Interconnect Technology Conference (IITC), 2010 International*, June 2010, pp. 1–3.

[4] A. Eghbal, P. M. Yaghini, and N. Bagherzadeh, "Tsv-to-tsv inductive coupling-aware coding scheme for 3d network-on-chip," in *Defect and Fault Tolerance in VLSI and Nanotechnology Systems (DFT), 2014 IEEE International Symposium on*, Oct 2014, pp. 92–97.

[5] P. Yaghini, A. Eghbal, M. Khayambashi, and N. Bagherzadeh, "Coupling mitigation in 3-d multiple-stacked devices," *Very Large Scale Integration (VLSI) Systems, IEEE Transactions on*, vol. PP, no. 99, pp. 1–1, 2015.

[6] C. Duan, V. Calle, and S. Khatri, "Efficient on-chip crosstalk avoidance codec design," *Very Large Scale Integration (VLSI) Systems, IEEE Transactions on*, vol. 17, no. 4, pp. 551–560, April 2009.

[7] K. N. Patel and I. L. Markov, "Error-correction and crosstalk avoidance in dsm busses," in *Proceedings of the 2003 International Workshop on System-level Interconnect Prediction*, ser. SLIP '03. ACM, 2003, pp. 9–14.

[8] C. Duan, B. J. LaMeres, and S. P. Khatri, "Preliminaries to on-chip crosstalk," in *On and Off-Chip Crosstalk Avoidance in VLSI Design*. Springer US, 2010, pp. 13–26.

[9] C. Liu, T. Song, J. Cho, J. Kim, J. Kim, and S.-K. Lim, "Full-chip tsv-to-tsv coupling analysis and optimization in 3d ic," in *Design Automation Conference (DAC), 2011 48th ACM/EDAC/IEEE*, June 2011, pp. 783–788.

[10] R. Weerasekera, M. Grange, D. Pamunuwa, and H. Tenhunen, "On signalling over through-silicon via (tsv) interconnects in 3-d integrated circuits," in *Design, Automation Test in Europe Conference Exhibition (DATE), 2010*, March 2010, pp. 1325–1328.

[11] K. Salah, A. El Rouby, H. Ragai, and Y. Ismail, "Tsv impact on circuit performance and recommended design methodologies," in *Microelectronics (ICM), 2012 24th International Conference on*, Dec 2012, pp. 1–4.

[12] R. Kumar and S. P. Khatri, "Crosstalk avoidance codes for 3d vlsi," in *Design, Automation Test in Europe Conference Exhibition (DATE), 2013*, 2013, pp. 1673–1678.

[13] Q. Zou, D. Niu, Y. Cao, and Y. Xie, "3dlat: Tsv-based 3d ics crosstalk minimization utilizing less adjacent transition code," in *Design Automation Conference, 2014 19th Asia and South Pacific*, 2014, pp. 762–767.

[14] PTM, "Predictive Technology Model," ptm.asu.edu.

[15] S. Itr, "ITRS 2012 Executive Summary," ITRS.

[16] C. Bienia and K. Li, "Parsec 2.0: A new benchmark suite for chip-multiprocessors," in *Proceedings of the 5th Annual Workshop on Modeling, Benchmarking and Simulation*, June 2009.

[17] Intel-cooperation, "Pin-A Dynamic Binary Instrumentation Tool," Intel. [Online]. Available: https://software.intel.com/en-us/articles/pin-a-dynamic-binary-instrumentation-tool

978-1-4799-7598-3/15 $31.00 © 2015 IEEE

2015 IEEE 33rd VLSI Test Symposium (VTS)

Improving Accuracy of On-chip Diagnosis via Incremental Learning

Xuanle Ren, Mitchell Martin and R. D. (Shawn) Blanton
Department of Electrical and Computer Engineering, Carnegie Mellon University
5000 Forbes Ave, Pittsburgh, PA, USA
{xuanler, mitchel1, blanton}@ece.cmu.edu

Abstract—On-chip test/diagnosis is proposed to be an effective method to ensure the lifetime reliability of integrated systems. In order to manage the complexity of such an approach, an integrated system is partitioned into multiple modules where each module can be periodically tested, diagnosed and repaired if necessary. The limitation of on-chip memory and computing capability, coupled with the inherent uncertainty in diagnosis, causes the occurrence of misdiagnoses. To address this challenge, a novel incremental-learning algorithm, namely dynamic k-nearest-neighbor (DKNN), is developed to improve the accuracy of on-chip diagnosis. Different from the conventional KNN, DKNN employs online diagnosis data to update the learned classifier so that the classifier can keep evolving as new diagnosis data becomes available. Incorporating online diagnosis data enables tracking of the fault distribution and thus improves diagnostic accuracy. Experiments using various benchmark circuits (e.g., the cache controller from the OpenSPARC T2 processor design) demonstrate that diagnostic accuracy can be more than doubled.

Key words—On-chip diagnosis, k-nearest-neighbor, machine learning, diagnostic accuracy, lifetime reliability

I. INTRODUCTION

Ensuring the lifetime reliability of integrated systems has become a central concern. A robust system should be able to continue acceptable operations over its intended lifetime even in the presence of failures [1]. Although manufacturing tests are performed to help ensure reliability, a chip may still degrade and even fail in the field due to various locations of failure. Early life failure, also called infant mortality, is caused by the defects that are not exposed during manufacturing tests. However, electrical and thermal stress during in-field use will eventually degrade the defect to a significant failure in functionality [2], [3]. Wear-out, also called aging, manifesting as progressive performance degradation, is induced by various mechanisms, e.g., negative-bias temperature instability (NBTI) and hot-carrier injection (HCI) [4].

Various methods have been proposed to detect and avoid failures. First, forward error control (FEC) method uses error correction codes (ECC) to detect and correct data faults during transmission by adding data redundancy to the packets [5]–[8]. However, FEC does not target permanent failures, e.g, early-life and wear-out [9]. On the other hand, most error detection/correction methods incur significant power, performance and area penalties [10]. Second, failure-prediction schemes provide an early warning of circuit aging before errors appear. Specifically, an aging sensor periodically checks the slack

of a critical path in order to avoid the occurrence of delay faults. This approach requires an aging sensor and a stability checker for each flip-flop which however incurs a large area overhead [11]. Third, on-chip self-test schemes test the system periodically for failure detection [10], [12]–[14]. To reduce the overhead, the DFT already on chip for manufacturing test is reused to perform the in-field testing based on vectors stored in off-chip flash memory.

To enable failure localization, diagnosis is performed when on-chip self-test detects a failure. There are generally two diagnostic approaches, namely, effect-cause and cause-effect [15]. In effect-cause, a complete model of the design, its test-vector set, and the test response from the failing circuit are analyzed by diagnostic software to identify possible fault locations. However, it requires both significant memory and run-time. Compared to effect-cause, cause-effect is more feasible. However, it too has challenges since it requires the generation, storage and usage of a fault dictionary that contains the test response for every fault of interest. The size of a full fault dictionary is measured in terabytes for modern designs [15]. In [14], an on-chip diagnosis scheme that performs diagnosis and repair however at the module level (i.e., sub-core/sub-uncore) results in a much more compact dictionary. Instead of storing the simulation response of each fault per test, the compact fault dictionary only stores a single bit per test that indicates the pass/fail status of a subset of faults that enable module-level diagnosis. Such an approach significantly reduces the size of a fault dictionary at the cost of diagnostic resolution (i.e., one module is assumed to be faulty, but multiple modules are diagnosed as fault candidates). After the faulty module is located, self-repair is performed to replace or bypass it [16]. However, if the diagnostic resolution is non-ideal and no further analysis is done to narrow down the fault location, all fault candidates have to be repaired, resulting in inefficient use of on-chip resources.

In this paper, a novel incremental-learning algorithm that we call dynamic k-nearest-neighbor (DKNN) is proposed to improve the accuracy of on-chip, module-level diagnosis. Here, accuracy is defined as the probability that the identified module is the one with the failure, assuming that a single module is faulty. It is assumed that on-chip testing is performed with a test clock that has a higher frequency than the system clock because it allows failure sources that slowly degrade system timing to be tracked over time. For example, delay

978-1-4799-7598-3/15 $31.00 © 2015 IEEE
165

degradation due to NBTI [4] can be monitored, enabling system adjustments (e.g., task scheduling, sleep scheduling, etc.) that mitigate adverse effects on system lifetime. Another consequence of using a faster clock for test, means failures will be much more frequent, thus creating sufficient data for learning a model for improving diagnostic accuracy. Different from the conventional KNN, DKNN employs online data to update the learned classifier, enabling the classifier to evolve as new data becomes available. Consequently, DKNN is able to track the fault distribution, especially when the fault distribution is non-stationary[1]. In [17]–[19], pattern recognition methods are used to make diagnostic decisions in analog circuits. In particular, possible circuit defects are identified through inductive fault analysis. Then a set of classifiers, trained offline using data from fault simulation, is used to map each defect to a score according to its likelihood of occurrence. In [20], an incremental KNN algorithm that also revises the composition of data set by exploiting a "correct-error" teacher is proposed. In their approach, each instance in the data set is associated with a dynamically-evolving weight, that requires additional overhead if implemented on chip. Moreover, the memory required increases as more data is collected, making it difficult to determine the required amount of memory a priori. Compared to the incremental KNN in [20], DKNN is more hardware-friendly because it maintains a fixed-size data set, and only requires little additional logic for data replacement.

Finally, the effectiveness of DKNN is validated using two benchmark circuits, L2B (the L2 cache bank controller of the OpenSPARC T2 processor [21]) and c7552 (an ISCAS benchmark circuit [22]), details of which will be elaborated upon in Section IV.

This work has two main contributions:

• DKNN copes with non-stationary distributed data by updating the classifier incrementally using online data.

• DKNN can be implemented on chip for improving diagnostic accuracy, using little additional logic and a fixed-size data set.

The rest of this paper is organized as follows. Section II provides details of the DKNN algorithm and demonstrates how it improves the accuracy of on-chip diagnosis. Section III presents the on-chip implementation of DKNN while Section IV evaluates the performance using the benchmark circuits and the UCI repository databases [23]. Section V draws conclusions.

II. METHODOLOGY

The flow for improving the accuracy of on-chip diagnosis is depicted in Figure 1. On-chip test/diagnosis generates a coarse diagnosis result which is then refined, if necessary, by the DKNN classifier.

A. Module-level Diagnosis

For the purpose of on-chip test/diagnosis, the core/uncore to be tested is partitioned into a set of interconnected modules,

---

[1]If the probability that a fault is located in a specific module changes over time, then the fault distribution is non-stationary; otherwise, it is stationary.

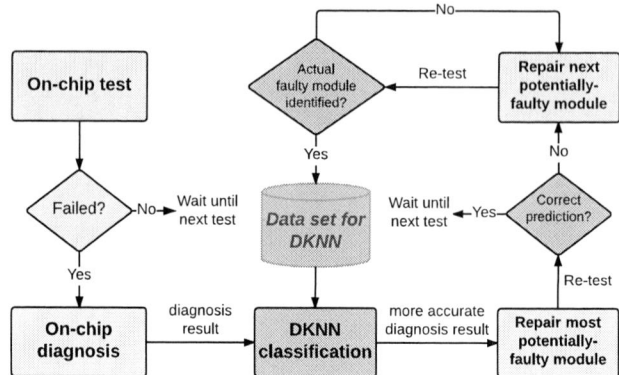

Figure 1: An on-chip diagnosis result is fed into the on-chip DKNN classifier. The module predicted by the DKNN classifier is repaired. If the predicted label proved to be incorrect, the DKNN data set is updated.

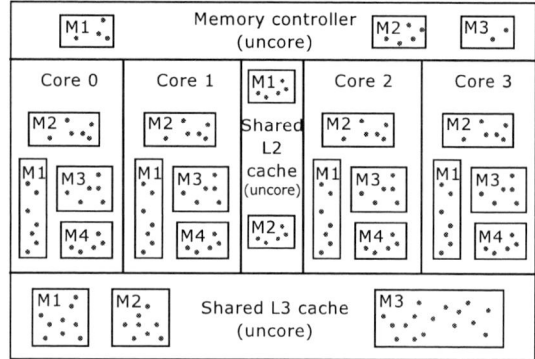

Figure 2: A processor with four cores and several uncores is shown hierarchically. Each core/uncore is partitioned into multiple modules, each of which is assumed to be independently repairable. Within each module lies a number of faults (shown as blue dots) that allow faulty modules to be distinguished via testing [14].

each of which is independently repairable [14]. Figure 2 shows a fictional multicore processor that contains four cores and several uncores. Uncores are defined as system components that are neither processor cores nor co-processors (e.g., memory controller, interrupt handlers, etc.) [13]. In Figure 2, each core/uncore is partitioned into multiple modules, and particular faults identified within each module capture those early-life failure and wear-out locations that allow faulty modules to be distinguished through diagnosis.

Periodic on-chip test/diagnosis starts with testing a core/uncore using a method such as the CASP (Concurrent Autonomous chip self-test using Stored test Patterns) technique [12]. If the core/uncore is faulty, then diagnosis is initiated to locate the faulty module. Specifically, diagnosis generates a list of potentially-responsible delay faults on a per module basis [14]. Further, counting the number of potentially-responsible delay faults for each module provides an indication of the likelihood that the corresponding module is indeed the location of failure. Predicting the faulty module to be the one with the maximum fault-count is called "select-max".

Table 1 shows three diagnosis results for L2B through simulation. Each simulation first injects a delay fault (which mimics early-life failure or wear-out) to a specific gate, and then runs diagnosis that results in a fault-count associated with each module. The first diagnosis result has ideal resolution, i.e., only one module has non-zero fault-count. Thus, M_9 is most likely to be faulty. However, most diagnosis results do not have ideal resolution. The second diagnosis result has three modules associated with non-zero fault-counts, either of which may be the faulty one. In this case, M_5 is deemed to be faulty using the select-max strategy. The third diagnosis result has nine modules associated with non-zero fault-counts. The select-max strategy would deem M_7 as the faulty module but the actual faulty module is M_9.

The examples from Table 1 demonstrate that select-max is not always correct. Based on the simulation results (described in Section IV), the diagnostic accuracy of select-max is 71% for L2B, and 20% for c7552. Here, a diagnostic outcome is deemed accurate if the module with the maximum fault-count is indeed the location of the failure. If select-max is incorrect (i.e., the wrong repaired), the benefit of on-chip test/diagnosis is hindered. Thus, improving diagnostic accuracy is a critical goal of on-chip test/diagnosis.

B. Dynamic k-nearest-neighbor

k-nearest-neighbor (KNN) is an instance-based non-parametric machine-learning algorithm used for classification. Specifically, an unlabeled instance[2] is classified based on the class of its k nearest neighbors [24]. KNN classification depends on the local-similarity nature of data, i.e., two instances belonging to the same class are supposed to be close to each other in the hyperspace that captures the data. This assumption is also applicable to module-level diagnosis. Figure 3 provides an example illustrating that failures in different modules may result in different diagnosis results based on how the modules are interconnected. For example, Figure 3(a) shows two modules of a core/uncore, namely M_1 and M_2, that are connected to each other. The failure effects stemming from M_1 can propagate to M_2, but the converse is obviously not true. Thus, as shown in Figure 3(b), a failure in M_1 (red dots) may cause both M_1 and M_2 to report non-zero fault-counts, but a failure in M_2 (blue dots) cannot cause M_1 to report a non-zero fault-count. In this case, it is easy to find the classification boundary in the two-dimensional space between the faults residing in M_1 and M_2, which indicates that the conventional KNN is capable of identifying the similarities among faults residing in the same module.

DKNN is an incremental-learning algorithm [25]–[27] (Figure 4). An incremental classifier can evolve using new instances without having to re-process past instances. It requires

---

[2] A data instance is considered to be "labeled" if its class is known, otherwise it is "unlabeled".

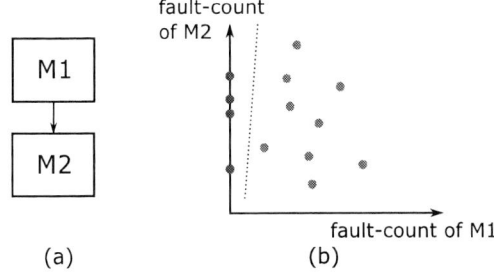

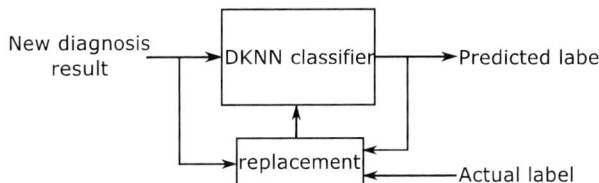

Figure 3: (a) Two modules M_1 and M_2 are connected in an core/uncore as shown. (b) A failure in M_1 may cause both M_1 and M_2 to report non-zero fault-counts (red dots), while a failure in M_2 cannot cause a non-zero fault-count for M_1 as illustrated by the blue dots that lie along the y-axis.

Figure 4: The incremental learning scheme of DKNN. If the predicted label is different from the actual label, then the DKNN classifier is updated.

less data to build the initial classifier and consumes less computational resources. In the DKNN algorithm, a new data instance (i.e., a diagnosis result) is fed into the classifier, resulting in a predicted label. After the actual label is known, both the predicted label and the actual label are fed into the replacement block for comparison. If the predicted label is correct, the data set used by DKNN is left intact. Otherwise, the nearest neighbor that is responsible for predicting the wrong label is replaced with the instance. In other words, the nearest neighbor whose label is the same as the instance is replaced in case of wrong classification. By employing replacement, the size of the data set is constrained to fit the on-chip resources allocated for learning.

The replacement of data instances gradually revises the composition of the data set so that it is more reflective of the instances now being generated by the chip. For example, if the distribution of the data changes, then the DKNN may make a wrong prediction, but then it gradually adapts the distribution of the DKNN data set by incorporating new data. In addition, the replacement can refine the data set by removing noisy data that has an adverse effect on classification. Specifically, if a noisy data causes a wrong label, it is very likely to be replaced. Moreover, although data replacement may cause the classifier to overfit the faults appearing in a specific period of time, the overfitting can be mitigated by dynamically updating the data set, which is also an advantage of DKNN over the conventional KNN.

Faulty gate	Faulty module	M_1	M_2	M_3	M_4	M_5	M_6	M_7	M_8	M_9	M_{10}
U6525_temp0_Y	M_9	0	0	0	0	0	0	0	0	2	0
U7423_temp1_Y	M_5	0	0	3	0	15	0	4	0	0	0
n11747	M_9	77	21	32	34	63	0	83	4	57	11

Table 1: L2B, partitioned into ten modules, is diagnosed at the module level through simulation. The first step of simulation is to inject a delay fault into a specific gate (the first column). The second step is to run on-chip diagnosis that results in a fault-count associated with each module (the third column). The second column is the module that contains the faulty gate.

C. DKNN for Improving the Accuracy of On-chip Diagnosis

Algorithm 1 describes the use of DKNN for improving the accuracy of on-chip diagnosis. The DKNN data set is initialized using fault simulation data, i.e., injecting faults into the gate-level netlist and then generating a fault-count for each module using simulation tools.

A periodic test/diagnosis starts with testing a core/uncore. Only if the test fails, then diagnosis is applied to identify which module of the core/uncore is most likely the reason for the observed test failure. The diagnosis result is reported as a n-dimensional vector of integers, where n is the number of modules. For each diagnosis result, a check for ideal resolution (i.e., only one fault-count is non-zero) is performed. If the resolution is ideal, then prediction via DKNN is obviously not needed. If the resolution is non-ideal, then DKNN is performed to predict which of the n modules is the faulty one. Specifically, the module is labeled using the outcome of a majority vote provided by its k nearest neighbors, and then data replacement is performed if the predicted label is incorrect. It is noted that only "relevant data" is considered when the classifier is searching for nearest neighbors. For example, in Table 1, the second diagnosis result has three modules (i.e., M_3, M_5 and M_7) associated with non-zero fault-counts, and therefore only data instances whose labels are 3, 5, or 7 are considered relevant. In addition, if a tie occurs during majority vote (i.e., two or more modules contribute the same number of neighbors), then either one module is selected as the label.

When a module is predicted to be faulty, it will be repaired and the system will be re-tested to determine if it was the actual faulty module. If the system passes, then the prediction is assumed to be correct; otherwise, the second-potentially faulty module will be repaired. Test, diagnosis, and repair continue until the actual faulty module is identified. Finally, after the actual faulty module is identified, the data replacement is performed.

III. ARCHITECTURE

Figure 5 shows the block diagram for improving the accuracy of on-chip diagnosis. DKNN takes as input the diagnosis result from the on-chip diagnosis scheme described in [14], performs classification using the labeled data that is stored in the on-chip buffer, and updates the data set if the classification is later deemed to be incorrect. The architecture of the KNN classifier is shown in Figure 6. The DKNN classifier hardware employs an IP-core design proposed in [28]. The fully pipelined KNN architecture shown in Figure 6 has $m+k$

Algorithm 1 DKNN for improving the accuracy of on-chip diagnosis.

1: partition a core/uncore into n modules, i.e., $M_1, M_2, ..., M_n$
2: initialize the data set S using simulation data
3: **for** each periodic test/diagnosis **do**
4:    run on-chip test
5:    **if** test passes **then**
6:      wait until next test
7:    **else**
8:      run on-chip diagnosis
9:      collect fault-counts, i.e., $v = (c_1, c_2, ..., c_n)$
10:      **if** v has only one non-zero fault-count **then**
11:        l_{pred} (i.e., predicted module) $\leftarrow i, s.t.\ c_i > 0$
12:      **else**
13:        find k nearest neighbors of v from S, i.e., $(v_1, l_1), ..., (v_k, l_k)$
14:        $l_{pred} \leftarrow$ majority vote of $l_1, ..., l_k$
15:        (if a tie occurs, select either one)
16:      repair module l_{pred} and run on-chip test
17:      l_{actl} (i.e., actual faulty module) $\leftarrow l_{pred}$
18:      **while** test fails **do**
19:        repair next potentially-faulty module l_{next}
20:        run on-chip test
21:        $l_{actl} \leftarrow l_{next}$
22:      **if** $l_{pred} \neq l_{actl}$ **then**
23:        **for** $i \leftarrow 1$ to k **do**
24:          **if** $l_i = l_{pred}$ **then**
25:            replace (v_i, l_i) with (v, l_{actl}) in S
26:            break

---

stages, where m is the number of features, and k is the number of nearest neighbors. The white nodes compute the distances, the dark gray nodes find the labels of the k nearest neighbors, and the light gray nodes find the majority of the KNN labels to be used for prediction. Assuming that the size of the data set is n, the architecture can classify a new data instance in $n + m + k$ clock cycles.

DKNN is described using Verilog and synthesized using Synopsys Design Compiler [29]. The area, critical path, and power are $45,305\mu m^2$, $5.3ns$, and $700\mu W$, respectively, for the following parameter values: $k = 5$, the number of features is 16, and the size of data set is 200. In addition, the latency of predicting the faulty module ($\sim 1.3\mu s @ 166MHz$) is much smaller than that of on-chip test, diagnosis and repair [13], [14], [16].

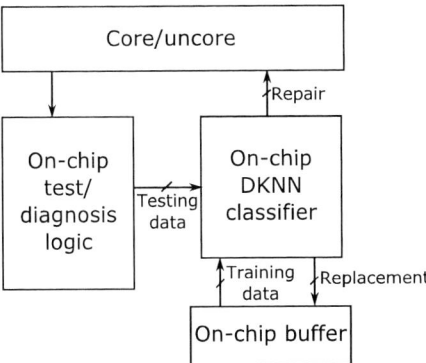

Figure 5: A DKNN classifier is used to improve accuracy of on-chip diagnosis. The block diagram shows the communication of data between various blocks within the system.

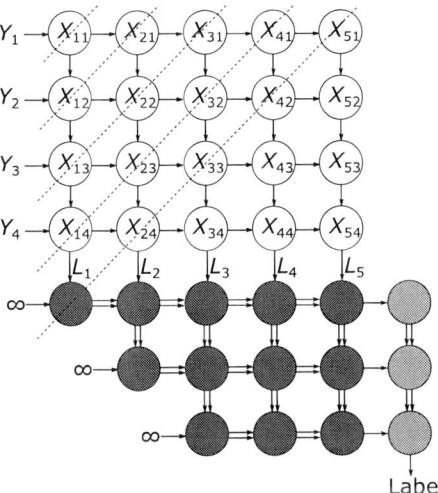

Figure 6: A pipelined architecture for KNN from [28]. The architecture classifies a data instance in $n+m+k$ clock cycles, where n is the size of the data set, m is the number of features, and k is the number of nearest neighbors. In this case, $n =5$, $m =4$ and $k =3$.

IV. EXPERIMENT

To validate the effectiveness of DKNN, two benchmark circuits, namely L2B [21] and c7552 [22], are used for simulation. L2B is the L2 cache bank controller of the OpenSPARC T2 processor, which is partitioned into 10 modules, and c7552 is one of the ISCAS benchmark circuits, which is partitioned into 12 modules.

The benchmark circuits, with delay faults injected into them, are tested/diagnosed, and then the faulty module is predicted using DKNN. The conventional KNN and select-max are also performed to the benchmark circuits for comparison. Table 2 shows the accuracy which is defined as the probability that the predicted module is actually faulty under the assumption that a single module is faulty. The result demonstrates that DKNN

Bench-mark circuits	No. of injected faults	Fault distribution	Select-max	Conventional KNN	DKNN
L2B	997	stationary	71%	79%	83%
		non-stationary	71%	51%	82%
c7552	2,694	stationary	21%	44%	46%
		non-stationary	21%	38%	46%

Table 2: The diagnostic accuracies for select-max, the conventional KNN, and DKNN are compared using the benchmark circuits L2B and c7552.

performs better than both the conventional KNN and select-max. Compared to select-max, DKNN improves the accuracy from 71% to 83% for L2B, and from 21% to 46% for c7552. After all injected faults are classified by DKNN, the amount of data replacement is 165 for L2B, and 1,212 for c7552, representing 16.5% and 45.0% of the initial data sets for the two benchmark circuits, respectively.

To simulate non-stationary fault distribution, we assume that the DKNN data set is initially constructed without any faults from M_9 and M_{10}, but the DKNN classifier has to deal with the faults from M_9 and M_{10} later in time. The result demonstrates that the DKNN is able to maintain high accuracy because it can dynamically incorporate new diagnosis data from M_9 and M_{10}. However, the conventional KNN performs even worse than select-max because it cannot predict any fault from M_9 and M_{10} (Table 2).

Figure 7(a) shows the overall diagnostic accuracy of DKNN versus the size of the DKNN data set. The accuracy is improved as the size of the DKNN data set increases. However, the accuracy saturates if the size is larger than 200. Figure 7(b) shows the overall diagnostic accuracy of DKNN versus k (i.e., the number of nearest neighbors). It shows that the highest overall accuracy is achieved when $k =5$. Thus, 200 is selected as the size of the DKNN data set, and 5 is selected as the value of k for KNN classification.

Finally, the performance of DKNN, as a generic machine-learning algorithm, is evaluated. Specifically, the accuracy of DKNN is compared with the conventional KNN and the incremental KNN algorithm proposed in [20], using seven data sets from the UCI repository databases [23]. These data sets are commonly used for verifying the performance of classifiers. The classification results demonstrate that DKNN performs slightly better than both the conventional KNN and the incremental KNN in [20] (Figure 8).

V. CONCLUSION

In this paper, we have developed a dynamic k-nearest-neighbor (DKNN) algorithm for improving the accuracy of on-chip, module-level diagnosis. Different from the conventional KNN, DKNN described here employs online data to update the learned classifier dynamically, so that the learned classifier can evolve as new data becomes available. Specifically, if the classification for a data instance is proved to be wrong, the nearest neighbor that is responsible for predicting the wrong label is replaced by the data instance itself. When used to

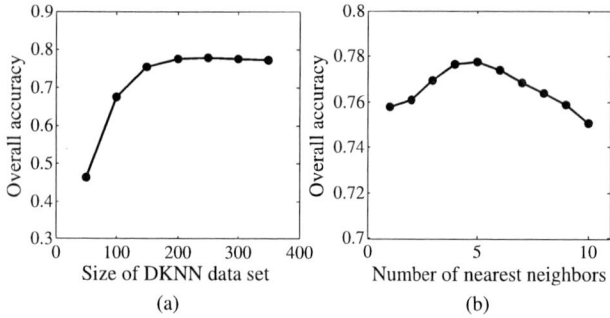

Figure 7: (a) The diagnostic accuracy of DKNN is improved as the size of the DKNN data set increases given $k = 5$. (b) The diagnostic accuracy of the DKNN varies for different values of k (i.e., the number of nearest neighbors) given the size of the DKNN data set is 200.

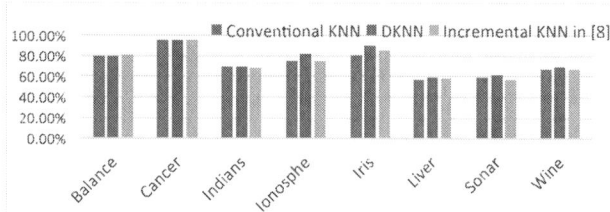

Figure 8: The performance of the conventional KNN, DKNN, and the incremental KNN in [20] are compared using seven randomly selected datasets from the UCI repository databases. 10-fold cross-validation is used, and size of data set is 200.

improve the accuracy of on-chip diagnosis, the experiments demonstrate that DKNN significantly improves the accuracy for benchmark circuits, L2B and c7552. DKNN can achieve acceptable accuracy with limited memory, or in case of non-stationary fault distribution, which makes it suitable for on-chip use. Finally, experiments using the UCI repository [23] show that DKNN performs as well or better than conventional KNN and another incremental KNN algorithm [20].

REFERENCES

[1] S. Borkar, "Designing Reliable Systems from Unreliable Components: the Challenges of Transistor Variability and Degradation," in *Micro*, 2005.

[2] T. Mak, "Infant Mortality-The Lesser Known Reliability Issue," in *International On-Line Testing Symposium*, 2007.

[3] Y. M. Kim, Y. Kameda, H. Kim, M. Mizuno, and S. Mitra, "Low-cost Gate-oxide Early-life Failure Detection in Robust Systems," in *Symposium on VLSI Circuits*, 2010.

[4] N. C. Laurenciu, Y. Wang, and S. D. Cotofana, "A Direct Measurement Scheme of Amalgamated Aging Effects with Novel On-chip Sensor," in *International Conference on Very Large Scale Integration*, 2013.

[5] S. Mitra, N. Seifert, M. Zhang, Q. Shi, and K. S. Kim, "Built-in Soft-error Resilience for Robust System Design," in *Integrated Circuit Design and Technology*, 2005.

[6] M. Zhang, S. Mitra, S. Member, T. M. Mak, N. Seifert, N. J. Wang, Q. Shi, K. S. Kim, N. R. Shanbhag, and S. J. Patel, "Sequential Element Design with Built-in Soft Error Resilience," *Very Large Scale Integration (VLSI) Systems*, vol. 14, no. 12, pp. 1368–1378, 2006.

[7] S. Shamshiri, A. Rani, K. Cheng, and S. Barbara, "End-to-End Error Correction and Online Diagnosis for On-Chip Networks," in *International Test Conference*, 2011.

[8] D. Gizopoulos, M. Psarakis, S. V. Adve, P. Ramachandran, S. K. S. Hari, D. Sorin, A. Meixner, A. Biswas, and X. Vera, "Architectures for Online Error Detection and Recovery in Multicore Processors," in *Design, Automation and Test in Europe*, 2011.

[9] A. Ghofrani, R. Parikh, S. Shamshiri, A. Deorio, K. Cheng, and V. Bertacco, "Comprehensive Online Defect Diagnosis in On-Chip Networks," in *VLSI Test Symposium*, 2012.

[10] Y. Li, Y. M. Kim, E. Mintarno, D. Gardner, and S. Mitra, "Overcoming Early-life Failure and Aging for Robust Systems," in *Design and Test of Computers*, 2009.

[11] M. Agarwal, B. C. Paul, M. Zhang, and S. Mitra, "Circuit Failure Prediction and Its Application to Transistor Aging," in *VLSI Test Symposium*, 2007.

[12] Y. Li, S. Makar, and S. Mitra, "CASP: Concurrent Autonomous Chip Self-Test Using Stored Test Patterns," in *Design, Automation and Test in Europe*, 2008.

[13] Y. Li, D. S. Gardner, O. Mutlu, and S. Mitra, "Concurrent Autonomous Self-Test for Uncore Components in System-on-Chips," in *VLSI Test Symposium*, 2010.

[14] M. Beckler and S. Blanton, "On-chip Diagnosis for Early-life and Wear-out Failures," in *International Test Conference*, 2012.

[15] D. Lazarevski, "VLSI Fault Diagnosis Problems and Decisions," in *ARSA-Advanced Research in Scientific Areas*, 2012.

[16] Y. Li, E. Cheng, S. Makar, and S. Mitra, "Self-repair of Uncore Components in Robust System-on-Chips: An OpenSPARC T2 Case Study," in *International Test Conference*, 2013.

[17] K. Huang, H.-G. Stratigopoulos, S. Mir, C. Hora, Y. Xing, and B. Kruseman, "Diagnosis of Local Spot Defects in Analog Circuits," *Instrumentation and Measurement, IEEE Transactions on*, vol. 61, no. 10, pp. 2701–2712, 2012.

[18] C. Yang, S. Tian, Z. Liu, J. Huang, and F. Chen, "Fault Modeling on Complex Plane and Tolerance Handling Methods for Analog Circuits," *Instrumentation and Measurement, IEEE Transactions on*, vol. 62, no. 10, pp. 2730–2738, 2013.

[19] M. A. El-Gamal, A.-K. S. Hassan, and A. A. Ibrahim, "Analog Fault Diagnosis Using Conic Optimization and Ellipsoidal Classifiers," *Journal of Electronic Testing*, vol. 30, no. 4, pp. 443–455, 2014.

[20] K. Forster, S. Monteleone, A. Calatroni, D. Roggen, and G. Troster, "Incremental k-NN Classifier Exploiting Correct-Error Teacher for Activity Recognition," in *Machine Learning and Applications*, 2010.

[21] "OpenSPARC T2," http://www.oracle.com/technetwork/systems /opensparc/opensparc-t2-page-1446157.html, Oracle.

[22] "ISCAS Benchmarks," http://web.eecs.umich.edu/ jhayes/iscas.restore/.

[23] "UC Irvine Machine Learning Repository," http://archive.ics.uci.edu/ml/, University of California, Irvine.

[24] L. Kozma, "K Nearest Neighbors Algorithm (kNN)," Tech. Rep., 2008.

[25] R. Polikar, L. Udpa, S. Member, S. S. Udpa, and V. Honavar, "Learn++ : An Incremental Learning Algorithm for Supervised Neural Networks," *Systems, Man, and Cybernetics*, vol. 31, no. 4, pp. 497–508, 2001.

[26] G. Cauwenberghs and T. Poggio, "Incremental and Decremental Support Vector Machine Learning," in *Advances in Neural Information Processing Systems*, 2001.

[27] R. Elwell, R. Polikar, and S. Member, "Incremental Learning of Concept Drift in Nonstationary Environments," *Neural Networks*, vol. 22, no. 10, pp. 1517–1531, 2011.

[28] E. S. Manolakos and I. Stamoulias, "IP-cores Design for the kNN Classifier," in *International Symposium on Circuits and Systems*, 2010.

[29] "Design Compiler," http://www.synopsys.com/Tools/Implementation/ RTLSynthesis/DesignCompiler/Pages/default.aspx, Synopsys.

Resiliency Challenges in sub-10nm Technologies
(*Special Session Paper*)

Rob Aitken

ARM
San Jose, CA
rob.aitken@arm.com

Ethan H. Cannon

The Boeing Company
Seattle, WA
ethan.cannon@boeing.com

Mondira Pant

Intel
Hudson, MA
mondira.pant@intel.com

Mehdi B. Tahoori

Karlsruhe Institute of Technology
Karlsruhe, Germany
mehdi.tahoori@kit.edu

Abstract— **Improvements in chip manufacturing technology, driven by high degree of integration due to small device sizes and additional complex functionalities enabled by heterogeneous integration, have propelled an astonishing growth of computing systems. While the pervasiveness of these systems enables emerging application domains, however, this trend is facing serious challenges, both at device and system levels. As the minimum feature size continues to shrink, a host of vulnerabilities influence the robustness, reliability, and resiliency of embedded and critical systems. Some of these factors are caused by the stochastic nature of the nanoscale manufacturing process, while other factors appear because of high frequencies and nanoscale features. This paper overviews the vision by some of the key industrial players regarding the emerging resiliency challenges faced at the extreme nanoscale technologies.**

I. INTRODUCTION

With down-scaling of CMOS technology into deep nanometer dimensions, reliability has become a major issue [1]. The main sources of unreliability in current technology nodes are mainly due to variability both during manufacturing and runtime, as well as transient soft errors.

Due to variability, the device (e.g. transistor, gate, circuit and processor) characteristics are different from the desired values. The variability could be due to *"time-zero" variation (process variation)* or *runtime variation* [2]. Process variation is a natural device parameter variation which makes the properties of fabricated devices different from those of designed ones. In other words, due to process variation similarly designed devices will behave differently after fabrication. Due to runtime variation, the device properties will change (degrade) during the chip operational lifetime. Runtime variations are routed in various sources such as supply voltage fluctuations, temperature variation and aging. The voltage and temperature variations are temporal or spatial according to the place of the transistor/gate and also the workload. Therefore, they cause a non-uniform variation on the properties of different transistors/gates at different location of the circuit and at different time points during the chip operational lifetime.

Device aging is caused by different wearout effects such as *Bias Temperature Instability* (BTI), *Hot Carrier Injection* (HCI), *Random Telegraph Noise* (RTN) as well as *Time Dependent Dielectric Breakdown* (TDDB) in case of transistors. All these effects degrade the transistors properties

(e.g. threshold voltage), and hence the switching delay of the transistor increases which can eventually lead to a timing failure if the delay of the circuit does not meet the timing constraint. In addition, the interconnects are degraded due to *ElectroMigration* (EM).

The common remedy to deal with variations is guardbanding, in which a timing margin is added to the designed clock cycle in order to guarantee the correct operation of the circuit in the presence of process and runtime variations. A pessimistic guardbanding leads to either yield or performance loss and an optimistic guardbanding affects Defects per Million (DPM) or field failures. Therefore, the required timing margin needs to be accurately predicted.

By further downscaling of transistor dimensions into the deep nanometer range, the aging effects become nondeterministic meaning that similar devices exposed to the same stress conditions may degrade differently. In other words, the aging behavior becomes *stochastic* [3]. As a result, the delay degradation becomes also stochastic and the timing guardband needs to be obtained according to the delay degradation distribution to guarantee reliable operation in the field, rather than just considering the mean value as in older technology nodes. Accurate stochastic timing analysis of the circuit becomes very important in this case since over and under margining can lead to significant performance or yield loss (timing failure), respectively. The matter is aggravated when it is combined with process variation, adding another degree of non-determinism. Thus, the circuit performance analysis becomes even more challenging, in particular considering that the sensitivity of the circuit properties to process and runtime variability increases with downscaling.

The other category of reliability issues is the transient soft errors caused by radiation-induced particles. Transient soft errors do not cause permanent degradations or faults, rather transient computational errors [4]. However, since its nature is random, the detection and correction of this type of errors is very challenging [4]. The soft error can affect memory cell and sequential elements of the circuit. Traditionally, only single errors caused by *Single Event Upsets* (SEUs) were considered as the target of detection and correction methods [5]. However, by continuous scaling of transistor dimensions, the probability that multiple nodes of the circuit are affected simultaneously by a strike, i.e., *Multi-Bit Upsets* (MBUs)

becomes larger, which makes the detection and correction even more challenging [6]. Already in a 14 nm FinFET technology node, the MBU/SEU ratio is around 40% and it will further increase with technology scaling beyond 10 nm nodes [7]. Soft errors can also affect the combinational logic part of the circuit by generating a transient pulse which may be propagated to sequential elements. Although the combinational logic soft error was not a big issue in the previous technology nodes, its contribution is rapidly increasing by scaling in deep nano-meter regime [5,8].

Besides the acceleration of these reliability issues due to transistor scaling [8], they become more pronounced in ultra low power paradigms such as *Near and Sub Threshold Computing* (NTC/STC) which has several applications in *Internet of Things* (IoT) domains. In such paradigms, the sensitivity of the device properties to supply voltage, temperature and threshold voltage is much higher, and hence any variation in these parameters due to process and runtime variability leads to significant circuit performance variations. For instance, in NTC the performance uncertainty is around 5X, while it is only 30% in the traditional regime for the same technology node [9]. In addition, the functional failure rate increases exponentially with supply voltage scaling, and as a result the failure rate of an SRAM cell is approximately 5 orders of magnitude larger [9]. Since in NTC the noise margin is very low, the susceptibility to soft errors is also extremely pronounced.

The next sections present the perspectives from the industry experts regarding their views on upcoming resiliency challenges in future technology nodes, and what are the outlooks to tackle these challenges. Section II focuses on discussing resiliency challenges in the nano-era, the responses and various steps we can take towards circumventing and mitigating a seemingly insurmountable problem ahead using sensor-driven runtime adaptation. Section III discusses technology, design, and market challenges of resiliency at 10nm and beyond. Section IV presents resiliency issues of integrated circuits used in the aerospace industry. Finally Section V concludes the paper.

II. RESILIENCY- THE TOUGH ROAD AHEAD

According to Wikipedia, in computer systems, "Resiliency" is the ability to provide and maintain an acceptable level of service in the face of faults and challenges to normal operation. Faults could be a result of numerous factors including static and dynamic variations in the computing infrastructure, triggered by inherent existing conditions or by environmental changes. This is true whether one is looking at the computer networking level, at the platform level, at the system level or at the chip/die level. In order to increase resiliency at any of these levels, it is important to understand the propensity to failures, identify resiliency metrics, set targets and design subsequently with a goal to meet/exceed them. Today the pace of change is like never before. Not only are we rapidly approaching scaling dimensions the size of an atom (0.1nm), our technology is heralding a new era of smart computing where generic devices all around considered to be "inanimate" are now appearing to

"spring into life" , be it the refrigerator, the thermostat, or the doorbell as they enter the mold of the Internet of Things. With the dramatic increase in the number of components, higher than ever performance requirements, lower than ever energy consumption and smaller than ever form factors, the susceptibility to failure has gotten exponentially higher and the challenges will only compound as we move ahead.

A widely referenced paper [10] discusses sources of chip variations, their impact and developing variation-tolerant designs. Our era of multi-voltage, multi-core, turbo [11], Advanced Vector Extensions (AVX) [12] multiprocessor designs for delivering top performance while staying within the defined power envelope is also an era of extreme variations, where common countering techniques like redundancy and guardbanding no longer suffice. Adaptive designs which typically use sensors to detect changes and adjust/respond to avoid errors are becoming increasingly widespread. Sensors are the heart of monitoring systems used to monitor CPU variations on the die. A well-tuned adaptive response requires the presence of a higher level of abstraction that understands and aggregate sensor data across a wide variety of sensors. Testing is a big component of making sure the system will be resilient when deployed. Test techniques need to keep up with the increasingly complicated systems to ensure resiliency. Big-data techniques can now allow for deep learning and analysis of phenomena affecting the reliability of systems in the field [13]. These techniques can be heavily used to benefit and improve our ability to optimize for and predict reliability issues examples include use of C-state and P-state residency analysis to reduce turbo guardbands, use of temperature cycling and time-in-state data to modify reliability guardbands, and so on. Moving forward intelligent data mining and analytics of data gathered will allow for design of systems that remain smart in light of extreme variations.

III. THE CHALLENGES AND META-CHALLENGES OF RESILIENCY AT 10NM AND BEYOND

Despite regular predictions of its demise, Moore's law appears poised to continue through the 10nm node and beyond. At each step, resiliency has become more challenging, due to new issues at each node with the fundamental technology (e.g. copper wiring, high-K metal-gate devices, FinFETs, and so on). Technology changes lead to both design challenges and opportunities (e.g. the area overhead of ECC is much less of an issue now than in the past). These in turn must be considered as part of an overall electronics market, which has its own challenges and meta-challenges relating to resiliency.

A. Technology challenges

For advanced CMOS manufacturers, the 10nm node is the second or third since FinFETs were introduced. As a result, there is already some experience in FinFET resilience, including material reliability, susceptibility to soft errors, aging behavior, etc. In many ways 10nm represents a brief return to classic scaling, similar to the transition to, say, 65nm, where trends that have already been observed can be expected to continue. On the other hand, dimensions are significantly

978-1-4799-7598-3/15 $31.00 © 2015 IEEE

smaller than they have been, leading to the need to consider atomic level effects.

For example, as shown in [14], manufacturing variation in fin dimensions lead to substantially different current flow within the fins, and this in turn leads to differential heating, which can induce thermal damage in the neighborhood of the fin. Similarly, differences in shape lead to differences in trapped charge that can lead to significantly different NBTI or PBTI.

Random telegraph noise (RTN) is also expected to be increasingly challenging in advanced nodes, since it scales with device length and width [15], rather than the square root of length and width that is more common for transistor random dopant variation [16]. At the moment its effects are restricted to SRAM variation and can be accommodated with careful design and margining, but it may become more critical at 10nm and certainly will in the future.

Finally, while the introduction of FinFETs led to a boost in power and performance, maintaining that gain through scaling is difficult, requiring complex device engineering, multiple work functions, stress and strain, and so on. Each of these introduces resiliency challenges: FinFETs can crack under strain, material boundaries introduce thermal and physical compatibility issues, etc. New materials can help, although there is some evidence that previously heralded solutions such as Germanium will be unable to provide expected gains [17]. While the industry has been lucky so far, it is certainly possible that a surprise is lurking in the future with respect to a systematic reliability failure of an exotic material.

B. Design challenges

It is unlikely that the 10nm node will require any new design for resiliency techniques, but the existing design challenges are amplified. For example, SRAM bit cell performance has not been keeping up with technology for many generations, due to the difficulties in getting cells to yield, while balancing read and write capability, all the while dealing with exponential increases in the number of bits per chip and corresponding increases in the likelihood of extreme tail bits [18]. A standard approach to resilience, such as applying single bit correction dual bit detection error correcting codes, faces increasing timing overhead as memory access increasingly forms part of a critical path. Designers can sometimes accommodate this delay either through additional memory access latency (which has ripple effects at system level), but sometimes they cannot and must either overdesign circuitry to hide the ECC cost, or ignore ECC for critical arrays (e.g. L1 caches), or add error detection at a higher level, through time redundancy, lockstep processing, or similar approaches.

C. Market challenges and meta-challenges

The major market drivers at advanced nodes are mobile computing (e.g. applications and graphics processors for smart phones and tablets) and enterprise computing (e.g. servers, networking). In the future, automotive electronics are also likely to move to more advanced nodes as processing needs increase due to both consumers demanding that their vehicles match their other devices in capability and connectivity (e.g. apps etc.), as well as the need to manage ever more complex systems of power delivery, driver assistance via video, radar, etc. In the so-called Internet of Things world, sensors and home automation are unlikely to move to advanced nodes, but some smart wearables might, in order to take advantage of better power/performance tradeoffs available with FinFETs.

This variety of applications and markets leads to a plethora of requirements for resiliency. Automotive systems require high reliability with lifetimes of a decade or more. Enterprise hardware must meet uptime requirements. Mobile devices are updated on timescales of 2 years or so. Wearables, on the other hand, are somewhat of an unknown quantity – will customers update them regularly or keep them for many years? Both are common in fashion. A given SoC might be used across one or two of these applications, which presents its own challenges, but a manufacturing process or an IP block could well span all of them. How are designers to know what resilience concerns to target? Several approaches are possible. Some circuitry can be optional; e.g. an IP block could have an option to include or not include ECC. For those challenges which affect performance and power, such as aging, the questions are more difficult. Previous research has shown that processor aging is workload dependent [19], and this means that simplistic "5 year aging" models and the like are not adequate to margin for failures across long lifetimes, but can also lead to overmargining and unnecessary loss of capability for devices that will be replaced before aging affects them significantly.

IV. REALIZING THE FUTURE OF AEROSPACE THROUGH RESILIENT IC DESIGN

Scaling enables truly revolutionary performance in aerospace applications. Small feature sizes and CMOS transistors with Radio Frequency (RF) performance enable integrating the functionality of several Integrated Circuits (IC) into a single die, greatly reducing *Size, Weight, and Power* (SWaP). *Radiation Hardening By Design* (RHBD) techniques have been demonstrated to provide robust performance at sub-100 nm process nodes where radiation hardened processes do not exist. Consequently, RHBD enables unprecedented designs for high radiation environments, such as satellite applications, that capitalize on the benefits of scaling, such as RHBD multi-core processors [20].

However, aerospace applications have stringent resilience requirements, including extended temperature ranges (such as the military standard -55° to 125°C), long mission life (as long as 150,000 power on hours), and harsh radiation environments. We highlight some of the challenges to achieving resilient IC designs that realize the performance promise of advanced CMOS process technologies.

A. Transistor Degradation

Over the IC lifetime, aging effects such as BTI and HCI cause transistor parameters to shift. In turn, transistor degradation impacts circuit performance. Transistor degradation is modeled at a transistor-level, but transistor-level simulations of a full System-On-a-Chip (SOC) are too time consuming, so an efficient method to assess the effect of aging on a digital IC is required.

The standard method for analyzing timing performance of large digital ICs is *Static Timing Analysis* (STA), where the propagation delay and output slew rate of each logic gate is modeled as a function of input slew rate and output load. STA tools ensure that the propagation delay through all timing arcs meet timing requirements. BTI increases the magnitude of the transistor threshold voltage. This BTI degradation can increase the propagation delay in digital circuits, eventually causing a timing error.

We developed an Age-Aware timing library, where the propagation delay and output slew rate for digital cells are aged based on the cell use conditions [21]. The duty cycle determines the amount of BTI in NMOS and PMOS transistors, while the toggle rate determines the magnitude of HCI degradation. At 45nm, propagation delay shifts on the order of ps were obtained. While these delays are quite acceptable, transistor degradation must be evaluated at each node to ensure acceptable lifetime.

B. Radiation Effects

Radiation effects are a particular concern for aerospace applications. The neutron flux at avionics altitudes is 200-300x higher than at ground level. Space environments include large fluxes of electrons and protons that cause cumulative *Total Ionizing Dose* (TID) effects. Heavy ions and protons cause *Single Event Effects* (SEE).

With scaling, transistors become less sensitive to TID, while circuits become more sensitive to SEE. TID effects are best mitigated by process changes to harden oxide layers, while SEE effects can be effectively mitigated by circuit design and fault tolerant architectures. Many RHBD techniques depend on spatial redundancy. Since the area of influence of a radiation event does not scale, a single radiation particle impacts more transistors, and spatial redundancy techniques become more challenging to implement.

C. Future Technologies

In the quest for every-increasing performance, novel design techniques are being implemented. Near-threshold voltage operation offers the promise of extreme energy efficiency, while demanding careful design techniques to protect against process variation. Heterogeneous 3D packaging brings ICs of disparate technologies in close proximity, reducing SWaP, while improving electrical performance. Hot spots are a particular concern in the vacuum of space, where heat dissipation is less efficient.

D. Summary

Process scaling brings benefits of greater functionality and higher performance with smaller SWaP for aerospace applications. Careful design is required to maintain reliability.

V. Conclusion

Resilience is complicated, and depends not just on technological factors, but also on architectural and structural factors, and must be viewed through a lens of target market and even application within that market. The 10nm node will exacerbate some recent resiliency issues, but the biggest challenges likely await the introduction of gate-all-around or similar novel structures.

References

[1] S. Guertin and M. White, "CMOS reliability challenges the future of commercial digital electronics and NASA," in NEPP Electronic Technology Workshop, 2010.

[2] International Technology Roadmap for Semiconductors, in ITRS 2013 Edition – Process Integration, Devices, and Structures, 2014.

[3] B. Kaczer et al., "Atomistic approach to variability of bias-temperature instability in circuit simulations," in Reliability Physics Symposium (IRPS), 2011 IEEE International. IEEE, 2011, pp. XT–3.

[4] M. A. Alam, K. Roy, and C. Augustine, "Reliability-and process-variation aware design of integrated circuits- a broader perspective," in Reliability Physics Symposium (IRPS), 2011 IEEE International. IEEE, 2011, pp. 4A–1.

[5] S. Mitra et al, "Robust system design to overcome cmos reliability challenges," Emerging and Selected Topics in Circuits and Systems, IEEE Journal on, vol. 1, no. 1, pp. 30–41, 2011.

[6] E. Ibe et al, "Impact of scaling on neutron-induced soft error in srams from a 250 nm to a 22 nm design rule," Electron Devices, IEEE Transactions on, vol. 57, no. 7, pp. 1527–1538, 2010.

[7] G. Hubert, L. Artola, and D. Regis, "Impact of scaling on the soft error sensitivity of bulk, fdsoi and finfet technologies due to atmospheric radiation," Integration, the VLSI Journal, 2015.

[8] H. Nguyen, "Resiliency challenges in future communications infrastructure," in Proceedings of the Communications Quality and Reliability Workshop, May 2014.

[9] R. G. Dreslinski et al, "Nearthreshold computing: Reclaiming moore's law through energy efficient integrated circuits," Proceedings of the IEEE, vol. 98, no. 2, pp. 253–266, 2010.

[10] S Borkar, "Designing reliable systems from unreliable components: the challenges of transistor variability and degradation" - Micro, IEEE, 2005

[11] http://www.intel.com/content/www/us/en/architecture-and-technology/turbo-boost/turbo-boost-technology.html

[12] http://en.wikipedia.org/wiki/Advanced_Vector_Extensions

[13] T. Siddiqua, A. Papathanasiou, A. Biswas, S. Gurumurthi, "Analysis and Modeling of Memory Errors from Large-Scale Field Data Collection", IEEE Workshop on Silicon Errors in Logic - System Effects, March 2013.

[14] A. Asenov et al, "Variability Aware Simulation Based Design-Technology Co-Optimisation (DTCO) Flow in 14 nm FinFET/SRAM Co-Optimisation", IEEE Trans. Elect. Devices 2014.

[15] K. Ito et al, "The impact of RTN on performance fluctuation in CMOS logic circuits", International Reliability Physics Symp., 2011.

[16] M. Pelgrom et al, "Matching Properties of MOS Transistors", IEEE J. of Solid State Circuits, 1989.

[17] L. Shifren et al, "Predictive Simulation and Benchmarking of Si and Ge pMOS FinFETs for Future CMOS Technology", IEEE Trans. Elect. Devices, 2014.

[18] A. Singhee and R. Rutenbar, "Extreme Statistics in Nanoscale Memory Design", Springer 2010.

[19] E. Mintarno et al, "Workload dependent NBTI and PBTI analysis for a sub-45nm commercial microprocessor", International Reliability Physics Symp., 2013.

[20] J. Ballast et al, "A Method for Efficient Radiation Hardening of Multicore Processors," to be published in IEEE Aerospace Conf., Big Sky, MT, 2015.

[21] M. Katoozi et al, "An Age-Aware Library for Reliability Simulation of Digital ICs," in IEEE IRPS, Monterrey, CA, 2013, pp. 3A.3.1-3A.3.5.

Innovative Practices Session 7C:
Mixed Signal Test and Debug

Organizer:
Suriya Natarajan (Intel)

Moderator:
Suriya Natarajan (Intel)

Presenters & Abstracts:

Presenter #1: *Srinivas Modekurty and Prashant Goteti (Intel, USA)*

Title: What's more important for high-speed IO? Test or Debug?

Abstract: The traditional focus of work in test has been innovation and efficiency for high volume manufacturing test. However, in reality, a significant amount of effort is expended on design validation and debug prior to deeming analog and high speed serial IO test stimuli content production worthy. In this presentation we emphasize the importance of widening the post-silicon envelope to include early design validation and debug in addition to manufacturing test. The impact on DFX architecture is considered in which in order to be efficient the architecture has to be scalable across a wider set of post-silicon stages. We also discuss the need for establishing quick correlation between design validation and manufacturing test measurements and the time-to-market savings it brings for analog and IO content designed essentially on a digital process technology.

Presenter #2: *Ashraf Takla (Mixel, USA)*

Title: MIPI D-PHY RX+, Optimized test configuration

Abstract: With the proliferation of the mobile platform, the accelerating adoption of MIPI® beyond the traditional mobile platform and into safety related applications, testability of MIPI® PHY is becoming a key requirement. The D-PHY is the MIPI® PHY with the widest adoption in the industry today. The RX+ is a D-PHY receiver configuration optimized for full-speed production testing. The presentation will cover an overview of testability challenges in safety related applications, examine the advantages of the RX+ configuration, and compare this configuration to traditionally used D-PHY configurations.

Presenter #3: *Shahin Toutounchi, Andrew Taylor, Seng Lao and Ekarat Laohavaleeso (Xilinx, USA)*

Title: Application of On-chip, C based post analysis to the test of mixed signal IPs

Abstract: How do you test the ability to calibrate? In todays mixed signal IPs, whether a die is good or bad is decided by its ability to calibrate. Calibration typically hinges on a few key sub-blocks like delay elements or phase interpolators. Rarely do these sub-blocks work in isolation. Designs often require these sub-blocks be well behaved and also behave similar to their brethren. We explore the significance of on-chip post analysis of mixed signal sub-block measurements and how it can be used to test calibration. We start with the uniformity of delay elements which have relevance to DDR calibration as a first example. Next we will cover the correction & post analysis of duty cycle measurements forming the basis for testing phase interpolators.

Panel: Analog/RF BIST: Are we there yet?

Organizers:
Sule Ozev (Arizona State University)

Moderator:
Linda Milor (Georgia Institute of Technology)

Abstract:
BIST for analog and RF circuits has been proposed many years ago and we are still chasing it. One school of thought is to have generic BIST components for input stimulus generation and output analysis and to use them in a plug-and-play fashion. Another school of thought is to develop dedicated circuits for each functionality and re-use the same blocks for the same functionality. A third approach is designing completely circuit-specific BIST for each primary circuit. The truth is ad-hoc examples of BIST have been around for years. However, there is no standardized way of implementing or inserting BIST for analog and RF circuits. The panelists, all experts in this domain, will share their view of the best way of implementing BIST for analog and RF circuits, if there is such a thing...

Panelists:
- *Abhijit Chatterjee (Georgia Institute of Technology)*
- *Steve Sunter (Mentor Graphics Inc.)*
- *Karim Arabi (Qualcomm, Inc.)*
- *Bertan Bakkaloglu (ASU)*
- *Haralampos Stratigopoulos (TIMA)*

No Fault Found: The Root Cause

Erik Larsson, Bill Eklow, Scott Davidsson, Rob Aitken, Artur Jutman, and Christophe Lotz

Abstract—No Trouble Found (NTF) has been discussed for several years [1]. An NTF occurs when a device fails at the board/system level and that failure cannot be confirm by the component supplier. There are several explanations for why NTFs occur, including: device complexity; inability to create system level hardware/software transactions which uncover hard to find defects; different environments during testing (power, thermal, noise). More recently a new concept, No Fault Found (NFF), has emerged. A NFF represents a defect which cannot be detected by any known means so far. The premise is that at some point the defect will be exposed - most likely at a customer site when the device is in a system. Given that we looking for a defect that we know nothing about and are theoretically undetectable it will be interesting to see what the panel has to say about the nature of these defects and how we intend to find them.

The panel is organized by Erik Larsson and moderated by Bill Eklow. The statements from the panel are:

1 SCOTT DAVIDSSON

In my experience NFFs have many root causes [2]. The biggest is that board and system test exercise the part in ways a chip tester cant. Another cause is poor fault coverage. It sometimes pays to test a returned part with the test program it was tested to before shipment, and then retest with the latest. The reduction in NFFs with the newest one proves that the cause was coverage. There are other reasons. The diagnostic process is imperfect, and some returned parts are good. There might be psychology involved. If your last revision had problems, repair people will tend to replace your part just to be safe, even if the quality has improved. But a low NFF rate on the original test program is not something to be wished for. If the part passed and now fails, you have a reliability problem and none of us wants that.

2 ROB AITKEN

Broadly speaking, NFFs fall into three categories: test escapes, marginal chips, and type 1 errors (working parts mistakenly believed to be failures). The relative abundance of each category depends on the type and complexity of the product, the maturity of the manufacturing process, and the robustness of the test flow. A recent consultant's study [3] determined that the vast majority of NFF returns for consumer products appear to be type 1 errors. Specifically, 68% of NFF parts met all specs but somehow don't meet consumer expectations, 27% were "buyer's remorse" (falsely reported as bad in order to return part), leaving a surprisingly low 5% as test escapes or marginal parts. This report also found that a 1% reduction in the rate of NFF parts led to a 4% reduction in return and repair costs. This leads to two interesting areas for further discussion: How many of the 68% of parts could potentially be found by improved screening, and for chip vendors, are there failure modes that are only observable at a higher level of the system, and if so, how can they be mitigated or avoided? Providing good answers to these questions can have a substantial effect on perceived quality while at the same time helping to control costs.

3 ARTUR JUTMAN

A certain portion of NFFs could be due to the lack of Board-level defect or fault coverage. The incompleteness of existing board-level test coverage metrics could simply appear e.g. due to the combination of the following two factors. First, the inability of the classical structural test techniques to apply test patterns at-speed, hence limiting the covered fault spectrum to static faults leaving the delay and performance fault domain traditionally to the functional tests, which in its turn do not produce measurable coverage of structural faults (e.g. delays on the board). As a result, the quality of the functional test sets depends highly on the human factor and statistics (experience, field returns, yield learning) providing no guaranteed quality. The latter fact represents the second factor contributing to the potentially missing fault coverage at the board level, while also creating problems for technology transfer and production outsourcing.

4 CHRISTOPHE LOTZ

Test strategy and defect occurrence should be tied together: (1) High test coverage should be applied for defects that occur frequently, (2) Lack of coverage on defects that never occur, has no consequence to the final quality. It is usually unknown, precisely, where the defect really occurs. We will have to qualify the test strategy against the true defects. In the BASTION project [4] we are developing the tool QuadDPMO to extract the true Defect Per Million Opportunities from a traceability database, with data collected from the production line and throughout the product life. The analysis combines repair information, board modelization and test coverage data. Linking test coverage and true DPMO provides new business opportunities such as: test cost reduction, improved tests that target the true defects, new opportunities for adaptive test, test overlap reduction, lower escape rates and culminating in fewer NFF.

REFERENCES

[1] Z. Conroy, G. Richmond, X. Gu, W. Eklow, "A Practical Perspective on Reducing ASIC NTFs, International Test Conference, 2005.

[2] Davidson, S., Understanding NTF Components from the Field, 2005 International Test Conference, Austin, TX, November 2005.

[3] "Big Trouble with No Trouble Found: How Consumer Electronics Firms Confront the High Cost of Customer Returns", Accenture Report, 2009

[4] BASTION (http://fp7-bastion.eu)

- E. Larsson, Lund University, Sweden, e-mail: erik.larsson@eit.lth.se.
- B. Eklow, Cisco, California, USA, e-mail: beklow@cisco.com
- S. Davidsson, Oracle, California, USA, e-mail: scott.davidson@oracle.com
- R. Aitken, ARM, California, USA, e-mail: Rob.Aitken@arm.com
- A. Jutman, Testonica, Estonia, e-mail: artur@testonica.com
- C. Lotz, Aster Tech., France, e-mail: christophe.lotz@aster-technologies.com

Special Session 8C: E.J. McCluskey Doctoral Thesis Award Semi-Final

Organizers: M. Portolan (TIMA Laboratory) & K. Huang (San Diego State University)
Moderator: K. Huang (San Diego State University)

Named after Prof. E.J. McCluskey, a key contributor to the field of test technology, the 2015 TTTC's Doctoral Thesis Award serves the purpose to i) promote the most impactful doctoral student work, ii) provide the students with the exposure to the community and the prospective employers, and iii) support interaction between academia and industry in the field of test technology. TTTC's E.J. McCluskey Best Doctoral Thesis Award will be given to the winning student of the doctoral student contest and his or her advisor. The award consists of a certificate, an honorarium and an invitation to submit a paper on the presented work to the IEEE Design & Test magazine.

The contest is held in two stages: semi-finals and finals: In 2015, semi-finals will be held at the IEEE VLSI Test Symposium (VTS), the IEEE European Test Symposium (ETS), the IEEE Latin American Test Workshop (LATW) and the Asian Test Symposium (ATS). At each semi-final, a jury composed of industrial and academic experts will determine the winner, and semi-final winners will compete against each other in the finals, held at the International Test Conference (ITC) 2015.

In this session, each contestant is given 9 minutes for presentation. After the end of all presentations, a poster session follows, Q&A from a panel of industry experts and the audience. The panel of experts will judge the presented doctoral theses with regards to theoretical advancement, industrial relevance and presentation. The grades submitted by the industrial panel will be combined with the grades given by the academic jury, consisting of distinguished professors. The winner will be announced during the VTS 2015 social event.

Contestants:
1. Abhishek Basak (Case Western Reserve University), Advisor: Swarup Bhunia
Thesis title: Low-cost Design for Security Approaches for IC Integrity Validation

2. Navankur Beohar (Arizona State University), Advisor: Bertan Bakkaloglu
Thesis title: System Identification, Diagnosis, and Built-In-Self-Test of High Switching Rate DC-DC Converters

3. Sergej Deutsch (Duke University), Advisor: Krishnendu Chakrabarty
Thesis title: Test and Debug Solutions for 3D-Stacked Integrated Circuits

4. Ashkan Eghbal (University of California, Irvine), Advisor: Nader Bagherzadeh
Thesis title: Three-Dimensional NoC Reliability Evaluation Automatic Tool (TREAT)

5. Bahar Farahani (University of Tehran), Advisor: Saeed Safari
Thesis title: Cross-layer Resilient Processor Design to Address Soft Error and PVTA Variations

6. Amirali Ghofrani (University of California, Santa Barbara), Advisor: Kwang-Ting (Tim) Cheng
Thesis title: Memristive Memories: Test & Reliability Challenges, Application Opportunities

7. Shahrzad Mirkhani (University of Texas at Austin), Advisor: Jacob A. Abraham
Thesis title: Statistical Methods for Rapid System Evaluation under Transient and Permanent Faults

8. Lu Wang (The University of Texas at San Antonio), Advisor: Bao Liu
Thesis title: VLSI Dynamic Statistical Performance Verification and Timing/Soft Error-Resilient Design

9. Li Xu (Iowa State University), Advisor: Degang Chen
Thesis title: Efficient and Accurate Testing of High Performance ADCs without Precise Instrumentation

10. Fangming Ye (Duke University), Advisor: Krishnendu Chakrabarty
Thesis title: Knowledge-Driven Board-Level Functional Fault Diagnosis

Abstraction-based Relation Mining for Functional Test Generation

Kelson Gent and Michael S. Hsiao

Bradley Department of Electrical and Computer Engineering
Virginia Tech, Blacksburg, VA 24061, USA
{kelsong, mhsiao}@vt.edu

Abstract—**Functional test generation and design validation frequently use stochastic methods for vector generation. However, for circuits with narrow paths or random-resistant corner cases, purely random techniques can fail to produce adequate results. Deterministic techniques can aid this process; however, they add significant computational complexity. This paper presents a Register Transfer Level (RTL) abstraction technique to derive relationships between inputs and path activations. The abstractions are built off of various program slices. Using such a variety of abstracted RTL models, we attempt to find patterns in the reduced state and input with their resulting branch activations. These relationships are then applied to guide stimuli generation in the concrete model. Experimental results show that this method allows for fast convergence on hard-to-reach states and achieves a performance increase of up to 9× together with a reduction of test lengths compared to previous hybrid search techniques.**

I. INTRODUCTION

Today, functional test generation represents a critical portion of the total effort in modern hardware designs, in particular for design validation. Due to the high costs associated with design errors caught during production, ensuring correctness in the design is critical for keeping costs low while managing the short time-to-market window. As a result, a significant effort has been invested in automated tools for aiding designers in validation efforts. Generation of functional tests at the Register Transfer Level (RTL) is a significant step in ensuring the correctness of the design. Testing at the RTL gives the benefit of automating test generation at a higher level of design abstraction than at the lower levels. This abstraction level yields structural information that can be utilized for heuristics to guide the test generation process. Recently, software metrics have been employed to evaluate the effectiveness of RTL test generation. A commonly used metric is branch coverage or line coverage [1, 2]. The base method for generating these tests is random vector generation. Although generation and application of random test vectors is very fast, these vectors tend to fail to reach corner cases and states that require a narrow activation path. To counteract this shortcoming, directed tests are generated to directly reach uncovered corner cases.

Recently, several significant advances have been made in test generation at the RTL utilizing branch coverage as a metric. HYBRO [3] utilizes a hybrid mechanism that extracts execution paths from concrete simulation and attempts to reach new branches by mutating these paths. These mutations are passed to a formal Satisfiability Modulo Theory (SMT) solver as constraints, which attempts to find inputs that satisfy the mutation.

This approach can be limited due to the computational costs of calling the SMT solver multiple times. Following HYBRO, BEACON [4] was developed as a purely stochastic method of RTL test generation utilizing a combination of evolutionary and swarm intelligence techniques. Bounded model checking was augmented in [5] to help in reaching critical states and trim unreachable branches from the search space. Most recently, PACOST [6] utilizes formal methods to build an onion ring abstraction from the RTL specification for different branch points. This abstraction is highly effective for state justification and allows PACOST to quickly reach difficult corner cases.

In this paper, we present a constraint based abstraction for improving branch coverage based test generation by utilizing mined relations from multiple HDL program slices in order to generate more effective inputs during runtime. By slicing the HDL, we obtain a reduced model of the circuit behavior across a few state state variables. Through random simulation of this model, we may find relationships between state variables, inputs and specific branch activations. By applying these relationships during test generation, we theorize that our method can reach high levels of coverage with greater efficiency. While the concept of abstraction is not new to design validation, we offer the simultaneous use of multiple abstractions to achieve our goal.

The high-level description of the algorithm and abstraction is as follows. The algorithm follows a similar framework to BEACON[4]; however, the base heuristic for input guidance focuses on mined relations for guiding the target towards uncovered states. During preprocessing, a set of executable HDL program slices is generated and instrumented for branch coverage. Then, each program slice is simulated for a set number of random vectors. Based on the results of this simulation, a set of relations is obtained that characterize the inputs and circuit state(s) within particular single-cycle execution paths. The concrete model is instrumented for branch coverage, compiled into a static simulation library, and the controlling FSMs are extracted. A set of simulation units is created using the simulation library and random simulated to obtain a base level of branch coverage. Following initialization, each simulation unit is assigned an unreached branch as a target and we begin constrained vector generation. Our abstraction is used to attempt to remain on the dominating FSM path to the target by constraining the inputs based on the mined relations. Stimuli generation continues until each simulation unit reaches its target or generates the current maximum number of vectors. Then, if no targets are reached, the search space is expanded,

supported in part by an NSF grant 1422054

otherwise, the vectors are saved and new targets are generated. After a set number of rounds with no targets reached, the algorithm terminates.

The contributions can be summarized as follows:

- We propose a simulation-based multiple abstraction model for mining relationships between inputs and internal states.
- The relations mining from multiple slices, combined with FSM extraction allows for the specific targeting of branches in the concrete model during test generation.
- No formal tools are invoked to save computational costs, but the method is still able to reach corner cases with the help of mined relations.
- Significant speedup over previous techniques is achieved while maintaining high coverage and reducing test set lengths in difficult benchmarks.

The rest of the paper is organized as follows. Section II discusses relevant past work and fundamental theories. Section III covers the integration of the abstraction in the test framework Section IV discusses the performance of the algorithm compared to previous works and the purely stochastic algorithm. Finally, Section V provides the concluding summary.

II. BACKGROUND AND PRELIMINARIES

In this section, we highlight the relevant previous work as well as covering necessary theories used in this work.

A. Control Flow Graphs

The Control Flow Graphs (CFG), proposed by Allen [7], provides the basis for many compiler level optimizations and static analysis tools. The CFG is represented as a directed graph $G(V, E)$ with vertices representing basic blocks and edges representing the flow of execution between basic blocks. Each basic block is the maximal number of program statements such that it meets the following conditions:

1) Each block can only be entered through the first statement.
2) Each block may only contain one statement that leads to the execution of another basic block.
3) All statements must execute sequentially within the block.

Graph edges are created based on the execution targets of the final statement in the block to form the CFG. Based on this analysis, loop optimization and unreachable program segments can be determined at compilation.

B. Dominators and Dominator Trees

Dominators [8] play a critical role in analyzing flow diagrams. Dominance has been utilized for many applications related to code analysis, including single static assignment form, and loop identification and reduction during code compilation. The definition of dominance for a given node can be expressed as follows:

$Dominator(x)$: A node $n \in CFG$ dominates x if n lies on every path from the entry node of the CFG to x. The set $Dominators(x)$ contains every node n that dominates x. Dominance is reflexive and transitive.

Several other definitions are useful for the calculation and use of dominance in graph analysis. Strict dominance asserts that if x dominates y and $x \neq y$, then x is a strict dominator of y. The immediate dominator of y, is the dominator x such that x is a strict dominator of y and no other node in the flow graph. The dominance tree is a tree structure where the children of a node x are all immediately dominated by x. The root of this tree is the entry node of the flow graph. An example of the dominator node is shown below in Figure 1. The dominators of the target node, 5, are $(1, 2, 3)$ highlighted in gray.

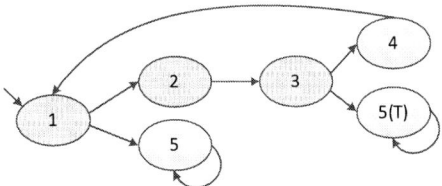

Fig. 1. Dominator example

The dominator tree structure was utilized in [9] to create a space efficient, fast algorithm for the calculation of dominators in a graph. This method was used in this work for necessary dominance calculations.

III. METHODOLOGY

Previous methods in functional test generation rely on distributing the computational load of navigating corner cases between a deterministic solver and a stochastic search engine. However, for large circuits, the computation cost associated with a single call to a deterministic solver may be significant. Additionally, some conditions may require specific decisions to be made many cycles before the target condition, which further increases the cost of of using a deterministic solver and limits the effectiveness of the algorithm. This paper introduces a method for improving performance of stochastic functional test generation algorithms without the invocation of any formal tools for generating vectors. By utilizing mined relationships between branch activation and circuit state, our algorithm attempts to stay in states that dominate the target, applying inputs that were derived during the mining process. This section describes the proposed abstraction and its integration into the test generation framework.

Test generation takes place across two stages, a preprocessing and mining stage, and a vector generation stage. During the preprocessing stage, program slices are taken of the HDL under test and are compiled using Verilator [10] and instrumented with a database of counters for branch coverage. Additionally, the HDL under test is compiled and a mapping is created between instrumentation points in the sliced circuits and the complete circuit. Then, each slice is analyzed for relationships corresponding to branch activations. Finally, the concrete circuit's CFG is extracted to generate the finite state machine(FSM) for use during test generation.

The flow of the test pattern generation is shown in Figure 2. At initialization, a set of K single simulation units, is initialized. Each unit is given an initialization vector to start their search at a reset condition S_0. Following the application

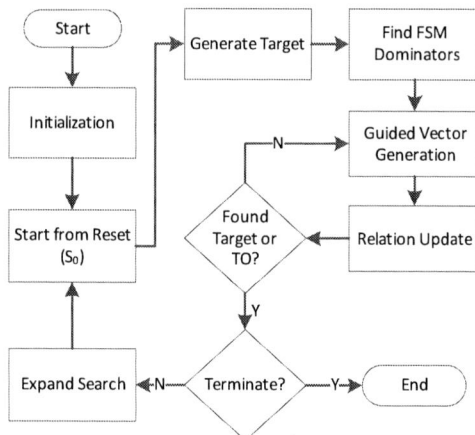

Fig. 2. Algorithm Flow

of the initialization vector, each unit simulates a randomly generated vector to eliminate easily reachable branches from target selection. Next, each search chooses an unreached branch as a target and the dominators to this node are calculated from the extracted FSM. Once a target is analyzed, stimuli generation begins and relies on information gained during the mining process to generate inputs that will advance towards a target state. Vector generation terminates after a unit reaches its target branch or after generating a maximum number of vectors, L_{max}. Vector generation terminates when no unit reaches a target for N_r rounds or all branches are covered. The pseudo-code is shown below in Algorithm 1.

Algorithm 1 Test Generation Framework

1: extract FSM
2: **while** $S_{uncov} \neq \emptyset$ **do**
3:     set the initial states to S_0
4:     $count_{ss} = 0$
5:     set $Branch_{globalcov} = \emptyset$
6:     **for** simulation units 1 to K **do**
7:         obtain target, t_k
8:         $Branch_{cov} = S_{cov} \cup$ vector generation(t_k, L_{max})
9:     **end for**
10:     **if** $S_{cov} \setminus S_{globalcov} = \emptyset$ **then**
11:         $count_{ss}$++
12:         increase L_{max}
13:     **else**
14:         $count_{ss} = 0$
15:         $Branch_{globalcov} = S_{cov} \cup Branch_{globalcov}$
16:     **end if**
17:     **if** $count_{ss} == N_r$ **then**
18:         RETURN
19:     **end if**
20: **end while**

A. Relation Mining

During preprocessing, we generate a set of slices such that each slice satisfies the constraints of an executable slice as defined by [11]. This implies that for any program point p

and a variable x included in the slice, x will have the same value as in the corresponding program point p' of the concrete model. Therefore, relationships and properties mined from the slice will still hold in concrete model. Each slice is generated based on variables used in control statements at the RTL. For example, the controlling variable of a switch-case block.

Each slice is compiled using Verilator[10], and randomly simulated from an initialization for N_{slice} cycles by y simulation units. The execution path and circuit state are saved at each cycle and added to a database. For each unique path within a single cycle, we attempt to find relations using a template matching system [12]. We have implemented templates for constant values, variable equality and basic boolean logical operations such as AND, OR, NOT, etc. For each instance of a particular path, we attempt to match the state to these templates. Since each template has distinct, distinguishing behavior, this cycle analysis is used as either a support or counter example to existing relations. Once all cycles have been processed, the relations are added to a database that can be accessed via their associated path in the CFG.

The database generated provides a set of constraints for the stochastic search during the input generation phase of simulation. Each constraint in the database is represented by the pair $(P, f(x))$ where P is the single cycle execution path and $f(x)$ is the function of x that represents the mined constraint. Consider the code and CFG shown in Figure 3. For the path $P = 48, 50$ (covering coverage points 50 and 48 in lines 2 and 5), we see that $k\&0x01 == 1$ (k[0] is KEY_ON) and $data\_out == 0$ must be true to activate the path, which yields the two relations $(\{48, 50\}, k\&1 = 1)$ and $(\{48, 50\}, data\_out = 0)$. The application of this abstraction in vector generation is described fully in Section III-C.

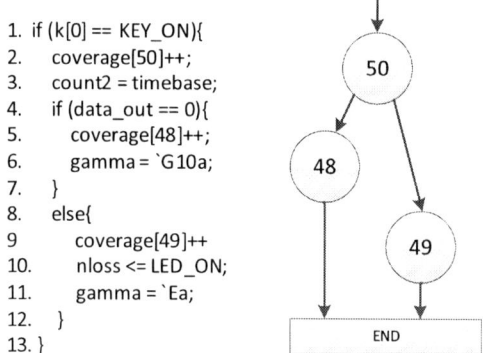

Fig. 3. Relation Example

B. FSM Extraction

At the RTL, the FSM for the circuit is often explicitly declared. Additionally, each state in the machine is typically associated with a single branch coverage statement. In order to associate the branches, we utilize a static analysis method on the CFG extracted by Verilator.

Generally, at the RTL, the FSM is represented by a set of branches controlled by a single variable which are mutually exclusive within a single cycle. Based the execution of these

978-1-4799-7598-3/15 $31.00 © 2015 IEEE

branches, the controlling variable will be updated for the next cycle. In order to extract relationships between these branches, each exclusive branch within the CFG is established as a node in the FSM and examined for potential assignments to the controlling variable. If an explicit constant assignment to the controlling variable exists, the path is saved as an edge in the FSM and the new variable is symbolically executed to determine the next states corresponding branch. Exploring each mutually exclusive path in this manner generates a representation of the FSM using branch coverage instrumentation points.

Each state in the stored graph is represented by a single branch coverage point s and each edge is represented as a tuple (s_{curr}, P, s_{next}) where P is a set of branches executed during a single cycle to transition between s_{curr} and s_{next}.

C. Vector Generation

Vector generation for each simulation unit is heavily influenced by relations mined during slice analysis. Compared to previous methods, which walk randomly, our method uses the relations as generate constrained intelligent inputs that activate branches in the path of the target branch.

During each simulation units generation, a new input vector is created at each cycle. The pseudo-code for this generation is shown below in Algorithm 2. For each unit, an unreached target branch is passed from the general flow. From this target, t_k, the dominator set, $doms(t_k)$, is generated. Following the calculation of dominators, vector generation begins. At each cycle, based on the execution path in the prior cycle, we calculate the expected state branch activation in the current cycle, s_{curr}. Then, all relations, $R(s_{curr})$, whose identifying path contains s_{curr} are fetched from the database. If there exists any relation in $R(s_{curr})$ which defines a relationship based on an input variable, then we attempt to find a matching state. For each unique path in the relation database that contains branch s_{curr}, the concrete state is compared to the relations associated with that path. If all relations hold a potential match is found, the FSM is then queried to determine if s_{next} is also on the dominating path to the target. If this set of relations leads to another dominating state, then, we apply the applicable input relations to the generated vector. If not match is found, and there exists a path in the FSM that leads to a dominating node, then generate an input such that it does not match any relation with a unique P that yields the same s_{next}. If vector has been generated that leads to a dominating node, then we generate input randomly because there is either no node known relation that leads to the target or all possible next states exist along the path to the target.

Following the generation of a vector, it is simulated and checked for consistency with the expected results. If our simulated path differs from the result expected based on the given relation, they are invalidated and removed from the guidance pool, as to not mislead the algorithm. Additionally, when new branch execution paths are reached, the state is observed and new potential input relations are identified by passing the inputs into the relation engine.

Algorithm 2 Abstraction Constrained Vector Generation

1: calculate $doms(t_k)$
2: **for** 1 to N_{max} **do**
3:    get current state in FSM, s_{curr}
4:    Gen_input = false
5:    **if** $s_{curr} \in doms(t_k)$ **then**
6:       Get all relations $R(s_{curr})$ whose path includes s_{curr}
7:       **if** $R(s_{curr})$ contains input relationships **then**
8:          **for** each unique path, P, in $R(s_{curr})$ **do**
9:             **if** state variables match relations $\wedge$
                P leads to $s_{next} \mid s_{next} \in doms(t_k)$ **then**
10:               constrain $V_{constraint}$ to input relations
11:               $V_{sim} = V_{constraint}$
12:               $gen\_input = true$
13:            **end if**
14:         **end for**
15:         **if** no matches $\wedge$
                $\exists e(s_{curr}, s_{next}) \in E(FSM) \mid s_{next} \in doms(t_k)$
             then
16:            constrain $V_{constraint}$ to not match other relations in s_{curr}
17:            $gen\_input = true$
18:         **end if**
19:      **end if**
20:   **end if**
21:   **if** $\neg gen\_input$ **then**
22:      $V_{sim} = V_{rand}$
23:   **end if**
24:   simulate(V_{sim})
25:   update relations
26: **end for**

TABLE I
BENCHMARK CHARACTERISTICS

benchmark	#Lines	#Branches	#PIs	#POs	#FFs
b07	92	19	1	8	49
b10	167	32	11	6	17
b11	118	32	7	6	31
b12	697	105	5	6	121
b12word	877	140	17	18	153
b14	509	211	32	54	245
OR1200-0	943	19	79	111	75
OR1200-1	982	25	144	109	77
OR1200-2	478	19	111	106	4
OR1200-3	579	47	171	230	96
OR1200	26959	684	161	303	2777

IV. EXPERIMENTAL RESULTS

The relation mining and targeted ATPG algorithm were developed on an Intel Core i7-3770k@3.5Ghz with 8 cores and 16GB of memory running Ubuntu 14.04. Experiments utilized a single core and were conducted on a set of ITC99 [13] circuits to assess their performance relative to prior works. Additionally, four modules of the OpenRISC1200 [14] as well as the complete OR1200 circuit are also tested. The four different OR1200 modules are the instruction cache controller(0), data cache controller(1), Wishbone bus interface(2) and exception(3). As representatives of general sequential

TABLE II
BRANCH COVERAGE AND COMPARISON WITH PRIOR TECHNIQUES

Circuit	Mining Time (s)	Branch Coverage %				Unreachable [5]	Vector Generation Run Time(s)				Speedup Over	
		HYBRO	BEACON	[5]	Ours		HYBRO	BEACON	[5]	Ours	HYBRO	[5]
b07	0.11	NA	90.00	**95.00**	**95.00**	1	NA	0.0024	0.37	0.148	NA	2.5
b10	0.54	96.77	93.75	**100.00**	**100.00**	1	52.14	11.40	3.12	0.29	179.8	10.7
b11	1.12	91.30	96.88	96.88	**100.00**	1	326.85	11.95	14.2	2.41	135.62	5.89
b12	0.41	NA	98.09	97.1	**99.04**	1	NA	111.42	69.6	7.73	NA	9.00
b12word	19.2	NA	NA	NA	**98.58**	0	NA	NA	NA	52.15	NA	NA
b14	1.17	83.50	91.94	**93.4**	**93.4**	2	301.69	204.65	94.2	11.73	25.72	8.03
OR1200-0	0.06	**93.75**	89.47	89.47	89.47	0	37.73	4.38	4.38	1.15	32.81	3.89
OR1200-1	0.36	**96.30**	92.00	92.00	92.00	0	21.9	4.87	2.75	1.34	16.34	2.05
OR1200-2	0.43	**100.00**	**100.00**	**100.00**	**100.00**	0	302.67	2.85	2.52	1.27	1.98	1.98
OR1200-3	0.66	96.61	**97.87**	**97.87**	**97.87**	0	287.62	27.35	6.4	2.26	101.63	2.83
OR1200	14.63	NA	92.84	NA	**94.15**	0	NA	300.4	NA	87.59	NA	NA

logic, many of the ITC99 [13] benchmarks circuits have hard-to-reach states (*control states*), making them a useful case study for design validation. Additionally, b12word, introduced in [6], is an extension of b12 to expand the search space. B12 is an implementation of a 1 player game where the computer generates a sequence of numbers and the user must hit the correct button associated with each generated number, in order, for a sequence of 512 guesses. B12word expands the number of inputs from 4 to 16, greatly increasing the difficulty of reaching the winning state. The characteristics of each circuit are shown in Table I, including the number of lines in RTL, number of branches, number of primary inputs, number of primary outputs, and the number of flip-flops for each circuit. The complete OR1200 with 2777 state elements is also included to show that our method is able to process the entire circuit while formal methods will have difficulty with such a circuit.

A. Algorithm Settings

The stochastic search algorithm uses the following parameters: the number of simulation units is set to $K = 50$. The maximum number of rounds is set to $R = 10$. For each round of local search, the length of random vectors N_c is chosen based on the circuit size with a max vector length of 3000. Local search steady state is assumed after $N_r = 5$ rounds with no new branches activated.

B. Branch Coverage

Our algorithm maintains the high level of coverage seen in previous works, but at a significant speedup due to the new feedback techniques employed in this work. The results compared to HYBRO, BEACON [4], and the technique in [5] are shown in Table II. For each circuit, the mining time is first reported. Columns 3 to 6 report the branch coverage for HYBRO, BEACON, [5] and our method, respectively. The known unreachable branches are given in column 7 and the run time comparison is given in columns 8-11, with speedups reported in the last two columns. Note that the mining time is short. Most are within 1 second except for b12word and OR1200, which took under 20 seconds. The initial data mining process is a one-time cost which does not add much overhead to the final result.

Compared to previous methods, our approach achieves a significant speedup for many benchmarks. For example, in circuit b11, our method achieves higher coverage than HYBRO, maintains the same coverage as [5], but it achieves a speedup of 135.62× and 5.89× over these two techniques, respectively. This speedup comes from two primary factors. First, the use of purely stochastic methods drastically reduces the computational cost compared to calls to deterministic solvers on long sequences. Second, the feedback generated by the proposed path-activation abstraction allows for significantly improved convergence time of the algorithm compared to previous stochastic methods. This effect is particularly apparent in circuit b12, which is highly resistant to random stimuli, because following initialization there are many branches which are uncovered. With our method, those uncovered branches become explicit targets and vectors are found much more quickly compared to prior approaches. This convergence factor leads to the 9.00× speedup seen over [5], which was already considered fast for this challenging circuit. Additionally, We are able to quickly uncover many branches in the extended circuit b12word providing a high level of coverage in the case of a significantly expanded input space. In OR1200 with 2777 state elements, we utilize the mined information from the 4 modules as well as slices taken from OR1200 with "fsm" and "ctrl" labels. With our approach, we achieve 94.15% coverage, an improvement compared to BEACON [4] which achieves a lower 92.84% coverage. Additionally, the proposed algorithm completes the search in 87.59 seconds, a 3.42× speedup over BEACON. This improvement comes from significantly higher levels of activation for some FSM states, which leads to additional coverage of previously unseen conditional branches.

We also report the test vector lengths generated by our algorithm and compare them to previous works, shown in Table III. [5] does not list vector lengths, so we compare them to the preceding work, BEACON, and HYBRO. Our algorithm produces quality compact vectors due to the guidance of the path-activation abstraction. For example, in b10, HYBRO generated 6450 vectors without reaching 100% coverage, BEACON generated 3547 vectors that achieved 100% coverage, and the proposed method, which also achieves 100% coverage, uses only 1973 vectors. By using knowledge about branch activation paths across cycles, certain conditions are excited earlier in operation leading to the improvements in applied test length. The notable exception is OR1200, which required more vectors to uncover several additional hard-to-reach branches.

TABLE III
TEST VECTOR LENGTHS

Circuit	Vector Length		
	HYBRO	BEACON	Ours
b06	NA	1731	**41**
b07	NA	759	**171**
b10	6450	3547	**1973**
b11	4530	1235	**1111**
b12	NA	37006	**33617**
b12word	NA	NA	**49622**
b14	NA	4381	**3707**
OR1200-0	1170	642	**357**
OR1200-1	NA	2146	**423**
OR1200-2	NA	1261	**706**
OR1200-3	11410	4615	**3322**
OR1200	NA	7946	*9388*

C. State Justification

Instead of targeting all branches in the circuit, PACOST [6] is a recent state justification method also utilizing formal methods for generating vectors to reach a targeted branch condition. Of particular interest for state justification is the reaching of properties c1 and c2 for circuit b12, defined by [4]. Unlike PACOST, which targets individual branches, our method targets all branches. We report the time when the target property is first reached. In Table IV, the minimum-length sequences found by PACOST are compared to our method.

The performance of our method compares favorably to PACOST, despite targeting branch coverage for the entire circuit. In the case of property c1, our generated sequence is slightly longer (141 vs. 109) than PACOST, due to the effects of stochastic search and that our method did not target only property c1. However, this indicates that our abstraction is effective in guiding the search toward the target without expending many more vectors. Our execution time 1.38 seconds, as opposed to 1.87 seconds in PACOST. The performance gain is likely from the exclusion of expensive formal techniques in our technique. For the more challenging property c2, our method shows significant improvement over PACOST when compared to their best reported time. Our sequence is slightly shorter (31900 vs. 33148 vectors). In addition, our method achieves an $11.29\times$ speedup over PACOST even when we are targeting the circuit as a whole. Additionally, our method is able to justify both hard-to-reach properties, c1 and c2, in b12word with a relatively low computational effort, even though PACOST reported results only for c1. These results indicate that the proposed approach is effective at reaching corner states that require traversing through narrow activation paths. These properties all timed out utilizing the bounded model checker in [5]

TABLE IV
STATE JUSTIFICATION FOR B12

Property	PACOST		Ours	
	length	time(s)	length	time(s)
b12(c1)	109	1.87	141	1.38
b12(c2)	33148	54.22	31900	4.8
b12word(c1)	109	1.92	137	2.1
b12word(c2)	N/A	N/A	33681	30.86

V. CONCLUSIONS

In this paper, we presented a novel model for guiding a search for functional test generation at the RTL. This guidance mechanism is based on utilizing mined relations from multiple HDL program slices to control the execution path in the concrete model. This guidance allows us to avoid the invocation of expensive formal engines to navigate narrow activation paths. The algorithm yields a significant performance again showing improvements of up to $9\times$ over previous methods. Additionally, the method yields significantly smaller vector lengths while maintaining the same level of coverage and is able to justify previous uncovered states in some benchmarks.

REFERENCES

[1] S. Devadas, A. Ghosh, and K. Keutzer, "An observability-based code coverage metric for functional simulation," in *Proc. Int. Conf. Computer-Aided Design*, 1996, pp. 418–425.

[2] B. Beizer, *Software testing techniques (2nd ed.)*, Van Nostrand Reinhold Co., New York, NY, USA, 1990.

[3] Lingyi Liu and S. Vasudevan, "Efficient validation input generation in rtl by hybridized source code analysis," in *Proc. Design Automation & Test Europe Conf.*, march 2011, pp. 1–6.

[4] Min Li, K. Gent, and M.S. Hsiao, "Design validation of rtl circuits using evolutionary swarm intelligence," in *Proc. Int. Test Conf.*, 2012.

[5] K. Gent and M.S. Hsiao, "Functional test generation at the rtl using swarm intelligence and bounded model checking," in *Proc. Asian Test Symp.*, Nov 2013, pp. 233–238.

[6] Y. Zhou, T. Wang, T. Lv, H. Li, and X. Li, "Path constraint solving based test generation for hard-to-reach states," in *Proc. Asian Test Symp.*, Nov 2013, pp. 239–244.

[7] F.E. Allen, "Control flow analysis," *SIGPLAN Not.*, vol. 5, no. 7, pp. 1–19, July 1970.

[8] R.T. Prosser, "Applications of boolean matrices to the analysis of flow diagrams," in *Eastern Joint IRE-AIEE-ACM Computer Conference*, New York, NY, USA, 1959, IRE-AIEE-ACM '59 (Eastern), pp. 133–138, ACM.

[9] K.D. Cooper, T.J. Harvey, and K. Kennedy, "A Simple, Fast Dominance Algorithm," .

[10] "Verilator," http://www.veripool.org/wiki/verilator.

[11] E.M. Clarke, M. Fujita, S.P. Rajan, T. Reps, S. Shankar, and T. Teitelbaum, "Program slicing of hardware description languages," in *Correct Hardware Design and Verification Methods*, vol. 1703 of *Lecture Notes in Computer Science*, pp. 298–313. Springer Berlin Heidelberg, 1999.

[12] O. Guzey, *Data Mining in Constrained Random Verification*, Ph.D. thesis, Santa Barbara, CA, USA, 2008, AAI3319881.

[13] S. Davidson, "Itc99 benchmark circuits - preliminary results," in *Proc. Int. Symp. Circuits & Systems*, 1999, p. 1125.

[14] "OpenRISC web page," http://www.opencores.org.

Random Pattern Generation for Post-Silicon Validation of DDR3 SDRAM

Hao-Yu Yang*, Shih-Hua Kuo*, Tzu-Hsuan Huang*, Chi-Hung Chen†, Chris Lin† and Mango C.-T. Chao*

*Dept. of Electronics Engineering & Institute of Electronics, National Chiao Tung University, Hsinchu, Taiwan

†Winbond Electronics Corporation, Hsinchu, Taiwan

{max0327.eecs94@nctu.edu.tw}

Abstract—Due to the demand of pursuing a main memory with larger data bandwidth, higher data density, and lower power, the specification of DRAM has been constantly evolved in the past decade. The new DRAM specifications support multiple operating modes with multiple timing settings. It then becomes computationally infeasible to exhaustively validate all the combinations of different operating modes, timing settings and address/data with pure simulation before silicon. In this paper, we propose a framework to generate proper random patterns for validating a newly designed DDR3 SDRAM based on its first silicon chips. The proposed framework needs to not only guarantee the correctness of the generated patterns according to the state diagram and timing constraints defined in the specification but also provide the flexibility of exploring various design corners for the targeted DDR3 SDRAM. We will also show some successful silicon-validation cases of applying the proposed framework to identify the design errors based on real DDR3 SDRAM products.

I. Introduction

Due to the advantage of structural simplicity, DRAM has been the mainstream of the main memory market since its invention by Dr. Dennard in 1966 [1]. In order to achieve higher bandwidth and higher data density, the DRAM industry has put significant research efforts on not only improving the DRAM process [2][3][4][5] but also evolving new DRAM specifications [6][7][8][9][10][11]. The earliest version of DRAM utilizes an asynchronous interface to communicate with the system, where the system needs to wait for the response of DRAM before applying the next operation. Its successor, the *synchronous DRAM (SDRAM)* [6], utilizes an interface synchronized with the system clock and hence enables the use of the pipeline technique to increase the system performance. Also, SDRAM divides its storage data into multiple banks and hence can provide data interleaving to further increase the data bandwidth.

The successor of SDRAM is the *double-data-rate SDRAM (DDR SDRAM)* [7], which doubles the original data rate by using the double pumping technique to transfer data on both the rising and falling edges of the clock signal. DDR SDRAM also introduces 3 read latencies. The next DRAM specification is the *DDR2 SDRAM* [8], which increases the number of banks to 4/8 and the number of prefetch data to 4. DDR2 SDRAM also introduces 5 additive latencies, 5 latencies for write recovery (auto-precharge), 2 latencies for active power-down exit time, and RTT (input termination). *DDR3 SDRAM* [9] further increases the number of banks to 8/16 and the number of prefetch data to 8. DDR3 SDRAM also introduces 9 read latencies, 8 write latencies, 8 latencies for write recovery, 8 settings for partial array self-refresh, asynchronous reset, ZQ calibration, and write leveling.

Compared to DDR SDRAM supporting only 1G data and 6 mode registers with only 18 combinations of mode settings, DDR3 SDRAM supports up to 8G data and total 20 mode registers with more than 300k combinations of mode settings, which derives various combinations of cycle latencies associated with 45 different timing constraints between commands. As a result, it is computationally infeasible to exhaustively validate all the combinations of different operating modes, timing

constraints and address/data for DDR3 SDRAM with pure simulation before silicon. In addition, certain design parameters, such as dynamic power, thermal profile, and process variations, are also difficult to explicitly simulate for the entire DRAM chip. Therefore, the coverage of the conventional simulation-based design validation is insufficient for DDR3 SDRAM. Validating the DRAM design with random patterns based on silicon chips then becomes an efficient and practical solution to cover the unexplored design space left by simulation.

Another challenge on design validation for DRAM providers is that the system integrators usually validate a DRAM product with their own system benchmarks, which are not accessible for DRAM providers. When the DRAM product fails the system validation, some friendly system integrators may feed back the failure condition to the DRAM providers. However, most of the time, the DRAM providers will not receive any failure report and still have no clue how to fix the failed DRAM product after losing a business opportunity. Applying random patterns to DRAM chips for validation can be like a mock test for a DRAM product, which may run for days. If any bug is found, DRAM providers can have the complete information for debugging.

In this paper, we propose a framework to generate the random cycle-based patterns used for the silicon validation of DDR3 SDRAM. The proposed framework can support any configuration of DDR3 SDRAM with the adjustable probabilities of transferring one DRAM state to another. The generated patterns are guaranteed to follow the state diagram and the timing constraints defined in DDR3 SDRAM specification for each operating mode. The proposed framework also tries to create worst-case power by extensively toggle the data and address. The proposed framework also supports user-defined patterns, which can cover some specialized patterns that is difficult to generate based on pure random scheme. The generated cycle-based random patterns are initially in the format of CSV [12] and can then be transferred to the patterns of any designated ATE (or even simulation patterns). In our case, the ATE in use is the Adventest Verigy 93K [13]. We will later show three silicon-validation cases based on real DDR3 SDRAM products where the design errors were successfully identified with the help of the proposed random pattern generator.

II. Background of DDR3 SDRAM

A. Architecture

The specification of DDR3 SDRAM supports the storage size from 512M to 8G (bits). In Figure 1, we use a 2G DDR3 SDRAM as an example to illustrate the overall architecture of a DDR3 SDRAM, where the 2G data is divided into 8 banks and each bank contains 256M bit. The input/output bit-width is 16 bits. The row address and column address share the same address pins. The row address uses 14 pins while the column address uses 10 of 14 pins only.

Following lists the function of each pin.

- *CK, CKn*: Positive-edge trigger clock pin and negative-edge trigger clock pin.
- *CKE*: Clock enable pin. The DRAM macro is only received the commands when CKE pin is high.
- *CSn*: Chip select pin with negatively enabled.

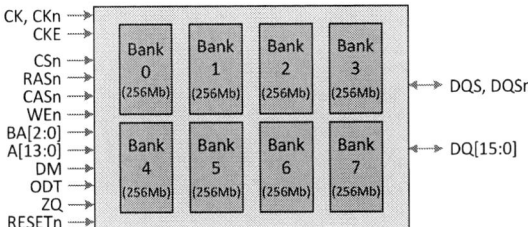

Fig. 1. **Architecture of a 2Gb DDR3.**

- *RASn, CASn, WEn*: Pins for row address strobe, column address strobe, and write enable. All these three pins are negatively enabled and the DRAM commands are composed by the combination of these three pins.
- *BA[2:0]*: Bank address pins.
- *A[13:0]*: Address pins for both row and column address. During read/write commands, only 10 pins are used for column address and the rest pins are used for other functions, such as auto-precharge and burst chop.
- *DM*: Pin for data mask, which masks the written-in data and keeps original stored data.
- *ODT*: Pin for on-die termination, which is used for improving signal integrity.
- *ZQ*: Reference pin for ZQ calibration.
- *RESETn*: Reset pin, which is negatively enabled.
- *DQ[15:0]*: Data pins, which are the bidirectional for written-in and read-out data.
- *DQS, DQSn*: Data strobe pins, one for positively enabled and another for negatively enabled.

B. State Diagram

Figure 2 shows the state diagram of DDR3 SDRAM [9]. In Figure 2, a circle represents a DRAM state and a rectangle represents an action. A solid edge represents a DRAM command to perform and a dash edge represents an automatic transfer after the designated action is done. The upper part of Figure 2 is denoted as the *common part* of the state diagram. The commands in the common part can be performed only when all the banks are in the idle state. The lower part of Figure 2 is denoted as the *bank part* of the state diagram, which is associated with a designated bank only and each bank has its own copy of the bank-part state diagram.

In Figure 2, the DRAM is initialized with all banks in the idle state. In the common part of the state diagram, this initial state can then be transferred to the self-refresh state or the precharge power-down (PD) state. When entering the self-refresh state, the DRAM can save power by turning off its peripheral circuitries while maintaining the stored data by frequent precharge. When entering the precharge power-down state, the DRAM can save power by turning off only the input/output buffers. Through an ACT command, the DRAM can select one designated bank and enter into the bank-active state in the bank-part state diagram. Once any bank is in the bank-active state, we can enter into the active power-down state, which turns off the input/output buffers just like the precharge power-down state. Its difference to the precharge power-down state is that when leaving the active power-down state, all the original active banks remain in the bank-active state.

Following is the brief description of each command.

- *REF* activates the refresh function.
- *MRS* updates the mode register (MR) with the desired settings.
- *ZQC* calibrates the on-die resistor with the ZQ pin.
- *SRE, SRX* enter and exit the self refresh mode.
- *PDE, PDX* enter and exit the power-down mode.
- *ACT* activates a row in a given bank.
- *PRE* denotes precharge, which deactivates the active row in a given bank.

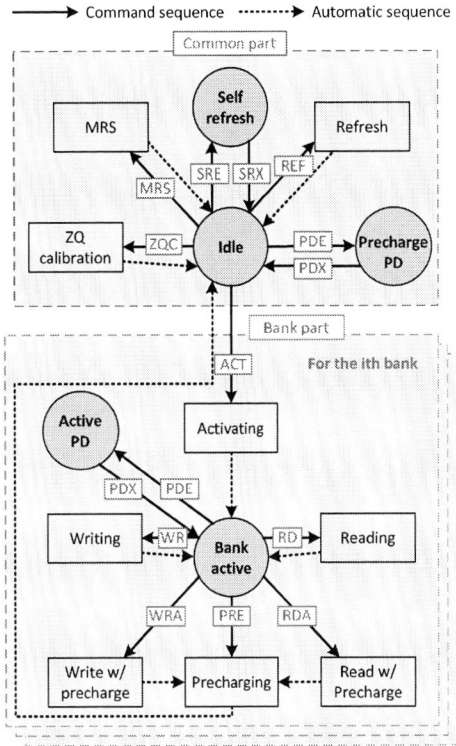

Fig. 2. **State diagram of DDR3 SDRAM.**

- *WR, WRA* denote the write command and write with auto-precharge command.
- *RD, RDA* denote the read command and read with auto-precharge command.

Note that the MRS command is used to set the content of the mode registers. DDR3 SDRAM divides its total 20 mode registers into three sets, *MR0* for 7 register modes, *MR1* for 7 register modes, and *Mr2* for 6 register modes. One MRS command can only update the content of one set of the register modes.

C. Timing Specification

The timing specification defines the minimum timing constraint and the maximum timing constraint between a pair of commands. The minimum timing constraint represents the minimum waiting time before applying the next command given a current command. Different pairs of a current command and a next command require different minimum timing constraints. Also, different settings of the mode registers may result in different minimum timing constraints between a pair of commands. Note that during the waiting time, other commands may be applied as long as the applied commands can satisfy the minimum timing constraint associated with each of the commands applied previously.

Figure 3 illustrates an example of applying a ACT command followed by a RD command. After applying the ACT command, we have to wait for the time of tRCD before applying the RD command. Then we need to wait for another time of tRL such that the data can be sent out at DQ, where tRL can be adjusted by different settings of the mode registers.

On the other hand, the maximum timing constraint between a current command and a next command specifies the time within which the next command must be applied after the current command is applied. In the specification of DDR3 SDRAM, only the refresh command needs

978-1-4799-7598-3/15 $31.00 © 2015 IEEE   187

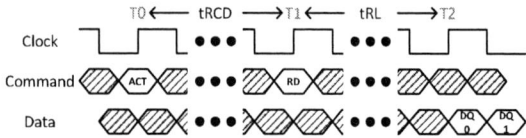

Fig. 3. **Example of the timing constraint between ACT and RD.**

to use a maximum timing constraint. All the other timing constraints defined in the specification are the minimum timing constraints.

III. OVERVIEW OF RANDOM PATTERN GENERATOR

·A. Overall Flow

The proposed random pattern generator needs to achieve the following three main objectives. First, the sequence of the applied commands must satisfy the state diagram defined in the specification of DDR3 SDRAM. Second, the generated cycle-based commands must satisfy all the timing constraints defined in the specification. Third, the frequency of the occurrence of the applied patterns can be adjusted so that different combinations of the commands, operating modes, timing constraints, and data/addresses can be examined.

Figure 4 illustrates the overall flow of the proposed random pattern generator. Its input files include (1) the configuration of the targeted DRAM, (2) the state diagram with a specified probability of each transfer between commands, (3) the timing constraints defined in the specification, (4) the setting for address & data inversion, and (5) the user-defined patterns. The setting for the address & data inversion specifies the probability of completely inverting the address or data in the command sequence, which can help to create higher power consumption for validation purpose. The user-defined patterns specify some command sequences that are useful in DRAM validation based on the experience but difficult to be generated under a random scheme.

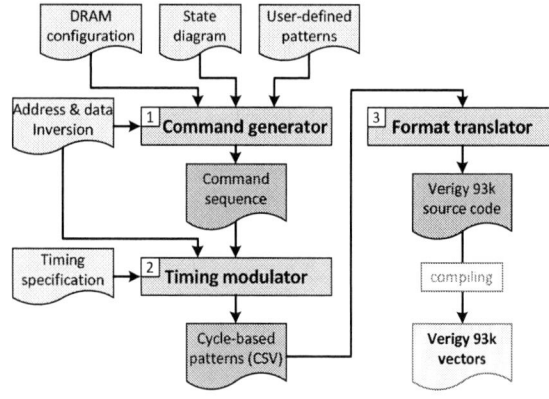

Fig. 4. **Overall flow of the proposed random pattern generator.**

The proposed framework contains three main procedures: the *command generator*, the *timing modulator*, and the *format translator*. The command generator will generate a random command sequence that satisfy the specified state diagram based on the given state-transfer probabilities for the targeted DRAM. Also, the value at each pin is randomly assigned for each command. Then the timing modulator will take the command sequence generated by the command generator, insert proper no-operations (NOPs) to satisfy the specified timing constraints, and output the cycle-based commands in the CSV format [12], which is a platform-independent text file. In our silicon-validation environment, we apply the random patterns through the ATE, Advantest Verigy 93k [13]. Next, the format translator will convert the cycle-based patterns in CSV format into Verigy 93k source code and

then be compiled into the Verigy 93k vectors, which can be directly applied on Verigy 93k.

In the proposed framework, the command generator and timing modulator can guarantee that the generated cycle-based command sequence must satisfy the specified state diagram and timing constraints. We can also specify several different combinations of the command-transfer probabilities in the input file of the state diagram to generate different sets of the random patterns that cover different corners of the specifications. Also, with the settings of the address/data inversion and the user-defined patterns, we can include some useful pre-defined command sequences, address sequences, and data sequences in the generated random patterns, which further improves the flexibility of the proposed framework. In the following subsections, we will introduce more details about each of the input files.

B. DRAM Configuration

This input file defines the configuration of the targeted DRAM, including the number of banks, the number of bits for row and column addresses, and the input/output width. It also defines the initial data background and the initial setting of each mode register.

C. State Diagram

Instead of directly specifying the state transfers shown in Figure 2, we represent the state diagram by specifying the valid transfer between a pair of commands with a 2-dimension array. If there exist a valid transfer between two commands, the corresponding entry in the array will list the probability of executing this transfer. This probability can be defined by the users or randomly generated. Table I shows an small exemplary state diagram, where the transfer probability between a current command and a next command is listed and the symbol '-' represents no valid transfer exists between the two commands. Note that we need to provide a 16-command by 16-command array for the common-part state diagram and a 13-command by 13-command array for the bank-part state diagram in this input file.

TABLE I
Example of representing a state diagram using a 2D array with transfer probabilities.

Transfer probability		Next command			
		ACT	RD	WR	PRE
Current command	ACT	-	40%	40%	20%
	RD	-	60%	40%	0%
	PRE	100%	-	-	-

In addition, this input file also defines some other probabilities used in the command generator, such as the probability of selecting the bank-part commands or the common-part command, the probability of selecting a bank, and the probability of the length of the consecutive commands to be applied to a bank once the bank is selected.

D. Timing Constraints

The minimum timing constraints is specified by a 29-command by 29-command array. The 29 commands include both the common-part and bank-part commands since a minimum timing constraint may also exist between a common-part command and a bank-part command. Table II shows a small exemplary array for representing the minimum timing constraints between two commands, where each minimum timing constraint is denoted by a timing label, which follows the naming defined in the specification of DDR3 SDRAM. Note that a timing label can be represented by the number of clock cycles or an equation associated with other timing labels or MRS settings. In Table II, the symbol '-' means there is no minimum timing constraint between the corresponding two commands. As to the maximum timing constraints, we used a one-dimension list to record since it only involves the refresh command.

978-1-4799-7598-3/15 $31.00 © 2015 IEEE   188

TABLE II
Example of representing minimum timing constraints.

Minimum timing constraint		Next command			
		ACT	RD	WR	PRE
Current command	ACT	-	tRCD	tRCD	tRAS
	RD	-	tCCD	tRTW	tRTP
	PRE	tRP	-	-	-

E. Address & Data Inversion

The address or data inversion means the address or data bits of the next command is completely opposite to those of the current command, which in our experience is an effective way to create higher power consumption. We apply the address/data inversion in both the command generator and the timing modulator. In the command generator, we can toggle the address and data directly applied to the peripheral circuits of a bank, such as the address decoders and write drivers. In the timing modulator, we need to further toggle the address and data for those newly inserted no-operations (NOP), which can consume higher power at the address and data bus of the chip. The probability of applying the address/data inversion and the probability of the length of the applied period needs to be specified for both the command generator and the timing modulator in this input file.

F. User-defined Pattern

Under the above random scheme, some useful command sequences, such as consecutive read (or write) commands and the address sequence of GALPAT, are difficult to be obtained. This input file defines the command sequence of each of those specialized patterns along with the activation condition and the probability for applying the defined pattern once its activation condition is satisfied. A loop representation is supported in this input file to help the users to define the command sequence.

IV. IMPLEMENTATION OF RANDOM PATTERN GENERATOR

A. Command Generator

The command generator only generates the command sequence without timing information based on given state diagram and transfer probability. Figure 5 shows the flow chart of the command generator, which starts with all banks in the idle state. The command generator first checks whether the status of the previous command satisfies the activated conditions of any user-defined pattern. If yes, we will determine whether to apply the user-defined pattern based on its defined applying probability. If no user-defined pattern is selected, we jumped into the normal procedure of generating random commands.

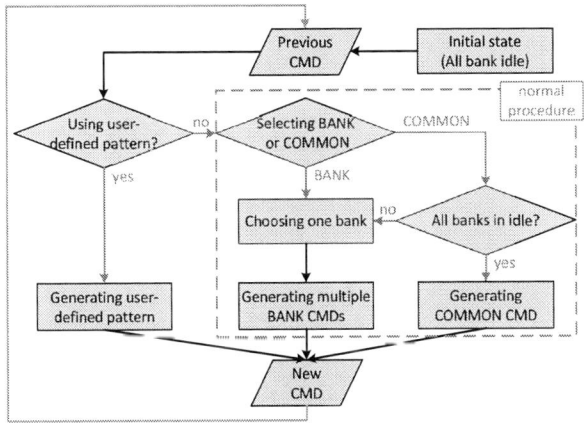

Fig. 5. **Flow chart of the command generator.**

In the normal procedure, we first need to randomly determine whether to generate random commands in the bank part or the common part of the state diagram. If the bank part is selected, we randomly select a bank and the number of consecutive commands to be applied to the bank. In this way, we could create the situations of staying in one bank for a long period or switching to another bank quickly. If the common part is selected, we need to check whether all banks are in the idle state before randomly generating any command of the common part. Note that all probabilities used to the normal procedure are defined in the input file of the state diagram.

B. Timing modulator

In the timing modulator, we needed to resolve the minimum and maximum timing constrains for the command sequence generated by command generator. For the minimum timing constraints, a command may be constrained by N previous commands. If we directly check the minimum timing between each current command and each of its N relating previous commands, the required data structure and time complexity to maintain can be large since the minimum timing associated with each of the N previous commands need to be recorded and updated for each command. Instead of maintaining a series of previous commands, we maintain the minimum applicable cycles (from the current cycle) for each potential following command constrained by the current command. This information is recorded by a proposed data structure, called Checking_List.

Following are the four steps of using Checking_List to determine the applied cycle of each next command in the command sequence.

1) Find the minimum applicable cycles of the next command in Checking_List.
2) Determine the applied cycle of the next command by adding the minimum applicable cycles to the current cycle and output each DRAM pin's value of the next command.
3) Adjust Checking_List to the applied cycle by subtracting the applied minimum applicable cycles from each entry.
4) Update Checking_List with the new minimum timing constraints induced by the just applied next command. An entry in Checking_List will only be updated when the new minimum timing constraint is larger than its original value.

After determining the applied cycle of each command, NOPs will be inserted into the cycles between to the applied cycles of two adjacent commands. Figure 6 shows an example of how the timing modulator determines the applied cycle of each command with the help of Checking_List. The upper part of Figure 6 lists the minimum timing constraints used in this example. The applied command sequence is ACT, RD, and PRE. At the beginning, Checking_List is initialized as a null list. For the first ACT command, no timing constraint limits the applied cycle of it and hence its applied cycle is 0. The ACT command will then induce three minimum timing constraints for RD, WR, and PRE to the originally empty Checking_List. For the second RD command, its minimum applied cycle is 12 as listed in Checking_List and hence its applied cycle is 12 (0+12). Then we adjust Checking_List to cycle 12 by subtracting 12 from each entry. The RD command will also induce three minimum timing constraints, but only the minimum applicable cycles for RD and WR in Checking_List are updated since only their induced minimum timing constraints are larger than the original value. This means that the following PRE command is actually constrained by the previous commands instead of this RD command. For the third PRE command, the applied cycle is 28 (12+16) using a similar manner.

Since the maximum timing constraints involve only the refresh commands, we need to constantly check before which cycles in the commands a refresh command is required and then insert one. In order to insert a refresh command, several other commands may need to be inserted before it. After inserting a new command, Checking_List needs to be updated as well.

978-1-4799-7598-3/15 $31.00 © 2015 IEEE

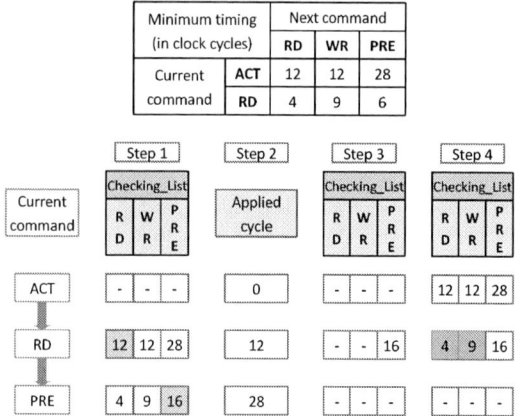

Fig. 6. **Example of determining the applied cycles of commands using** `Checking_List` **in the timing modulator.**

C. Runtime

The proposed random pattern generator is programmed in C++ and performed on a computer with a 2.4GHz Intel Xeon E5620 CPU and 16GB main memory. Table III lists the runtime of each part of the random pattern generator for generating 64M random patterns of a 2Gb DDR3 SDRAM. The part of initializing memory space is used for allocating the memory space to store the expected read-out data. As the result shown, the overall runtime is around 46 minutes and most of the runtime is on the timing modulator.

TABLE III
Runtime of each part of the random pattern generator.

	Initializing memory space	Command generator	Timing modulator	Total time
Runtime	17.6s	4m28.5s	41m32.6s	46m18.7s

V. EXPERIMENTAL RESULTS

When using the proposed random pattern generator for silicon validation, we first randomly sample 1000 state-diagram files all with different transfer probabilities and generate a set of 64M random patterns for each of the 1000 sampled state-diagram files. Then we will apply each set of 64M random patterns to obtain its tAC-vs-tCK shmoo plot for the targeted DDR3 SDRAM chip. It takes around 20s to get one shmoo plot and less than one day to finish applying all 1000 sets of random patterns.

In this sections, we will show three successful cases of using the proposed random-pattern framework in silicon validation to identify the bug of a failed DDR3 SDRAM product. All three cases have different data sizes and targeted frequencies. One of the cases is actually the product rejected from a system integrator without providing any information of the failure. Note that each of these three DDR3 SDRAM products passes all the conventional validation patterns on ATE, including the functional patterns used for simulation, all the March patterns used for production test, and some empirical patterns that has successfully detected a previous bug.

A. Case 1: tRFC Timing Bug

The first case of the failed products is a 1Gb DDR3 SDRAM with 8 banks, 16-bit width of the input/output port, 13-bit row address, and 10-bit column address. The targeted clock period (tCK) is 1.35ns. This failed DDR3 SDRAM can be detected by the random patterns (denoted as PAT-1) where a REF command is applied followed by an ACT command with the minimum timing in between. Figure 7(a) first shows the tAC-vs.-tCK shmoo plot by applying the PAT-1 patterns to

the failed product, where tAC means the data access time starting from the clock's positive edge and tCK means the clock period. The quantity of tAC is defined as a ratio of tCK.

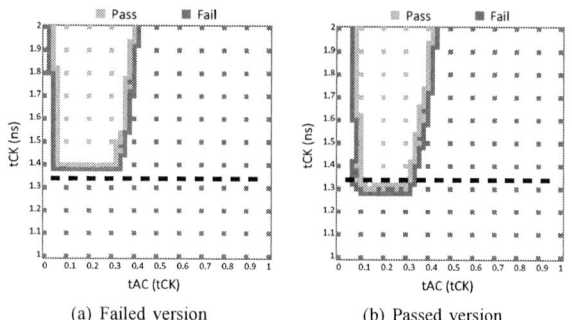

(a) Failed version   (b) Passed version

Fig. 7. **tAC-vs.-tCK shmoo plots of applying PAT-2 to (a) failed version with tRFC timing bug and (b) passed version.**

Note that tAC is a timing that can be dynamically calibrated at the system when first accessing the DRAM chip. Therefore, as long as there exists a valid tAC that corresponds to a tCK smaller than the specified value (1.35ns in this case) in the shmoo plot, the DRAM chip is considered passed. As shown in Figure 7(a), the tCK corresponding to each tAC in the shmoo plot is larger than 1.35ns, which is clearly a case of failed products. We later applied the same PAT-1 patterns to a passed design and its shmoo plot is shown in Figure 7(b), where we can find some valid tAC that corresponds to a tCK smaller than 1.35ns. This result demonstrates that the generated random patterns can effectively detect this failed product.

In fact, the minimum timing between a REF command and a ACT command is defined by tRFC in the specification. In this case, the failed chip cannot satisfy the minimum timing constraint of tRFC. After further examination to the design, we found that this bug results from an error in the refresh controller, which reports the wrong refresh period and let the DRAM accidentally block the next ACT command.

B. Case 2: tRAS Timing Bug

The second case is a 2Gb DDR3 SDRAM product with 8 banks, 16-bit width of the input/output port, 14-bit row address, and 10-bit column address. The targeted clock period (tCK) is 1.25ns. This failed DDR3 SDRAM can be detected by the random patterns (denoted as PAT-2) which satisfies the following three conditions: (1) an ACT command is applied followed by an PRE command to the same bank with the minimum timing in between, (2) the mode register of CAS latency is set to 8, and (3) the mode register of the additive latency is set to 6. Figure 8(a) shows the tAC-vs.-tCK shmoo plot by applying the PAT-2 patterns to the failed product, where no resulting tCK is smaller than the specified value 1.25ns. Figure 8(b) shows the shmoo plot by applying the same PAT-2 patterns to a passed design, where we can find some valid tAC that corresponds to a tCK smaller than 1.25ns. This result again demonstrates the effectiveness of our random generated patterns on detecting design errors.

The minimum timing between an ACT command and a PRE command is defined by tRAS in the specification and the value of tRAS may vary depending on the setting of the mode registers. The function of an ACT command is to activate a bank by turning on a targeted row and then the data on the row will automatically write back. The function of a PRE command is to turn off the active row, perform precharge, and then going back to the idle state. The bug in this case will turn off the active row too early with the PRE command while the automatic write-back of the ACT command has not finished, which in turn results in the data lose.

Note that in the normal functional operation of a DDR3 SDRAM, an ACT command is seldom immediately followed by a PRE command

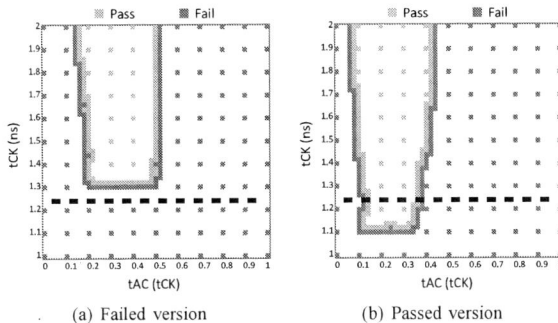

(a) Failed version   (b) Passed version

Fig. 8. **tAC-vs.-tCK shmoo plots of applying PAT-2 to (a) failed version with tRAS timing bug and (b) passed version.**

because a bank turned on by an ACT command will usually perform some read or write operations before being turned off by a PRE command. This also a reason why this command sequence is usually not covered by the functional simulation patterns. Also, the conditions of detecting this bug requires the setting of proper CAS latency and additive latency, which are specified in two sets of mode registers with two different commands. Therefore, compared to the previous case, it is statistically more difficult to randomly generate the command sequence that can detect the tRAS bug in this case. The PAT-2 patterns are generated by a state-diagram file with a high transfer probabilities from ACT to PRE and a high probability from any state to MR0 (for setting CAS latency) and MR1 (for setting additive latency).

C. Case 3: Dummy Read Bug

The third case is a 4Gb DDR3 SDRAM product with 8 banks, 16-bit width of the input/output port, 15-bit row address, and 10-bit column address. The targeted clock period (tCK) is 1.25ns. This failed DDR3 SDRAM can be detected by the random patterns (denoted as PAT-3) where a RDA command is applied followed by a RD command to the same bank. Figure 9 shows the shmoo plot by applying the PAT-3 patterns to the failed product, where the product fails on all the plotted tCKs for each tAC. Note that a RDA command will perform a read operation and then auto-precharge, which transfers the designated bank into idle mode. As a result, the followed RD command cannot perform any action to the idle bank. Such a dummy-read command sequence (RDA immediately followed by RD) usually will not occurs in the normal functional operation of a DDR3 SDRAM since the later RD command is a dummy. This is also the reason why the functional simulation patterns and the general March patterns cannot detect this bug.

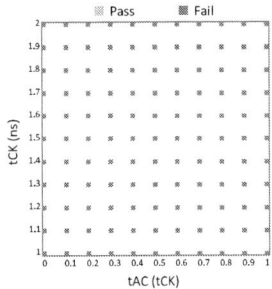

Fig. 9. **tAC-vs.-tCK shmoo plot of applying PAT-3 to a failed version with dummy read bug.**

Figure 10 further shows the timing diagram at targeted specification tCK=1.25ns and the failed cycles are highlighted. As Figure 10 shows, the output data of the first read command is not the same as we expected. The upper half bits and the lower half bits are exchanged at the failed cycles due to an error on data bus controller.

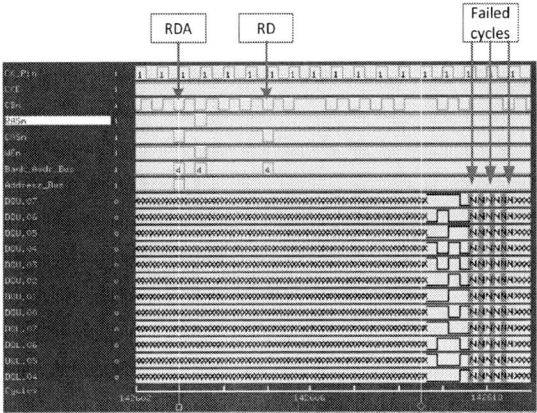

Fig. 10. **Failed timing diagram associated with the dummy read bug.**

VI. CONCLUSION & FUTURE WORK

In this paper, we have developed a framework to generate random cycle-based patterns for silicon validation of DDR3 SDRAM. The proposed framework can not only provide the randomness of exploring various design corners but also guarantee that the generated patterns satisfy the specified state diagram and timing constraints. Also, a the proposed framework provides the flexibility of including those useful pre-defined command/address/data sequences that are difficult to be obtained through a random mechanism. We further demonstrated three successful cases of applying this random-pattern framework to identify the design errors of three failed DDR3 SDRAM products.

Even though DDR3 SDRAM is currently the mainstream of the DRAM market. Some more advanced specifications of SDRAM, such as low-power DDR3 SDRAM [10] and DDR4 SDRAM [11], have already be announced and used. Our next step is to extend the current random-pattern generator for validating low-power DDR3 SDRAM and DDR4 SDRAM.

REFERENCES

[1] R.H. Dennard. Field-effect Transistor Memory, 1968. US Patent 3,387,286.

[2] H.K. Kang, K.H. Kim, Y.G. Shin, et al. Highly Manufacturable Process Technology For Reliable 256 Mbit and 1 Gbit DRAMs. In *IEEE International Electron Devices Meeting*, pages 635–638, 1994.

[3] J. Alsmeier and R.J. Stengl. Deep Trench DRAM Process on SOI for Low Leakage DRAM Cell, May 6 1997. US Patent 5,627,092.

[4] K.-W. Song, J.-Y. Kim, J.-M. Yoo, et al. A 31 ns Random Cycle VCAT-based 4f DRAM with Manufacturability and Enhanced Cell Efficiency. *IEEE Journal of Solid-State Circuits*, 45(4):880–888, 2010.

[5] H.J. Park, I.C. Ryu, Y.P. Song, et al. The Results of Self-annealing Process for a Copper Interconnection on the 4xnm DRAM Products. *Advanced Materials Research*, 816:101–105, 2013.

[6] JEDEC Standard Memory Configurations. JESD21-C. *JEDEC Solid State Technology Association*, January 2004. No. 3.11.

[7] JEDEC Standard Double Date Rate (DDR) SDRAM. JESD79F. *JEDEC Solid State Technology Association*, pages 1–84, February 2008.

[8] JEDEC Standard DDR2 SDRAM. JESD79-2F. *JEDEC Solid State Technology Association*, pages 1–128, November 2009.

[9] JEDEC Standard DDR3 SDRAM. JESD79-3F. *JEDEC Solid State Technology Association*, pages 1–226, July 2012.

[10] JEDEC Standard Low Power DDR3 SDRAM. JESD209-3B. *JEDEC Solid State Technology Association*, pages 1–116, August 2013.

[11] JEDEC Standard DDR4 SDRAM. JESD79-4. *JEDEC Solid State Technology Association*, pages 1–214, September 2012.

[12] CSV. http://en.wikipedia.org/wiki/comma-separated_values.

[13] ADVENTEST. https://www.advantest.com/us/index.htm.

UPF-based Formal Verification of Low Power Techniques in Modern Processors

Reza Sharafinejad[1], Bijan Alizadeh[1,2] and Masahiro Fujita[3]

[1]School of Electrical and Computer Engineering, College of Engineering, University of Tehran, P.o.Box 14395-515, Tehran, Iran
[2]School of Computer Science, Institute for Research in Fundamental Sciences (IPM), P.o.Box 19395-5746, Tehran, Iran
[3]VLSI Design and Education Center, The University of Tokyo, P.o.Box 113-8654, Tokyo, Japan
E-mails: r.sharafi@ut.ac.ir, b.alizadeh@ut.ac.ir, fujita@ee.t.u-tokyo.ac.jp

Abstract— **Ensuring from the correctness of system on a chip (SoC) designs after the insertion of high level power management strategies that are disconnected from low level controlling signals, is a serious challenge to be addressed. This paper proposes a methodology for formally verifying dynamic power management strategies on implementations in modern processors. The proposed methodology is based on correspondence checking between a golden model without power features as a specification and a pipelined implementation with various power management strategies. Our main contributions in this paper are: 1) extracting Power Management Unit (PMU) from Unified Power Format (UPF) and Global Power Management (GPM), 2) automatically integrating PMU into the implementation and 3) checking the correspondence between two models with efficient symbolic simulation. The experimental results show that our method enables the designers to verify the designs with different power management strategies up to several thousands of lines of Register Transfer Level (RTL) code in minutes. In comparison with existing methods such as [7], our method reduces the number of state variables, the number of clauses, the number of symbolic simulation steps, and the CPU time by 11.04×, 17.57×, 2.08× and 13.71×, respectively.**

Keywords—Formal verification; global power management (GPM); unified power format (UPF); power management unit (PMU)

I. INTRODUCTION

High demand of portable devices such as laptops, mobile phones, personal digital assistance, smart cards, including Internet of Things (IoT), utilize intensively low-power design strategies and techniques such as clock gating, multi-supply multi voltage (MSMV), and power gating with state retention [1]. On the other hand, such low power strategies should not be adopted in real processor designs unless their correctness on the target processor has been fully verified. Hence, power-aware verification techniques especially, formal ones are highly desirable. Such techniques must ensure that not only the power intents have been completely and correctly implemented, but also the design continues correct operations after the insertion of power management strategies [2].

In this regard, efforts have been spent to establish a standard that represents the power features of a system. The first version of the standard was presented by Accellera organization in 2007, and finally IEEE institute published the IEEE 1801 standard as Unified Power Format (UPF) in 2013 [3]. This standard enables the defining and verifying of the power intent that represents the power-management architecture (including set of power domains, power switches, supply networks, etc.) and strategy specifications (legal system power modes and power transitions). By providing this standard many Electronic Design Automation (EDA) tools try to support this power format in different phases of digital system design flow (from simulation and synthesis to physical design) [4].

Works have been performed to use this standard for verifying power aware designs at different levels of abstraction such as Register Transfer Level (RTL) and Transaction Level Modeling (TLM). Authors of [5] proposed assertion based structure for power management in TLM that presents a generic framework to abstract relevant power concepts specified by UPF. The authors of [6] presented a method to extract assertions from Unified Power Format (UPF) and then verify them with high level global power intent, including power down and power up states.

The authors of [7] proposed a method to verify power gating strategy in processors. In this method, first of all, the processor is abstracted in such a way that its components just work in a single power domain. In contrast to it, our proposed method works in multiple power domains. Moreover, it manually performs the power management control process while in our proposed method; the power management controller has been integrated into the implementation that automatically controls the power domains. Finally, using UCLID 1.0 [8], they have tried to check the correspondence between the abstracted processor, including some power aware features with its specification. Although the authors of [7] have tried to formally verify a power gating strategy in processors, the proposed verification technique may not be scalable enough to handle more complex power management strategies which are usually used in modern processors due to the large number of symbolic simulation steps. It should be noted that using UCLID 1.0 to verify modern processors is a well-known approach utilized by [14] to verify interrupts in modern processors.

In this paper, we propose a verification methodology for low-power modern processors in two different viewpoints: 1) make sure that processor's functionality doesn't change after the insertion of low power strategies and 2) check to see whether each power domain of given processor performs its function properly as defined in UPF. The proposed methodology is based on correspondence checking between a specification without power features and a pipelined implementation with various power management strategies. Hence, our main contributions in this paper are as follows:

- Automatic extraction of Power Management Unit (PMU) from UPF and Global Power Management (GPM) which makes our verification technique efficient in terms of the number of symbolic simulation steps and CPU time.
- Integrating the PMU into the implementation so that low-level controlling signals for each domain can be automatically adjusted.

The rest of the paper is organized as follows. In Section II, a brief review on power management protocol is presented. In Section III, the main idea of our proposed modeling of low power processors using UCLID and our verification methodology are

explained in details. Section IV shows the experimental results, and Section V concludes the paper.

II. POWER MANAGEMENT PROTOCOL

Before describing our proposed power-aware verification methodology, we first explain power management protocol used in this work. There are four fundamental blocks in power management protocol: 1) RTL design, 2) UPF block, 3) Local Power Management (LPM) and 4) Global Power Management (GPM) that will be explained in the following subsections.

A. UPF block and its power control circuitries

Unified Power Format (UPF) is a low-power specification standard [3]. This standard defines the low-power intent in the portable form so that power management components can be implemented in different stages of the design from RTL to physical design. Since the Hardware Description Language (HDL) code is not adequate to specify the power information in the design, UPF provides a consistent format without impacting the existing HDL code. The UPF code is a side file separated from the HDL file, and therefore, changes to the HDL code do not require to rewriting the UPF code. The UPF block consists of several components such as power domains, supply networks, retention registers, isolation cells and level shifter cells that are explained in the following subsections.

a) Power domain

A basic and fundamental unit in the UPF is the power domain, and chips today have 20 to 50 power domains and hundreds of power modes. In the power management terminology, power domain means a collection of instances in the design that share power supply set and have the same power strategy. A power domain may be composed of some other power domains called subdomain. In order to create a power domain in the UPF create_power_domain command is used.

b) Supply networks

Supply networks consist of supply ports, supply nets, and power switches. The UPF supply network defines how supply nets are distributed to the instances or power domains and how that distribution is controlled. Supply nets provide a connection between supply ports and power domains whereas power switches control these connections with their input signals. The UPF provides the connect_supply_net command for creating a connection between a supply port and supply net [3].

c) Isolation

Isolation is a cell to protect active power domains from off power domains that have floating values in their outputs. These floating inputs might cause high current consumption or improper logic behavior in active power domains. In order to solve these problems, isolation cell is placed between power domains that AND, OR, NOR, and NAND gates are good choices of isolation cells. For creating an isolation cell between two power domains, set_isolation command is used [3].

d) Level shifter

Between domains with different supply nets, a level shifter translates signal values from an input voltage swing to a different output voltage swing. Typically, level shifters convert the high voltage signal in the transmitter domain to the low voltage signal in the receiver domain. The UPF provides the set_level_shifter command for creating a level shifter between two power domains with different supply nets.

e) Retention cell

When volatile memories or sequential elements are power off, their states are lost and when power is on again, their states must be restored. To ensure this, always-on cells called retention cells are used to retain internal states and values of design components even when primary power supplies are turned off. There are two types of retention cells: retention flips flop and retention latches.

B. Local power management (LPM)

Power domains contain different power modes such as IDLE, ON (Active) and OFF. In each power mode, power domains have specific voltage and frequency. For example, the ON mode is characterized using highest voltage and frequency pairs, while the OFF mode is characterized using ground voltage. The Local Power Management (LPM) adjusts power control circuitries (i.e., isolation cell, level shifter, etc.) of each power domain based on its power mode. For example, if a power domain wants to be OFF, then by adjusting the inputs of power switch, isolation cell, and in some cases with enabling of retention register (RR), LPM prepares the domain circuitries in such a way that the power domain is turned off. An important point to be noted here is that the LPM controls power domains independently.

C. Global power management (GPM)

Low power applications and tasks (receiving, processing, storing and transmitting of data) have several power states. In each power state, the status of each power domain mode is defined. Global Power Management (GPM) specifies the legal and illegal power states and also identifies how the transition between power states can occur. In other words, GPM controls LPMs to implement the power management scenario according to the target applications.

In order to have a better understanding, let us consider the design shown in Fig. 1. Suppose we are given a RTL design which has three modules, namely *Module1*, *Module2* and *Module3*. Each of these modules is in an individual power domain PD1, PD2 and PD3. We also assume that power networks of PD2 and PD3 are a subset of PD1, which means that if PD1 is in OFF mode, then PD2 and PD3 have to be in OFF mode. However, power domains PD2 and PD3 can be independently ON or OFF. Obviously, in the power management protocol, the UPF bridges the gaps between the RTL design and low level power control circuitries (PCC).

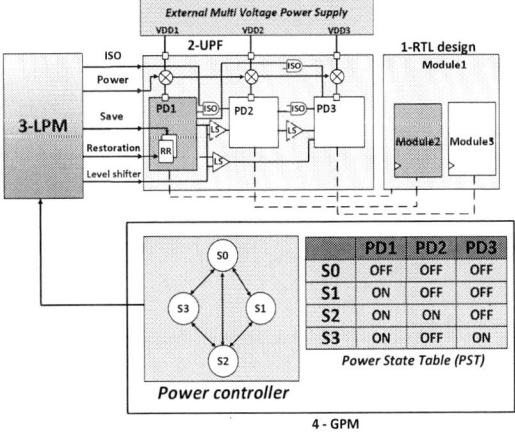

Fig. 1. An Example of power management protocol

UPF contains lower level circuits such as power switches, rails, supply ports, supply nets, retention, level shifters and isolation cells. Supply nets in UPF are connected to external power supply or ground voltage according to the logical state of control inputs in power switches. The output of domain PD2 is clamped from inputs of PD3 using isolation cells, when the domain PD2 is not in normal mode operation. On the other hand, GPM includes power controller and power state table as shown in Fig. 1. The GPM specifies the legal system power modes and power transitions. In Fig. 1, the GPM consists of four power states called S0 (initial state), S1, S2 and S3. In each state, the status of power domains is defined using Power State Table (PST). For example, in S2, PD1 and PD2 are in ON mode while PD3 is OFF. Note that in the power controller, there is no state in which PD2 and PD3 can be ON, simultaneously. Therefore, this state would be illegal. Furthermore, there is not a direct transition between S1 and S3 and for going to S3 from S1 we must travel to S0 or S2. Another interesting point in the power management protocol is the existence of an interface, i.e. LPM, between high level GPM and lower level UPF. The inputs of LPM module that comes from GPM are power states; however its outputs are low level power control signals such as the isolation, the power supply switch, the restoration and the level shifter. These low-level signals control the operation of PCC defined in the UPF. In other words, the LPM is defined as a power controller for each domain of RTL design.

III. PROPOSED METHODOLOGY FOR LOW POWER PROCSESSOR VERIFICATION

Most of today's processors provide at least one power management strategy with multi power domains, built-in privileged data isolation and state retention. These features make the modern processors complicated so that their verification is not straightforward at all. In this section, we discuss how to verify low power processors with UCLID 3.0 [9] where correspondence checking has been implemented as a formal verification technique. Fig. 2 shows our proposed verification methodology consisting of four steps that are explained in the following subsections.

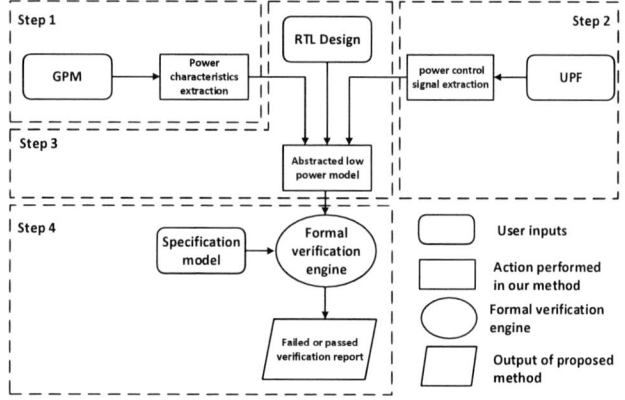

Fig. 2. Proposed power verification methodology

A. Step1: Power characteristics extraction from GPM

The Global power management (GPM) which expresses general strategies for low-power circuits is taken into account as one of inputs of our methodology that is provided by designers as shown in Fig. 2. These strategies that define the relationship among power domains are usually expressed using system-level languages such as SystemC. Power characteristics extracted from the GPM specify the number of power states, the status of each power domain, those power domains that are not allowed to be active simultaneously, and the transition conditions from a power state to another one. For example, consider LEON3 processor [10], [11] which has six power domains: integer unit (IU), instruction & data cache (CACHE), multiplication unit (MUL), division unit (DIV), storage element unit (STORAGE) and cache and memory controller (MEM). A part of GPM for LEON3 is shown in Fig. 3 where the power states are defined in lines 1-4. In each state, the power mode of each domain should be specified. Based on the instructions of the processor, transition conditions from one state to another are defined. For example, in the state S0, if the instruction

type be *load* or *store*, then the GPM drives a transition from S0 to S1 else the next state be S0 (lines 10-14 of Fig. 3). Moreover, IU and CACHE domains will be OFF and the MEM domain will return to ON mode.

```
1:    enum STATE { S0,          //IU-ON, DIV-OFF MUL-OFF, CACHE-ON
2:    S1,                       //IU-OFF, MEM-ON, DIV-OFF, MUL-OFF
3:    S2,                       //IU-OFF, MEM-OFF, DIV-ON, MUL-OFF
4:    S3 };                     //IU-OFF, MUL-ON, DIV-OFF, MEM-OFF
5:    SC_MODULE (GPI-LEON3){
6:    const enum STATE Pstate ;
7:    while(1){
8:    switch (Pstate){
9:    case (S0):
10:        if (instr.read ( ) == load | instr.read ( ) == store )
11:        Pstate = S1 ;          MEM-ON = 1;
12:        else
13:            Pstate = S0;
14:            CACHE-ON = 1;   IU-ON = 1;        ... };
```

Fig. 3. A part of LEON3 processor's GPM described by SystemC

Step 2: Power control signals extraction from UPF

From UPF a set of useful information, including power domains and power control signals such as the isolation, save and restoration ones, are extracted. In order to clarify it, let us consider a UPF of LEON3 processor. A part of such a UPF is shown in Fig. 4. This file contains power domains (lines 3-4), supply networks (lines 6-9), power switches (lines 11-13), isolation cells (lines 15-18) and retention cells (lines 20-23).

```
1:    set_scope top
2:    ## Creating Power Domains ##
3:    create_power_domain TOP -include_scope top
4:    create_power_domain IU -elements {/top/iu}
5:    ## Creat Supply Port ##
6:    create_supply_port VCC -domain TOP
7:    create_supply_port GND -domain TOP
8:    ## Connect Supply Ports to Supply nets##
9:    connect_supply_net VCC -ports { VCC }
10:   ## Power Switch Creation ##
11:   create_power_switch IU_power_switch
12:   -domain MULT
13:   -input supply port {IU_input_port1 /top/VDD_high_top}
14:   ## Isolation Definition ##
15:   set_isolation IU_isolation_0 -domain IU
16:   -isolation_power_net /top/VDD_low_top
17:   -isolation_ground_net VSS_top
18:   -clamp value 0
19:   ## Retention Definition ##
20:   set_retention IU_retention_strategy -domain IU
21:   -retention_power_net VDD_low_top
22:   -retention_ground_net VSS_top
23:   -save_signal {save_IU_posedge} -restore_signal {restore_IU_negedge}
```

Fig. 4. A part of LEON3 processor's UPF

Fig. 5 shows how to extract power control signals from UPF. First of all, different power domains (i.e., MULT, IU, STORAGE, CACHE, MEM and DIV) are determined. After that, we specify power modes in which power domains work. For example, the IU domain works in three modes, IDLE, ON and OFF, while the DIV domain operates in two modes, ON and OFF. Finally, the status of all power control signals is defined. For example, when the DIV domain is in ON mode, ISO, PWR and RES should be 0, 1, and 0, respectively.

One important point to be noted here is the fact that by transitioning from a power mode to another one, a specific sequence of power signals must occur. Otherwise, it results in functional errors. For example, let us consider the DIV domain of the LEON3 processor. Fig. 6(a) shows the related power signal sequence to transition from OFF mode to ON mode (i.e., power on state). In the first step, the PWR signal becomes 1 and then, RES signal becomes 0. Finally, the isolation signal ISO is set to 0. In the transition from ON mode to OFF mode (i.e., power off state), the DIV domain has been isolated by ISO = 1, then the state of the DIV domain has been saved in the retention register (SAVE = 1) and in the last step, the PWR signal is disabled (PWR = 0). As another example, let us consider the CACHE domain of the LEON3 processor. Fig. 6(b) shows its state machine to transition from FULL_ON mode to PARTIAL_ON mode and vice versa. As shown in this figure, in transition from FULL_ON mode to PARTIAL_ON mode (power down state), the PWR_LOW signal is set to 1 whereas in power up state (transition from PARTIAL_ON mode to FULL_ON mode), the PWR_HIGH signal is enabled.

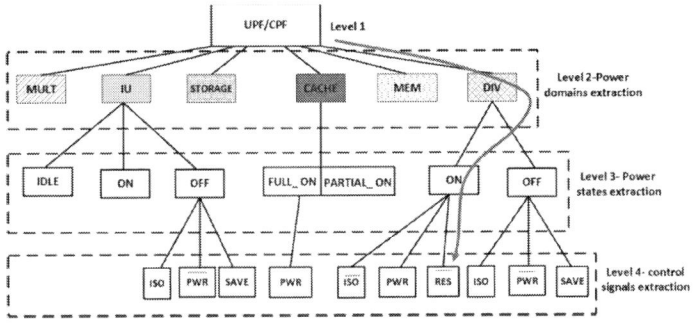

Fig. 5. Extraction of power control signals from UPF of LEON3 processor

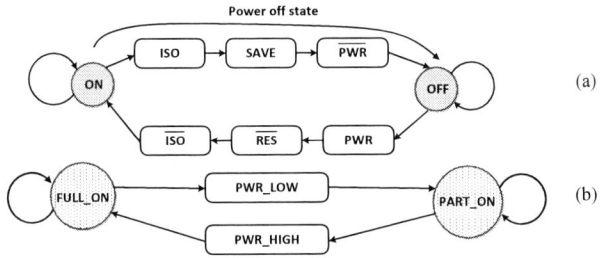

Fig. 6. Power state machine for: (a) DIV domain (b) CACHE domain

B. Step 3: Abstracted low power implementation model

In order to reduce the complexity of processor verification, we have abstracted the RTL implementation of the processor in such a way that it contains a detailed controller plus an abstracted data path. Abstraction of the data path's components is processed in such a way that the details of their operations are modeled as un-interpreted or predicate functions, and also their operands are abstracted to word level (TERM) and bit level (TRUTH) symbolic values. An un-interpreted function is a black box of the exact function which only satisfies functional consistency. Fig. 7 shows the abstracted low-power pipelined model of the LEON3 processor that consists of two main parts: 1) abstracted data path, and 2) Power Management Unit (PMU).

In the abstracted data path, nine types of instructions are existed: the division instruction (DIV), the multiplication instruction (MUL), store word instructions (ST), load word instructions (LD), conditional branch instructions (BR), trapping and interrupt instructions (TR), ALU register instructions (RR), ALU immediate instructions (RI) and the null control instruction (NULL). While its pipeline stages are: fetch (FE), decode (DE), access to register file (RA), execution of instruction (EX), memory and data cache access (MA), traps and interrupts resolving (XC) and write-back to register file (WR). As shown in Fig. 7, each power domain, before going to its power off mode, saves its values and states in the external memory and when its power is on, it restores its values and states from the external memory.

The PMU is a part of the abstracted low power model of the LEON3 processor which controls the processor operations according to power strategies defined by the GPM (explained in subsection III.A) and UPF (explained in subsection III.B). Each power domain (such as IU, MULT, DIV, etc.) can contain power control signals including PWR, ISO, RES, and SAVE (RET),

which can be automatically enabled or disabled by PMU based on the signal sequences extracted from UPF and the instructions to be executed. For example, suppose a multiplication should be performed by the processor while the MULT domain is in the OFF mode. When the MUL instruction is decoded in the DE-RA stage, the PMU actives the PWR signal in the MULT domain. After that, when MUL instruction is in the RA-EX stage, the PMU restores the state of the MULT domain from the external memory by RES=0. Finally, the PMU removes the isolation cell of the MULT domain by ISO = 0, thereby the MULT domain can transfer its result to another component, e.g. the Register File, with a different power domain.

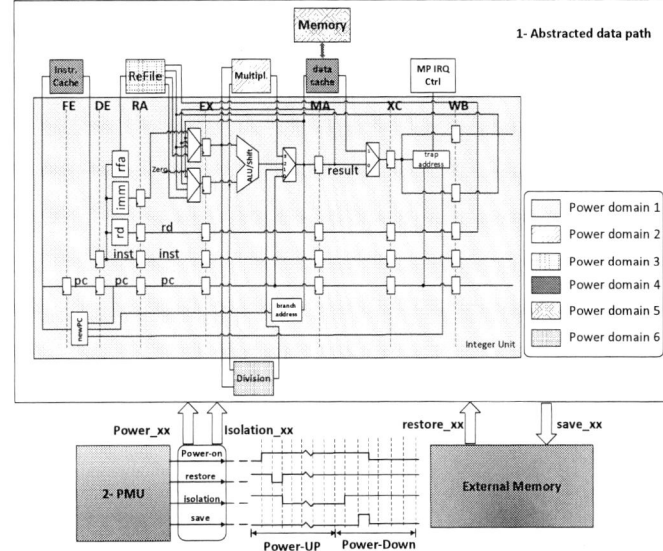

Fig. 7. Abstracted low power model of LEON3

C. Step 4: Formal verification

A flushing-based technique is used to check whether the state of a pipelined processor with low-power strategies is equivalent to its Instruction Set Architecture (ISA) by completing the partially executed instructions in the pipeline [12]. The abstracted low-power processor described in the previous subsection and the ISA without power features are taken into account as pipelined implementation model and golden model of the processor, respectively. As shown in Fig. 8(a), based on our proposed methodology with PMU, the abstracted model of low-power processor is symbolically simulated with arbitrary input combinations for one cycle. Then the pipeline is flushed for N cycles until all partially executed instructions can be completed, and the programmer-visible parts of the design, i.e. Program Counter ($Q^N_{impl}.PC$), Register File ($Q^N_{impl}.RF$), Data Cache ($Q^N_{impl}.dCache$), and Data Memory ($Q^N_{impl}.dMem$), are saved. It should be noted that N is the number of pipeline stages (for LEON3 is seven). In order to have a corresponding state of the implementation in the specification, first the state of the implementation after one cycle simulation, i.e. Q^1_{impl}, is projected into that of the specification ($Q^1_{impl} \rightarrow Q_{spec}$). Then the non-pipelined specification is symbolically simulated with the same instructions and the programmer visible components are saved as $Q^1_{spec}.PC$, $Q^1_{spec}.RF$, $Q^N_{spec}.dCache$ and $Q^1_{ispec}.dMem$. Finally, the correspondence property in (1) is checked to see whether Q^N_{impl} — Q^1_{spec} or not. All of these have been implemented with UCLID 3.0 [9].

$$(Q^N_{imp}.PC = Q^1_{imp}.PC) \& (Q^N_{imp}.RF = Q^1_{imp}.RF) \& \quad (1)$$
$$(Q^N_{imp}.dCache = Q^1_{imp}.dCache) \& (Q^N_{imp}.dMem = Q^1_{spec}.dMem)$$

In order to see why our proposed method reduces the number of symbolic simulation steps, let us explain how UCLID based correspondence checking without PMU works. Fig. 8(b) shows the steps in correspondence checking when power control signals should be manually adjusted where N, PS, and M are the number of pipeline stages, the average number of power control signals in different domains, and the number of power domains, respectively. For example, let us consider the functionality of the MEM domain in the LEON3 processor. Suppose this power domain is in ON mode (Q^1_{impl} in Fig. 8(b)). Therefore, in the initial step, the power signals including the power supply (PWR_MEM), the isolation (ISO_MEM), and the retention (RET_MEM) are manually set to 1, 0, and 0, respectively. In the next step (i.e., Q^i_{impl}), ISO_MEM is manually set to 1 in order to clamp MEM domain from other domains. Then, RET_MEM is manually set to 1 in order to make MEM domain save its value and state in the external memory before going to power off mode (Q^{i+1}_{impl}). At the final step (Q^{i+2}_{impl}), the power supply is disconnected by enabling inputs of the power switch, i.e., PWR_MEM = 0. In addition, to return from OFF mode to ON mode, the four other steps should be performed. Finally, it is checked whether the implementation (Q^F_{impl}) is equivalent to the specification (Q^1_{spec}) or not. Moreover, in the case of multiple power domains, this process must be repeated for each power domain according to its low-power strategy. Hence, the number of symbolic simulation steps would be $4 \times (PS \times M) + 2 \times N + 3$.

IV. EXPERIMENTAL RESULTS

In order to show the effectiveness of the proposed verification methodology, three experiments with two processors, the 5-stage MIPS [13] and the 7-stage LEON3 [11] have been conducted. The number of lines of MIPS and LEON3 RTL codes are 8750 and 10809, respectively. All experiments were carried out on a 2.4 GHz Intel core i7 Haswell with 6GB main memory running VMware Linux, where MiniSAT has been used as the SAT solver within UCLID 3.0 [9]. In the first experiment, the necessity of using PMU to adjust power control signals will be discussed. The second experiment shows the effect of power domains on the performance of the proposed method. In the third experiment, we compare the results of the proposed method with those of the method explained in [7].

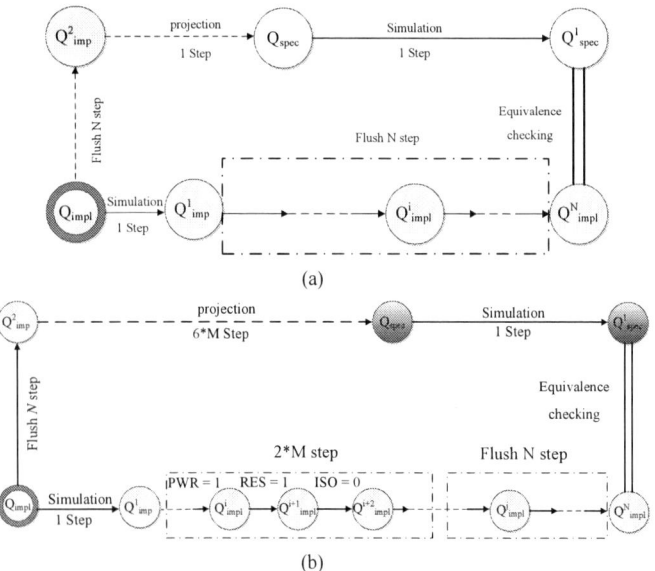

Fig. 8. UCLID based correspondence checking: (a) with PMU (b) without PMU

A. Experiment 1: advantage of generating PMU

As explained before, one of our contributions in this work is to automatically generate PMU by extracting power characteristics and power control signals from GPM and UPF, respectively. In order to clarify the advantage of generating PMU, we have performed this experiment on the LEON3 processor which has six power domains, including MULT, IU, STORAGE, CACHE, MEM, and DIV. Each power domain may have its own power strategy and usually contains power control signals such as PWR, ISO, RES, and SAVE (RET).

One way to verify LEON3 processor with six power domains is to manually control and check each power domain independently. For doing so, power signals should be manually set and then check the correspondence between the specification and implementation. We refer to such a verification technique as manual power management (without PMU) in the rest of the paper. For example, let us consider the MEM domain in the LEON3 processor. First, it is assumed that this domain is ON. Therefore, in the initial step, power signals including the power supply (PWR_MEM), isolation (ISO_MEM), and retention (RET_MEM) are manually set to 1, 0, and 0, respectively. In the next step, ISO_MEM is manually set to 1 in order to clamp MEM domain from other domains. Then, RET_MEM is set to 1 in order to make MEM domain save it value and state in the external memory before going to power off mode. In the final step, power supply is disconnected by enabling the inputs of power switches, i.e. PWR_MEM=0. Finally, it is checked whether the implementation is equivalent to the specification or not. The main problem of such a verification technique is the fact that the number of symbolic simulation cycles exponentially increases with respect to the number of power domains and power control signals. This results in increasing in the size of the model and therefore the verification time also increases exponentially. In the case of multiple power domains, this process should be performed in each domain according to the related strategy.

In order to alleviate this issue, we have generated the PMU to automatically control power control signals of each domain according to the instruction types in pipeline stages and the power strategy defined in the GPM. Hence, such an automatic control process reduces the complexity of the verification by reducing the number of symbolic simulation steps. TABLE I shows the results of LEON3 processor verification for two cases: 1) manual power management (without PMU) and 2) automatic power management (with PMU). In this table, columns #Domain, #Step, #Var, #Clause and CPU Time (in seconds) give the number of power domains, the number of symbolic simulation steps, the number of variables, the number of CNF variables, the number of clauses, and the verification time in seconds, respectively. As you can see in this table, the verification technique based on PMU could reduce the symbolic simulation steps by 5.23× and the verification time by 31.95× in comparison with the manual power management technique, i.e. without PMU. Since the PMU block has been in implementation model and manages the power domains automatically, therefore the number of symbolic simulations in the verification methodology based on PMU is less than without PMU.

TABLE I. FORMAL VERIFICATION RESULTS OF THE 7-STAGE LEON3

Method	#Domain	#Step	#Var	#Clause	CPU Time (seconds)
without PMU	6	89	1659668	4597174	16874
with PMU	6	17	104978	309691	528
Improvement with PMU		5.23×	15.8×	14.85×	31.95×

B. Experiment 2: effect of the number of power domains

In this experiment, we show the effect of increasing the number of power domains on the performance of the proposed verification methodology in two cases: 1) without PMU, and 2) with PMU. Fig. 9 shows how the number of CNF clauses and verification time increase when the number of power domains in LEON3 processor increases.

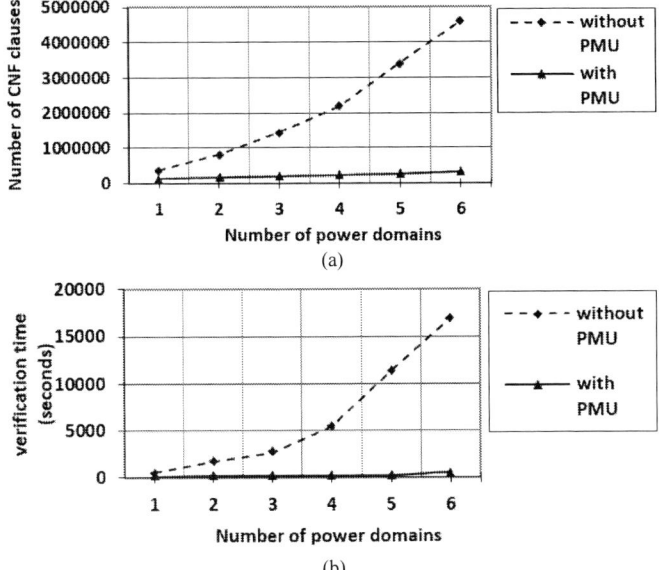

(a)

(b)

Fig. 9. The effect of increasing the number of power domains on: (a) the number of clauses and (b) the verification time in two cases (without PMU and with PMU)

C. Experiment 3: comparison with existing method

In the final experiment, in order to show the effectiveness of the proposed verification methodology, we have compared our results with those of the method proposed in [7]. In this paper, the authors tried to verify power gating strategy in a MIPS processor which only has a single power domain while in our proposed methodology, the processor can be empowered by various power management techniques such as power gating and clock gating in multiple domains. The power gating technique makes the processor go to the power off state immediately after saving the internal values in the external memory and when the processor is going back to the power on mode, first internal values contents are loaded from the external memory. When the processor is going to power on mode, the execution starts from the fetch state. It is verified that the processor successfully goes to power down and returns to power up, and all the instructions before and after the power gating are finished their executions correctly.

In the method of [7], power off and power on processes are performed manually by using external signals that indicate start and end of power gating for the processor. These signals are manually activated at cycle i and deactivated at cycle $i+1$. In our method, in contrast to [7], these signals are automatically activated based on the PMU integrated into the abstracted RTL implementation. TABLE II shows the results of our proposed verification methodology for two cases (with and without PMU) in comparison with those of the method in [7]. It is obvious that the proposed verification methodology is scalable enough to handle complex power management strategies, which are usually used in modern processors. Moreover, adding PMU to the abstracted implementation results in fewer symbolic simulation steps, in comparison with [7], which reduce the number of clause and the verification time by 17.57× and 13.71×, respectively.

V. CONCLUSION AND FUTURE WORK

We have proposed a verification methodology to effectively formally verify dynamic power management techniques in modern processors in two aspects of functionality and automatically controlling of the power management in the verification process. For doing so, the PMU is extracted from UPF and the GPM and integrated into the implementation so that low-level controlling signals for each power domain can be automatically adjusted. The results show that the verification time is reduced by 13.71× in comparison with the existing methods such as [7]. As a future work, we are going to extend our methodology to verify multicore processors.

TABLE II. FORMAL VERIFICATION RESULTS OF THE 5-STAGE MIPS

Method	#Domain	#Step	#Var	#Clause	CPU Time (seconds)
without PMU	1	25	277108	823441	132
	3	49	276841	902241	1452
Method in [7]	1	25	434837	1292158	924
	3	29	606104	1802971	1804
with PMU	1	13	25964	75898	67
	3	13	34178	100249	132
Improvement in comparison with the method in [7]	2.08×	11.04×	17.57×	13.71×	

REFRENCES

[1] S. Ahuja, A. Lakshminarayana, and S. Kumar Shukla, "Low power design with high level power estimation and power aware synthesis," *Springer*, 1st edition, pp. 1-170, 2012.

[2] S. Baily, G. Chidolue, Allan Crone, and Mentor Graphics, "Low power design and verification techniques," *Mentor graphics's white paper*.

[3] IEEE Std 1801-2013: IEEE Standard for Design and Verification of Low-Power Integrated Circuits, *IEEE Press, 2013*.

[4] http://www.mentor.com/products/fv/questa-power-aware-simulator.

[5] O. Mbarek, A. Pegatoquet, and M. Auguin, "Using unified power format standard concepts for power-aware design and verification of systems-onchip at transaction level," *IET Circuits, Devices & Systems.*, vol. 6, no. 5, pp. 287–296, 2012.

[6] A. Hazra, R. Mukherjee, P. Dasgupta, A. Pal, K. M. Harer, A. Banerjee, and S. Mukherjee, "POWER-TRUCTOR: An integrated tool flow for formal verification and coverage of architectural power intent," *in IEEE Transactions On Computer-Aided Design Of Integrated Circuits And Systems (TCAD)*, vol 32, no 11, pp. 1801-1813, 2013.

[7] A. M. Gharehbaghi, M. Fujita, "Specification and formal verification of power gating in processors," *in 15th International Symposium on Quality Electronic Design (ISQED)*, pp. 604 - 610, 2014.

[8] Uclid ver. 1.0. [online]: http://www.cs.cmu.edu/ uclid.

[9] Uclid ver. 3.0. [online]: http://uclid.eecs.berkeley.edu/downloads.html.

[10] M. Dan°ek, L. Kafka, L. Kohout, J. S´ykora, R. Bartosi´nski, "UTLEON3: Exploring fine-grain multi-threading in FPGAs," *Springer*, 1st edition, pp. 1-216, 2013.

[11] GRLIB IP Core User's Manual, Version 1.3.7-B4144, 2014. [online]: http://www. gaisler.com/

[12] J. Burch and D. Dill, "Automatic verification of pipelined microprocessor control," *in proceedings of the International Conference on Computer-Aided Verification (CAV'94)*, pp. 68–80, 1994.

[13] MIPS32 architecture. [online]: http://www.imgtec.com/mips/mips32-architecture.asp.

[14] B. Alizadeh, "Formal verification and debugging of precise interrupts on high performance microprocessors," *in ACM Transactions on Design Automation of Electronic Systems*, vol. 17, no. 4, pp. 371-378, 2012.

2015 IEEE 33rd VLSI Test Symposium (VTS)

MBIST and Statistical Hypothesis Test for Time Dependent Dielectric Breakdowns due to GOBD vs. BTDDB in an SRAM Array

Woongrae Kim, Chang-Chih Chen, Soonyoung Cha, and Linda Milor
School of Electrical and Computer Engineering, Georgia Institute of Technology, Atlanta, GA, USA
wrkim@gatech.edu

Abstract— **In this paper we present Memory Built-In Self-Test (MBIST) diagnosis methodologies for failure analysis of time-dependent breakdown due to gate oxide breakdown (GOBD) and backend time-dependent dielectric breakdown (BTDDB) in an SRAM array. First, a Built-In Self-Test (BIST) system and algorithm detect the breakdown mechanisms and identify the locations of the faulty sites in SRAM cells. Then, probabilities of failure are estimated for BTDDB and GOBD by matching the observed failure rate from BIST and the expected failure distribution functions from system simulations under realistic use scenarios, with different simulated failure rates for BTDDB and GOBD.**

Keywords—Built-In Self-Test (BIST); statistical hypothesis test; dielectric breakdown; wearout mechanisms; BTDDB; GOBD; wearout distribution

I. INTRODUCTION

Reliability of SRAM systems, embedded in CPUs, GPUs, and application processors, is considered as the one of barriers for process technology scaling. Wearout of devices and interconnects occur more quickly since process technology scaling leads to the reduction of interconnect and transistor dimensions without the reduction of the supply voltage in proportion. When there is insufficient redundancy or the failure rate is faster than predicted, wearout can lead to system failure even with the use of the error correcting codes.

When failing chips are detected and returned from the field, the manufacturer diagnoses the cause of wearout in the returned chips. The standard methodology for diagnosis is physical failure analysis which includes deprocessing to determine the nature of the defect visually. However, since the success rate of physical failure analysis is not always high, but has a high cost, there is a need to devise another diagnosis methodology to achieve cost reduction and a higher success rate. In this work, we propose the combination of built-in electrical tests and statistical analysis using volume data to estimate probabilities of wearout in SRAMs due to gate oxide breakdown (GOBD) and backend time-dependent dielectric breakdown (BTDDB).

Time-dependent dielectric breakdown problems from process technology scaling include frontend wearout due to GOBD and backend wearout due to BTDDB. In prior work, the impact and detection of dielectric breakdown has been discussed in [1][2]. However, there is no study that separates GOBD and BTDDB, both of which create leakage paths in the SRAM cell. GOBD is modeled as a leakage path through the

gate oxide of transistors [3] and BTDDB is modeled as a resistive bridge between unconnected metals in the same layer [2]. Some of the resistive short locations in an SRAM cell overlap, and hence these faults result in exactly the same failure signature. Hence, it is hard to separate them with only electrical tests. Nevertheless it is important to separately determine the failure rate for these mechanisms so that the manufacturing process can be improved.

In this paper, we present a diagnosis methodology that combines BIST and statistical estimation using volume data to determine separate wearout distributions for GOBD and BTDDB. First, our BIST system detects resistive short defects due to GOBD and BTDDB in an SRAM cell. The lifetime and failure rates for GOBD and BTDDB depend on geometric components and system activity [3]. These geometric components and activities vary significantly in a cell. Hence, variation in the failure rate of specific sites can be utilized to diagnose the failure rates due to GOBD and BTDDB. Our BIST methodology diagnoses the site of failure within an SRAM cell. By matching the failure rates from BIST and the failure distributions from a reliability simulator, described in [4][5], we estimate the probability of failure from GOBD vs. BTDDB. This work is similar to our prior work where the probability of resistive open wearout failures is estimated, distinguishing electromigration vs. stress-induced voiding [6].

The proposed work is not the first attempt at electrical diagnosis of the cause of failure in an SRAM array. In prior work, diagnosis methodologies have aimed at identifying the locations of failure to aid physical failure analysis [7]-[9]. However, prior studies mainly focus on fault detection and the repair algorithm. Unlike the previous studies, our work focuses on the estimation of wearout distributions for specific wearout mechanisms in the field.

Our work has been implemented and simulated with commercial-grade IBM 90nm PDK, thereby ensuring practicality and credibility.

II. MODELING GOBD AND BTDDB IN AN SRAM CELL

Gate oxide breakdown (GOBD) is modeled as a leakage current which flows through the gate oxide in an SRAM cell [3]. Although the leakage path appears between the gate and substrate, the gate-to-substrate leakage is usually neglected since it has a little effect on performance [1]. Hence, in this work, we model only the dominant leakage paths: gate-to-source leakage and gate-to-drain leakage. The leakage paths due to GOBD that are modeled in an SRAM cell are shown in

---

The authors thank the Defense Advanced Research Projects Agency under grant HR0011-11-1-0011 for support.

978-1-4799-7598-3/15 $31.00 © 2015 IEEE

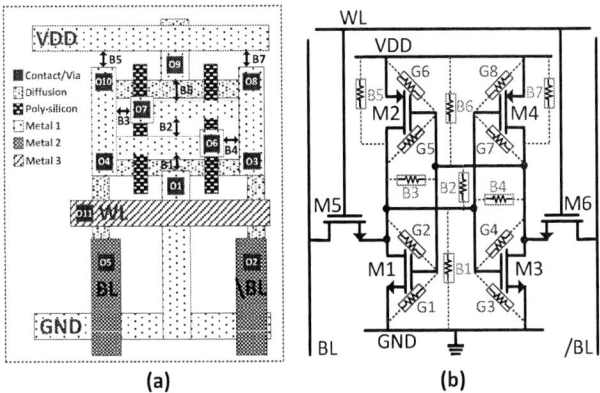

Fig. 1. (a) Possible BTDDB locations in a physical layout of a 6T SRAM cell (B1–B7). The locations of potential via/contact failures are also noted (O1-O11). (b) Modeling of the GOBD (G1-G8) and BTDDB effects (B1-B7).

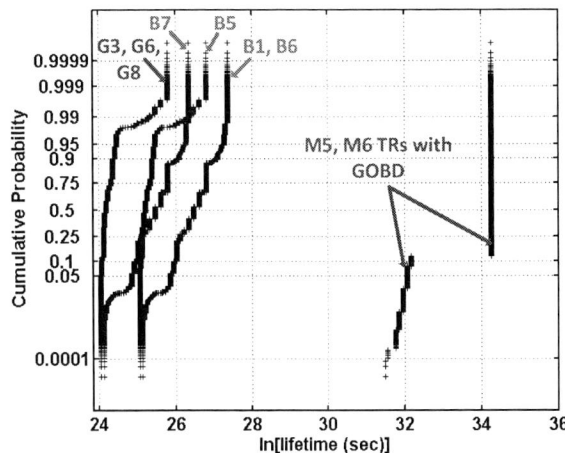

Fig. 2. Cumulative probability distribution of lifetime for 32Kb SRAM cells for both mechanisms and each site within the cell, including the access transistors.

Fig. 1(b).

We consider only GOBD in the four cell transistors since stress for access transistors is almost negligible, especially when the system frequency is high. When the system frequency is high, read/write operations are faster, and the access transistors are mostly turned off. Fig. 2 shows that the lifetimes for the access transistors (M5, M6) due to GOBD are much larger than those for other transistors in an SRAM cell.

The higher electric fields in the backend insulators caused by process technology scaling lead to leakage paths through the dielectrics. Fig. 1(a) presents the BTDDB locations in a physical layout of a 6T SRAM cell. Seven possible leakage paths due to BTDDB exist in this SRAM cell. Fig. 1(b) shows the modeling of the leakage paths induced by BTDDB in a schematic for an SRAM cell. Unlike resistive-bridging models in previous studies [10], our fault model is based on resistive-short faults for GOBD and BTDDB, considering only possible vulnerable dielectric segments in the physical layout of the SRAM presented in Fig. 1.

Since most short sites due to BTDDB and GOBD which can be detected by BIST are the same, there is a need to do additional statistical analysis using lifetime data from a reliability simulator [4][5] to distinguish them.

Both GOBD and BTDDB are modeled with Weibull distributions with two parameters, a characteristic lifetime (η) and a shape parameter (β). The characteristic lifetime, η_{GOBD}, for GOBD is as follows [3],[4]:

$$\eta_{GOBD} = A_{GOBD} \left(\frac{1}{WL}\right)^{\frac{1}{\beta_{GOBD}}} e^{-\frac{1}{\beta_{GOBD}}} exp\left(\frac{c}{T} + \frac{d}{T^2}\right)/\alpha \quad (1)$$

where W and L are the device width and length, respectively, β_{GOBD} is the Weibull shape parameter, α is the fraction of time that the gate is under stress, T is temperature, V is the gate voltage, and a, b, c, d, and A_{GOBD} are fitting parameters. The characteristic lifetime for GOBD is a function of the location of the failure since all failure sites do not experience the same stress. Stress depends on workload.

The characteristic lifetime for BTDDB is [3]-[5]:

$$\eta_{BTDDB} = A_{BTDDB} L_{BTDDB}^{\frac{1}{\beta_{GOBD}}} exp(-\gamma E^M - E_a/k_B T)/\alpha \quad (2)$$

which is a function of the vulnerable length, L_{BTDDB}, its associated line space, S_{BTDDB}, the corresponding electric field,

$E = V/S_{BTDDB}$, where V is the supply voltage, the Weibull shape parameter, β_{BTDDB}, the field acceleration factor, γ, the activation energy, E_a, Boltzmann's constant, k_B, and the probability that the adjacent nets to the dielectric segment are at opposite voltages, α', and fitting parameters, A_{BTDDB} and M [3]-[5].

The reliability simulator [4][5] estimates lifetimes of each fault site due to GOBD and BTDDB. The 32Kb bit SRAM is generated by a memory generator [11]. Then, by running a variety of standard benchmarks [12], the simulator extracts the detailed electrical stress and temperature of each SRAM cell in the embedded memory within the microprocessor [13], from which the lifetime of each bit is estimated.

Fig. 2 shows that the cumulative probability distributions of the lifetimes of the resistive short sites due to GOBD are not the same as those due to BTDDB, even if these faults result in exactly the same electrical failure signature.

III. BIST FOR ON-CHIP MONITORING OF GOBD AND BTDDB

A. Overview of the BIST System and the Test Structures

The BIST system detects GOBD and BTDDB in an SRAM array. The BIST system consists of a BIST controller, a sensing circuit (SC), and an output response analyzer (ORA).

The BIST controller contains a test pattern generator (TPG). The TPG generates test input vectors to implement the BIST algorithm for wearout detection. The test vectors consist of the write enable (WE), data inputs (W_data), the sense amplifier enable (SAE), the precharge enable (PRE), and test Row/Column addresses.

Fig. 3(a) presents the test area in half of the SRAM system where each bank contains 16 Kb cells. Normally, the column decoder for the 16 Kb bank uses four bit addresses as inputs to choose eight global data line pairs from 128 bitline pairs. To run our BIST algorithm, we activate each individual SRAM cell for each test pattern. Hence, we slightly revised the column decoder so that it uses seven bit column addresses to disconnect the other seven global data line pairs from the 127 inactive bitline pairs.

The current test circuit in SC, which was proposed to detect

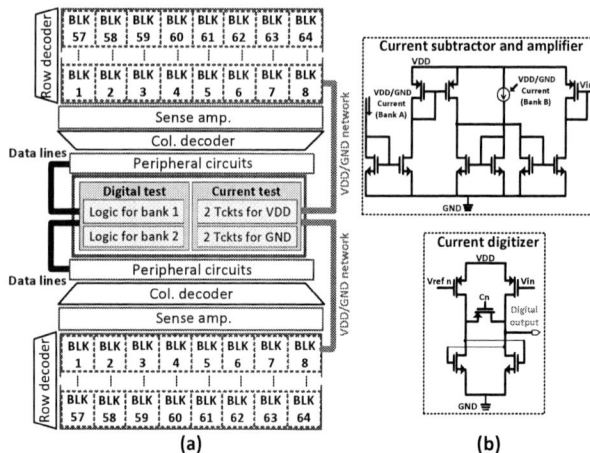

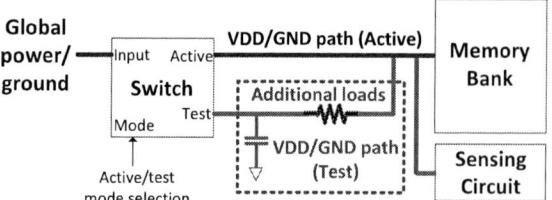

Fig. 5. Additional test structure for VDD/GND variation test.

Fig. 3. (a) Test structures in the built-in self-test area, including digital test and current test, where the current test involves four (b) sensing circuits for analysis of current variations in power/ground networks [1].

TABLE I. TEST PATTERN FOR BIST ALGORITHM

Test group	Detection	Test pattern	Test point	Test Name
Proper	Reference	(w1,r1,w0,r0)	Data lines	Refer
Short 1	B6,B7,G6	(w1,r1)	VDD paths	TVDD1
Short 2	B5,G8	(w0,r0)	VDD paths	TVDD2
Short 3	B1,G3	(w1,r1)	GND paths	TGND1
Short 4	G1	(w0,r0)	GND paths	TGND2
Short 5	B2,B3,B4, G2,G4,G5,G7	(w0,w1, pre [1.2V],r1)	Data lines	TF

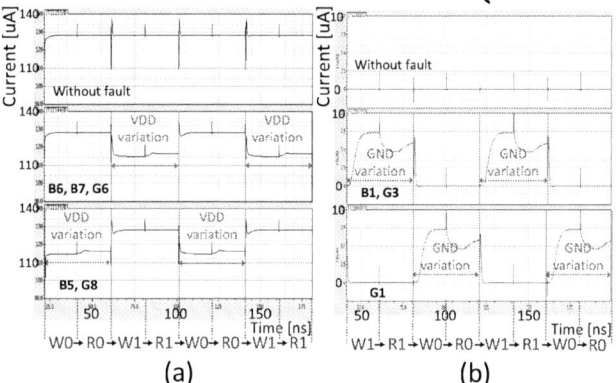

Fig. 4. Simulation results for current variation at an input of the current sensing circuit: (a) current variation in the VDD network due to B5-B7, G6, and G8, and (b) current variation in the GND network due to B1, G1, and G3.

GOBD [1] and BTDDB [2], is used to detect current on the power/ground networks in the SRAM array. The circuit is shown in Fig. 3(b). Through a pairwise comparison of current on the power/ground network from a cell from a bank with another cell from another bank, the short defects between VDD and a signal node or between GND and a signal line can be detected.

Four current test circuits detect the current variations in the power/ground networks of two banks. The VDD path for the upper bank is connected to the "Bank A" input of the first current subtractor and the "Bank B" input of the second current subtractor (see Fig. 3(b)). Another VDD network from the lower bank is connected to the "Bank B" input of the first current subtractor and the "Bank A" input of the second current subtractor. The ground paths for both banks are connected to two more current test circuits similarly. The test results are digitized with the current digitizer in Fig. 3(b). Fig. 4 presents the current variations at an input of the current test circuit due to short defects.

The current test circuit in Fig. 3(b) requires a minimum input voltage level to detect current variation on power/ground network. To make the variation more visible, an additional test

structure is placed between a global power/ground network and the SRAM bank (see Fig. 5). In test mode, a switch switches the VDD/GND paths for active mode to another test path with a larger resistance. The larger noise enables VDD/GND variations from bridging faults to be detected with better test coverage.

For the short defects between internal nodes in an SRAM cell due to GOBD and BTDDB, the digital test logic detects the voltage patterns on bitline pairs (see Fig. 3(a)).

The output response analyzer (ORA) contains a fault analyzer to store diagnosis results when the fault detect signal is sent from the test area.

B. BIST Algorithm

Reference test: The BIST algorithm detects faulty cells by comparing the currents in power/ground networks from two cells, one in an upper and the other in a lower bank. The BIST algorithm first has to find a set of fault-free reference cells to identify whether the cell under test from the paired reference cell is a faulty cell [2],[6]. We find several reference cells to minimize the impact of mismatch in the power/ground path length, as mismatch can lead to a fault detection error. Hence we divide an SRAM bank into 64 sub-blocks, as shown in Fig. 3(a) [6].

The BIST algorithm can select any cell in a sub-block of a bank as the reference cell for any cell in the same sub-block of another paired bank. To find the 64 reference cells in the 64 sub-banks of each bank, the power/ground currents from a pair of cells are compared with the test pattern in Table I. When there is no fault-signal from the SC, both cells are set as reference cells for their sub-bank. On the other hand, when at least one of the cells is faulty cell, a search for a nearby test cell in the same sub-block continues until a fault-free reference cell is detected.

When the first test step for proper reference cell selection is complete, the cell under test is located together with the reference cell in the complementary bank for a pairwise comparison of current in their power/ground networks. Then the diagnosis steps begin with the test patterns in Table I.

978-1-4799-7598-3/15 $31.00 © 2015 IEEE   200

TABLE II. VDD/GND VARIATION SIMULATION RESULTS

Test group	VDD current variation (max) at input of SC	GND current variation (max) at input of SC
Proper	0 uA	0 uA
Short 1-2	13.2 uA	0.2 uA
Short 3-4	3.1 uA	7.3 uA
Short 5	2.8 uA	0.1 uA
O1	0.3 uA	Less than 0.1 uA
O2-O5	2 uA	0.5 uA
O6-O7	3.0 uA	Less than 0.1 uA
O8-O10	Less than 0.1 uA	Less than 0.1 uA
O11	1.3 uA	Less than 0.1 uA

We have set the resistance to 10Ω for resistive-short defects in our simulations. Undetectable faults exist if the leakage currents from faulty cells are exactly the same. However, since the leakage currents depend on the degree of wearout, undetectable faults from matched leakage currents are highly unlikely and have little impact on test coverage.

Current variation analysis of the VDD network: The BIST system next starts the current variation tests on the VDD networks of the SRAM array to detect resistive-bridging defects between VDD and a signal node (B5, B6, B7, G6, G8 in Fig. 1(b)). The two inputs for the current sensing circuit in Fig. 3(b) are connected to the VDD networks for both the upper and lower banks, respectively. Then, the BIST controller sends the test address of the cell under test, its paired reference cell, and the test vector ((w1, r1) for TVDD 1 and (w0, r0) for TVDD 2 (see Table I)). If the current test circuits detect current variation in the VDD networks, the address and fault type are stored in the ORA's registers.

During write '1' and read 1' operations for the TVDD 1 pattern, the leakage path induced by short group 1 (B6, B7, G6) becomes the bridge which leads current in the VDD line to flow to ground. Also, the leakage path due to short group 2 (B5, G8) enables current in the VDD line to flow to GND during write '0' and read '0' operations. Fig. 4(a) shows the reduction of VDD current caused by short group 1 for pattern (w1, r1) and by short group 2 during (w0, r0).

Table II shows that short groups 1 and 2 can be detected with signature analysis of VDD variation. The 11 possible open via/contacts (O1-O11) due to other wearout mechanisms, including electromigration and stress-induced voiding, can be confounded in an SRAM array [6] (see Fig. 1(a)). With our BIST system, which includes VDD variation analysis, short groups 1 and 2 due to GOBD and BTDDB are distinguished from other wearout mechanisms (see Table II). To distinguish them from other faults, we set the current trigger level to 7.4uA by using a weighted reference current generator [1].

Current variation analysis of the GND network: The BIST system conducts current variation tests on the GND network to detect resistive-bridging faults between ground and a signal path (B1, G1, G3) with the test patterns TGND1 and TGND2 in Table I.

When the signal line is shorted to GND through the leakage path due to short group 3 (B1, G3), the GND level temporarily goes up during the write '1' and read '1' operations. Also, the leakage path induced by short group 4 (G1) leads the ground current to spike up during write '0' and read '0' operations

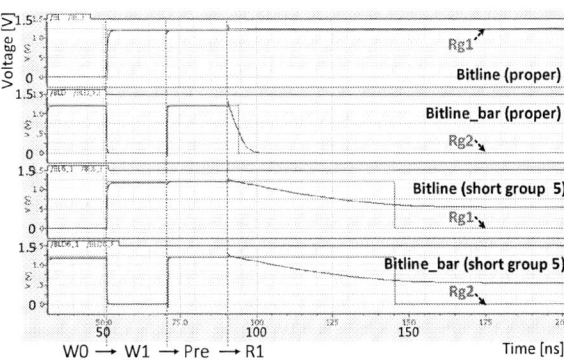

Fig. 6. Simulation results for an SRAM cell with test pattern TF: bitline pair voltages/their digitized values from a proper cell and a cell with a short from group 5.

TABLE III. SIMULATION RESULTS FOR DATA LINE VOLTAGES AND DIAGNOSIS LOGIC FOR TF PATTERN (DURING READ OPERATION)

Fault	From BL	From \BL	Diagnosis
Proper	1.24 V / logic 1	0 V / logic 0	logic 1
Short 5	0.546 V / logic 0	0.546 V / logic 0	logic 0
O1	1.25 V / logic 1	1.21 V / logic 1	logic 1
O2-O3	0 V / logic 0	1.24 /logic 1	logic 1
O4-O10	1.24 / logic 1	0 / logic 0	logic 1
O11	1.25 V / logic 1	1.25 V / logic 1	logic 1
Reg.	Rg_1	Rg_2	$Rg_1 \cup Rg_2$

(see Fig. 4(b) and Table II).

The short defects connected to GND are distinguished from other wearout mechanisms through GND current variation analysis. We set the current trigger level to 4.1uA. A larger resistance in the VDD/GND test path in Fig. 5 can be used to enhance the trigger values, which allows the variation to be more visible.

Digital test for resistive shorts between internal nodes: For the detection of resistive bridging faults for short group 5 (B2, B3, B4, G2, G4, G5, G7), the BIST controller generates the transient fault (TF) test pattern in Table I. TF consists of (w0, w1, precharge a bitline pair to 'high', r1). The test patterns include the precharge enable signal (PRE) to pull up the bitline pair before the read operation. All access transistors are inactive during the precharge operations and sense amplifiers are turned off during the read operation in the TF pattern.

Since the two internal nodes are shorted by GOBD or BTDDB, both voltages on a bitline pair go up to an intermediate voltage (0.546 V) at the capturing point in Fig. 6 during the read '1' operation. This voltage is less than the trigger point for digitization, and, hence, the digitized values associated with both bitlines are logic '0'. Table III shows that both digitized logic values from a bitline pair with the defects in short group 5 are '0'. The test logic stores digitized values from the bitlines in register-type circuits (Rg_1-Rg_2). Hence, the shorts in short group 5 can be distinguished from proper cells and other causes of wearout.

Detectable range for each leakage path: Fig. 7 presents detectable ranges of the inserted resistances for BTDDB and GOBD (short groups 1-4) using power/ground variation test. BIST can detect the leakage path resistance up to 15.4 KΩ for short groups 1-2 and up to 1.51KΩ for short groups 3-4,

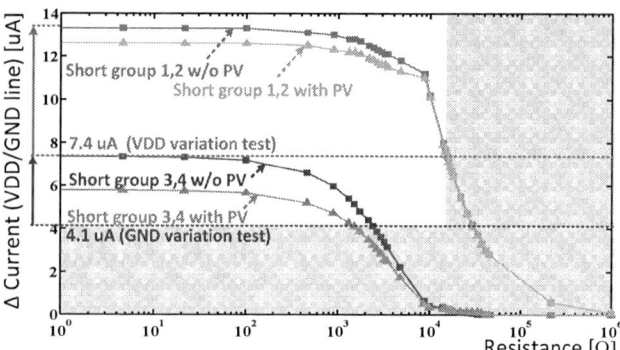

Fig. 7. Detectable ranges of an inserted resistance in an SRAM cell using the current test (VDD and GND variation tests) for short groups 1-4.

TABLE IV. DETECTABLE RANGE OF INSERTED RESISTANCES

Fault	Range 1 [Ω]	Range 2 [Ω] (with 10% PV)	The worst range [Ω]
Short 1-2	0~15.4K	0~15.4K	0~15.4K
Short 3-4	0~2.41K	0~1.51K	0~1.51K
Short 5	0~29.6K	0~26.8K	0~26.8K

TABLE V. FAULT GROUP AND INDEX FOR EACH MECHANISM

Group	GOBD	BTDDB
Short 1 (k=1)	G6 (j=1)	B6 (j=1), B7 (j=2)
Short 2 (k=2)	G8 (j=1)	B5 (j=1)
Short 3 (k=3)	G3 (j=1)	B1 (j=1)
Short 5 (k=4)	G2 (j=1), G4 (j=2), G5 (j=3), G7 (j=4)	B2 (j=1), B3 (j=2), B4 (j=3)

TABLE VI. SOLUTION OF EQUATIONS (3)~(7) USING SIMULATION DATA FOR A 32KB EMBEDDED SRAM (N = 32768)

Group 1	G6 (j=1)	$\eta_{k=1,GOBD}$	B6 (j=1)	B7 (j=2)	$\eta_{k=1,BTDDB}$
$\eta_{j,1}$ Ln (sec)	16.09	16.09 for $\sum P_{j,1}=1$	20.84	20.84	20.13 for $\sum P_{j,1}=1$
$P_{j,1}$	1.00		0.5	0.5	
Group 2	**G8 (j=1)**	$\eta_{k=2,GOBD}$	**B5 (j=1)**		$\eta_{k=2,BTDDB}$
$\eta_{j,2}$ Ln (sec)	14.93	14.93 for $\sum P_{j,2}=1$	19.75		19.75 for $\sum P_{j,2}=1$
$P_{j,2}$	1.00		1.00		
Group 3	**G3 (j=1)**	$\eta_{k=3,GOBD}$	**B1 (j=1)**		$\eta_{k=3,BTDDB}$
$\eta_{j,3}$ Ln (sec)	16.09	16.09 for $\sum P_{j,3}=1$	20.84		20.84 for $\sum P_{j,3}=1$
$P_{j,3}$	1.00		1.00		
Group 5	**G2 (j=1)**	**G4 (j=2)**	**G5 (j=3)**	**G7 (j=4)**	$\eta_{k=4,GOBD}$
$\eta_{j,4}$ Ln (sec)	14.93	16.09	16.09	14.93	14.14 for $\sum P_{j,4}=1$
$P_{j,4}$	0.396	0.102	0.102	0.396	
Group 5	**B2 (j=1)**	**B3 (j=2)**	**B4 (j=3)**		$\eta_{k=4,BTDDB}$
$\eta_{j,4}$ Ln (sec)	18.69	19.7	19.7		18.12 for $\sum P_{j,4}=1$
$P_{j,4}$	0.58	0.215	0.215		

even in the presence of both 10% process variation corners and the maximum allowed path length mismatch (110um). Leakage resistances up to 26.8KΩ can be detected for short group 5 with the digital test procedure (see Table IV).

The worst case test time for the 32Kb SRAM with 60 faulty cells (28 shorts due to BTDDB and 32 shorts due to GOBD) is 0.0114s.

IV. STATISTICAL ANALYSIS TO SEPARATE WEAROUT DISTRIBUTIONS FOR GOBD AND BTDDB

For short groups 1,2,3, and 5 in Table V, the cause of a fault cannot be identified using only electrical tests since both mechanisms can cause the short. Hence, we need additional analysis to identify the cause of wearout. We propose the statistical analysis of volume field data, combined with reliability simulation data. By matching the failure rate of each fault site from our BIST to statistical data from simulation, the fraction of failures from GOBD vs. BTDDB can be estimated.

The embedded SRAM system in the microprocessor is composed of n SRAM cells and 5n short groups which can be detected by BIST. The lifetime distributions for each wearout mechanism are modeled with Weibull distributions. $\eta_{i,j,k}$ is the Weibull characteristics lifetime and $\beta_{i,j,k}$ is the Weibull shape parameter. k is an index for the short group (1-3, and 5) in ith cell and j is for each possible wearout location within the short groups (see Table V). $\eta_{31000,2,4\ for\ GOBD}$ is the characteristics lifetime for G4 (j=2) of the short group 5 (k=4) in 31,000th cell (i=31000) (see Fig. 1(b) and Table V).

The characteristic lifetimes, $\eta_{i,j,k}$, of resistive short defects due to GOBD and BTDDB are computed with the reliability simulator [4][5] for all possible wearout sites induced by GOBD and BTDDB in the embedded SRAM system.

The characteristic lifetimes of each mechanism are determined by solving for $\eta_{j,k}$ with [3]:

$$1 = \sum_{i=1}^{n=32768} P_{i,j,k} \qquad (3)$$

where

$$P_{i,j,k} = \left(\eta_{j,k}/\eta_{i,j,k}\right)^{\beta_{i,j,k}}. \qquad (4)$$

For a single mechanism, $\beta_{i,j,k}$ is usually assumed as a constant for each mechanism. With this assumption, a closed form solution can be derived for GOBD and BTDDB:

$$\eta_{j,k\ for\ GOBD} = \left(\sum_{i=1}^{n=32768} \frac{1}{\eta_{i,j,k}^{\beta_{GOBD}}}\right)^{1/\beta_{GOBD}} \qquad (5)$$

and

$$\eta_{j,k\ for\ BTDDB} = \left(\sum_{i=1}^{n=32768} \frac{1}{\eta_{i,j,k}^{\beta_{BTDDB}}}\right)^{1/\beta_{BTDDB}}. \qquad (6)$$

When $\eta_{j,k}$ for both GOBD and BTDDB are determined, we can compute the characteristic lifetime of each short group, η_k, and the entire chip, η_{chip}, for both mechanisms by solving:

$$1 = \sum_j \left(\eta_k/\eta_{j,k}\right)^\beta \qquad (7)$$

$$1 = \sum_k \left(\eta_{chip}/\eta_k\right)^\beta \qquad (8)$$

Table VI presents the solution of equations (3)-(7) to estimate the lifetime of each fault group. The lifetimes of entire chip due to both mechanisms are determined in Table VII by solving equation (8) with data in Table VI.

The failure rate of the entire SRAM chip, P_{chip}, relies on the relative frequency of GOBD vs. BTDDB, since the failure rate for each wearout mechanism is different. The relative frequency of different short groups is a function of the relative frequency of each mechanism, which can be determined by

TABLE VII. SOLUTION OF EQUATIONS (8) FOR ENTIER CHIP

GOBD	$\eta_{k=1,GOBD}$	$\eta_{k=2,GOBD}$	$\eta_{k=3,GOBD}$	$\eta_{k=4,GOBD}$	$\eta_{chip,GOBD}$
$\eta_{k,GOBD}$ Ln (sec)	16.09	14.93	16.09	14.14	13.73 for $\sum P_k =1$
$P_{k,GOBD}$	0.063	0.245	0.063	0.619	
BTDDB	$\eta_{k=1,BTDDB}$	$\eta_{k=2,BTDDB}$	$\eta_{k=3,BTDDB}$	$\eta_{k=4,BTDDB}$	$\eta_{chip,BTDDB}$
$\eta_{k,BTDDB}$ Ln (sec)	20.13	19.75	20.84	18.12	17.76 for $\sum P_k =1$
$P_{k,BTDDB}$	0.10	0.145	0.050	0.71	

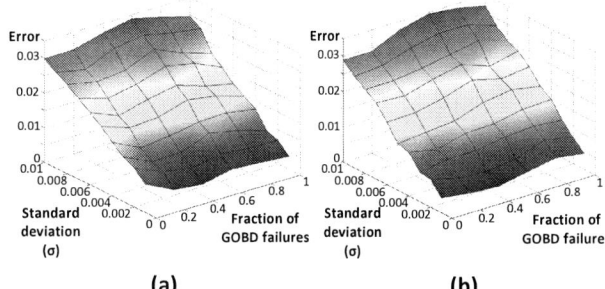

Fig. 9. The error, $|\gamma - \gamma'|$, with (a) test bench 1 and (b) test bench 2.

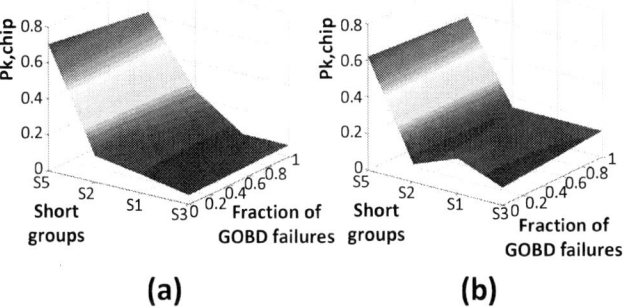

Fig. 8. Failure rate distribution for GOBD and BTDDB using a simluator which determines the stress distribution of SRAM cells inside a microprocessor with (a) test bench 1 and (b) test bench 2.

by the relative frequency of GOBD (γ) and BTDDB ($1 - \gamma$).

The relative frequency of the short groups, $P_{k,chip}$, is determined with:

$$P_{k,chip} = \gamma P_{k,GOBD} + (1 - \gamma)P_{k,BTDDB} \qquad (9)$$

$P_{k,chip}$ for two test benches [12] is presented in Fig. 8 by varying the fraction of GOBD failures, γ.

We suppose that $P_{k,chip}$ can be obtained from the observed fraction of failures in each short group, using BIST. When equation (9) is solved for γ with regression, γ is:

$$\gamma = \frac{\sum_{k=1}^{4}(P_{k,GOBD}-P_{k,BTDDB})(P_{k,chip}-P_{k,BTDDB})}{\sum_{k=1}^{4}(P_{k,GOBD}-P_{k,BTDDB})^2} \qquad (10)$$

For the error analysis for the relative frequency, γ, we assume that $P_{k,GOBD}$ and $P_{k,BTDDB}$ are modeled as normal distributions with standard deviation, σ. First, we compute $P_{k,chip}$ for the samples of γ using equation (9). Then, by varying σ for the normal distribution for $P_{k,GOBD}$ and $P_{k,BTDDB}$, equation (10) is solved for γ' with the computed $P_{k,chip}$. Fig. 9 presents the error, $|\gamma - \gamma'|$, with the two different test benches [12].

When the short groups 1-3, 5 due to GOBD or BTDDB in Table V are detected by the BIST system, the number of failed chips and failures due to each short group are counted. Before the failure of the SRAM occurs, there can be several failures from which the system can recover by error correcting codes (ECCs) and redundancy [14]. These failures can be used to estimate γ more accurately by increasing the sample size of the estimate.

V. CONCLUSIONS

We have proposed a methodology involving BIST and statistical analysis of volume data to detect and separate wearout distributions for GOBD vs BTDDB in SRAMs. The BIST methodology provides an electrical method to determine

the location of the fault within a cell, and volume data analysis is needed because both GOBD and BTDDB cause resistive shorts in SRAM cells. Therefore, we match the observed failure rate from the test data and the simulation data to estimate the failure rate distribution for each mechanism.

As we move to more scaled technologies, leakage currents increase, making the leakage current tests less effective. Nevertheless, if the reference cell selection controls for the initial leakage currents, then it is likely that the proposed diagnosis methods will work for scaled technologies as well.

REFERENCES

[1] F. Ahmed and L. Milor, "Analysis of on-chip monitoring of gate oxide breakdown in SRAM cells," IEEE Trans. VLSI, vol. 20, no. 5, pp. 855-864, May 2012.

[2] W. Kim and L. Milor, "Built-in self-test methodology for diagnosis of backend wearout mechanisms in SRAM cells," Proc. IEEE VLSI Test Symp., 2014.

[3] C.-C. Chen and L. Milor, "System-level modeling and reliability analysis of microprocessor systems," Proc. IEEE Int. Workshop on Advanced in Sensors and Interfaces, 2013, pp. 178-183.

[4] C.-C. Chen, F. Ahmed, and L. Milor, "A comparative study of wearout mechanisms in state-of-art microprocessors," Proc. IEEE Int. Conf. Computer Design, 2012, pp. 271-276.

[5] C.-C. Chen and L. Milor, "Microprocessor aging analysis and reliability modeling due to back-end weraout mechanisms," IEEE Trans. VLSI, 2015.

[6] W. Kim, C.-C. Chen, and L. Milor, "Diagnosis of Resistive-Open Defects due to Electromigration and Stress-Induced Voiding in an SRAM Array," Proc. IEEE International Integrated Reliability Workshop (IIRW), 2014.

[7] S. Naik, F. Agricola, and W. Maly, "Failure analysis of high-density CMOS SRAMs," IEEE Design & Test of Computers, vol. 10, no. 2, pp. 13-23, June 1993.

[8] J.B. Khare, W. Maly, S. Griep, and D. Schmitt-Landsiedel, "Yield-oriented computer-aided defect diagnosis," IEEE Trans. Semiconductor Manufacturing, vol. 8, no. 2, pp. 195-206, May 1995.

[9] H. Balachandran and D.M.H. Walker, "Improvement of SRAM-based failure analysis using calibrated Iddq testing," Proc. VLSI Test Symp., 1996, pp. 130-136.

[10] R. Alves Fonseca, L. Dilillo, A. Bosio, P. Girard, S. Pravossoudovitch, A. Virazel, and N. Badereddine., "Analysis of resistive-bridging defects in SRAM core-cells: A comparative study from 90nm down to 40nm technology nodes," Proc. IEEE European Test Symp., 2010, pp. 132-137.

[11] Memory compiler: www.arm.com.

[12] M.R. Guthaus, J.S. Ringenberg, D. Ernst, T.M. Austin, T. Mudge, and R.B. Brown, "MiBench: A free, commercially representative embedded benchmark suite," Proc. IEEE Int. Workshop on Workload Characterization., 2001, pp. 3-14.

[13] LEON3 processor: www.gaisler.com

[14] B. Sklar and F.J. Harris, "The ABCs of Linear Block Codes," IEEE Signal Processing Magazine, pp. 14-35, July 2004.

An Early Prediction Methodology for Aging Sensor Insertion to Assure Safe Circuit Operation due to NBTI Aging

Andres Gomez[1], Leticia Poehls[2], Fabian Vargas[2], Victor Champac[1]

[1]National Institute for Astrophysics, Optics and Electronics - INAOE, Mexico
[2]Catholic University of Rio Grande do Sul - PUCRS, Brazil

Abstract—This paper proposes an early resilience methodology to identify circuit output nodes where aging sensors should be inserted for an error prediction framework. The methodology is based in a pre-layout statistical estimation of the signal paths likely to become critical due to NBTI and/or Process Variations. To handle the fact that spatial correlation information is not available at early steps of the design flow, a statistical approach maximizing critical paths coverage is proposed. The results obtained with the early prediction methodology are compared with those obtained with spatial correlation information. The proposed methodology provides a good prediction of the set of critical paths to be monitored. Furthermore, location and number of aging sensors required to be inserted at critical paths output nodes are closely predicted.

I. INTRODUCTION

Negative Bias Temperature Instability (NBTI) is a major reliability concern in nanometer technology nodes [1]. NBTI gradually increases PMOS threshold voltage (V_{th}) over time, which translates into an increase of circuit delay. Due to NBTI effect, a circuit may fail its timing specification before the expected lifetime. Therefore, the impact of NBTI on circuit delay needs to be considered early in the design phase to assure circuit correct functionality over the entire lifetime.

The error prediction technique [2] allows to counteract NBTI by performing corrective actions like to decrease clock frequency, to increase supply voltage, and to redistribute computer tasks, before a faulty behavior occurs. Aging sensors [2][3][4][5][6] are inserted at circuit output nodes to keep under surveillance delay degradation and to detect if corrective actions are needed. However, the aging sensors insertion can introduce a large power and area overhead for large circuits. Therefore, the error prediction technique requires a cost−effective methodology to shorten the number of aging sensors to be inserted while critical paths coverage is assured. Moreover, to make the error prediction technique suitable to be applied in a conventional design flow, critical paths identification and the aging sensors insertion should be performed at early steps of the design flow.

In [7] aging-aware Static Timing Analysis is performed to identify the critical paths set. However, NBTI effect strongly interacts with process parameters variations producing shifts on both the mean and the variance of circuit timing responses [8]. Therefore, static timing analysis (even considering NBTI effect) may lacks of accuracy.

In [9] and [10] the effect of process variations in the critical paths selection procedure has been considered by performing Statistical Static Timing Analysis (SSTA). Those approaches require gates placement information to compute spatial correlation between process parameters. However, this information is not usually available at early design stages. Other statistical approach is proposed in [11]. A small set of representative critical paths is selected to be monitored and to statistically infer the aged delay of a larger set of critical paths. Although this approach allows to reduce the required number of aging sensors to be inserted, uncertainties introduced by process variations and aging may require to be considered.

For timing prediction early in the design flow, SSTA approaches using correlation bounds becomes atractive. Approaches in [12][13] compute bounds for the circuit delay distribution, but this is not the concern here. In an error prediction framework the aim is to minimize the probability that a critical path affecting circuit reliability is unmonitored.

This paper proposes an early prediction methodology to identify the circuit output nodes where aging sensors should be inserted for an error prediction framework. The combined effect of NBTI and process variations is considered. A statistical approach maximizing path coverage is proposed to cope with the unknown spatial correlation information. The results show that our approach allows to identify most of the critical paths and nodes to be monitored at the early design stages.

The rest of this paper is organized as follows: Section II presents the NBTI statistical model and the maximum path delay degradation criterion used in this work. Section III presents an overall description of the proposed methodology for cost-effective aging sensor insertion. Section IV presents our approach to select the critical paths set without availability of spatial correlation information. Section V presents the results and the validation of the proposed methodology in some ISCAS85 benchmark circuits. Finally, Section VI presents the conclusions of this work.

II. NBTI PHENOMENOM AND MAXIMUM DEGRADATION CRITERION

A. NBTI Impact in Threshold Voltage

NBTI phenomenom occurs in PMOS transistors biased in inversion. During the stress condition, $Si - H$ bonds at the gate oxide-channel interface are broken, leaving unsatisfied bonds that act like hole traps. When stress condition is removed, a partial recovery of interface traps take place. As a result, NBTI gradually increases V_{th} of PMOS devices along life operation. NBTI effect is aggravated in scaled CMOS, due to higher electric fields and higher temperatures [14]. The amount of V_{th} shift due to NBTI depends on time, technology parameters, supply voltage, temperature and stress probability, which can be defined as the average time the device is at stress condition. A simplified model that accounts for process variations and NBTI effect in V_{th} was proposed in [15]:

$$\Delta Vth_{NBTI} = (1 - S_v \cdot \Delta Vth_{PV}) \cdot A \cdot \alpha^n \cdot t^n \quad (1)$$

where A and S_v are fitted constants related to technology and operation conditions, ΔVth_{PV} is the initial variation in V_{th} due to process variations, α is the stress probability and n is a constant with a value of $1/6$. It should be highlighted that ΔVth_{PV} is composed of both correlated and independent (due to Random Dopant Fluctuations) variations.

The total ΔV_{th} a device may experience is obtained by adding the contributions of process variation (ΔVth_{PV}) and aging due to NBTI (ΔVth_{NBTI}). After rearranging the terms the following expression is obtained [15]:

$$\Delta V_{th} = A \cdot \alpha^n \cdot t^n + (1 - S_v \cdot A \cdot \alpha^n \cdot t^n) \cdot \Delta Vth_{PV} \quad (2)$$

It can be observed at $t = 0$ that total ΔV_{th} is due to process variation only. However, as time increases the first term increases the V_{th} mean value, while the second term indicates the rate of V_{th} degradation is reduced over time.

B. Maximum Path Delay Degradation Criterion

For a signal path, its delay degradation depends on the specific ΔVth of each device activated for the performed transition. However, devices stress probabilities (α) are hard to be estimated for random logic circuits, because the specific circuit workload is unknown. A maximum path delay degradation criterion is assumed to assure a conservative analysis for the early resilience methodology.

The worst case stress probabilitiy at the main input of a path, which produces the maximum delay degradation can be found as shown in [16]. Consider the five inverter chain of Figure 1. The α value at the path input (α_{in}) to obtain its maximum delay degradation should set α maximum at each activated PMOS transistor along the chain. Thus, a low value of α_{in} (0.1 in our work) is considered for the 0 to 1 transition at the path input, and a high value of α_{in} (0.9 in our work) is considered for the 1 to 0 transition at the path input. It should be highlighted that in this work the worst case value of α_{in} was found for each single path transition (t_{plh} or t_{phl}).

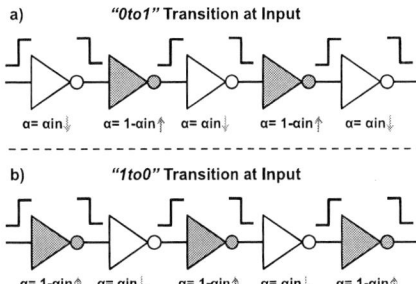

Figure 1. Stress probability conditions for worst case degradation

III. GLOBAL METHODOLOGY FOR AGING SENSOR INSERTION

Figure 2 shows the flow diagram of the early prediction methodology to identify the set of paths and nodes to be monitored. The input file is the circuit netlist at gate level (Pre-Layout Level). Path pre-filtering using static timing analysis based on process corners is performed. A coarse selection condition is used to define the set of pre-filtered paths. This allows to reduce the overall topological paths set in a reasonable amount of computational time. Then, the set of pre-filtered paths is analyzed with SSTA. Two statistical selection criteria are evaluated to identify the paths likely to become critical due to NBTI and/or Process Variations. Once the set of critical paths is obtained, the output nodes (primary outputs) of these paths are identified. The aforementioned methodology has been implemented in C++ code. In the following, a more detailed description of each step in the proposed methodology is given.

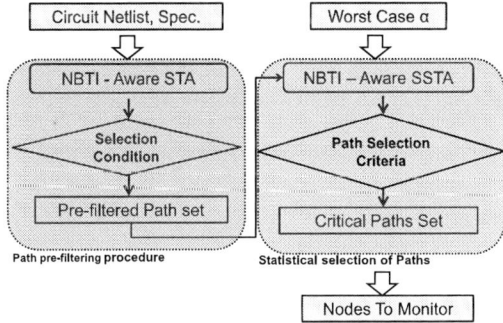

Figure 2. Flow diagram of the proposed early prediction methodology

A. Path pre-filtering

Figure 3 illustrates the path pre-filtering step, which allows to identify those paths having certain probability to become critical in a reasonable amount of computational time. The paths are ranked according to their nominal delay values as P_1, P_2, P_3, P_4, where P_1 states for the Longest Critical Path (LCP) and P_4 represents the path with the shortest delay. Delay information of each path using Fast-Fast (FF) and Slow-Slow (SS) process corners are obtained by Static Timing

Analysis. The time delay of $P1$ at FF corner (D_{FF1}) is taken as the delay threshold to obtain the *pre-filtered paths set*.

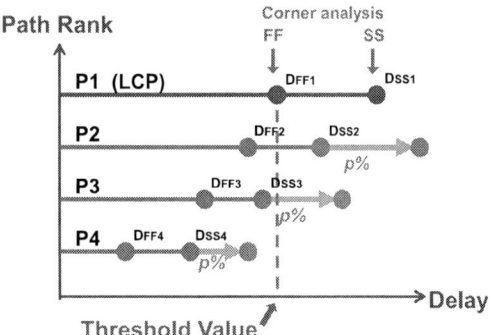

Figure 3. Path pre-filtering procedure.

Selection condition for path-prefiltering: a path is selected, if its delay in the SS corner (D_{SS}) plus some worst case percentage of aging degradation ($p\%$) is greater than the delay threshold value: $D_{SS}(1 + p\%) > D_{FF1}$. Paths that met this condition could exhibit a delay greater than the LCP delay during operational lifetime.

As can be seen in Figure 3, $P2$ and $P3$ meet previous condition, then those paths are included in the pre-filtered path set. On the other hand, $P4$ does not take delays greater than the LCP delays under any process corner and any aging condition. Then, $P4$ does not require to be further analyzed. In this work, a maximum possible percentage of degradation $p = 20\%$ is assumed [17]. This conservative value of possible aging degradation assures that any path endangering correct circuit operation will not be left out.

B. Statistical Identification of Paths to be monitored

The pre-filtered paths set is next analyzed using SSTA. Gate delays are modeled as linear functions of process parameters. Variations in channel width, channel length, oxide thickness and threshold voltage have been considered. Applying properties of linear models, paths delays distributions and covariance between them can be computed.

1) Statistical Selection Criteria: A Statistical Selection Criteria allows to shorten the number of paths that really need to be under surveillance. In such way, the number of aging sensors to be inserted is minimized while critical paths coverage is assured. The two following statistical criteria are proposed to establish whether a path should be monitored.

Criterion 1: The first criterion establishes that a path should be monitored if under NBTI and Process Variations effects (at $t = 10$ years) its probability to have delays greater than the LCP delays without aging (at $t = 0$ years) is greater than an user-defined threshold (ε),

$$P(D_{i,a} > D_{LCP,f}) > \varepsilon \qquad (3)$$

where $D_{i,a}$ stands for the delay distribution of the path under analysis (taken from the pre-filtered paths set), and $D_{LCP,f}$ stands for the fresh (without aging) LCP delay distribution. The path under analysis is assumed under maximum

degradation conditions (See Section II) while the LCP is considered fresh, to assure that a discarded path does not become critical under any aging condition.

The threshold ε defines the acceptable degree of probability that a path under analysis becomes more critical than the LCP. The paths that exceed this threshold should be monitored. As ε value decreases, a higher number of critical paths are selected to be monitored, which means that circuit reliability increases. Using a small value of ε assures that a path not passing this criterion is quite unlikely to become critical in spite of process variations and aging conditions.

Criterion 2: Even if a path is selected with *Criterion 1*, it could exist another already selected path whose delays cover the delays of the path under analysis. Therefore, a second criterion is applied to the set of paths already selected with *Criterion 1*. This criterion establishes that a path does not need to be monitored if its probability of having delays greater than the delays of any other selected path with *Criterion 1* is lower than ε (See Equation 4).

$$P(D_{i,a} > D_{j,f}) < \varepsilon \qquad (4)$$

Similar to *Criterion 1* the path i under analysis is assumed under maximum degradation conditions while the other path (j) is considered fresh.

2) Computing Selection Criteria: For two path delay distributions D_A and D_B, where D_A is for the LCP and D_B is for the path under analysis, *Criterion 1* can be translated to the probability that the *delay difference random variable DD* ($= D_B - D_A$) takes values greater than zero by the threshold value ε. This can be expressed as: $P(DD > 0) > \varepsilon$. *Criterion 1* is evaluated by obtaining the mean and variance of DD, which are given by equations 5a and 5b, respectively.

$$\mu_{DD} = \mu_{D_B} - \mu_{D_A} \qquad (5a)$$

$$\sigma_{DD}^2 = \sigma_{D_A}^2 + \sigma_{D_B}^2 - 2 \cdot COV(D_A, D_B) \qquad (5b)$$

The mean (μ_{DD}) and the variance (σ_{DD}^2) of DD can be computed by means of SSTA. At early steps of the design flow, the mean value μ_{DD} can be estimated by a pre-characterization of nominal delays of the gates library. However, σ_{DD}^2 estimation requires to compute the *paths variance* and the *inter-path covariance*, which depend on the spatial correlation between gates. Since actual layout gate placement information is not usually available at early design stages, an approach to estimate σ_{DD}^2 without spatial correlation information will be presented in the next section.

Criterion 2 can be computed similarly to *Criterion 1*. In this case, DD is obtained between the path under analysis and each previously selected path with *Criterion 1*.

C. Nodes to be monitored

Only one single aging sensor per group of those selected paths converging at the same output node is required. This minimizes the total number of aging sensors that require to be placed at the output nodes.

Our cost−effective methodology allows the critical paths to be effectively covered by the inserted aging sensors and corrective actions can take place if a near-faulty behavior occurs in those paths.

IV. EARLY SELECTION OF CRITICAL PATHS TO BE MONITORED DUE TO NBTI AGING

This section describes in detail our approach to evaluate the statistical selection criteria without spatial correlation information.

A. Computing the Delay Diference Probability Density Function

For two path delay distributions D_A and D_B, where D_A is for the reference path (i.e. the LCP) and D_B is for the path under analysis, the mean value of the delay difference probability density function μ_{DD} is given by Equation 6.

$$\mu_{DD} = \sum_{i\epsilon B}^{N} \mu_i - \sum_{i\epsilon A}^{M} \mu_i \qquad (6)$$

where M and N are the number of gates in the paths A and B, and μ_i is the mean delay for gate i.

The variance of the delay difference distribution can be computed as shown in Equation 7.

$$\sigma_{DD}^2 = \sum_{i\epsilon A}^{M}\sum_{j\epsilon A}^{M} \rho_{ij} \cdot \sigma_i \cdot \sigma_j + \sum_{i\epsilon B}^{N}\sum_{j\epsilon B}^{N} \rho_{ij} \cdot \sigma_i \cdot \sigma_j$$
$$- 2\sum_{i\epsilon B}^{N}\sum_{j\epsilon A}^{M} \rho_{ij} \cdot \sigma_i \cdot \sigma_j \qquad (7)$$

where ρ_{ij} is the spatial correlation between gates i and j, and σ_i and σ_j are the gate delay standard deviations for gates i and j, respectively. Note that each term of Equation 7 corresponds to one term in Equation 5b. Also note that ρ_{ij} in the first and the second sumation relates two gates in the same path while ρ_{ij} in the third sumation relates one gate in path A and one gate in path B.

Analytical models for spatial correlation can be found in literature, which considers it as an exponentially decreasing function of the distance between two gates [18]. However, placement information is not usually available at early design stages to compute spatial correlation.

B. Correlation Estimation at Early Design Stage

The values of spatial correlation required to compute the delay difference variance (See Equation 7) can be represented by a correlation matrix as in Equation 8. The term ρ_{ij} represents the spatial correlation between gates i and j. Matrix indexes highlighted in orange correspond to the spatial correlation between gates in the same path. Those values are used to compute each path variance (First two summations of Equation 7). Matrix indexes highlighted in blue corresponds to the spatial correlation between a gate in one path with a gate in the other path. Those values are used to compute the inter-path covariance (Third summation of Equation 7). It

should be noted that if the gate i is the same gate j, the value of spatial correlation becomes the unit because it corresponds to the correlation of a gate with itself.

$$\rho_{ij} = \begin{pmatrix} \rho_{1_a1_a} & \rho_{1_a2_a} & \cdots & \rho_{1_aN_a} & \rho_{1_a1_b} & \rho_{1_a2_b} & \cdots & \rho_{1_aM_b} \\ \rho_{2_a1_a} & \rho_{2_a2_a} & \cdots & \rho_{2_aN_a} & \rho_{2_a1_b} & \rho_{2_a2_b} & \cdots & \rho_{2_aM_b} \\ \vdots & \vdots & \ddots & \vdots & \vdots & \vdots & \ddots & \vdots \\ \rho_{N_a1_a} & \rho_{N_a2_a} & \cdots & \rho_{N_aN_a} & \rho_{N_a1_b} & \rho_{N_a2_b} & \cdots & \rho_{N_aM_b} \\ \rho_{1_b1_a} & \rho_{1_b2_a} & \cdots & \rho_{1_bN_a} & \rho_{1_b1_b} & \rho_{1_b2_b} & \cdots & \rho_{1_bM_b} \\ \rho_{2_b1_a} & \rho_{2_b2_a} & \cdots & \rho_{2_bN_a} & \rho_{2_b1_b} & \rho_{2_b2_b} & \cdots & \rho_{2_bM_b} \\ \vdots & \vdots & \ddots & \vdots & \vdots & \vdots & \ddots & \vdots \\ \rho_{M_b1_a} & \rho_{M_b2_a} & \cdots & \rho_{M_bN_a} & \rho_{M_b1_b} & \rho_{M_b2_b} & \cdots & \rho_{M_bM_b} \end{pmatrix}$$
$$(8)$$

In order to evaluate the selection criteria under conservative assumptions due to spatial correlation uncertainty, σ_{DD}^2 should be approximated in such way that its value is maximum, because it increases the probability that the Delay Difference Distribution takes values greater than zero.

The approximation of σ_{DD}^2 can be made assuming spatial correlation bounds. For example, the spatial correlation to compute each path variance becomes the unit (gates are assumed fully correlated), while the spatial correlation to compute covariance between paths becomes zero (gates are assumed completely uncorrelated) [13]. However, this first approach would result in a significant overestimation of the number of paths to be monitored. Figure 4 plots the terms of Equation 5b obtained for each pre-filtered path (B_i) and the LCP (A) of ISCAS $C1908$ circuit. Each dot represents a different path in the circuit. The data for this figure has been obtained directly using spatial correlation information from layout. Figure 4 shows that there is a behavior relationship between the sum of the first two terms of Equation 5b and the third term. This trend was also observed for other ISCAS benchmark circuits. Hence, previous spatial correlation bounding approach would result in a significant overestimation of the critical paths set to be monitored. Based on this observation, a first approximation is made:

Approximation 1: *A same correlation value to compute each path variance and inter−path covariance is assumed.*

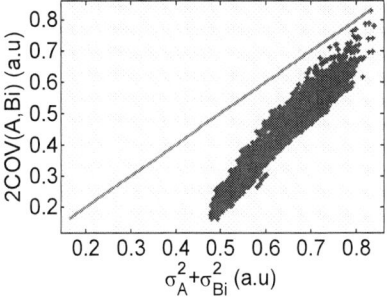

Figure 4. Behavior between addition of paths delay variances and their covariance computed for the LCP and each analyzed path of C1908 circuit.

This means that all the elements in matrix 8 that are not the unit are replaced by the global correlation ρ_g. This assumption introduces a trade-off between the added terms (path variances) and the substracted term (inter-path covariance) because both increase as ρ_g increases.

A second issue is to determine the ρ_g value that should be used to maximize σ_{DD}^2 under these global correlation assumption. Figure 5 shows two inverter chains that were analyzed to solve this issue. The path in the top is assumed the LCP with $M = 20$ inverters, and the path in the bottom is the path under analysis with N stages going from 2 to 20 inverters. σ_{DD} was computed for different ρ_g values as shown in Figure 6. It can be observed that for a short path ($N << M$), σ_{DD} is maximized by $\rho_g = 1$. However, for a long path ($N \approx M$), $\rho_g = 0$ provides an upper bound for σ_{DD}^2. In general, it would be expected that logic depths of the critical paths, including the LCP, are not dramatically different from each other. Similar results have been found for others types of gates structure, such as $NANDs$ and $NORs$. Based on this observation, a second approximation is made:

Approximation 2: *A value of $\rho_g = 0$ is used.*

Some logic paths could miss the previous maximization property, but if this set is reduced it would not significantly impact our proposed approach. It must be noted that an exception of *Approximation* 2 are those gates with structural correlation where $\rho_g = 1$ is used.

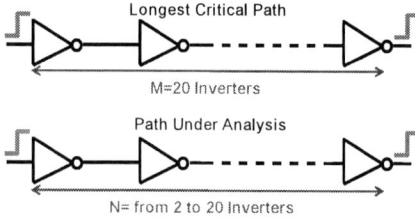

Figure 5.   Inverter chains to analyze the impact of ρ_g on the standard deviation of the delay difference (σ_{DD}) as function of the path depth.

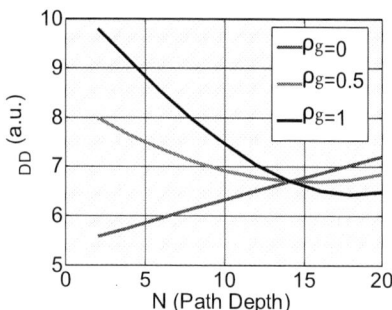

Figure 6.   Impact of ρ_g on σ_{DD} as function of the depth of the path under analysis.

V. SIMULATION RESULTS

The proposed methodology has been applied to some ISCAS85 benchmark circuits implemented in a commercial $65nm$ technology. Mentor Graphics suite of layout, synthesis, simulation, and DFT tools were used. Table I shows main characteristics of the analyzed circuits. It can be seen that the number of pre-filtered paths do not significantly decrease with respect to the number of topological paths. This is because corned-based static timing analysis with conservative aging gives a very pessimistic selection of critical paths.

Table I
CIRCUITS CHARACTERISTICS OF THE ISCAS85 BENCHMARK CIRCUITS

Circuit	Gates Number	Topological Paths	Output Nodes	Pre-filtered Paths
C432	168	82364	7	81884
C499	220	9440	32	9408
C880	226	4935	26	4566
C1908	244	15638	25	15330
C2670	393	3381	50	1896
C3540	748	670373	22	659495
C5315	1139	24662	123	23112
C7552	1365	43614	108	40512

The results obtained with our proposed early design methodology for different probability thresholds ε are shown in Table II. The number of selected critical paths and the number of nodes to be monitored are given. The set of paths to be monitored selected statistically (See Table II) is significantly smaller than the set of pre-filtered paths (See Table I). A high number of critical paths can be covered by monitoring few output nodes (See Table II). The results are given for three ε values, which set the degree of circuit reliability as defined in Section 4b (See equations 3 and 4). Circuit reliability increases as lower values of ε are used. This is because a more stringent selection criterion allows to select more critical paths. It can also be observed that the number of nodes to be monitored remain the same as ε reduces from 1% to 0.5% in spite that more critical paths were selected. This is because the additional selected critical paths converge at the already selected output nodes with ε=1%. However, the number of output nodes increase in ISCAS $C7552$ for ε=0.25% (from 12 to 13). An aging sensor also must be inserted at the new output node to cover new selected critical paths do not covered previously by ε values of 1% and 0.5%.

Table II
RESULTS WITH OUR PROPOSED EARLY METHODOLOGY.

Circuit	$\varepsilon = 1\%$		$\varepsilon = 0.5\%$		$\varepsilon = 0.25\%$	
	Selected Paths	Nodes to Monitor	Selected Paths	Nodes to Monitor	Selected Paths	Nodes to Monitor
C432	4720	3	5298	3	5883	3
C499	2304	32	2384	32	2432	32
C880	80	3	86	3	93	3
C1908	283	7	333	7	370	7
C2670	84	1	92	1	100	1
C3540	1052	3	1174	3	1280	3
C5315	166	4	184	4	200	4
C7552	95	12	114	12	126	13

In order to validate our proposal, the results obtained with the proposed methodology have been compared against those obtained using SSTA with layout information [9]. The analytical model proposed in [18] was used to calculate the

Table III
RESULTS USING LAYOUT INFORMATION.

Circuit	$\varepsilon = 1\%$		$\varepsilon = 0.5\%$		$\varepsilon = 0.25\%$	
	Selected Paths	Nodes to Monitor	Selected Paths	Nodes to Monitor	Selected Paths	Nodes to Monitor
C432	4676	3	5248	3	5816	3
C499	2272	32	2315	32	2400	32
C880	78	3	82	3	92	3
C1908	270	7	317	7	347	7
C2670	81	1	89	1	95	1
C3540	1030	3	1153	3	1242	3
C5315	186	4	209	4	220	4
C7552	95	12	116	13	126	13

spatial correlation. Table III shows the number of selected critical paths and nodes to be monitored obtained from layout-based analysis. The comparison of the proposed methodology result (See Table II) against the layout-based results (See Table III) makes evident that our approach approximates quite well the number of critical paths and nodes to be monitored. Differences between the selected paths in Table II with those selected in Table III may contain some wrong selected paths (paths considered as critical but actually they are not) and some unselected paths (paths not identified as critical but that actually they are). However, the fact that many paths could converge at the same output node makes that even if a path is not actually selected by the proposed methodology, it is still covered if it shares its output node with at least one selected path. As can be seen, for almost all cases, the correct number of nodes to be monitored was identified. Only for $C7552$ with $\varepsilon = 0.5\%$, an output node is not selected with the early design methodology. However, using in the early methodology an ε value lower than that used in the layout-based analysis makes less likely that a true critical path is unselected. Thus, some nodes and paths do not selected can be included in the predicted critical path set (compare columns 4 and 5 from Table III with columns 6 and 7 from Table II).

VI. CONCLUSIONS

An early prediction methodology for aging sensors insertion in an error prediction framework has been proposed. The combined effect of process variations and aging due to NBTI were taken into account. To approximate the locations where aging sensors should be inserted, paths that are likely to become critical are identified first. Two statistical criteria are used to select the critical paths set to be monitored. The set size depends on the desired degree of circuit reliability.

The proposed methodology is suitable to be applied at an early design step, which reduce design complexity since computation of spatial correlation between each pair of gates in the circuit is not required. To handle the lack of spatial correlation information a statistical non-layout dependent approach that maximizes the probability of a critical path to be identified was proposed. A comparison of the proposed methodology with that based in availability of layout information shows that our proposal accurately predicts the set of critical paths and nodes to be monitored.

ACKNOWLEDGMENT – The work has been partially supported by CONACYT (Mexico) through the PhD scholarship number 420129/264560.

REFERENCES

[1] Seyab, Nor Zaidi Haron, Said Hamdioui, "CMOS scaling impacts on reliability, What do we understand?", Delft University of Technology, Computer Engineering Laboratory.

[2] Agarwal, M.; Paul, B.C.; Ming Zhang; Mitra, S, "Circuit Failure Prediction and Its Application to Transistor Aging," VLSI Test Symposium, 2007. 25th IEEE, vol., no., pp.277,286, 6-10 May 2007

[3] Junyoung Park; Abraham, J.A.,"An aging−aware flip−flop design based on accurate, run−time failure prediction," VLSI Test Symposium (VTS), 2012 IEEE 30th , vol., no., pp.294,299, 23-25 April 2012

[4] Vazquez, J.C.; Champac, V.; Ziesemer, A.M.; Reis, R.; Teixeira, I.C.; Santos, M.B.; Teixeira, J.P., "Built-in aging monitoring for safety−critical applications," On−Line Testing Symposium, 2009. IOLTS 2009. 15th IEEE International , vol., no., pp.9,14, 24−26 June 2009.

[5] Vazquez, J.C.; Champac, V.; Teixeira, I.C.; Santos, M.B.; Teixeira, J.P., "Programmable aging sensor for automotive safety − critical applications," Design, Automation & Test in Europe Conference & Exhibition (DATE), 2010 , vol., no., pp.618,621, 8-12 March 2010

[6] Khan, S.; Haron, N.Z.; Hamdioui, S.; Catthoor, F., "NBTI Monitoring and Design for Reliability in Nanoscale Circuits," Defect and Fault Tolerance in VLSI and Nanotechnology Systems (DFT), 2011 IEEE International Symposium on , vol., no., pp.68,76, 3-5 Oct. 2011

[7] Jifeng Chen , Shuo Wang , Mohammad Tehranipoor, Efficient selection and analysis of critical-reliability paths and gates, Proceedings of the great lakes symposium on VLSI, May 03-04, 2012, Salt Lake City, Utah, USA

[8] Wenping Wang; Reddy, V.; Yang, B.; Balakrishnan, V.; Krishnan, Srikanth; Yu Cao, "Statistical prediction of circuit aging under process variations," Custom Integrated Circuits Conference, 2008. CICC 2008. IEEE , vol., no., pp.13,16, 21-24 Sept. 2008

[9] Vazquez, J.C., Champac, V., Semio, J., Teixeira, I.C., Santos, M.B., Teixeira, J.P., "Process Variations-Aware Statistical Analysis Framework for Aging Sensors Insertion", Journal of Electronic Testing, 2013.

[10] Dominik Lorenz, Martin Barke, Ulf Schlichtmann, "Monitoring of aging in integrated circuits by identifying possible critical paths", Microelectronics Reliability, Volume 54, Issues 6-7, June-July 2014, Pages 1075-1082.

[11] Firouzi, F.; Fangming Ye; Chakrabarty, K.; Tahoori, M.B., "Representative critical-path selection for aging-induced delay monitoring," Test Conference (ITC), 2013 IEEE International , vol., no., pp.1,10, 6-13 Sept. 2013

[12] Wei-Shen Wang; Orshansky, M., "Path-Based Statistical Timing Analysis Handling Arbitrary Delay Correlations: Theory and Implementation," Computer-Aided Design of Integrated Circuits and Systems, IEEE Transactions on , vol.25, no.12, pp.2976,2988, Dec. 2006

[13] Heloue, Khaled R.; Najm, Farid N., "Early Statistical Timing Analysis with Unknown Within-Die Correlations," In IEEE TAU Workshop, Austin 2007.

[14] Khan, S.; Hamdioui, S., "Temperature dependence of NBTI induced delay," On-Line Testing Symposium (IOLTS), 2010 IEEE 16th International , vol., no., pp.15,20, 5-7 July 2010

[15] Song Jin; Yinhe Han; Huawei Li; Xiaowei Li, "P^2CLRAF: An Pre- and Post-Silicon Cooperated Circuit Lifetime Reliability Analysis Framework," Test Symposium (ATS), 2010 19th IEEE Asian , vol., no., pp.117,120, 1-4 Dec. 2010

[16] Wenping Wang; Zile Wei; Shengqi Yang; Yu Cao, "An efficient method to identify critical gates under circuit aging," Computer−Aided Design, 2007. ICCAD 2007. IEEE/ACM International Conference on , vol., no., pp.735,740, 4-8 Nov. 2007

[17] Wenping Wang; Shengqi Yang; Bhardwaj, S.; Vattikonda, R.; Vrudhula, S.; Liu, F.; Yu Cao, "The Impact of NBTI on the Performance of Combinational and Sequential Circuits," Design Automation Conference, 2007 DAC '07. 44th ACM/IEEE , vol., no., pp.364,369, 4-8 June 2007

[18] Jinjun Xiong; Zolotov, V.; Lei He, "Robust Extraction of Spatial Correlation," Computer-Aided Design of Integrated Circuits and Systems, IEEE Transactions on , vol.26, no.4, pp.619,631, April 2007.

Integral Impact of BTI and Voltage Temperature Variation on SRAM Sense Amplifier

Innocent Agbo   Mottaqiallah Taouil   Said Hamdioui
Delft University of Technology
Faculty of Electrical Engineering, Mathematics and CS
Mekelweg 4, 2628 CD Delft, The Netherlands
{I.O.Agbo, M.Taouil, S.Hamdioui}@tudelft.nl

Halil Kukner   Pieter Weckx   Praveen Raghavan   Francky Catthoor
IMEC vzw
Kapeldreef 75, 3001 Leuven, Belgium
{Halil, Pieter.Weckx, Ragha, Francky.catthoor}@imec.be

Abstract—With the continuous downscaling of CMOS technologies, ICs become more vulnerable to transistor aging mainly due to Bias Temperature Instability (BTI). A lot of work is published on the impact of BTI in SRAMs; however most of the work focused mainly on the memory cell array. An SRAM consists also of peripheral circuitries such as address decoders, sense amplifiers, etc. This paper characterizes the combined impact of BTI and voltage temperature fluctuations on the memory sense amplifier for different technology nodes (45nm up to 16nm). The evaluation metric, the sensing delay (SD), is analyzed for various workloads. In contrast to earlier work, this paper thoroughly quantifies the increased impact of BTI in such sense amplifiers for all the relevant technology scaling parameters. The results show that the BTI impact for nominal voltage and temperature is 6.7% for 45nm and 12.0% for 16nm when applying the worst case workload, while this is 1.8% for 45nm technology and 3.6% higher for 16nm when applying the best case workload. In addition, the results show that the increase in power supply significantly reduces the BTI degradation; e.g., the degradation at $-10\% V_{dd}$ is 9.0%, while this does not exceed 5.3% at $+10\% V_{dd}$ at room temperature. Moreover, the results that the increase in temperature can double the degradation; for instance, the degradation at room temperature and nominal V_{dd} is 6.7% while this goes up to 18.5% at 398K.

Index Terms—BTI, NBTI, PBTI, SRAM sense amplifier

I. INTRODUCTION

In recent decades, CMOS technology has been sustained with aggressive downscaling that severely impacts the reliability of devices [1,2,28]. These trends are due to advancements in the fabrication technology, introduction of novel materials and evolution of architecture designs. Bias Temperature Instability (BTI) (i.e., Negative BTI in PMOS transistors and Positive BTI in NMOS transistors) is a reliability failure mechanism which affects the performance of MOS transistors by increasing their threshold voltage and reducing their drain current (I_d) over the operational lifetime [5,11]. However, studies of such individual devices or small composites do not allow extrapolation of these effects on larger circuits like SRAMs.

Static Random-Access Memories (SRAM) occupy a large fraction of semiconductor chip and play a major role in the silicon area, performance, and critical robustness [12]. An SRAM system consists of an array of cells, its peripherals circuits such as row and column address decoders, control circuits, write drivers, and sense amplifiers. Designing an optimal reliable memory system requires the consideration of all its sub-parts, i.e., their degradation rate depends on the application (e.g. workload, temperature, etc); for instance, the

aging rate of the sense amplifier may differ from that of the memory array and other peripheral circuits.

Many publications analyzed the BTI impact on SRAM cell array, while very limited work is published on the SRAM peripheral circuitry. For instance, Binjie *et al.* [17] investigated NBTI impact on Static Noise Margin (SNM) and Write Noise Margin (WNM) degradation of 6T SRAM cell. Kumar *et al.* [18] Analyzed the impact of NBTI on the read stability and SNM of SRAM cells. Andrew [19] investigated the mechanism of NBTI degradation on SRAM metrics such as SNM. Bansal et al [20] presented insights on the stability of an SRAM cell under the worst-case conditions and analyzed the effect of NBTI and PBTI, individually and collectively. Rodopoulos *et al.* [22] investigated the atomistic pseudo-transient BTI simulation with built-in workloads. Khan et al [23] investigated BTI analysis of FinFET based SRAM cell.

On the other hand, few authors have focused on reliability analysis of the SRAM peripheral circuit. Khan *et al.* [21] investigated the impact of partial opens conjuction BTI in SRAM address decoders. Menchaca *et al.* [24] analyzed the BTI impact on different sense amplifier designs implemented on 32nm technology node by using failure probability (i.e., flipping a wrong value) as a reliability metric. Agbo *et al.* [25] investigated the BTI impact on SRAM drain-input latch type sense amplifier design implemented on 90nm, 65nm, and 45nm for different supply voltages by using sensing delay and sensing voltage as reliability metrics. However, quantitative analysis of BTI impact of peripheral circuits (including sense amplifiers) while considering different workloads, temperatures, supply voltages and how they correlate with technology scaling is still to be explored. It is worth noting that understanding and quantifying the aging rate of each memory part is needed for optimal reliable memory design; this is because the different parts may degrade with different rates depending e.g. on the workload (application).

This paper focuses on standard latch-type sense amplifier design due to its superior performance [30] and analyzes the BTI impact for different temperatures and supply voltages, and different workloads. The main contributions of the paper are as follows:

- Investigation of BTI impact on the sense amplifier's sensing delay. In contrast to previous work, we analyze the BTI stress and relaxation cycles for each transistor individually to obtain more accurate results. These stress and relaxation cycles are workload dependent. In this

978-1-4799-7598-3/15 $31.00 © 2015 IEEE

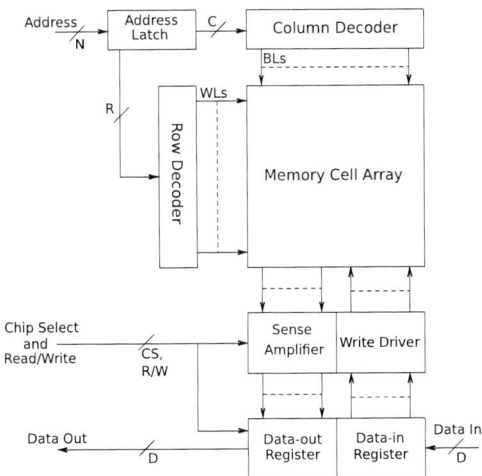

Fig. 1. Functional model of SRAM system

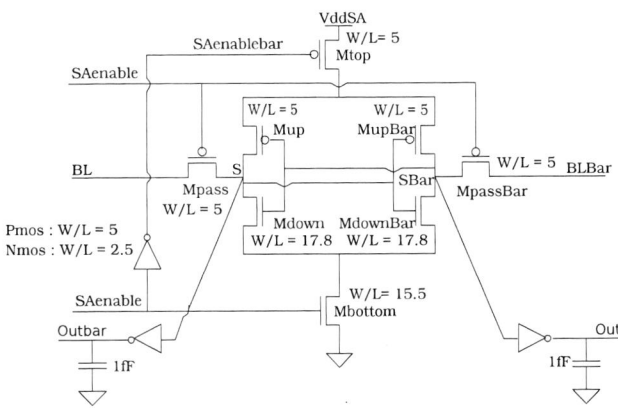

Fig. 2. Standard latch-type Sense Amplifier

work we define eight realistic workloads.

- Investigation of BTI impact on different workloads for different technology nodes.
- Thorough quantitative analysis of the BTI impact on the sense amplifier for deeply nano-scaled technology nodes, i.e., 45nm, 32nm, 22nm, and 16nm.
- Analysis of BTI impact under different supply voltages and temperatures on the SRAM sense amplifier sensing delay using different workloads.

The rest of the paper is organized as follows: Section II introduces the SRAM model, standard latch-type sense amplifier, BTI mechanism and its model. Section III provides our analysis framework, analysis metric, and the performed experiments. Section IV analyzes the result for different technology nodes, workloads, varying supply voltages and temperatures. Finally, Section V concludes the paper.

II. BACKGROUND

This section presents first the functional SRAM model. Afterwards, it focuses on the behavior of the standard latch-type sense amplifier. Finally, it explains the BTI mechanisms and its model analyzed in this paper.

A. Memory model

Figure 1 depicts a functional model of the SRAM system [28]. A memory system is comprised of a memory cell array, row and column address decoders, read/write circuitry, input/output data registers and control logic. The main focus of the paper is the sense amplifier.

SRAM Sense Amplifier

Several implementations of sense amplifiers have been proposed. In this paper, the standard latch-type SRAM strobed sense amplifier will be addressed which is representative for industrial SA designs [30].

The structure of the Standard latch-type Sense Amplifier is depicted in Fig. 2. The width length ratio of each transistor is presented by W/L.

The operation of the sense amplifier consists of two phases. In the first phase, when SAenable is low, the access transistors Mpass and MpassBar connect to the BL (BLBar) with the internal nodes S (SBar). In this phase, Mtop and Mbottom transistors are switched off. In the second phase, when SAenable is high, the pass transistors disconnect the BL (BLBar) input from the internal nodes. The cross coupled inverters get their current from Mtop and Mbottom and subsequently amplify the difference between S and SBar and produce digital outputs on Out and Outbar. S (SBar) node is actively pulled down when SBar (S) exceeds the threshold voltage of Mdown. The positive feedback loop ensures low amplification time and produces the read value at its output. Moreover, all current paths are disabled when S (SBar) is at 0V and SBar (S) is at V_{ddSA} or vice versa. This process is repeated for each read operation.

B. Bias Temperature Instability

BTI mechanism takes place inside the MOS transistors and causes a threshold voltage shift that impacts the delay negatively; its mechanism is described below.

BTI Mechanism

BTI increases the absolute V_{th} value in MOS transistors. For PMOS, negative BTI (NBTI) reduces the V_{th} while for NMOS, positive BTI (PBTI) the V_{th} increases. Recently, exhaustive efforts have been put to understand NBTI [5,10,11]. Kaczer et al. in [10] have analyzed NBTI using an atomistic model. Alam et al. [5] have modeled NBTI and presented the overall dynamics of NBTI as a reaction diffusion process. The model is usable at a higher level such as circuit level. In Kukner et al. in [26], both the atomistic and RD models have been compared. The authors conclude that for the long-term simulation time, RD model is lightweight than the atomistic model. For this reason we select this model [5]. For a MOS transistor, there are two BTI phases, i.e., the stress phase and the relaxation phase.

Stress Phase: In the stress phase, the Silicon Hydrogen bonds (≡Si-H) break at Silicon-Oxide interface. The broken Silicon bonds (≡Si-) remain at the interface (known as interface traps), and the released H atoms/molecules diffuse towards the gate oxide. The number of interface traps (N_{IT}) generated after applying a stress of time (t) is given by [5]:

$$N_{IT}(t) = \left(\frac{N_o \cdot k_f}{k_r}\right)^{2/3} \cdot \left(\frac{k_H}{k_{H_2}}\right)^{1/3} \cdot (6 \cdot D_0 \cdot t)^{1/6} \quad (1)$$

where N_o, k_f, k_r, k_H, and k_{H_2}, represent initial $\equiv$Si-H density, $\equiv$Si-H breaking rate, $\equiv$Si- recovery rate, H to H_2 conversion rate, and H_2 to H conversion rate inside the oxide layer, respectively. $D_0 = D_{H_2} \cdot \exp(-E_A/kT)$ [6] is the diffusion coefficient of the produced H_2 species and E_A is the activation energy, k is the boltzman constant, and T is the temperature in Kelvin.

Relaxation Phase: In the relaxation phase, there is no $\equiv$Si-H breaking. However, the H atoms/molecules diffuse back towards the interface and anneal the $\equiv$Si- bonds. The number of interface traps that *do not* anneal by the approaching H atoms during the relaxation phase is given by [18]:

$$N_{IT}(t_o + t_r) = \frac{N_{IT}(t_o)}{1 + \sqrt{\frac{\xi \cdot t_r}{t_o + t_r}}} \quad (2)$$

where $N_{IT}(t_o)$ is the number of interface traps at the start of the relaxation, ξ is a relaxation coefficient with $\xi=0.5$ [18], t_o is the duration of the previous stress phase and t_r is the relaxation duration.

Threshold voltage increment: The N_{IT} oppose the gate voltage which result in a threshold voltage increment (ΔV_{th}). The relation between N_{IT} and ΔV_{th} is given by [4]:

$$\Delta V_{th} = (1 + m) \cdot q \cdot N_{IT}/C_{ox} \cdot \chi, \quad (3)$$

where m, q, and C_{ox} are the holes/mobility degradation that contribute to the V_{th} increment [16], electron charge, and oxide capacitance, respectively. χ is a BTI coefficient with a value $\chi=1$ for NBTI and $\chi=0.5$ for PBTI [14].

III. ANALYSIS FRAMEWORK

In this section, the analysis framework of the standard latch type sense amplifier circuit is described. Furthermore, the workloads used in performing the experiment are explained. Thereafter, the output performance metric is presented. Finally, the conducted experiments are presented.

A. Framework Flow

Figure 3 depicts a flexible and generic BTI framework for the standard latch type sense amplifier circuit. The framework evaluates the BTI impact for different designs, technologies, workload under normal conditions and considering VT (i.e., voltage, and temperature) variations. The framework consist of a MATLAB and HSPICE working environment. The MATLAB environment typically is used for pre-processing and post-processing, it prepares BTI augmented files to run in HSPICE. The results of HSPICE that simulates the BTI augmented netlist, are subsequently post-processed in MATLAB. Furthermore, the framework analysis is divided into three parts (i.e., input, processing, and output blocks) and they are

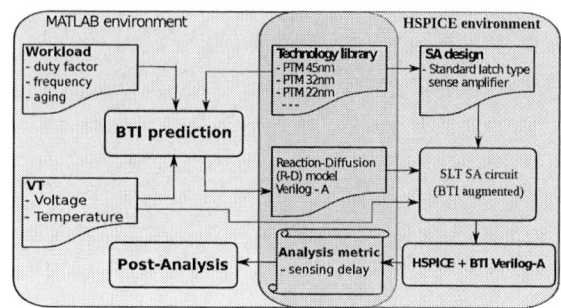

Fig. 3. Analysis framework for the standard latch sense amplifier circuit.

explained next.

Input: The general input blocks of the framework are the SA design, technology library, workload, and voltage temperature (VT). They are explained as follows.

- SA design: Generally, all sense amplifier design can be used. In this paper we focus only on the standard latch type sense amplifier. The SA design is described by an HSPICE netlist.

- Technology library: Different technology nodes are considered in this work, they are 45nm, 32nm, 22nm, and 16nm and are obtained from PTM library cards [31].

- VT: This block specifies the temperatures and voltages. In this paper, we restrict ourselves to temperatures $T_1 = 298K$, $T_2 = 348K$, and $T_3 = 398K$ and supply voltages $V_1 = -10\%V_{dd}$, $V_2 = V_{dd}$, and $V_3 = +10\%V_{dd}$. Note that each technology has its own nominal voltage.

- Workload: The shift in threshold voltage is a function of stress and relaxation durations of the transistors. This implies that BTI degradation depends on the amount of ON and OFF (idle) states of the input patterns which translates to workload. To perform this analysis, we assume that today's application consists of 10% - 90% memory instructions and the percentage of read instructions is typically 50% - 90%. Furthermore, we derive from this the following cases: best case with stress period of 0.1 * 0.1 = 0.01, worst case with 0.9 * 0.9 = 0.81, and mid case: 0.5 * 0.5 = 0.25. They lead to the following workload sequences: S1: $R0I^{99}$, S2: $R0R1I^{198}$, S3: $R0^4I^1$, S4: $(R0R1)^4I^2$, S5: $(R0)I^{24}$, S6: $(R0R1)I^{24}$, S7: $(R0)I^{50}$, and S8: $(R0R1)I^{50}$. In these sequences, $R0$ stands for read 0, $R1$ stands for read 1, I for idle operation (which includes memory write operations). For example, $S1$: $R0I^{99}$ is workload where read 0 is followed by 99 idle operations. The best and worst case will be analyzed in most detail.

The workload inputs are typically characterized by their duty factor, frequency, and aging (or stress time). The BTI impact sensitivity is highly dependent on the input stimulus clock cycle (i.e., frequency), its aging and duty factor (DC stress or AC stress) w.r.t., the affected device or circuitry.

i **Frequency** The BTI induced degradation depends on the signal frequency to the sense amplifier design. In this experiment, the frequencies considered for the SA design are 1.32GHz, 1.89GHz, 2.70GHz, and 3.86GHz for 45nm,

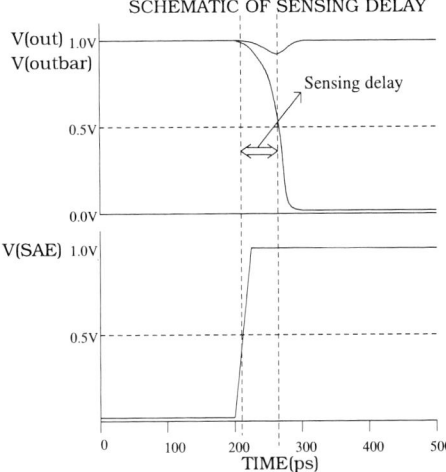

Fig. 4. Metric diagram of Sensing delay.

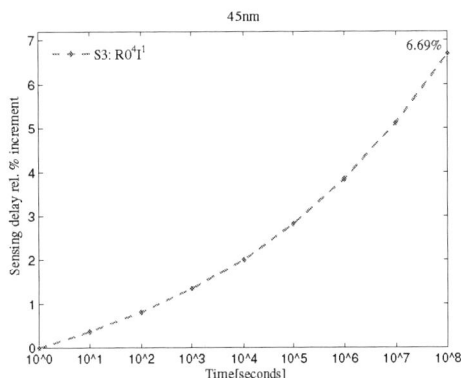

Fig. 5. BTI impact on Sensing delay.

32nm, 22nm, and 16nm technology nodes, respectively.

ii **Aging** The BTI induced degradation depends strongly on the stress time. The stress time defines how long the workload sequence is being applied. A workload sequence is assumed to be repeated until the age time is reached.

iii **Duty Factor** The input signal is a function of the duty factor that affects the BTI induced degradation of the sense amplifier design. The duty factor of the input SA enable signal (see Figure 4.) is approximately 0.48. This applies only to read operations. During write or idle operations, SAenable signal is disabled. From the waveform analysis, we extract for all transistors individually their stress and relaxation cycles; thereby is able to obtain accurate duty cycles for each transistor. This enhances the accuracy of our simulation results. Based on the duty factor and age time, interface traps (Eqns. 1 and 2) or threshold voltage increments (Eqn. 3) can be attributed to all transistors in an accurate manner.

Processing: There are two processing blocks, the BTI predictor and the HSPICE simulation unit. The long term BTI predictor uses the duty factor, frequency, and aging to predict the interface traps/ threshold voltage increment of each device in the sense amplifier. In addition to workload inputs, inputs are required from the reaction-diffusion model (such as K_f, K_r, D_H, etc.), technology parameters, and voltage temperature (VT). Once the BTI induced V_{TH} increments are calculated per transistor, the original BTI free netlist will be updated. This new netlist is simulated in HSPICE/Verilog-A.
Output: Finally, post-analysis of the results are performed for varying voltages and temperatures in MATLAB environment.

B. Output Analysis Metrics

In this section, the sensing delay metric used for analyzing BTI impact on sense amplifier is described.
Sensing delay: The sensing delay metric is determined when the trigger signal (i.e., sense amplifier enable input signal) reaches 50% of the supply voltage and the target (i.e., either out or outbar falling output signal) reaches 50% of the supply voltage. The difference between the target and the trigger

results in sensing delay as shown in Fig. 4. Furthermore, the relative variation of the sensing delay due to BTI is the difference between the measured sensing delay when BTI is added and referenced sensing delay when BTI is not added.

C. Experiments Performed

In this paper, four sets of experiments are performed to analyze BTI impacts. These experiments are described below:

1 **BTI Impact Experiments:** BTI impact on sensing delay of the SRAM sense amplifier is investigated.

2 **Workload Dependent Experiments:** BTI impact on the sensing delay of the SRAM sense amplifier for different workloads on different technology nodes is investigated.

3 **Technology Dependent Experiments:** BTI impact on sensing delay using different technology nodes is investigated.

4 **Supply Voltage and Temperature Dependent Experiments:** BTI impact on sensing delay of the SRAM sense amplifier for varying supply voltages (i.e., $-10\% V_{dd}$, V_{dd} and $+10\%$ V_{dd}) and temperatures (i.e., $298K$, $348K$ and $398K$) for different technology nodes are explored.

IV. EXPERIMENTAL RESULTS

In this section, we present the analysis results of the experiments mentioned in the previous section.

A. Temporal BTI Impact

The BTI in MOS transistors affect the sensing delay of the sense amplifier, i.e., the time required to amplify the input from BL and BLBar to outputs Out and Outbar (see Figure 2). In order to quantify this delay, we simulate the initial BTI-free SA design, for each technology node and take their sensing delays as references. To obtain proper sensing delays, appropriate values of BL and BLBar should be selected. For 45nm, we assume the differential input to be 100mV (V_{dd}) [30]. Subsequently, we modify this differential voltage in such a way to meet up with the frequencies of the lower technology nodes. Note that the transistors are scaled with each technology.

Figure 5 shows the relative increment of the sensing delay w.r.t. the stress time (aging) for workload $S3$ using 45nm technology. The figure shows a quadratic delay increment with respect to the stress time. For example, after 10^8 sec the delay increments equals 6.69% due to BTI.

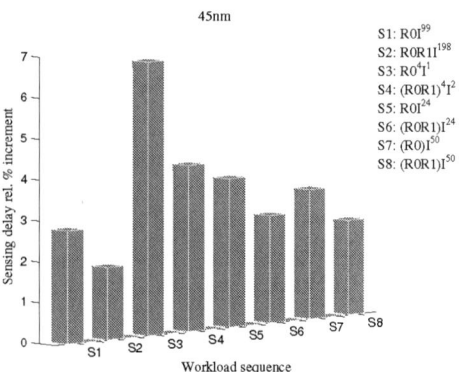

Fig. 6. Workload dependent Sensing delay.

S1: $R0I^{99}$
S2: $R0R1I^{198}$
S3: $R0^4I^1$
S4: $(R0R1)^4I^2$
S5: $R0I^{24}$
S6: $(R0R1)I^{24}$
S7: $(R0)I^{50}$
S8: $(R0R1)I^{50}$

B. Workload Dependency

The BTI induced degradation is sensitive to the workload. Hence, this workload-dependent behaviour is one of the main contributions of this study. A better understanding of this behaviour will strongly help to select the proper mitigation schemes and to reduce the cost for highly dynamic cost-sensitive embedded systems. The workload defines when and how long each transistor is stressed. Figure 6 shows the relative BTI induced sensing delay for sequences $S1$, $S2$, $S3$, $S4$, $S5$, $S6$, $S7$, and $S8$ respectively. There is a significant variation in the relative sensing delay increment. For instance, workload sequence, $S2$ ($R0R1I^{198}$) has a lower impact as it is activated (stressed) only 1% of the signal duration than the other workloads, whereas workload sequence, $S3$ ($R0^4I^1$) has the highest impact as it is activated (stressed) 80% of the signal duration. The remaining workloads result in a delay increment between the extreme cases $S2$ and $S3$. The same trends are observed for the other technology nodes.

C. Technology Dependency

CMOS technology scaling results in an apparent oxide field increment which speeds up the BTI covalent bond breaking phenomenon and thus the BTI induces threshold voltage. Therefore, it is essential to evaluate the reliability of different technology nodes. Experiments are performed at nominal supply voltage and temperature (T_1 = 298K) for different technology nodes with their corresponding supply voltages 1.0V, 0.9V, 0.8V, and 0.7V for 45nm, 32nm, 22nm, and 16nm, respectively. Here, we focus only on the best and worst case workloads $S2$ and $S3$ respectively. Experiments are performed to investigate the impact of BTI on sensing delay for different technologies for worst and best case workloads, i.e., workloads S3 and S2, respectively. Figure 6 depicts the sensing delay of these experiments for a stress time of 10^8s. The figure shows for both the worst case and best case workloads that the relative delay increment increases with advanced technology nodes. However, this increment is larger for the worst case workload. For instance, the variation for the worst case workload increases from 6.7% for 45nm to 12.0% for 16nm, while for the best case the increment is 1.8% for 45nm and 3.6% for 16nm only.

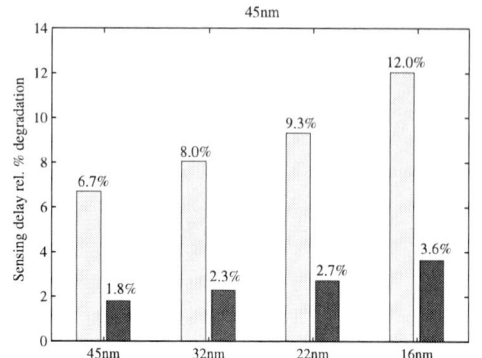

Fig. 7. BTI impact on Sensing delay for all technology nodes.

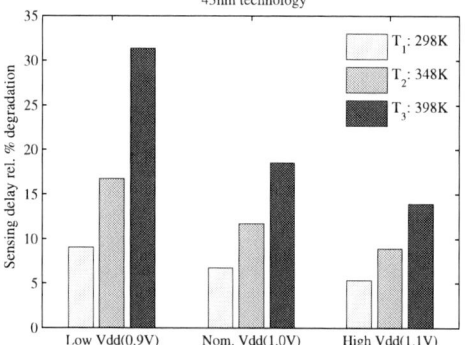

Fig. 8. Worst case sensing delay for supply voltage and temperature variations.

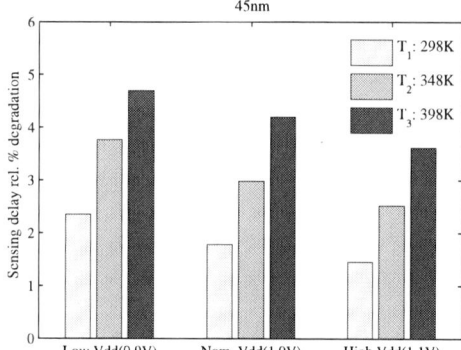

Fig. 9. Best case sensing delay for supply voltage and temperature variations.

D. Supply Voltage and Temperature Dependency

Conventionally, supply voltage and temperature fluctuations impact the operating condition of MOS transistors. Supply voltage variations in a transistor can impact the sensing delay significantly as it impacts the operational speed. In addition, variation in supply voltage also affects the oxide field (capacitance) and subsequently the BTI impact (C_{ox} in Eqn. 3). The analysis of the supply voltage variation is performed on two workloads, (i.e., worst-case (S3) and best-case (S2)) for 45nm technology. The supply voltage is varied between -10% and +10% of nominal V_{dd}, i.e., between 0.9V and 1.1V. Figures 8 and 9 depict the BTI induced sensing delay for various supply voltages and temperatures for the worst-case and best-case workloads, respectively. The figures show for different voltages, i.e., low V_{dd} (Low V_{dd} = 0.9V), nominal V_{dd} (Nom. V_{dd} = 1.0V), and high V_{dd} (High V_{dd} = 0.9V) and temperatures (i.e., $T_1 = 298$K, $T_2 = 348$K, and $T_3 = 398$K) the impact on the sensing delay (vertical axis). From the figures we conclude the following:

- Increasing the temperature leads to a higher BTI induced degradation. For example, in Figure 8, for a fixed supply voltage, e.g., $V_{dd} = 1.1$V, the BTI induce degradation is 5.3% for T_1, and while 8.9% and 13.9% for T_2 and T_3, respectively. In addition to that, the same trend is observed at $V_{dd} = 1.0$V and $V_{dd} = 0.9$V. In Figure 9, for considering $V_{dd} = 1.1$V as a fixed reference, the relative sensing delay degradation becomes 1.5% for T_1 while 2.5% and 3.6% for T_2 and T_3, respectively.

- Increasing the supply voltage reduces the sensing delay degradation. For instance, in Figure 8, for a fixed temperature, e.g., T_1, the BTI induced degradation on the sensing delay at 10^8s is 9.0% for low V_{dd}, while 6.7%, and 5.3% for nominal V_{dd} and high V_{dd}, respectively. Moreover, the same trend is observed at T_2 and T_3. In Figure 9, for fixed T_1, the relative sensing delay degradation is 2.3% for $V_{dd} = 0.9$V, while 1.8% and 1.5% for $V_{dd} = 1.0$V and $V_{dd} = 1.1$V, respectively.

- The performance degradation is much higher for the worst-case workload (S3) as compared to the best-case workload (S2). For instance, in Figure 8, at low V_{dd} and $T_3 = 398$K, the sensing degradation is 31.3% while in Figure 9, for the same voltage and temperature, the BTI induced degradation is only 4.7%.

In conclusion, based on the experiments, it is extremely important for designers to include proper design margins to guarantee the life-time operation of the considered SA circuitry under given operating conditions (such as voltage and temperature). In order to design a reliable memory system, designers need to understand the degradation of each sub-component. Based on such information, monitoring and mitigation schemes might be considered for SAs. For example, a sensing delay monitoring circuit could be used as sensor to adaptively control the supply voltage. Another approach could focus on redesigning more robust SA, by increasing the drive strength (or device width) of critical transistors [29].

V. CONCLUSION

This paper investigated the combined impact of Bias Temperature Instability (BTI), voltage and temperature variation and different workloads on the standard latch type memory sense amplifier for different technologies. The results show that the sensing delay degradation is strongly workload dependent and reaches up to 12.8%. Both the scaling and increase in temperature severely impact the BTI degradation. Increasing the supply voltage reduces the BTI induced degradation leading to more reliable and robust sense amplifiers, but at the cost of a higher power consumption. Optimizing a reliable memory system requires the consideration of all its sub-parts, i.e., their degradation rate depends on the application (e.g. workload, temperature, etc); for instance, the aging rate of the sense amplifier may differ from that of the memory array and other peripheral circuits.

REFERENCES

[1] ITRS, "International Technology Roadmap for Semiconductor 2004" "www.itrs.net/common/2004 update/2004update.htm.".

[2] S. Borkar, et al "Micro architecture and Design Challenges for Giga scale Integration", *Pro. of Intl. Sympos. Micro architecture*, 2004.

[3] S. Hamdioui et al, "Reliability Challenges of Real-Time Systems in Forthcoming Technology Nodes", *DATE*, 2013.

[4] B.C. Paul et al, "Impact of NBTI on the Temporal Performance Degradation of Digital Circuits", *IEEE Electron Device Letter*, Vol. 26, No.8, Aug. 2005.

[5] M. A. Alam et al, "A Comprehensive Model of PMOS NBTI Degradation", *Microelectronics Reliability* , Vol:45, 2005.

[6] D. Varghese, et al, "On the Dispersive versus Arrhenius Temperature Activation of NBTI Time Evolution in Plasma Nitrided Gate Oxides: Measurements, Theory, and Implications", *IEDM*, Dec. 2005, pp. 1-4.

[7] K. Kang, et al, "Estimation of Statistical Variation in Temporal NBTI Degradation and Its Impact on Lifetime Circuit Performance", *ICCD*, 2005, pp. 730-734.

[8] R. Wang, et al., Threshold Voltage Variation with Temperature in MOS Transistors, *IEEE Transaction on Electron Devices*, pp: 386- 388, 1971.

[9] S. Sapatnekar, et al., Overcoming Variatins in Nano-scale Technologies, IEEE Transaction on Emerging and Selected Topics in Circuits and Systems, pp: 5-18, 2011.

[10] B. Kackzar, et al., "Disorder-Controlled-Kinetics Model NBTI and its Experimental Verification", *IPRS* , pp: 381-387, 2005.

[11] S. Zafar, et al, "A comparative study of NBTI and PBTI in SiO2/HfO2 stacks with FUSI, TiN gates", *Pro. of VLSI Technology symp.*, 2006.

[12] P. Pouyan, et al, "Process Variability-Aware Proactive Reconfiguration Technique for Mitigating Aging effects in Nano Scale SRAM lifetime", *IEEE 30th VLSI Test Symposium.*, 2012.

[13] D. Rodopoulos, et al, "Time and Workload Dependent Device Variability in Circuit Simulations" *Proc. Intl. Conf on IC Design and Technology*, pp: 1-4, 2011.

[14] M. T. Luque, et al, "From Mean Values to Distribution of BTI Lifetime of Deeply scaled FETs through Atomistic Understanding of the Degradation" *Sym. on VLSI Technology*, pp: 152-153, 2011.

[15] T. Sakurai, "Alpha-Power law MOSFET model and its applications to CMOS delay and other formulas", *IEEE JSSC*, Vol.25, No.2, April 1990,

[16] A. T. Krishnan, et al., "NBTI impact on transistor and circuit: Models, mechanisms and scaling effects", *IEDM*, 2003.

[17] B. Cheng, A. R. Brown, "Impact of NBTI/PBTI on SRAM Stability Degradation", *IEEE ELECTRON DEVICES LETTERS*, 2011.

[18] S. Kumar et al, "Impact of NBTI on SRAM Read Stability and Design for Reliability", *ISQED*, pp: 212-128, 2006.

[19] A. Carlson, "Mechanism of Increase in SRAM VMIN Due to Negative-Bias Temperature Instability", *IEEE TDMR*, 2007.

[20] A. Bansal et al, "Impact of NBTI and PBTI on SRAM static/dynamic noise margins and cell failure probability", *JMR*, 2009.

[21] S. Khan, et al, "Impact of Partial Resistive Defects and Bias Temperature Defects and Bias Temperature Instability on SRAM Decoder Reliability", *Pro. of 7th IDT*, 2012.

[22] D. Rodopoulos, et al, "Atomistic Pseudo-Transient BTI Simulation with Inherent Workload Memory", *IEEE TDMR*, June 2014.

[23] S. Khan, et al., "Bias Temperature Instability Analysis of FinFET based SRAM cells", *DATE*, 2014.

[24] R. Menchaca et al, "Impact of Transistor Aging Effects on Sense Amplifier Reliability in Nano-Scale CMOS", *13th ISQED*, 2012.

[25] I. Agbo et al, "BTI Impact on SRAM Sense Amplifier", *8th IDT*, 16-18 Dec. 2013.

[26] H. Kukner et al, "Comparison of Reaction-Diffusion and Atomistic Trap-based Models for Logic Gates", *IEEE TDMR*, 2013.

[27] V. Chandra, R. Aitken, "Impact of Voltage Scaling on Nanoscale SRAM Reliability", *DATE*, 2009.

[28] S. Hamdioui, "Testing Static Random Access Memories: Defects, Fault Models and Test Patterns", *Kluwer Academic Press Publishers, The Netherlands*, 2004.

[29] H. Kukner et al, "BTI reliability from Planar to FinFET nodes: Will the next node be more or less reliable", *MEDIAN*, 2014.

[30] S. Cosemans, "Variability-aware design of low power SRAM memories", *Ph.D Thesis Katholieke Universiteit Leuven*, 2009.

[31] Predictive Technology Model "http://ptm.asu.edu/".

978-1-4799-7598-3/15 $31.00 © 2015 IEEE

2015 IEEE 33rd VLSI Test Symposium (VTS)

A Robust Digital Sensor IP and Sensor Insertion Flow for In-Situ Path Timing Slack Monitoring in SoCs

M. Sadi [1], L. Winemberg [2] and M. Tehranipoor [1]

[1]Dept. of Electrical & Computer Engineering, University of Connecticut, Storrs, USA

[2]Freescale Semiconductor, Austin, TX, USA

Emails: mehdi.sadi@engr.uconn.edu, [2]leroyw@freescale.com, tehrani@engr.uconn.edu

ABSTRACT

Because of process variations, the post-silicon critical or near-critical paths differ from those identified in the pre-silicon stage. Thus, it has become necessary to extract timing slack information from circuit paths in the post-silicon phase. In this paper, we present a robust digital sensor IP for in-situ timing slack monitoring on actual circuit paths from SoCs. The timing slack data is converted into a digital format and stored in a dedicated scan register chain for easy extraction at any point in time during test and functional modes. A novel layout-aware and netlist-level sensor insertion flow is proposed. The sensor IP has been designed with 32/28nm standard cell library and its performance is demonstrated in the physical design of several benchmark circuits.

1 INTRODUCTION

With aggressive technology scaling the transistor density per unit chip area has increased significantly over the past few years. This paved the way for many-core processors and highly integrated System-on-Chips (SoCs). The increased variability in transistor parameters and workload dependent fluctuations in operating conditions have made harnessing the full benefits of scaling a challenging task [1] [2]. Variations can be categorized into three major classes - 1) One-time static process variations due to manufacturing imperfections that cause transistor and interconnect parameters to drift from their designed values. 2) Run-time dynamic variations - power supply noise and temperature fluctuations - resulting shifts in operating conditions. 3) Aging variations, induced by stress from prolonged operations, that cause parametric degradation over time [2]. In terms of spatial locations, the variation can be segmented into die-to-die and within-die components. Variations impact transistor length, width, threshold voltage, oxide thickness etc., all of which directly contribute to path delay fluctuations. As a result, a certain path may show significant discrepancy in the delay between pre- and post-fabrication stages [3][4] and the actual speed limiting paths might be masked in the simulation phase [5]. Further, the workload-dependent aging variations can cause an initially non-critical path to be critical anytime in the lifetime of the chip [6].

Significant amount of research effort is being spent on ensuring that the flip-flops/latches located at path ends capture the correct data despite fluctuations in timing margins due to Process, Voltage and Temperature (PVT) variations, and, Bias Temperature Instability (BTI) and Hot Carrier Injection (HCI) induced dynamic aging effects. There are two approaches towards this goal. The first method almost eliminates the guard-band or timing slack and uses error correction methods to recover from timing failures [7]-[9]. Razor II flip-flops use in-situ detection and architectural correction against variation-induced delay errors [7][8]. In [9] error detection sequential circuits are used to replay failing instructions at lower clock frequency to guarantee correct functionality. The second method assigns a conservative guard-band with the nominal path delay as a safety margin against variations and aging [10][11]. The timing slack or guard-band is defined as the delay, between the data arrival time and active edge of the clock, minus the flip-flop setup time. The guard-band assignment is done at the pre-silicon design phase based on the target clock frequency. In [7]-[9] timing error correction comes at the expense of greater cycles per instruction (CPI) overhead and these proprietary circuits, that require architectural replay for correct operation, may not be applicable to general purpose SoCs. As a result, majority of SoCs still assign conservative timing slacks at the path ends as guard-bands against timing uncertainties. As the guard-band is estimated based on the worst-case design corner, it is often pessimistic. In [10] the authors reported experimental data to corroborate the fact that guard-band reduction improves performance in terms of area and power. Variations and workload-dependent aging complicates pre-silicon fixed guard-band assignment. As a result, optimization of guard-band has become a significant research problem. Hierarchically focused and application adaptive [11] guard-banding techniques have been proposed that will utilize the on-chip sensor data to adjust the path slack at run-time. Light-weight on-chip structures that can acquire accurate timing slack information from critical and near-critical paths in the production test, in Built-in-Self-Test (BIST) and in in-field functional mode can aid in optimization of guard-band for lifetime reliability.

Online monitoring of path slack is necessary at initial speed binning phase as well as in run-time for reliable operation of the chip. The monitoring sensor circuits can be broadly classified into two groups - in-situ [12]-[17] and replica [18]. In-situ monitors inspect on actual circuit paths at the expense of adding minimal capacitive load to the path under test. On the contrary, replica circuits offer non-intrusive monitoring, but they may fail to detect PVT variations and aging that are local to the actual circuit path. In [15] the authors use a canary flip-flop that generates a warning signal on impending path delay failure to activate adaptive body biasing techniques. An all-digital self-calibrating slack measurement architecture that utilizes a Vernier chain based time-to-digital converter is reported in [14]. Although the system offers a very high resolution of 5ps, it incurs high area overhead and requires complex calibration. [12] proposes a method to monitor path slacks for impending failure by probing intermediate nodes. This strategy requires a critical control of path-dependent delay margins representing a group of paths. In [17] an on-chip scan-based method for quantifying path delay using special signature registers and variable clock generator is proposed. The selected paths are required to be single path sensitizable. The critical path monitor (CPM) sensors of Power7+ constitute a mixture of synthesized paths which are expected to mimic all the actual critical paths in the design [19]. The CPM sensor data are used to control the DPLL for dynamic frequency adjustment.

In addition to the concerns of variations, aging (BTI, HCI etc.) induced guard-band degradation and consequent timing failure is dependent on the dynamic workload, and hence, cannot be accurately predicted at the post-fabrication design phase [6][18]. In [20] the circuit path is transformed into a ring oscillator to convert the path delay into a frequency. This frequency is compared against a reference clock to quantify the delay degradation due to aging. This method offers a high resolution at the expense of considerable performance and area overhead.

978-1-4799-7598-3/15 $31.00 © 2015 IEEE   216

The shortcomings of existing slack monitors, however, are: (i) Most of the sensors act on a timing failure to activate error recovery mechanisms through architectural replay rather than quantifying the guard-band continuously in production test as well as in functional mode. (ii) The existing sensors incur high area overhead that deters them from deployment in a large scale at the end of many critical paths across the layout. (iii) Most of the reported sensors do not accurately handle the spurious glitches in the circuit.

In light of the above, our objectives and contributions in this work are:

- Designing a compact, robust and fully-digital embedded sensor IP that can monitor slack of actual critical or near-critical paths in-situ throughout the manufacturing test, system bring-up, BIST, and lifetime of the SoC in the field.
- Among the many paths that terminate on a certain capture flip-flop, some are shorter or less-critical than others. As these paths can be activated in any order depending on the workload, the recorded timing slack of the longest or most-critical path might be masked by the slack of a subsequently activated less-critical path. A circuit technique is proposed to avoid this masking effect.
- Easy storage and extraction of sensor results using scan-based JTAG or iJTAG features.
- Development of a novel layout-aware and netlist-level sensor insertion flow to insert sensor IPs in the synthesized netlist and physically place those at strategic locations across the layout.

The rest of the paper is organized as follows. In Section 2 the architecture of the sensor IP is presented. Section 3 describes the various features and modes of the sensor. Sensor insertion algorithm is stated in Section 4. Experimental results are presented in Section 5 followed by conclusions in Section 6.

2 THE SENSOR ARCHITECTURE

The circuit diagram of the proposed sensor IP is depicted in Fig. 1. When activated, the sensor monitors and records the worst-case timing slack at the capture flip-flop for all the paths terminating at that particular flip-flop. The sensor probes path ending capture flip-flop's D and Q ports through minimum size buffers. The small size of the buffers ensure that minimum amount of load capacitance is added to the main circuit path, thus impacting the path delay minimally. The components of the sensor and their task are briefly described below.

2.1 Rise Transition Detector

The flip-flop inside the rise transition detector stores the data sampled by the capture flip-flop in the previous CLK cycle. This stored value is compared to the data sampled at the capture flip-flop by the current CLK edge. A pulse is generated at the ACT output in the event that the current data is logic one and previous a logic zero, indicating a rising transition in the incoming data. The latency between the active edge of the CLK and the pulse at the ACT is the sum of the propagation delays of a buffer and an AND gate. This small latency is irrelevant since the purpose of the ACT pulse is to latch the states from the Monitor Unit (MU) flip-flops to the Capture and Result Storage Unit's (CRSU) scan flip-flop chain.

2.2 Delay Line and Monitor Unit

The delay line comprises of a chain of minimum size buffers. The number of required buffer stages depend on the desired slack detection range. A flip-flop is attached at the end of each buffer to capture the states at the active CLK edge. The combination of delay line and the flip-flops in the Monitor Unit (MU) convert the timing slack into a corresponding digital data.

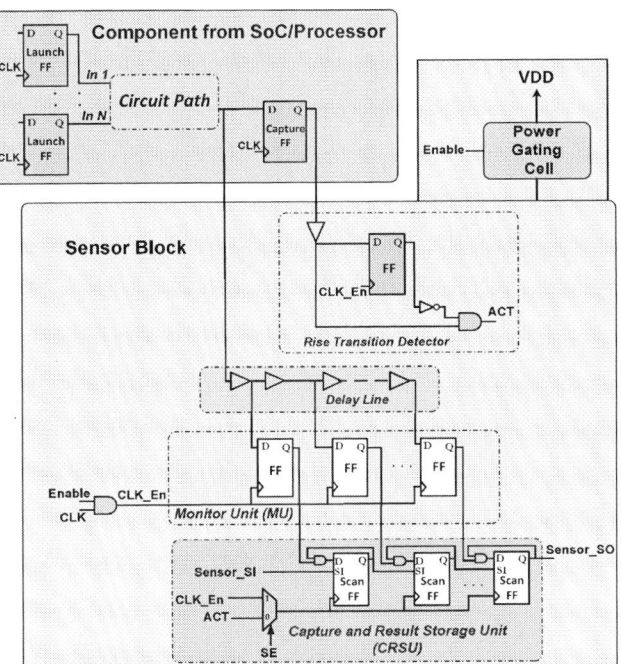

Fig. 1. The sensor architecture.

2.3 Capture and Result Storage Unit

Each flip-flop in the scan chain of CRSU samples the corresponding MU flip-flop's output. The flip-flops in this unit are clocked by the ACT pulse rather than the system CLK for achieving glitch immunity. The AND gate in front of each scan flip-flop make it sticky, in the sense that if it ever records a zero, it will hold on to the zero until reset or it is scan loaded with logic one. At the initialization stage all the scan flip-flops are connected in a scan chain by asserting the Scan Enable (SE) signal and a logic one is loaded in each of the flip-flops. During the slack evaluation phase SE is set to zero and the flip-flops inside CRSU record the worst-case slack observed at the monitored capture flip-flop. The sensor data extraction process is initiated by setting SE pin to logic high which connects the flip-flops in a scan chain clocked by the main clock and finally the stored slack data is extracted through the Senor_SO pin. Any meta-stability in the storage flip-flops is automatically resolved in our architecture as we are using two flip-flops in series - forming a synchronizer circuit [21] - with the output of each buffer in the delay line.

3 FEATURES AND MODES OF OPERATION

The sensor IP is designed to be immune against spurious glitches that occur before the data settles down. The immunity is achieved as a result of using two layers of sampling flip-flops, one in the MU and the other in the CRSU. For example, if there is a narrow rising glitch before the data settles to a stable high value, the flip-flops in the MU will respond to the glitch as well as the stable data that arrives after the glitch. But the flip-flops in the CRSU will only respond to the stable data, because, these flip-flops are clocked by the ACT trigger pulse which is generated by sampling the capture flip-flop's stable response. This glitch immunity is demonstrated in Fig. 2, where a narrow glitch had occurred between 7.6n to 7.8n seconds before the data settled to a stable value at 7.9n second. The ACT pulse was generated after the rising CLK edge had appeared at 8n second, this effectively masked the glitch from interfering with slack results.

Power and clock gating techniques used in the sensor IP make it robust against aging effects. The gating architecture also reduces energy consumption of the sensor block in the field.

978-1-4799-7598-3/15 $31.00 © 2015 IEEE   217

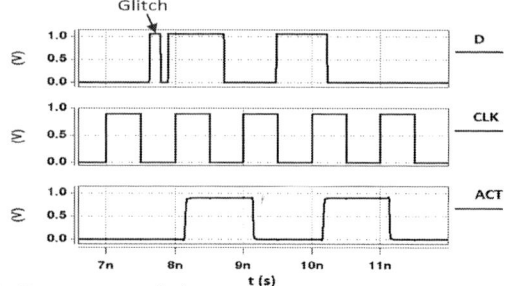

Fig. 2. Response to glitch

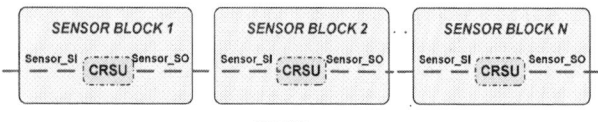

Fig. 3. Multiple sensors connected in scan chain

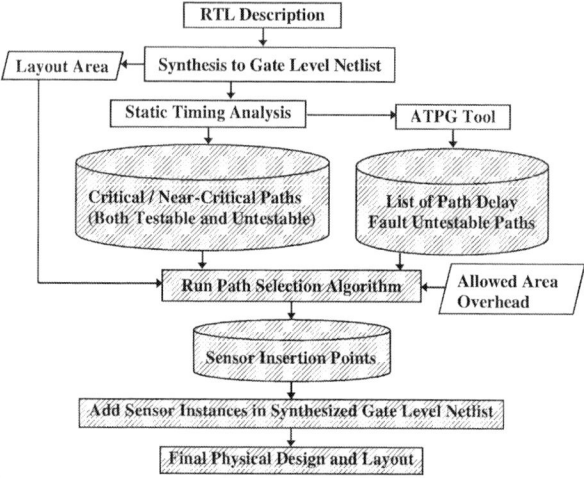

Fig. 4. Sensor insertion flow

Since Vernier delay line based Time-to-Digital Converters (TDC) are most often used for timing slack digitization, we compared our proposed architecture with it. Vernier TDC might offer a very high resolution, but at the expense of complex calibration with delay locked loops, and they also suffer from dead zone issues [14]. On the contrary, our proposed sensor IP's resolution is limited by a minimum size buffer delay. It does not require complex calibration steps and hence is area efficient.

For slack data extraction from the multiple instances of the sensor IP placed in the layout, the flip-flops in each sensor's CSRU will be connected in a long scan chain as depicted in Fig. 3. The data stored in the scan chain can be recovered using JTAG or iJTAG based test data access features.

At post-silicon validation and production test modes multiple test patterns are applied to exercise the critical or near-critical paths. The embedded sensor IPs record the worst-case timing slacks at the monitored capture flip-flops. The effect of noise on timing integrity can be quantified at this stage. During in-field functional mode the sensors would be activated when required and workloads would be run to detect timing slack degradation considering noise and aging effects.

4 SENSOR INSERTION FLOW

In modern SoCs the number of near-critical paths are extensive. As a result it is not practical to place sensor monitors at the end of each of these paths. Our proposed sensor insertion flow, shown in Fig. 4, addresses this concern. The flow starts with the synthesis of hardware from RTL to the gate level netlist. At this stage, based on the the standard cell library used in synthesis and the gate-level netlist, an estimate of the layout area is obtained from the synthesis CAD tool. After that, static timing analysis is performed on the synthesized netlist and critical/near-critical paths (considering both single and multi-cycle paths) are sorted in order of their respective path slacks. At this stage the netlist is also taken to the Automatic Test Pattern Generator (ATPG) tool to identify those paths that are not testable for Path Delay Fault (PDF) at the current design specifications. The rationale behind the PDF coverage analysis is that, those paths that are not ATPG testable, and, at the same time critical/near-critical, are more important to be monitored by sensors than those paths that are PDF testable. Next the Path Selection () algorithm is executed. The algorithm takes as input the sorted critical/near critical path list, PDF test ability information, an estimated area-overhead - which can be 2% to 5% of the main design depending on area-overhead budget - and

outputs the senor insertion points located inside the synthesized netlist of the SoC . In Algorithm 1, Lines 1 to 4 calculate the total number of sensors allowed for the given area-overhead. Lines 5 to 6 identify logical module-aware unique capture flip-flop sets. Since one or more paths can terminate on the same capture flip-flop, only the unique capture flip-flops are identified. In Lines 7 to 10, the total sensors are distributed among the logical modules in the ratio of the modules' share of the total critical/near-critical paths. Lastly, in Lines 11 to 13, the algorithm reports capture flip-flop nodes in the netlist where the sensors will be inserted. The algorithm ensures that the sensors cover a wide range of paths and those paths are spatially distributed across the layout. Spatial distribution of the sensors allow monitoring PVT variation effects on the timing critical paths. Finally, the sensor instances are added inside the synthesized netlist of the SoC at those identified insertion points and the physical design is completed.

Algorithm 1 Path Selection Algorithm

1: Specify area of the sensor module, $Area_{Sensor}$
2: Specify estimated total area of the chip, $Area_{Chip}$
3: Specify allowed area-overhead, $Area_{Overhead}$
4: Total sensors, $Total_{Sensor} = (Area_{Chip} * Area_{Overhead})/Area_{Sensor}$
5: Read static timing analysis report and create a set of unique capture flip-flops where one or more paths terminate
6: Create a path group for each of the logical modules in the netlist and assign the respective capture flip-flops into that path group
7: **for** (each path group k) **do**
8:     Identify the number of critical/near-critical paths, N_k
9:     Identify contribution to the total number of critical or near-critical paths, $P_k = N_k / \sum N_k$;
10:     Allotted sensor for each path group, $S_k = P_k * Total_{Sensor}$
11:     Select top S_k critical or near-critical paths. ATPG Untestable path is given priority when two paths have the same slack
12:     Report the list of capture flip-flops for the selected S_k paths
13: **end for**
       Report the list of sensor insertion points in netlist for each path group k

---

5 EXPERIMENTAL RESULTS

The sensor IP was written in Verilog and synthesized with Design Compiler using Synopsys 28/32nm standard cell library [25]. Each sensor occupied a layout area of $136\mu m^2$. A post layout SPICE netlist was extracted and simulated with HSPICE [25]. The sensor calibration

TABLE 1
CALIBRATION RESULTS FOR 28/32nm STANDARD CELL LIBRARY

Slack (ps)	Nominal VDD 1.05 V			95% VDD 0.9975 V			90% VDD 0.9450 V		
	25°C	50°C	75°C	25°C	50°C	75°C	25°C	50°C	75°C
120	1 1 1 1 1 0	1 1 1 1 1 0	1 1 1 1 1 0	1 1 1 1 1 0	1 1 1 1 1 0	1 1 1 1 1 0	1 1 1 1 0 0	1 1 1 1 0 0	1 1 1 1 0 0
100	1 1 1 1 0 0	1 1 1 1 0 0	1 1 1 1 0 0	1 1 1 1 0 0	1 1 1 1 0 0	1 1 1 1 0 0	1 1 1 0 0 0	1 1 1 0 0 0	1 1 1 0 0 0
80	1 1 1 0 0 0	1 1 1 0 0 0	1 1 1 0 0 0	1 1 1 0 0 0	1 1 1 0 0 0	1 1 1 0 0 0	1 1 0 0 0 0	1 1 0 0 0 0	1 1 0 0 0 0
60	1 1 0 0 0 0	1 1 0 0 0 0	1 1 0 0 0 0	1 1 0 0 0 0	1 1 0 0 0 0	1 1 0 0 0 0	1 1 0 0 0 0	1 1 0 0 0 0	1 1 0 0 0 0
40	1 0 0 0 0 0	1 0 0 0 0 0	1 0 0 0 0 0	1 0 0 0 0 0	1 0 0 0 0 0	1 0 0 0 0 0	1 0 0 0 0 0	1 0 0 0 0 0	1 0 0 0 0 0
20	0 0 0 0 0 0	0 0 0 0 0 0	0 0 0 0 0 0	0 0 0 0 0 0	0 0 0 0 0 0	0 0 0 0 0 0	0 0 0 0 0 0	0 0 0 0 0 0	0 0 0 0 0 0

results for 28/32nm standard cell library are given in Table 1, where Column 1 reports the slack obtained by subtracting the flip-flop setup time from the delay between the data arrival time and the active clock edge. Columns 2 to 4 show the effects of temperature variations on the sensor response at nominal VDD. Since only 6 buffers were used in the delay line, it was observed - for the selected 28/32nm technology library - that the sensor response was invariant of any shift in temperature within 25°C to 75°C, the typical operating temperature range of chips. As mentioned in Section 2, the sensors are initialized with a sensor reading of 111111, before the path slack measurements commence. At the upper detection limit the sensor code is 111110 and as the slack decreases, the sensor code changes from 111110 to 111100, then to 111000 and eventually to the lower detection limit at 000000, where each zero represents slack reduction by an amount equal to the delay of a buffer. Hence, the sensor reading of 111111 and all other states not mentioned in the calibration data of Table 1, indicate that the monitored slack was either negative or beyond the detection range. The slack detection range is determined by the number of buffer stages in the delay line. The detection resolution is limited by the delay of a unit buffer. The resolution at our selected 28/32nm standard cell library is 20ps. The expected resolution at 14nm and 22nm nodes are reported in Table 2 considering minimum size buffers.

TABLE 2
SENSOR RESOLUTION

Technology	Sensor Resolution
28/32nm [25]	20ps
20nm [26]	10ps
14nm [26]	6ps

Since the speed of the buffers, used in the delay line, is sensitive to power supply noise and voltage droop, we analyzed the power supply variation sensitivity of the sensor IP and tabulated the results in Table 1. The effect of voltage droop - slowing down of the sensor as shown for the 90% VDD case in Columns 8 to 10 - can be easily decoupled from slack data with the aid of power supply noise sensors such as in [22][23].

As the sensor modules are power gated, the impact of aging degradation are minimal. To assess the impact of process variations on the sensitivity of the sensor we conducted a 100 sample Monte Carlo simulation on the sensor IP. The selected parameters to vary from nominal were transistor length (L) by 15%, width (W) by 15%, threshold voltage (Vth) by 10% and oxide thickness (Tox) by 5%. All simulations were done within the statistical range of 3 standard deviations (3σ). The Monte Carlo simulation results are shown in Fig. 5, where X-axis depicts calibration points (timing slacks) and Y-axis shows sample counts. It was observed that out of 100 samples, most of the responses matched the expected calibration results as in Table 1, in rest of the cases the sensor response was either slower or faster by an unit buffer delay - as denoted by 'Slower than expected' and 'Faster than expected', respectively, in Fig 5. These results imply that worst-case observation error is limited to one buffer delay.

For implementing our sensor insertion flow presented in Section 4, we selected two circuits - the Execution Unit (EXU) and the Floating Point and Graphics Unit (FGU) from the OpenSPARCT2 SoC's SPARC core [27]. The circuits were synthesized to gate level netlist with 28/32nm

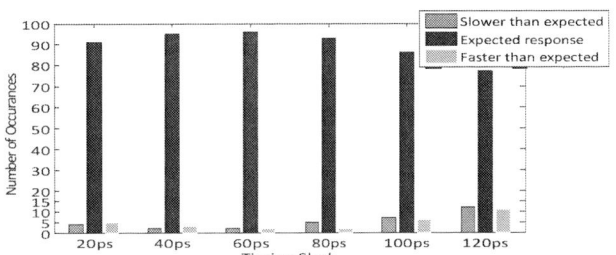

Fig. 5. Monte Carlo process variation simulation results on sensor IP at nominal VDD

standard cell library using Design Compiler [25]. For both of these circuits, from an initial post synthesis timing analysis with PrimeTime [25], the list of all the timing paths sorted in the descending order of nominal delay were obtained. Since one or more paths terminated on each capture flip-flop, the paths were grouped according to the unique capture flip-flops where they terminated and the circuit modules which contained the flip-flops. All the unique capture flip-flops were tested with ATPG tool Tetramax [25] for their Path Delay Fault (PDF) testability (without adding any extra test point). The layout areas of EXU and FGU were estimated from the synthesis tool [25]. Since the register file macro modules of EXU and FGU are supplied as hard macro blocks, in calculating the allowed area-overhead we only considered the areas pertaining to the standard cells. The statistics from the sensor insertion flow are shown in Column 2 for EXU and in Column 3 for FGU in Table 3. The area-overhead budget was set to 5% of the area of the main circuit layout as an estimate. Based on the area-budget and the area of each sensor IP, 28 and 75 capture flip-flops were selected, respectively, for EXU and FGU modules for sensor insertion in Table 3. As described in Section 4, to ensure that the sensors were distributed spatially across the layout, the sensors are placed evenly across the different logical modules of the netlist. In both Tables 4 and 5, Column 1 reports the different logical modules, Column 2 reports the number of unique capture flip-flops in each logical module, Column 5 depicts the ATPG testability (without adding any extra test point) for PDF and Column 6 presents the number of allotted slack sensor for each module in EXU and FGU blocks, respectively. The cut-off margin in Column 3 of Tables 4 and 5 were set as 50% of the maximum slack as an estimate and can be adjusted if required.

After finalizing the candidate flip-flops for sensor insertion, the sensor netlist was added to those points inside the synthesized gate level netlist of the EXU and FGU modules. The netlist-level sensor insertion flow was automated with Perl scripts. Finally, the full physical designs of the EXU and FGU, with embedded sensor network, were completed with IC Compiler [25]. Fig. 6 shows the sensor inserted layouts. From Fig. 6(b) and 6(d), it can be observed that the sensor blocks are distributed across the layout as those were placed inside the different logical modules of the main netlist according to our proposed flow.

The sensor IP is designed in a way that it monitors the delays of all the paths ending at the target capture flip-flop and records the worst-case slack due to the most-critical of the paths. We extracted the detail transistor-level netlist of the most-critical paths monitored by each sensor and simulated those with HSPICE [25] at t=0 and t=3

TABLE 3
LAYOUT STATISTICS

	Execution Unit (EXU) of OpenSPARCT2	Floating Point and Graphics Unit (FGU) of OpenSPARCT2
Cell count	15200	58802
Flip-flop count	1662	7514
Estimated layout area (non macro)	76711 μm^2	204304 μm^2
5% area overhaed	3835 μm^2	10215 μm^2
Each sensor area	136 μm^2	136 μm^2
Sensor count	28	75

TABLE 4
SENSOR DISTRIBUTION IN EXU OF OPENSPARCT2

Logical Module Name	Number of Unique Capture Flip-flops	Flip-flops with Worst Slack Below Cut-off		Number of ATPG Untestable Flip-flops	Number of Allotted Slack Sensors
		Number	Percent of Total		
ECC	32	16	1.8	13	1
ECT	196	53	6.12	10	2
EDP	726	528	60.96	185	17
RML	391	269	31.06	27	8

TABLE 5
SENSOR DISTRIBUTION IN FGU OF OPENSPARCT2

Logical Module Name	Number of Unique Capture Flip-flops	Flip-flops with Worst Slack Below Cut-off		Number of ATPG Untestable Flip-flops	Number of Allotted Slack Sensors
		Number	Percent of Total		
FAC	495	161	2.98	103	3
FAD	693	590	10.9	309	9
FDC	76	61	1.13	21	1
FDD	1040	740	19.29	340	14
FEC	84	33	0.61	2	1
FGD	1102	709	13.15	109	9
FIC	65	49	0.9	14	1
FPC	699	271	5	84	4
FPE	97	67	1.24	28	2
FPF	731	506	9.38	111	7
FPY	1931	1902	35.29	425	27

years to account for aging effects. We performed a 100 point Monte Carlo simulation by varying transistor W and L by 15%, Vth by 10% and Tox by 5% of the nominal to mimic chip-to-chip variations. The results of the top 10 sensor-monitored paths of EXU are presented in Table 6. For aging (t = 3 years), BTI and HCI were considered using HSPICE MOSRA [25]. In terms of nominal delay, Path 1 is the most critical or speed-limiting followed by Path 2, Path 3 and so on. But if we consider the impact of process variations on the chip-to-chip path delay distribution - as revealed by the statistical 3σ ranges - any of these 10 paths might be the most critical or speed-limiting after fabrication. For example, the nominal delay at t=0, as identified in the pre-silicon stage timing analysis, of Path 10 is 51ps less than that of Path 1, making Path 10 non-critical. But for some chips Path 10 might exhibit a delay of 620ps, whereas Path 1 exhibited 539ps, making Path 10 the critical speed-limiting path. Similar scenario can be observed for in-field functional mode at t = 3 years in Columns 6 to 9 in Table 6 . These results reiterate the fact that because of process variations it is hard to pinpoint the actual speed limiting critical paths at the pre-silicon design phase [24]. For these type of paths, the post layout path slack can be extracted with the aid of our proposed sensor IP for different samples of the same chip.

Next we evaluated the slack measurement errors stemming from the use of the sensor IP. For this purpose, we extracted the transistor-

level netlist of each sensor, the corresponding capture flip-flop and the monitored most-critical path for that particular flip-flop. The results are demonstrated in Table 7 for the top 10 sensors from the EXU module at t=0 and t = 3 years. The CLK period was set to 750ps and $t_{CLK-to-Q}$ for flip-flops were observed to be 80ps. All the paths are single cycle. Form HSPICE [25] simulations, with accuracy level set to maximum, the actual path slacks were recorded by subtracting the flip-flop setup-time from the delay between data arrival time and the respective capture CLK edge as reported in Column 2. Column 3 shows the extracted codes from the respective slack sensor modules. Using the calibration results from Table 1, the sensor codes were converted to equivalent timing slacks in Column 4. Finally the measurement errors were estimated by taking the absolute difference between the path slack measured directly from SPICE and that obtained from the sensor readings. As shown in Columns 5 and 9 of Table 7, the measurement errors were observed to be within the delay of a unit buffer as expected. The results in Columns 6 to 9 of Table 7 indicate that the sensors can capture aging induced path delay degradation in functional mode.

To assess the impact of chip-to-chip process variations, we chose a particular path-sensor combination and performed a 100 point Monte Carlo simulation on it. The distribution of slack extracted from the sensor readings for Path 5 are shown in Fig. 7(a). This observation implies that it is possible to extract local and global process variation profiles using multiple sensor data form different chips. The measurement errors for each of the Monte Carlo runs are depicted in Fig. 7 (b). In some samples the measurement errors exceeded the unit buffer delay of 20ps. This is because in Monte Carlo simulation the process variations impacted both the circuit path and the sensor itself.

Since circuit delay is a strong function of supply voltage, path slack deteriorates significantly with voltage droop and power supply noise. The power supply droop effect on Path 10 is reported in Table 8. With

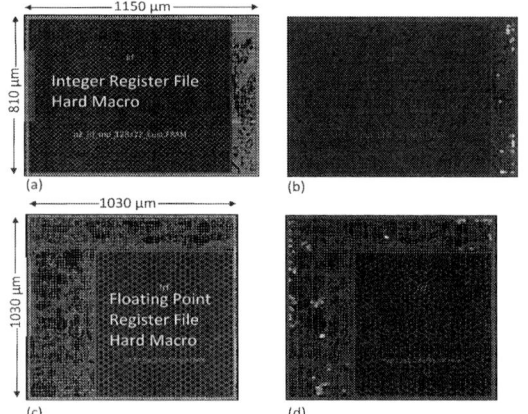

Fig. 6. (a) The layout of EXU with embedded sensor network. (c) The layout of FGU with embedded sensor network. (b),(d) Cells of sensor modules are highlighted.

TABLE 6
MONTE CARLO SIMULATION RESULTS FOR DELAY VARIATIONS OF THE TOP 10 MONITORED PATHS OF EXU

Path ID	Path Delay at t = 0 (No Aging)				Path Delay at t = 3 years (Aged)			
	Mean μ (ps)	σ (ps)	3σ Range		Mean μ (ps)	σ (ps)	3σ Range	
			μ-3σ (ps)	μ+3σ (ps)			μ-3σ (ps)	μ+3σ (ps)
1	605	22	539	671	643	22	577	709
2	590	21	527	653	620	23	551	689
3	585	21	522	648	615	24	543	687
4	580	19	523	637	610	22	544	676
5	578	21	515	641	618	23	549	687
6	576	22	510	642	608	22	542	674
7	573	21	510	636	612	21	549	675
8	567	21	504	630	601	23	532	670
9	556	19	499	613	589	21	526	652
10	554	22	488	620	585	23	516	654

TABLE 7
SENSOR DATA EXTRACTION FROM ACTUAL PATHS OF EXU

Path ID	No Aging (t = 0)				With Aging (t = 3 years)			
	Actual Slack (ps)	Sensor Reading	Slack from Sensor Reading (ps)	Measurement Error (ps)	Actual Slack (ps)	Sensor Reading	Slack from Sensor Reading (ps)	Measurement Error (ps)
1	69	1 1 0 0 0 0	60	9	29	0 0 0 0 0 0	20	9
2	82	1 1 1 0 0 0	80	2	43	1 0 0 0 0 0	40	7
3	83	1 1 0 0 0 0	60	17	47	1 0 0 0 0 0	40	7
4	94	1 1 1 0 0 0	80	14	53	1 1 0 0 0 0	60	7
5	91	1 1 1 0 0 0	80	11	52	1 1 0 0 0 0	60	8
6	98	1 1 1 0 0 0	80	18	57	1 1 0 0 0 0	60	3
7	111	1 1 1 1 0 0	100	11	51	1 0 0 0 0 0	40	11
8	112	1 1 1 1 0 0	100	12	73	1 1 1 0 0 0	80	7
9	114	1 1 1 1 0 0	100	14	78	1 1 1 0 0 0	80	2
10	125	1 1 1 1 1 0	120	5	82	1 1 1 0 0 0	80	2

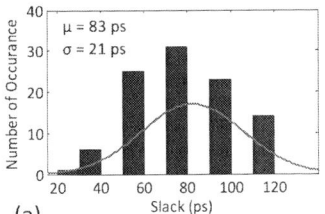
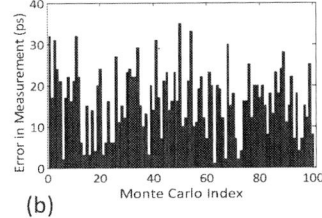

(a)   (b)

Fig. 7. Monte Carlo simulations for Path 5. (a) The extracted slack distribution (c) Measurement errors for each sample.

the aid of local voltage fluctuation information from power supply noise sensor [22][23] and the calibration data from Table 1, the sensor codes would be decoded into corresponding path slacks.

TABLE 8
EFFECT OF VOLTAGE FLUCTUATIONS ON PATH 10

Mean Voltage	Actual Slack (ps)	Sensor Reading	Slack From Sensor (ps)	Measurement Error (ps)
Nominal VDD	125	1 1 1 1 1 0	120	5
95% VDD	72	1 1 1 0 0 0	80	8
90% VDD	17	0 0 0 0 0 0	20	3

6 CONCLUSIONS

We have proposed a compact digital sensor IP and the corresponding layout-aware netlist-level sensor insertion flow to continuously monitor timing slack of critical/near critical paths in the post-silicon phase. We demonstrated the robustness of the sensor IP against variation and aging effects. The netlist-level sensor insertion flow has been implemented in the layout of selected benchmark circuits. Our future plan is to collect and analyze silicon results from fabricated SoCs.

REFERENCES

[1] S. Ghosh and K. Roy, "Parameter variation tolerance and error resiliency: New design paradigm for the nanoscale era," Proc IEEE, vol. 98, pp. 1718-1751, 2010.

[2] S.S. Sapatnekar, "Overcoming variations in nanometer-scale technologies," Emerging and Selected Topics in Circuits and Systems, IEEE Journal on, vol. 1, pp. 5-18, 2011.

[3] P. Das and S.K. Gupta, "Extending pre-silicon delay models for post-silicon tasks: Validation, diagnosis, delay testing, and speed binning," in VLSI Test Symposium (VTS), 2013 IEEE 31st, pp. 1-6, 2013.

[4] J. Zeng, R. Guo, W. Cheng, M. Mateja and J. Wang, "Scan-based Speed-path Debug for a Microprocessor," IEEE Design & Test of Computers, 2011.

[5] E.J. Jang, A. Gattiker, S. Nassif and J.A. Abraham, "Efficient and product-representative timing model validation," in VLSI Test Symposium (VTS), 2011 IEEE 29th, pp. 90-95, 2011.

[6] M. Ebrahimi, F. Oboril, S. Kiamehr and M.B. Tahoori, "Aging-aware logic synthesis," in Proceedings of the International Conference on Computer-Aided Design, pp. 61-68, 2013.

[7] S. Das et al., "RazorII: In situ error detection and correction for PVT and SER tolerance," Solid-State Circuits, IEEE Journal of, vol. 44, pp. 32-48, 2009.

[8] M. Fojtik et al., "Bubble Razor: Eliminating Timing Margins in an ARM Cortex-M3 Processor in 45 nm CMOS Using Architecturally Independent Error Detection and Correction," Solid-State Circuits, IEEE Journal of, vol. 48, pp. 66-81, 2013.

[9] K.A. Bowman et al., "Energy-efficient and metastability-immune resilient circuits for dynamic variation tolerance," Solid-State Circuits, IEEE Journal of, vol. 44, pp. 49-63, 2009.

[10] K. Jeong, A.B. Kahng and K. Samadi, "Impact of guardband reduction on design outcomes: A quantitative approach," Semiconductor Manufacturing, IEEE Transactions on, vol. 22, pp. 552-565, 2009.

[11] A. Rahimi, L. Benini and R. Gupta, "Application-adaptive guard-banding to mitigate static and dynamic variability," Computers, IEEE Transactions on, vol. 63, pp. 2160-2173, 2014.

[12] L. Lai, V. Chandra, R.C. Aitken and P. Gupta, "SlackProbe: A Flexible and Efficient In Situ Timing Slack Monitoring Methodology," Computer-Aided Design of Integrated Circuits and Systems, IEEE Transactions on, vol. 33, pp. 1168-1179, 2014.

[13] J. Li and M. Seok, "Robust and In-Situ Self-Testing Technique for Monitoring Device Aging Effects in Pipeline Circuits," in Proceedings of the The 51st Annual Design Automation Conference on Design Automation Conference, pp. 1-6, 2014.

[14] D. Fick et al., "In situ delay-slack monitor for high-performance processors using an all-digital self-calibrating 5ps resolution time-to-digital converter," in ISSCC, pp. 188-189, 2010.

[15] H. Fuketa, M. Hashimoto, Y. Mitsuyama and T. Onoye, "Adaptive performance compensation with in-situ timing error predictive sensors for subthreshold circuits," Very Large Scale Integration (VLSI) Systems, IEEE Transactions on, vol. 20, pp. 333-343, 2012.

[16] X. Wang et al., "Radic: A standard-cell-based sensor for on-chip aging and flip-flop metastability measurements," in Test Conference (ITC), 2012 IEEE International, pp. 1-9, 2012.

[17] K. Katoh, K. Namba and H. Ito, "An on-chip delay measurement technique using signature registers for small-delay defect detection," Very Large Scale Integration (VLSI) Systems, IEEE Transactions on, vol. 20, pp. 804-817, 2012.

[18] F. Firouzi, F. Ye, K. Chakrabarty and M.B. Tahoori, "Representative critical-path selection for aging-induced delay monitoring," in Test Conference (ITC), 2013 IEEE International, pp. 1-10, 2013.

[19] A.J. Drake et al., "Single-cycle, pulse-shaped critical path monitor in the POWER7 microprocessor," in Low Power Electronics and Design (ISLPED), 2013 IEEE International Symposium on, pp. 193-198, 2013.

[20] X. Wang, M. Tehranipoor, S. George, D. Tran and L. Winemberg, "Design and analysis of a delay sensor applicable to process/environmental variations and aging measurements," Very Large Scale Integration (VLSI) Systems, IEEE Transactions on, vol. 20, pp. 1405-1418, 2012.

[21] R. Ginosar, "Metastability and Synchronizers: A Tutorial," Design & Test of Computers, IEEE, vol. 28, pp. 23-35, 2011.

[22] M. Sadi, Z. Conroy, B. Eklow, M. Kamm, N. Bidokhti and M.M. Tehranipoor, "An All Digital Distributed Sensor Network Based Framework for Continuous Noise Monitoring and Timing Failure Analysis in SoCs," in Test Symposium (ATS), 2014 IEEE 23rd Asian, pp. 269-274, 2014.

[23] R. Petersen, P. Pant, P. Lopez, A. Barton, J. Ignowski and D. Josephson, "Voltage transient detection and induction for debug and test," in Test Conference, 2009. ITC 2009. International, pp. 1-10, 2009.

[24] J. Zeng, J. Wang, C. Chen, M. Mateja and L. Wang, "On evaluating speed path detection of structural tests," in Quality Electronic Design (ISQED), 2010 11th International Symposium on, pp. 570-576, 2010.

[25] http://www.synopsys.com/COMMUNITY/UNIVERSITYPROGRAM/

[26] http://ptm.asu.edu/

[27] http://www.oracle.com/technetwork/systems/opensparc/index.html

Scalability Study of PSANDE: Power Supply Analysis for Noise and Delay Estimation

Sushmita Kadiyala Rao, Bharath Shivashankar, Ryan Robucci, Nilanjan Banerjee and Chintan Patel

CSEE Department, University of Maryland, Baltimore County

Abstract

Variations in the power-distribution network are exacerbated because of scaled supply voltages and smaller noise margins in sub-nanometer designs, which adversely affect performance and yield. Power-Supply noise incurred by excessive simultaneous switching of multiple paths negatively impacts the timing of a circuit. Supply noise is a major issue especially during transition and delay test where test vectors cause increased switching as compared to functional operation resulting in increase in path delays. Test rejects due to excessive noise-induced failures during delay and transition testing negatively impacts yield. Hence there is a need to accurately characterize the resistive and inductive voltage drop caused by excessive switching. To our knowledge, inductive drop has been excluded to simplify noise analysis. In our previous work, we have presented a convolution-based dynamic method (herein referred to as PSANDE) to estimate both IR and Ldi/dt drop on small combinational and sequential circuits. In this paper we show that the effectiveness of the design partitioning technique makes the framework feasible for a larger design. Our dynamic approach involves selectively simulating only extracted switching logic which makes the run-time tractable as compared to prohibitive full-chip SPICE simulations. We also present data to show that PSANDE can accurately predict the power-supply noise due to clock tree switching. Data presented in this paper for power supply noise is based on a large ITC'99 sequential benchmark b17 circuit. with a maximum error of 8.2% in comparison to full-chip SPICE results.

Keywords: Power-Supply Noise, Transition & Delay Testing, IR-drop, *Ldi/dt* drop, PSANDE

1.0 Introduction

Power-Supply noise continues to plague deep sub-micron technologies especially during delay testing. Simultaneous switching of multiple paths can degrade the supply voltages affecting circuit performance. The variations in supply voltages increases path delays that can sometimes lead to timing failures. In particular, during test mode extensive switching is incurred by the test vectors that may not occur in functional mode. These test conditions cause a larger voltage droop that results in false failures increasing yield loss. Techniques are needed to accurately model supply noise and the ill-effects it has on path delays so that test vector generation and compaction can be based on it that can alleviate test-related yield loss due to false positives.

Power supply noise (PSN) refers to the variations caused in the supply voltages in the power-distribution network due to considerable switching activity in the circuit. PSN comprises the IR-drop caused by parasitic

resistors of signal wires in a circuit and the voltage drop caused by package and parasitic inductance known as *Ldi/dt* drop. Inductive drop is often excluded to simplify noise analysis leading to inaccuracies in supply noise estimation. In this work we show that the dynamic approach, PSANDE [1][2][3] can be scaled to estimate power supply noise for large circuits. We also predict the contribution of power supply noise from the off-path clock elements.

A survey highlighting the causes and effects of PSN in delay testing is presented in [6]. In [7] a statistical analysis is used to study the impact of IR-drop on critical path delays. Authors in [8] propose an event-driven simulation technique to estimate the worst-case noise where they use genetic algorithms to optimize the noise problem.

The authors in [9] claim that on-chip inductance can be ignored for most of the frequencies since at high-frequencies the effects of inductance are localized in nature and comparatively smaller. But in [10] the authors show that the on-chip self inductance caused the voltage drop to deviate significantly, by more than 50% over that of the IR predicted drop. In [11] the effect of both self and mutual inductance is illustrated from the simulation results of a 4-bit coupled bus. Results show that error for delay predictions can be more than 100% and that for rise time computation can be greater than 70% when the effect of inductance is ignored.

Test pattern generation flow has also been affected by PSN. Authors in [12] predict that the test industry will move towards newer ATPG methods, one being the usage of partition-aware ATPG techniques, by the year 2020 due to the design sizes that increase with shrinking technology. The overkill induced by random filling of don't-care bits during test compaction has been addressed by authors in [13]. Lee et al. [14] proposed a layout-aware test compaction technique such that transition test patterns cause switching activity that is evenly distributed across the chip. In [15] Ma et al. use a layout-aware test pattern generation method to increase switching in the neighborhood cells of the critical path gates to maximize supply noise effects on the critical path under test. The authors in [16] present a fast technique to validate test vectors in the presence of noise caused by IR-drop. In [17] the authors have proposed a virtual circuit partitioning technique in which only test patterns that cause high power dissipation are applied to the subcircuits while the low power test vectors are applied at the full circuit level.

In [18] the authors use genetic algorithms to generate a small set of test patterns that cause higher power-supply noise and use a lower-level simulator to validate them and find the worst case test vector. The authors in [19] propose iterative algorithms to solve linear equations of the power grid that are far more accu-

rate and faster than the random-walk-based algorithm. In [20] the authors have proposed multigrid-like techniques where the grid is reduced for power grid analysis. PSANDE is unique in its focus on prediction of PSN caused by simultaneous switching activity. Using this framework we can accurately predict the increase in path delays due to power consumption in the vicinity of critical paths (or any path) and therefore we can supplement the techniques described above.

2.0 Power Supply Analysis for Noise and Delay Estimation (PSANDE)

In previous work [1][2][3], we have established a fast and accurate convolution-based approach to estimate power-supply noise caused by simultaneous switching of multiple paths. We have also predicted the increase in path propagation delays induced by PSN with SPICE-like accuracy. The basis of PSANDE framework is separating the power grid from the core logic circuitry in any CMOS chip. Analysis of each of these subsytems is explained briefly below.

2.1 Power-Grid Characterization

Since the power grid can be modeled as a linear system, it can be characterized by its Impulse Response (IR) functions. Grid simulations are carried out to compute IRs between input and output locations of the power grid. Inputs are points where standard cells connect to the power ground network and outputs are either the power pads of the chip or any point where standard cells connect to the grid. The Current-to-Current impulse responses (C2C IR) provide the relationship between the current source applied at any input and the corresponding currents measured at any output location on the power grid. For the purpose of this paper we aim to estimate the total transient current at the power pads, though as described in our previous work we can characterize both current and voltage IR responses to any input/output node in the system. Therefore the current responses are measured only at the power pads or C4 bumps. The IR computation is shown in Figure 1.

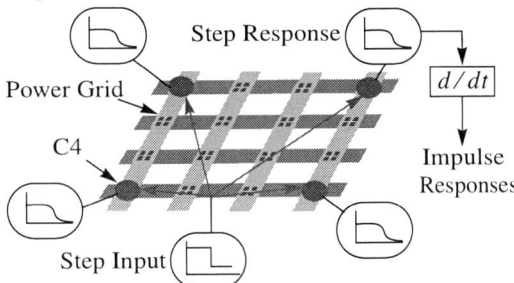

Figure 1. C2C Impulse response [1]

As seen in the figure, a current step input is applied to an input and the output step responses are measured at the power pads are differentiated to compute the C2C IR functions. This pre-characterization of the power grid is carried out only once for a design and accelerates computation.

2.2 Core-Logic Circuit Evaluation

In our work, we carry out isolated path simulations that are much faster in comparison to full-chip simulations accelerating computation. The Value Change Dump (VCD) file obtained from the Verilog simulation is used to identify gates of a switching path. These gates are then extracted from the full-chip circuit netlist with RLC parasitics. Next, the isolated paths are simulated with ideal V_{DD} voltages and the corresponding current transients at the power supply pins of each logic gate in the path are saved. This transient current process is repeated for every test pattern.

To estimate the power supply waveforms at the power pads, the current transients saved during path simulations are convolved with the impulse responses computed during the grid characterization process. Previously, we have published results for estimating PSN and delays for small combinational and sequential benchmark circuits that consisted of about 2000 logic gates.

In this paper, we have applied the PSANDE framework to a much larger sequential benchmark circuit. We have shown that the power supply noise scales well for large designs using the partitioning scheme. Previously we used test circuits that consisted of multiple instances of the c6288 design. Hence, the design partitioning technique divides the design into identical copies of the c6288 circuit. In this work, the partitioning scheme results in several non-identical partitions. These unique partitions are evaluated to estimate power supply noise. We also show that the convolution-based PSN estimation method can be used efficiently to predict the current transients contributed by the on and off-path sequential elements of the clock tree.

This paper presents three novel research contributions that are described in the following sections

1. *Macro-level (full-chip) Design Partitioning into micro-level subsections towards fast full-chip PSN estimation*

2. *Superposition and redistribution techniques to predict macro-level currents using only micro-level simulations*

3. *Accurately quantifying the effect of clock tree switching on full-chip PSN*

3.0 Macro-Design Partitioning Strategy

In this work, we use the ITC'99 b17 sequential benchmark circuit to study the scalability of the PSANDE framework. The b17 circuit is a combination of three 80386 processors (subset) that communicate with each other and to primary inputs/outputs. The layout is implemented in 180-nm technology using a place-and-route tool. The circuit netlist with extracted RLC parasitics consisting of 318,006 transistors; 2,828,786 resistors; 22,721,081 capacitors; and 338,056 parasitic inductors is representative of a large circuit block by today's standards. The circuit is subdivided into nine sections u1-u9 based on the power pad placement as shown in Figure 2. A section of the chip encompassed by power pads or C4s colored in dark orange is

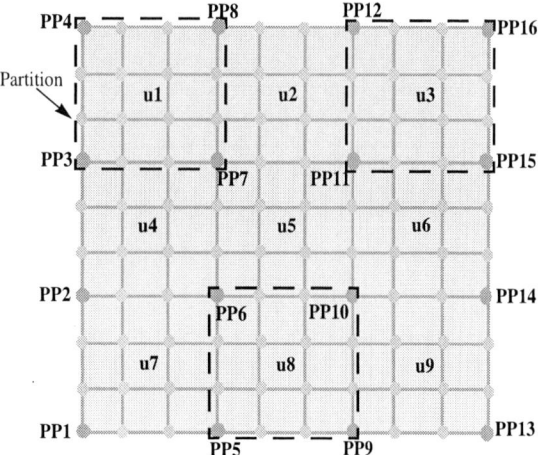

Figure 2. b17 Partitioning

defined as a partition. Current measurements are taken at 16 power pads named PP1-PP16. The SPICE netlist for each of the partition is created using the DEF (Design Exchange Format) file and the full-chip netlist of the b17 design with RLC parasitics.

4.0 Micro-Design Partitioning

The b17 design is broken into the power grid and core logic subsystems. The partitioning is performed at the metal connections between the local V_{DD}/GND and global supply rails called follow-pins. The two independent subsystems are characterized using the grid impulse response and gate current transient computations.

4.1 Grid Impulse Response Computation

In our previous work, we characterized the power grid as a single entity. However, for large circuits even a linear characterization of the power grid would be infeasible. One of the major contributions of this work is to demonstrate the scaling capabilities of PSANDE by partitioning the power grid (which is already separated from the non-linear core logic subsystem) for C2C IR analysis. The number of inputs for the power grid IR computation for each partition is shown in Table 1. Grid simulations are run for each of the partition to obtain the C2C impulse responses for each input in the partition. To reiterate, this pre-characterization process is performed only once for a given design and can be run in parallel. This one time characterization of the power grid can be used to predict full-chip response for *any given input from the core logic circuit* i.e switching paths.

Partition u_X and Number of Inputs					
u1	7370	u4	6908	u7	7093
u2	6799	u5	5724	u8	4658
u3	6212	u6	4697	u9	2866

Table 1: Inputs to the Linear Power Grid for each Partition

4.2 Gate Current Transients Computation

A switching path is identified from the VCD output obtained from the Verilog simulation of the b17 design. The current transient from each switching logic gate is an input to the pre-characterized linear power grid circuit. As the design is partitioned using the placement of power pads, a switching path is also divided into several subpaths across all partitions. This step identifies subpaths that are encompassed by a given partition. Each subpath is then individually simulated in parallel to produce the current waveforms at the power supply pins of the logic gates which are the inputs to the power grid. These subpath simulations take only a few minutes, in contrast to full-chip SPICE simulations which are run-time intensive and therefore prohibitive.

5.0 Comprehensive PSN Estimation

Using the power grid IR characterization and subpath simulations we discuss the technique for estimating full-chip PSN. This involves using logic gate power supply inputs per partition to compute current transients at micro-level power ports and then superimposing and redistributing these transients to predict macro-level full-chip PSN. The gate-level power current transients saved during simulations of each subpath are convolved with the C2C IRs and this step is repeated for all subpaths switching in a given partition. The cumulative sum of current waveforms estimated for all subpaths of a partition result in the micro-level power pad current waveforms for each partition. These partition current waveforms are then superimposed accordingly to generate the current waveforms at the macro-level power pads of the b17 design. For example, to estimate the total macro-level current transient for PP6 shown in Figure 2, we need to superimpose and redistribute currents from each partition. For this estimation, at the micro-level we superimpose currents from upper-left PP of u8, lower-left PP of u5, lower-right PP of u4 and upper-right PP of u7. Additionally we need to redistribute the superimposed micro-level currents to determine the final macro-level PP6 current transients as explained in Section 7.0.

6.0 Clock Tree Transient Estimation

Partitioning the chip based on the power pad placement also necessitates characterizing a fragmented clock tree. In this work we demonstrate a novel technique for predicting the current transients attributed to the global clock signal. It enhances our preliminary method that was previously described in [1], that shows results for clock-tree current transient estimation. Our new approach addresses the scalability of predicting the PSN contributed by the global clock tree as it is impractical to simulate the entire clock distribution circuit of a large design. Therefore we divide the clock tree based on our micro-level partitions and specifically differentiate between clocked elements within a switching path and non-switching sequential elements.

The transients caused by the sequential elements that switch are included in the path extractions and captured during path simulations. Also we must account for the current transients generated by non-switching off-path

Figure 3. Clock-tree Highlighted (in white) in b17

flip-flops and clock buffers as they are not included in the switching path simulations. We estimate the current transients from non-switching sequential elements and include it along with micro-level computations in the PSANDE framework to estimate the full-chip PSN.

Highlighted in Figure 3 is the entire clock tree of the design.2

7.0 Signal Redistribution Technique

The current waveforms computed at each power pad at the micro-level are superimposed to get an initial estimate of the current transients at the macro-level PPs. Superposition of micro-level current transients only encompasses the effect of switching activity from partitions that a particular PP belongs to. However other power pads on the chip provide current to the PP under

consideration that we need to account for. This is accomplished by estimating contributions from other PPs (not included at the micro-level) using a redistribution technique described in [4] and [5]. The technique involves applying an AC current step input at the PP under consideration, measuring the response at all other PPs, creating a transfer-function matrix and using it to redistribute the current contributions. This process is repeated for each PP (under consideration). Each row in the transfer-function matrix refers to one measurement (i.e. a current source applied to one PP at a time) and the columns represent the current measured at all other PPs. Each element in the matrix is normalized by the average of the DC levels of the measured current transient.

8.0 Results

As mentioned in Section 3.0, the ITC'99 b17 benchmark circuit, is used for all the simulations in this work using the Cadence UltraSim simulator. Results for the three major contributions of this work are presented in this section that demonstrate the scalability of the PSANDE framework.

Test Pattern	Combinational Elements	Sequential Elements	Total
Path1	6274	258	6532
Path2	6277	266	6535

Table 2: Switching Gates Count

Test patterns were generated using a commercial ATPG tool and PSN was estimated. The number of both sequential and combinational elements that switch in

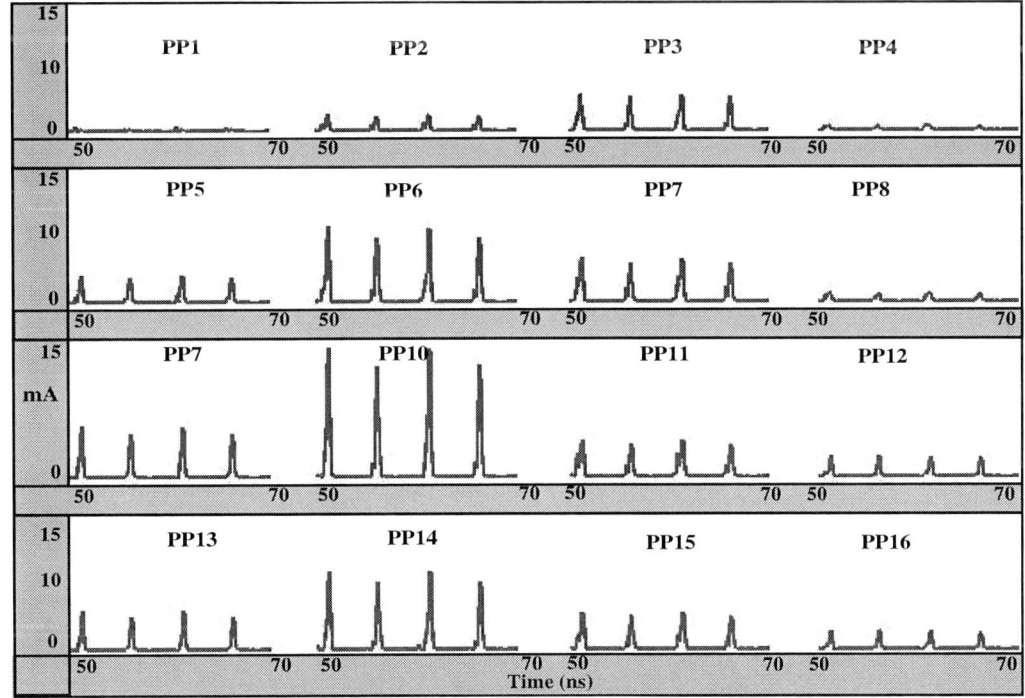

Figure 4. Estimated Path1 Clock Current Transients

978-1-4799-7598-3/15 $31.00 © 2015 IEEE

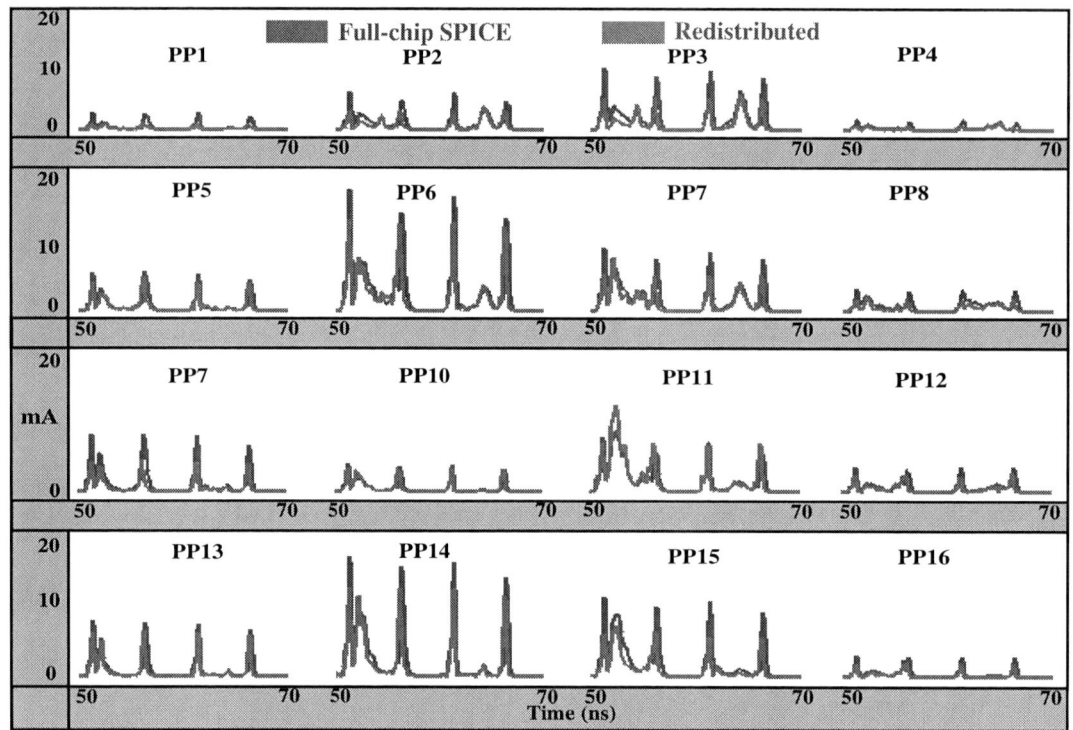

Figure 5. Full-Chip vs. Estimated Path1 Current Transients

each path (or test pattern) are shown in Table 2. Sub-paths (on-path sequential and combinational logic gates) in each partition are identified and simulated to obtain the gate current transients which are then convolved with the C2C IR's to estimate the power pad current transients. The same process is repeated for all the off-path sequential elements in the clock tree to predict the current transients contributing to the total power transients measured at the sixteen PPs of the b17 design. The estimated current waveforms for both the clock tree and switching logic gates are presented.

8.1 Clock Current Transients

path	u1	u2	u3	u4	u5	u6	u7	u8	u9
1	76	30	139	168	111	158	52	183	247
2	73	35	137	167	113	156	52	181	245

Table 3: Off-path sequential elements and clock buffer count

The clock transients are estimated for 1164 and 1159 for path 1 and 2 respectively off-path clock buffers and flip-flops in each test pattern spread across the partitions as shown in Table 3. The partition based divide and simulate strategy speeds up the current estimation process since each of the clock path simulation takes less than 5 min. and can be run in parallel. The current waveforms estimated using our method is shown in Figure 4 for one representative path.

8.2 Combinational Current Transients

The off-path clock tree waveforms are superposed with the current waveforms estimated for the switching sequential and combinational logic gates and then redistributed using the transfer-function matrix. These redistributed waveforms are then compared to the full-chip SPICE current waveforms as shown in Figure 5 for this particular path. The figure shows that the estimated waveforms follow the expected full-chip SPICE results closely.

$$\text{NME} = \frac{\left(\dfrac{1}{N} \sum_{n=0}^{N-1} |(I_{\text{Full Chip}}[n] - I_{\text{Estimated}}[n])| \right)}{max(I_{\text{Full Chip}}[n])} \times 100\% \quad (1)$$

To quantify the results, we compute the Normalized Mean Error (NME) to compare the full-chip and estimated waveforms as defined in equation 1 shown below. In the equation $I_{\text{Full Chip}}$ is the full-chip transient current, $I_{\text{Estimated}}$ is the estimated transient current and N is the number of datapoints in the waveforms. The NME recorded for two paths is shown in Figure 6. A maximum error of 8.2% is observed.

The scalability and efficiency of PSANDE is illustrated in Table 4. The table shows the comparison of worst-case simulation time required for SPICE simulations of 1) full-chip b17 circuit; 2) each subpath; 3) power grid to compute the C2C IR's per input and; 4)

978-1-4799-7598-3/15 $31.00 © 2015 IEEE

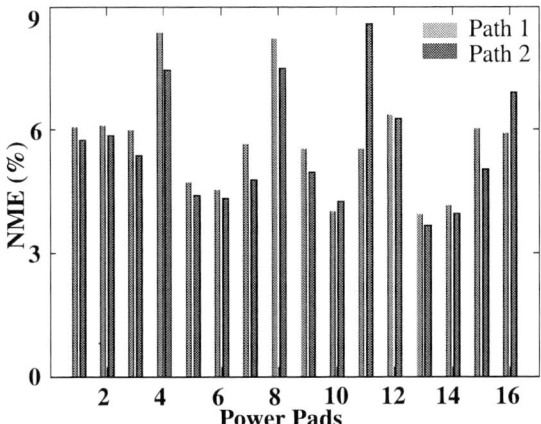

Figure 6. Normalized Mean Error for Two Switching Paths is shown on the Y-axis. The X-axis represents 16 different PPs in the design.

off-path sequential elements to estimate clock-tree transients. In the worst case each subpath and grid simulation take the time shown in the table, where 1415 subpath simulations were performed for the longest path in parallel.

Component Simulated	Time
Full-Chip SPICE	147h 28m
One Subpath	2m 13s
One Grid	1m 42s
One Clock path	4m 16s

Table 4: Simulation Time Comparison

9.0 Conclusion

This paper shows major enhancements to PSANDE that include, complete implementation of macro and micro-level design partitioning, superposition and redistribution techniques to predict macro-level currents using only micro-level simulations and more importantly PSN induced by the clock network. This research specifically demonstrates the scalability of PSANDE. Effect of both switching (on-path) and non-switching (off-path) clocked sequential elements is accurately modeled in the enhanced framework. The accuracy of the PSANDE is also presented and the worst-case error is 8.2%. The run-time for accurately predicting PSN shows improvement from several hours of full-chip simulation to a few minutes using the framework.

10.0 References

[1] Rao, S.K.; Robucci, R.; Patel, C.; , "Simulation Based Framework for Accurately Estimating Dynamic Power-Supply Noise and Path Delay," Journal of Electronic Testing., February 2014, Volume 30, Issue1,pp.125-147.

[2] Rao, S.K.; Robucci, R.; Patel, C.;, "Scalable Dynamic Technique for Accurately Predicting Power-Supply Noise and Path Delays," VLSI Test Symposium (VTS), 2013 IEEE 31st, vol., no., pp.1-6, April 29 - May 02 2013.

[3] Rao, S.K.; Sathyanarayana, C; Kallianpur,A.; Robucci, R.; Patel, C.; , "Estimating Power Supply Noise and its impact on path delay", VLSI Test Symposium (VTS), 2012

IEEE 30th, vol., no., pp.276-281, 23-25 April 2012.

[4] Abhishek Singh, Jim Plusquellic, Dhananjay Phatak and Chintan Patel, "Defect Simulation Methodology for i_{ddt} Testing", J.Electron Test., 22:255-272, June 2006.

[5] Reza M. Rad and Jim Plusquellic, "A Novel Fault Localization Technique based on Deconvolution and Calibration of Power Pad Transient Signals", J.Electron. Test./ 25:169-185, June 2009.

[6] M.Tehranipoor and K.M. Butler, "Power Supply Noise: A Survey on Effects and Research", IEEE Design & Test of Computers, 27(2):51-67,2010.

[7] Chunsheng Liu, Yang Wu and Yu Hang, "Effect of IR-drop on Path Delay Testing using Statistical Analysis", In Proc. 16th Asian Test Symposium ATS '07, pages 245-250, 2007

[8] S.Zhao, K.Roy and C.K.Koh, "Estimation of Inductive and Resistive Switching Noise on Power Supply Network in Deep Sub-micron CMOS Circuits", In Proc. Int Computer Design Conf, pages 65-72,2000.

[9] Pant, S.; Chiprout, E.; Blaauw, D, "Power Grid Physics and Implications for CAD," Design & Test of Computers, IEEE , vol.24, no.3, pp.246,254, May-June 2007.

[10] Andersson, D.A.; Nilsson, B.; Pihl, J.; Svensson, L.; Larsson-Edefors, P.; , "Supply voltage drop study considering on-chip self inductance of a 32-bit processor's power grid," Signal Propagation on Interconnects, 2009. SPI '09. IEEE Workshop on , vol., no., pp.1-4, 12-15 May 2009

[11] Ismail, Y.I., "On-chip inductance cons and pros," Very Large Scale Integration (VLSI) Systems, IEEE Transactions on , vol.10, no.6, pp.685,694, Dec. 2002.

[12] Galivanche, R.: Kapur, R.: Rubio, A.: , "Tesing in the Year 2020," Design, Automation & Test in Europe Conference & Exhibition, 2007. DATE '07 , vol., no., pp.1-6, 16-20 April 2007.

[13] Wang, L.-C.; Walker, D.M.; Xiang Lu; Majhi, A.; Kruseman, B.; Gronthoud, G.; Villagra, L.E.; van de Wiel, P.J.A.; Eichenberger, S., "Modeling Power Supply Noise in Delay Testing," Design & Test of Computers, IEEE , vol.24, no.3, pp.226,234, May-June 2007.

[14] Lee, J.; Narayan, S.; Kapralos, M.; Tehranipoor, M.; , "Layout-Aware, IR-Drop Tolerant Transition Fault Pattern Generation," Design, Automation and Test in Europe, 2008. DATE '08 , vol., no., pp.1172-1177, 10-14 March 2008

[15] Junxia Ma; Lee, J.; Tehranipoor, M.; , "Layout-Aware Pattern Generation for Maximizing Supply Noise Effects on Critical Paths," VLSI Test Symposium, 2009. VTS '09. 27th IEEE , vol., no., pp.221 226, 3 7 May 2009.

[16] Kokrady, A.; Ravikumar, C. P., "Fast, layout-aware validation of test-vectors for nanometer-related timing failures," VLSI Design, 2004. Proceedings. 17th International Conference on , vol., no., pp.597,602, 2004.

[17] Qiang Xu; Dianwei Hu; Dong Xiang; , "Pattern-directed circuit virtual partitioning for test power reduction," Test Conference, 2007. ITC 2007. IEEE International , vol., no., pp.1-10, 21-26 Oct. 2007.

[18] Yi-Min Jiang; Kwang-Ting Cheng; , "Vector generation for power supply noise estimation and verification of deep submicron designs," Very Large Scale Integration (VLSI) Systems, IEEE Transactions on , vol.9, no.2, pp.329-340, April 2001.

[19] Yu Zhong; Wong, M.D.F.; , "Fast algorithms for IR drop analysis in large power grid," Computer-Aided Design, 2005. ICCAD-2005. IEEE/ACM International Conference on , vol., no., pp. 351- 357, 6-10 Nov. 2005.

[20] Kozhaya, J.N.; Nassif, S.R.; Najm, F.N.;,"A multigrid-like technique for power grid analysis," Computer-Aided Design of Integrated Circuits and Systems, IEEE Transactions on , vol.21, no.10, pp. 1148- 1160, Oct 2002.

978-1-4799-7598-3/15 $31.00 © 2015 IEEE

Robust Counterfeit PCB Detection Exploiting Intrinsic Trace Impedance Variations

Fengchao Zhang, Andrew Hennessy, and Swarup Bhunia
Department of Electrical Engineering and Computer Science
Case Western Reserve University, Cleveland, OH 44106, USA
Emails: {fxz67, ajb200, skb21}@case.edu

ABSTRACT

The long and distributed supply chain of printed circuit boards (PCBs) makes them vulnerable to different forms of counterfeiting attacks. Existing chip-level integrity validation approaches cannot be readily extended to PCB. In this paper, we address this issue with a novel PCB authentication approach that creates robust, unique signatures from a PCB based on process-induced variations in its trace impedances. The approach comes at virtually zero design and hardware overhead and can be applied to legacy PCBs. Experiments with two sets of commercial PCBs as well as a set of custom designed PCBs show that the proposed approach can obtain unique authentication signature with inter-PCB hamming distance of 47.94% or higher.

Keywords: Printed Circuit Board (PCB), Piracy, Counterfeiting, Trust, PUF, Authentication

I. INTRODUCTION

A counterfeit electronic component is one that has a discrepancy in functionality, performance, and reliability - but is sold as an authentic one. Counterfeiting of Integrated Circuits (ICs) is a global issue and a growing multi-billion dollar industry [1]. There have been numerous studies on how to prevent, identify, or mitigate IC-level counterfeiting. Similar to integrated circuits, Printed Circuit Boards (PCBs), which are common to all electronic products, also share a long globally distributed supply chain involving multiple untrusted parties, as illustrated in Fig. 1 (a). Hence, PCBs are also vulnerable to different forms of counterfeiting attacks. The relative ease of PCB reverse engineering and piracy of a PCB design make it highly vulnerable to cloning attacks. Hence, counterfeiting of a PCB has emerged as a prevalent practice [2]. However, there has been a dearth of study on the prevention, identification, or mitigation of counterfeit PCBs.

Counterfeit PCBs can be categorized into three major classes. The first and the most obvious one is outright cloning of the entire PCB, often times with the identification of the Bill-of-Materials (BoM) included in the process, so that a functioning counterfeit product can be quickly brought to market. By acquiring a sample PCB, some manufacturer can duplicate it without the original design and layout through a reverse-engineering process [3]. The second class consists of legitimate PCBs that do not meet the standards of the target

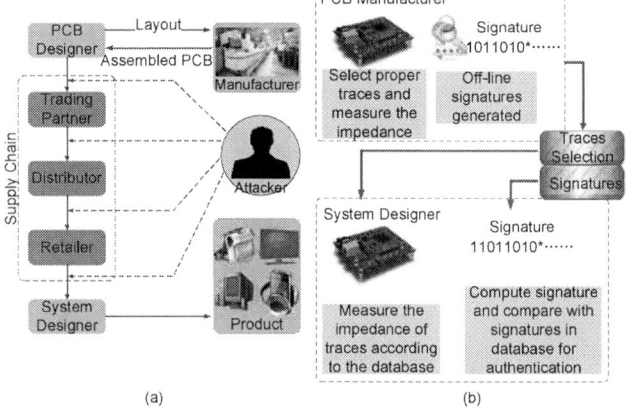

Fig. 1. (a) Typical stages in a PCB supply chain, which are vulnerable to counterfeit PCB insertion; and (b) overall steps of the proposed impedance based PCB authentication procedure.

customer and hence discarded. These PCBs can then by picked up by ghost shift workers in a factory; filled with components; and then sold to customers as real products. The final class of counterfeit PCBs is comprised of legitimate PCBs, which are bought, used, refurbished, and then sold as new involving multiple parties in the process. The quality of these counterfeit PCBs may be poor causing early failures, performance degradation, or potential damage and loss of information to the end users due to the unreliable board material, poor layout, or construction. Furthermore, counterfeit PCBs can potentially have additional undesired functionalities or malicious circuits (i.e. Hardware Trojans) [11], which largely compromises the integrity of the resultant system.

Several counterfeit PCB detection approaches have been used in practice or explored in the research community. A method trademarked as "DNA marking" [4] was developed by Applied DNA Sciences, which embeds unique and unclonable botanical DNA on the products. This mark can be detected by laser readers down the supply chain. Such a technology, however, is simply able to detect the individual components (specifically chips) on the PCBs. This means that while the components soldered onto the PCB can be verified, the actual PCB itself cannot be. Radio Frequency Identification (RFID) and its variants constitute another technology platform, which

is becoming popular in order to provide authentication to electronic products [5]. It relies on the wireless non-contact use of radio-frequency electromagnetic fields to transfer data for the purposes of identifying and tracking tags attached to objects. Though RFID is more robust over the traditional authentication mechanisms, it still suffers from the cloning problem – cloned RFID tags are indistinguishable from authentic ones. A final class of solutions for detecting counterfeit products is based on a security primitive called Physical Unclonable Function (PUF) [6] [7]. A PUF can extract a unique signature from each production unit of a design due to the random intrinsic manufacturing process variations. However, open literatures on the subject focus on a PUF embedded into a die or the IC package. They require special circuit structures (such as ring oscillators [6], memory array [7], or scan chain [8]) and hence, cannot be readily extended to PCB level.

We note that the metal traces on a PCB, typically numbering in the hundreds to thousands in a PCB of moderate complexity, can be powerful resources for PCB authentication. These metal traces, commonly made with copper (Cu) lines of different thickness, are subject to random intrinsic manufacturing process variations, such as random shift in length/width or contamination of Cu. Such variations reflect into variations of the DC resistance, AC impedance, and signal propagation delay through these lines. They vary from board to board and can be measured by a test equipment. Impedances from multiple traces in a board can collectively construct unique signature from each board, essentially acting as a PUF, and hence can be used for PCB integrity validation or authentication. A hallmark of modern PCB production processes is the set of automated test fixtures. Some of them use flying probes that securely make contact with test points in a design to provide quality assurances to the manufacturer and system designer. Some of the existing probes or extra probes added to the text fixtures can be used to automatically measure impedances and resistances of pre-defined traces for the purpose of PCB authentication. We observe that the DC resistance measurement is usually very sensitive to the exact contact condition between the probe and a PCB copper trace. However, measurement of the AC impedance of the copper traces is more robust. Moreover, since impedance $Z = \sqrt{R^2 + X_L^2}$, where $X_L = 2\pi f L$, L is the inductance and R is the resistance, it can capture variations in both resistance and inductance in the measured values.

In this paper, we propose a novel counterfeit PCB detection approach that utilizes the intrinsic impedance variations in metal traces on a PCB to create unique signature for authentication. It requires no design modification or hardware overhead. The overall approach is illustrated in Fig. 1 (b), which is separated into two stages. In the first stage, PCB manufacturers select apprpriate wire traces on the board and measure their impedances under a stable frequency on all authentic PCBs. Signatures are produced off-line based on the impedance measurements. The selection of traces and the corresponding signatures are stored in a database. In the second stage, system designers or end users who bought the PCBs from the market need to measure the same selected

traces for each PCB and compute the signature, which is then compared to the signatures stored in the database. A PCB is determined to be counterfeit if the produced signature does not match with the ones in the database. In particular, the paper makes the following key contributions:

1) It presents a novel methodology for PCB authentication, which does not require physical storage of key. Instead, it generates unique authentication signatures from each PCB exploiting random process variations that change the PCB trace impedances from board to board. Such an approach is low-cost requiring virtually no design changes or hardware overhead; and robust against invasive attacks (since no key is physically stored). To the best of our knowledge, this is the first key-less PCB integrity validation approach, and the first PCB PUF, which exploits intrinsic variations in PCB manufacturing process.

2) It evaluates the proposed authentication approach with two sets of widely used commercial PCBs. The experimental measurements show very high level of uniqueness and robustness of the signature for both sets of boards. In the experiment, 16 double-layer boards (Arduino UNO R3 SMD Edition) with the layout shown as Fig. 2 (a) [9] and 25 four-layer boards (Terasic DE0) were measured. We selected 10 traces on each board and each trace on the board was measured several times to minimize the impact of random measurement noise. For the Terasic DE0, 84 bits of unique signature were generated with the average inter-PCB Hamming Distance (HD) of 50.24% and intra-PCB HD of 2.14%. For the Arduino UNO, 120 bits of signature were generated with the average inter-PCB HD of 47.94% and intra-PCB HD of 1.06%.

3) To enhance the level of security as well as the ease of impedance measurements, it also presents a novel design-for-security (DfS) approach for PCB, which inserts carefully crafted trace patterns in a PCB design for the purpose of signature generation and authentication. We custom designed and fabricated such a PCB (30 copies) and measured all the trace impedances for all the copies. We observe very high level of impedance variations in these traces, which are suitable for signature generation with high entropy.

Remainder of the paper is organized as follows. Section II provides background on PCB wire impedance and motivation for the proposed solution. Section III describes the methodology of wire impedance based authentication. Section IV presents the measurement results and analysis. Section V presents a DfS approach for new PCBs. Section VI describes test apparatus that can be used in production to implement the proposed authentication. We conclude in Section VII.

II. BACKGROUND AND MOTIVATION

PCB copper traces have resistive, inductive and capacitive effects distributed throughout them. Two basic trace types of PCB are the microstrip and stripline. On a single layer PCB,

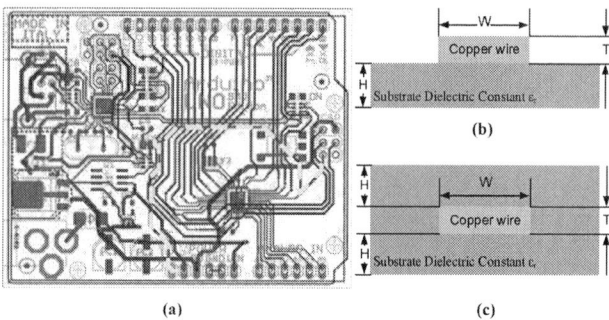

Fig. 2. (a) The layout of the Arduino UNO R3 SMD Edition with a selected trace (highlighted in yellow dash line); (b) microstrip Trace in single layer or multilayer PCB; and (c) stripline Trace in a multilayer PCB.

the microstrip trace is the dominant type of trace for the underlying pattern of copper wire. However, in a multilayer PCB, both types of traces are used. Thus, different PCBs may have different wire impedance models when considering the copper trace and substrate dielectric. Cross-sections of these trace types are shown in Fig. 2(b) and Fig. 2(c).

Impedance of a microstrip trace can be calculated as [10]:

$$Z_0 = \frac{87}{\sqrt{\epsilon_r + 1.41}} \ln\left(\frac{5.98H}{0.8W + T}\right) \quad (1)$$

Impedance of a stripline trace can be calculated as [10]:

$$Z_0 = \frac{60}{\sqrt{\epsilon_r}} \ln\left(\frac{4(2H + T)}{0.67\pi(0.8W + T)}\right) \quad (2)$$

Where Z_0 is the unit length characteristic impedance. From equations (1) and (2), the impedance of unit length is determined by width and thickness of the copper trace, thickness and dielectric constant of the substrate. During the PCB manufacturing processes, the dimensions of the traces cannot be exactly uniform in both width and height as well as the dielectric constant of the substrate varying over the area of the PCB. These factors will result in process-induced variations of the unit length impedance of a trace.

A flying probe test provides the feasibility to automatically and precisely measure the impedance of metal traces on a PCB. The test harness provides support for any number of probes to perform the analogue measurements of resistance, impedance, and inductance. In our experiments, a scaled down harness was built to fulfill the testing needs.

III. METHODOLOGY

In order to achieve a trace impedance based PCB authentication, or PCB PUF, first, we need to select a set of appropriate traces from a PCB design. Most PCBs are made using an FR4 substrate with so-called "One Ounce Thick Copper". This is defined as one ounce of copper spread over a square foot. Furthermore, some PCBs are bathed in molten tin or gold after fabrication (known in the industry as plating). This will result in difference in related parameters. For example, under room temperature, gold has a resistivity of 2.44×10^{-8} ohm·meter,

while tin has a resistivity of 1.09×10^{-7} ohm·meter, almost an order of magnitude higher. If the original specifications for a PCB called for gold plating and a counterfeit PCB used cheaper tin plating, the counterfeit one will have a higher trace resistance. The difference cannot however be detected visually because during PCB assembly, solder covers up the gold plating and a PCB with gold plating will look indistinguishable from one with tin plating.

A measurable way to determine the authenticity of a PCB is through measuring the impedance of a trace that passes through multiple vias. A via is a small hole drilled in a circuit board that, when plated with metal, connects the top layer of copper to the bottom layer of copper. For PCB with more than two layers, there are two more types of vias. A blind via is a via that connects one of the outer layers of the PCB to one of the inner layers, while a buried via is a via that connects two of the inner layers together.

Each manufacturer of PCB starts with a similar piece of copper clad and they use their knowledge and skill to make the finished product. Each board house has a different process for etching the copper off of the substrate as well as drilling and plating the vias. These different methods have different intrinsic resistances associated with them. For example, a via that is electrochemically plated onto the FR4 will have a lower resistance than a via that is riveted on. Finding a good path for impedance measurement needs to consider many variables and factors. First, we need to ensure that two probes can make good contact with the path. Often times PCB designers coat their PCB in solder mask (a typically green substance that helps keep solder where it belongs) and silkscreen (a typically white paint that helps designate areas of the PCB). Both silkscreen and solder mask have a very high electrical resistivity, preventing an accurate measurement of the trace impedance. Furthermore, for multilayer PCBs such as Terasic DE0, many traces ran exclusively in the inner layers of the PCB, using blind and buried vias to travel through the PCB without ever touching the outer layers. These traces cannot be used for measurement, because they are hard to probe.

Once the traces are selected, the impedance can be measured by commercial instruments, such as an Keysight 4263B LCR Meter [13]. The LCR meter needs to be self-calibrated before measurement. Additionally open connection correction needs to be done (leaving the two probes disconnected from everything) along with short connection correction (shorting the two probes together). As shown in Fig. 3 (a), probes need to be carefully placed on the pad of a PCB to avoid unwanted contact with other pads on the PCB. Gold-plated probes should be used to obtain the lowest possible parasitic resistance and to maintain the contact between the probes and the pads.

A possible setup for trace measurement is shown in Fig. 3 (b). The impedance of each trace needs to be measured and data collected by averaging over a number of measurements (five in our case) in order to mitigate the effect of random measurement noise. In Step 1, we select total n paths and measure their impedances on PCB c to create a set $d^{(c)} = [d_1^{(c)}, d_2^{(c)}, ..., d_n^{(c)}]$. In Step 2, we compare

978-1-4799-7598-3/15 $31.00 © 2015 IEEE

Fig. 3. (a) Schematic showing probe placement on a PCB substrate; (b) Terasic DE0 PCB measurement setup.

the impedances between any two paths on the same PCB by computing the distance between them. We then compute the vector $\overline{\Delta d^{(c)}}$ including $n(n-1)/2$ distance values (such as, $d_1^{(c)} - d_2^{(c)}$ and $d_1^{(c)} - d_3^{(c)}$). Thus, we get $\Delta d^{(c)} = [\Delta d_1^{(c)}, \Delta d_2^{(c)}, ..., \Delta d_{\frac{n(n-1)}{2}}^{(c)}]$. In Step 3, the normalization is done as: $\overline{d^{(c)}} = (\Delta d_i^{(c)} - min\Delta d^{(c)})/(max\Delta d^{(c)} - min\Delta d^{(c)})$. In Step 4, we select k specific bits (between i-th and j-th bit position) of the normalized values, to obtain $dig_{i,j}(\overline{\Delta d^{(c)}})$. We discard few least significant bits, which are highly vulnerable to environmental variations, as well as few most significant bits, which have poor variations. Finally, in Step 5, we combine the select bits from all normalized distances to generate a PCB signature of $k*n*(n-1)/2$ bits.

IV. RESULTS AND ANALYSIS

We have evaluated the proposed approach with two widely used commercial PCBs - in particular, sixteen Arduino UNO R3 Edition double-layer boards and twenty five Terasic DE0 four-layer boards. For both set of boards, first we judiciously selected a set of total 10 traces. Next we performed impedance measurements at room temperature, $25°C$, for each trace for five times and averaged the values to eliminate random noise. Based on the measured impedances for each PCB, we generated the authentication signatures. Similar to other PUFs, the uniqueness and robustness were evaluated by the Hamming Distance (HD) metric. Assuming $HD_{i,j}$ stands for the Inter-PCB HD between PCB_i and PCB_j, the average inter HD for m PCBs, denoted by HD_{avg}, was calculated as:

$$InterHD_{avg} = \frac{2}{m(m-1)} \sum_{i=1}^{m-1} \sum_{j=i+1}^{m} HD_{i,j} \quad (3)$$

In our experiment, we observed HD_{avg} of 50.24% based on 25 Terasic DE0 boards. Fig.4 (a) shows the histogram plot. The inter-PCB HD centers at around 50%, and ranges from 25% to 75%. As a result, the authentication of each PCB

can be completed successfully with good uniqueness of their signatures.

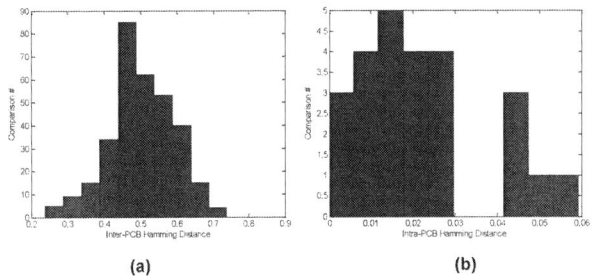

Fig. 4. (a) Inter-PCB HD; and (b) intra-PCB HD for Terasic DE0.

The robustness of the signature was evaluated under room temperature at two different times, which means the environment conditions, such as temperature, humidity and noise floor as well as the accuracy of the measurement setup were likely to vary between the two measurements. Assuming $HD_{p,q}$ stands for the intra HD of all boards between the pth measurement and qth measurement, the average intra-PCB HD for n times measurements on m PCBs, denoted by $IntraHD_{avg}$ was calculated as:

$$IntraHD_{avg} = \frac{2}{mn(n-1)} \sum_{1}^{m} \sum_{p=1}^{n-1} \sum_{q=p+1}^{n} HD_{p,q} \quad (4)$$

The distribution of intra-PCB HD is shown in Fig. 4 (b) with an average of 2.14%. The Arduino UNO R3 boards were evaluated in the same way. The histogram of inter-PCB HD is shown in Fig. 5 (a) with an average of 47.94% in the range between 25% to 70% and the histogram of intra-PCB HD is shown in Fig. 5 (b) with an average of 1.06%.

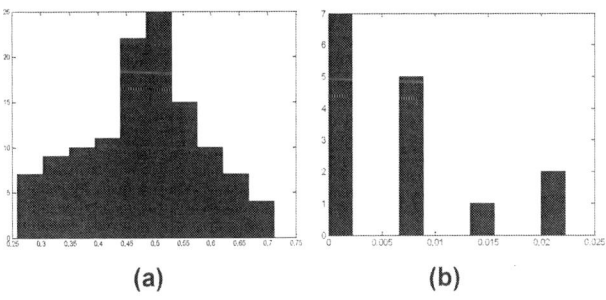

Fig. 5. (a) Inter-PCB HD; and (b) intra-PCB HD of Arduino UNO R3 SMD.

The average value of intra-PCB and inter-PCB Hamming Distance for both Arduino UNO R3 SMD and Terasic DE0 are shown in Table. I.

TABLE I
MEASURED RESULTS ON INTER AND INTRA-PCB HAMMING DISTANCE

PCB Type	Avg. Intra-PCB HD	Avg. Inter-PCB HD
Terasic DE0	2.14%	50.24%
Arduino UNO R3 SMD	1.06%	47.94%

The signature is generated from the copper traces on the PCB with statistical variations inherent in the manufacturing processes. Considering the large number of traces which can be used for authentication, it's practically infeasible to fully clone the signatures in cloned PCB instances. Even if the distributions are copied, if the number of traces for authentication is large enough, the signature is still practically unclonable. Therefore, the trace impedance based authentication is an effective and secure method. The end user can identify a cloned PCB by producing its signature and compare it with the manufacturer's database.

V. DESIGN FOR SECURITY

The methods described thus far in this paper have been geared towards identifying counterfeit, particularly cloned PCBs, already in production. However, it is possible to design a new PCB with a set of additional traces (and access points to facilitate probing) built in to help identify cloned boards. The impedance values for these additional traces can be easier to measure. They provide new and potentially better source of entropy as well. Since these traces are not used for signal propagation, they can be designed to suffer from increased variations. Furthermore, since they are not used for PCB operation, they are less likely to suffer from wearing or aging effects. Hence, signatures generated from these additional traces can be more robust against aging effects.

It is very likely that a cloned PCB will be made using a cheaper method than the original. If the copper in the original circuit board was milled away while the cloned circuit boards copper was etched out, then a properly designed board would have traces that were well designed for a mill while being poorly designed for an etching process. For example, both processes of fabricating a PCB will have issues in making a large obtuse angle. A milling process will have issues making the inner section of the angle while an etching process will have issues making the outer portion of the angle. This will affect the measurement properties of the trace when viewed as a micro strip.

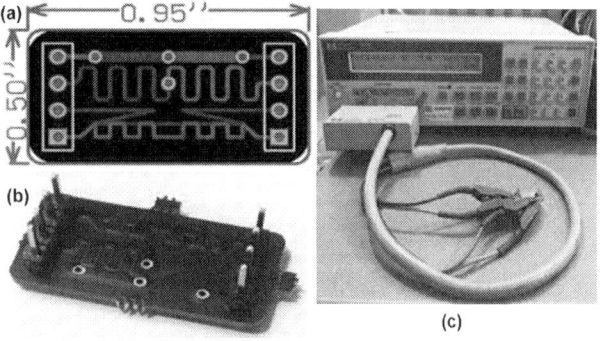

Fig. 6. (a) Suggested PCB authentication trace patterns in 0.5 square inches of PCB substrate; (b) photograph of the fabricated PCB with authentication patterns; and (c) measurement setup for these additional patterns.

Fig. 6 (a) shows a sample pattern that can be deployed to PCBs taking up less than 0.5 square inches. This pattern

consists of four independent traces. Each trace was designed to take advantage of limitations in cheaper PCB manufacturing processes. The top trace, Trace 1, uses four differently sized micro strip transmission lines, each one transitioning between sizes (and layers) with the use of a via. The second trace, Trace 2, simulates a controlled impedance transmission line. The third trace, Trace 3, utilizes sharp angles to expose the afore-mentioned weaknesses in the manufacturing process. The final trace, Trace 4, continues to expose weaknesses in the manufacturing process by having numerous sharp right angle bends in the trace path. If tested at a high frequency, then variances in the process can be effectively exposed. Finally, the Trace 3 and Trace 4 partially overlap on different layers of the PCB, enabling the characterization of crosstalk.

Using the design in Fig. 6 (a) 30 PCBs were manufactured by OSHPark to test the uniqueness and robustness of the signatures generated by the judiciously designed traces. A photograph of one of the fabricated PCBs is provided in Fig. 6 (b).

Each of the four traces were tested using the same methodology described in Section III. The testing setup is shown in Fig. 6 (c). The average trace impedance for Traces 1 – 4 are 9.93 mΩ, 39.76 mΩ, 15.26 mΩ, and 22.38 mΩ with a standard deviation of 0.82 mΩ, 2.44 mΩ, 1.13 mΩ, and 2.44 mΩ, respectively. It can be observed in Fig. 7 (a) to (d), which show the trace impedance histograms for the four traces.

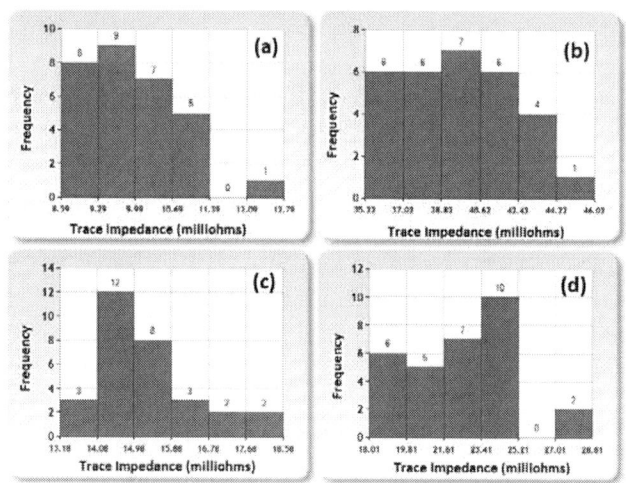

Fig. 7. OSHPark trace impedance histogram for: (a) Trace 1; (b) Trace 2; (c) Trace 3; and (d) Trace 4.

We observed that each of the four traces in the custom designed test PCB had a distinct signature that was easily measured by the test equipment. This is especially evident with the difference between Traces 2 and 3 (as shown in Fig. 7 (b) and Fig. 7 (c)) which have a similar histogram profile, however, Trace 2's resistance is almost three times greater than Trace 3's measured values. From these designs, it can be inferred that an individual trace can be identified from a group of traces and these traces provide high signature entropy.

VI. IMPEDANCE MEASUREMENT

In modern automated PCB production systems, one of the most important aspects is testing. This can range from test engineers physically handling the product to visually inspect for defects to robots probing the inner workings of the product for imperfections. In this section, we describe a system to fully implement the method of PCB identification and authentication described in Section III using automatic test equipment.

A common and effective PCB testing method is known as "flying probe" method of testing. Because the probes are on what is essentially an X-Y table, they can be moved into any position at any angle, allowing the probing of components such as ones depicted in Fig. 3 (a). Additionally, the probe heads are replaceable and four-wire probe heads, commonly called Kelvin Probes, are able to be attached to the "flying probe" system [12].

If we replace the heads of any common "flying probe" system with the heads described in [12] and attach them to any production-grade Micro-Ohm Meter, such as the Keysight 34420A, then very accurate and precise measurements of trace impedance can be taken. Next, through the use of the IEEE 488 General Purpose Interface Bus to communicate with a computer, the signature of a PCB under test can be automatically generated and verified without any user input [13].

Fig. 8. An example of a commercial flying-Probe based In-Circuit Tester (ICT) [15].

One example of the aforementioned solution that is used in production today is the Keysight Technologies, Inc. *Medalist* i3070 Series 5 In-Circuit Test System [14]. This is an integrated all-in-one flying probe-based tester, an example of which is shown in Fig. 8. One of its key features is a unique system where an user can add-in custom measurement cards and software solutions. The i3070 supports finding points on the Device-Under-Test (DUT) either manually or by programmed knowledge of the Device-Under-Test [14]. By combining these two features, accurate impedance measurements can be preformed automatically. Using these measurements a signature for a specific PCB can be created. Finally, by

communicating with a central server the signature can be saved for future verification by an end-user.

VII. CONCLUSION

We have presented a low-cost and robust approach to check PCB integrity in presence of counterfeiting attacks, in particular, cloning attacks. It relies on intrinsic manufacturing variations in metal traces of a PCB to create unique authentication signature from each PCB instance. Through detailed experimental analysis with common commercial boards, we have shown that the approach can be highly effective in robust PCB authentication. A major advantage of the proposed approach is that it does not require design modification or hardware overhead. We have shown that impedance measurements from 10 or more traces can provide adequate entropy for authentication. Existing production PCB testing setup can be used to automatically measure trace impedances for this purpose. We have also presented a DfS solution, where a PCB designer inserts carefully crafted additional wire traces, which are easier to measure and hidden from a board's surface. They simplify the authentication process and enhance the entropy.

VIII. ACKNOWLEDGEMENT

This work is funded in part by National Science Foundation grants #1245756 and #1054744.

REFERENCES

[1] "Defense Industrial base Assessment: Counterfeit Electronics", Bureau of Industry and Security, U.S. Department of Commence, Jan. 2010.
[2] "Integrated circuits china Manufacturer", BLD Electronic Co., Ltd. Available: http://dlbld.en.alibaba.com/product/954988019-215979726/integrated_circuits_china_Manufacturer.html
[3] "PCB Clone, PCB Copy, PCB Cloning, PCB Copying, PCB duplicating — PCB Reverse", HuaLan Technology. Available: http://www.hualantech.com/pcb-clone
[4] "DNA Marking and Authentication: A unique, secure anti-counterfeiting program for the electronics industry", *Applied DNA Sciences*, Stony Brook, NY, USA.
[5] S. Devadas, *et al.* "Design and implementation of PUF-based "unclonable" RFID ICs for anti-counterfeiting and security applications", *EEE International Conference on RFID*, 2008.
[6] G.E. Suh and S. Devadas, "Physical unclonable functions for device authentication and secret key generation", *DAC*, 2007.
[7] A.R. Krishna, S. Narasimhan, X. Wang, and S. Bhunia, "MECCA: A Robust Low-Overhead PUF Using Embedded Memory Array", *CHES*, 2011.
[8] Y. Zheng, A.R. Krishna, and S. Bhunia, "ScanPUF: Robust ultralow-overhead PUF using scan chain", *ASPDAC*, 2013.
[9] "ArduinoBoardUno". Available: http://arduino.cc/en/Main/ArduinoBoardUno
[10] "Design Guide for Electronic Packaging Utilizing High-Speed Techniques", 4th Working Draft, IPC-2251, February 2001.
[11] S. Ghosh, et al., "How Secure Are Printed Circuit Boards Against Trojan Attacks?", *IEEE Design & Test of Computers*, 2014.
[12] J.E. Boyette et al., "Dual-contact probe tip for flying probe tester". *US Patent 6023171*, February 8, 2000.
[13] Keysight 34420A NanoVolt/Micro-Ohm Meter, Available: http://literature.cdn.keysight.com/litweb/pdf/5968-0161EN.pdf
[14] Keysight *Medalist* i3070 In-Circuit Test System, Available: http://www.keysight.com/en/pc-1041067
[15] "ICT without Expensive Fixtures ACDI Expands Capabilities with In-House Flying Probe Tester", American Computer Development, Inc., 2011.

Field, Experimental, and Analytical Data on Large-scale HPC Systems and Evaluation of the Implications for Exascale System Design

Nathan DeBardeleben[1], Sean Blanchard[1], David Kaeli[2], and Paolo Rech[3]

[1]Ultrascale Systems Research Center, Los Alamos National Laboratory, NM, USA
[2]Northeastern University, Boston, MA, USA [3] UFRGS, Porto Alegre, Brazil
Email: ndebard@lanl.gov, seanb@lanl.gov, kaeli@ece.neu.edu, prech@inf.ufrgs.br

Abstract— Reliability is an issue for today's large scale computing systems designers, producers, and users. As we approach exascale, the resilience challenge will become critical due to increase in system-scale. It is then fundamental to understand the nature of errors, evaluate their probability of occurrence, and improve the design to reduce their impact on the overall system. In the paper we will present experimental, field, and analytical data to characterize and quantify errors on accelerators, providing a thorough understanding of errors impact on today and future large-scale systems.

I. INTRODUCTION

As we approach exascale, the resilience challenge will become critical due to increase in system-scale. Parallel accelerators like GPUs are anticipated to be a part of the projected path to exascale due to their ability of offer more FLOPS compared to the traditional CPUs. Therefore, understanding the nature of errors in both traditional CPUs and parallel computing system is critical for operating today's large-scale HPC systems as well as designing future extreme-scale systems. In fact, lack of understanding about the resilience of these emerging compute system components may lead to lower scientific productivity, lower operational efficiency, and significant monetary loss. This led us to conduct a detailed study to characterize and quantify errors on accelerators and processors, providing a thorough understanding of errors impact on large-scale systems. Our study aims at providing an updated, pragmatic, and analytic view on radiation-induced errors impact and effect on large-scale HPC systems. The presented results serve as an overview of the experience gathered on a GPU cluster at the Los Alamos National Lab. and on IBM BlueGene/Q processors. Moreover, we will take advantage of fault injection on GPUs to give insights on error propagation and characterize algorithms vulnerability.

II. RELIABILITY OF HPC SYSTEM HARDWARE

HPC systems, especially those at the U.S. Department of Energy laboratories, are by their very nature cutting edge. These systems often represent extremely early models of next generation hardware and as such can often be the harbingers of things to come. Furthermore, due to their large size most often hardware vendors are unable to fully test their products at scale before delivery to a supercomputing center. DoE HPC systems drive national policy, make scientific breakthroughs, and provide national security. These areas require a high level of confidence in the results of the computations performed. It is therefore essential to understand the reliability of HPC systems today and use that to drive improvements on future systems.

Resilience is considered one of the challenge areas for exascale[8] and that is largely due to a number of factors. These include a requirement to drive down computer power consumption (which may increase error rates), a need for first-of-its-kind hardware designs (which often are unproven), and a need for extreme system scale (which often results in considerably more components to fail). An early example of a technology that might be similar to the kind used in exascale is the GPGPU. At Los Alamos National Laboratory (LANL) in 2011 we deployed a cluster of several hundred M2090 NVIDIA GPGPUs with memory error protection. This system was a testbed to explore the technology and determine if it was appropriate for some of the workloads of the laboratory. Not uncommon with early hardware, we encountered problems including performance variability, bandwidth issues, inconsistent error logging, detectable uncorrectable errors, and silent data corruption [2], [9]. Many of the problems were mitigated over the lifetime of the cluster through a variety of changes including hardware replacements and driver upgrades. In a small testbed follow-on system with K20 NVIDIA GPGPUs we were unable to observe any of the previous issues. This story is indicative of the kinds of challenges seen deploying cutting edge technology and the HPC community should expect more of these as it moves towards exascale.

III. RESILIENCY CHARACTERIZATION AND IMPROVEMENT ON GPUS

We tested the raw sensitivity of the GPU memory structures for two GPUs: Fermi architecture based C2050 (similar to the Moonlight's GPU) and Kepler architecture based K20 (similar to the Titan's GPU). Since our goal for this part of the study is to measure the raw sensitivity of the SRAM structures of the GPUs, we disabled built-in ECC mechanism.

To measure the raw sensitivity of these devices, we store a particular pattern (all $1s$ or all $0s$) in the structure, expose the device to the controlled high-energy neutron flux available at LANSCE, LANL and, finally, check if radiation corrupted the initially stored values.

Figure 1 shows the *per-device* cross section for L2 cache and register file normalized to the cross section of K20 L2

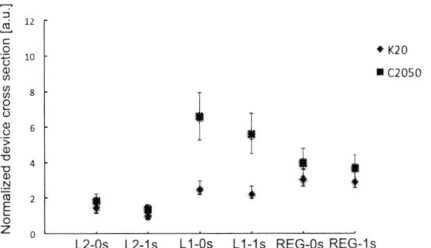

Fig. 1: Normalized cross sections for the K20 and C2050 structures. The K20 is less prone to bit corruptions than C2050.

cache obtained with the $1s$ pattern. The 95% confidence intervals for our results, which is a combination of neutron counts uncertainty and statistical error, are lower than 12% for all the reported values. Note that even if the K20 structures are significantly larger than the C2050 structures, the *total-area* cross section for the K20 is lower than the C2050 for L2, L1 caches and register file. This is an encouraging result, because at lower feature sizes (newer process technology), the reliability can get worse. The observed improvement can be attributed to various factors, including a better cell design in Kepler architecture. Unfortunately, the details of cell design are considered business-sensitive. However, this indicates that architects have improved the cell design to combat the danger of increased sensitivity at lower feature sizes. A more detailed analysis of GPUs application and parallel management resources reliability can be found in [9], [7], respectively

IV. FAULT INJECTION VS. AVF MODELING/SIMULATION IN GPUS

In previous device generations, Graphics Processing Units (GPUs) were designed for graphics rendering (written in DirectX and OpenGL). Graphics workloads tend to be fault insensitive as compared to emerging applications that leverage the GPU for computation. Graphics Processing Units (GPUs) have become the *compute accelerator of choice* for many general purpose and supercomputing applications. As more and more applications start to leverage the power of a GPU for compute acceleration, GPU reliability becomes a first-rate concern [11]. In this contribution, we consider soft-error vulnerability of next generation GPU designs.

Graphics processors are equipped with a large number of compute units, each provided with local memory to support vast context switching and a shared multi-level memory hierarchy to support high-bandwidth memory access. While spatial redundancy (e.g., parity or ECC) has been used to protect these devices, many bit flips in the data path can turn into silent data corruptions (SDCs). We need to develop better methods to evaluate design tradeoffs to address reliability and quantify vulnerability in next generation GPU devices. Previous studies on CPU have used statistical fault injection [4] and Architecturally Correct Execution (ACE) analysis to measure architectural vulnerability factors (AVFs) of a design [5], [1]. These two methods help designers analyze the vulnerability of an architecture in various stages of the design process.

In this work we consider the benefits and tradeoffs associated with fault injection and AVF analysis on next gener-

ation graphics processors. Each of these techniques presents different kinds of useful information when considering design reliability. Using the Multi2Sim [10] simulation infrastructure, we demonstrate how each of these techniques can be used to evaluate design tradeoffs as they impact the vulnerability and reliability of GPU designs [3], [12], [6]. Equipped with these tools, we consider how applications can be designed more robustly to reduce vulnerability.

V. CONCLUSION

The presented data serve as a pragmatic and precise estimation of the realistic error rate of modern large-scale systems exposed to natural radiation.

REFERENCES

[1] "Using hardware vulnerability factors to enhance avf analysis," in *Proceedings of the 37th Annual International Symposium on Computer Architecture*, ser. ISCA '10. New York, NY, USA: ACM, 2010, pp. 461–472.

[2] N. DeBardeleben, S. Blanchard, L. Monroe, P. Romero, D. Grunau, C. Idler, and C. Wright, "Gpu behavior on a large hpc cluster," in *Euro-Par 2013: Parallel Processing Workshops, Resilience 2013*. Springer Berlin Heidelberg, 2014, vol. 8374, pp. 680–689. [Online].

[3] N. Farazmand, R. Ubal, and D. Kaeli, "Statistical Fault Injection-Based Analysis in GPU Architecture," in *Proceedings of SELSE-8: Silicon Errors in Logic - System Effects*, March 2012.

[4] M.-C. Hsueh, T. K. Tsai, and R. K. Iyer, "Fault injection techniques and tools," *Computer*, vol. 30, no. 4, pp. 75–82, Apr. 1997.

[5] S. S. Mukherjee, C. Weaver, J. Emer, S. K. Reinhardt, and T. Austin, "A systematic methodology to compute the architectural vulnerability factors for a high-performance microprocessor," in *Proceedings of the 36th Annual IEEE/ACM International Symposium on Microarchitecture*, ser. MICRO 36. Los Alamitos, CA, USA: IEEE Computer Society Press, 2003, pp. 29–40.

[6] F. Previlon, M. Wilkening, V. Sridharan, S. Gurumurthi, and D. Kaeli, "Examining the Impact of ACE Interference of MultiBit AVF Estimates," in *Proceedings of SELSE-8: Silicon Errors in Logic - System Effects, year = 2015, month = March, location = Austin, Tx*.

[7] P. Rech, L. L. Pilla, P. O. A. Navaux, and L. Carro, "Impact of GPUs Parallelism Management on Safety-Critical and HPC Applications Reliability," in *DSN 2014*, Atlanta, USA, 2014.

[8] M. Snir, R. W. Wisniewski, J. A. Abraham, S. V. Adve, S. Bagchi, P. Balaji, J. Belak, P. Bose, F. Cappello, B. Carlson, A. A. Chien, P. Coteus, N. A. DeBardeleben, P. C. Diniz, C. Engelmann, M. Erez, S. Fazzari, A. Geist, R. Gupta, F. Johnson, S. Krishnamoorthy, S. Leyffer, D. Liberty, S. Mitra, T. Munson, R. Schreiber, J. Stearley, and E. V. Hensbergen, "Addressing failures in exascale computing," *International Journal of High Performance Computing Applications*, vol. 28, no. 2, pp. 129–173, 2014. [Online]. Available: http://hpc.sagepub.com/content/28/2/129.abstract

[9] D. Tiwari *et al.*, "Understanding GPU Errors on Large-scale HPC Systems and the Implications for System Design and Operation," *In HPCA*, 2015.

[10] R. Ubal, B. Jang, P. Mistry, D. Schaa, and D. Kaeli, "Multi2sim: A simulation framework for cpu-gpu computing," in *Proceedings of the 21st International Conference on Parallel Architectures and Compilation Techniques*, ser. PACT '12. New York, NY, USA: ACM, 2012, pp. 335–344.

[11] R. Ubal, D. Schaa, P. Mistry, X. Gong, Y. Ukidave, Z. Chen, G. Schirner, and D. Kaeli, "Exploring the heterogeneous design space for both performance and reliability," in *Proceedings of the 51st Annual Design Automation Conference*, ser. DAC '14. New York, NY, USA: ACM, 2014, pp. 1–6.

[12] M. Wilkening, V. Sridharan, S. Li, F. Previlon, S. Gurumurthi, and D. Kaeli, "Calculating architectural vulnerability factors for spatial multi-bit transient faults," in *Proceedings of the 47th Annual International Symposium on Microarchitecture*. Los Alamitos, CA, USA: IEEE Computer Society Press, Dec 2014, pp. 293–305.

Multi-Cycle Circuit Parameter Independent ATPG for Interconnect Open Defects

Dominik Erb*, Karsten Scheibler*, Matthias Sauer*, Sudhakar M. Reddy[†], Bernd Becker*

*University of Freiburg, Germany, {erb|scheibler|sauerm|becker}@informatik.uni-freiburg.de

[†]University of Iowa, USA, reddy@engineering.uiowa.edu

Abstract—**Interconnect opens are known to be one of the predominant defects in nanoscale technologies. Generating tests to detect such defects is challenging due to the need to accurately determine the coupling capacitances between the open net and its aggressors and fix the state of these aggressors during test. Process variations cause deviations from assumed values of circuit parameters thus potentially invalidating tests generated with assumed circuit parameters. Additionally, recent investigation using test chips showed that the steady state voltage on open nets may drift slowly with the application of circuit inputs and can be different at different nets.**

Recently we proposed a class of tests called Circuit Parameter Independent (CPI) tests to detect interconnect opens and reported on an implementation of a test generator for them. CPI tests detect opens independent of the values of coupling capacitances and the initial trapped charge on the open net and hence are robust against process variations affecting these parameters. Yet, this work did not address the effects of leakage currents on open nets.

In this work we extend the validity of CPI tests by introducing so-called multi-cycle CPI tests and single-value CPI tests. By doing so, we significantly improve the coverage of open defects and ensure their detection whilst including the additional effect of leakage currents on opens. Experimental results for circuits with over 500k non-equivalent faults and several thousand aggressors show the effectiveness of the newly proposed CPI tests as well as the high efficiency of a new ATPG algorithm to generate these new classes of CPI tests.

Index Terms—**interconnect opens, SAT, test generation, ATPG, circuit parameter independent tests**

I. INTRODUCTION

Interconnect open defects constitute a high percentage of defects in nanoscale VLSI circuits [1] since interconnects account for a very high percentage of chip area. Interconnect opens could be partial, resistive or full opens. In this work we study tests to detect full opens. An open isolates driven gates from the driving sources – the affected gate inputs are *"floating"*. The voltage on a floating interconnect caused by an open may depend on the state of the neighboring interconnects (called aggressors), the values of the coupling capacitances to these aggressors as well as to power and ground lines and to the substrate. Also initial trapped charge, leakage currents and the internal capacitances of the gates driven by the floating interconnect may affect the voltage [2–11]. Leakage currents affect the open net in a temporal way such that the steady state voltage on an open net changes slowly and in an unpredictable manner with the application of inputs to the circuit under test [12]. The values of the capacitances, the trapped charge and the effect of leakage currents may be difficult, if not impossible, to accurately determine. For example, the value of the trapped charge is unknown and its polarity could be both positive and negative [4]. Thus, it is critical to develop methods to detect opens that avoid the need to know the above circuit parameters.

Earlier approaches to generate tests to detect opens can be classified into two categories: The first category of approaches requires the knowledge of the circuit parameters such as coupling capacitances, initial trapped charge etc. [2–9, 13]. Yet, coupling capacitance values may differ from assumed values due to process variations and more importantly the unknown initial trapped charge and the leakage current induced voltage on open nets may invalidate such tests. The other category of tests we refer to as

parameter independent tests, only require the neighboring nets whose state may affect the voltage on the floating net [10, 11, 14].

The test generated in [10] called maximal favorable neighborhood tests do not guarantee detection of open net faults. The tests proposed in [11] require several stuck-at tests while setting the aggressors of an open net to all possible combinations of states. Thus the number of tests generated could be extremely large. Furthermore, these tests may also be invalidated due to the time varying nature of the voltage induced on the open net by leakage currents. The method in [14] proposed to derive a stuck-at 0 and a stuck-at 1 test for each floating net with its aggressors in the same state for the two tests. It was proved that if a net is floating then one of the two tests will detect the open fault. In an earlier work we presented the first complete ATPG algorithm which generates circuit parameter independent (CPI) tests for the model proposed in [14] and thus are robust against process variations affecting the influence of aggressors and trapped charge [15]. Furthermore, oscillating behavior in the context of CPI tests was handled for the first time. However, the presented approach assumed two independent tests and consequently requires two scan loads for each CPI test. Moreover, *leakage current effects* [12] were not considered. Specifically, the CPI tests proposed in[14, 15] use two separate scan loads which may be separated by several clock cycles and hence may not be effective since detection of net opens by these tests is based on the assumption that the initial voltage on the open net due to trapped charge and leakage current is the same during the application of a related pair of stuck-at-0 and stuck-at-1 tests.

In this paper we propose new classes of CPI tests to extend the approach proposed in [14, 15]. Specifically we present:

- Multi-cycle CPI tests which are similar to [14, 15] but use a single scan load and launch off capture *(LOC)* test method with two capture cycles containing a stuck-at 0 and a stuck-at 1 test for the faulty net.
- A novel type of CPI tests (single-value CPI tests) to detect interconnect opens by two related tests for stuck-at 0 or two related tests for stuck-at-1 at the faulty net. This new class of CPI tests improve the coverage of CPI tests by detecting opens not detectable by tests of [14, 15].
- A novel ATPG system generating normal and single-value CPI tests within independent cycles and in multi-cycle scenario using two LOC connected cycles to achieve high fault coverage.

The newly proposed multi-cycle CPI tests use the LOC test application method and hence require only a single scan load instead of two scan loads for the earlier CPI tests. More importantly, they address the effect of leakage current on the open interconnect [12] – as the two related stuck-at-0 and stuck-at-1 tests are applied using a single scan load and two consecutive cycles and thus can expect the initial voltage on the open net to be the same when either of the tests is applied.

Experimental results for the largest ISCAS89 circuits and in total over 1 500 000 non-equivalent faults show that the proposed single-value CPI tests increase significantly the number of faults for which a CPI test exists. In addition, for most of the faults a

multi-cycle CPI test exists which guarantees robustness against leakage current effects.

The rest of the paper is organized as follows. Section II introduces the terminology and the CPI test model and Section III discusses the newly proposed multi-cycle and single-value CPI tests. Section IV explains the algorithm for generating multi-cycle CPI as well as single-value CPI tests in detail, Section V discusses experimental results and Section VI concludes the paper.

II. TERMINOLOGY AND CPI TEST MODEL

An interconnect consists of a source, segments and one or multiple sinks as shown in Figure 1. A segment (called RC-element) is a combination of a resistor r and zero or more capacitances CC. The capacitances represent the influence of surrounding interconnects (*aggressors*) on the output of a segment. According to [16], all open defects on a given piece of an interconnect are mapped to an open fault at the output of the corresponding RC-element and break the interconnect in two parts: the first starting from the source is the stable part and fault-free; the second is disconnected from the source and affected by numerous electric parameters. This part is also called the *floating part*.

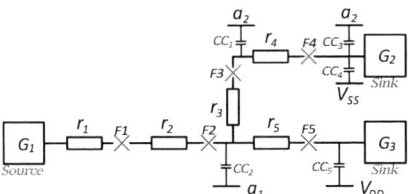

Fig. 1: **Example of an interconnect.**

Several models exist to describe open defects based on layout information. Sato et al. [9] introduced the *Aggressor Victim model* and stated that the floating part is mainly influenced by aggressors. The value of the coupling capacitance (CC_i) determines the influence of the aggressor i on the floating part. V_{DD} and V_{SS} can occur as aggressors, but their value is constant.

Figure 2 shows a full-scan sequential circuit with a fault located at the source of the interconnect (I_1) influenced by four aggressors: S_1, I_2, V_{SS} and S_3. For the pattern $S_1 = 1, I_1 = 0, I_2 = 1$ a logic 1 will be stored in the flip-flop in a fault free circuit and the aggressors S_1, I_2 and S_3 show logic 1, while V_{SS} shows logic 0. In [9] a fault is testable, if one or more aggressors induce a parasitic coupling capacitance to at least one RC-element of the floating part. C_0 (C_1) represents the cumulative coupling capacitance of all aggressors showing logic 0 (logic 1). If the cumulative coupling capacitance C_1 exceeds C_0, all gates driven by the floating part interpret the voltage as logic 1 – or logic 0 otherwise. Hence, for the example in Figure 2 the pattern $S_1 = 1, I_1 = 0, I_2 = 1$ detects the fault – as gate G_1 and G_2 interpret the voltage as logic 1, resulting in the faulty value logic 0 to be stored in the scan flip-flop.

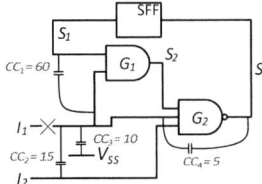

Fig. 2: **Open fault with full knowledge of circuit parameters.**

It is important to note that if initial trapped charge and/or the leakage currents are introduced into the fault model, then an unknown voltage has to be added to the voltage on the floating net. This unknown voltage may have a time dependency [12] when leakage current effects are included and dependent on its

influence may invalidate tests based on assumed circuit parameters. In contrast, the proposed multi-cycle CPI tests detect opens irrespective of the precise influence of trapped charge and/or the leakage currents.

A. Circuit Parameter Independent test model

In order to detect interconnect open defects independent of the actual influence of each aggressor affecting the floating part, [14, 15] proposed *circuit parameter independent* (CPI) tests. CPI tests do not assume knowledge of the circuit parameters such as coupling capacitances and initial trapped charge and hence remain valid for any value of these parameters. As proposed by [9] for interconnect open defects, each affected gate is assumed to interpret the voltage the same.

The central idea for CPI tests is the following: If for the same fault but two different patterns the logic values of all aggressors are equal, then the voltage of the floating part is equal too – and consequently all gates will interpret it in both tests the same (as either logic 0 or logic 1). Therefore, two patterns are generated per CPI test – one, assuming the voltage to be interpreted as logic 0 (logic 1) and the other assuming the voltage to be interpreted as the opposite logic value while forcing all aggressors to hold their value.

Furthermore, these conditions could be relaxed if prior to each test one is selected as *leading frame* enforcing constraints to the non-leading test. Figure 3 shows an example for such a CPI test and the fault shown in Figure 2. Within this test, the first pattern is selected as leading frame and assumes a segment stuck-at 0 fault to be present at the fault site (in the following called F_{SA0}) in order to handle the case that the voltage of the floating part is interpreted as logic 0. The second pattern assumes a segment stuck-at 1 fault (F_{SA1}) in case the voltage is interpreted as logic 1. Within F_{SA0} the aggressors S_3 and V_{SS} show a logic 0 in a fault-free circuit while S_1 and I_2 show a logic 1. Now two possible scenarios may occur: (1) either S_3 and V_{SS} induce such a voltage to the floating part, that it is indeed interpreted as logic 0 and F_{SA0} will detect the open fault or (2) the assumption is wrong and S_1 and I_2 force the voltage to be interpreted as logic 1. Thus, all aggressors showing logic 1 within the first test need to keep their value in the second test while all other aggressors may be freely assigned – if they stay at logic 0 then nothing would change and if they switch to logic 1 the already present logic 1 would be "strengthened". Hence, regarding our example, S_1 and I_2 need to show logic 1 within the second test while S_3 and V_{SS} may be freely chosen. The same argument applies for the opposite direction if F_{SA1} is selected as leading frame. Either the aggressors showing logic 1 within F_{SA1} are sufficient and the voltage is indeed interpreted as logic 1 – or all aggressors showing logic 0 in F_{SA1} are stronger and the voltage needs to be interpreted as logic 0 – and thus F_{SA0} will detect the open fault.

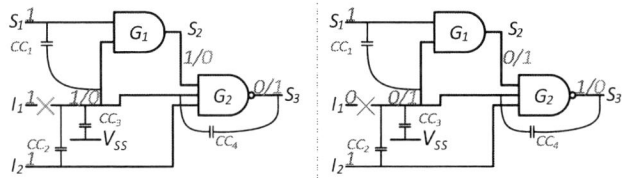

Fig. 3: **CPI test for the fault in Figure 2.**

As stated above, in [14, 15] it is assumed that both patterns are completely independent and consequently for full-scan sequential circuits the consideration of the combinational core is sufficient. In the following, we refer to these tests as *independent CPI tests*.

978-1-4799-7598-3/15 $31.00 © 2015 IEEE   237

However, for full-scan sequential circuits the use of independent CPI tests may result in a lot of scan operations between the application of the pair of tests constituting a CPI test. The time lapse caused by scan operations may invalidate the assumption that the steady state voltage determined by the initial trapped charge and leakage currents remains the same at the application of the test pair. This could invalidate the CPI tests. Hence, for full-scan sequential circuits, *multi-cycle CPI tests* (ref. to Section III-A) may be desirable to reduce the amount of scan-operations and to achieve robustness against leakage current effects.

B. Oscillating behavior within CPI tests

While the original model of [14] did not handle oscillation, [15] distinguished between static CPI tests (guaranteed to be oscillation free) and dynamic CPI tests which may show oscillating behaviour – depending on the actual influence of each aggressor.

Consider output S_3 in Figure 3. Because S_3 itself depends on the fault site, an oscillating behavior might be observed [17][1]. As stated in Section II-A, a fault is detected by an independent CPI test iff either F_{SA1} is leading and all aggressors showing logic 0 within F_{SA1} also show logic 0 within F_{SA0} – or F_{SA0} is leading and aggressors with logic 1 also show logic 1 in F_{SA1}. Additionally, an independent CPI test with F_{SA0} leading is *guaranteed* to be oscillation free (static CPI test) if (1) all aggressors showing logic 0 in F_{SA0} keep this value in F_{SA0} and (2) all aggressors showing logic 1 in F_{SA0} show logic 1 in F_{SA1} no matter they are affected by the fault or not. Similar requirements exist if F_{SA1} is leading. Hence, the independent CPI test shown in Figure 3 only detects the fault dynamically (as the value of aggressor S_3 differs in the fault affected case).

Yet, a dynamic CPI test does not imply that oscillation occurs in a later test – as this highly depends on the influence of each aggressor. Thus, if we consider the actual capacitance values of each aggressor (cf. Figure 2) the non-static independent CPI test of Figure 3 will indeed detect the fault statically (as the influence of S_3 is too weak).

III. IMPROVING THE CPI TEST MODEL

This section describes new classes of CPI tests which extend the CPI tests of [15] in two directions: (1) *multi-cycle CPI tests* allow to half the number of scan operations per test and guarantee robustness against leakage current, (2) *single-value independent CPI* and *single-value multi-cycle CPI tests* additionally increase the overall fault coverage.

A. Multi-cycle CPI test model

For full-scan sequential circuits it may be beneficial to prefer *multi-cycle CPI tests* considering two or more connected launch off capture (LOC) or launch off shift (LOS) cycles instead of independent CPI tests as proposed in [15]. As described in Section II-A, each CPI test consists of two patterns: one assumes a segment stuck-at 0 (F_{SA0}) and one assumes a segment stuck-at 1 (F_{SA1}). Furthermore, as we don't know which assumption reflects the actual behavior of the considered open fault both fault-effects need to be propagated to a primary output or to scan cells.

Consequently, we propose to use two LOC-connected cycles for each multi-cycle CPI test – as this is sufficient to incorporate F_{SA0} and F_{SA1}[2]. This scheme (1) allows the detection with

as few as possible cycles without intervening shifts and (2) guarantees each multi-cycle CPI test to be robust against leakage-current-induced variations in the voltage on the open net.

Yet, several problems need to be addressed in order to transfer the CPI model described in Section II-A to a multi-cycle scenario considering two LOC-connected cycles. Thus, both assumptions (F_{SA0} and F_{SA1}) need to be incorporated into the same test and propagated to at least one (possibly the same) output – irrespective if in the real circuit the segment stuck-at 0 and/or the segment stuck-at 1 is observable. Furthermore, for multi-cycle CPI tests both cycles could be selected as *leading frame* enforcing constraints to the non-leading frame. Therefore, the definition needs to be extended in order to reflect this. Finally, in case both assumptions are true and both fault-effects are present, the static constraints need to be adapted to guarantee validity.

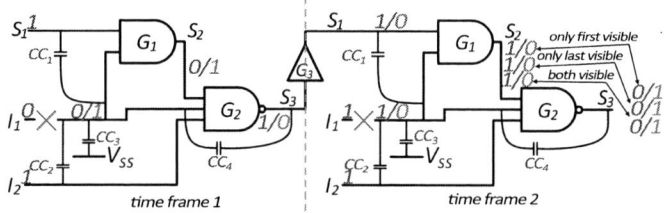

Fig. 4: **Multi-cycle CPI test for the fault in Figure 2.**

Figure 4 shows an example for such a multi-cycle CPI test and the circuit presented in Figure 2. Within this example, the flip-flop is used (1) as input (signal S_1) of the first time frame, (2) as a buffer (G_3) connecting the two time frames, and (3) as output of time frame 2 (S_3). Furthermore, we selected the last cycle as leading frame and assumed a segment stuck-at 0 (F_{SA0}). Thus, the first cycle needs to represent a segment stuck-at 1 (F_{SA1}) and all aggressors showing logic 1 in F_{SA0} (i.e. S_1 and I_2) need to show logic 1 in the first cycle as well – while no constraints for all other aggressors are required (S_3 and V_{SS}). Furthermore, the fault-effects of F_{SA0} and F_{SA1} need to be propagated no matter if one or both of them reflect the actual behavior of the considered open fault. This is represented by the three differently colored numbers. Red, in case F_{SA1} reflects the actual behavior of the circuit. Green, in case of F_{SA0} and light blue in case the aggressors force the voltage to be interpreted as logic 1 in time frame 1 and at the same time as logic 0 in time frame 2. For all three cases a faulty logic 1 is present at signal S_3 instead of a fault-free logic 0 and hence the fault is at least dynamically detected.

For a *static multi-cycle CPI test* the constraints described in Section II-B can be directly transfered to multi-cycle tests. For the example shown in Figure 4 (assuming the last frame is leading and a segment stuck-at 0 (F_{SA0})), all aggressors showing logic 0 in F_{SA0} need to keep this value in this cycle irrespective F_{SA0} is visible or not and all aggressors showing logic 1 in a fault free circuit in F_{SA1} need to show logic 1 in F_{SA1} no matter the assumed stuck-at 1 is present or not. Hence, the values of aggressors influenced by F_{SA1} (which is inserted in time frame 1) do not have to be considered within time frame 2. This is because F_{SA1} may only influence F_{SA0} in case a segment stuck-at 1 was already present in time frame 1 and consequently F_{SA1} already allows to detect the fault – as we demand propagation irrespective of which reflects the actual behavior. Regarding Figure 4, the shown test only allows to detect the fault dynamically – because aggressor S_3 shows logic 0 in a fault free circuit but not in case F_{SA0} reflects the actual behavior.

Earlier in [18], a class of tests called Double Observation

---

[1]As it may happen, that depending on the logic value of aggressor S_3, the affected gates interpret the induced voltage differently and as a result the logic value of output S_3 toggles between logic 0 and logic 1.

[2]For a two-cycle scenario the use of classical LOS is disadvantageous as within shift-cycles the fault-effect could only be propagated to primary outputs and not captured within a scan flip-flop.

978-1-4799-7598-3/15 $31.00 © 2015 IEEE

was proposed. They use an LOC test with two capture cycles to detect a stuck-at 0 and a stuck-at 1 fault at an open net whose voltage cannot be influenced by aggressors. The difference with the proposed multi-cycle tests is that the Double Observation requires fault-effect propagation of stuck-at 0 and stuck-at 1 tests to different outputs, which is a stronger but not a necessary condition for our ATPG. Additionally, the tests did not consider the most common case of opens whose node voltages can be influenced by aggressors.

B. Single-value CPI tests

Several reasons could lead to the problem that no independent CPI and consequently also no multi-cycle CPI test exists: for some faults it is impossible to adjust both, a segment stuck-at 0 and a segment stuck-at 1 even within independent cycles. Furthermore, especially in the multi-cycle scenario the propagation of both fault-effects to the outputs – irrespective if only one or both are observable – might prohibit a valid CPI test. Finally, if many aggressors affect the faulty interconnect, the large amount of constraints to the aggressors might hinder the propagation of one or both fault-effects.

In all these cases a single-value CPI test (in an independent or multi-cycle fashion) may help to increase the overall fault coverage (note, single-value CPI test could not guarantee robustness against trapped charge and hence it should only be used in case no other CPI test exists). Similar to normal CPI tests we assume for single-value CPI tests a segment stuck-at 1 (segment-stuck-at 0) to be present at the fault site and select it as leading frame. Now two possible scenarios could occur: (1) either the assumption was correct and the test will detect the fault or (2) we know that all aggressors which show the opposite logic value are stronger. In contrast to normal CPI we demand the fault site to keep its value while all aggressors which show the opposite logic value are required to show the value of the fault site (while no constraints for the remaining aggressors exist). Hence, the main difference between normal CPI tests (as explained in the Sections II-A, II-B and III-A) and single-value CPI tests is that the aggressors need to change their value and not the fault site – while all other requirements are similar despite the restriction that a single-value CPI test is not possible in case V_{DD} and V_{SS} occur as aggressors at the same time. This is because these aggressors are unable to show the opposite logic value and consequently a single-value CPI test assuming a segment-stuck-at 1 (segment stuck-at 0) in both cycles is only possible in case the interconnect is not affected by V_{SS} (V_{DD}).

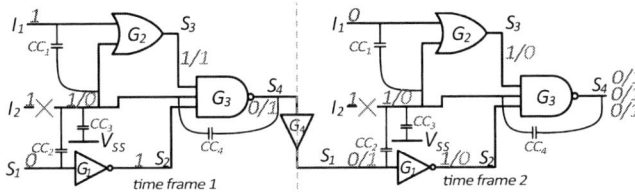

Fig. 5: **Example for a single-value multi-cycle CPI test.**

Figure 5 shows an example for a fault for which no normal multi-cycle CPI test exists. On the one hand, time frame 2 is not allowed to assume a segment stuck-at 1 – because this prevents the propagation of the fault assumed in time frame 1. On the other hand, time frame 1 is not allowed to assume a segment stuck-at 1 because it always violates the requirements for aggressor S_1. Yet, a single-value CPI test exists in case time frame 1 is leading and a segment stuck-at 0 is assumed in both cycles – because both fault-effects are visible at the output (irrespective of which

one reflects the actual behavior) and aggressor I_1 which showed logic 1 in time frame 1 changes to logic 0 in time frame 2.

IV. CPI MULTI-CYCLE ATPG

This section provides an overview of the proposed SAT-based ATPG algorithm and the main constraints for generating multi-cycle CPI tests guaranteeing static or dynamic detection and using two LOC-connected cycles.

The proposed ATPG procedure highly exploits the capabilities of incremental SAT solving and consequently already the generation of tests guaranteeing only dynamic detectability is divided in two parts: (1) the encoding of the circuit (Section IV-B) and (2) the encoding of dynamic constraints (Section IV-C). This incremental strategy allows a fast abort of untestable faults (e.g. in case when the circuit structure already prohibits a valid test) and helps reducing the runtime of the overall algorithm.

When a multi-cycle CPI test detecting the fault dynamically exists and static detection is desired, the instance guaranteeing dynamic detection is extended by the constraints to ensure static detection which are described in Section IV-D.

Finally, Section IV-E describes all required adjustments to generate single-value independent CPI tests and single-value multi-cycle CPI tests.

A. SAT formula and incremental solving

Combinatorial circuits can also be represented as Boolean formulas. SAT solvers can be used to determine if a Boolean formula has a solution – i.e. whether an assignment to the variables exists such that the formula evaluates to true. In this paper we use Boolean formulas to encode ATPG instances. If a SAT solver finds an assignment, a test is found – if the formula is unsatisfiable the fault is untestable.

Usually, SAT solvers do not operate on arbitrary Boolean formulas – instead they expect a so-called *conjunctive normal form* (CNF). A CNF consists of a conjunction of clauses with each clause being a disjunction of literals. Each literal is either a boolean variable or the negation thereof. The Tseitin-transformation [19] allows the conversion of arbitrary Boolean formulas into CNF. When considering CNF formulas with a common set of clauses *incremental solving* can be used to further speed-up the solving process. For more details on SAT solvers and incremental solving refer to [20, 21].

B. Boolean encoding of the circuit

As described in Section III-A and Section III-B, each multi-cycle test consists of two patterns. In the following, we use F_{TF1} and F_{TF2} for the value assumed at the fault site in the first and second cycle. In case normal multi-cycle CPI tests are considered, the first cycle needs to show logic 0 ($F_{TF1} = 0$) or logic 1 ($F_{TF1} = 1$) and the second cycle needs to show the opposite logic value ($F_{TF2} = \neg F_{TF1}$). In case single-value multi-cycle CPI tests are considered, the second cycle needs to show the same logic value ($F_{TF2} = F_{TF1}$).

Furthermore, we need to ensure detection of the considered fault irrespective of whether the assumption made in the first, the second or both cycles reflects the behavior of the fault within a later test. Thus, we need to consider four different instances of the circuit: (1) a fault-free version, (2) a faulty version with segment stuck-at F_{TF1}, (3) a faulty version with segment stuck-at F_{TF2} and (4) a faulty version with both F_{TF1} and F_{TF2} being present. In fact, we do not consider the whole circuit – but only those signals and gates (sub-circuits) relevant for adjustment and propagation of the currently examined open fault. We denote these sub-circuits with SC_{FF} (fault-free), SC_{F1} (F_{TF1} present), SC_{F2} (F_{TF2} present), and SC_{BF} (both faults present). Additionally, we exploit the fact that some gates and signals are shared between the different sub-circuits. For example SC_{F1} will reference gates and signals from SC_{FF} if they are not affected by F_{TF1}. Similar examples for SC_{F2} and SC_{BF} exist as well.

The Tseitin-transformation [19] is used to transform SC_{FF}, SC_{F1}, SC_{F2}, and SC_{BF} into a CNF. Additionally, we encode D-Chains [22] into the CNF in order to speed-up the ATPG. Further clauses are added to: (1) describe the relation between F_{TF1} and F_{TF2} as described above and (2) guarantee a difference at at least one output in case F_{TF1}, F_{TF2} or both are visible.

C. Constraints to ensure dynamic detection

As described in Section III-A, the constraints which need to be fulfilled by a multi-cycle CPI test in order to guarantee it detects a fault at least dynamically, highly depend on which frame is leading and which value is inserted in this frame. In the following, we write L_{TF1} in case time frame 1 is leading and otherwise $\neg L_{TF1}$ which implies that not time frame 1 and therefore time frame 2 is leading. Furthermore, only the values of each aggressor in SC_{FF} need to be considered (e.g. $A_{FF,i}^{TF1}$ for aggressor i of time frame 1), as we do not handle oscillation.

In case time frame 1 is leading and represents a segment stuck-at 0 ($F_{TF1} = 0$), all aggressors showing logic 1 in time frame 1, need to show logic 1 in time frame 2:

$$C1 : \bigwedge_{\forall i}(L_{TF1} \rightarrow (F_{TF1} \vee \neg A_{FF,i}^{TF1} \vee A_{FF,i}^{TF2}))$$

In case time frame 2 is leading and represents a segment stuck-at 1 ($F_{TF2} = 1$ implying $F_{TF1} = 0$), all aggressors showing logic 0 in time frame 2, need to show logic 0 in time frame 1:

$$C2 : \bigwedge_{\forall i}(\neg L_{TF1} \rightarrow (F_{TF1} \vee A_{FF,i}^{TF2} \vee \neg A_{FF,i}^{TF1}))$$

Obviously, $C1$ and $C2$ are equivalent to:
$$\bigwedge_{\forall i}(F_{TF1} \vee \neg A_{FF,i}^{TF1} \vee A_{FF,i}^{TF2})$$

Similarly, the both other cases yield:
$$\bigwedge_{\forall i}(\neg F_{TF1} \vee A_{FF,i}^{TF1} \vee \neg A_{FF,i}^{TF2})$$

D. Additional constraints to ensure static detection

To ensure static detection, only constraints for aggressors which are themselves dependent on the fault site need to be added – for all other aggressors the constraints described in section IV-C are sufficient. Furthermore, as described in section III-A, to guarantee static detection for the fault assumed in the second cycle (F_{TF2}), the value of the aggressors of SC_{F2} in time frame 2 needs to be considered (e.g.$A_{F2,j}^{TF2}$ for aggressor j) while for F_{TF1} the consideration of the value in SC_{F1} in time frame 1 is sufficient. Similar to dynamic constraints, we write L_{TF1} if time frame 1 is leading. Otherwise time frame 2 is leading implying $\neg L_{TF1}$. In contrast, we use F_{TF1} and F_{TF2} directly as no easy simplifications are possible with the help of $F_{TF2} = \neg F_{TF1}$.

If time frame 1 is leading, we get the following constraints for an aggressor j which is itself affected by the fault:

$$L_{TF1} \rightarrow$$
$$((F_{TF1} \vee A_{FF,j}^{TF1} \vee \neg A_{F1,j}^{TF1}) \wedge (F_{TF1} \vee \neg A_{FF,j}^{TF1} \vee A_{F2,j}^{TF2}) \wedge$$
$$(\neg F_{TF1} \vee \neg A_{FF,j}^{TF1} \vee A_{F1,j}^{TF1}) \wedge (\neg F_{TF1} \vee A_{FF,j}^{TF1} \vee \neg A_{F2,j}^{TF2}))$$

If time frame 2 is leading, the constraints are:

$$\neg L_{TF1} \rightarrow$$
$$((F_{TF2} \vee A_{FF,j}^{TF2} \vee \neg A_{F2,j}^{TF2}) \wedge (F_{TF2} \vee \neg A_{FF,j}^{TF2} \vee A_{F1,j}^{TF1}) \wedge$$
$$(\neg F_{TF2} \vee \neg A_{FF,j}^{TF2} \vee A_{F2,j}^{TF2}) \wedge (\neg F_{TF2} \vee A_{FF,j}^{TF2} \vee \neg A_{F1,j}^{TF1}))$$

E. Required adjustments for generating single-value independent CPI and single-value multi-cycle CPI tests

To generate single-value multi-cycle CPI tests with the algorithm described above, only a few adjustments are necessary. First, the computation of F_{TF2} is adjusted as explained in Section IV. Furthermore, in case V_{DD} or V_{SS} occur as aggressors, F_{TF1} needs to be restricted to not show the corresponding value. At last, the constraints introduced in Section IV-C, IV-D need to be adjusted to demand the aggressors of the non-leading frame to change their values and not to stay the same. Thus, e.g. the constraints ensuring dynamic detectability change to:

$$\bigwedge_{\forall i}((F_{TF1} \vee \neg A_{FF,i}^{TF1} \vee \neg A_{FF,i}^{TF2}) \wedge (\neg F_{TF1} \vee A_{FF,i}^{TF1} \vee A_{FF,i}^{TF2}))$$

Restricted by the space limitations, we omit the details for static constraints here. Yet, the adjustments are similar to the ones explained for dynamic constraints.

The generation of single-value independent CPI tests is similar to the generation of single-value multi-cycle CPI tests. However, as the name independent CPI tests suggests, for these tests the two cycles are considered to be completely independent and not connected regarding fault adjustment and propagation – only the aggressors need to be constrained.

V. EVALUATION

We evaluated the proposed approach using full-scan versions of ISCAS89 benchmarks and the flow explained in [23] for the circuit layout, and the extraction of the aggressors. Note, as we generate CPI tests, no parasitic coupling capacitances are required. We always used the union of intra-layer (via-open) and inter-layer open faults and merged equivalent faults. All experiments are conducted on a single core of an Intel Xeon CPU running at 3.3 GHz. As solver back-end iSAT3 [21] is used with a timeout of one second – without having any aborts.

A. Comparison of different open defect ATPG algorithms

The first experiment compares an ATPG having full knowledge of circuit parameters and assuming zero trapped charge and no effect of leakage currents, with the CPI ATPG algorithms proposed in this paper. Figure 6 shows the achieved fault coverage for circuit s38417 with 557 895 non-equivalent faults for five ATPG approaches. It shows, that ATPG assuming full knowledge allows to classify over 99% of the faults as detected. Furthermore, for 90.00% of the faults, an independent CPI test exists (ind. CPI ATPG) which guarantees robustness against variations affecting the influence of aggressors as well as trapped charge. The generation of single-value tests increases it slightly to 90.33% (ind. CPI ATPG + single). Finally, the generation of multi-cycle CPI tests (including single-value tests) allows to achieve robustness against the leakage current effects for 80.24% (82.20%) of the faults.

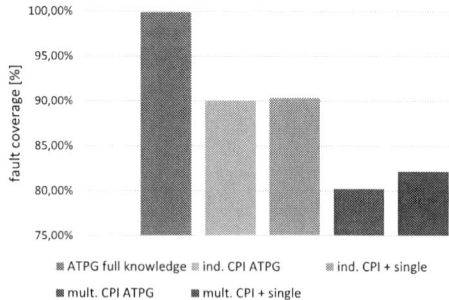

Fig. 6: Comparison of different open defect ATPG algorithms.

978-1-4799-7598-3/15 $31.00 © 2015 IEEE

To give some information on the impact of including single-value tests for all circuits shown in Table I, the generation of single-value tests in addition to independent CPI tests allowed to increase fault coverage on average by 1.22% . If included into the multi-cycle CPI approach, the impact is even larger and thus for 4.47% more faults a multi-cycle CPI test is generated.

B. Results of the proposed CPI ATPG

In the second experiment we first used the independent CPI ATPG proposed in [15] extended by single-value CPI tests (Section III-B) to measure for how many open faults an independent CPI test exists. Whenever the ATPG was able to generate such a test, afterwards the newly proposed multi-cycle ATPG (Section III-A) using two LOC-connected cycles and including single-value CPI tests was used to achieve robustness against leakage current.

Table I shows the achieved results for the largest ISCAS89 benchmarks. The first three columns contain the circuit name along with the number of gates and non-equivalent faults. Columns 4 and 5 contain the number of faults for which a static independent CPI test exists in absolute numbers as well as in %. Columns 6 and 7 additionally show the absolute number of faults for which a multi-cycle CPI test exists and the percentage of independent CPI tests the multi-cycle approach was able to generate a test for. Columns 8 to 11 contain the same information for dynamic CPI tests and columns 12 and 13 contain the number of faults for which no independent CPI test exists – and consequently also no multi-cycle CPI test. Finally, column 14 contains the number of aborts and column 15 the overall runtime of the ATPG including independent CPI ATPG + single-value CPI tests and multi-cycle CPI ATPG + single-value CPI tests.

The results show, that overall the proposed ATPG was able to generate an independent CPI test for 1 290 361 out of 1 507 380 faults (i.e. 85.60%) while all other faults are CPI untestable. Furthermore, for 1 123 641 faults for which an independent CPI test was found (i.e. 87.08%), a multi-cycle CPI test using connected two cycle LOC exists, which ensures robustness against effects of leakage currents and initial trapped charge. Additionally, for most faults a static test was found and consequently only 13.73% may show oscillating behavior within a later test. Yet, this does not imply that oscillating behavior will occur for these tests (refer to Section II-B) – as this highly depends on the actual influence of each aggressor which was not considered here.

The total runtime of around 9.5h for classifying over 1 500 000 faults is acceptably low in our opinion considering the hardness of the problem and the many different tests which are included in one ATPG run.

VI. Conclusions

This paper presented new classes of CPI tests for interconnect open defects. These new tests allow to improve the overall fault coverage of CPI tests and reduce scan loads. Furthermore, the generated tests are robust against leakage current. In addition,

a novel SAT-based ATPG system generating normal and single-value CPI tests in single and multi-cycle scenario was presented.

Experimental results showed the efficiency of this framework allowing the generation of tests achieving robustness against process variations affecting the influence of aggressors, trapped charge, oscillating behavior and leakage current at the same time.

Acknowledgments

The authors thank Linus Feiten from the University of Freiburg as well as Joan Figueras and Daniel Arumi from the University of Catalonia for supporting this work. This work was supported by the German Research Foundation (DFG) under grant SFB/TR14 AVACS.

References

[1] A. Sreedhar, A. Sanyal, and S. Kundu, "On modeling and testing of lithography related open faults in nano-CMOS circuits," in *DATE*, 2008.
[2] W. Maly, P. Nag, and P. Nigh, "Testing oriented analysis of CMOS ICs with opens," in *ICCAD*, 1988.
[3] R. Rodriguez-Montanes and J. Figueras, "Electrical and topological characterization of interconnect open defects," in *Current and Defect Based Testing*, May 2005, pp. 42–46.
[4] H. Konuk and F. Ferguson, "An unexpected factor in testing for CMOS opens: the die surface," in *VLSI Test Symposium*, Apr 1996.
[5] V. Champac and A. Zenteno, "Detectability conditions for interconnection open defects," in *VLSI Test Symposium*, 2000.
[6] D. Arumi, R. Rodriguez-Montane, and J. Figueras, "Defective behaviours of resistive opens in interconnect lines," in *ETS*, May 2005.
[7] R. Gomez, A. Giron, and V. Champac, "Test of interconnection opens considering coupling signals," in *International Symposium on DFT 2005*.
[8] S. Rafiq, A. Ivanov *et al.*, "Testing for floating gates defects in CMOS circuits," in *Asian Test Symposium, 1998.*, 1998, pp. 228–236.
[9] Y. Sato, L. Yamazaki *et al.*, "A persistent diagnostic technique for unstable defects," in *Test Conference, 2002. Proceedings. International*, 2002.
[10] R. Blanton, K. Dwarakanath, and A. Shah, "Analyzing the effectiveness of multiple-detect test sets," in *ITC*, 2003.
[11] J. Nelson, J. Brown *et al.*, "Multiple-detect ATPG based on physical neighborhoods," in *Design Automation Conference*, 2006, pp. 1099–1102.
[12] D. Arumi, R. Rodriguez-Montaes *et al.*, "Gate Leakage Impact on Full Open defects in Interconnect Lines," in *TVLSI*, 2011.
[13] D. Erb, K. Scheibler *et al.*, "Efficient SMT-based ATPG for interconnect open defects," in *DATE*, 2014.
[14] S. Reddy, I. Pomeranz, and C. Liu, "On tests to detect via opens in digital CMOS circuits," in *Design Automation Conference*, 2008, pp. 840–845.
[15] D. Erb, K. Scheibler *et al.*, "Circuit parameter independent test pattern generation for interconnect open defects," in *to be published in ATS14*, 2014.
[16] S. Spinner, I. Polian *et al.*, "Automatic test pattern generation for interconnect open defects," in *26th IEEE VTS 2008.*, 2008.
[17] H. Konuk and F. Ferguson, "Oscillation and sequential behavior caused by interconnect opens in digital CMOS circuits," in *ITC*, 1997.
[18] X. Lin and J. Rajski, "Test Generation for Interconnect Opens," in *ITC*, 2008.
[19] G. Tseitin, "On the complexity of derivation in propositional calculus," *Studies in constructive mathematics and mathematical logic*, vol. 2, 1968.
[20] H. K. Büning and U. Bubeck, *Handbook of Satisfiability*, scr. Frontiers in Artificial Intelligence and Applications 185. IOS Press, 2009, ch. Theory of quantified Boolean formulas.
[21] K. Scheibler, S. Kupferschmid, and B. Becker, "Recent improvements in the SMT solver iSAT," in *MBMV'13*, 2013, pp. 231–241.
[22] T. Larrabee, "Test pattern generation using Boolean satisfiability," *IEEE Trans. CAD*, vol. 11, no. 1, pp. 4–15, jan 1992.
[23] S. Hillebrecht, I. Polian *et al.*, "Extraction, simulation and test generation for interconnect open defects based on enhanced aggressor-victim model," in *Test Conference, 2008. ITC 2008. IEEE International*, 2008, pp. 1–10.

TABLE I: RESULTS OF THE PROPOSED CPI ATPG INCLUDING SINGLE-VALUE CPI TESTS AND MULTI-CYCLE CPI TESTS.

circuit	gates	faults	statically detected				dynamically detected				CPI untestable		aborts	overall runtime
			indep. CPI		multi-cycle CPI		indep. CPI		multi-cycle CPI					
			num	[%]	num	[%] ind. CPI	num	[%]	num	[%] ind. CPI	num	[%]		[s]
s09234	5 597	61 915	41 229	66.59	31 251	75.80	10 888	17.59	8 737	80.24	9 798	15.82	0	683
s13207	8 027	124 727	87 322	70.01	72 030	82.49	19 394	15.55	16 829	86.77	18 011	14.44	0	1 162
s15850	9 786	127 890	96 047	75.10	82 860	86.27	18 278	14.29	16 401	89.73	16 565	10.61	0	2 803
s35932	17 793	171 300	109 953	64.19	107 882	98.12	23 084	13.48	23 079	99.98	38 263	22.34	0	2 548
s38584	19 407	463 653	318 044	68.60	256 899	80.77	62 161	13.41	49 101	78.99	83 448	18.00	0	10 888
s38417	22 397	557 895	411 240	73.71	365 712	88.93	92 721	16.62	92 860	100.15	53 934	9.67	0	13 514
$\sum$	83 007	1 507 380	1 063 835	70.58	916 634	86.16	226 526	15.03	207 007	91.38	217 019	14.40	0	34 274

2015 IEEE 33rd VLSI Test Symposium (VTS)

Test Vector Omission with Minimal Sets of Simulated Faults

Irith Pomeranz
School of Electrical and Computer Engineering
Purdue University
West Lafayette, IN 47907, U.S.A.
E-mail: pomeranz@ecn.purdue.edu

Abstract—Test vector omission is a static test compaction procedure for functional test sequences that removes unnecessary test vectors from a sequence. The test vector omission procedure requires fault simulation for every test vector (or subsequence) that it considers for omission. It was noted earlier that it is possible to reduce the set of simulated faults based on the clock cycles where the faults are detected. However, this reduction is effective only for the later test vectors of a sequence. This paper defines a minimal set of faults that need to be simulated for the omission of a test vector by considering, in addition to detection clock cycles, also clock cycles where test subsequences start. The former are computed by a conventional sequential fault simulation process. For the latter, the paper introduces a sequential reverse order fault simulation process, and an approximation with a reduced computational complexity. Experimental results show significant reductions in the run time for test vector omission without affecting the level of compaction.

Index Terms—Finite-state machines, functional test sequences, reverse order fault simulation, test compaction.

I. INTRODUCTION

Test compaction procedures are important for reducing the test data volume and the test application time. Dynamic test compaction procedures are test generation procedures that include test compaction heuristics. Static test compaction procedures are applied following test generation. Considering functional test sequences [1]-[16], in general, static test compaction procedures are effective at reducing the length of a functional test sequence even if a dynamic test compaction procedure was used for generating it. In addition, static test compaction procedures that consider a single functional test sequence can produce shorter sequences than procedures that produce a set of sequences. After compacting a set of sequences, if the sequences are concatenated into a single sequence, their total length can be reduced further.

The possibility of applying static test compaction to a single functional test sequence was first demonstrated in [5]. One of the procedures described in [5] attempts to omit test vectors from a functional test sequence one at a time, or in subsequences of consecutive vectors. After a test vector (or subsequence) is omitted, the procedure performs fault simulation to check whether the fault coverage of the sequence is maintained. If the fault coverage is reduced, the vector (or subsequence) is reintroduced into the sequence. Otherwise, the omission is accepted. The results in [5] showed that test sequences, which are generated by sequential test generation

procedures, contain large numbers of test vectors that can be omitted without reducing the fault coverage.

The set of faults that needs to be simulated in order to decide whether a test vector can be omitted is defined in [5] as follows. Let t_u be the test vector at clock cycle u of a test sequence T. If a fault f is detected by T for the first time before clock cycle u, then f will be detected by T even if t_u is omitted. To decide on the omission of t_u, it is necessary to simulate the set of faults that are detected by T for the first time at clock cycle u or later. Figure 1 illustrates the use of the detection clock cycle for a fault f_0. The figure shows a sequence $T = t_0 t_1 ... t_{L-1}$ of length L. If f_0 is detected at clock cycle v of T, then f_0 is detected by the subsequence $t_0 ... t_v$ of T. With $v < u$, t_u can be omitted without simulating f_0.

Since very few faults are typically detected by the first test vectors of a sequence, all or most of the target faults need to be simulated when the first test vectors of a sequence are considered for omission. The use of detection clock cycles is effective in reducing the sets of simulated faults only for the later test vectors of a sequence. To address the fault simulation effort of the test vector omission procedure, restoration based compaction was introduced in [8]. In a restoration based procedure, all or most of the test vectors are initially omitted from the sequence. Test vectors are then restored in order to restore the detection of target faults. Since only one fault needs to be simulated in order to decide on the restoration of a test vector, the fault simulation effort is reduced significantly.

However, a test vector omission procedure performs a more thorough search than a restoration based procedure for test vectors that can be omitted. Specifically, a test vector omission procedure identifies any single test vector that can be omitted. In contrast, a restoration based procedure restores test vectors until it detects a fault before considering the next fault. Suppose that the procedure restores a subset of test vectors U_0 in order to detect a fault f_0, and then a subset U_1 in order to detect f_1. It is possible that restoring U_1 will make some of the vectors from U_0 unnecessary. The restoration based procedure will not identify such cases. This applies to all the variations of the restoration based procedure such as [8], [11], [12] and [14].

Motivated by this discussion, this paper defines the minimal set of faults that needs to be simulated in order to decide on

978-1-4799-7598-3/15 $31.00 © 2015 IEEE

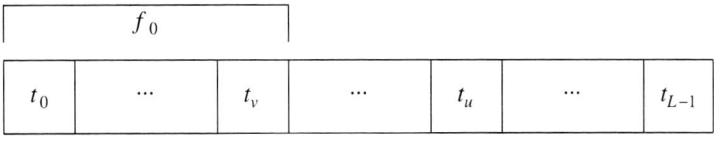

Fig. 1.   Detection Clock Cycle

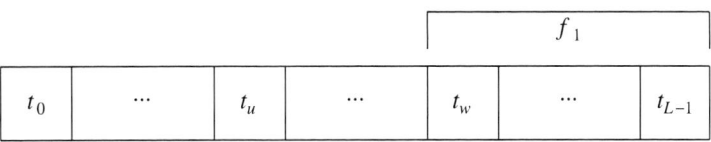

Fig. 2.   Starting Clock Cycle

the omission of a test vector. By simulating a minimal set of faults, the computational effort of the test vector omission procedure is reduced significantly.

To obtain the minimal set of faults that needs to be simulated for the omission of a test vector t_u, this paper complements the use of detection clock cycles, as illustrated by Figure 1, with the concept of a starting clock cycle for a subsequence that detects a fault. Figure 2 illustrates the use of the starting clock cycle for a fault f_1. In Figure 2, f_1 is known to be detected by the subsequence $t_w...t_{L-1}$ of T when it is applied starting from the all-unspecified state. With $w > u$, the omission of t_u does not affect the detection of f, and t_u can be omitted without simulating f_1.

The paper introduces a sequential reverse order fault simulation procedure that can find the highest clock cycle where a subsequence for a fault starts. The procedure drops faults similar to a conventional fault simulation procedure that computes detection clock cycles. However, the procedure simulates larger numbers of clock cycles. The paper discusses the tradeoffs related to this procedure, and describes an approximation whose computational complexity is similar to that of conventional fault simulation with fault dropping. Using this approximation it describes a test vector omission procedure with reduced computational effort compared with a procedure that considers only detection clock cycles.

In general, the approach described in this paper can be used for reducing the computational effort of test compaction procedures that consider subsequences for modification or removal.

The paper is organized as follows. Section II discusses the minimal set of faults that needs to be simulated when attempting to omit a test vector. Section III describes the sequential reverse order fault simulation procedure. Section IV describes the test vector omission procedure. Section V presents experimental results.

II. MINIMAL SETS OF FAULTS

This section discusses the definition of a minimal set of faults that needs to be simulated when attempting to omit a test vector. The same discussion applies to the omission of

a subsequence. For simplicity of discussion, only single test vectors are considered for omission.

Let the test sequence under consideration be $T = t_0 t_1...t_{L-1}$. The length of the sequence is L. For $0 \leq u < L$, t_u is the test vector at clock cycle u of T. A subsequence of T that starts at clock cycle u_0 and ends at clock cycle u_1 is denoted by $T[u_0, u_1]$. We have that $T[u_0, u_1] = t_{u_0} t_{u_0+1}...t_{u_1}$. Let F be the set of target faults that are detected by T.

The omission of t_u from T does not affect the detection of a fault $f \in F$ in one of two cases as discussed next.

The first case occurs when the subsequence $T[0, u-1]$ is known to detect f. The omission of t_u does not affect this subsequence. Therefore, this subsequence continues to detect f after t_u is omitted.

The information needed for determining which faults are detected by $T[0, u-1]$ can be computed by performing conventional sequential fault simulation with fault dropping of F under T. For every fault $f \in F$, let $u_{det}(f)$ be the first clock cycle where f is detected by T. After $u_{det}(f)$ is found, f is dropped from further simulation. The subsequence $T[0, u-1]$ is guaranteed to detect f if $u_{det}(f) < u$.

This is the only case considered in [5]. Thus, a fault $f \in F$ is simulated when t_u is considered for omission in every case where $u_{det}(f) \geq u$.

The second case, which was not considered in [5], occurs when the subsequence $T[u+1, L-1]$ is known to detect f when it is applied starting from the all-unspecified state. The omission of t_u does not affect this subsequence. Therefore, this subsequence continues to detect f after t_u is omitted.

The information needed for determining which faults are detected by $T[u+1, L-1]$ can be computed by performing sequential reverse order fault simulation of F under T. This process is described in the next section. It computes, for every fault $f \in F$, a clock cycle denoted by $u_{start}(f)$. This is the highest clock cycle such that the subsequence $T[u_{start}(f), L-1]$ detects f if it is applied starting from the all-unspecified state. The subsequence $T[u+1, L-1]$ is guaranteed to detect f if $u_{start}(f) > u$.

Using both detection and starting clock cycles, a fault $f \in F$ needs to be simulated when t_u is considered for

978-1-4799-7598-3/15 $31.00 © 2015 IEEE   243

TABLE I
Example based on $s1423$

u	u_{det}	u_{start}
0	100.00	6.36
10	99.86	6.36
20	98.80	6.36
30	84.23	6.22
40	83.45	6.22
50	71.78	4.60
60	65.06	4.60
70	61.74	4.60
80	61.53	4.60
90	53.96	4.60
100	52.12	4.60
200	43.85	5.02
300	15.70	0.71
400	12.66	5.73
500	6.58	2.12
600	5.23	1.49
700	4.46	1.06
800	2.62	2.55
900	1.98	1.98
1000	0.35	0.35

omission if $u_{det}(f) \geq u$ and $u_{start}(f) \leq u$. In this case, the omission of t_u affects both subsequences $T[0, u_{det}(f)]$ and $T[u_{start}(f), L-1]$ that detect f. Therefore, in order to omit t_u, it is necessary to verify by simulation that f continues to be detected. If $u_{det}(f) < u$ or $u_{start}(f) > u$, f does not need to be simulated.

The set of simulated faults that is based on $u_{det}(f)$ and $u_{start}(f)$ is minimal since $u_{det}(f)$ is the lowest clock cycle such that $T[0, u_{det}(f)]$ detects f, and $u_{start}(f)$ is the highest clock cycle such that $T[u_{start}(f), L-1]$ detects f. Therefore, a fault in the set must be simulated in order to determine whether or not the omission of t_u affects its detection.

The effects of using $u_{start}(f)$ in addition to $u_{det}(f)$ in order to define sets of simulated faults for test vector omission are illustrated next. The circuit under consideration is ISCAS-89 benchmark $s1423$ with a test sequence of length 1024. Table I shows the percentages of faults that need to be simulated when the test vector t_u is considered for omission, for $u = 0, 10, 20, ..., 100, 200, ..., 1000$ (these percentages are computed using the test vector omission procedure described in Section IV). Using only the variables $u_{det}(f)$, the procedure needs to simulate the percentages of faults shown in Table I under column u_{det}. Using the variables $u_{start}(f)$ as well, the procedure simulates the percentages of faults shown in Table I under column u_{start}.

Table I demonstrates that the use of $u_{det}(f)$ alone is effective in reducing the fault simulation effort for the higher values of u. A higher value of u implies that more faults have detection clock cycles that are lower than u. Therefore, there are more faults that do not need to be simulated. For the lower values of u, all or most of the faults need to be simulated, and the use of $u_{det}(f)$ alone is not effective in reducing the sets of simulated faults.

The use of $u_{start}(f)$ balances this effect. With a lower value of u, more faults have detecting subsequences that start later than u. Such faults do not need to be simulated. Thus, with

both $u_{det}(f)$ and $u_{start}(f)$, there are large percentages of faults that do not need to be simulated for both the lower and higher values of u.

Overall, the significant reduction in the percentage of faults that need to be simulated for the lower values of u results in a significant reduction in the computational effort of the test vector omission procedure.

III. Sequential Reverse Order Fault Simulation

This section describes the sequential reverse order fault simulation procedure. It first describes its basic form. It then describes an approximation with a reduced computational complexity.

The basic sequential reverse order fault simulation procedure starts by including all the target faults in the set F. For $u = L-1, L-2, ..., 0$, the procedure simulates F under the subsequence $T[u, L-1]$ starting from the all-unspecified initial state. If a fault $f \in F$ is detected by $T[u, L-1]$, the procedure assigns $u_{start}(f) = u$, and removes f from F.

The procedure simulates F (with fault dropping) under the subsequences $T[L-1, L-1], T[L-2, L-1], ..., T[0, L-1]$ of lengths $1, 2, ..., L$, respectively. The total length of the subsequences it simulates is $1 + 2 + ... + L = L(L+1)/2$, or $O(L^2)$. Compared with conventional fault simulation with fault dropping of F under T, the procedure simulates $O(L^2)$ instead of $O(L)$ clock cycles. To bring the computational complexity of the sequential reverse order fault simulation procedure closer to that of a conventional fault simulation procedure, a parameter denoted by λ is introduced as follows.

With $1 \leq \lambda \leq L$, the sequential reverse order fault simulation procedure considers only λ subsequences instead of the L subsequences that it considers in its basic form. The subsequences start $\lfloor L/\lambda \rfloor$ clock cycles apart. Thus, instead of considering the subsequence $T[u, L-1]$ for $u = L-1, L-2, ..., 0$, the procedure considers $T[u, L-1]$ for $u = L - \lfloor L/\lambda \rfloor, L - 2\lfloor L/\lambda \rfloor, ..., L - \lambda \lfloor L/\lambda \rfloor$. The total length of the subsequences that the procedure considers is $(1 + 2 + ... + \lambda)\lfloor L/\lambda \rfloor = (1 + \lambda)(\lambda/2)\lfloor L/\lambda \rfloor$. With a constant value for λ, the procedure simulates F under $O(L)$ clock cycles, as in the case of conventional fault simulation with fault dropping.

It should be noted that $\lambda = L$ results in the basic case where the procedure considers $u = L-1, L-2, ..., 0$. In addition, $\lambda = 1$ results in a single value of u, $u = 0$. Since all the faults in F are known to be detected by $T = T[0, L-1]$, this case does not require any fault simulation, and it yields $u_{start}(f) = 0$ for every $f \in F$. In this case, the variables $u_{start}(f)$ are not effective in reducing the sets of faults that need to be simulated during test vector omission.

The effect of using $1 < \lambda < L$ on the values of $u_{start}(f)$ is the following. Suppose that $u_{start}(f)$ is obtained for a fault f when $\lambda = L$ causes the procedure to consider all the clock cycles of T. This implies that the subsequence $T[u_{start}(f), L-1]$ detects f starting from the all-unspecified state. In addition, $T[u, L-1]$ detects f for every $u \leq u_{start}(f)$. Therefore, when the procedure considers $u = L - r\lfloor L/\lambda \rfloor$, for the lowest value

978-1-4799-7598-3/15 $31.00 © 2015 IEEE

of r such that $L - r\lfloor L/\lambda \rfloor \leq u_{start}(f)$, it will find that f is detected. It will therefore assign $u_{start}(f) = L - r\lfloor L/\lambda \rfloor$. As a result, $u_{start}(f)$ will be lower than its highest value by at most $\lfloor L/\lambda \rfloor - 1$. This will cause the test vector omission procedure to simulate the fault for at most $\lfloor L/\lambda \rfloor - 1$ clock cycles unnecessarily.

IV. TEST VECTOR OMISSION PROCEDURE

This section describes the test vector omission procedure and the updating of the variables $u_{det}(f)$ and $u_{start}(f)$ as they change when test vectors are omitted during the procedure.

The test vector omission procedure considers the clock cycles of T in the order $u = 0, 1, ..., L - 1$. While different orders are possible, this order reduces the need to update the variables $u_{start}(f)$ during the procedure as will be clarified later. When clock cycle u is considered, the procedure omits t_u from T tentatively. It performs fault simulation to determine whether T continues to detect all the faults in F. If it does, the omission of t_u is accepted. Otherwise, t_u is reintroduced into T.

To keep track of the test vectors that are omitted from T, the procedure uses a variable denoted by $omit(u)$. Initially, $omit(u) = 0$ for $0 \leq u < L$. If the procedure decides to omit t_u, it assigns $omit(u) = 1$. In this case, the omission of additional test vectors is considered with t_u omitted from T. After considering all the clock cycles of T, the procedure omits every test vector t_u such that $omit(u) = 1$ to obtain the compacted test sequence.

Suppose that only the variables $u_{det}(f)$ are available. When t_u is considered for omission, the procedure assigns $omit(u) = 1$ and simulates the set of faults $F_u = \{f : u_{det}(f) \geq u\}$ under T. If any fault from F_u remains undetected, the omission of t_u is not accepted, and $omit(u) = 0$ is assigned. Otherwise, $omit(u) = 1$ remains, and the procedure updates the detection clock cycle $u_{det}(f)$ for every $f \in F_u$ based on the results of the fault simulation process. In this case, F_u includes all the faults whose detection clock cycles may be affected by the omission. Therefore, the variables $u_{det}(f)$ remain up-to-date throughout the procedure.

Suppose next that both $u_{det}(f)$ and $u_{start}(f)$ are available. When t_u is considered for omission, the procedure simulates the set of faults $F_u = \{f : u_{det}(f) \geq u \text{ and } u_{start}(f) \leq u\}$. As in the earlier case, if any fault from F_u remains undetected, the omission of t_u is not accepted. Otherwise, the fault simulation process updates the detection clock cycle $u_{det}(f)$ for every $f \in F_u$. However, in this case, not all the variables $u_{det}(f)$ for $f \in F$ are up-to-date. This issue is discussed next.

If f is such that $u_{det}(f) \geq u$ but $u_{start}(f) > u$, the detection clock cycle of f may be affected by the omission of t_u. The reason f is not simulated is that it is guaranteed to be detected by $T[u_{start}(f), L - 1]$, which is not affected by the omission of t_u. The detection clock cycle of f does not have to be updated as long as the procedure considers for omission a test vector t_u such that $u_{start}(f) > u$. However, when the procedure considers the omission of t_u for $u = u_{start}(f)$, the condition $u_{start}(f) > u$ is not satisfied any more. Therefore,

$u_{det}(f)$ needs to be up-to-date in order to decide whether or not f needs to be simulated. To update $u_{det}(f)$, the procedure simulates f under T. Thus, f is excluded from F_u for $u_{start}(f)$ clock cycles, and it is simulated only before $t_{u_{start}(f)}$ is considered. In this discussion, $u_{det}(f) \geq 0$ and $u_{start}(f) > 0$.

To ensure that the updating of $u_{det}(f)$ is performed only when necessary, the procedure uses the initialization $u_{start}(f) = 0$ for every $f \in F$. In the case where $\lambda = 1$ and sequential reverse order fault simulation is not used, the procedure keeps $u_{start}(f) = 0$. Updating of $u_{det}(f)$ is done before a test vector t_u is considered for omission only for a fault f such that $u_{start}(f) = u > 0$. After updating $u_{det}(f)$, the procedure assigns $u_{start}(f) = 0$ in anticipation that the omission of additional test vectors will render $u_{start}(f)$ incorrect. This does not affect the accuracy of the procedure since $u_{start}(f)$ is not useful in avoiding the simulation of the fault when the next test vectors are considered for omission.

With this updating of $u_{start}(f)$, the procedure does not consider cases where $u_{start}(f)$ increases because of the omission of test vectors t_u such that $u \geq u_{start}(f)$. These cases are expected to be rare, and they do not justify the computational effort of recomputing the variables $u_{start}(f)$.

Another speed-up technique that the procedure uses is to associate with every fault $f \in F$ the number of times it remains undetected when a test vector is omitted, thus causing the test vector to remain in T. This number is denoted by $n_{fail}(f)$. Faults are simulated in decreasing order of $n_{fail}(f)$ based on the observation that the same faults typically cause the omission of test vectors to fail. Such faults have high values of $n_{fail}(f)$, and they are simulated earlier.

The test vector omission procedure is summarized next.

Procedure 1: Test vector omission

1) Let $T = t_0 t_1 ... t_{L-1}$ be a functional test sequence of length L that detects a set of faults F. Assign $omit(u) = 0$ for $0 \leq u < L$.
2) Perform conventional fault simulation with fault dropping of F under T and record the first detection clock cycle $u_{det}(f)$ for every $f \in F$.
3) Perform sequential reverse order fault simulation of F under T and record the starting clock cycle $u_{start}(f)$ for every $f \in F$.
4) Assign $u = 0$.
5) For every $f \in F$, if $u_{start}(f) = u > 0$:

    a) Simulate f under T and update its detection clock cycle $u_{det}(f)$.

    b) Assign $u_{start}(f) = 0$.
6) Assign $omit(u) = 1$.
7) Define $F_u = \{f : u_{det}(f) \geq u \text{ and } u_{start}(f) \leq u\}$. Simulate F_u under T. If not all the faults are detected, assign $omit(u) = 0$.
8) Assign $u = u + 1$. If $u < L$, go to Step 5.

The worst-case computational complexity of the procedure is determined as follows. The procedure considers L test vectors for omission. When a test vector t_u is considered, the procedure simulates the set of faults F_u under a sequence of length $O(L)$. With $|F_u| \leq |F|$, the worst-case computational complexity is that of simulating the circuit for $O(L^2|F|)$ clock cycles. The use of the variables $u_{det}(f)$ and $u_{start}(f)$ results in sets of faults F_u that are significantly smaller than $|F|$ as illustrated by Table I.

V. EXPERIMENTAL RESULTS

This section presents the results of the test vector omission procedure.

The test sequences to which the procedure is applied were generated by sequential test generation and compacted by a restoration based procedure from [8] targeting single stuck-at faults. The sequences are applied starting from the all-unspecified state (the procedure can also be applied in the case where reset is available, and to different fault models). The test vector omission procedure was implemented using a fault simulation procedure that considers one fault at a time.

The value of λ needs to balance the run time of the sequential reverse order fault simulation process with the run time of the test vector omission process. The value of λ does not affect the length of the compacted test sequence. The results using $\lambda = 1, 2, 4, 8, 16, 32$ and 64 are shown for three benhcmark circuits in Table II to illustrate the tradeoffs related to λ.

The results in Table II point to $\lambda = 8$ as the first value that yields the lowest or close to the lowest run time overall. A lower value of λ means a lower run time for the sequential reverse order fault simulation process. The results for additional circuits using $\lambda = 8$ are shown in Tables III and IV. For the circuits in Table III, the results using $\lambda = 1$ are also shown for comparison. For $b15$ in Table IV, the run time with $\lambda = 1$ was too high, and the results for this value are not reported. Tables II, III and IV are organized as follows.

Column λ shows the value of the corresponding parameter.

Column len shows the following information related to test sequence lengths. Let T_{init} be the test sequence to which the test vector omission procedure is applied. Let T_{omit} be the compacted test sequence obtained by applying the test vector omission procedure to T_{init}. Subcolumn $omit$ shows the length of the compacted test sequence T_{omit}. Subcolumn $ratio$ shows the length of T_{omit} divided by the length of T_{init}.

Column $ntime$ shows the following information related to run times. Let rt_{init} be the run time for conventional fault simulation with fault dropping of T_{init}. Let $rt_{reverse}(\lambda)$ be the run time for sequential reverse order fault simulation of T_{init} with parameter λ. Let $rt_{omit}(\lambda)$ be the run time of the test vector omission procedure with parameter λ (including the run time for sequential reverse order fault simulation when $\lambda = 8$). Subcolumn $reverse$ shows the normalized run time $rt_{reverse}(\lambda)/rt_{init}$. Subcolumn $omit$ shows the normalized run time $rt_{omit}(\lambda)/rt_{init}$. Normalization provides an indication of the computational effort in terms of fault simulation time of T_{init}. In addition, subcolumn $ratio$ shows the ratio $rt_{omit}(\lambda)/rt_{omit}(1)$.

TABLE II
EXPERIMENTAL RESULTS ($\lambda = 1, 2, 4, ..., 64$)

| circuit | λ | len | | ntime | | |
		omit	ratio	reverse	omit	ratio
s382	1	489	0.95	0.00	130.91	1.00
s382	2	489	0.95	0.98	74.62	0.57
s382	4	489	0.95	1.18	51.56	0.39
s382	8	489	0.95	1.82	43.02	0.33
s382	16	489	0.95	3.18	42.71	0.33
s382	32	489	0.95	6.02	45.04	0.34
s382	64	489	0.95	11.64	49.89	0.38
b08	1	350	0.84	0.00	129.34	1.00
b08	2	350	0.84	0.87	50.72	0.39
b08	4	350	0.84	1.17	29.17	0.23
b08	8	350	0.84	2.32	28.40	0.22
b08	16	350	0.84	4.34	26.60	0.21
b08	32	350	0.84	8.38	28.81	0.22
b08	64	350	0.84	16.89	37.02	0.29
usb_phy	1	1296	0.94	0.00	200.57	1.00
usb_phy	2	1296	0.94	0.41	88.08	0.44
usb_phy	4	1296	0.94	0.83	52.56	0.26
usb_phy	8	1296	0.94	1.01	31.43	0.16
usb_phy	16	1296	0.94	2.05	30.47	0.15
usb_phy	32	1296	0.94	3.87	31.36	0.16
usb_phy	64	1296	0.94	7.81	34.94	0.17

The following points can be seen from Tables III and IV. The test vector omission procedure is able to reduce the lengths of the test sequences significantly even though they were already compacted by a restoration based procedure. This demonstrates the importance of the test vector omission procedure.

The same compacted test sequence is obtained for all the values of λ. Thus, the use of the variables $u_{start}(f)$ does not affect the length of the compacted test sequence.

With $\lambda = 1$, the sequential reverse order fault simulation procedure is not applied, and does not contribute to the run time. For $\lambda = 8$, based on the computational complexity of the sequential reverse order fault simulation procedure, it is expected that $rt_{reverse}(8)/rt_{init} \approx (1 + \lambda)/2 = 4.5$. Lower values are obtained because of faster fault dropping during sequential reverse order fault simulation.

Considering $rt_{omit}(8)/rt_{omit}(1)$, the ratio is typically lower than 0.4, indicating more than 60% reduction in run time.

The normalized run time of the test vector omission procedure does not always increase with the size of the circuit. This indicates that the procedure scales similar to a fault simulation procedure.

VI. CONCLUDING REMARKS

This paper defined a minimal set of faults that need to be simulated in order to accept the omission of a test vector from a functional test sequence. The set is defined by considering the first detection clock cycles of the faults, and the highest clock cycles where test subsequences for the faults start. The former are computed by a conventional sequential fault simulation process, and were used in earlier implementations of test vector omission. For the latter, the paper introduced a sequential reverse order fault simulation process. It also

978-1-4799-7598-3/15 $31.00 © 2015 IEEE

TABLE III
EXPERIMENTAL RESULTS ($\lambda = 1$ AND 8)

circuit	λ	len omit	len ratio	ntime reverse	ntime omit	ntime ratio
s298	1	81	0.69	0.00	28.80	1.00
s298	8	81	0.69	1.80	11.20	0.39
s382	1	489	0.95	0.00	134.90	1.00
s382	8	489	0.95	1.83	44.27	0.33
s526	1	934	0.93	0.00	214.47	1.00
s526	8	934	0.93	1.07	45.74	0.21
s641	1	83	0.82	0.00	26.14	1.00
s641	8	83	0.82	2.50	12.93	0.49
s820	1	362	0.74	0.00	208.68	1.00
s820	8	362	0.74	2.66	33.78	0.16
s1196	1	227	0.95	0.00	158.07	1.00
s1196	8	227	0.95	1.96	16.78	0.11
s1423	1	847	0.83	0.00	314.98	1.00
s1423	8	847	0.83	2.63	61.85	0.20
s5378	1	615	0.95	0.00	61.56	1.00
s5378	8	615	0.95	0.94	10.53	0.17
s35932	1	138	0.92	0.00	25.56	1.00
s35932	8	138	0.92	1.63	5.81	0.23
b03	1	103	0.79	0.00	36.47	1.00
b03	8	103	0.79	0.73	6.20	0.17
b04	1	129	0.77	0.00	40.47	1.00
b04	8	129	0.77	1.18	7.48	0.18
b05	1	237	0.74	0.00	31.94	1.00
b05	8	237	0.74	0.30	2.69	0.08
b07	1	330	0.87	0.00	114.42	1.00
b07	8	330	0.87	0.19	4.88	0.04
b08	1	350	0.84	0.00	130.36	1.00
b08	8	350	0.84	2.36	28.26	0.22
b10	1	116	0.61	0.00	57.95	1.00
b10	8	116	0.61	1.65	12.90	0.22
b11	1	350	0.63	0.00	146.90	1.00
b11	8	350	0.63	3.49	52.25	0.36
b14	1	5363	0.96	0.00	822.06	1.00
b14	8	5363	0.96	1.67	76.56	0.09
aes_core	1	779	0.84	0.00	402.21	1.00
aes_core	8	779	0.84	2.18	22.86	0.06
i2c	1	906	0.41	0.00	220.18	1.00
i2c	8	906	0.41	0.96	22.02	0.10
des_area	1	209	0.87	0.00	94.76	1.00
des_area	8	209	0.87	3.76	16.10	0.17
simple_spi	1	10299	0.73	0.00	7688.16	1.00
simple_spi	8	10299	0.73	0.94	532.73	0.07
usb_phy	1	1296	0.94	0.00	201.48	1.00
usb_phy	8	1296	0.94	1.01	31.68	0.16

TABLE IV
EXPERIMENTAL RESULTS ($\lambda = 8$)

circuit	λ	len omit	len ratio	ntime reverse	ntime omit
b15	8	11518	0.81	1.18	349.76

ACKNOWLEDGMENT

This work was supported in part by NSF Grant No. CCF-1320263.

REFERENCES

[1] R. K. Roy, T. M. Niermann, J. H. Patel, J. A. Abraham and R. A. Saleh, "Compaction of ATPG-Generated Test Sequences for Sequential Circuits", in Proc. Intl. Conf. on Computer-Aided Design, 1988, pp. 382-385.

[2] I. Pomeranz and S. M. Reddy, "On Generating Compact Test Sequences for Synchronous Sequential Circuits" in Proc. EURO-DAC, 1995, pp. 105-110.

[3] T. J. Lambert and K. K. Saluja, "Methods for Dynamic Test Vector Compaction in Sequential Test Generation", in Proc. Intl. Conf. on VLSI Design, 1996, pp. 166-169.

[4] A. Raghunathan and S. T. Chakradhar, "Dynamic Test Sequence Compaction for Sequential Circuits", in Proc. Intl. Conf. on VLSI Design, 1996, pp. 170-173.

[5] I. Pomeranz and S. M. Reddy, "On Static Compaction of Test Sequences for Synchronous Sequential Circuits", in Proc. Design Autom. Conf., 1996, pp. 215-220.

[6] E. M. Rudnick and J. H. Patel, "Simulation-based Techniques for Dynamic Test Sequence Compaction", in Proc. Intl. Conf. on Computer-Aided Design, 1996, pp. 67-73.

[7] F. Corno, P. Prinetto, M. Rebaudengo and M. Sonza Reorda, "New Static Compaction Techniques of Test Sequences for Sequential Circuits", in Proc. European Design and Test Conf., 1997, pp. 37-43.

[8] I. Pomeranz and S. M. Reddy, "Vector Restoration Based Static Compaction of Test Sequences for Synchronous Sequential Circuits", in Proc. Intl. Conf. on Computer Design, 1997, pp. 360-365.

[9] E. M. Rudnick and J. H. Patel, "Putting the Squeeze on Test Sequences in Proc. Intl. Test Conf., 1997, pp. 723-732.

[10] M. S. Hsiao and S. T. Chakradhar, "State Relaxation Based Subsequence Removal for Fast Static Compaction in Sequential Circuits", in Proc. Design Autom. and Test in Europe, 1998, pp. 577-582.

[11] S. K. Bommu, S. T. Chakradhar and K. B. Doreswamy, "Static Compaction using Overlapped Restoration and Segment Pruning", in Proc. Intl. Conf. on Computer-Aided Design, 1998, pp. 140-146.

[12] X. Lin, W.-T. Cheng, I. Pomeranz and S. M. Reddy, "SIFAR: Static Test Compaction for Synchronous Sequential Circuits Based on Single Fault Restoration", in Proc. VLSI Test Symp., 2000, pp. 205-212.

[13] I. Pomeranz and S. M. Reddy, "Vector Replacement to Improve Static Test Compaction for Synchronous Sequential Circuits", IEEE Trans. on Computer-Aided Design, Feb. 2001, pp. 336-342.

[14] R. Guo, S. M. Reddy and I. Pomeranz, "Reverse Order Restoration Based Static Test Compaction for Synchronous Sequential Circuits", IEEE Trans. on Computer-Aided Design, March 2003, pp. 293-304.

[15] M. Dimopoulos and P. Linardis, "Efficient Static compaction of Test Sequence Sets through the Application of Set Covering Techniques", in Proc. Design, Autom. and Test in Europe Conf., 2004, pp. 194-199.

[16] I. Pomeranz, "Restoration Based Procedures with Set Covering Heuristics for Static Test Compaction of Functional Test Sequences", IEEE Trans. on VLSI Systems, April 2014, pp. 779-791.

described an approximation with a computational complexity that is similar to that of conventional sequential fault simulation with fault dropping. Experimental results showed significant reductions in the run time for test vector omission without affecting the level of test compaction.

Test Compaction by Test Cube Merging for Four-Way Bridging Faults

Irith Pomeranz

School of Electrical and Computer Engineering
Purdue University
West Lafayette, IN 47907, U.S.A.
E-mail: pomeranz@ecn.purdue.edu

Abstract—Test compaction that accommodates the constraints of test data compression can be achieved by generating test cubes for target faults, and then merging the test cubes. This paper describes an improved test cube merging procedure for four-way bridging faults. The procedure is motivated by the prevalence of bridging defects and the fact that test sets for bridging faults are larger than test sets for single stuck-at faults. A four-way bridging fault $g_i/a_i/h_i$ models the case where a value a_i of a line h_i dominates the value of a line g_i. A basic test cube merging procedure considers a set of test cubes C_{det} that detects target faults. The paper extends the set of test cubes to include, in addition to C_{det}, a set of test cubes C_{dom} that assign values to dominating lines. Test cubes from C_{dom} have significantly fewer specified values than test cubes from C_{det}. When test cubes from C_{dom} are merged with test cubes from C_{det}, each resulting test cube detects more faults, and fewer test cubes are needed for detecting the same set of target faults.

Index Terms—Bridging faults, static test compaction, test cubes, test data compression.

I. INTRODUCTION

Test generation that produces test cubes for target faults, followed by merging of test cubes to achieve test compaction, fits well with the need to accommodate the constraints of a test data compression method [1]-[5]. Test cubes that are generated for single target faults have the smallest possible numbers of specified values. Merging of test cubes can take into consideration the constraints of the test data compression method on the specified values of a test cube.

Test compaction procedures based on test cube merging were described in [6]-[9]. The procedures described in [6]-[7] merge only compatible test cubes. Two test cubes are said to be compatible if, for every input where they are both specified, they are specified to the same value. When two compatible test cubes t_i and t_j are merged into a test cube $t_{i,j}$, $t_{i,j}$ is guaranteed to detect all the faults that are detected by t_i and t_j alone. Therefore, a test cube merging procedure that merges only compatible test cubes does not require fault simulation during the merging process. Fault simulation followed by reverse order fault simulation are applied after the test cube merging process is complete in order to identify test cubes that can be removed from the test set without reducing the fault coverage. Such test cubes exist since a test cube $t_{i,j}$, which is obtained by merging of test cubes t_i and t_j, may detect faults that are not detected by t_i or t_j individually.

The procedures described in [8]-[9] merge test cubes even if they are not compatible. In [8], conflicts between merged test cubes are recorded, and they imply that more than one test will be produced based on a test cube. In [9], fault simulation is used to determine whether incompatible test cubes can be merged without losing fault coverage. To avoid these overheads, only merging of compatible test cubes is considered in this paper.

Test compaction by merging of test cubes can be applied to any fault model. This paper focuses on bridging faults. Bridging faults are important to target because of the prevalence of bridging defects. Test compaction is important for bridging faults since the number of bridging faults that need to be considered is typically larger than the number of single stuck-at faults, and test sets for bridging faults are larger than test sets for single stuck-at faults.

Several bridging fault models were proposed for modeling physical defects [6], [10]-[14]. The bridging fault model that is considered in this paper is the four-way bridging fault model from [13]-[14]. Under this model, a pair of lines is associated with four faults corresponding to the two options for the line whose value is dominating, and the two options for the dominating value. A four-way bridging fault is denoted by $g_i/a_i/h_i$, where g_i is the dominated line, h_i is the dominating line, and a_i is the value of h_i that causes it to dominate the value of g_i. The fault is detected by a test that assigns $g_i = a_i'$, $h_i = a_i$, and propagates the fault-free/faulty value a_i'/a_i from g_i to an output. Thus, the test assigns $h_i = a_i$ and detects the fault g_i stuck-at a_i.

The paper describes an improved test cube merging procedure for four-way bridging faults. The extended procedure produces smaller test sets than a basic test cube merging procedure. This is achieved by taking advantage of the following observations.

Let t_{i0} be a test cube that detects a four-way bridging fault $g_i/a_i/h_{i0}$, and let t_{i1} be a test cube that detects a four-way bridging fault $g_i/a_i/h_{i1}$. If t_{i0} and t_{i1} conflict on an input, they will not be merged by a test cube merging procedure that considers only compatible test cubes. Nevertheless, it may be possible to specify t_{i0} such that it would satisfy the condition $h_{i1} = a_i$. In this case, t_{i0} would detect the fault g_i stuck-at a_i (since it detects the four-way bridging fault $g_i/a_i/h_{i0}$), and it would assign $h_{i1} = a_i$. Therefore, t_{i0} will detect the

four-way bridging fault $g_i/a_i/h_{i1}$. The possibility of selecting tests for single stuck-at faults such that they would detect several bridging faults is the basis for the procedures described in [15]-[16]. These procedures improve the quality of a test set (without changing the number of tests) by increasing its coverage of bridging faults.

In the context of a test cube merging procedure, the purpose of ensuring that test cubes will detect additional bridging faults is to reduce the number of test cubes that are needed for detecting a given set of faults. This is achieved in this paper as follows.

Let C_{det} be a set of test cubes for a set of four-way bridging faults BR. A basic test cube merging procedure considers only the test cubes in C_{det}. The improved test cube merging procedure uses, in addition to C_{det}, a set of test cubes denoted by C_{dom}. The test cubes in C_{dom} assign values to dominating lines. For a four-way bridging fault $g_i/a_i/h_i$ that is detected by C_{det}, let $c_{det(i)}$ be the first test cube in C_{det} that detects the fault. The procedure extracts from $c_{det(i)}$ a minimally specified test cube $c_{dom(i)}$ that assigns $h_i = a_i$ (without necessarily detecting the bridging fault). The test cube $c_{dom(i)}$ is included in C_{dom} for every fault $g_i/a_i/h_i$ that is detected by C_{det}.

The extended test cube merging procedure merges test cubes from $C_{det} \cup C_{dom}$. It merges test cubes from C_{det} as much as possible since these test cubes guarantee the detection of target faults. In addition, to the extent possible, the procedure merges test cubes from C_{dom} with test cubes from C_{det} in order to increase the numbers of four-way bridging faults that the resulting test cubes detect. Since test cubes from C_{dom} have significantly fewer specified values than test cubes from C_{det}, merging with test cubes from C_{dom} is possible even when it is not possible with test cubes from C_{det}. An increase in the number of detected faults occurs when a test cube already detects some bridging faults, and test cubes from C_{dom} that are merged with it satisfy the conditions on the dominating line values that are needed for detecting additional bridging faults.

Similar to a basic test cube merging procedure, the extended procedure does not perform any fault simulation until the merging process is complete. Test cubes that can be removed without losing fault coverage are identified after the test cube merging process is complete by performing fault simulation followed by reverse order fault simulation of the test set. Experimental results demonstrate that merging of test cubes from C_{dom} contributes to the ability to remove test cubes without reducing the bridging fault coverage. The reduction is significant for some benchmark circuits.

Only four-way bridging faults are considered in this paper. However, improved test cube merging procedures that produce smaller test sets can be designed for other fault models. For other bridging fault models, a set of test cubes C_{dom} can be defined similar to the definition for four-way bridging faults. For other fault models, an extended procedure can use a set of test cubes C_{det} that detect target faults, and a set of test cubes C_{prop} that satisfy a simpler property of tests for target faults. By merging test cubes from C_{prop} with test cubes from C_{det},

more target faults are detected by each test cube, and fewer test cubes are needed for detecting the faults.

Test cube merging is described in this paper without considering the constraints of a particular test data compression method. Such constraints can be incorporated into the merging procedures by avoiding the merging of test cubes t_i and t_j if the resulting test cube $t_{i,j}$ does not satisfy the constraints.

The paper is organized as follows. Section II describes a basic test cube merging procedure. Section III describes the extended test cube merging procedure for four-way bridging faults. Section IV presents experimental results.

II. BASIC PROCEDURE

This section describes a basic test cube merging procedure. The procedure can be applied to any fault model. It is applied here to a set of four-way bridging faults BR with a set of test cubes C_{det}.

The basic test cube merging procedure is given below as Procedure 1. The test set that the procedure produces from C_{det} is denoted by T. Initially, $T = C_{det}$. After this initialization, the test cubes in T are arranged by order of decreasing number of specified values. This order gives precedence to the merging of test cubes with larger numbers of specified values, which are more difficult to merge. It thus results in a reduced number of test cubes.

The procedure considers every pair of test cubes $t_i \in T$ and $t_j \in T$ such that $i < j$. If t_i and t_j are compatible, the procedure computes a test cube $t_{i,j}$ that contains all the specified values of t_i and t_j. It replaces t_i with $t_{i,j}$, and marks that t_j has been merged with another test cube by assigning $merged_j = 1$. With $merged_j = 1$, the test cube t_j is not considered again for merging.

After attempting to merge all the possible pairs of test cubes, the procedure removes from T every test cube t_i such that $merged_i = 1$. This is followed by fault simulation and forward-looking reverse order fault simulation to remove unnecessary test cubes from T.

Procedure 1: Basic test cube merging

1) Let $C_{det} = \{c_0, c_1, \ldots, c_{n-1}\}$ be a set of test cubes for a set of target bridging faults BR. Assign $T = C_{det}$. Arrange the test cubes in T by order of decreasing number of specified values. Let $T = \{t_0, t_1, \ldots, t_{n-1}\}$. For $0 \leq i < n$, assign $merged_i = 0$.

2) For $0 \leq i < n$, if $merged_i = 0$:
   a) For $i < j < n$, if $merged_j = 0$:
      i) If t_i and t_j are compatible:
         A) Replace t_i with $t_{i,j}$.
         B) Assign $merged_j = 1$.

3) For $0 \leq i < n$, if $merged_i = 1$, remove t_i from T.

4) Perform fault simulation followed by forward-looking reverse order fault simulation of BR under T, and remove unnecessary test cubes from T.

III. EXTENDED PROCEDURE

This section describes the extended test cube merging procedure. The procedure improves the ability of the basic

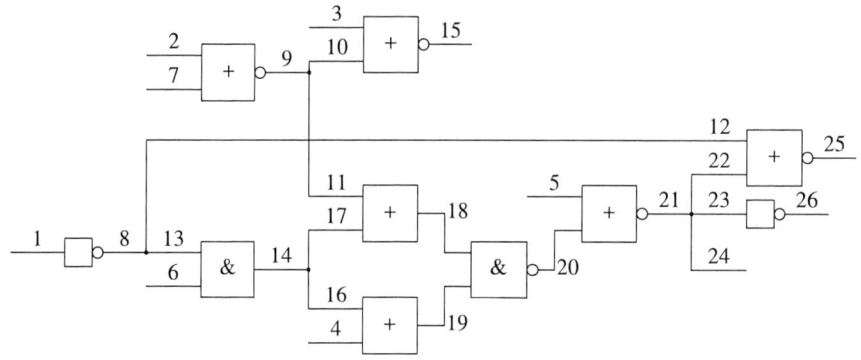

Fig. 1.   ISCAS-89 benchmark s27

procedure to reduce the number of test cubes by using two sets of test cubes. The set C_{det} includes test cubes that detect the four-way bridging faults in BR. This is the same set that is used by the basic procedure. The set C_{dom} is added to increase the number of four-way bridging faults that each test cube detects after test cube merging, thus allowing the number of test cubes in the final test set to be reduced. This section first describes the derivation of C_{dom}. It then describes the extended test cube merging procedure.

A. Computing C_{dom}

Fault simulation with fault dropping of BR under C_{det} yields, for every fault $g_i/a_i/h_i \in BR$, the first test cube in C_{det} that detects it. Let this test cube be $c_{det(i)}$.

Initially, $C_{dom} = \emptyset$. The procedure considers every four-way bridging fault $g_i/a_i/h_i \in BR$ that is detected by C_{det}. Starting from the test cube $c_{det(i)}$ that detects the fault, the procedure finds a test cube $c_{dom(i)}$ that only assigns $h_i = a_i$. The test cube $c_{dom(i)}$ is added to C_{dom} (if it is not already included in C_{dom}).

The derivation of $c_{dom(i)}$ starts by assigning $c_{dom(i)} = c_{det(i)}$. The procedure considers every specified value of $c_{dom(i)}$. Let $c_{dom(i)}(k)$ be the value of input k under $c_{dom(i)}$. If $c_{dom(i)}(k)$ is specified, the procedure assigns an unspecified value to $c_{dom(i)}(k)$, and checks whether $h_i = a_i$ under the modified test cube. If h_i is unspecified, the procedure restores the previous specified value of $c_{dom(i)}(k)$. Otherwise, it accepts the unspecified value of input k under $c_{dom(i)}$.

The procedure considers all the inputs in a random order. The final test cube it obtains is added to C_{dom}.

For illustration, ISCAS-89 benchmark $s27$ is shown in Figure 1. A set of test cubes C_{det} for $s27$ is shown in Table I under column C_{det}. The test cubes are generated for a set BR of 64 four-way bridging faults.

Column C_{dom} of Table I shows the set of test cubes C_{dom} obtained for $s27$. The test cube $c_{dom(i)} \in C_{dom}$ is generated for the fault $g_i/a_i/h_i$ based on $c_{det(i)}$ whose index is shown under subcolumn $det(i)$. The resulting test cube $c_{dom(i)}$ is shown in the last subcolumn. Additional test cubes that are generated based on other four-way bridging faults are equal to the ones shown in Table I.

The test cubes in C_{dom} have significantly fewer specified values than the test cubes in C_{det}. This is important since it allows the extended test cube merging procedure to specify relatively small numbers of additional input values of every test cube in order to detect additional four-way bridging faults. Thus, even when it is not possible to merge a test cube $t_i \in C_{det}$ with other test cubes from C_{det}, it is possible to merge t_i with test cubes from C_{dom}, and thus increase the number of detected four-way bridging faults.

TABLE I
SETS OF TEST CUBES

C_{det}		C_{dom}			
i	c_i	i	$g_i/a_i/h_i$	$det(i)$	$c_{dom(i)}$
0	01x1000	23	2/0/1	0	0xxxxxx
1	110xxx0	24	2/0/8	1	1xxxxxx
2	0x1xxx1	25	3/0/11	2	xxxxxx1
3	10x10x0	26	5/0/7	4	xxxxxx0
4	0xxx110	27	6/1/4	0	xxx1xxx
5	10x11x0	28	7/0/3	7	xx0xxxx
6	0xxx001	29	8/0/15	2	xx1xxxx
7	x00xxx1	30	8/1/6	9	xxxxx1x
8	000xxx1	31	8/1/11	5	x0xxxx0
9	1xx0x1x	32	9/0/19	10	1xx0xxx
10	1000xx0	33	9/1/16	11	0xxxx1x
11	0x0xx11	34	12/0/4	13	xxx0xxx
12	00x1000	35	15/0/22	1	11xxxxx
13	0xx01xx	37	16/0/2	18	x0xxxxx
14	11x101x	38	18/1/15	19	x10xxxx
15	10x0010	39	24/0/25	3	x0x10x0
16	0xxx011	40	25/1/26	0	x1xxx0x
17	0xx001x	41	26/0/10	0	x1xxxxx
18	00x001x	42	26/1/15	22	xx0xxx1
19	11010xx				
20	00x0000				
21	xx1xx01				
22	0x0x011				

B. Procedure

The extended test cube merging procedure is given below as Procedure 2. The procedure computes C_{dom} in Step 1. For initialization of the test set T, it places C_{det} followed by C_{dom} in T. A test cube $t_i \in T$ that is obtained from a test cube in C_{det} has an index $0 \leq i < n$. A test cube $t_i \in T$ that is obtained from a test cube in C_{dom} has an index $n \leq i < m+n$.

978-1-4799-7598-3/15 $31.00 © 2015 IEEE

The procedure considers the merging of test cubes t_i and t_j only for $0 \leq i < n$. For j it considers all the possible values, $i < j < m + n$. This implies that t_i is initially a test cube from C_{det}, and not from C_{dom}. In addition, by placing C_{det} before C_{dom} in T, the procedure merges test cubes from C_{det} with t_i before it considers test cubes from C_{dom}. As a result, Procedure 2 merges all the pairs of test cubes that Procedure 1 merges. In addition, it may merge a test cube from C_{dom} with one of these test cubes.

It is possible to select the test cubes from C_{dom} that will be merged with a test cube t_i so as to ensure that additional four-way bridging faults will be detected. For example, suppose that t_i detects a single stuck-at fault g_j stuck-at a_j, and BR contains a four-way bridging fault $g_j/a_j/h_j$. Suppose also that C_{dom} contains a test cube t_j for $h_j = a_j$ that does not conflict with t_i. In this case it is possible to merge t_i and t_j into a test cube $t_{i,j}$ that detects $g_j/a_j/h_j$. However, such computations would require fault simulation to be carried out during the test cube merging process, increasing the computational effort of this process.

In addition, even with fault simulation, it is difficult to predict which faults are important to detect so that reverse order fault simulation will remove test cubes from the final test set. Instead, the extended test cube merging procedure merges as many test cubes from C_{dom} as possible in order to increase the numbers of detected four-way bridging faults as much as possible without focusing on specific faults.

If a test cube $t_j \in C_{dom}$ is merged with a test cube t_i, $merged_j$ is not changed to 1. Since t_j may contribute to the detection of different faults when it is merged with different test cubes, the procedure allows it to be merged multiple times.

A fixed order of the test cubes from C_{dom} in T implies that the first test cubes from C_{dom} will be merged more times. To avoid such a bias, the procedure permutes the test cubes from C_{dom} randomly before starting to merge test cubes with a test cube t_i. This is done in Step 3(a) of Procedure 2. It ensures that different test cubes from C_{dom} will be merged with every test cube t_i, potentially contributing to the detection of different four-way bridging faults.

After the test cube merging process is complete, the procedure removes from T the test cubes that were originally included in C_{dom}. It also removes every test cube t_i such that $merged_i = 1$. Finally, the procedure performs fault simulation followed by forward-looking reverse order fault simulation of BR under T in order to remove unnecessary test cubes from T.

Procedure 2: Extended test cube merging

1) Let $C_{det} = \{c_0, c_1, ..., c_{n-1}\}$ be a set of test cubes for a set of target bridging faults BR. Assign $C_{dom} = \emptyset$. For every bridging fault $g_i/a_i/h_i \in BR$ that is detected by C_{det}:
   a) Extract a test cube $c_{dom(i)}$ that assigns $h_i = a_i$.
   b) Add $c_{dom(i)}$ to C_{dom}.
2) Let $C_{dom} = \{c_n, c_{n+1}, ..., c_{m+n-1}\}$. Assign $T = C_{det}$. Arrange the test cubes in T by order of decreasing

number of specified values. Add C_{dom} to T following C_{det}. Let $T = \{t_0, t_1, ..., t_{n-1}, t_n, ..., t_{m+n-1}\}$. For $0 \leq i < m + n$, assign $merged_i = 0$.
3) For $0 \leq i < n$, if $merged_i = 0$:
   a) Permute $\{t_n, t_{n+1}, ..., t_{m+n-1}\}$ randomly.
   b) For $i < j < m + n$, if $merged_j = 0$:
      i) If t_i and t_j are compatible:
         A) Replace t_i with $t_{i,j}$.
         B) If $j < n$, assign $merged_j = 1$.
4) For $0 \leq i < n$, if $merged_i = 1$, remove t_i from T. For $n \leq i < m + n$, remove t_i from T.
5) Perform fault simulation followed by forward-looking reverse order fault simulation of BR under T, and remove unnecessary test cubes from T.

In the example of $s27$, column $Procedure1$ subcolumn $merged$ of Table II shows the indices of the test cubes from C_{det} that are merged when Procedure 1 is applied. Column $Procedure2$ subcolumn $merged$ of Table II shows the indices of the test cubes from $C_{det} \cup C_{dom}$ that are merged when Procedure 2 is applied. The indices of the test cubes are different from the ones in Table I since both C_{det} and C_{dom} are permuted in T. However, test cubes from C_{det} have indices between 0 and 22, and test cubes from C_{dom} have indices between 23 and 42. For C_{dom}, Table II shows only test cubes that increase the number of specified values. Table II demonstrates that the procedures merge the same test cubes from C_{det}. They differ in that Procedure 2 also merges test cubes from C_{dom}.

To demonstrate the effects of this difference, subcolumns det of Table II show the numbers of four-way bridging faults that are detected after each test cube is simulated using fault simulation with fault dropping. Although the test sets contain the same number of test cubes and detect the same set of faults, the test cubes that are produced by Procedure 2 detect more faults than the corresponding test cubes that are produced by Procedure 1. This contributes to the ability to remove test cubes when forward-looking reverse order fault simulation is applied. In the case of Procedure 1, the final number of test cubes is ten. In the case of Procedure 2, the final number of test cubes is nine. Larger differences in the numbers of test cubes are obtained for larger circuits.

TABLE II
TEST CUBE MERGING

Procedure1		Procedure2	
merged	det	merged	det
0	7	0 25	12
1	15	1 23	20
2 6 20	28	2 6 20	32
3	31	3 38	36
4	33	4 28 42	38
5	35	5 23 39	40
7 9 11	41	7 9 11	43
8 10 14 15 16 17 19	51	8 10 14 15 16 17 19	52
12 21	52	12 21 24 29	53
13 18 22	57	13 18 22 27 33	57

IV. EXPERIMENTAL RESULTS

This section presents experimental results for ISCAS-89, ITC-99 and IWLS-05 benchmark circuits comparing the basic and the improved test cube merging procedures, Procedures 1 and 2.

Four-way bridging faults are selected randomly for the experiments. For every line g_i and value a_i, four lines are selected randomly and used as h_i to define four different four-way bridging faults of the form $g_i/a_i/h_i$. It is also possible to consider realistic bridging faults [10]-[12] or hard-to-detect bridging faults [17] using the same test cube merging procedures.

The results of test cube merging are shown in Tables III-V and Figure 2 as follows. The first row for every circuit in Tables III-V shows the results of Procedure 1. The second row shows the results of Procedure 2.

Column *proc* shows which procedure is applied, where $merge1$ refers to Procedure 1 and $merge2$ refers to Procedure 2. Column *cubes* shows the number of test cubes that the procedure considers (the test cubes in C_{det} or $C_{det} \cup C_{dom}$). Column *tests* shows the number of test cubes in the final test set that the procedure produces. Column *ratio* shows the number of test cubes as a fraction of the number of test cubes produced by Procedure 1. Column *f.c.* shows the four-way bridging fault coverage. Column *n.time* shows the run time as a fraction of the run time of Procedure 1. The run time does not include the test generation process that is required for computing test cubes to detect target faults. The difference in run time is mainly because of the need to compute C_{dom} as part of Procedure 2.

Figure 2 shows the numbers of test cubes in the following format. For a circuit where Procedure 1 produces m_1 test cubes and Procedure 2 produces m_2 test cubes, there is a circle at coordinates m_2, m_1. The dashed line shows the case where $m_1 = m_2$. Above the line, $m_2 < m_1$.

The following points can be seen from Tables III-V and Figure 2. The use of Procedure 2 instead of Procedure 1 produces a smaller test set for all the circuits considered. In several cases the difference in the number of test cubes is significant. Figure 2 demonstrates that these cases occur for the circuits with the larger test sets.

When the set of test cubes C_{det} does not detect all the detectable bridging faults in BR, Procedure 2 detects more faults accidentally and achieves a higher four-way bridging fault coverage. This is possible since Procedure 2 merges test cubes that Procedure 1 does not.

The number of test cubes in C_{dom} is significantly smaller than the number of test cubes in C_{det}. This can be seen by comparing the numbers of test cubes that the procedures consider.

V. CONCLUDING REMARKS

This paper considered the process of test compaction by merging of test cubes for four-way bridging faults. It extended a basic test cube merging procedure to consider a set of test cubes C_{dom} that assign values to dominating lines. By merging

TABLE III
EXPERIMENTAL RESULTS (ISCAS-89)

circuit	proc	cubes	tests	ratio	f.c.	n.time
s1423	merge1	3498	141	1.00	96.11	1.00
s1423	merge2	4651	120	0.85	96.19	6.84
s5378	merge1	12310	329	1.00	97.69	1.00
s5378	merge2	14524	298	0.91	97.69	4.86
s9234	merge1	16768	543	1.00	90.16	1.00
s9234	merge2	20791	477	0.88	90.25	3.26
s13207	merge1	30360	496	1.00	95.87	1.00
s13207	merge2	35832	459	0.93	95.97	6.73
s15850	merge1	33374	503	1.00	94.83	1.00
s15850	merge2	41134	441	0.88	94.88	5.31
s35932	merge1	106545	120	1.00	84.91	1.00
s35932	merge2	128165	103	0.86	84.91	10.87
s38417	merge1	104583	790	1.00	99.11	1.00
s38417	merge2	130261	650	0.82	99.14	4.97
s38584	merge1	107462	462	1.00	89.15	1.00
s38584	merge2	128131	446	0.97	89.19	3.43

TABLE IV
EXPERIMENTAL RESULTS (ITC-99)

circuit	proc	cubes	tests	ratio	f.c.	n.time
b04	merge1	2620	102	1.00	94.97	1.00
b04	merge2	3511	91	0.89	94.97	6.95
b05	merge1	1891	186	1.00	90.74	1.00
b05	merge2	2957	160	0.86	90.75	5.71
b07	merge1	1849	107	1.00	89.86	1.00
b07	merge2	2664	96	0.90	89.86	5.95
b11	merge1	1117	107	1.00	85.79	1.00
b11	merge2	1777	103	0.96	85.79	4.89
b14	merge1	16061	683	1.00	86.60	1.00
b14	merge2	24682	576	0.84	86.69	2.46
b15	merge1	36907	1572	1.00	87.23	1.00
b15	merge2	53231	1302	0.83	87.36	2.01
b20	merge1	46420	1439	1.00	90.85	1.00
b20	merge2	70202	1092	0.76	90.93	3.97

TABLE V
EXPERIMENTAL RESULTS (IWLS-05)

circuit	proc	cubes	tests	ratio	f.c.	n.time
aes_core	merge1	147748	2004	1.00	98.99	1.00
aes_core	merge2	181560	1847	0.92	99.00	1.62
des_area	merge1	17240	613	1.00	94.82	1.00
des_area	merge2	27795	456	0.74	94.96	2.45
i2c	merge1	6003	151	1.00	96.56	1.00
i2c	merge2	7186	147	0.97	96.58	6.81
pci_spoci_ctrl	merge1	2402	231	1.00	86.66	1.00
pci_spoci_ctrl	merge2	3247	223	0.97	86.68	4.84
simple_spi	merge1	6054	109	1.00	97.42	1.00
simple_spi	merge2	7358	102	0.94	97.43	6.37
spi	merge1	14629	1016	1.00	93.11	1.00
spi	merge2	19867	850	0.84	93.20	1.62
systemcaes	merge1	58460	685	1.00	98.06	1.00
systemcaes	merge2	80623	479	0.70	98.11	4.81
systemcdes	merge1	15393	465	1.00	97.32	1.00
systemcdes	merge2	24714	300	0.65	97.38	2.88
tv80	merge1	34258	1891	1.00	93.17	1.00
tv80	merge2	50359	1498	0.79	93.30	1.82
wb_dma	merge1	29204	228	1.00	98.34	1.00
wb_dma	merge2	35616	210	0.92	98.37	7.11

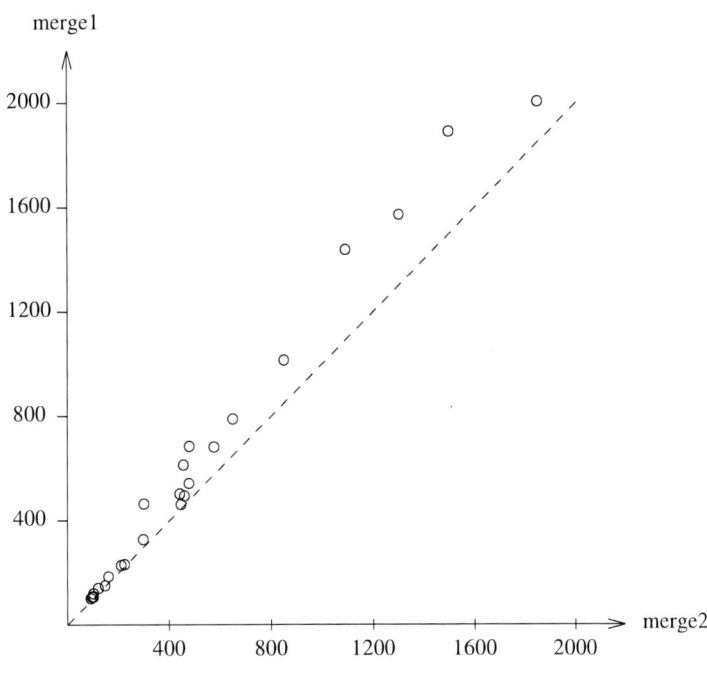

Fig. 2. Numbers of test cubes

test cubes from C_{dom} with test cubes that detect bridging faults, the procedure increases the number of bridging faults that are detected by the resulting test cubes. As a result, fewer test cubes are needed for detecting the same set of target faults. Experimental results were presented to demonstrate the levels of additional test compaction that can be achieved by this procedure.

REFERENCES

[1] B. Koenemann, "LFSR-Coded Test Patterns for Scan Designs", in Proc. Europ. Test Conf., 1991, pp. 237-242.

[2] K.-J. Lee, J. J. Chen and C. H. Huang, "Using a Single Input to Support Multiple Scan Chains", in Proc. Intl. Conf. on Computer-Aided Design, 1998, pp. 74-78

[3] C. Barnhart, V. Brunkhorst, F. Distler, O. Farnsworth, B. Keller and B. Koenemann, "OPMISR: The Foundation for Compressed ATPG Vectors", in Proc. Intl. Test Conf., 2001, pp. 748-757.

[4] J. Rajski, J. Tyszer, M. Kassab, N. Mukherjee, R. Thompson, K.-H. Tsai, A. Hertwig, N. Tamarapalli, G. Mrugalski, G. Eide and J. Qian, "Embedded Deterministic Test for Low Cost Manufacturing Test", in Proc. Intl. Test Conf., 2002, pp. 301-310.

[5] N. A. Touba, "Survey of Test Vector Compression Techniques", IEEE Design & Test, Apr. 2006, pp. 294-303.

[6] M. Abramovici, M. A. Breuer and A. D. Friedman, *Digital Systems Testing and Testable Design*, IEEE Press, 1995.

[7] A. H. El-Maleh and Y. E. Osais, "Test Vector Decomposition-Based Static Compaction Algorithms for Combinational Circuits", ACM Trans. on Design Autom. of Elect. Sys., Oct. 2003, pp. 430-459.

[8] D. Czysz, G. Mrugalski, N. Mukherjee, J. Rajski, P. Szczerbicki and J. Tyszer, "Deterministic Clustering of Incompatible Test Cubes for Higher Power-Aware EDT Compression", IEEE Trans. on Computer-Aided Design, Aug. 2011, pp. 1225-1238.

[9] I. Pomeranz, "On Test Compaction of Broadside and Skewed-Load Test Cubes", IEEE Trans. on VLSI Systems, Sept. 2013, pp. 1705-1714.

[10] F. J. Ferguson and J. P. Shen, "A CMOS Fault Extractor for Inductive Fault Analysis", IEEE Trans. on Computer-Aided Design, Nov. 1988, pp. 1181-1194.

[11] S. T. Zachariah and S. Chakravarty, "A Scalable and Efficient Methodology to Extract Two Node Bridges from Large Industrial Circuits", in Proc. Intl. Test Conf., 2000, pp. 750-759.

[12] Z. Stanojevic and D. M. H. Walker, "FedEx - A Fast Bridging Fault Extractor", in Proc. Intl. Test Conf., 2001, pp. 696-703.

[13] S. Sengupta et. al., "Defect-Based Tests: A Key Enabler for Successful Migration to Structural Test", Intel Technology Journal, Q.1, 1999.

[14] V. Krishnaswamy, A. B. Ma, P. Vishakantaiah, "A Study of Bridging Defect Probabilities on a Pentium (TM) 4 CPU", in Proc. Intl. Test Conf., 2001, pp. 688-695.

[15] H. Tang, G. Chen, S. M. Reddy, C. Wang, J. Rajski and I. Pomeranz, "Defect Aware Test Patterns", in Proc. Design Autom. and Test in Europe Conf., 2005, pp. 450-455.

[16] J. E. Nelson, J. G. Brown, R. Desineni and R. D. Blanton, "Multiple-Detect ATPG Based on Physical Neighborhoods", in Proc. Design Autom. Conf., 2006, pp. 1099-1102.

[17] I. Pomeranz, S. M. Reddy and S. Kundu, "On the Characterization of Hard-to-Detect Bridging Faults", in Proc. Design Autom. and Test in Europe Conf., 2003, pp. 1012-1017.

Panel: Is Design-for-Security the new DFT?

There has been a lot of interest recently in design-for-security, while at the same time research opportunities in conventional DFT are waning. Is DFS the new DFT, since many similar techniques can be applied, or is DFS something new, requiring new approaches to be successful, or is DFS merely a distraction? The panel will discuss the issues and give the audience the opportunity to decide.

Moderator and Organizer: Rob Aitken, ARM

Panelists
Subhasish Mitra, Stanford
Ronald Perez, Cryptography Research
Mark Tehranipoor, U Connecticut
Steve Trimberger, Xilinx

Innovative Practices Session 11C:
Advanced Scan Methodologies

Organizer: Janusz Rajski (Mentor Graphics)
Moderator: Nilanjan Mukherjee (Mentor Graphics)

Presenters and Abstracts

Presenter #1: *Vivek Chickermane (Cadence)*

Title: Practical Aspects of Hierarchical Test Implementation

Abstract: In this presentation we discuss several practical considerations of implementing a traditional style of hierarchical compression using IEEE 1500 wrapped cores and using a parallel test access mechanism (TAM) to supply compressed patterns to the core inputs and gather compressed responses from the embedded cores. We will also describe briefly the diagnostic approaches needed to support hierarchical compression. We will also address two other emerging trends that can be exploited to create more efficient hierarchical compression. The first is increased use of serialization and de-serialization of scan patterns to match a limited number of tester pins to the large number of core-level scan pins. The second trend is the standardization of embedded instrument access in hierarchical designs using the IEEE 1687 standard.

Presenter #2: *Kamlesh Pandey (Broadcom)*

Title: ASIC Test cost reduction using advanced scan techniques

Abstract: Ever increasing integration scale of modern SOCs poses two key test challenges. These challenges are (1) exploding test data volume to support multiple fault models and (2) massive test development efforts. The problem is further compounded by shrinking design cycles to achieve aggressive marketing goals. Good scan compression technique can cut down the test data volume by factor of 100 to 200. It is possible to further reduce test data volume by factor of 3 to 4 by using efficient test point insertion scheme. Similarly hierarchical test implementation can significantly cut down the test development time and efforts. Both these test challenges were very effectively addressed using EDT scan compression, EDT test point insertion and Tessent pattern retargeting solutions.

Presenter #3: *Tassanee Payakapan (AMD)*

Title: An IJTAG (IEEE1687) based Method to Automate ATPG Setup for an SoC with Complex Embedded Test Access Infrastructure

Abstract: IEEE 1149.1-based top level access to IEEE 1500-compliant IP cores is commonly used in industrial designs as the underlying infrastructure to provide test access, control, instrumentation, and ease of use. Validating the test infrastructure and its usage in the early design stages is critical to the success of the project. The new Internal Joint Test Action Group (IJTAG or IEEE 1687-2014) standard is a valuable component of this test infrastructure and is designed to promote efficient embedded instrument access. This presentation discusses our experience using an IJTAG-based method to a state-of-the-art Server Microprocessor to automate ATPG test pattern setup procedures.

2015 IEEE 33rd VLSI Test Symposium (VTS)

Testing Cross Wire Opens within Complex Gates

Chao Han and Adit D. Singh
{chaohan,singhad@auburn.edu}
Department of Electrical and Computer Engineering
Auburn University, Auburn AL, 36849

Abstract—**Recent test studies on volume production data suggest that a significant number of CMOS open defects remain undetected by commonly applied TDF timing tests, potentially leading to high defectivity in the shipped parts. This has focused attention on developing tests that explicitly target open faults, in particular transistor stuck open faults (TSOFs). However, while TSOFs cover all open faults in circuits implemented from primitive logic gates, they do not model a type of open fault found only in complex CMOS gates. We refer to these as cross wire open (CWO) faults. In this paper, we develop the first tests that target CWOs. Although we observe that the fault list of potential CWOs can be significantly reduced if the layouts of complex gate cells used in the design are available, we present test generation methodologies both with and without this layout information. CWO fault coverage results for scan based tests are presented for ISCAS89 and ITC99 benchmark circuits that have been resynthesized using an open source cell library containing complex gates.**

Keywords — open defects, complex gates, cross wire opens

I. INTRODUCTION

With the continuing shrinking of CMOS technology feature sizes and decrease in transistor threshold voltages, open defects are becoming increasingly significant during test due to their complex timing behavior caused by the presence of relatively large circuit leakage currents. Transition delay fault (TDF) testing has been widely used by the industry for targeting circuit timing faults and delay defects. It has been commonly assumed that since stuck-open defects in the pull up or pull down networks of CMOS gates behave like transition faults exhibiting infinite delays (or at least relatively large delays even in the presence of significant leakage current), they are detected by the TDF timing tests. However, increasing evidence indicates that high quality TDF tests do not detect many open defects that may cause circuit malfunction. *A recent industrial study [1] on 32nm microprocessor parts has shown that performing detailed "cell aware" testing resulted in the detection of additional defect level of as much as 885 DPPM beyond what was screened by industrial grade stuck-at and N detect (N=5) TDF tests. Furthermore, 87% of these additional defects missed by the TDF tests but found by the cell aware tests required two pattern tests, suggesting opens. This has motivated interest in explicitly targeting open defects during test.*

There has been research on testing open defects reported in the literature going back many decades. Some more recent papers [3,4] have been proposed enhancing the open defect coverage of TDF test to also target the transistor stuck-open faults (TSOFs) by adding the constraint to the ATPG to additionally require gate output initialization when targeting TDFs at gate input nodes. However, the open fault coverage improvement from this approach appears limited, as a large number of TSOFs remain undetectable. Unfortunately, these open faults are not all benign; many can actually be activated by common circuit hazards (glitches) to cause errors. To address this issue earlier work has developed hazard initialized tests [5,6] and a DFT scheme [14] to enhance TSOF coverage beyond that achieved by traditional TSOF tests.

For primitive gates, TSOFs are the only type of open defect that needs to be considered because they can model any break in the pull up or pull down path. However, modern industrial circuits consist of not only primitive gates, but also a rich set of complex CMOS gates. *An additional type of open fault is found in complex gates which behaves differently to TSOFs – we call this the cross wire open (CWO) fault. This important fault type has not been studied as a target for test generation so far.* The likely reason is that commonly available implementations of the benchmark circuits used in most published studies are based on primitive logic gates and do not employ complex gates.

Figure 1 shows a cross wire open fault d in the schematic of an AOI22 complex gate. It is obvious that the two pattern tests targeting the four PMOS TSOFs alone may miss the cross wire open fault shown. To detect this fault, in addition to a V1 vector that initializes the Z output to low, we need a V2 vector that either activates only the pull up path a1, b2 or only the pull up path a2, b1, both of which contain the open defect. Thus for V2 (a1, a2) = 01 and V2 (b1, b2) = 10, or V2 (a1, a2) = 10 and V2 (b1, b2) = 01, if the cross wire open exists the gate output will stay at its initialized low value and not be pulled up, indicating a fault. Existing TDF timing tests and TSOF tests supported by the commercial ATPG cannot generate test vectors to satisfy the required conditions for V2 described above, and generally depend on the chance detection of such open defects by traditional TDF tests.

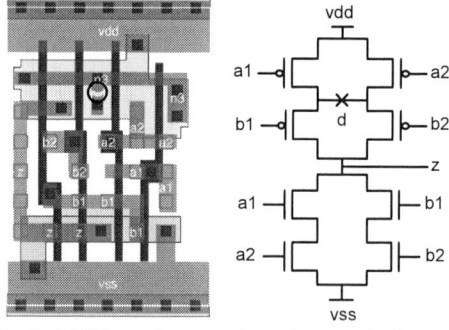

Fig. 1 AOI22 gate layout and transistor level schematic

The recently proposed defect-oriented cell aware tests [1] target all possible defects within the library cells and therefore

978-1-4799-7598-3/15 $31.00 © 2015 IEEE

256

cross wire open faults are assumed to be included. However, these tests are not explicitly focused on such opens, and target very many other possible defects in the cell layout, leading to a large increase in the test set. Furthermore, detailed cell level layout information is not always available at test generation. *There is, therefore, a need to develop compact and effective tests from a gate level model of the circuit. This is a key motivation for this work.* Note further that the proposed cell aware tests[1] currently only target open detection under timing unaware Boolean analysis that ignores hazards. They are unable to generate test for the many open faults that appear undetectable from functional states but can actually be activated by hazards and cause circuit errors [5,6].

In the paper, we for the first time present a test methodology that explicitly targets and generates tests for all the CWO faults within complex gates. Furthermore, we also present a DFT scheme to target those CWO faults that are not detectable by traditional timing unaware Boolean tests but remain a reliability concern because of potential activation by hazards.

The rest of the paper is organized as follows. Section 2 discusses the background and prior work. Section 3 analyzes the cross wire fault list for different complex gates, both when cell layout information is available, and when it is not. We introduce a modified circuit structure, *to be used only during test generation*, to force commercial ATPG tools to yield TDF tests that meet the requirements for the detection of the target cross wire opens (CWOs) in Section 4. Section 5 presents CWO test coverage results obtained for some benchmark circuits. The DFT scheme approach from [14] is employed for targeting the CWO faults that are undetectable by traditional timing unaware scan tests in Section 6, with experimental results showing the improvement in coverage presented in Section 7. We conclude in Section 8.

II. BACKGROUND AND PRIOR WORK

A cross wire open defect (circled in black) is shown in the layout of an AOI22 gate in Figure 1. This open defect can cause the pull up network to electrically behave like a true open, or more generally, result in a large pull up delay fault on account of some leakage current through the off transistors. In either case, detection of the fault generally requires a two vector test. The first vector initializes the gate output to 0; the second vector turns on a pull up path containing only two PMOS transistors connected by the target cross wire (PMOS at input a1 and b2, or input a2 and b1). Furthermore, the fault effect must also be sensitized to the circuit outputs by the second vector. It is obvious that traditional TSOF tests targeting the four PMOS transistors alone may miss the open fault since the tests may turn on PMOS at a1and b1, or a2 and b2 such that the open fault at d is never activated.

It is possible that because of the structural limitations of scan which restrict the V2 vector that can be applied, no such test exists in either LOC or LOS mode to detect this fault, where the V1 vector initializes output z to 0, and V2 vector activates and sensitizes the fault effect to circuit outputs. In such a case, while the fault appears undetectable based on stable Boolean analysis from functional states (which are a subset of the LOC test states), it is still possible for it to be activated due to the common hazards (glitches) in normal functional operation and

result in output errors. Consider two sets of signal waveforms for the AOI22 gate in Figure 3. For Case 1, the two (V1,V2) values are: a1 = 10, a2 = 01, b1 = 11 and b2 = 10. Under timing unaware Boolean analysis, when input signals transit from V1 to V2 as in Case 1, output z remains 1 and the open fault cannot be activated. However, if in fact the 1 to 0 transition at a1 occurs later than the 0 to 1 transition at a2 and generates a hazard at the z output as shown, z can be initialized to 0 by the hazard, which would initialize the fault. Similarly, in Case 2, steady state (V1V2) values are a1 = 00, a2 = 01, b1 = 11 and b2 = 10. Timing unaware analysis suggests output z cannot be initialized to 0. But if we assume there is a 0(1)0 hazard received at a1 in between (V1) 0 and (V2) 0, and at the time a1 receives this hazard a2 is already 1, output z will be initialized to 0 by this propagating hazard arriving at input a1. For both cases, output z can be initialized to 0, and the open fault will be activated at gate output z when (V2) values are: a1 = b2 = 0 and a2 = b1 = 1. Malfunction occurs if this error can be sensitized to circuit outputs.

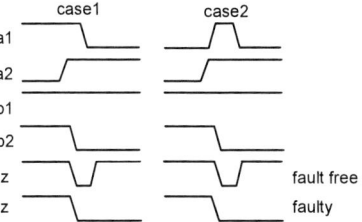

Fig. 2 Signal waveforms for AOI22 gate that generate output hazards

Note that besides hazard generation caused by a1 and a2 inputs, hazards from b1 and b2 inputs can also cause the open fault to be activated in a similar way. And even though the fault appears undetectable from timing unaware analysis, there is a significant possibility for hazards to activate the open fault and cause circuit errors. Hazards are very common during circuit normal operation and they consume as much as 40% of switching power. In the presence of hazards, virtually all open faults need to be targeted for detection. Cell aware tests [1] target cross wire open faults, as one of many possible defects, based only on traditional timing unaware analysis; there is no reported work on hazard initialized detection of cross wire faults. In addition to targeting cross wire opens with traditional LOC and LOS scan tests, in this paper we also employ the DFT methodology in [14] to target the remaining undetectable faults. It involves the use of multiple scan enable signals to support a mix of simultaneous LOC and LOS tests that allow more possibilities for the V2 vector in launching a two pattern test, thereby increasing test coverage.

Observe that cross wire open (CWO) faults only exist in complex gates, while most available netlists for ISCAS-89 and ITC99 benchmark circuits only contain primitive gates. We have re-synthesized the ISCAS-89 and ITC99 benchmark circuits to include complex gates using an open source TSMC 250nm standard cell library to conduct the test generation and fault simulation experiments reported here.

III. CROSS WIRE FAULT LIST FOR COMPLEX GATES

To target the CWO faults within the complex gates independent of library cell layout, we need to identify all the

This research was supported in part by the National Science Foundation under Grants No. EECS-0903449 and CCF-1319529.

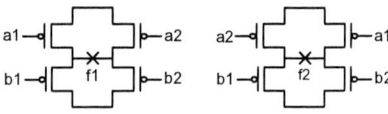

Fig. 3 PMOS networks for an AOI22 gate

possible open faults for all inputs configurations of the complex gates. This is because without layout information, it is impossible to tell how inputs from the netlist map to physical transistor locations in silicon. For example, Figure 3 shows the PMOS networks of an AOI22 gate. It is obvious that by exchanging input a1 and a2 we have two possible network configurations which create two possible open faults f1 and f2. To activate fault f1 the V2 vector needs to turn on PMOS at input a1 and b2 (a1, b2) only or a2 and b1 (a2, b1) only. Similarly for fault f2 we need PMOS at (a2, b2) or (a1, b1) to be turned on. If the layout information is unavailable, both these CWOs must be targeted during test generation to guarantee fault detection.

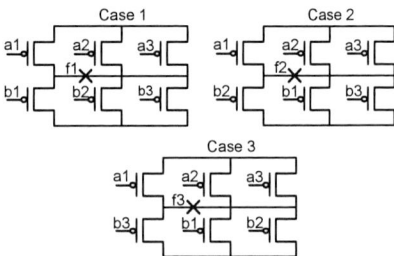

Fig. 4 PMOS networks for an AOI33 gate

Now consider the PMOS network of an AOI33 gate. Figure 4 shows the different input configurations possible. As can be seen in Case 1, the CWO fault f1 separates the upper and lower level of inputs into two groups respectively: input a1 on the left side and input a2, a3 on the right side; input b1 on the left side and inputs b2, b3 on the right side. If we keep the input configuration for the upper level inputs the same and list all the possible configurations for the lower level inputs based on different grouping, we have three possible PMOS networks that for Case 1 fault f1 separates input b1 and b2, b3; for Case 2 fault f2 separates input b2 and b1, b3; for Case 3 fault f3 separates input b3 and b1, b2. To activate fault f1 in Case 1, the V2 vector should turn on PMOS at a1 and b2 or b3 (a1, (b2 + b3)), or PMOS at a2 or a3 and b1 ((a2 + a3), b1). Similarly, for fault f2 in Case 2 we need (a1, (b1 + b3)) or ((a2 + a3), b2), and for fault f3 in Case 3 we need (a1, (b1 + b2)) or ((a2 + a3), b3). Since we also have three possible configurations for the upper level inputs, totally we have a total of 3x3 = 9 faults for all possible input configurations of the AOI33 gate. Similarly, for the AOI333 gate, when we consider the open faults at the first and second level of inputs, we have 3x3 = 9 faults; when we consider the second and third level, we have another 9 faults. Thus for the AOI333 gate we have 18 possible faults to target. Figure 5 shows the different PMOS network configurations of an AOI44 gate. For fault f0 which is at the side of the network, the analysis is similar to our discussion of AOI33 gate; for each level of inputs we have 4 possible configurations and totally we have 4x4 = 16 possible side open faults for the AOI44 gate. Now

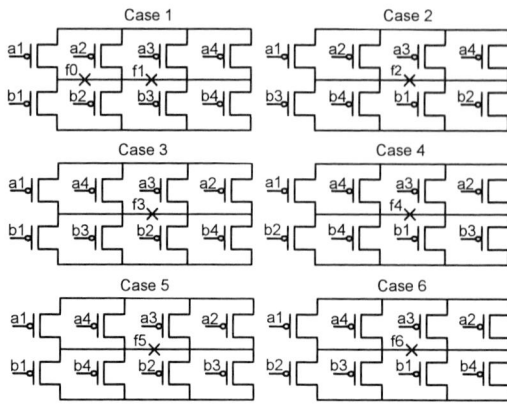

Fig. 5 PMOS networks for an AOI44 gate

consider fault f1 which is at the center of the network, it separates the upper and lower inputs into two groups respectively: input a1, a2 and a3, a4 for the upper level; input b1, b2 and b3, b4 for the lower level. If we keep the upper level network the same and interchange the lower level network, we have six possible configurations (As shown in Figure 5) based on different groups of inputs: input b1, b2 and b3, b4 separated by fault f1; input b3, b4 and b1, b2 separated by fault f2; input b1, b3 and b2, b4 separated by fault f3; input b2, b4 and b1, b3 separated by f4, input b1, b4 and b2, b3 separated by f5, input b2, b3 and b1, b4 separated by f6. As there can be three possible configurations for the upper level network (the reason there is three instead of six configurations as in the lower level is we interchange the two groups in the lower level to have six configurations, and the interchanging of input groups in the upper level network is redundant since the network is symmetrical), totally we have 6x3 = 18 possible faults at the center of the AOI44 network. Thus the AOI44 gate have 18 + 16 = 34 possible CWO faults.

AOI22, AOI221		AOI222		AOI32, AOI321		AOI322		AOI33	
2	1	4	2	6	2	8	3	9	2
AOI332		AOI333		AOI422		AOI43		AOI44	
15	4	18	4	16	4	30	3	34	3

Table 1. Number of possible cross wire open (CWO) faults for sample AOI complex gates with and without layout information

Table 1 shows the number of cross wire open faults possible for varying sizes of AOI complex gate cells. The number on the left is for total possible faults from all layout configurations, while the number on the right is for the faults from a particular configuration assuming we have the cell layout information.

IV. TARGETING LOC AND LOS DETECTABLE OPENS

We use a commercial ATPG tool to generate deterministic tests for the timing un-aware cross wire opens in complex gates. This is done by transforming the complex gates into another circuit structure *for the purposes of test generation*. This converts the detection of open faults into the detection of corresponding transition delay faults with additional test activation conditions. Figure 6 shows an example of AOI33 gate transformation. The function of the transformed structure is the same as that of the AOI33 gate. In the figure we target

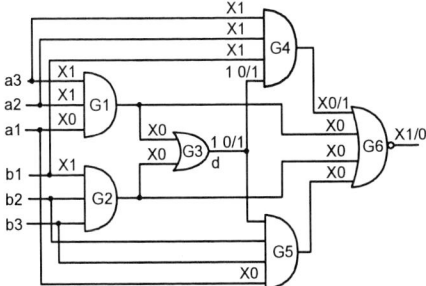

Fig. 6 Gate transformation for cross wire open detection

node d (output of gate G3) slow-to-fall TDF in the transformed circuit. Since (V1) d = 1 and d = (a1 AND a2 AND a3) OR (b1 AND b2 AND b3), we have either (V1) a1 = a2 = a3 =1 or b1 = b2 = b3 = 1 to initialize the output y to 0. The fault effect has two paths (gate G4 or G5) to sensitize to output y. Suppose it sensitizes through gate G4, then we have (V2) a2 = 1, a3 = 1 and b1 = 1. Since (V2) d = 0, we have gate G1's output = 0 and G2's output = 0, which give us a1 = 0 and b2 or b3 = 0. Thus the V2 vector will try to turn on PMOS at input a1 and input b2 or b3 only. This is exactly a test for the cross wire open fault f1 in Figure 5. Similarly, if the fault effect sensitizes through gate G5, we have the V2 vector to set a1 = 1, a2 = 0 or a3 = 0 and b1 = 0, b2 = 1, b3 = 1, which is the other test for fault f1. Thus by using ATPG to target the TDF fault, we have the test for the corresponding cross wire open (CWO) fault. The transformed sub-circuits circuits for other complex gates can be created in a similar way. We use a small script to automate the transformation before the circuit is given to ATPG.

Using the transformed circuit we can only generate the test targeting one cross wire open fault in each of the complex gates. To target all the possible faults within all the complex gates, we need to run test generation on multiple transformed copies of the CUT. The total number of test generation runs needed equals the total number of possible cross wire opens faults in the largest complex gate. This results in a somewhat large fault list and test data volume for the CUTs, which can be greatly reduced if layout information is available.

Note that while we have worked with a traditional circuit structure based commercial ATPG tool, adding the required constraints on two pattern TDF tests to target opens can perhaps be more easily implemented in SAT based ATPG, and also using new commercial tools that support user defined fault models.

V. EXPERIMENTAL RESULTS FOR LOC AND LOS DETECTABLE OPENS

As explained earlier in the paper, the experiments reported in this section were all conducted using a commercial ATPG tool for the re-synthesized ISCAS-89 and ITC-99 benchmark circuits using an open source 250nm technology standard cell library containing a rich set of complex gates. We initially employ LOC TDF tests to conduct the cross wire open fault simulation using the transformed circuit structure discussed in Section 4. We then incrementally allow LOC and LOS cross wire open fault test generation to target the CWO faults undetected by LOC TDF tests. The rationale for this order is

CUT	Total CWOs	CWO % of total TSOFs	CWO FC of TDF Tests	ΔLOC CWO FC	ΔLOS CWO FC	Total FC
s1423	82	4.75%	69.51%	19.51%	0%	89.02%
s9234	148	4.91%	45.27%	48.65%	2.03%	95.95%
s13207	580	7.61%	54.83%	22.41%	18.97%	96.21%
s15850	734	7.37%	43.19%	37.87%	11.04%	92.10%
s35932	1674	4.96%	97.85%	2.15%	0%	100%
s38417	2080	6.99%	63.94%	31.54%	1.20%	96.68%
s38584	2653	7.77%	69.05%	21.75%	2.86%	93.67%
b14	2347	12.45%	47.59%	34.73%	8.27%	90.58%
b15	3976	15.04%	31.34%	40.34%	12.60%	84.28%
b20	4395	11.39%	53.11%	32.38%	7.87%	93.36%
b21	4469	10.99%	51.31%	31.17%	9.64%	92.12%
b22	6839	11.50%	49.42%	33.73%	8.77%	91.93%
b17	14837	16.28%	32.04%	39.81%	11.59%	83.45%
b18	34670	14.13%	32.32%	39.95%	17.55%	89.82%
average	NA	9.72%	52.91%	31.14%	8.03%	92.08%

Table 2. Incidental CWO coverage of LOC TDF Tests and coverage increase from targeting undetected CWOs in LOC and LOS modes.

to maximize the test speed with which most of tests are applied. LOC tests can always be run at speed, so the LOC tests are used to target as many CWOs that can be detected by such tests. The remaining faults are still targeted by LOS tests because even when LOS tests must be run much slower because of a slow scan enable signal, they can still detect many opens that display very large delay fault behavior.

CUT	Total CWOs	CWO % of total TSOFs	CWO FC of TDF Tests	ΔLOC CWO FC	ΔLOS CWO FC	Total FC
s1423	31	1.79%	77.42%	22.58%	0%	100.00%
s9234	66	2.19%	50.00%	48.48%	0%	98.48%
s13207	228	2.99%	30.26%	28.51%	37.28%	96.05%
s15850	294	2.95%	37.76%	46.60%	13.61%	97.96%
s35932	558	1.65%	95.88%	4.12%	0%	100.00%
s38417	806	2.71%	60.42%	34.37%	0.12%	94.91%
s38584	907	2.65%	70.23%	20.07%	2.76%	93.05%
b14	991	5.26%	53.28%	27.45%	8.48%	89.20%
b15	1774	6.71%	29.59%	50.34%	6.76%	86.70%
b20	1961	5.08%	54.51%	29.63%	9.43%	93.57%
b21	1985	4.88%	51.28%	29.87%	10.18%	91.34%
b22	2811	4.73%	47.95%	33.48%	11.10%	92.53%
b17	6296	6.91%	34.55%	40.71%	6.40%	81.66%
b18	14377	5.86%	31.01%	41.12%	17.96%	90.10%
average	NA	4.03%	51.73%	32.67%	9.54%	93.25%

Table 3. Incidental CWO coverage of TDF tests and improvement from targeted LOC and LOS tests for a fixed layout configuration.

The experimental results for all possible CWO faults are shown in Table 2. Column 2 shows the total number of possible CWO faults in the absence of layout information. Column 3 shows the percentages of total number of possible CWO faults to the total number of TSOFs for each CUT. Note the percentages are relatively large since we consider all possible inputs configurations of the complex gates. Column 4 shows the CWO fault coverage (as percentage of the total number of possible CWO faults) of the TDF LOC tests for the

978-1-4799-7598-3/15 $31.00 © 2015 IEEE   259

original circuits with complex gates. Column 5 and 6 show the incremental coverage improvement over TDF LOC tests by LOC and LOS cross wire open tests. As we can see, by applying TDF tests alone, on average only 50% of the CWO faults are detected. By employing LOC and LOS cross wire open tests to target the TDF undetected faults, we can get close to a 40% improvement in CWO fault coverage on average. For the results in Table 3 we have repeated the test generation and fault simulation experiments in Table 2 based on a fixed and known transistor configuration in each complex gate (which assumes the cell layout information is available). While the number of target CWOs are 2-3X smaller, the coverage results and trends can be seen to be quite similar.

VI. TARGETING OPENS UNDETECTABLE BY TRADITIONAL TIMING UNAWARE SCAN TESTS

Observe in Tables 2 and 3 that the CWO fault coverage for the larger circuits ranges between 80 and 95%, which may lead to some test escapes in production tests. To boost this test coverage beyond what is reachable by conventional LOC and LOS scan tests, we investigate the DFT scheme in [14] that employs multiple control lines to distribute the scan enable signals to different set of flip-flops. This allows some flip flops to operate in LOC mode while others in LOS mode, thereby increasing the available scan states for launching the tests beyond those reachable by LOC and LOS test modes alone.

Note that papers have proposed the idea of using multiple scan enable signals to simultaneously have flip-flops operating in the LOC and LOS mode, but in a more restrictive way. To minimize the overhead of implementing a high speed scan enable signal to support LOS timing tests, the hybrid TDF test generation methodology in [11] and similar methods selected only a very small number of critical flip-flops to be operated in the LOS mode. [12] avoided the need for an at-speed scan enable by using the LOS flip-flops (along with LOC flip-flops) only to launch the TDF timing test; the test response was only captured and observed by the flip-flops operating in the LOC mode. While this combination of LOC and LOS modes led to some increase in the available test launch states, test output observability will be significantly impact by restricting the response capture to the LOC flip-flops.

Furthermore, unlike small delay timing defects, the target in our methodology are open faults which cause delay faults of much longer durations. Therefore we are not significantly constrained by the need for a fast scan enable to support LOS tests, although some low cost methods such as pipelining to speedup the scan enable signals will help support faster launch to capture cycle and can potentially prevent the test escape of a few opens with very high leakage. We therefore assume here that we can independently combine LOC and LOS test modes using multiple partitions and the overhead of routing multiple scan enable signals is still manageable.

In the experiments in the next section, we report simulation results for tests conducted using 2, 4 and 8 scan enable signals (each enable signal independently controls the same number of flip-flops), with each partition operated in either the LOC or LOS mode during any test. For example, for two partitions controlled by two independent scan enable lines, the circuit can be tested in LOC-LOC, LOC-LOS, LOS-LOC and LOS-LOS

test modes. For 4 partitions we have 16 test mode combinations. We modify the scan structure such that the ATPG determines whether each partitions of flip-flops would be under LOC mode or LOS mode. This leads to the availability of a larger set of test launch states than from LOC and LOS test modes alone and results in the improved fault coverage. In this study, we first of all randomly assign all the flip-flops to different scan enable signals. And then we re-assign them in a simple greedy manner such that the cases where any gate is directly driven by flip flops controlled by the same scan enable signal is minimized. This can help to reduce some of the shift dependence compared to a purely random assignment. More effective assignment of the scan enable lines to flip-flops may be possible [13] and will be the subject of future research.

VII. EXPERIMENTAL RESULTS FOR UNDETECTED OPENS

CUT	Δ Multiple Scan_en Tests FC			ΔEnhanced Scan FC	Enhanced Scan Undet
	2 scan_en	4 scan_en	8 scan_en		
b14	0.72%	0.77%	0.77%	1.36%	8.05%
b15	5.01%	5.51%	6.24%	7.75%	7.97%
b20	0.98%	1.00%	1.02%	1.34%	5.30%
b21	1.03%	1.12%	1.16%	1.41%	6.47%
b22	1.02%	1.08%	1.13%	1.27%	6.80%
b17	3.67%	4.25%	4.62%	7.82%	8.73%
b18	3.13%	3.58%	3.74%	4.47%	5.70%

Table 4. CWO coverage improvement with multiple scan enable tests. Results for enhanced scan show the best coverage possible

Table 4 shows the experimental results for the LOC and LOS undetectable cross wire open faults for all possible configurations. Table 5 shows the same results for a fixed configuration, which assumes that we have the cell layout information. Columns 2, 3 and 4 show the fault coverage improvement using 2, 4 and 8 scan enable tests. Column 5 in Table 4 shows the fault coverage improvement beyond combined LOC and LOS tests achievable by Enhanced Scan tests. Recall that the Enhanced Scan approach allows all possible V2 vectors to be applied to the circuit, completely removing any structural restrictions of scan in applying two pattern tests. Thus the CWO coverage achieved by Enhanced Scan tests is the highest achievable. We can see that with an 8 scan enable design, the fault coverage improvement for most of the circuits is close to what is achievable from the Enhanced Scan tests. For some large circuits (b15, b17) we see more than 4% FC improvement. Most of this improvement is achieved with just 2 scan enables. Column 6 shows the percentage of CWOs that are undetectable even with Enhanced Scan. Recall that enhanced scan removes all restrictions and arbitrary two-pattern tests can be apply to the combinational logic of a sequential design. It is obvious that an enhanced scan undetectable fault might still be activated by hazards, but the fault effect never propagates to the circuit outputs, and therefore causes no errors in functional operation. First consider the V2 pattern needed for detection. It must attempt to turn on the transistors path that containing the cross wire open fault, turn off any other parallel paths (so that the gate output floats in the presence of the fault), and at the same time sensitize a path from the gate output to a circuit output. For complex gates with multiple inputs, satisfying all the above conditions simultaneously may not always be possible,

CUT	Δ Multiple Scan_en Tests FC			ΔEnhanced Scan FC	Enhanced Scan Undet
	2 scan_en	4 scan_en	8 scan_en		
b14	1.41%	2.02%	2.22%	2.72%	8.07%
b15	3.10%	3.78%	4.11%	7.05%	6.26%
b20	1.73%	1.78%	1.89%	1.89%	4.54%
b21	2.07%	2.27%	2.27%	2.57%	6.10%
b22	1.74%	1.92%	1.99%	2.28%	5.19%
b17	3.76%	4.27%	4.75%	9.50%	8.85%
b18	3.67%	4.17%	4.51%	5.65%	4.26%

Table 5. CWO coverage improvement with multiple scan enable tests. Enhanced scan shows the best coverage achievable.

making the corresponding cross wire open fault functionally redundant. Note that the V1 vector can always independently set the faulty gate output to the desired pre-charged logic value (unless the gate output only takes on a single value, in which case the logic itself is redundant) such that an incorrect output value is presents if the output floats for V2. Thus the undetectability of any cross wire open fault in the Enhanced Scan mode is entirely due to the fault being functionally redundant and not because of a failure to initialize the fault.

CUT	# LOC TDF tests	# LOC Open tests	# LOS Open tests	# 2 Scan enable tests	# 4 Scan enable tests	# 8 Scan enable tests
b14	866	269	80	16	17	14
		116	38	14	16	12
b15	953	332	122	113	131	134
		211	37	48	57	65
b20	1292	447	105	37	35	36
		245	59	26	29	34
b21	1621	469	127	28	33	36
		264	69	27	32	28
b22	1307	565	131	41	51	59
		267	71	25	37	39
b17	1441	464	320	235	294	370
		258	72	85	103	137
b18	1435	586	424	258	347	423
		320	211	110	142	178

Table 6. Number of test patterns applied in each test mode

Table 6 shows the number of test patterns for each test type. Column 1 shows the number of traditional LOC TDF tests for each CUT. For Column 2 to Column 6, the upper and lower numbers are the number of tests for all possible cross wire open faults and the number of tests for the cross wire open faults within one configuration respectively.

VIII. CONCLUSION

Traditional transistor stuck open fault (TSOF) tests are not sufficient to screen out all possible CMOS open defects because another type of open - the cross wire open (CWO) fault is commonly found in complex gates. The detection of such faults has not been addressed so far. We show that commonly employed scan based LOC TDF timing tests fail to detect many, between 30% and 70%, of such open faults. We therefore present a methodology to explicitly target CWO faults using commercial ATPG tools. Our approach uses a circuit transformation that converts the detection of CWO faults into detection of a corresponding transition delay faults, which can then be conventionally targeted by the ATPG. Furthermore, since such open defects can be expected to exhibit long delays, first LOC and then also LOS mode tests

are generated to maximize CWO fault coverage, even if the LOS tests may need to be run at relatively slow speeds due to timing limitations on the scan enable. However, some CWOs remain undetectable even by such targeted test generation. These remaining open defects are not all benign since many can be activated by commonly occurring circuit hazards and can generate erroneous circuit outputs. We have investigated the detection of these CWOs, undetectable by traditional timing unaware scan tests, by a DFT approach that uses multiple independent scan enable signals to allow mixed LOC and LOS tests. This approach can detect nearly all the remaining non redundant CWO faults, albeit with some hardware overhead. Future work continues to focus on the efficient detection on open faults in CMOS with high coverage, a subject that has perhaps not received the attention it deserves because of the common misconception that most opens are detected by TDF timing tests.

REFERENCES

[1] F. Hapke, M. Reese, J. Rivers, A. Over, V. Ravikumar, W. Redemund, A. Glowatz, J. Schloeffel and J. Rajski, "Cell-aware Production test results from a 32-nm notebook processor", 2012 IEEE International Test Conference (ITC), pp. 1-9.

[2] R. Madge, B.R. Benware and W.R. Daasch, "Obtaining high defect coverage for frequency-dependent defects in complex ASICs", 2003 IEEE Design & Test of Computers, pp. 46-53.

[3] N. Devtaprasanna, A. Gunda, P. Krishnamurthy, S. M. Reddy and I. Pomeranz, "A Unified Method to Detect Transistor Stuck-Open Faults and Transition Delay Faults", Proc. ETS, 2006, pp. 185-192.

[4] N. Devtaprasanna, A. Gunda, P. Krishnamurthy, S. M. Reddy and I. Pomeranz, "Test Generation for Open Defects in CMOS Circuits", International Symposium on Defect and Fault Tolerance in VLSI systems, 2006, pp. 41-49.

[5] C. Han and A. D. Singh, "Improving CMOS Open Defect Coverage Using Hazard Activated Tests", Proc. VLSI Test Symposium, pp 2014, pp. 1-6.

[6] C. Han and A. D. Singh, "Hazard Initialized LOC Tests for TDF Undetectable CMOS Open Defects", Proc. ATS 2013, pp. 189-194.

[7] J. Saxena, K. M. Butler, J. Gatt, R. Raghuraman, S. P. Kumar, S. Basu, D. J. Campbell and J. Berech, "Scan-based transition fault testing - implementation and low cost test challenges", in Proc. International Test Conference, 2002, pp. 1120-1129.

[8] Brand, D. and Iyengar, V.S., "Identification of Redundant Delay Faults", IEEE Transactions on Computer-Aided Design of Integrated Circuits and Systems, Vol. 13, No. 5, May 1994, pp. 553-565.

[9] I. Pomeranz and S.M. Reddy, "Hazard-Based Detection Conditions for Improved Transition Fault Coverage of Scan-Based Tests", IEEE Trans. On VLSI Systems. Vol. 18, Issue: 2, pp. 333-337.

[10] S. M. Reddy and M. K. Reddy. "Testable realizations for FET stuck-open faults in CMOS combinational logic circuits", IEEE Transactions on Computers. Vol. C-35, Issue: 8, pp. 742-754.

[11] N. Ahmed and M. Tehranipoor, "Improving Transition Delay Fault Coverage Using Hybrid Scan-Based Technique", International Symposium on Defect and Fault Tolerance in VLSI systems, 2006, PP. 187-195.

[12] N. Devtaprasanna, A. Gunda, P. Krishnamurthy, S.M. Reddy and I. Pomeranz, "A Novel Method of Improving Transition Delay Fault Coverage Using Multiple Scan Enable Signals", Proc. International Conference on Computer Design (ICCD), 2005, pp. 471-474.

[13] F. J. Ferguson and J. P. Shen, "A CMOS Fault Extractor for Inductive Fault Analysis", IEEE Transactions. on CAD, Vol. 7, No. 11, pp. 1181-1194.

[14] C. Han and A. D. Singh, "On the Testing of Hazard Activated Open Defects", Proc. International Test Conference 2014.

A Definition of the Number of Detections for Faults with Single Tests in a Compact Scan-Based Test Set

Irith Pomeranz

School of Electrical and Computer Engineering
Purdue University
West Lafayette, IN 47907, U.S.A.
E-mail: pomeranz@ecn.purdue.edu

Abstract—Test quality metrics that use the numbers of detections of target faults are based on the premise that increasing the number of tests for a fault increases the likelihood of detecting defects around the site of the fault. This paper describes a new definition of the number of detections for faults that have only one test in a given test set. Such faults are prevalent in compact test sets. For a fault with a single test, metrics based on the number of detections yield the same value, one, for any test. The new definition associates different numbers of detections with different tests for the fault by considering the number of distinct test cubes that a test contains. It thus provides a target for the generation of a single test with a higher quality for the fault. The effectiveness of the definition is demonstrated by modifying a compact test set to increase the numbers of detections of single stuck-at faults with single tests, and comparing a bridging fault coverage of the test set before and after the modification.

Index Terms—Bridging faults, number of detections, single stuck-at faults, test generation, test quality.

I. INTRODUCTION

The quality of a test set T for a set of target faults F increases with the number of distinct tests in T that detect each fault in F. For a fault $f \in F$ this number is denoted by $n(f)$. A higher value of $n(f)$ makes it more likely that T will detect defects around the site of f. This is the basis for n-detection test generation procedures [1]-[12].

In general, increasing the numbers of detections for target faults goes together with an increase in the number of tests. For cases where an increase in the number of tests is not acceptable, the procedure described in [12] increases the values of $n(f)$ for the faults in F without increasing the number of tests in T relative to the number of tests in a conventional compact one-detection test set. However, it is not possible to increase the value of $n(f)$ for *every* $f \in F$ without increasing the number of tests. This can be seen as follows.

Let T be a conventional compact one-detection test set. It is expected that such a test set will not contain any redundant tests. A test $t \in T$ is redundant if every fault it detects is also detected by another test in T. In this case, t can be removed from T without reducing the fault coverage. In an irredundant test set T, every test $t \in T$ has at least one fault $f \in F$ such that t is the only test in T that detects f. Therefore, t cannot be removed from T without losing fault coverage. The procedure from [13] is a reverse order fault simulation procedure, with

the computational complexity of fault simulation with fault dropping, that yields an irredundant test set.

Considering an irredundant test set T, every test has at least one fault f such that $n(f) = 1$. With $|T|$ tests in T, there are at least $|T|$ such faults. Even if a compact test set has redundant tests, their number is expected to be small. It is thus expected to have a large number of faults with single tests. As a result, there is a large number of faults for which it is not possible to increase the number of tests beyond $n(f) = 1$ without increasing the number of tests.

The goal of this paper is to derive a definition of the number of detections that is applicable to the faults with $n(f) = 1$, and can be used for increasing the quality of a compact test set as related to these faults. This issue is not addressed by n-detection test generation procedures or by metrics that are based on the numbers of detections [1]-[12]. It is important because of the prevalence of faults with single tests in compact test sets, and because defects around the sites of faults with single tests are the least likely to be detected by T. Increasing the likelihood that they will be detected is more important than addressing defects around the sites of faults with $n(f) > 1$.

The definition described in this paper is based on the following considerations. Let f be a fault such that $n(f) = 1$. Let t be the single test in T that detects f. By unspecifying input values of t, it is possible to find a test cube c, with a minimal set of specified input values, that is sufficient for detecting f. In some cases the test cube c may be unique. In general, by unspecifying different input values of t it is possible to find different test cubes. Suppose that it is possible to extract from t a set of $m(f)$ distinct test cubes, $C(f) = \{c_0, c_1, ..., c_{m(f)-1}\}$, such that each test cube from $C(f)$ detects f, and uses for this purpose a minimal set of specified input values from t. A higher value of $m(f)$ implies that t allows f to be detected in more distinct ways. This can contribute to the detection of defects around the site of f similar to the way $m(f)$ different tests contribute to defect detection.

The test cubes based on t do not conflict in any input values. However, it is possible even with different tests that the test cubes they use for detecting a fault would be compatible. Moreover, without increasing the number of tests, it is important to allow compatible but distinct test cubes to be considered as different detections of a fault. The usefulness of

the definition will be verified by using the following procedure.

Given a conventional compact one-detection test set T_1 for single stuck-at faults, the procedure attempts to increase the numbers of detections $m(f)$ for single stuck-at faults that are detected by single tests in T_1 (faults with $n(f) = 1$). This is achieved by modifying the tests in T_1 so as to increase the numbers of test cubes that are embedded in the single tests that detect the faults with $n(f) = 1$. The target number of detections for faults with single tests is denoted by m.

The procedure also considers the following issue. For a single stuck-at fault f with $n(f) > 1$, the modification of T may cause $n(f)$ to decrease accidentally. When $n(f)$ is low, this may reduce the quality of T. For a constant n, the procedure ensures that, if a fault f is detected by $n(f)$ tests in T, and $1 < n(f) \leq n$, then none of the tests for f will be lost, and $n(f)$ will not decrease.

With a limit of n for single stuck-at faults where $n(f) > 1$, and a target of m for the number of detections considering single stuck-at faults where $n(f) = 1$, the resulting test set is denoted by $T_{n,m}$.

The quality of T_1 and $T_{n,m}$ is compared based on their bridging fault coverage. Since bridging faults are not considered during the generation of T_1 or $T_{n,m}$, the bridging fault coverage provides an independent metric for the quality of T_1 and $T_{n,m}$, and their ability to detect defects that are different from single stuck-at faults [1].

When test data compression is used, the definition of the number of detections for faults with $n(f) = 1$ can be applied to the test cubes that the test data compression method uses for compression, or to the fully-specified test set that it applies to the circuit. The modification procedure can be applied to the set of test cubes. The modification can be performed such that the constraints of the test data compression method will continue to be satisfied.

The paper is organized as follows. Section II defines the number of detections for faults with single tests in a given test set, and describes its computation. Section III describes the modification of a test set T_1 into a test set $T_{n,m}$, for $n \geq 1$ and $m \geq 2$. Section IV presents experimental results.

II. NUMBER OF DETECTIONS

This section describes the definition of the number of detections for target faults with single tests in a test set T. The set of target faults (single stuck-at faults) that are detected by T is denoted by F.

A. Faults with Single Tests

The faults in F with single tests in T are identified by performing two-detection fault simulation of F under T. During this process, a fault is dropped after it is detected by two different tests. The indices of the tests are stored in variables denoted by $d_0(f)$ and $d_1(f)$. If the second test does not exist, $d_1(f) = -1$

A fault $f \in F$ is detected by a single test $t_i \in T$ if $d_0(f) = i$ and $d_1(f) = -1$. The set of faults that are detected only by t_i is denoted by U_i. Thus, $U_i = \{f \in F : d_0(f) = i, d_1(f) = $

$-1\}$. The set of faults that are detected by single tests in T is $U = \cup \{U_i : t_i \in T\}$.

B. Sets of Test Cubes

The procedure for computing the numbers of detections for faults with single tests considers every fault $f \in U$. Let $f \in U_i$ be detected by t_i. The procedure extracts from t_i a set of test cubes that is denoted by $C(f)$. The number of test cubes in $C(f)$ is denoted by $m(f)$. This is the number of detections associated with f.

For given parameters m and M such that $m \leq M$, the procedure makes up to M attempts to extract from t_i a test cube for f. It stops if it obtains m distinct test cubes for f. Thus, $m(f) \leq m$ even if it is possible to extract from t_i additional test cubes for f. This is analogous to n-detection fault simulation, where a fault is dropped after it is detected by n different tests, even if there are additional tests that detect it. The procedure allows M to be larger than m since a particular attempt to extract a new test cube may result in a test cube that already exists in $C(f)$.

Procedures for extracting a test cube c for a fault f from a test t_i were described in [14]-[15]. Any one of these procedures can be used here after modifying it to produce different test cubes when it is called multiple times with the same fault f and test t_i. The procedure used in this paper proceeds as follows.

For a circuit with N inputs, let $t_i = t_i(0)t_i(1)...t_i(N-1)$. The procedure first finds the input cone of f. Tracing the circuit from the site of f to the outputs, the procedure finds all the outputs on which f can be detected. Tracing the circuit from these outputs to the inputs, the procedure finds the input cone that can affect the detection of f.

For $0 \leq j < N$, if input j is in the input cone of f, the procedure assigns $c(j) = t_i(j)$. Otherwise, the procedure assigns $c(j) = x$, where x denotes an unspecified value.

Next, the procedure considers the specified values of c one at a time in a random order. The random order is important for producing different test cubes in different attempts. When an input j is considered, the procedure assigns $c(j) = x$ and simulates f under $c(j)$. If f is not detected, the procedure assigns $c(j) = t_i(j)$ in order to restore the detection of f. Otherwise, the unspecified value of $c(j)$ is accepted.

The worst-case computational complexity of the procedure is determined as follows. The procedure considers at most $|F|$ faults. For every fault it computes at most M test cubes, where M is a constant. To compute every test cube, the procedure simulates a fault under at most N test cubes that are obtained by unspecifying input values one at a time. Overall, the worst-case computational complexity is that of logic simulation of the circuit under $O(N \cdot |F|)$ test cubes. In effect, the procedure considers fewer than N input values for every fault, and fewer than $|F|$ faults.

C. Test Set Quality

It is possible to compute a single metric value for the quality of T based on the numbers of detections for all the faults in F.

Specifically, $n(f)$ can be used for faults with more than one test in T, and $m(f)$ can be used for faults with single tests. For example, suppose that m is used for computing the number of detections for faults with single tests. In this case, $1 \leq m(f) \leq m$ for every $f \in F$ that has a single test in T. A fault f with more than one test can contribute $n(f)$ to the metric. A fault with a single test can contribute $1 + [m(f) - 1]/m$ to the metric. In this way, a higher value of $m(f)$ will result in a higher contribution, but the contribution is lower than the contribution of a fault with two tests.

This paper focuses on the faults with single tests. For $1 \leq m_0 \leq m$, let u_{m_0} be the number of faults with a single test and m_0 test cubes. The quality of the test set is represented by m values, $u_1, u_2, ..., u_m$.

III. TEST SET MODIFICATION

This section describes a procedure for modifying a test set T_1 that detects a set of target faults F. The goal of the modification is to increase the quality of the test set based on the faults that are detected by single tests, and using the number of detections $m(f)$ introduced in the previous section.

A. Overview

While modifying the test set, the procedure may accidentally reduce the number of tests for faults with more than one test in T. This may reduce the quality of the test set. To prevent this from occurring, the procedure uses a parameter denoted by n. If a fault f has $n(f)$ tests in T, and $1 < n(f) \leq n$, the procedure ensures that the number of tests for f will not decrease as T is modified. For faults that are detected by $n+1$ tests or more, the procedure allows the number of tests to be reduced. These faults are expected to have a lower impact on the quality of the test set since they have more tests.

For simplicity, n is assumed to be a constant. For any value of n, T_1 corresponds to $m = 1$. The procedure considers $m = 2, 3, ...$ For every value of m the procedure attempts to increase to m the numbers of detections for faults with single tests in T and $m(f) < m$. The test set obtained for given values of n and m is denoted by $T_{n,m}$.

Let $T_{n,1} = T_1$. For $m \geq 2$, the procedure starts the computation of $T_{n,m}$ by assigning $T_{n,m} = T_{n,m-1}$. The procedure considers the tests from $T_{n,m}$ one at a time. When a test $t_i \in T_{n,m}$ is considered, the procedure applies the steps described next.

B. Target Faults

The procedure performs $(n + 1)$-detection fault simulation of F under T. During this process, a fault is dropped after it is detected by $n+1$ different tests. The indices of the tests that detect a fault $f \in F$ are stored in variables denoted by $d_0(f)$, $d_1(f), ..., d_n(f)$. If f has $n(f)$ tests in T, then $d_j(f) \geq 0$ for $j < n(f)$, and $d_j(f) = -1$ for $n(f) \leq j \leq n$.

The procedure finds the set of faults U_i that are detected only by t_i. This is the set $U_i = \{f \in F : d_0(f) = i, d_j(f) = -1 \; for \; 1 \leq j \leq n\}$. In addition, the procedure finds the set of faults D_i that are detected by t_i and have $1 < n(f) \leq n$ tests in T. This is the set $D_i = \{f \in F : d_j(f) = i \; for \; some \; 0 \leq j < n, d_0(f) \geq 0, d_1(f) \geq 0, d_n(f) = -1\}$.

The numbers of tests in T for the faults in D_i should not be reduced. This is achieved by ensuring that t_i will continue to detect all the faults in D_i while it is being modified.

It is possible to obtain $U_i = \emptyset$ even if T was initially irredundant. This can occur as follows. The modification of other tests in $T_{n,m}$ can cause them to detect the faults that were initially detected only by t_i. If this happens for all such faults, $U_i = \emptyset$ will be obtained. In this case, t_i is not modified.

C. Test Modification

If $U_i \neq \emptyset$, the procedure computes the number of test cubes $m(f)$ based on t_i for every fault $f \in U_i$. It also assigns $m_{curr}(f) = m(f)$. During the modification of t_i, $m_{curr}(f)$ is the number of test cubes for f before a modification of t_i. It is used for deciding whether or not a modification of t_i is acceptable.

A fault $f \in U_i$ for which $m_{curr}(f) = m$ does not require t_i to be modified. The procedure considers t_i for modification only if there is at least one fault $f \in U_i$ for which $m_{curr}(f) < m$.

The modification of t_i is carried out by complementing its input values one at a time in a random order. The inputs under consideration are such that they are in the input cone of at least one fault from U_i. The input cones are defined as described earlier. For every input j that the procedure considers, it assigns $t_i(j) = \overline{t_i(j)}$. It then checks whether the complementation of input j is acceptable, as follows.

The procedure checks whether all the faults in D_i continue to be detected by t_i. If any one of the faults is not detected, the complementation of $t_i(j)$ is not accepted. In this case, the procedure complements $t_i(j)$ again in order to restore its previous value.

If all the faults in D_i are detected, the procedure computes the number of test cubes $m(f)$ based on the modified test t_i for every fault $f \in U_i$. To accept the complementation of $t_i(j)$, the procedure requires that $m(f) \geq m_{curr}(f)$ for every $f \in U_i$. If this condition is not satisfied for a fault $f \in U_i$, the procedure complements $t_i(j)$ again in order to restore its previous value. Otherwise, the complemented value is accepted, and the procedure assigns $m(f) = m_{curr}(f)$ for every $f \in U_i$.

After accepting a complemented value the procedure again checks whether there is a fault $f \in U_i$ with $m_{curr}(f) < m$. If no such fault exists, no additional input values are considered.

The worst-case computational complexity of the modification procedure is determined as follows. The procedure considers $|T|$ tests. For each test it attempts to complement at most N input values. To check whether a value can be complemented it computes test cubes for at most $|F|$ faults. The computation of test cubes has the worst-case computational complexity of logic simulation of $O(N \cdot |F|)$ test cubes. Overall, the worst-case computational complexity of the modification procedure is that of logic simulation of $O(|T| \cdot N^2 \cdot |F|)$ test cubes.

IV. EXPERIMENTAL RESULTS

This section describes the results of an experiment that studies the effectiveness of the definition and modification procedure described in the previous sections.

A. Parameter Values

The test set T_1 is a compact one-detection test set for single stuck-at faults. To ensure that the test set has as few faults with single tests as possible, it is obtained from a compact two-detection test set by removing tests one at a time. The test that results in the smallest number of faults with single tests and the same fault coverage is selected for removal at every step. This continues until it is not possible to remove any test without reducing the fault coverage.

The modification procedure was applied with various values of n in order to select an appropriate value. For a given value of n, the modification procedure was applied with $m = 2, 3, ..., m_{max}$ for a constant m_{max}. In all the cases, $M = 2m$. In general, the following considerations affect the selection of n and m_{max}.

A higher value of m is expected to yield a higher quality test set. However, beyond a certain value, it is not possible to increase the numbers of detections for a sufficient fraction of the faults with single tests in order to obtain a meaningful increase in quality.

Considering n, the modification procedure can cause the numbers of detections to be reduced for faults with more than n tests ($n(f) > n$). This can have a significant effect on the quality of the test set if n is too low.

If n is too high, the modification procedure needs to maintain the number of detections for a large number of faults with $1 < n(f) \le n$. This prevents it from increasing the numbers of detections for faults with single tests. Thus, it prevents it from increasing the quality of the test set.

To address these issues, and based on the results of experiments with different values of n and m_{max}, the procedure was run as follows. Beyond $m_{max} = 5$ the quality of the test set typically does not increase any further. The procedure was, therefore, run with $m_{max} = 5$.

With $n < 3$ there is a loss of test quality when faults with $n(f) = 3$ lose detections. Therefore, $n = 3$ was used for $m = 2$ and 3. It was then increased, and $n = 4$ was used for $m = 4$ and 5.

The test set $T_{3,2}$ was computed starting from T_1, $T_{3,3}$ was computed starting from $T_{3,2}$, $T_{4,4}$ was computed starting from $T_{3,3}$, and $T_{4,5}$ was computed starting from $T_{4,4}$. T_1 is also referred to as $T_{1,1}$.

For every test set, the numbers of detections are computed for faults with single tests using $m = 5$ and $M = 10$.

B. Assessing Test Set Quality

To assess the quality of a test set using an independent metric, bridging fault simulation was carried out. The bridging fault coverage measures the quality of a test set through its ability to detect faults that were not targeted during its generation. Non-feedback four-way bridging faults were selected randomly as follows.

A non-feedback bridging fault $g/a/h$ is associated with lines g and h and a value a. In the presence of the fault, the value a on line h dominates the value of line g. Thus, $h = a$ results in $g = a$. The fault is detected by a test that assigns $h = a$ and detects the single stuck-at fault g/a.

A bridging fault $g/a/h$ is less likely to be detected by T_1 if the single stuck-at fault g/a has a lower number of tests in T_1. For a constant n_{max}, let $F_{n_{max}}$ be the subset of single stuck-at faults with $n(f) \le n_{max}$ tests in T_1. For every single stuck-at fault $g/a \in F_{n_{max}}$, the set of bridging faults $BR_{n_{max}}$ includes 10 randomly selected faults of the form $g/a/h$.

The value of n_{max} was set to five. Beyond $n_{max} = 5$, the detectable bridging faults that are associated with a single stuck-at fault are likely to be detected.

When considering the bridging fault coverage it should be noted that it does not track the quality of the test set perfectly, especially since sets of randomly selected bridging faults are simulated. In addition, without increasing the number of tests, the possibility of increasing the bridging fault coverage is limited. As a result, only small increases in the bridging fault coverage can be expected.

C. Results

The results of the modification procedure are shown in Table I as follows. There is a row for every test set $T_{n,m}$ where the bridging fault coverage is increased. Every row contains the following information.

Column inp shows the number of inputs. Column n shows the value of n. Column m shows the value of m. Column $tests$ shows the number of tests in $T_{n,m}$. Column $s.a$ shows its single stuck-at fault coverage.

Column $single$ shows the following information. Subcolumn tot shows the number of faults with single tests in $T_{n,m}$ (the number of faults with $n(f) = 1$). Subcolumn $m(f) = m_0$, for $m_0 - 1, 2, ..., 5$, shows the number of faults with $n(f) = 1$ and $m(f) = m_0$. This number was denoted earlier by u_{m_0}.

Column $bridg$ shows the bridging fault coverage of $T_{n,m}$ with respect to BR_5. Column $ntime$ shows normalized run times. For normalization the run times are divided by the run time for fault simulation of T_1. For T_1, the run time is that required for the computation of the numbers of detections for faults with single tests using $m = 5$. For $T_{n,m}$, where $m \ge 2$, the run time is the cumulative run time required for the modification of the test sets $T_{3,2}, ..., T_{n,m}$.

D. Discussion

The following points can be seen from Table I. In general, the modification procedure is more effective for circuits with larger numbers of inputs. For such circuits there is more flexibility for embedding in a single test several distinct test cubes for a fault. More specifically, the ability to generate several distinct test cubes depends on the structure of the circuit and the number of inputs in the input cones of the faults with single tests.

Considering only faults with single tests, when the test modification procedure is applied with a certain value of m, it reduces the number of faults with $m(f) = m - 1$. It typically increases the number of faults with $m(f) = m$. More generally, it reduces the number of faults with $m(f) < m$, and increases the number of faults with $m(f) \geq m$. Alternatively, it may reduce the number of faults with single tests. This occurs accidentally, and typically only for a small number of faults.

For example, considering faults with single tests in the case of $s5378$, when the test modification procedure is applied with $m = 2$, it decreases the number of faults with $m(f) = 1$ from 85 to 61, and increases the number of faults with $m(f) = 2$ from 67 to 77. It also increases the number of faults with $m(f) = 3$ from 36 to 46.

The bridging fault coverage increases with m in almost all the cases. Within the constraints discussed earlier on the bridging fault coverage, the increase is consistent and significant. It indicates that the quality of the test set increases with m, and points to the effectiveness of the new definition.

The normalized run time does not increase with the size of the circuit or its number of inputs as fast as expected from the worst-case analysis. This is because the numbers of inputs in the input cones, and not the total number of inputs, determine the run time, and the numbers of inputs in the input cones may be significantly smaller than the total number of inputs.

V. CONCLUDING REMARKS

This paper described a new definition for the numbers of detections of faults that have only one test in a given test set. Compact test sets have large numbers of such faults. For these faults, existing metrics based on the number of detections do not distinguish between different tests. Thus, they cannot guide the generation of tests that increase the likelihood of detecting defects around the sites of these faults. The new definition is based on the number of distinct test cubes that can be extracted for a fault from the single test that detects it. A higher number of test cubes implies that the test allows the fault to be detected in more distinct ways. This can contribute to the detection of defects around the site of the fault. The effectiveness of the definition was demonstrated by modifying a compact test set to increase the numbers of detections for single stuck-at faults with single tests, and comparing a bridging fault coverage of the test set before and after the modification.

REFERENCES

[1] K. M. Butler and M. R. Mercer, "Quantifying Non-Target Defect Detection by Target Fault Test Sets", in Proc. Europ. Test Conf., 1991, pp. 91-100.

[2] S. C. Ma, P. Franco and E. J. McCluskey, "An Experimental Chip to Evaluate Test Techniques Experiment Results", in Proc. Intl. Test Conf., 1995, pp. 663-672.

[3] S. M. Reddy, I. Pomeranz and S. Kajihara, "Compact Test Sets for High Defect Coverage", IEEE Trans. on Computer-Aided Design, Aug. 1997, pp. 923-930.

[4] M. R. Grimaila, S. Lee, J. Dworak, K. M. Butler, B. Stewart, H. Balachandran, B. Houchins, V. Mathur, J. Park, L.-C. Wang and M. R. Mercer, "REDO - Random Excitation and Deterministic Observation - First Commercial Experiment", in Proc. VLSI Test Symp., 1999, pp. 268-274.

[5] I. Pomeranz and S. M. Reddy, "On n-Detection Test Sets and Variable n-Detection Test Sets for Transition Faults", IEEE Trans. on Computer-Aided Design, March 2000, pp. 372-383.

[6] I. Pomeranz and S. M. Reddy, "Definitions of the Numbers of Detections of Target Faults and their Effectiveness in Guiding Test Generation for High Defect Coverage", in Proc. Design Autom. and Test in Europe Conf., 2001, pp. 504-508.

[7] C.-W. Tseng and E. J. McCluskey, "Multiple-Output Propagation Transition Fault Test", in Proc. Intl. Test Conf., 2001, pp. 358-366.

[8] B. Benware, C. Schuermyer, N. Tamarapalli, K.-H. Tsai, S. Ranganathan, R. Madge, J. Rajski and P. Krishnamurthy, "Impact of multiple-detect test patterns on product quality", in Proc. Intl. Test Conf., 2003, pp. 1031-1040.

[9] S. Venkataraman, S. Sivaraj, E. Amyeen, S. Lee, A. Ojha and R. Guo, "An Experimental Study of n-Detect Scan ATPG Patterns on a Processor", in Proc. VLSI Test Symp., 2004, pp. 23-28.

[10] K. R. Kantipudi and V. D. Agrawal, "On the Size and Generation of Minimal N-Detection Tests", in Proc. VLSI Design Conf., 2006.

[11] J. E. Nelson, J. G. Brown, R. Desineni and R. D. Blanton, "Multiple-Detect ATPG Based on Physical Neighborhoods", in Proc. Design Autom. Conf., 2006, pp. 1099-1102.

[12] J. Geuzebroek, E. J. Marinissen, A. Majhi, A. Glowatz and F. Hapke, "Embedded Multi-Detect ATPG and Its Effect on the Detection of Unmodeled Defects", in Proc. Intl. Test Conf., 2007, pp. 1-10.

[13] I. Pomeranz and S. M. Reddy, "Forward-Looking Fault Simulation for Improved Static Compaction", IEEE Trans. on Computer-Aided Design, Oct. 2001, pp. 1262-1265.

[14] S. Kajihara and K. Miyase, "On Identifying Don't Care Inputs of Test Patterns for Combinational Circuits", in Proc. Intl. Conf. on Computer-Aided Design, Nov. 2001, pp. 364-369.

[15] A. El-Maleh and A. Al-Suwaiyan, "An Efficient Test Relaxation Technique for Combinational & Full-Scan Sequential Circuits", in Proc. VLSI Test Symp., 2002, pp. 53-59.

[16] S. Kajihara, I. Pomeranz, K. Kinoshita and S. M. Reddy, "Cost-Effective Generation of Minimal Test Sets for Stuck-at Faults in Combinational Logic Circuits", IEEE Trans. on Computer-Aided Design, Dec. 1995, pp. 1496-1504.

TABLE I
EXPERIMENTAL RESULTS

circuit	inp	n	m	tests	s.a	single tot	m(f)=1	m(f)=2	m(f)=3	m(f)=4	m(f)=5	bridg	ntime
s1423	91	1	1	33	99.08	122	54	31	12	6	19	73.30	10.66
s1423	91	3	2	33	99.08	105	42	29	15	5	14	74.07	119.93
s1423	91	3	3	33	99.08	101	38	23	20	5	15	75.25	291.69
s1423	91	4	5	33	99.08	101	38	23	16	2	22	75.42	763.79
s5378	214	1	1	111	99.13	286	85	67	36	37	61	72.78	3.79
s5378	214	3	2	111	99.13	273	61	77	46	35	54	73.52	90.93
s5378	214	3	3	111	99.13	272	60	29	64	53	66	73.70	218.95
s5378	214	4	4	111	99.13	271	57	27	35	67	85	73.88	381.23
s5378	214	4	5	111	99.13	271	57	22	34	51	107	74.03	584.56
s9234	247	1	1	135	93.47	646	236	207	45	51	107	68.24	3.39
s9234	247	3	2	135	93.47	602	175	232	49	49	97	68.91	109.85
s9234	247	3	3	135	93.47	580	166	177	56	83	98	69.03	260.37
s9234	247	4	4	135	93.47	577	161	166	36	79	135	69.70	436.04
s13207	700	1	1	240	98.46	973	810	17	40	40	66	62.89	24.30
s13207	700	4	4	240	98.46	966	790	4	16	85	71	62.92	7265.20
s13207	700	4	5	240	98.46	966	790	3	12	19	142	63.13	11305.16
s15850	611	1	1	108	96.68	577	153	265	87	25	47	70.23	9.66
s15850	611	3	2	108	96.68	547	138	240	96	39	34	71.19	932.89
s15850	611	4	4	108	96.68	541	112	181	118	82	48	71.26	4636.02
s35932	1763	1	1	18	89.81	2776	2776	0	0	0	0	77.54	2.50
s35932	1763	3	2	18	89.81	2622	2622	0	0	0	0	77.79	286.29
s35932	1763	3	3	18	89.81	2614	2614	0	0	0	0	77.89	690.28
s35932	1763	4	4	18	89.81	2608	2608	0	0	0	0	77.96	1099.06
s35932	1763	4	5	18	89.81	2604	2604	0	0	0	0	78.00	1497.29
s38417	1664	1	1	102	99.47	1305	122	65	64	99	955	73.94	4.12
s38417	1664	3	2	102	99.47	1189	79	174	131	131	674	75.10	506.97
s38417	1664	3	3	102	99.47	1154	74	36	107	160	777	75.29	1355.14
s38417	1664	4	4	102	99.47	1127	70	31	41	192	793	75.89	2334.87
s38417	1664	4	5	102	99.47	1116	68	28	36	75	909	75.98	3883.38
s38584	1464	1	1	147	95.85	1038	423	149	111	137	218	70.64	2.41
s38584	1464	3	2	147	95.85	975	383	264	102	112	114	71.16	228.37
s38584	1464	4	4	147	95.85	961	365	117	94	195	190	71.38	825.12
b14	280	1	1	321	94.87	768	235	161	59	54	259	70.06	3.42
b14	280	3	2	321	94.87	702	116	207	68	63	248	70.77	210.18
b14	280	4	4	321	94.87	684	103	162	27	76	316	71.09	884.40
b15	483	1	1	379	98.51	1532	676	174	133	120	429	61.29	7.17
b15	483	3	2	379	98.51	1390	476	193	124	141	456	62.55	579.36
b15	483	4	4	379	98.51	1337	446	92	99	157	543	62.85	2338.22
b20	527	1	1	303	94.08	931	133	90	85	117	506	74.53	2.63
b20	527	3	2	303	94.09	820	57	91	82	111	479	75.24	199.05
b20	527	3	3	303	94.11	789	53	40	74	102	520	75.84	462.99
b20	527	4	4	303	94.12	777	53	36	38	94	556	75.99	764.55
b20	527	4	5	303	94.13	776	53	35	31	35	622	76.07	1095.33
aes_core	788	1	1	214	100.00	5266	3976	891	323	69	7	67.16	3.96
aes_core	788	3	3	214	100.00	5187	2983	1493	572	122	17	67.17	944.73
aes_core	788	4	5	214	100.00	5187	2983	1493	572	122	17	67.19	3191.28
des_area	367	1	1	119	100.00	691	230	252	159	44	6	73.72	4.98
des_area	367	3	2	119	100.00	512	162	191	120	33	6	75.14	373.99
des_area	367	3	3	119	100.00	456	144	152	117	36	7	75.52	938.25
des_area	367	4	5	119	100.00	451	140	149	116	38	8	75.57	2724.02
spi	274	1	1	385	99.98	748	259	195	125	103	66	63.37	11.66
spi	274	3	2	385	99.98	707	108	295	122	122	60	63.71	498.06
spi	274	3	3	385	99.98	690	81	179	200	134	96	64.32	1205.67
spi	274	4	4	385	99.98	683	65	182	63	244	129	64.73	2214.31
spi	274	4	5	385	99.98	678	62	182	34	190	210	64.81	3575.15
systemcaes	928	1	1	142	99.82	467	150	63	58	65	131	78.79	6.27
systemcaes	928	3	2	142	99.82	404	90	64	54	64	132	80.57	502.35
systemcaes	928	3	3	142	99.82	397	77	58	52	77	133	80.84	1470.27
systemcaes	928	4	4	142	99.83	397	75	58	45	71	148	81.39	2855.72
systemcaes	928	4	5	142	99.83	387	74	56	39	59	159	81.54	4894.87
systemcdes	320	1	1	78	100.00	344	166	98	45	30	5	74.29	3.54
systemcdes	320	3	2	78	100.00	323	156	85	55	23	4	74.90	117.63
systemcdes	320	3	3	78	100.00	315	153	70	52	37	3	74.95	316.81
systemcdes	320	4	4	78	100.00	313	151	69	39	51	3	75.15	586.48
tv80	372	1	1	431	99.35	1244	494	163	181	127	279	67.46	2.34
tv80	372	3	2	431	99.35	1154	427	195	179	114	239	68.33	332.80
tv80	372	3	3	431	99.37	1110	419	127	197	141	226	68.49	749.38
tv80	372	4	4	431	99.38	1102	417	121	112	190	262	68.78	1273.13
tv80	372	4	5	431	99.38	1100	414	117	100	143	326	68.92	1847.11
wb_dma	738	1	1	64	100.00	283	130	48	33	21	51	75.97	8.73
wb_dma	738	3	2	64	100.00	257	109	69	23	9	47	76.41	310.20
wb_dma	738	3	3	64	100.00	257	106	40	35	28	48	76.69	774.49
wb_dma	738	4	4	64	100.00	252	106	34	31	36	45	76.76	1359.00
wb_dma	738	4	5	64	100.00	247	106	28	33	30	50	76.95	2061.16

Efficient Built-in Self Test of Regular Logic Characterization Vehicles

Ben Niewenhuis and R. D. (Shawn) Blanton
Department of Electrical and Computer Engineering
Carnegie Mellon University, Pittsburgh, PA 15213
http://www.ece.cmu.edu/~actl/

Abstract— Fast and efficient analysis of test chips is crucial for effective yield learning. Prior work proposed the Carnegie-Mellon logic characterization vehicle (CM-LCV) as an improved test chip for yield learning. The highly regular nature of the CM-LCV test chip is particularly appealing for BIST; the current work describes a BIST scheme that achieves 100% input-pattern fault coverage with an 86.9% reduction in test time for a reference design. Furthermore, all of these properties are achieved with a minimal hardware overhead.

I. INTRODUCTION

Test chips broadly encompass any design fabricated for purposes other than use or sale. Prior work has established a distinct category of product-like test chips, that is, integrated circuits that share features and some functionality with commercial products, but are fabricated to gather information about the design and manufacturing process. This current work builds on the Carnegie Mellon logic characterization vehicle (CM-LCV), which proposed removing the product functionality from product-like test chips while maintaining a similar physical structure [1]. This approach allows the CM-LCV to be designed with optimal test and diagnosis properties without compromising its ability to extract relevant information about the semiconductor design and manufacturing process.

In this paper the unique characteristics of the CM-LCV are used to create an efficient built-in self test (BIST) system based on the circular BIST concept [2]. Several proofs are supplied for existence of the crucial properties of the CM-LCV. Additionally, a reference design is constructed and simulated to support the claims made concerning fault coverage and test set size.

The remainder of the paper is organized as follows: Section II provides background on LCVs and BIST. Section III discusses the properties of the CM-LCV relevant to this paper and describes the implementation of the BIST system. Section IV describes in detail the experiments used to evaluate this new BIST architecture while Section V concludes the paper.

II. BACKGROUND

This section presents the background relevant to the current work. Specifically, Section II-A describes the current state-of-the-art for test chips. Section II-B then provides an overview of built-in self-test techniques, with particular emphasis on circular BIST.

A. Logic Characterization Vehicle

The concept of a test chip encompasses any design fabricated with the primary goal of gathering actionable feedback about the manufacturing process as opposed to commercial sale. The traditional test chip is a collection of specialized test structures that are optimized for gathering specific information with high precision. Examples include comb- and serpentine-shaped interconnect used for measuring defect size and density distributions [3][4]. However, the increasing complexity of the IC design and manufacturing process in the nanoscale regimes has resulted in the emergence of systematic defects. Systematic defects are failures caused by complex interactions between the design and the fabrication process which result in significant yield loss whenever specific features exist in a design. Traditional test chips are ill-suited for uncovering systematic defects because they do not include many of the design features of an actual product. This has led to the development of the product-like test chip, which is composed of portions of actual product designs, utilizes cells from standard-cell libraries, and is created using typical design flows (synthesis, place-and-route, etc.). While many of these product-like test chip designs are ad hoc in nature, Hess et. al. introduced a more systematic approach in their Logic Characterization Vehicle (LCV) [5]. Their methodology is primarily focused on a jig (essentially a ring of logic) that applies logical and parametric tests to any combinational circuit placed inside the jig. Their jig can be considered to be a form of DFT since it provides the tester different access modes to the circuit under test. The flexibility and extensibility of this jig enables easy creation of an LCV using any product sub-circuit.

The CM-LCV is centered on the concept of removing the product functionality from the test chip design; the key insight is that the manufacturing process is sensitive only to the physical features of a design, not its functionality. Although the functionality (logic) and physical features (layout) are related, they can be sufficiently separated to maximize testability and diagnosability through careful design of the logic function while at the same time ensuring that the layout mimics the targeted product or family of products. To achieve the goals of maximizing testability and diagnosability the following test-chip characteristics are adopted:

- Regularity - A set of functional unit blocks (FUBs) with regular connections can be designed to be C-testable

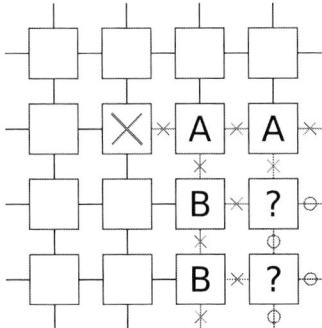

Fig. 1. VH-bijectivity demonstrated in the presence of a defective FUB in a two-dimensional array. Note that signal propagation moves left to right, top to bottom.

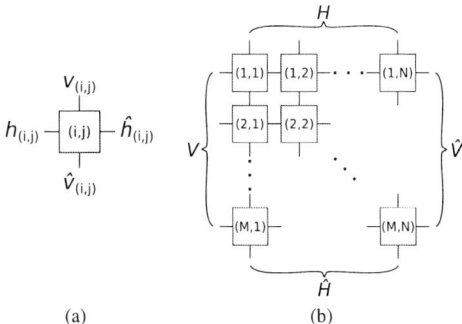

Fig. 2. Notation used for (a) the inputs and outputs of a single FUB and (b) the inputs and outputs of an entire array of FUBs.

[6][7][8], a property which provides strong guarantees about the test-set size and fault coverage over the entire array regardless of its size.

- Two-dimensional array - Arranging FUBs in a two-dimensional array with vertical and horizontal connections enables propagation of fault effects in two directions, allowing for better localization of defects within the array and within a defective FUB.
- Bijective FUBs - Constraining the FUB function to be bijective guarantees propagation of fault effects and simplifies test set construction.

In addition to these characteristics, the FUBs are designed to exhibit VH-bijectivity, a more constrained form of bijectivity that propagates an error residing exclusively on either the horizontal or vertical input of the FUB to both the vertical and horizontal FUB outputs. Figure 1 demonstrates how VH-bijectivity controls error propagation within a FUB array. Assume some FUB instance in the array of Figure 1 is defective (marked with the red "X"), and produces an erroneous value (represented by the red-crossed connection) for some test along its horizontal output. All of the FUBs in the same row as the defective module (marked with an "A") fall under the horizontal case of the VH-bijective property: an error only present at a horizontal input will propagate to both outputs. Furthermore, all FUBs in the first downstream column (marked with a "B") fall under the vertical case of the VH-bijective property: an error that appears only at a vertical input will propagate to both outputs. The remaining downstream FUBs may experience errors on both inputs; because they are bijective the error will propagate to at least one of the outputs, but this case is not covered by VH-bijectivity and thus the error propagation is unpredictable without precise knowledge of the defect behavior (represented by the blue-dotted connections). Regardless, the array location indicated by the first row/column errors observed must be either (a) the site of the defect, (b) horizontally adjacent to the defect, or (c) vertically adjacent to the defect, resulting in near-perfect diagnostic resolution for a single defective FUB.

B. Built-In Self Test

BIST is a circuit design methodology that seeks to reduce the cost of test by building some portion of the test flow into the circuit under test itself. BIST is thus a balancing act between the savings generated on the testing side with the incurred hardware and performance overhead. It is important to note that the output produced by most BIST schemes is not ideal for actual diagnosis; instead, BIST is typically used as a simple pass/fail indicator. Regular designs (e.g. memories [9], adders [10], multipliers [11], etc.) have always been appealing targets for BIST, since their regular test patterns simplify the hardware required to implement BIST.

The two-dimensional FUB array used in the CM-LCV is particularly appealing in this regard as it possesses an additional property: given careful test-vector selection, the output of the fault-free array is equivalent to another (related) test vector. This makes the CM-LCV a perfect candidate for a circular feedback approach for BIST wherein the circuit inputs are directly driven by the circuit outputs. Previous BIST architectures with circular feedback include the Circular Self-Test Path (CTSP)[2] and the Circular Celluar BIST (C^{2}BIST)[12]. However, both of these methods require additional circuitry along the feedback path, which is not necessary for the CM-LCV due to the aforementioned property.

III. BUILT-IN SELF TEST DESIGN

This section describes the BIST architecture. Specifically, Section III-A defines and formally establishes the relevant properties of the CM-LCV. Section III-B then details the changes to the circuit architecture required for circular BIST.

A. Theory

The core assumption of this work is that the output of the logic array used in the CM-LCV is equivalent to another test for some high-quality set of test vectors. This property is proven through three theorems. First, however, it is necessary to establish the notation used in these theorems.

Figure 2a summarizes the relevant notation for a single FUB. A FUB is a combinational circuit that implements function F with vertical and horizontal inputs, denoted as h and v, respectively, and horizontal and vertical outputs, denoted as $\hat{h}$ and $\hat{v}$, respectively. The input pattern of a FUB is denoted as $p = (h, v)$. Similarly, the output response of a FUB is represented as $r = (\hat{h}, \hat{v})$. Finally, a cycle C of function F is defined as a sequence of input patterns $\{p^{(0)}, p^{(1)}, \ldots, p^{(|C|-1)}\}$ such that $F(p^{(i)}) = r^{(i)} = p^{((i+1) mod\ |C|)}$ and $p^{(i)} \neq p^{(j)}\ \forall p^{(i)}, p^{(j)} \in C$.

Figure 2b summarizes the relevant notation for an array of FUBs. An array of FUBs is denoted as A and consists of M rows and N columns. The two tuple (i, j) is used to refer to the FUB location in the array, and is also used as a subscript to refer to the specific FUB inputs/outputs at that location, i.e. $h_{(i,j)}, v_{(i,j)}, \hat{h}_{(i,j)}, \hat{v}_{(i,j)}$. An array input pattern includes all of the logic values applied to the primary inputs of the array and is denoted as $P = (H, V)$, where $H = \{h_{(1,1)}, h_{(2,1)}, \ldots, h_{(M,1)}\}$, $V = \{v_{(1,1)}, v_{(1,2)}, \ldots, v_{(1,N)}\}$. Similarly an array output response is defined as $R = (\hat{H}, \hat{V})$ where $\hat{H} = \{\hat{h}_{(1,N)}, \hat{h}_{(2,N)}, \ldots, \hat{h}_{(M,N)}\}$, $\hat{V} = \{\hat{v}_{(M,1)}, \hat{v}_{(M,2)}, \ldots, \hat{v}_{(M,N)}\}$.

Given these definitions, Theorem 1 describes a method for constructing a test set for an array of FUBs from a cycle that exists within the FUB function F:

Theorem 1. *Given an $M \times N$ array of FUBs implementing bijective function F with cycle C of length k, there exists an array test set T of size k constructed from C such that all patterns in C are applied to all FUBs in the array when the tests of T are applied.*

Proof: Suppose a cycle C in F is of length k, that is, $C = \{p^{(1)}, p^{(2)}, \ldots, p^{(k)}\}$, where $p^{(i)} = (h^{(i)}, v^{(i)})$. Let T be composed of k array input patterns, that is, $T = \{P_1, P_2, \ldots, P_k\}$, and let each $P_s \in T$ be constructed according to the following:

$$H_s = \{h_{(1,1)}^{(s)}, h_{(2,1)}^{(s+1) mod\ k}, \ldots, h_{(M,1)}^{(s+M-1) mod\ k}\}$$
$$V_s = \{v_{(1,1)}^{(s)}, v_{(1,2)}^{(s+1) mod\ k}, \ldots, v_{(1,N)}^{(s+N-1) mod\ k}\}$$

Now consider all of the FUBs along the diagonal of the array, that is, all $F_{(i,j)}$ where $i + j = \lambda$. (see Figure 3 for an illustration.)

- For $\lambda = 2$ only $F_{(1,1)}$ meets the constraint. Observe that for test P_s, $h_{(1,1)} = h^{(s)}$ (determined by H_s) and $v_{(1,1)} = v^{(s)}$ (determined by V_s); thus $r_{(1,1)} = F(p^{(s)}) = p^{(s+1) mod\ k}$.

- For $\lambda = 3$ both $F_{(2,1)}$ and $F_{(1,2)}$ meet the constraint. Observe that for test P_s, $p_{(2,1)} = p^{(s+1) mod\ k}$ (determined by H_s and $\hat{v}_{(1,1)}$); thus $r_{(2,1)} = p^{(s+2) mod\ k}$. Similarly, for test P_s, $p_{(1,2)} = p^{(s+1) mod\ k}$ (determined by V_s and $\hat{h}_{(1,1)}$); thus $r_{(1,2)} = p^{(s+2) mod\ k}$.

- For $\lambda > 3$, $h_{(i,j)}$ will be determined by either H_s or a FUB on diagonal $\lambda - 1$; in both cases $h_{(i,j)} = h^{(s+\lambda-2) mod\ k}$. Similarly $v_{(i,j)}$ will be determined by either V_s or a FUB on diagonal $\lambda - 1$; in both cases $v_{(i,j)} = v^{(s+\lambda-2) mod\ k}$. Thus $p_{(i,j)} = p^{(s+\lambda-2) mod\ k}$, and therefore $r_{(i,j)} = F(p^{(s+\lambda-2) mod\ k}) = p^{(s+\lambda-1) mod\ k}$.

Thus for any test P_s, all FUBs along the diagonal defined by λ have input pattern $p^{(s+\lambda-2) mod\ k}$ applied. Given that s

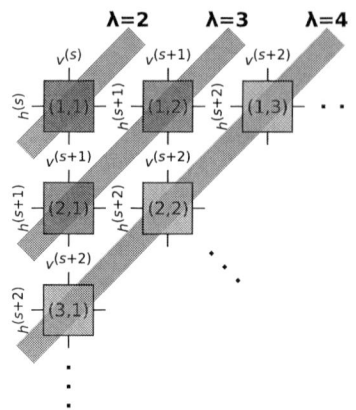

Fig. 3. Visualization of the FUB input patterns applied to the FUBs along the diagonals of the array by a test derived from a cycle in the FUB function F. Note that there is a $mod\ k$ (not shown due to space constraints) on all horizontal and vertical superscripts.

ranges from 0 to $(k-1)$ over the k tests in T, $p^{(s+\lambda-2) mod\ k}$ will cover all $p \in C$ irrespective of λ; thus, all FUBs along all diagonals will experience all input patterns in C over the application of the k tests in T. ∎

Theorem 1 is particularly useful given that error propagation is guaranteed if the FUBs are bijective, as is the case in the CM-LCV; thus, all input pattern faults [13] in a cycle C can be effectively tested across the entire array using a fixed test set of size $|C|$. Theorem 2 builds on this result by showing that the output of the FUB array is equivalent to another test constructed from this same cycle if the array is square (i.e., $M = N$).

Theorem 2. *Given an $N \times N$ square array of FUBs implementing bijective function F with cycle C of length k and the test set T constructed from C according to Theorem 1, the array output $R_s \in T$ for all test $P_s \in T$.*

Proof: Consider test $P_s \in T$ derived from cycle C. The proof for Theorem 1 demonstrated that for a fixed $\lambda = i + j$, $p_{(i,j)} = p^{(s+\lambda-2) mod\ k}$. Thus by definition $\hat{p}_{(i,j)} = p^{(s+\lambda-1) mod\ k}$. However, recall that $R_s = (\hat{H}_s, \hat{V}_s)$ where in this case:

$$\hat{H}_s = \{\hat{h}_{(1,N)}, \hat{h}_{(2,N)}, \ldots, \hat{h}_{(N,N)}\}$$
$$\hat{V}_s = \{\hat{v}_{(N,1)}, \hat{v}_{(N,2)}, \ldots, \hat{v}_{(N,N)}\}$$

Substituting according to λ these become:

$$\hat{H}_s = \{h^{(s+N) mod\ k}, h^{(s+N+1) mod\ k}, \ldots, h^{(s+2N-1) mod\ k}\}$$
$$\hat{V}_s = \{v^{(s+N) mod\ k}, v^{(s+N+1) mod\ k}, \ldots, v^{(s+2N-1) mod\ k}\}$$

This R_s is identical to $P_q \in T$ where P_q is constructed starting at $p^{(s+N) mod\ k}$; thus $R_s \in T$. ∎

Thus a cycle-based test set constructed according to Theorem 1 can be applied to a square array of bijective FUBs

978-1-4799-7598-3/15 $31.00 © 2015 IEEE   270

by driving the array input with the array output. While this result is useful, its utility is limited by the length of the cycles present in the bijective function. To circumvent this limitation, Theorem 3 is introduced.

Theorem 3. *An $M \times N$ array of FUBs implementing bijective function F is equivalent to a single bijective function G with inputs and outputs corresponding to the array inputs and outputs, respectively.*

Proof: Suppose two array inputs $P_i \neq P_j$ both evaluate to the same array output R_k. The difference between P_i and P_j can be represented by a set of errors on the array inputs. Because every FUB in the array is bijective, these errors are guaranteed to propagate through the array to some set of array outputs. Thus $G(P_i) \neq G(P_j)$ for all $P_i \neq P_j$, that is, G is one-to-one. Furthermore, because the array input space is the same size as the array output space (same number of array inputs/outputs), G must be onto. Therefore G is bijective. ∎

Thus, according to Theorem 3, any $m \times n$ sub-array within a FUB array can be represented as a single bijective function. Furthermore, if the dimensions of a FUB array are a scalar multiple of $m \times n$, the overall array can be considered a square array of FUBs implementing this sub-array bijective function. Thus a test set for this array can be derived using Theorem 1 and the cycles of the sub-array bijective function. Assuming some limited design freedom for the FUB array dimensions, it is expected that this process can be used to derive optimal test sets for the CM-LCV array by examining the cycles of various sub-array sizes. Exploiting Theorem 3 is further examined in Section IV.

B. Circuit Design

Implementation of circular BIST requires very little modification to the original CM-LCV design. The main components consist of the two dimensional array of FUBs and a single scan chain around its periphery. Circular feedback is added to connect the array outputs to the normal-mode inputs of the scan chains that drive the array input. Thus, for normal-mode operation (i.e., when the scan enable signal is not asserted), the scan chains feeding the array are updated with the array output values. Execution of the circular BIST test cycle is achieved through three steps: first, a seed vector is loaded into the scan chain. Second, a series of normal-mode clocks are applied to run the test cycle. Finally, the resulting values are unloaded from the scan chain and compared to the expected, fault-free signature.

Two variants of this circular BIST architecture based on the length of the scan chain are explored: Figure 4a uses the minimum scan chain length required, while Figure 4b extends the scan chain to capture the outputs in a second set of flip-flops. While the short-chain variant requires less hardware overhead and has a shorter scan chain length, the long-chain variant has several advantages:

- Diagnosability - the additional flip-flops in the long-chain implementation allow the feedback connections to be

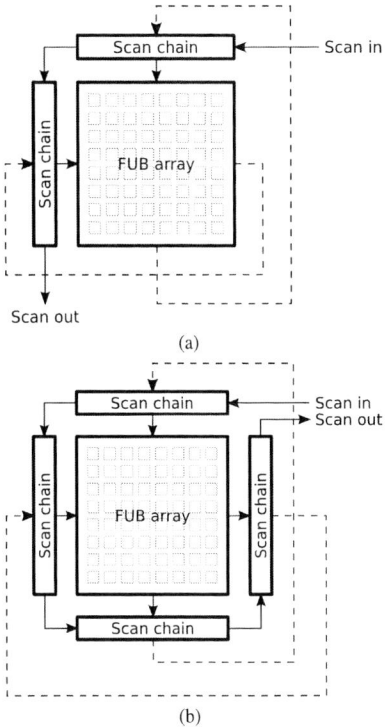

Fig. 4. Circuit diagrams for (a) the short-chain variant and (b) the long-chain variant of the proposed circular BIST design for the CM-LCV. The additional feedback paths required for the proposed BIST are represented by the dashed lines.

TABLE I
TRUTH TABLE FOR THE "GAMMA" 4-INPUT VH-BIJECTIVE FUNCTION.

Input (h, v)	Output $(\hat{h}, \hat{v})$	Input (h, v)	Output $(\hat{h}, \hat{v})$
(00, 00)	(00, 00)	(10, 00)	(10, 01)
(00, 01)	(01, 11)	(10, 01)	(11, 10)
(00, 10)	(10, 10)	(10, 10)	(00, 11)
(00, 11)	(11, 01)	(10, 11)	(01, 00)
(01, 00)	(01, 10)	(11, 00)	(11, 11)
(01, 01)	(00, 01)	(11, 01)	(10, 00)
(01, 10)	(11, 00)	(11, 10)	(01, 01)
(01, 11)	(10, 11)	(11, 11)	(00, 10)

tested independent of the logic array. This results in enhanced resolution because the short-chain implementation cannot distinguish between a failure in the FUBs located on the output edge of the array and a failure due to the feedback connections.

- Test independence - the long-chain implementation allows for two independent test vectors to be applied during BIST. Specifically, one vector can be applied to the logic array by the input scan chain while a second vector is simultaneously transferred from the output scan chain to the input scan chain via the feedback connections.

IV. EXPERIMENT

This work continues to use the "gamma" 4-input VH-bijective function identified in the CM-LCV work [1]. The truth table for this function is given in Table I. A test set and a fault model are needed to properly evaluate the effectiveness

TABLE II
CYCLE LENGTHS FOR FUNCTIONS CORRESPONDING TO SUB-ARRAYS OF
FUBs OF VARIOUS DIMENSIONS.

Sub-array dimensions	Cycle lengths
1×1	1, 15
1×2	1, 7($\times 9$)
2×1	1, 21($\times 3$)
2×2	1, 15($\times 17$)
2×3	1, 7, 127, 889
3×2	1, 31($\times 33$)
3×3	1, 5($\times 3$), 85($\times 48$)
3×4	1, 127($\times 129$)
4×3	1, 5461($\times 3$)
4×4	1, 15($\times 4369$)

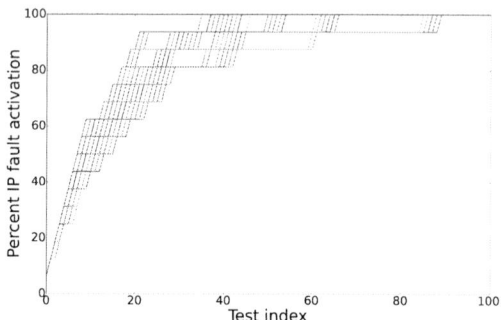

Fig. 5. Evolution of IP fault activation over test index for a 6×9 array using a test set derived from the 2×3 sub-array cycle of length 889. Each trace represents the IP fault activation for each FUB in the 6×9 array.

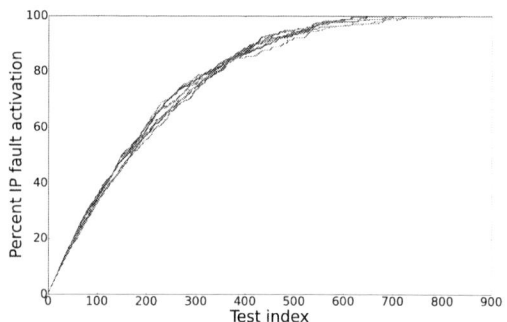

Fig. 6. Evolution of IP fault activation for 2×2 sub-arrays over test index for a 6×9 array using a test set derived from the 2×3 sub-array cycle of length 889. Each trace represents the IP fault activation for each unique 2×2 sub-array in the 6×9 array.

of the proposed circular BIST architecture. First, a high-quality test set is derived according to Theorem 1. Section III-A stated that an array can be considered a square array of FUBs implementing the sub-array bijective function (assuming the array dimensions are a multiple of the sub-array dimensions). Thus, given freedom over the size of the FUB array, the cycles of various sub-array sizes are all of interest as potential sources for a test set for the overall FUB array. Table II lists the cycle lengths for the sub-array bijective functions for various sub-array sizes; each entry under the column labeled "Cycle lengths" denotes a specific cycle length present in the sub-array function as well as the number of such cycles shown as an integer in parentheses. Note that the number of BIST cycles required for a single execution of BIST is at most the maximum cycle length for the corresponding sub-array.

The data shown in Table II support the previous assertion that test cycles of varying length can be found by considering various sub-array sizes. Of note are the single cycle of length 889 present in the 2×3 sub-array bijective function, and the three cycles of length 5,461 present in the 4×3 sub-array bijective function. Recalling Theorem 1, the existence of these cycles indicates that it is possible to create, for example, a design with a circular BIST test cycle of length 5,461 can be constructed if the array dimensions are a multiple of 4×3. The remainder of this section focuses on test sets derived from the 889 test cycle derived from the 2×3 sub-array bijective function for a 6×9 array.

Quantifying fault coverage for the proposed circular BIST architecture using fault simulation is computationally expensive since fault effects (i.e., errors) may be masked when errors are repeatedly fed back through a defective array. To mitigate this issue, fault coverage can instead be equated to the combination of (a) fault activation, and (b) the probability of error masking.

First the fault activation achieved by a test set is examined. A single-pattern input pattern (IP) fault model [13] is assumed for each FUB in the array. Thus, in order to achieve 100% fault activation, the test set must apply all 2^4 possible input patterns to each FUB in the array. Figure 5 shows how fault activation evolves as more tests are applied from the 889 tests derived from the 889-length cycle. Figure 5 indicates that the test set achieves 100% fault activation for all locations in the array by

the 89th test.

However, the fault activation results of Figure 5 is not indicative of all the capabilities of this test set. Activation for each 2×2 sub-array within the overall array translates to all 2^8 possible input patterns applied to the 8-input sub-array. From another perspective, fault activation for each 2×2 sub-array is equivalent to a 16-detect (or, rather, 16-activate) test set for each individual FUB. Figure 6 shows how fault activation evolves for all 2×2 sub-arrays for the same test set and 6×9 array. The test set achieves 100% IP fault activation for all 2×2 sub-arrays by the 835th test.

Activation of all faults is necessary but not sufficient for 100% fault detection; one way to check sufficiency is to consider the probability of error masking within the circular BIST architecture. Recalling the VH-bijectivity property, it is guaranteed that an error will propagate to both the vertical and horizontal outputs of the array for some test that activates the fault. These erroneous signals are guaranteed to be propagated by the non-faulty FUBs in the array during subsequent BIST cycles; the only way for an activated fault to escape detection is for these erroneous signals to converge on the faulty FUB to be masked before the final signature is observed. The probability of this form of masking is expected to be extremely

TABLE III
SUMMARY OF SIMULATION RESULTS FOR BOTH A SCAN TEST SET AND
CIRCULAR BIST APPLIED TO A 6×9 FUB ARRAY.

Metric	Scan test	Circular BIST
Clock cycles	7714	1013
SSL fault coverage	100%	100%
IP fault coverage	100%	100%
Simulation time (min.)	12.9	261.3

low, and moreover, it is expected to decrease exponentially as the size of the FUB array increases[1]. Thus the true fault coverage is expected to be equivalent to the level of fault activation achieved by the applied test set in the case of a single faulty FUB.

Empirical justification of the low likelihood for masking is investigated by simulating an implementation of a 6×9 FUB array using a commercial tool for both single-stuck line (SSL) and IP faults. The results are summarized in Table III for both the circular BIST scheme based on the 2×3 sub-array and a scan test set that achieves a similar level of fault detection. Note that 100% SSL and IP fault coverage is achieved by both the circular BIST and scan test for the given design.

Furthermore, the circular BIST achieves this perfect fault coverage with reduced test time compared to the scan test. The number of test cycles for the scan test is determined by (a) the length of the scan chain, and (b) the number of test vectors that need to be applied to achieve the desired fault coverage on all FUB blocks in the array. Note that in this discussion we are assuming the "long-chain" variant (Figure 4b), in which case only half of the scan chain needs to be observed for each scan test vector. Thus the number of test cycles for scan test can be expressed as: $TC\_SCAN = (vecs + 1) \times \frac{sc\_length}{2} + vecs + k$, where k is a small constant number of test cycles added to ensure the scan chain is functioning properly, and $vecs$ is determined by the FUB size (in the example design used here, 256 vectors are required to apply all input patterns to the 2×2 sub-arrays of the 4-input FUBs). However, circular BIST only requires two full scan-chain loads/unloads and a constant number of BIST cycles. Thus the number of test cycles required can be expressed as: $TC\_BIST = 2 \times sc\_length + BIST\_cycles + k$, where k is a small constant number of test cycles added to ensure the scan chain is functioning properly, and $BIST\_cycles$ is the number of BIST cycles. Thus, for the 6×9 FUB array discussed in this section, the circular BIST achieves an 86.9% reduction in the number of test clock cycles. This reduction will asymptotically approach $1 - \frac{4}{257} = 98.4\%$ for the 4-input FUBs as the array size increases (due to the dominance of the scan chain length factor in the two expressions). Note that

[1]In a larger FUB array, a single defective FUB has less impact on the overall functionality of the array (which, being composed of bijective FUBs, will continue to propagate errors indefinitely). Furthermore, the state space for the FUB array increases exponentially with the number of array inputs/outputs; if we assume that an injected fault causes the FUB array to output random states, the probability of masking decreases exponentially with the size of the state space. However, at this point the only means of verifying the probability of masking is simulation of the design.

this reduction is only in the number of test cycles, and fails to account for the fact that the circuit can be clocked faster for BIST as compared to scan; thus the actual test-time reduction is likely greater than the 86.9% presented for the 6×9 FUB array.

V. CONCLUSION

This work presented a circular BIST scheme for the CM-LCV design. The CM-LCV is an appealing target for circular BIST for reasons that include: (i) its high degree of regularity and (ii) unique test-set properties. The presented circular BIST implementation achieves 100% SSL and IP fault coverage for the CM-LCV design with an 86.9% reduction in test time. Furthermore, implementation of the proposed circular BIST incurs minimal hardware overhead cost, consisting exclusively of additional routing to/from the scan cells. The significant increase in simulation time (due to the fact that faults cannot be dropped until the BIST execution is complete) is a challenge, but may be addressed through fault sampling. Future work includes developing additional approaches to improving simulation time, proving fault detection properties under the circular BIST scheme, and measuring and assuring similarity between the test chip and customer products.

REFERENCES

[1] R. Blanton, B. Niewenhuis, and C. Taylor, "Logic characterization vehicle design for maximal information extraction for yield learning," in *International Test Conference*, Oct 2014.

[2] A. Krasniewski and S. Pilarski, "Circular self-test path: a low-cost bist technique for vlsi circuits," *IEEE Transactions on Computer-Aided Design of Integrated Circuits and Systems*, vol. 8, no. 1, pp. 46–55, Jan 1989.

[3] J. Nelson, T. Zanon, J. Brown, O. Poku, R. Blanton, W. Maly, B. Benware, and C. Schuermyer, "Extracting defect density and size distributions from product ics," *IEEE Design and Test of Computers*, vol. 23, no. 5, pp. 390–400, May 2006.

[4] C. Hess, D. Stashower, B. Stine, L. Weiland, G. Verma, K. Miyamoto, and K. Inoue, "Fast extraction of defect size distribution using a single layer short flow nest structure," *IEEE Transactions on Semiconductor Manufacturing*, vol. 14, no. 4, pp. 330–337, Nov 2001.

[5] C. Hess, B. Stine, L. Weiland, and K. Sawada, "Logic characterization vehicle to determine process variation impact on yield and performance of digital circuits," in *International Conference on Microelectronic Test Structures*, April 2002, pp. 189–196.

[6] A. D. Friedman, "Easily testable iterative systems," *IEEE Transactions on Computers*, vol. C-22, no. 12, pp. 1061–1064, Dec 1973.

[7] R. Blanton and J. Hayes, "Efficient testing of tree circuits," in *International Symposium on Fault-Tolerant Computing*, June 1993, pp. 176–185.

[8] H. Elhuni, A. Vergis, and L. Kinney, "C-testability of two-dimensional iterative arrays," *IEEE Transactions on Computer-Aided Design of Integrated Circuits and Systems*, vol. 5, no. 4, pp. 573–581, October 1986.

[9] S. Jain and C. Stroud, "Built-in self testing of embedded memories," *IEEE Design and Test of Computers*, vol. 3, no. 5, pp. 27–37, Oct 1986.

[10] D. Nikolos, D. Nikolos, H. Vergos, and C. Efstathiou, "An efficient bist scheme for high-speed adders," in *IEEE On-Line Testing Symposium*, July 2003, pp. 89–93.

[11] A. Paschalis, D. Gizopoulos, N. Kranitis, M. Psarakis, and Y. Zorian, "An effective bist architecture for fast multiplier cores," in *Design, Automation and Test in Europe Conference*, March 1999, pp. 117–121.

[12] F. Corno, N. Gaudenzi, P. Prinetto, and M. Reorda, "On the identification of optimal cellular automata for built-in self-test of sequential circuits," in *IEEE VLSI Test Symposium*, Apr 1998, pp. 424–429.

[13] R. Blanton and J. Hayes, "Properties of the input pattern fault model," in *IEEE International Conference on Computer Design: VLSI in Computers and Processors*, Oct 1997, pp. 372–380.

Special Session 12B:
Panel: IOT - Reliable? Secure? Or Death by a Billion Cuts?

Organizers:
Sreekumar V. Kodakara (Intel) and Suriya Natarajan (Intel)

Moderator:
Suriya Natarajan (Intel)

Abstract:

The era of Internet-Of-Things, or IOT - ubiquitous connected components, shared across an intelligent communication medium, and analyzed by massive storage and analysis "clouds" - has begun. With development of deep integration and sensor technologies, very high speed network connectivity, and huge server farms that can provide 24x7 service, the ability to monitor, process and optimize our work and personal life is at our door step.

As with any large scale decentralized system, this progress usually comes with its own massive price. The two major concerns that immediately arise are reliability and security. For dependable operation, each link of an IOT system has to be reliable in the field. Sensors which form the gateway to information sources collect data whose integrity is critical to the entire system performance. For example, health sensors that become unreliable due to field environmental factors or inherent manufacturing issues, can result in either false alarms leading to wasted resources, or more seriously, miss critical warning signs leading to loss of life. The communication medium also has to be reliable to ensure proper quality of service – since a service delayed in some situations can be tantamount to service denied. And finally, reliability of operation of server farms that process the information has to be assured with enough redundancy to ensure outages don't bring work and life to a standstill. As important as reliability is the aspect of security. Internet by itself has brought several security vulnerabilities at both a personal and national level. Malware and Trojans have known to wreak havoc even with compute nodes that are assumed to reasonably protected. Recently, there have been reports of hardware being compromised by leveraging mechanisms built-in for design-for-debug purposes, and memories being gateways of infiltration by exploiting electrical interference in adjacent memory cells. In the era of IOT, any sensor gateway is a potential gateway for malicious inputs and attacks. The more ubiquitous the connected components become, the more the vulnerability. For example, in a connected home, each appliance is a potential home invasion vulnerability. A malicious attack on a network can paralyze a city in ways not dreamt of before. An attack on a cloud service can result in cyber-crime of a scale that can be very hard to even detect and manage.

In this panel, experts debate reliability and security issues in each of the building blocks of an IOT system, and articulate the potential solution vectors that need to be in place to avoid a catastrophe in an IOT world. Experts in the intersection of test, validation, reliability, security and data analysis address the different challenges in this space and roadmaps to potential solutions.

Panelists:

- Mei Jiang, Hewlett-Packard
- Sridharan Ranganathan, Intel Corporation
- Rob Aitken, ARM Holdings PLC
- Bill Eklow, Cisco Systems
- Umit Ogras, Arizona State University

978-1-4799-7598-3/15 $31.00 © 2015 IEEE

AUTHOR INDEX

A-Bounouar, M 85
Abraham, Jacob A. 48
Adell, Philippe 29
Agarwal, Aditya 102
Agbo, Innocent 210
Ahmadi, Ali 79
Aitken, Rob 171, 177, 254
Ali, Sk Subidh 54
Alizadeh, Bijan 120, 192
Amyeen, Enamul 126
Azais, Florence 88
Bagherzadeh, Nader 159
Bakkaloglu, Bertan 29, 62
Bakliwal, Priyanka 29
Banerjee, Debashis 133
Banerjee, Nilanjan 222
Barragan, Manuel J. 88
Becker, Bernd 108, 236
Beohar, Navankur 29
Bertacco, Valeria 87
Bhattacharya, Bhargab B. 1
Bhunia, Swarup 228
Blanchard, Sean 234
Blanton, R.D. 88, 165, 268
Brisk, Philip 7
Cannon, Ethan H. 171
Carulli, John M. 79
Catthoor, Francky 210
Cha, Soonyoung 198
Chakrabarty, Krishnendu 1, 90
Champac, Victor 204
Chang Hao 96
Chang, Doohwang 62
Chao, Mango C.-T. 73, 186
Chatterjee, Abhijit 133
Chattopadhyay, Santanu 102
Chen, Chang-Chih 198
Chen, Chi-Hung 186
Chen, Degang 19, 24
Chen, Harry H. 73
Chen, Tao 24
Chen, Yong-Xiao 13
Cheng, Da 42
Coyette, Anthony 139
Crouch, Al 58
Davidsson, Scott 177
DeBardeleben, Nathan 234
Dobbelaere, Wim 139
Droniou, T. 153
Drouin, D 85
Droulers, G 85
Duan, Yan 19
Dworak, Jennifer 58
Ecoffey, S 85
Eghbal, Ashkan 159

Eklow, Bill 177
Erb, Dominik 236
Esen, Baris 139
Fei, R. 153
Fujita, Masahiro 192
Gent, Kelson 180
Gielen, Georges 139
Gomez, Andres 204
Grissom, Daniel 7
Guoliang Li 90
Gupta, Sandeep K. 36, 42
Hamdioui, Said 210
Han, Chao 256
Harutyunyan, G. 145
Hayashi, Taisuke 127
Hennessy, Andrew 228
Hsiao, Michael S. 180
Hsiung, Hsunwei 36
Hu, Kai 1
Huang, K. 178
Huang, Ke 79
Huang, Tzu-Hsuan 186
Huss, E. 153
Jaress, Christopher 7
Jun Qian 90
Jutman, Artur 177
Kaeli, David 234
Karmakar, Rajit 102
Kim, Woongrae 198
Kodakara, Sreekumar V. 67, 274
Kukner, Halil 210
Kuo, Shih-Hua 73, 186
Labalette, M 85
Larsson, Erik 177
Leger, Gildas 88
Li, Jin-Fu 13
Liang Huaguo 96
Lin, Chris 186
Lotz, Christophe 177
Makris, Yiorgos 79
Mandal, Debashis 29
Mandier, C. 153
Marcellin, A. 153
Martin, Mitchell 165
Milor, Linda 176, 198
Mir, S. 153
Mirkhani, Shahrzad 48
Miura, Noriyuki 127
Moreau, J. 153
Mukherjee, Nilanjan 255
Nagata, Makoto 127
Nahar, Amit 79
Natarajan, Suriya 175, 274
Navabi, Zainalabedin 120
Nicolaidis, Michael 149

AUTHOR INDEX

Niewenhuis, Ben 268
Orr, Bob 79
Ozev, Sule 29, 62, 176
Palmigiani, G. 153
Pant, Mondira 171
Papavramidou, Panagiota 149
Pas, Michael 79
Patel, Chintan 222
Pioro-Ladriere, M 85
Poehls, Leticia 204
Pomeranz, Irith 114, 242, 248, 262
Portolan, M 178
Raghavan, Praveen 210
Rajski, Janusz 255
Rao, Sushmita Kadiyala 222
Rech, Paolo 234
Reddy, Sudhakar 108
Reddy, Sudhakar M. 236
Refan, Fatemeh 120
Ren, Xuanle 165
Ricchetti, Mike 86
Riefert, Andreas 108
Robucci, Ryan 222
Roy, Sidhanto 29
Rui Li 90
Sadi, M. 216
Sagar, Mehul V. 67
Samynathan, Balavinayagam 48
Sauer, Matthias 108, 236
Scheibler, Karsten 236
Sharafinejad, Reza 192
Shivashankar, Bharath 222
Sinanoglu, Ozgur 54
Sindia, Suraj 66
Singh, Adit D. 88, 256
Souifi, A 85
Sunter, Stephen 88
Tahoori, Mehdi B. 171
Taniguchi, Kohki 127
Taouil, Mottaqiallah 210
Tehranipoor, M 216
Tracey, Paul 35
Tshagharyan, G. 145
Tung, Jonathan 73
Vanhooren, Ronny 139
Vargas, Fabian 204
Vermeire, Bert 29
Vitrou, P. 153
Wang, Ran 90
Wang, Xian 133
Weckx, Pieter 210
Winemberg, L. 216
Xu, Li 19
Yaghini, Pooria M. 159
Yang, Hao-Yu 186

Yuen, Joel 67
Zhang, Fengchao 228
Zorian, Y. 145